D0138601

Ergonomics

PRENTICE HALL INTERNATIONAL SERIES
IN INDUSTRIAL AND SYSTEMS ENGINEERING

W. J. Fabrycky and J. H. Mize, Editors

ALEXANDER *The Practice and Management of Industrial Ergonomics*
AMOS AND SARCHET *Management for Engineers*
AMRINE, RITCHEY, MOODIE, AND KMEC *Manufacturing Organization and Management*, 6/E
ASFAHL *Industrial Safety and Health Management*, 2/E
BABCOCK *Managing Engineering and Technology*
BADIRU *Expert Systems Applications in Engineering and Manufacturing*
BANKS AND CARSON *Discrete-Event System Simulation*
BLANCHARD *Logistics Engineering and Management*, 4/E
BLANCHARD AND FABRYCKY *Systems Engineering and Analysis*, 2/E
BUSSEY AND ESCHENBACH *The Economic Analysis of Industrial Projects*, 2/E
BUZACOTT AND SHANTHIKUMAR *Stochastic Models of Manufacturing Systems*
CANADA AND SULLIVAN *Economic and Multi-Attribute Evaluation of Advanced Manufacturing Systems*
CHANG AND WYSK *An Introduction to Automated Process Planning Systems*
CHANG, WYSK, AND WANG *Computer Aided Manufacturing*
CLYMER *Systems Analysis Using Simulation and Markov Models*
EBERTS *User Interface Design*
ELSAYED AND BOUCHER *Analysis and Control of Production Systems*, 2/E
FABRYCKY AND BLANCHARD *Life-Cycle Cost and Economic Analysis*
FABRYCKY AND THUESEN *Economic Decision Analysis*, 2/E
FRANCIS, MCGINNIS AND WHITE *Facility Layout and Location: An Analytical Approach*, 2/E
GIBSON *Modern Management of the High-Technology Enterprise*
HALL *Queuing Methods: For Services and Manufacturing*
HAMMER *Occupational Safety Management and Engineering*, 4/E
HUTCHINSON *An Integrated Approach to Logistics Management*
IGNIZIO *Linear Programming in Single- and Multiple-Objective Systems*
IGNIZIO AND CAVALIER *Linear Programming*
KROEMER, KROEMER, KROEMER-ELBERT *Ergonomics: How To Design for Ease and Efficiency*
KUSIAK *Intelligent Manufacturing Systems*
LANDERS, BROWN, FANT, MALSTROM, SCHMITT *Electronics Manufacturing Processes*
MUNDEL AND DANNER *Motion and Time Study: Improving Productivity*, 7/E
OSTWALD *Engineering Cost Estimating*, 3/E
PULAT *Fundamentals of Industrial Ergonomics*
SHTUB, BARD, GLOBERSON *Project Management: Engineering Technology and Implementation*
TAHA *Simulation Modeling and SIMNET*
THUESEN AND FABRYCKY *Engineering Economy*, 8/E
TURNER, MIZE, CASE, AND NAZEMETZ *Introduction to Industrial and Systems Engineering*, 3/E
WOLFF *Stochastic Modeling and the Theory of Queues*

Ergonomics

How to Design for Ease and Efficiency

K. H. E. Kroemer
Virginia Polytechnic Institute and State University

H. B. Kroemer
Burns Clinic Medical Center

K. E. Kroemer-Elbert
Howmedica Inc.

Prentice Hall International Series in Industrial and Systems Engineering
W.J. Fabrycky and J.H. Mize, Editors

PRENTICE HALL Englewood Cliffs, NJ 07632

Library of Congress Cataloging-in-Publication Data

Kroemer, K. H. E.
Ergonomics : how to design for ease and efficiency / K.H.E.
Kroemer, H.B. Kroemer, K.E. Kroemer-Elbert.
p. cm.
Includes bibliographical references and index.
ISBN 0-13-278359-2
1. Human engineering. 2. Engineering design. I. Kroemer, H. B.
II. Kroemer-Elbert, K. E. III. Title.
TA166.K77 1994
620.8'2--dc20 93-3143
CIP

Acquisitions Editor: Marcia Horton
Copy Editor: Bob Lentz
Cover Designer: Bruce Kenselaar
Prepress Buyer: Linda Behrens
Manufacturing Buyer: Dave Dickey
Supplements Editor: Alice Dworkin
Editorial Assistant: Dolores Mars

Simon & Schuster Company A Viacom Company
Upper Saddle River, New Jersey 07458

Printed in the United States of America

10 9 8 7 6 5 4 3

ISBN 0-13-278359-2

Prentice-Hall International (UK) Limited, *London*
Prentice-Hall of Australia Pty. Limited, *Sydney*
Prentice-Hall Canada Inc., *Toronto*
Prentice-Hall Hispanoamericana, S.A., *Mexico*
Prentice-Hall of India Private Limited, *New Delhi*
Prentice-Hall of Japan, Inc., *Tokyo*
Simon & Schuster Asia Pte. Ltd., *Singapore*
Editora Prentice-Hall do Brasil, Ltda., *Rio de Janeiro*

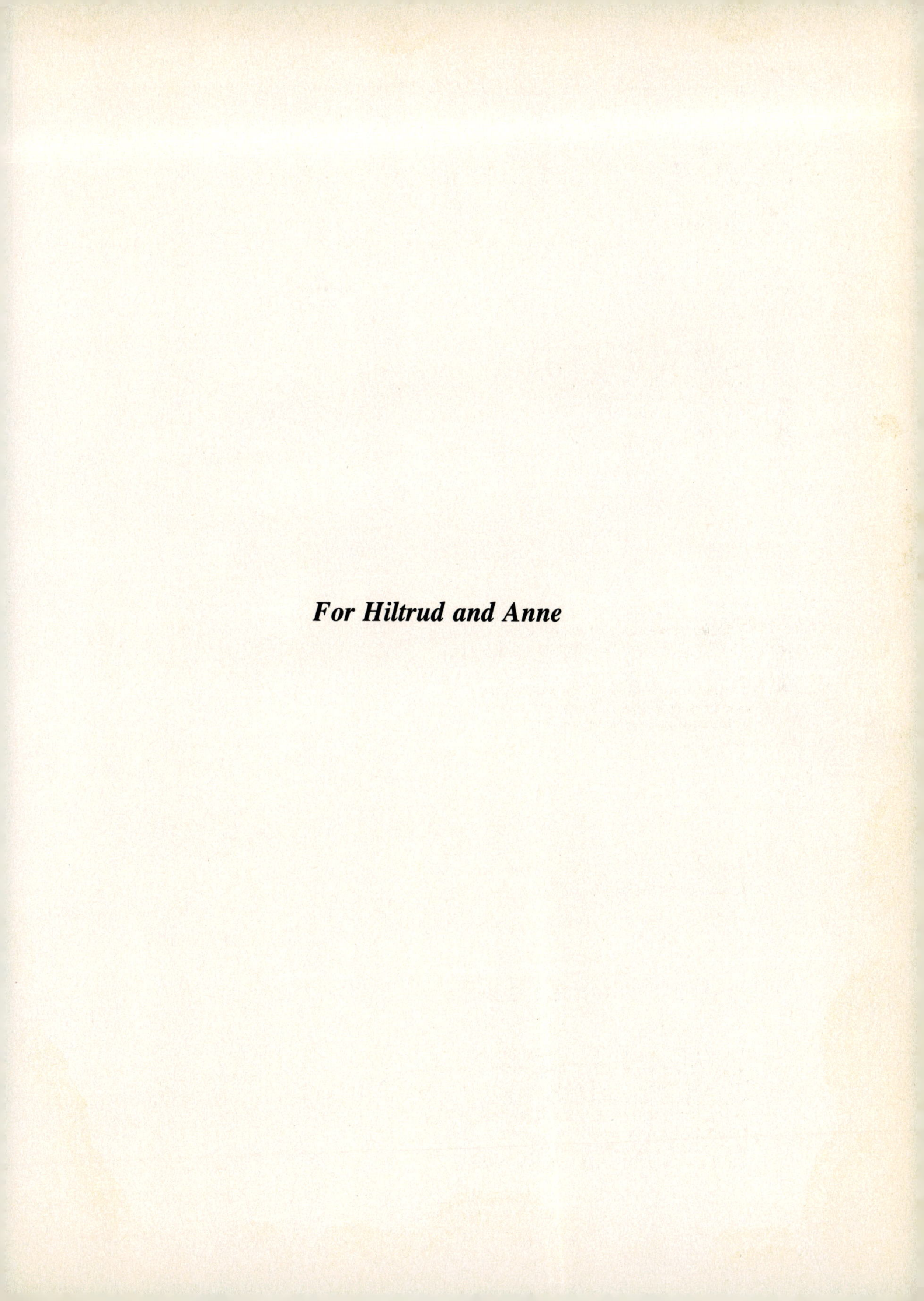

For Hiltrud and Anne

Contents

Preface

What do we know about the human body and mind at work? Given what we know, how then should we design the work task, tools, the interface with the machine, and work procedures so that the human can perform safely, efficiently, and with satisfaction—perhaps even enjoy working?

These challenges are the main themes of this book. The solutions are the WHY and HOW of ergonomics/human engineering.

Tell us of new developments to be considered in the future, and about old knowledge not to be forgotten.

K. H. E. Kroemer
Ergonomics Research Institute
P.O. Box 3019
Radford, Virginia 24143-3019 USA
Telephone 703•639•0514

Acknowledgments

We are grateful to colleagues and friends who helped us with material and suggestions, who read texts and improved them: M. M. Ayoub (Texas Tech University) John G. Casali (Virginia Tech, VPI & SU), Martin G. Helander (University of Buffalo), Paul T. Kemmerling (Virginia Tech), William S. Marras (Ohio State University), Rudolph G. Mortimer (University of Illinois, Champaign), Dennis L. Price

(Virginia Tech), Jerry D. Ramsey (Texas Tech University), John A. Roebuck, Jr. (Roebuck Research & Consulting), James L. Smith (Texas Tech University), Leo A. Smith (Auburn University), and Harry L. Snyder, Walter W. Wierwille, Jeffrey C. Woldstad (all at Virginia Tech).

We appreciate the helpfulness and expertise of the Prentice Hall professionals Marcia Horton, Elizabeth Kaster, Bob Lentz and Joe Scordato.

We owe thanks and admiration to Teresa Huff and Sandy Dalton for processing, correcting, and re-doing this text over and over with endurance, patience, and good humor.

This book would not exist without them.

Karl H. E. Kroemer
Henrike B. Kroemer
Katrin E. Kroemer-Elbert

Introduction: Goals of Ergonomics

Ergonomics is the application of scientific principles, methods, and data drawn from a variety of disciplines to the development of engineering systems in which people play a significant role. Among the basic disciplines are psychology, cognitive science, physiology, biomechanics, applied physical anthropometry, and industrial systems engineering. The engineering systems to be developed range from the use of a simple tool by a consumer to a multiperson sociotechnical system.

Ergonomic specialists involved in the system design process, declares the National Research Council, "are united by a singular perspective on the system design process: that design begins with an understanding of the user's role in overall system performance and that systems exist to serve their users, whether they are consumers, system operators, production workers, or maintenance crews. This user-oriented design philosophy acknowledges human variability as a design parameter. The resultant designs incorporate features that take advantage of unique human capabilities as well as build in safeguards to avoid or reduce the impact of unpredictable human error" (1983, pages 2 and 3).

Success according to the Council, is measured by improved productivity, efficiency, safety, and acceptance of the resultant system design.

☞☞☞ *There is a hierarchy of goals in ergonomics. The fundamental task is to generate "tolerable" working conditions that do not pose known dangers to human life or health. This basic requirement assured, the next goal is to generate "acceptable" conditions upon which the people involved can voluntarily agree according to current scientific knowledge and under given sociological, technological, and organizational circumstances. Of course, the final goal is to generate "optimal" conditions which are so well adapted to human characteristics, capabilities, and desires that physical, mental, and social well-being is achieved.*

To be more specific, we may define Ergonomics (also called Human Factors or Human Engineering in the United States) as the discipline to "study human characteristics for the appropriate design of the living and work environment." Its fundamental aim is that all man-made tools, devices, equipment, machines, and environment should advance, directly or indirectly, the safety, well-being, and performance of humans.

Thus, ergonomics has two distinct aspects: (1) study, research, and experimentation, in which we determine specific human traits and characteristics that we need to know for engineering design; (2) application and engineering, in which we design tools, machines, shelter, environment, work tasks, and job procedures to fit and accommodate the human. This includes, of course, the observation of actual performance of human and equipment in the environment to assess the suitability of the designed human-machine system and to determine possible improvements.

Ergonomics adapt the man-made world to the people involved because they focus on the human as the most important component of our technological systems. Thus, the utmost goal of ergonomics is "humanization" of work. This goal may be symbolized by the "E & E" of Ease and Efficiency, for which all technological systems and their elements should be designed. This requires knowledge of the characteristics of the people involved, particularly of their dimensions, their capabilities, and their limitations.

Ergonomic is "neutral": it takes no sides, neither of employers nor of employees. It is not for or against progress. It is not a philosophy, but a scientific discipline and technology.

GOALS OF THIS BOOK

In this book we discuss the human interaction with work task and technology.

Our intention in exploring this interaction is to build a knowledge-based understanding so that we can

- amplify human capabilities,
- utilize human abilities,
- facilitate human efficiency, and
- avoid overloading or underloading.

This understanding will benefit a variety of specialists and generalists who are concerned with people's performance and well-being at work. These include

Designer,
Engineer,
Architect,
Ergonomist, human factors specialist,
Industrial hygienist,
Industrial physician,
Occupational nurse,
Manager, whether supervising others or "just oneself,"
Student in any of these areas—and, of course,
Everybody interested in "humanizing work," i.e., making work safe, efficient, and satisfying.

HOW THIS BOOK IS ORGANIZED

The book has three major parts:

Part One, "The Ergonomics Knowledge Base," includes Chapters 1 through 6. Here, we explore the properties of the human body and mind, as manifested in people's interactions with the environment. The focus is on human *dimensions*, *capabilities*, and *limitations*—the human factors to be considered in E & E designing. For convenience, we use anatomy, physiology, and psychology as traditional disciplinary divisions while, in fact, the human functions wholistically and synergistically.

Part Two, "Design Applications," includes Chapters 7 through 12. Here, we discuss the design of task, equipment and environment in the light of the obtained knowledge about human size, strengths and weaknesses—the knowledge base developed in Part One. This information is brought to bear with the aim of matching human capabilities with demands.

Part Three, "Further Information," includes a listing of References, an extensive Glossary with concise descriptions and definitions, and a detailed Index which refers the reader to specific pages.

HOW TO USE THE BOOK

You may use the book in three ways: (1) Read it straight through from beginning to end, as in a university course; and work on the "challenges" listed at the end of each chapter. (2) Read a chapter of interest, absorb the background information, and proceed to design applications that make use of this information. (3) Start with the Index, pick a topic of interest, and look up the information in the book sections that the Index indicates.

THE DEVELOPMENT OF ERGONOMICS

The use of objects found in the environment for use as tools and weapons is an ancient and fundamental activity; this distinguishes humans from many other primates. Pieces of stone, bone, and wood were at first not shaped but simply selected for their fit to the human hand and their suitability as scrapers, pounders, and missiles. Purposeful shaping of these tools was the next step. Creating from raw materials and manufacturing in quantity followed. Fitting clothes and making shelters also were early and fundamental "ergonomic" activities.

As human society grew more complex, organizational and management challenges developed. Purposeful training of workers and soldiers, for example, became necessary, together with formation and control of behavior. For major projects, such as building the pyramids in old Egypt, assemblmg armies for warfare, sheltering the inhabitants of ancient cities, and supplying them with food and water, sophisticated knowledge of human needs and desires was required, and careful planning and complex logistics had to be mastered. The aims and means of training became sophisticated. Roman soldiers, for example, underwent well-organized training and conditioning until they could perform military exercises without sweat accumulating uselessly on their skin.

☞☞☞ *"Drying the legions" of the Roman Empire relied, consciously or by experience, on the principle of training and adapting the physiological capabilities of the recruits to the physical requirements.* ☜☜☜

Evolution of Disciplines

Artists, military officers, employers, and sports enthusiasts were always interested in body build and physical performance. Specialized "medicine men" and "herb women" treated illnesses and injuries. Anatomic and anthropological disciplines began to develop. About 400 B.C., Hippocrates described a scheme of four body types, which were supposedly determined by their fluids. The "moist" type was believed to be dominated by black gall; the "dry" type by yellow gall; the "cold" type by slime, and the "warm" type by blood.

Over the centuries, information accumulated into specialized disciplines. In the fifteenth to seventeenth centuries, however, gifted persons such as Leonardo da Vinci and Alfonso Giovanni Borrelli could still master the existing knowledge of anatomy, physiology, and equipment design; they were artist, scientist, and engineer in one.

In the eighteenth century, the sciences of anatomy and physiology diversified and accumulated specific detail knowledge. Psychology began to develop as a separate science. Well into the ninteenth century, these tended to be oriented toward theories, trying to understand the complex human being: the stereotype is the scientist in a white coat leading a research-devoted life in the laboratory. But an increasing interest in industrial work, together with the old interest in military employment of the human, brought forward "applied" aspects of the "pure" sciences. In the early 1800s, in France, Lavoisier, Duchenne, Amar, and Dunod researched energy capabilities of the working human body, Marey developed methods to describe human motions at work, and Bedaux made studies to determine work payment systems, before Taylor and the Gilbreths did similar work in the United States in the early 1900s. In England, the "Industrial Fatigue Research Board" considered theoretical and practical aspects of the human at work. In Italy, Mosso constructed dynamometers and ergometers to research fatigue. In Scandinavia, Johannsson and Tigerstedt developed the scientific discipline of "work physiology." A Work Physiology Institute was founded in Germany by Rubner in 1913. In the United States, Benedict and Cathcard (1913) described the efficiencies of muscular work. The Harvard Fatigue Laboratory was established in the 1920s (Lehmann, 1962; Brouha, 1967).

In the first half of the twentieth century, industrial physiology and psychology were well advanced and widely recognized, both in their theoretical research "to study human characteristics" and in the application of this knowledge "for the appropriate design of the living and work environment." Two distinct approaches to study human characteristics had developed: one was particularly concerned with physiological and physical traits of the human, the other was mainly interested in psychological and social properties. Although there was much overlap between these approaches, the anatomical-physical and physiological aspects were studied mainly in Europe, and the psychological and social aspects in North America.

Directions in Europe

Based on a broad fundament of anatomical, anthropological, and physiological research, applied or "work physiology" assumed great importance in Europe, particularly during the hunger years associated with the First World War. Marginal living conditions stimulated research on the minimal nutrition required to perform certain activities, the consumption of energy while performing agricultural, industrial, military, and household tasks, the relationships between energy consumption and heart rate, the use of muscular capabilities, suitable body postures at work, design of equipment and workplaces to fit the human body, and other related topics. Another development in the 1920s was "psychotechnology," which involved testing persons for their ability to perform physical and mental work, their vigilance and attention, their ability to carry mental workload, their behavior as vehicle drivers, their ability to read road signs, and related topics.

Directions in North America

"Most psychologists at this time [around 1900] were strictly scientific and deliberately avoided studying problems that strayed outside the boundaries of pure research" (Muchinsky, 1987, p. 13). Some investigators, however, had practical concerns, such as sending and receiving Morse code, perception and attention at work, using psychology in advertising, and promoting industrial efficiency. A particularly important step was the development of "intelligence testing," used to screen military recruits during the First World War, and later to screen industrial workers for assigning them to jobs appropriate to their mental capabilities. The terms *intelligence testing* and *industrial psychology* won acceptance. Gould (1981) provides a partly amusing, partly disturbing account of the early years of intelligence testing.

The term "industrial psychology" first appeared as a typographical error, actually meant to read "individual" psychology (Muchinsky, 1987).

Among the best-known findings in industrial psychology are those yielded by the experiments at the Hawthorne Works near Chicago in the mid-1920s. The study was designed to assess relationships between lighting and efficiency in work rooms where electrical equipment was produced (Roethlisberger and Dickson, 1943). The bizarre finding was that the workers' productivity increased or remained at a satisfactory level whether or not the illumination was changed, partly at least in response to the attention paid to the workers by the researchers (Jones, 1990; Parsons, 1974, 1990). This became known as the Hawthorne Effect.

Industrial psychology divided into special branches, including personnel psychology, organizational behavior, industrial relations, and engineering psychology. Under the pressures of the Second World War, the "human factor" as part of a "man-machine system" became of major concern. The technological development led to machines and to "operational systems" that put higher demands on attention, strength, and endurance of individuals and teams than many could muster. For example, operators had to observe radar screens over periods of many hours, with the intent of detecting some blips among others. In high-performance aircraft the pilot was subjected to forceful accelerations, such as in sharp turns, where he might be unable to operate hand controls properly, even black out. Crew members had to fit into tank and aircraft cockpits that were narrow and low, requiring that anthropometrically small crew members be selected. Stressful conditions made it difficult to maintain combat morale and performance. Thus, military activities and related efforts on the "home front" generated the need to consider human physique and psychology, purposefully and knowingly, in the design of task, equipment, and environment.

Names for the Discipline: "Ergonomics" and "Human Factors"

In Europe and in North America various terms were used to describe the activities of anthropologists, physiologists, psychologists, sociologists, statisticians, and engineers who studied the human and used this information in design, selection, and training. On January 13 and 14, 1950, British researchers met in Cambridge, England, to discuss the name of a new society to represent their activities. Among others, the term "ergonomics" was proposed. This word had been coined in late 1949 by K. F. H.

Murrell, who derived it from the Greek terms *ergon,* indicating work and effort, and *nomos,* law or rules, apparently reinventing a word already used in Poland a hundred years earlier (Monod and Valentin, 1979). The term was neutral, implying no priority of contributing disciplines, such as physiology or psychology or functional anatomy or engineering. It was easily remembered and recognized and could be used in any language. *Ergonomics* was formally accepted as the name of the new society at its council meeting on February 16, 1950 (Edholm and Murrell, 1974).

Several aspects are worthy of a brief note. The original proposal for the name included two alternative suggestions. One was the Ergonomic Society. Note that there is no "s" at the end of Ergonomic; apparently, it somehow slipped onto the ballot, and since that time has made derivating an adjective or adverb difficult. The alternately proposed term, Human Research Society, bears some similarity to the term Human Engineering.

In the United States, a group of persons convened in 1956 to establish a formal society. The name Ergonomics was rejected, and instead *Human Factors* was selected. Often, the word Engineering is added or substituted to indicate applications, as in Human (Factors) Engineering (Christensen, Topmiller, and Gill, 1988).

There has been some discussion of whether Human Factors differs from Ergonomics, whether one relies more heavily on psychology or physiology, or is more theoretical or practical than the other. Today, the two terms are usually considered synonymous, as exemplified by the naming practice of the Canadian Society: it uses "human factors" in its English name, and "ergonomie" in its French version. In 1992, the Human Factors Society decided to rename itself, adding the term Ergonomics.

THE ERGONOMIC KNOWLEDGE BASE

Ergonomics today is growing and changing. Development stems from increasing and improving knowledge about the human, and is driven by new applications and new technological developments.

As discussed earlier, a number of classic sciences provide the fundamental knowledge about the human. The anthropological basis consists of anatomy, describing the build of the human body; orthopaedics, concerned with the skeletal system; physiology, dealing with the functions and activities of the living body, including the physical and chemical processes involved; medicine, concerned with illnesses and their prevention and healing; psychology, the science of mind and behavior; and sociology, concerned with the development, structure, interaction, and behavior of individuals or groups. Of course, physics, chemistry, mathematics, and statistics also supply knowledge, approaches, and techniques.

From these basic sciences, a group of more applied disciplines developed into the core of ergonomics. These include primarily anthropometry, the measuring and description of the physical dimensions of the human body; biomechanics, describing the physical behavior of the body in mechanical terms; industrial hygiene, concerned with the control of occupational health hazards that arise as a result of doing work; industrial psychology, discussing people's attitude and behavior at work; management, dealing with and coordinating the intentions of the employer and the employees; and work physiology, applying physiological knowledge and measuring techniques to the body at work. Of course, many other disciplinary areas have developed that also are part of ergonomics, or contribute to it, or partly overlap, such as labor relations.

Several distinct application areas use ergonomics as components of their knowledge base, or of their work procedures. Among these are industrial engineering, by definition concerned with the interactions among people, machinery, and energies; bio-engineering, working to replace worn or damaged body parts; systems engineering, in which the human is an important component of the overall work unit; safety engineering and industrial hygiene, which focus on the well-being of the human; and military engineering, which relies on the human as soldier or operator. Naturally, other application areas are in urgent need of ergonomic information and data, such as computer-aided design, in which information about the human must be provided in computerized form. Oceanographic, aeronautical, and astronautical engineering also rely intensively on ergonomic knowledge.

The development from basic sciences to applied disciplines in ergonomics, and the use of ergonomic knowledge in specific areas, is depicted schematically in Figure 1. As more knowledge about the human becomes available, as new opportunities develop to make use of human capabilities in modern systems, and as needs arise for protecting the person from outside events, ergonomics changes and develops.

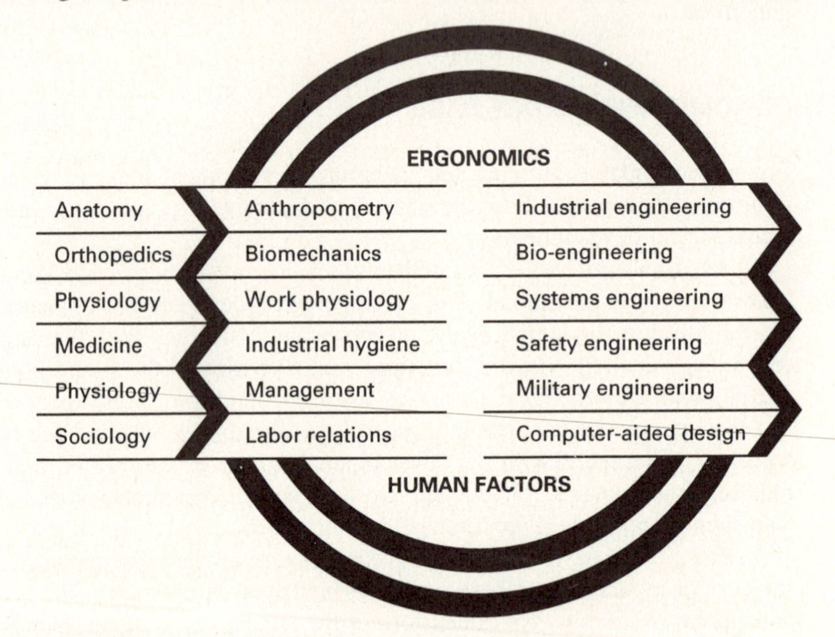

Figure 1. Origins, developments, and applications of ergonomics.

Professional Organizations

The *Ergonomics Society* is located in the United Kingdom (see the Addresses Section). Founded in 1950, it is the oldest ergonomics organization. The Ergonomics Society had about 830 members in 1989, mostly in the United Kingdom. The Society supports two journals, *Ergonomics* and *Applied Ergonomics,* and organizes an annual professional congress.

The largest single (national) professional organization in ergonomics/human factors is the Human Factors and Ergonomics Society, in the United States, with various special technical interest groups. The society publishes its own journal, *Human Factors,* and organizes an annual professional congress.

More than two dozen national societies exist, many carrying the term Ergonomics in their name. Most are members of the International Ergonomics Association, founded in 1959. In 1991, over 14,000 ergonomists were working, mostly in North America, Europe, Japan, and Australia/New Zealand, with fairly large contingents also in Southeast Asia and South America, as well as in Asia and China—see Table 1.

Sources of Ergonomic Information

After the initial development from basic "theoretical sciences" to applications of ergonomic knowledge, several major consumers and generators of human-factors information emerged. First among them is the military (like it or not), which for its success must "integrate" people (soldiers) and equipment (weapons). Thus, military regulations and standards (such as MIL STD 1472 in the United States) are still major sources of ergonomic information. Industries that supply the military are immediately affected, because they must comply with human engineering requirements.

Industries and occupations (such as mining, forestry, agriculture, construction, steel, tire manufacturing, and nurses and their aides) that require much physical work are in need of physiological and biomechanical information. Newer "hi-tech" industries (such as manufacture of automobiles, aircraft, spacecraft, and computers) need similar knowledge with a shift in emphasis toward sensori-motoric, perception, decision-making, and sociopsychological aspects. Modern management combines economical, psychological, and engineering aspects with much consideration of safety and health.

Of course, many journals provide information about recent findings, most associated with national professional organizations. In the United States, well-established human engineering standards are available from the military (MIL STD), the automotive industry (SAE), and NASA. Furthermore, government agencies issue guidelines (NIOSH) and requirements (OSHA). Professional groupings provide specialized guides (e.g., ASHRAE, ACGIH, National Safety Council), as do industry-wide associations (ANSI). Similar conditions exist in many countries. The International Standards Organization (ISO) publishes standards for worldwide use.

TABLE 1. ERGONOMIC SOCIETIES

Country/region	Founding date	Membership	Population (m)	Members per m
Australia	1964	613 (1991)	17	36
Austria	1976	42 (1990)	7.5	6
Belgium	1986	140 (1991)	9.9	14
Brazil	1983	244 (1988)	143.3	2
Canada	1968	366 (1991)	25.6	14
China	1989	300 (1989)	1096	0.3
CIS	1989	434 (1990)	288.8	2
Colombia	1987		30.7	
Denmark		1063 (1991)	5.1	208
Finland	1985	110 (1991)	5	22
France	1963	531 (1987)	55.2	10
Germany	1958	700 (1991)	76.6	9
Hungary	1987	90 (1988)	4.6	20
India	1987	100 (1991)	900	0.1
Indonesia	1988	120 (1988)	176.8	0.7
Israel	1982	120 (1992)	4.3	28
Italy	1968	215 (1990)	57.2	4
Japan	1964	1558 (1989)	121.4	13
Korea (South)	1982	250 (1988)	43.9	6
New Zealand	1986	119 (1992)	3.3	35
Netherlands	1963	604 (1992)	15	40
Norway		147 (1991)	4.2	35
Poland	1977	100 (1990)	37.5	3
Singapore	1988	36 (1990)	2.6	14
South Africa	1984	107 (1988)	34.3	3
South East Asia	1984	86 (1990)	250	3
Spain	1988	187 (1992)	39.6	5
Sweden		305 (1991)	8.5	36
United Kingdom	1949	1030 (1992)	56.5	18
USA	1957	4658 (1990)	242.2	19
Yugoslavia	1973	50 (1989)	23.2	2
TOTAL		14335		

SOURCE: Adapted from *Ergonomics* 35(11), page 1410, 1992.

Addresses

ACGIH American Conference of Government Industrial Hygienists
6500 Glennway Avenue
Cincinnati, OH 45221, USA

ANSI American National Standards Institute
11 West 42nd Street, 13th Floor
New York, NY 10036, USA

ASHRAE American Society of Heating, Refrigerating, and Air-conditioning Engineers
1791 Tullie Circle
Atlanta, GA 30329, USA

CSERIAC Crew System Ergonomics Information Analysis Center
AL/CFH/CSERIAC
Wright-Patterson AFB, OH 45433-6573, USA

Ergonomics Society
Devonshire House
Devonshire Square
Loughborough, Leics., 11 3DW, UK

Human Factors and Ergonomics Society
P.O. Box 1369
Santa Monica, CA 90406-1369, USA

Military (USA)
U.S. Military and Federal Standards, Handbooks, and Specifications are available from:

National Technical Information Service, NTIS
5285 Port Royal Road
Springfield, VA 22161, USA

or

Naval Publications and Forms Center, NPODS
5801 Tabor Avenue
Philadelphia, PA 19120-5099, USA

or

Standardization Division, U.S. Army Missile Command, DRSMI-RSDS
Redstone Arsenal, AL 35898, USA

or

U.S. Air Force Aeronautical Systems Division, Standards Branch
ASD-ENESS
Wright-Patterson AFB, OH 45433, USA

NASA National Aeronautics and Space Administration SP 34-MSIS
LBJ Space Center
Houston, TX 77058, USA

National Safety Council
1121 Spring Lake Drive
Itasca, IL 60143-3201, USA

NIOSH National Institute of Occupational Safety and Health
4676 Columbia Parkway
Cincinnati, OH 45226, USA

OSHA Occupational Safety and Health Agency
200 Constitution Avenue, NW, N3651
Washington, DC 20210, USA

SAE Society of Automotive Engineers
400 Commonwealth Drive
Warrendale, PA 15096-0001, USA

ISO International Organization for Standardization
1 rue Varembe
Case Postale 56
CH 1211 Genève 20, Switzerland

1

The Anatomical and Mechanical Structure of the Human Body

OVERVIEW

Body dimensions of many people on earth are only estimated. For Europeans and North Americans, the measurements of adult civilians must be derived from data taken on soldiers. Correlations among measurements allow us to establish design rules for things that must "fit" the body.

Biomechanically, one can describe the human body as a basic skeleton whose parts are linked in joints; the members have volumes and mass properties and are moved by muscles. Understanding the properties, capabilities, and limitations allows us to design equipment and tools that use and enhance human strengths.

DEVELOPMENT

Human species development may be compared to the growth of a bush and its branches. Some branches develop and die while others grow more and more twigs, some of which vanish while others flourish. In the process, variations occur in body dimensions of different groups of people. Ergonomics must take these variations into account.

The development of the human race can be traced by fossils and by reconstruction of mitochondrial DNA over several million years in Africa, and for hundreds of thousands of years in Europe and Asia (Diamond, 1988; Gould, 1988; Asimov, 1989).

Current theory holds that the Austrolopithecine was a predecessor of the genus homo about 3 million years ago in Africa, where *homo erectus* then developed. One humanoid branch started about 250 thousand years ago and remained in Africa. Another branch developed 60 or 70 thousand years later. Some of its members stayed in Africa and others spread all over the earth.

Remains of anatomically modern humans who lived about 130 to 180 thousand years ago have been found in South Africa and in the Levant. There and in central Europe, about 150 thousand years ago, the *Neanderthal* emerged. Neanderthals apparently were stocky, heavy-set, and cold-adapted with a brain as big as ours. They lived tens of thousands of years side-by-side with *Cro-Magnons*, but vanished about 30 thousand years B.C. Cro-Magnons lived in the area of what is today Israel around 90 thousand years ago; their stock probably grew into *homo sapiens*.

Popular notions about the different appearances of Cro-Magnons and Neanderthals are mostly based on conjecture, often Hollywood movie style. For example, there is no indication that the Cro-Magnons were dark skinned, or the Neanderthals light. Furthermore, there is no evidence of violent struggles for superiority between the two races.

Bushmen and Pygmies probably occupied most of subequatorial Africa until about two thousand years ago. Bantu-speaking people living in the area of Cameroon and Congo learned to use iron, developed agriculture, and domesticated animals. The flourishing Bantu then drove Bushmen and Pygmies into areas unsuitable for agriculture. Subsequently, 60 million Bantu occupied half the African continent.

From Africa, homo spread over the earth. About 50 thousand years ago, Australia was settled by early humans who arrived from eastern Indonesia. Their descendants became the Aboriginal population. Most of the current inhabitants of Indonesia, the Philippines, and parts of Southeast Asia are descendants of a population that emigrated from Taiwan about six thousand years ago.

The Americas were settled by emigrants from Asia who crossed what was then the Bering land bridge to Alaska, probably about 15 thousand years ago. It is believed that bands of hunters passed through today's Canada into the area of the present United States. Their descendants populated the entire hemisphere, becoming the ancestors of North and South American Indians.

Europe, after its long history of pre-Neanderthals, Neanderthals, and Cro-Magnons, has been reconstituted twice fairly recently: about eight thousand years ago by farmers from the Near East, and about two thousand years later by Indo-Europeans from southern Russia.

In 1776, the German anthropologist Johann Friedrich Blumenbach (1753–1840) divided groups of humans into "races": Caucasians, Mongolians, Malayans, Ethiopians, and (native) Americans (Asimov, 1989).

Thus, the human stock with its many current branches appears African in origin, and about a quarter-million years old. Today, the number of people is growing fast; "population explosions" are occurring in some parts of the earth. The total number of humans was about 10^9 (one thousand million, i.e., one billion) around 1800. The second billion was reached by 1930. The third billion was present in 1960, the fifth in 1987. Just before the year 2000, the earth's population is likely to reach six billion. If current birth and death rates continue, 80 billion people will live on earth in 2100, and 150 billion crowd it in 2125.

In the late 1980s, about 90 million people were added each year to the earth's population. The current projection is an increase of approximately one billion people every ten years. Most will be born in third-world countries where food supplies are insufficient even now; of the five billion people on earth in 1987 about 1.2 billion lived in industrialized countries.

Emigration from certain areas and immigration to others are on a much smaller scale than population growth but can be locally of great importance. In North America, for example, during the last few centuries, waves of immigrants from certain geographical areas have been changing the composition of the inhabitant population, replacing most native Indians by Europeans. In today's United States, the influx of Cubans and Haitians is strongly felt in Florida, the arrival of South Americans affects southwestern states, and Asians are very evident along the Pacific coast.

Marco Polo (1254–1324) traveled to the Far East and stayed in China for 20 years. There he found a nation far in advance of Europe in population, wealth, technology, and the civilized amenities. He returned in 1292 to Italy, where he was taken prisoner in the war between Venice and Genoa. While detained, he began to dictate his reminiscences of China, published in 1298. Although largely disbelieved, his book was immensely popular, and stimulated much interest in the study of other countries (Asimov, 1989).

ANTHROPOLOGY AND ANTHROPOMETRY

Anthropology, the study of mankind, was primarily philosophical and esthetical in nature until about the middle of the nineteenth century. Yet, the size and proportions of the human body have always been of interest to artists, warriors, and physicians. Physical anthropology is that scientific subgroup in which the body, particularly bones, is measured and compared. In the middle of the nineteenth century, the Belgian statistician Adolphe Quételet first applied statistics to anthropological data. This was the beginning of modern *anthropometry*, the measurement of the human body. By the end of the nineteenth century, anthropometry was a widely applied scientific discipline, used both in measuring the bones of early people and in assessing the body sizes and proportions of contemporaries. A new offspring, *biomechanics,* had already developed. Engineers have become highly interested in the application of anthropometric and biomechanical information.

In 1316, Mondino D. Luzzi, professor at the medical school of Bologna, Italy, published the first book devoted entirely to anatomy. The Flemish anatomist Andreas Vesalius, in the early sixteenth century, upset many traditional but false Greek and Egyptian notions about anatomy of the human body in his book De Corporis Humani Fabrica, *concerning the structure of the human body. His book contained careful illustrations of anatomical facts, drawn by a student of Titian (Asimov, 1989).*

Standardization of measuring methods became necessary, achieved primarily by conventions of anthropologists in Monaco, 1906, and in Geneva, 1912. Bony landmarks were established on the body, to and from which measurements were taken. In 1914 an authoritative textbook was published, Martin's *Lehrbuch der Anthropologie*, editions of which shaped the discipline for several decades. Beginning in the 1960s, new engineering needs for anthropometric information, newly developing measuring techniques, and advanced statistical considerations stimulated the need for updated standardization. In the 1980s, the International Standardization Organization (ISO) began efforts to normalize anthropometric measures and measuring techniques worldwide.

Measurement Techniques

Body measurements are usually defined by the two endpoints of the distance measured. For example, forearm length is often measured as elbow-to-fingertip distance; stature (height) starts at the floor on which the subject stands, and extends to the highest point on the skull.

EXAMPLE

There are four customary positions of the subject for measurement of stature: (1) standing naturally upright; (2) standing stretched to maximum height; (3) leaning against a wall with the back flattened and buttocks, shoulders, and back of the head touching the wall; or (4) lying on one's back. The difference between measures when the standing subject either stretches or just stands upright can easily be 2 cm or more. Lying supine results in the tallest measure. This example shows that standardization is needed to assure uniform postures and comparable results.

Specific terminology and measuring conventions have been described by Garrett and Kennedy (1971); Gordon, Churchill, Clauser, Bradtmiller, McConville, Tebbetts, and Walker, (1989); Hertzberg (1968); Kroemer, Kroemer, and Kroemer-Elbert (1990); Lohman, Roche, and Martorel (1988); NASA/Webb (1978); Pheasant (1986); Roebuck (1993); and Roebuck, Kroemer, and Thomson (1975). These publications provide exhaustive information about traditional measurement procedure and techniques.

APPLICATION

The following terms are used in classical anthropometry:

Height—is a straight-line, point-to-point vertical measurement.

Breadth—is a straight-line, point-to-point horizontal measurement running across the body or a segment.

Depth—is a straight-line, point-to-point horizontal measurement running fore-aft the body.

Distance—is a straight-line, point-to-point measurement between landmarks on the body.

Curvature—is a point-to-point measurement following a contour; this measurement is usually neither closed nor circular.

Circumference—is a closed measurement that follows a body contour; hence this measurement is not circular.

Reach—is a point-to-point measurement following the long axis of the arm or leg.

For most measurements, the subject's body is placed in a defined upright straight posture, with body segments either in line with each other or at 90 degrees. For example, the subject may be required to "stand erect; heels together; buttocks, shoulder blades and back of head touching a wall; arms and fingers straight and vertical." This is similar to the so-called "anatomical position." The head is often positioned in the Frankfurt Plane: the pupils are on the same horizontal level; the right ear hole (tragus) and the lowest point of the socket (orbit) of the right eye are also on a horizontal plane. When measurements are taken on a seated subject, the flat and horizontal surfaces of feet and foot support are so arranged that the thighs are horizontal, the lower legs vertical, and the feet flat on their horizontal support. The subject is nude, or nearly so, and does not wear shoes. The standard reference planes are the medial (midsagittal), the frontal (or coronal), and the transverse planes, usually thought to meet in the center of mass of the whole body.

Figure 1-1 shows reference planes and descriptive terms. Figures 1-2 and 1-3 illustrate anatomical landmarks on the human body.

Classical measuring techniques. The conventional measurement devices are quite simple. In the Morant technique, one uses a set of *grids,* best attached to the inside corner of two vertical walls meeting at right angles. The subject is placed in front of the grids, and projections of bony landmarks onto the grids are used to determine anthropometric values. Other boxlike jigs with grids are used to provide references for the measurement of head and foot dimensions.

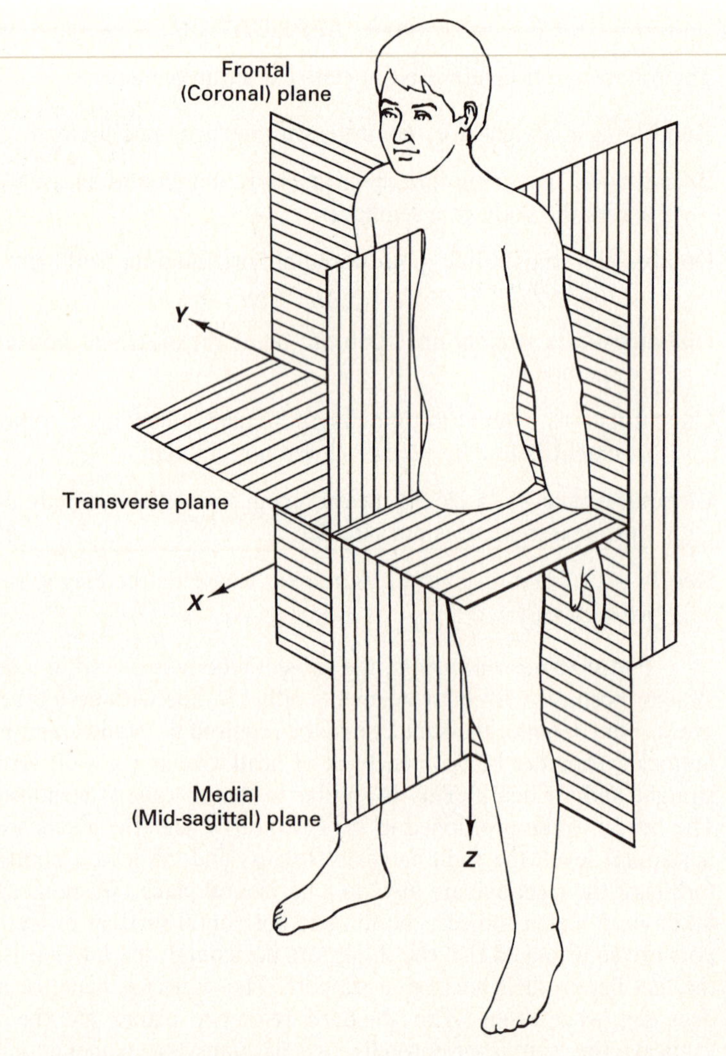

Figure 1-1. Terms and measuring planes used in anthropometry.

Many bony landmarks, however, cannot be projected easily onto grids. In this case, special instruments are used. The most important is the *anthropometer*, a graduated rod with a sliding edge at right angle. The rod can be disassembled for transport and storage, but put together is 2 meters long. (Anthropometric data are traditionally recorded in metric units.) The *spreading caliper* consists of two curved branches joined in a hinge. The distance between the tips of the branches is read from a scale. A small *sliding caliper* can be used for short measurements, such as finger thickness or finger length. A special caliper is used to measure the thickness of skinfolds. A cone is em-

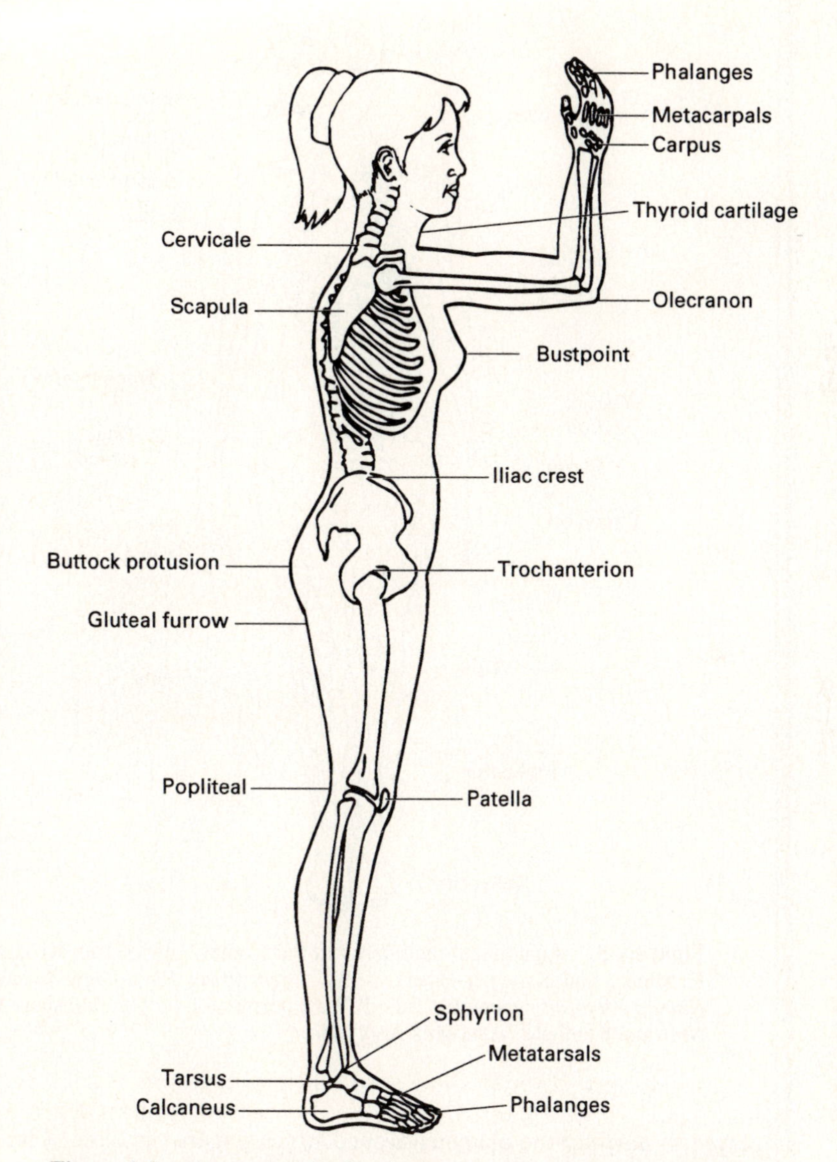

Figure 1-2. Anatomical landmarks in the sagittal view. (From Kroemer, Kroemer, and Kroemer-Elbert, 1990, *Engineering Physiology: Bases of Human Factors/Ergonomics,* 2d ed. With permission by the publisher, Van Nostrand Reinhold. All rights reserved.)

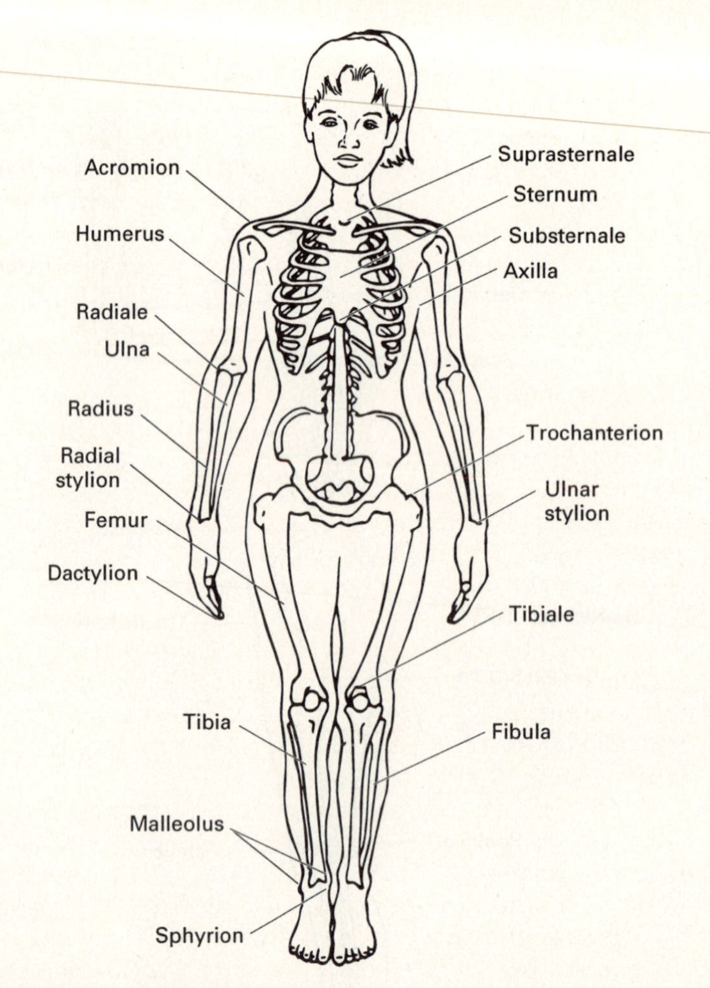

Figure 1-3. Anatomical landmarks in the frontal view. (From Kroemer, Kroemer, and Kroemer-Elbert, 1990, *Engineering Physiology: Bases of Human Factors/Ergonomics,* 2d ed. With permission by the publisher, Van Nostrand Reinhold. All rights reserved.)

ployed to measure the diameter around which fingers can close. Circular holes of increasing sizes drilled in a thin plate serve to measure external finger diameter. Circumferences and curvatures are measured with tapes. A scale is used to measure the weight of the body (Kroemer, Kroemer, and Kroemer-Elbert, 1990). Many other measuring methods can be applied in special cases, such as the shadow technique, use of templates, or casting; they are explained by Gordon, Churchill, Clauser et al. (1989) and by Roebuck, Kroemer, and Thomson (1975).

Most traditional measurement instruments are applied by the hand of the measurer to the body of the subject. This is simple but time consuming. Also, it requires that each measurement and tool be selected in advance, and what was not measured in the test session remains unknown.

A major shortcoming of the classical techniques is that they leave many of the body dimensions unrelated to each other in space. For example, as one looks at a subject from the side, stature, eye height, and shoulder height are located in different undefined frontal planes. Another shortcoming is that contact measurements cannot be made on certain parts of the body, such as the eyes, that are very sensitive.

New measurement techniques. Photographs can record all three-dimensional aspects of the human body. They allow the recording of practically infinite numbers of measurements, which can be taken from the record at one's convenience. Photographs also have drawbacks, however: the equipment (particularly for data analyses) is expensive; the body is depicted in two dimensions; a scale may be difficult to establish; parallax distortions occur; and bony landmarks under the skin cannot be palpated on the photograph.

For these and other reasons photographic anthropometry has not been widely used, in spite of many recent technical improvements, such as stereophotometry with several cameras or mirrors, holography, and the use of film and videotape instead of still photography. These methods can extract a large number of datum points from the recorded picture of the body (Coblentz, Mollard, and Ignazi, 1991).

Many techniques for acquiring three-dimensional anthropometric data have been proposed in the past, as described in some detail by Roebuck, Kroemer, and Thomson (1975) and NASA/Webb (1978). Some techniques rely on projecting a regular geometric grid onto the irregularly shaped human body. The projected grid remains regular when viewed along its axis of projection, but appears distorted if viewed at an angle. The displacements of projected grid points from their regular positions can be used to determine the shape of the surface.

The laser can be used as a distance measuring device to determine the shape of irregular bodies (Coblentz, Mollard, and Ignazi, 1991; Zehner, 1986). The body to be measured is rotated, or the sending and receiving units of the laser device rotate around the body.

With the increased use of computer models of the human body, requirements for anthropometric data have become much more complex now than they were just a few decades ago. Most such models represent the long bones of the human body as links, connected in simplified joints, and powered by muscles spanning one or two joints. Computer programs can integrate the recorded data to generate models of the human body, from which desired dimensions and contour identifiers can be extracted (see, e.g., Baughman, 1982; Drerup and Hierholzer, 1985; and Herron, 1973).

The "stick-person" is the seventeenth-century concept of Giovanni Alfonso Borelli, taken up two centuries later by Weber and Weber in their discussion of the mechanics of the legs, by Harless (1860) and von Meyor (1863) in their considerations of body mass properties, and by Braune and Fischer (1889) in their analysis of the biomechanics of a gun-firing infantryman. In 1873 von Meyer modeled body segments as ellipsoids and spheres. This biomechanical model was refined and expanded by Dempster in the 1950s. The Simons and Gardner model of 1960 still depicted body segments as uniform geometric shapes: cylinders for the appendages, neck, and torso, and a sphere for the head. Using equations developed by Barter in 1957, inertial parameters were computed for the geometric forms and the moment for the total body of inertia. This elementary work still is the basis for much of the present biodynamic modeling.

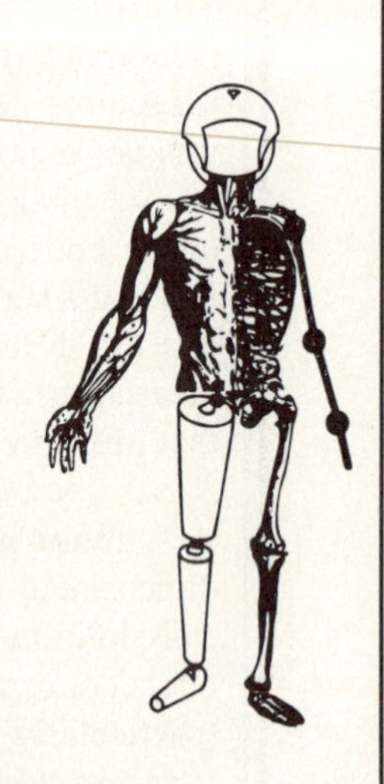

Mathematical-statistical techniques are available to locate specific landmarks and describe three-dimensional surfaces in terms of "facet algorithms," which provide a complete topographic description. The facet approach defines flats, peaks, pits, ridges, ravines, saddles, and hillsides on the human body in ways similar to those used in describing earth contours (Watson, Laffey, and Haralick, 1985).

NEED

It is desirable to replace the traditional heights, lengths, breadths, etc. by three-dimensional coordinates that provide exact point locations from a common origin. This approach should describe static body as well as a moving one. Before such a "system anthropometry" (a term coined by Reynolds — see Kroemer, Snook, Meadows and Deutsch, 1988) can be developed and an effective anthropometric database created, fundamental interrelated problems need solutions. We need an effective data-collection system which supplies accurate and reproducible results, allows immediate digital outputs as well as rapid data transfer to storage, and is fairly inexpensive and easy to use.

Available Anthropometric Information

In the past, interest in the body build of populations other than one's own group was based mostly on curiosity and general "wish to know." More recently, as industry and marketing reach around the globe, body size has become a matter of practical interest to designers and engineers. In the early 1970s, a conference on "ethnic variables in human-factors engineering" first attempted a compilation of world-wide ergonomic information (Chapanis, 1975). A thorough collection of data, available in the mid-1970s, was published in the NASA/Webb (1978) anthropometric sourcebook. Since then, an increasing number of publications describing national populations has appeared in the literature. For example, body sizes of southeast Asians are becoming well known, reflecting both scientific interest and economic concern. Wang, Whin, and Shi (1990) and Li, Hwang, and Wang (1990) demonstrated how proper use of modern technology and statistics allows anthropometric surveys (in this case, of Taiwan) to be performed rapidly and exactly.

Juergens, Aune, and Pieper (1990) attempted to classify the total population of the earth into 20 area groups and to estimate 19 of their main anthropometric dimensions. Because of many voids, much of the data had to be "guesstimated" and certain subgroups (e.g., pygmies) are not represented in this global survey. An excerpt from their global estimates is given in Table 1-1.

TABLE 1-1. AVERAGE ANTHROPOMETRIC DATA (IN CM) ESTIMATED FOR 20 REGIONS OF THE EARTH

	Stature		Sitting height		Knee height, sitting	
	Females	Males	Females	Males	Females	Males
NORTH AMERICA	165.0	179.0	88.0	93.0	50.0	55.0
LATIN AMERICA						
Indian population	148.0	162.0	80.0	85.0	44.5	49.5
European and Negroid population	162.0	175.0	86.0	93.0	48.0	54.0
EUROPE						
Northern	169.0	181.0	90.0	95.0	50.0	55.0
Central	166.0	177.0	88.0	94.0	50.0	55.0
Eastern	163.0	175.0	87.0	91.0	51.0	55.0
Southeastern	162.0	173.0	86.0	90.0	46.0	53.5
France	163.0	177.0	86.0	93.0	49.0	54.0
Iberia	160.0	171.0	85.0	89.0	48.0	52.0
AFRICA						
North	161.0	169.0	84.0	87.0	50.5	53.5
West	153.0	167.0	79.0	82.0	48.0	53.0
Southeast	157.0	168.0	82.0	86.0	49.5	54.0
NEAR EAST	161.0	171.0	85.0	89.0	49.0	52.0
INDIA						
North	154.0	167.0	82.0	87.0	49.0	53.0
South	150.0	162.0	80.0	82.0	47.0	51.0
ASIA						
North	159.0	169.0	85.0	90.0	47.5	51.5
Southeast	153.0	163.0	80.0	84.0	46.0	49.5
SOUTH CHINA	152.0	166.0	79.0	84.0	46.0	50.5
JAPAN	159.0	172.0	86.0	92.0	39.5	51.5
AUSTRALIA (European population)	167.0	177.0	88.0	93.0	52.5	57.0

SOURCE: Juergens, Aune, and Pieper, 1990.

TABLE 1-2. RECENT ANTHROPOMETRIC DATA ON INTERNATIONAL POPULATION SAMPLES: AVERAGE AND STANDARD DEVIATION (ALL IN CM BUT WEIGHT IN KG)

	Sample size N	Stature	Sitting height	Knee height, sitting	Weight
Algerian females (Mebarki and Davies, 1990)	666	157.6 (5.56)	79.5 (5.01)	48.7 (3.61)	61.3 (12.9)
Brazilian males (Ferreira, 1988; cited by Al-Haboubi, 1991)	3076	169.9 (6.7)	—	—	—
Chinese females (Singapore) (Ong, Koh, Phoon, and Low, 1988)	46	159.8 (5.8)	85.5 (3.1)	—	—
Cantonese males (Evans, 1990)	41	172.0 (6.3)	—	—	60.0 (6.2)
Egyptian females (Moustafar, Davies, Darwich, and Ibraheem, 1987)	4960	160.6 (7.18)	83.8 (4.30)	49.9 (2.51)	62.6 (4.37)
Indian males (farmers) (Nag, Sebastian, and Mavlankar, 1980)	13	157.6 (1.7)	—	—	44.6 (1.4)
Indonesian females	468	151.6 (5.4)	71.9 (3.4)	—	—
Indonesian males (Sama'mur, 1985; cited by Intaranont, 1991)	949	161.3 (5.6)	87.2 (3.7)	—	—
Irish males (Gallwey and Fitzgibbon, 1991)	164	173.1 (5.83)	91.1 (3.03)	50.8 (2.77)	73.9 (8.66)
Italian females	753	161.0 (6.4)	85.0 (3.4)	49.5 (3.0)	58.0 (8.3)
Italian males (Coniglio, Fubini, Masali, Masiero, Pierlorenzi and Sagone, 1991)	913	173.3 (7.1)	89.6 (3.6)	54.1 (3.0)	75.0 (9.6)
Jamaican females	30	174.9	85.6	—	67.6
Jamaican males (Camey, Aghazadeh, and Nye, 1991)	123	164.8	83.2	—	61.4

Malay females (Ong, Koh, Phoon, and Low, 1988)	32	155.9 (6.6)	83.1 (3.9)	—	—
Saudi-Arabian males (Dairi, 1986; cited by Al-Haboubi, 1991)	1440	167.5 (6.1)	—	—	—
Sri Lankan females	287	152.3 (5.9)	77.4 (2.2)	—	—
Sri Lankan males (Abeysekera, 1985; cited by Intaranont, 1991)	435	163.9 (6.3)	83.3 (2.7)	—	—
Sudanese males					
Villagers	37*	168.7 (6.3)	—	—	57.1 (7.6)
City dwellers	16*	170.4 (7.2)	—	—	62.3 (13.1)
	48**	166.8	—	—	51.3
Soldiers	21*	173.5 (7.1)	—	—	71.1 (8.4)
	104**	172.8	—	—	60.0
*(Elkarim, Sukkar, Collins, and Dore, 1981) **(Ballal et al., 1982; cited by Intaranont, 1991)					
Thai females	250*	151.2 (4.8)	—	—	—
	711**	154.0 (5.0)	81.7 (2.7)	—	—
Thai males*	250*	160.7 (2.0)	—	—	—
	1478**	165.4 (5.9)	87.2 (3.2)	—	—
*(Intaranont, 1991) **(NICE; cited by Intaranont, 1991)					
Turkish females					
Villagers	47	156.6 (5.2)	79.2 (3.8)	48.6 (2.7)	69.1 (13.8)
City dwellers (Goenen, Kalinkara, and Oezgen, 1991)	53	156.3 (5.5)	78.6 (0.5)	47.1 (0.5)	65.9 (13.0)
Turkish males (soldiers) (Kayis and Oezok, 1991)	5108	170.2 (6.0)	88.8 (3.4)	51.3 (2.8)	63.3 (7.3)

More exact data on specific population samples, taken in recent years, are compiled in Table 1-2.

Variability. Anthropometric data show considerable variability stemming from four sources:

Measurement Variability. Differing care is exercised in selecting population samples, using measurement instruments, storing the measured data, and applying statistical treatments which may yield quite variable information.

Intraindividual Variability. The size of the same body segment of a given person changes from youth to age, depending also on nutrition, physical exercise, and health. Such changes become apparent in "longitudinal" studies, in which an individual is observed over years and decades. Most (but not all) such changes with age follow the scheme shown in Figure 1-4. During childhood and adolescence, body dimensions such as stature change rapidly. From the early 20s into the 50s, little change occurs in general, with stature remaining almost steady. From the 60s on, many dimensions decline, while others—for example, weight or bone circumference—often increase.

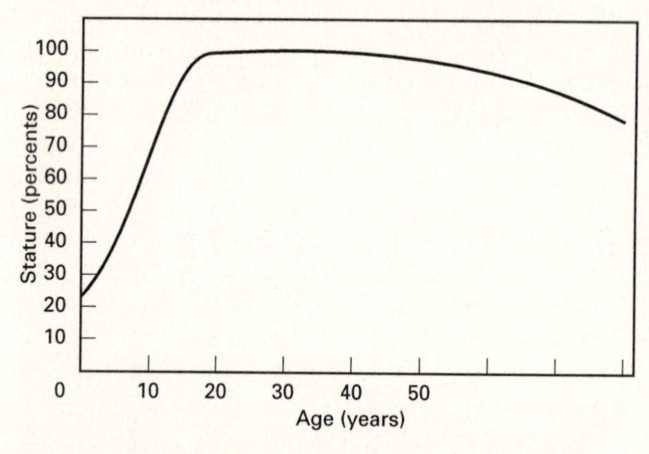

Figure 1-4. Approximate changes in stature with age.

Interindividual Variability. Individuals differ from each other in arm length, weight, height, and other measurements. Data describing a population sample are usually collected in a "cross-sectional" study, in which every subject is measured at the same moment in time. This means that people of different ages, nutrition, fitness, and so on, are included in the sample set. The anthropometric data found in most textbooks, including this one, are gathered in cross-sectional studies.

Secular Variations. There is some factual and much anecdotal evidence that people nowadays are larger, on average, than their ancestors. Reliable anthropometric information on this development is available only for about the last hundred years. During the last five decades stature has increased in North America and in Europe by about 1 cm per decade, on the average, while body weight has increased about 2 kg per decade. The reason is probably that improved nutrition and hygiene have allowed persons to achieve more of their genetically determined body-size potential. (If this explanation is correct, then the rate of increase should slowly taper off until a final body size is reached.) Data from Japan indicated initially a much faster average growth than found among Caucasians, but the rate seems to be slowing down now (Roebuck, Smith, and Raggio, 1988).

The most reliable data for observation of such secular trends are from military surveys. Measurements of U.S. soldiers have been minutely recorded since the Civil War. The military, however, is a selected sample of the general population, excluding, for example, people older than about 50 years, and people who are "unusual" in their body dimensions, such as extremely short or tall; also, only fairly healthy persons are included.

The anthropometric secular trends in 22 body dimensions of white, black, Hispanic, and Asian, female and male U.S. Army soldiers were investigated in 1990 by Greiner and Gordon. They found that some dimensions change only minimally, while others show fairly clear trends. The increases in stature and in sitting height seem to be slowing down: it now takes about 20 years before the gains are measurable with current techniques, by which time they are approximately another centimeter. Leg length (measured as crotch height), in contrast, is not changing appreciably. Yet, body weight is still increasing fairly fast, by 2 kg to 3 kg per decade. Shoulder breadth and chest circumference are increasing at rates of about 1 cm or more per decade. Altogether, white, black, and Hispanic U.S. Army soldiers show similar changes, while U.S. soldiers of Asian extraction exhibit quite different trends, which can be explained by recent immigration of Asians to the United States.

Population samples. Body dimensions of soldiers have long been of interest for a variety of reasons, among them the necessity to provide uniforms, armor, and equipment. Armies have medical personnel willing and capable to perform body measurements on large samples, available on command. Hence, anthropometric information about soldiers has a long history and is rather complete. For example, the anthropometric data bank of the U.S. Air Force Aerospace Medical Research Laboratory (CSERIAC) contains the data of approximately 100 surveys from many nations, though most on U.S. military personnel. Similarly, the "Human Biometry Data Bank Ergodata" at the University René Descartes in Paris contains European anthropometric information.

Soldiers are certainly a subsample of the general population, but they are a biased sample because they are youngish, healthy, and neither extremely small nor big. Thus, their body dimensions may not truly represent the adult civilian population (although it appears that there are no major differences in head, hand, and foot sizes). This problem was investigated by McConville, Robinette, and Churchill (1981). They selected several surveys done in the United States: for males, the 1965 HES (Health Examination Survey), the 1967 U.S. Air Force and the 1966 U.S. Army surveys; for females, the 1968 U.S. Air Force and 1977 U.S. Army surveys. Their underlying assumption was that if good pairing can be achieved in two "core" dimensions between civilian and military individuals, then the means and standard deviations of other dimensions should be well matched also. (This is a reasonable but arguable assumption.)

The procedure used was to match individuals from the civilian and military surveys on the basis of stature and weight, in intervals of plus-minus one inch and plus-minus five pounds. Thus, a new military sample was created which represented the civilians in height and weight. From this new matched military sample, dimensions other than height and weight were selected and compared to the equivalent data measured in the civilian surveys.

For the males (but not the females), an excellent fit was achieved: 99 percent of all civilian subjects could be matched with at least one military subject. The mean differences in stature and weight with regard to mean and standard deviation were nearly negligible. A comparison of six linear dimensions measured both in the military and civilian surveys provided similar good matches in means and standard deviations.

U.S. Civilians' Body Sizes

In earlier publications (e.g., Kroemer, 1981; Kroemer, Kroemer, and Kroemer-Elbert, 1986, 1990) the authors relied on *estimates* for the body dimensions of U.S. civilians based on data measured in the 1960s and 70s. In 1988, a thorough anthropometric survey of U.S. Army personnel was conducted (Gordon, Churchill, Clauser, Bradtmiller, McConville, Tebbetts, and Walker, 1989). In it, 2,208 female and 1,774 male soldiers were measured, who were subsets of soldiers sampled to match the proportions of age categories and racial/ethnic groups found in the active-duty army of June 1988. *Their anthropometric data are used in this book to represent the U.S. adult population.*

The decision to use these data was based on the following reasoning: among the U.S. military services, the Army is the largest and anthropometrically least biased sample of the total U.S. adult population. The measured sample in the 1988 survey is a mix of older and younger subjects: among the men, 30 percent were aged 31 and over, 25 percent between 25 and 30, and the others younger. Sixty-six percent were white, 26 percent black, 4 percent Hispanic, and the remaining 4 percent other racial/ethnic groups. Among the women, 22 percent were aged 31 and over, 32 percent between 25 and 30 years, and the others younger. Nearly 52 percent of the subjects were white, 42 black, 3 percent Hispanic, and 4 percent other racial/ethnic groups. Altogether, this survey is a reasonably good mix of ages and persons of various origins. Thus, the 1988 Army survey provides better information about the anthropometry of the civilian U.S. adult population than decade-old estimates.

We also compared specifically estimated civilian data, used previously, with those measured on the U.S. Army personnel. The largest absolute longitudinal differences were found in stature and sitting height, with the 1988 military population up to 2 cm taller. This appears to reflect the gains expected from the "secular" increase (discussed earlier) of about 1 cm per decade, but the majority of the data, particularly in widths and breadths, was similar in the two data sets. However, extremely heavy persons are more likely in the civilian population than in the army sample.

APPLICATION

Altogether, it is plausible that the 1988 U.S. Army survey constitutes, at present, the best estimate for the U.S. adult population. The 1988 data set contains 180 measurements (including 48 head and face dimensions) and 60 derived dimensions calculated from the measured data. The data are correlated in various ways. Thus, for information on data not reported here, the publication by Gordon, Churchill, Clauser et al. (1989) and the associated reports, listed there, should be consulted. Table 1-3 provides an excerpt of anthropometric data which should describe the adult U.S. civilian population well enough until better information becomes available.

NEED

Reliable information on body sizes of nonmilitary populations is missing—nationally and worldwide.

For standardization purposes, anthropometric measurements are done on persons standing or sitting erect with body joints at 0, 90, or 180 degrees—body postures not usually maintained at work. For the design of workstations and equipment, "functional" data are needed. Such data are often reported and used in engineering design guidelines (see, e.g., the sections on body posture, controls, and office design in this book), but they are dependent on stated or implied assumptions. A typical example is that of reach contours—see Figure 1-5.

Anthropometric Statistics

Fortunately, anthropometric data are usually distributed in a reasonably normal (or Gaussian) distribution (with the occasional exception of muscle-strength data). Hence, regular parametric statistics apply in most cases. The distribution of anthropometric information is, for practical purposes, well described by the *average* (mean), *standard deviation* SD, and *sample size* N. The *range* indicates extreme smallest to largest values.

TABLE 1-3. BODY DIMENSIONS OF U.S. CIVILIAN ADULTS, FEMALE/MALE, IN CM

	Percentiles			
	5th	50th	95th	SD
HEIGHTS				
(*f* above floor, *s* above seat)				
Stature ("height")^f	152.78 / 164.69	162.94 / 175.58	173.73 / 186.65	6.36 / 6.68
Eye height^f	141.52 / 152.82	151.61 / 163.39	162.13 / 174.29	6.25 / 6.57
Shoulder (acromial) height^f	124.09 / 134.16	133.36 / 144.25	143.20 / 154.56	5.79 / 6.20
Elbow height^f	92.63 / 99.52	99.79 / 107.25	107.40 / 115.28	4.48 / 4.81
Wrist height^f	72.79 / 77.79	79.03 / 84.65	85.51 / 91.52	3.86 / 4.15
Crotch height^f	70.02 / 76.44	77.14 / 83.72	84.58 / 91.64	4.41 / 4.62
Height (sitting)^s	79.53 / 85.45	85.20 / 91.39	91.02 / 97.19	3.49 / 3.56
Eye height (sitting)^s	68.46 / 73.50	73.87 / 79.20	79.43 / 84.80	3.32 / 3.42
Shoulder (acromial) height (sitting)^s	50.91 / 54.85	55.55 / 59.78	60.36 / 64.63	2.86 / 2.96
Elbow height (sitting)^s	17.57 / 18.41	22.05 / 23.06	26.44 / 27.37	2.68 / 2.72
Thigh height (sitting)^s	14.04 / 14.86	15.89 / 16.82	18.02 / 18.99	1.21 / 1.26
Knee height (sitting)^f	47.40 / 51.44	51.54 / 55.88	56.02 / 60.57	2.63 / 2.79
Popliteal height (sitting)^f	35.13 / 39.46	38.94 / 43.41	42.94 / 47.63	2.37 / 2.49
DEPTHS				
Forward (thumbtip) reach	67.67 / 73.92	73.46 / 80.08	79.67 / 86.70	3.64 / 3.92
Buttock-knee distance (sitting)	54.21 / 56.90	58.89 / 61.64	63.98 / 66.74	2.96 / 2.99
Buttock-popliteal distance (sitting)	44.00 / 45.81	48.17 / 50.04	52.77 / 54.55	2.66 / 2.66
Elbow-fingertip distance	40.62 / 44.79	44.29 / 48.40	48.25 / 52.42	2.34 / 2.33
Chest depth	20.86 / 20.96	23.94 / 24.32	27.78 / 28.04	2.11 / 2.15
BREADTHS				
Forearm-forearm breadth	41.47 / 47.74	46.85 / 54.61	52.84 / 62.06	3.47 / 4.36
Hip breadth (sitting)	34.25 / 32.87	38.45 / 36.68	43.22 / 41.16	2.72 / 2.52
HEAD DIMENSIONS				
Head circumference	52.25 / 54.27	54.62 / 56.77	57.05 / 59.35	1.46 / 1.54
Head breadth	13.66 / 14.31	14.44 / 15.17	15.27 / 16.08	0.49 / 0.54
Interpupillary breadth	5.66 / 5.88	6.23 / 6.47	6.85 / 7.10	0.36 / 0.37
FOOT DIMENSIONS				
Foot length	22.44 / 24.88	24.44 / 26.97	26.46 / 29.20	1.22 / 1.31
Foot breadth	8.16 / 9.23	8.97 / 10.06	9.78 / 10.95	0.49 / 0.53
Lateral malleolus height^f	5.23 / 5.84	6.06 / 6.71	6.97 / 7.64	0.53 / 0.55
HAND DIMENSIONS				
Circumference, metacarpale	17.25 / 19.85	18.62 / 21.38	20.03 / 23.03	0.85 / 0.97
Hand length	16.50 / 17.87	18.05 / 19.38	19.69 / 21.06	0.97 / 0.98
Hand breadth, metacarpale	7.34 / 8.36	7.94 / 9.04	8.56 / 9.76	0.38 / 0.42
Thumb breadth, interphalangeal	1.86 / 2.19	2.07 / 2.41	2.29 / 2.65	0.13 / 0.14
WEIGHT (in kg)	39.2* / 57.7*	62.01 / 78.49	84.8* / 99.3*	13.8* / 12.6*

*Estimated (from Kroemer, 1981).

Note: In this table, the entries in the 50th percentile column are actually "mean" (average) values. The 5th and 95th percentile values are from measured data, not calculated (except for weight). Thus, the values given may be slightly different from those obtained by subtracting 1.65 SD from the mean (50th percentile), or by adding 1.65 SD to it.

SOURCE: Adapted from U.S. Army data reported by Gordon, Churchill, Clauser, Bradtmiller, McConville, Tebbetts, and Walker (1989).

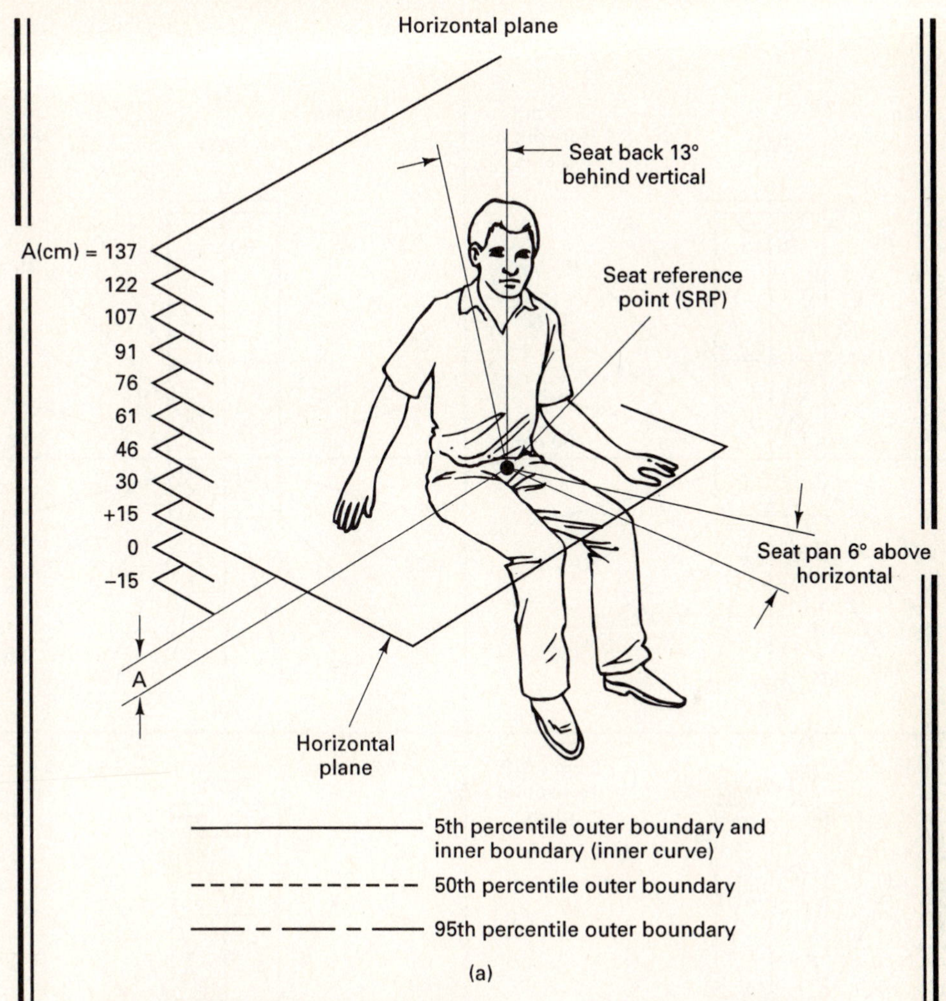

Figure 1-5(a). Definition of reach contours of U.S. Air Force males and females (adapted from NASA, 1989).

One easy way to check on data diversity is to divide the standard deviation of the data in question by their mean to get the *coefficient of variation,* CV. In most body dimensions, taken cross-sectionally, the CV is in the neighborhood of 5 percent; in most strength data, around 10 percent. Larger CV's are suspect and should prompt a thorough examination of the data.

Anthropometric data often are best presented in *percentiles*. They provide a convenient means of describing the range of body dimensions to be accommodated, making it easy to locate the percentile equivalent of a measured body dimension. Also, the use of percentiles avoids the misuse of the average in design (as discussed later).

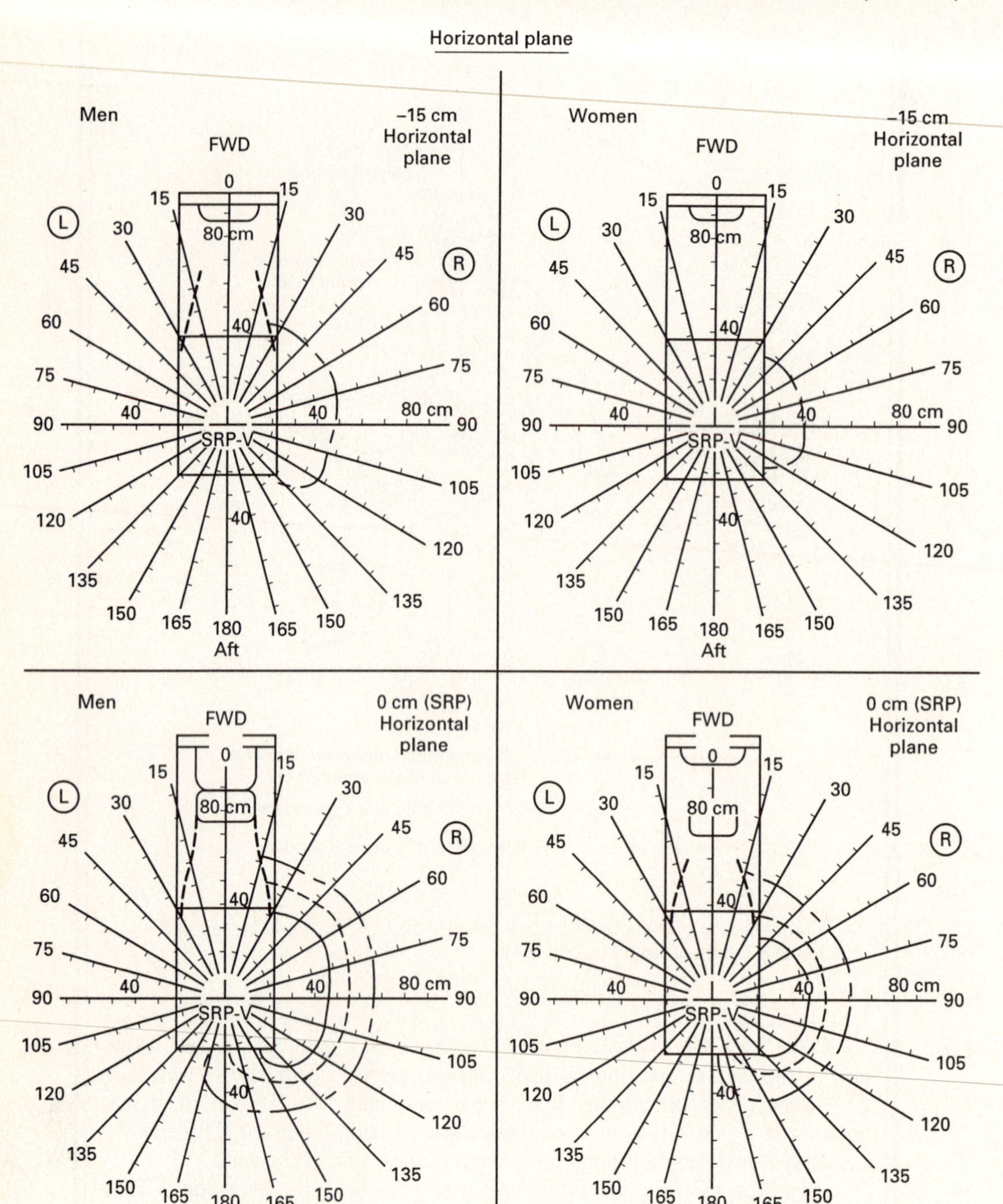

Figure 1-5(b). Reach contours in planes 15 cm below and at seat (SRP) height.

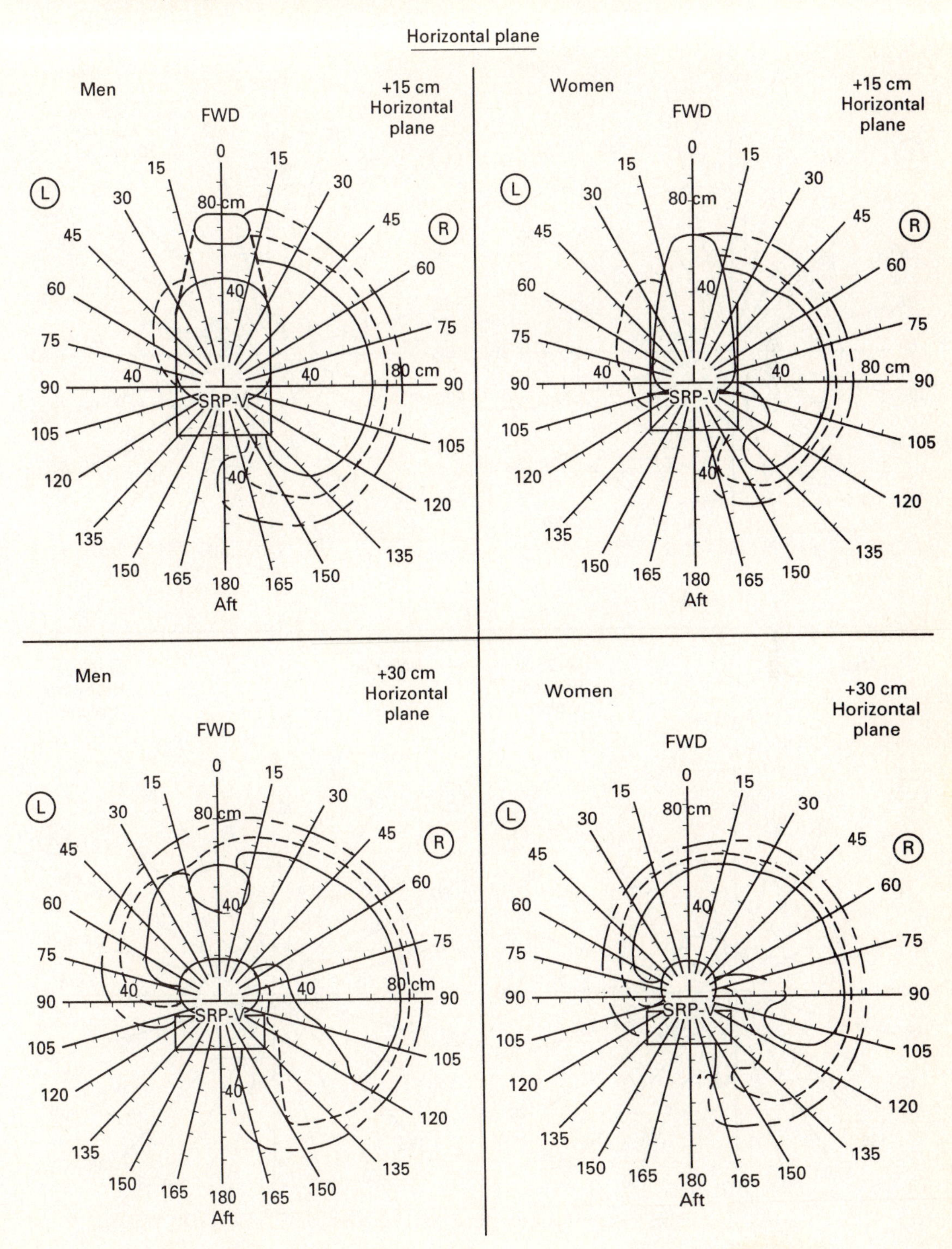

Figure 1-5(c). Reach contours in planes 15 and 30 cm above seat (SRP) height.

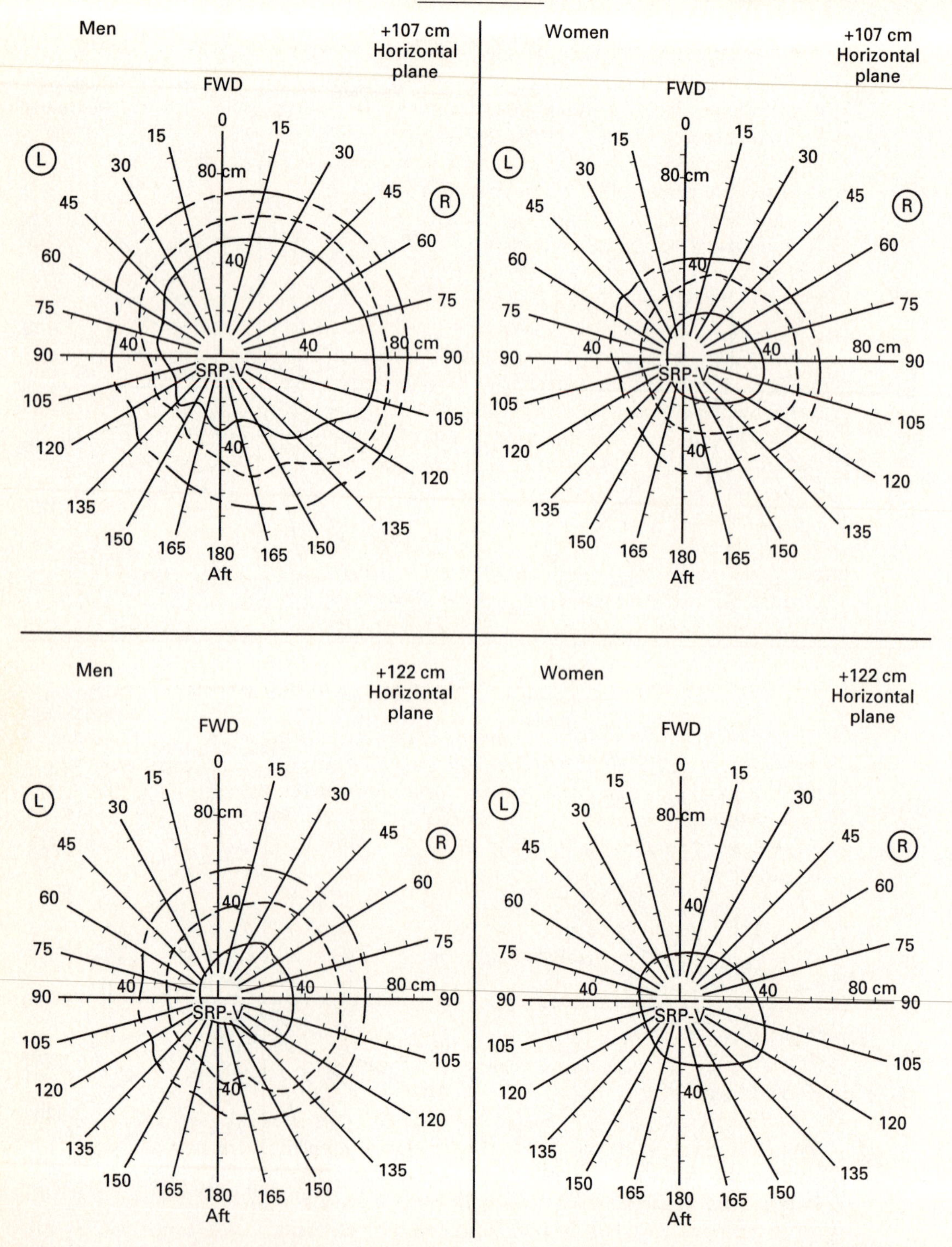

Figure 1-5(d). Reach contours in planes 107 and 122 cm above seat (SRP) height.

EXAMPLE

Percentiles serve the designer in several ways. First, they help to establish the portion of a user population that will be included in (or excluded from) a specific design solution. For example: a certain product may need to fit everybody who is taller than 5th percentile or smaller than 95th percentile in a specified dimension, such as grip size or arm reach. Thus, only the 5 percent having values smaller than 5th percentile, and the 5 percent having values larger than 5th percentile, will not be fitted. The central 90 percent of all users will be accommodated.

Second, percentiles are easily used to select subjects for fit tests. For example: if the product needs to be tested, persons having 5th or 95th percentile values in the critical dimensions can be employed for use tests.

Third, any body dimension, design value, or score of a subject can be exactly located. For example: a certain foot length can be described as a given percentile value of that dimension, or a certain seat height can be described as fitting a certain percentile value of popliteal height (a measure of lower leg length), or a test score can be described as being a certain percentile value.

Finally, the use of percentiles helps in the selection of persons to use a given product. For example: if a cockpit of an airplane is designed to fit 5th to 95th percentiles, one can select cockpit crews whose body measures are between the 5th and 95th percentile in the critical design dimensions.

For a normal distribution, percentiles are easily calculated from the mean and standard deviation. One need only multiply the standard deviation by a factor K, selected from Table 1-4, and then either deduct the result from the average to arrive at a certain percentile value below the 50th, or else add it to the average (which coincides with the 50th percentile) to arrive at a value above.

Body proportions. We often judge the human body by how body components "fit" together; our images of the beautiful body are affected by aesthetic codes, canons, and rules founded on often ancient (e.g., Egyptian, Greek, Roman) concepts of the human body. A more recent example is Leonardo da Vinci's drawing of the body within a frame of graduated circles and squares; it has been adopted, in simplified form, as the emblem of the U.S. Human Factors Society.

Categorization of body builds into different types is called *somatotyping*, from the Greek *soma* for body. Hippocrates developed, about 400 B.C., a scheme that included four body types, supposedly determined by their fluids. In 1921, the psychiatrist Ernst Kretschmer published a system of three body types intended to relate body build to personality traits ("Koerperbau and Charakter"). Kretschmer's typology consisted of the asthenic, pyknic, and ath-

TABLE 1-4. FACTOR K FOR COMPUTING PERCENTILES FROM MEAN $\bar{X}$ AND STANDARD DEVIATION S

	Percentile p associated with X	
K	$X = \bar{X} - KS$	$X = \bar{X} + KS$
2.576	00.5	99.5
2.326	1	99
2.06	2	98
1.96	2.5	97.5
1.88	3	97
1.65	5	95
1.28	10	90
1.04	15	85
1.00	16.5	83.5
0.84	20	80
0.67	25	75
0	50	50

Examples:

To determine 95th percentile, use $K = 1.65$.

To determine 20th percentile, use $K = 0.84$.

letic body builds. (The "athletic" type referred to character traits, not sports performance capabilities.) In the 1940s, the anthropologist W. H. Sheldon established a system of three body types, intended to describe (male) body proportions. Sheldon rated each person's appearance with respect to ecto-, endo-, and mesomorphic components—see Table 1-5. Sheldon's typology was originally based on intuitive assessment, not on actual body measurements; these were introduced into the system later by his disciples. Based on earlier research, Heath and Carter published in 1967 a revised procedure of somatotyping, which has been widely employed since.

Unfortunately, these and other attempts at somatotyping have not provided reliable predictors of human performance in technological systems. Hence, somatotyping is of little value for engineers or managers.

TABLE 1-5. BODY TYPOLOGIES

	Stocky, stout, soft, round	Strong, muscular, sturdy	Lean, slender, fragile
Kretschmer's terms	Pyknic	Athletic	Asthenic (leptosomic)
Sheldon's and Heath-Carter's terms	Endomorphic	Mesomorphic	Ectomorphic

❑

Body image. *Body image* is a person's mental picture of the physical appearance of his/her body. This mental image affects lifestyle behaviors. Often, one's body image does not agree with the anatomical appearance: this results in *phantom body size*. In general, men tend to underestimate their sizes while women tend to overestimate. Even following much reduction of body weight, some individuals perceive themselves as having lost almost no weight: they still overestimate their body size. Anorexic patients often do this. On the other hand, some obese individuals who have lost weight tend to underestimate their body size. The extent of body-image distortion an individual has may be reflected in the likelihood of weight-regain (recidivism) after an initially successful weight-loss program; that probability is 75 to 95 percent after a quick weight reduction.

Anthropometric surveys in the United States and Europe have shown that short people tend to overestimate their stature, while heavy people often underestimate their weight. If one applies appropriate *multipliers* to counteract these tendencies, simply asking people (instead of measuring them, which takes more effort) for their height and weight can lead to fairly reliable information.

"Desirable" body weight? People who are severely overweight have a higher risk of health problems and of early death than their slimmer contemporaries. The more overweight, the higher the risk (National Institutes of Health, 1985).

Adipose (fat-containing) tissue is a normal part of the human body. It stores fat energy for use under metabolic demands. *Obesity* is an excess of such tissue. The reasons for obesity may be both behavioral and genetic. They include too much caloric intake, too little physical activity, and metabolic and endocrine malfunctions.

However, determination of a "healthy" or normal body weight is a difficult enterprise. Since there are no given cut-off points, any quantitative definition of normality or obesity is arbitrary. In 1985, a specially called committee of experts agreed that 20% or more above "desirable" body weight should be called obese. For that definition it is necessary to establish a desirable reference weight. Several methods are in use in the United States: "relative weight" is the measured body weight divided by the midpoint of the recommended weight (for "medium frame") in the 1983 Metropolitan Life Insurance Tables (Metropolitan Life Foundation, 1983). The 1990 USDA Weight Table (U.S. Dept. of Agriculture, 1990) uses the "body mass index," calculated by dividing the body weight (in kilograms) by the square of body height (stature, in meters). Of course, all these measures are only approximate, because body composition varies among individuals of the same height and weight (Andres, 1985; National Institutes of Health, 1985); in the general U.S. population, body weight correlates with stature only moderately.

APPLICATION

Correlations. Some body dimensions are closely related with each other. For example, stature is very highly correlated with eye height, but not with head length, waist circumference, or weight. Table 1-6 shows selected

TABLE 1-6. SELECTED CORRELATION COEFFICIENTS FOR ANTHROPOMETRIC DATA ON U.S. AIR FORCE PERSONNEL: WOMEN ABOVE THE DIAGONAL, MEN BELOW

	1	2	3	4	5	6	7	8	9	10
1. Age		.223	.048	−.023	.039	−.055	.091	−.072	.233	.287
2. Weight	.113		.533	.457	.497	.431	.481	.370	.835	.799
3. Stature	−.028	.515		.927	.914	.849	.801	.728	.334	.257
4. Chest height	−.028	.483	.949		.897	.862	.673	.731	.271	.183
5. Waist height	−.033	.422	.923	.930		.909	.607	.762	.308	.238
6. Crotch height	−.093	.359	.856	.866	.905		.467	.788	.264	.190
7. Sitting height	−.054	.457	.786	.681	.580	.453		.398	.312	.239
8. Popliteal height	−.102	.299	.841	.843	.883	.880	.485		.230	.172
9. Shoulder circumference	.091	.831	.318	.300	.261	.212	.291	.182		.810
10. Chest circumference	.259	.832	.240	.245	.203	.147	.171	.114	.822	
11. Waist circumference	.262	.856	.224	.212	.142	.132	.167	.068	.720	.804
12. Buttock circumference	.105	.922	.362	.334	.278	.217	.347	.149	.744	.766
13. Biacromial breadth	.003	.452	.378	.335	.339	.282	.349	.316	.555	.401
14. Waist breadth	.214	.852	.287	.260	.215	.195	.216	.133	.715	.801
15. Hip breadth	.105	.809	.414	.380	.342	.283	.376	.221	.632	.647
16. Head circumference	.110	.412	.294	.251	.233	.188	.287	.194	.327	.340
17. Head length	.054	.261	.249	.218	.208	.170	.244	.175	.201	.196
18. Head breadth	.122	.305	.133	.097	.089	.066	.132	.075	.245	.271
19. Face length	.119	.228	.275	.220	.226	.199	.253	.193	.162	.172
20. Face breadth	.233	.453	.190	.160	.142	.099	.185	.098	.401	.421

	11	12	13	14	15	16	17	18	19	20
1. Age	.234	.219	.149	.146	.194	.095	.118	.190	.189	.089
2. Weight	.824	.886	.495	.768	.770	.403	.304	.290	.264	.358
3. Stature	.279	.360	.456	.329	.348	.331	.318	.136	.267	.199
4. Chest height	.216	.289	.412	.266	.276	.284	.284	.085	.222	.162
5. Waist height	.238	.336	.409	.293	.318	.306	.297	.123	.225	.200
6. Crotch height	.221	.246	.380	.277	.225	.294	.280	.089	.205	.172
7. Sitting height	.236	.383	.384	.277	.379	.294	.275	.136	.248	.146
8. Popliteal height	.186	.201	.327	.249	.181	.235	.253	.087	.185	.189
9. Shoulder circumference	.775	.717	.581	.719	.606	.330	.248	.252	.217	.313
10. Chest circumference	.796	.674	.370	.706	.551	.273	.204	.255	.176	.273
11. Waist circumference		.722	.382	.886	.600	.281	.149	.267	.174	.310
12. Buttock circumference	.852		.396	.668	.893	.310	.214	.238	.180	.269
13. Baicromial breadth	.288	.355		.401	.361	.311	.239	.178	.266	.211
14. Waist breadth	.936	.849	.327		.576	.292	.168	.263	.182	.296
15. Hip breadth	.724	.895	.340	.760		.265	.183	.188	.155	.215
16. Head circumference	.309	.330	.251	.310	.288		.692	.430	.273	.299
17. Head length	.158	.195	.179	.164	.166	.779		.115	.311	.113
18. Head breadth	.265	.252	.188	.268	.227	.521	.058		.174	.497
19. Face length	.129	.186	.187	.151	.161	.315	.289	.148		.144
20. Face breadth	.412	.394	.278	.410	.364	.464	.131	.660	.206	

SOURCE: From NASA/Webb, 1978.

correlation coefficients among body dimensions of U.S. Air Force personnel, male and female. (More detailed tables are contained in publications by NASA/Webb, 1978; Cheverud, Gordon, Walker, Jacquish, Kohn, Moore, and Yamashita, 1990; and Kroemer, Kroemer, and Kroemer-Elbert, 1990.)

Given the varying correlations among body measures, the attempt is futile to express all body dimensions as a portion of stature. For several years a scheme was used by designers which supposedly expressed body heights, body breadths, and segment lengths in terms of fixed percent of stature. For instance, hip breadth was said to be 19.1 percent of height—misleading nonsense, of course, because hip breadth varies widely among individuals and between males and females as groups, and furthermore nothing can be designed for a fixed "average" hip breadth.

☞☞☞ *Some people say they weigh too much for their height. Others say they are too short for their weight. But there is little correlation between stature and weight.* ☜☜☜

A useful phenomenon is the correlation of certain anthropometric data with each other in such a way that as one increases, another (or several others) increases as well, or conversely, that as one increases, others decrease. In statistics this is called *covariation*.

The simple correlation coefficient r (also called Pearson product-moment correlation) is a measure of the strength of the linear relationship between two variables.

The correlation coefficient between the variables x and y can be defined as

$$r_{x,y} = \frac{\text{COV}\,(x, y)}{\sqrt{V_x V_y}} = \frac{\text{COV}\,(x, y)}{S_x S_y}$$

where COV(x,y) is the covariance of x and y, and V their variance. The covariance measures the extent to which two variables vary in concert.

The covariance can be calculated from

$$\text{COV}(x, y) = \sum \frac{(x_i - \bar{x})(y_i - \bar{y})}{(N - 2)}$$

The *variance* V measures the extent of differences among individuals in x and y. It can be calculated from

$$V_x = \sum \frac{(x_i - \bar{x})^2}{N - 1} \quad \text{and} \quad V_y = \sum \frac{(y_i - \bar{y})^2}{N - 1}$$

or

$$V_x = \overline{(x^2)} - (\bar{x})^2 \quad \text{and} \quad V_y = \overline{(y^2)} - (\bar{y})^2$$

where $\bar{x}$ and $\bar{y}$ are the averages (means) of the values of x and y, respectively; and $\overline{(x^2)}$ is the average of the squared values of x, and $\overline{(y^2)}$ the average of the squared values of y.

A *bivariate* regression expresses the linear relationship between a dependent variable y and a single independent variable x according to the equation

$$y = a + bx$$

with a the *intercept* and b the *slope*. They can be calculated from

$$b = \frac{\mathrm{COV}\,(x, y)(N - 1)}{V_x(N - 2)} \qquad \text{and} \qquad a = \bar{y} - b\bar{x}$$

The *standard error of the estimate* for y, S_y, indicates the extent of variation in y for any given value of x. For large sample sizes N, a *95% confidence interval* for the estimated values of y can be calculated from

$$\bar{y} \pm 1.96S_y$$

meaning that 95% of all data fall within this range. (See Table 1-4 for other range factors.)

The coefficient of determination, R^2, measures the proportion of variation in the dependent variable y associated with the independent variable x, i.e., it measures the strength of association represented by the regression. R^2 is the square of the correlation coefficient between the two variables used in a bivariate regression equation, or among more variables in multiple regression equations.

It is common practice in engineering anthropometry (in fact in ergonomics altogether) to require a correlation coefficient of at least 0.7 as a basis for design decisions. The reason for this "0.7 convention" is that one should be able to explain at least 50 percent of the variance of the predicted value from the predictor variable: this requires r^2 to be at least 0.5, so r is at least 0.7075. (Note that r depends on sample size N.)

EXAMPLE

Clothing tariffs are examples of use, misuse, and non-use of correlations. In the United States, sizing of clothes for men is a fairly well organized and standardized procedure. Most men's jacket sizes run from "38" to "56", meaning that they should fit men with chest circumferences between 38 and 56 inches, in increments of one or two inches. Chest circumference, then, is used as the primary "predictor variable" for other design variables, such as coat length, shoulder width, and sleeve length. Similarly, trousers are ordered by waist circumference, and shirts by neck circumference.

In men's shirts, a given neck circumference is associated with a given chest circumference, while sleeve length may vary by one- or two-inch increments. This is an attempt to cover various body dimensions with a few shirt sizes, but it has obvious shortcomings: if a person needs a large neck size (e.g., size 17) such shirt also usually comes with ballooning chest and waist circumferences, which the buyer may not need. There is a trend to further consolidation of size ranges, providing shirts only in three neck sizes, "small," "medium," and "large," having only one sleeve length associated with each. Production variability is cut down very much in this simplified tariff, but fewer customers are fitted.

The situation for women's clothing is much less unified in the United States. There appears to exist only one ill-defined prototype "size 12" (based mostly on half-century-old data by O'Brian and Shelton, 1941), from which larger and smaller sizes are derived in nonstandard manners, as deemed suitable by each manufacturer. Hence, a woman well fitted by clothes of size 10 made by one producer may need a size 12, or 8, in clothing tailored by another company. This situation has allowed several manufacturers to become specialized in catering to "petite" or "mature" customers.

APPLICATION

How to Get Missing Data

Europe and North America have, anthropometrically, the best-known populations of the earth. Yet even here, the civilian populations are not assessed exactly, and current data on subgroups are sparse. The ergonomist may be interested in new information such as on Italians visiting swimming beaches (Coviglio, Fubini, Masali, Masiero, Pierlorenzi and Sagone, 1991), Irish workers (Gallery and Fitzgibbon, 1991), American farmers (Casey, 1989), pregnant American women (Culver and Vialo, 1990), U.S. hand sizes (Greiner, 1991), or Turkish schoolchildren (Vayis and Oezok, 1991). In many cases, the exact body dimensions needed for a design are not available in the literature.

Several routes exist to obtain the needed information. One is to actually measure a sufficiently large and well-selected sample of the population to be fitted. This is a time-consuming task, which should be done by anthropometrists or other specialists (although one can simply measure a few co-workers to get a rough estimate for the missing data). Another approach is to take the data of a population of known dimensions, if one has good reason to believe that population is similar to the one on which data are missing. (Yet: are Taiwanese similar in size to all Chinese?) In this case, it might also be highly advisable to seek help from an anthropologist or other well-informed person. The literature provides some help in discussing important aspects, such as sample selection, sample size, and composite populations (Chapanis, 1975; Kroemer, Kroemer, and Kroemer-Elbert, 1990; Lohman, Roche, and Martorell, 1988; Pheasant, 1986).

A rather interesting task is the prediction of future body dimensions, which are needed when equipment must be designed for use in decades to come. In the 1960s, for example, NASA was concerned about the body sizes of

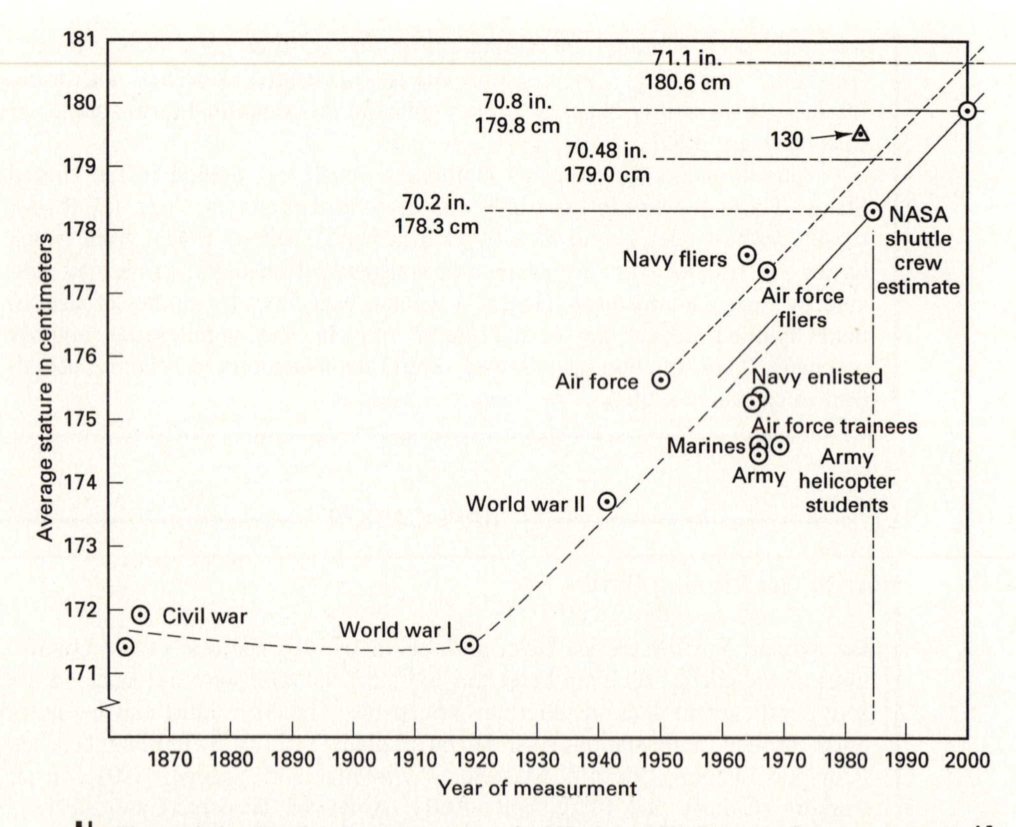

Figure 1-6. Predicted average stature for USAF and NASA male flying personnel (with permission from Roebuck, Smith, and Raggio, 1988).

astronauts in the 1980s and 90s. In 1988, Roebuck, Smith, and Raggio used a large variety of sources, military and civilian, U.S. and foreign, together with regression equations to forecast the body dimensions of astronauts in the year 2000. Figure 1-6 shows their predicted values for male U.S. Air Force and NASA flight crews in terms of stature. Figure 1-7 shows their predictions for American and Asian women.

☛☛☛ *Phantoms, Ghosts, and the "Average Person" Homunculus*

Several misleadingly simple body-proportion templates have been used in the past (e.g., Drillis and Contini, 1966). In fact, all "fixed" design templates fall in that category, if they assume that all body dimensions, such as lengths, breadths, and circumferences, can be represented as given fixed proportions (percentages) of one body dimension, for example, stature. Obviously, such a simplistic assumption contradicts reality: the relationships among body dimensions are neither necessarily linear, nor the same for all persons. In spite of the obvious fallacy of the model, "single-percentile constructs" have been gener-

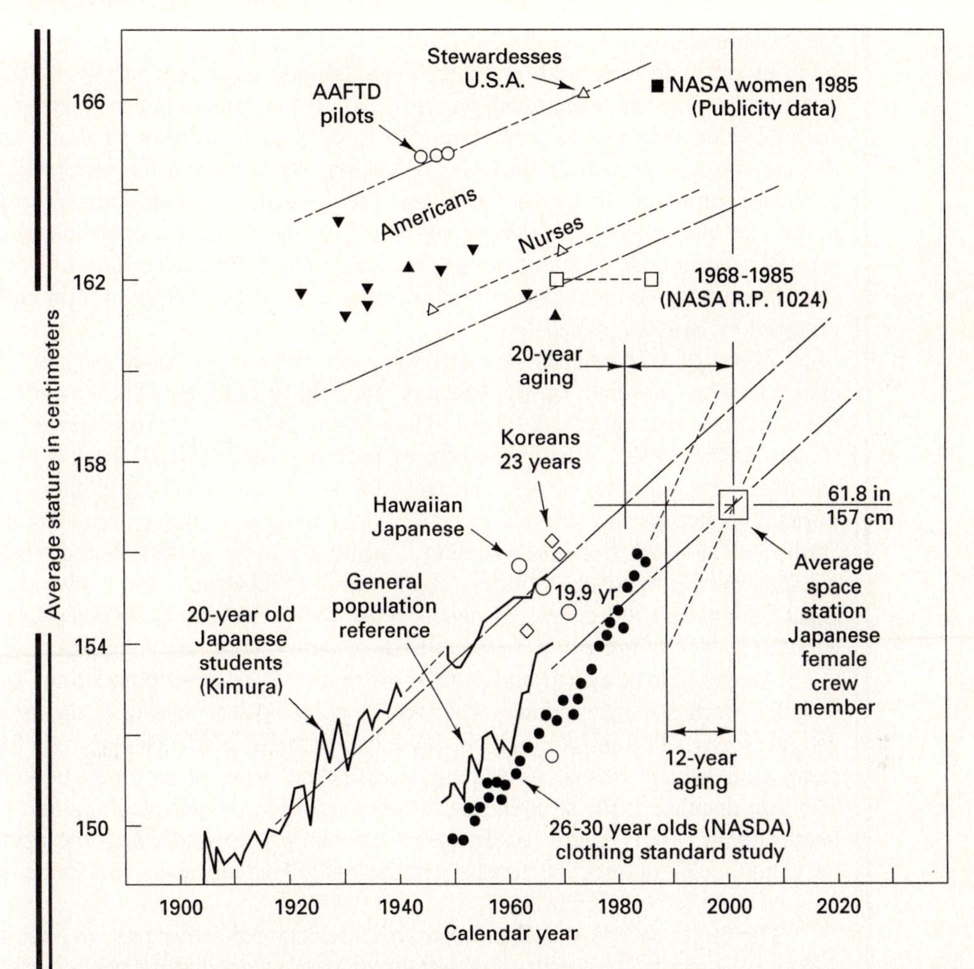

Figure 1-7. Predicted average stature for American and Asian women space crews (with permission from Roebuck, Smith, and Raggio, 1988).

ated, assuming that homunculi exist whose body segments are all of the same percentile value. Not only the 50th-percentile phantom (the "average person") has been used as a design template, but other ghostly figures have been created that have, for example, all 5th- or 95th-percentile values. Of course, designs for these figments do not fit actual users.

"Fitting" Design Procedures

Information about body size is needed when an object must fit the human body, such as tool handles to hold, protective equipment and clothing to wear, chairs to sit on, windows to look through, and workstations in general. Several of these applications are discussed elsewhere in this book (see, for example, the sections on hand tools and computer workstations). Different fitting methods are used, such as choosing exact percentiles on the continuum of the measuring scale to determine ranges (say, from 5th to 67th percentile) or to assure that the largest persons will fit through an opening, or that even the smallest can use the equipment. In this context, one often speaks of "functional" (or dynamic) anthropometry, meaning body data that depend on the coordinated efforts of several body segments to achieve a desired posture or perform an activity. These may define zones of convenience, of expediency, of minimally required or largest covered space.

Zones of convenience or expediency are difficult to define because the criterion is not absolute (in the sense of minimal or maximal) but depends on the situation, the subject, the task. The various "normal working areas" first shown in the 1940s, usually in form of partial spheres around the elbow or shoulder, are examples of plausible yet ill-defined convenience contours. It is difficult to accept that a male person should have a working area within a "radius of 394 mm from the shoulder," while a female worker should have a working area limited by a "radius of 356 mm" (Nicholson, 1991): why those exact dimensions? Of course, it makes sense that work should be done within easy reach—see Figure 1-8—one just has to define what "easy" means.

An example of a clear and defined procedure is the determination of "safe distance from a danger point." The danger point is that edge of a hazardous gadget (such as of a press mold, cutting edge, or pinch point) closest to the operator from which the operator's body (usually the finger or toe) must be kept. The safe distance is the straight line distance between the danger point and the barrier (wall, safety guard, enclosure of an opening) beyond which the operator's body cannot proceed toward the hazard. That distance should be increased by a safety margin.

For finger safety, the distance may be determined either past an opening which allows only the finger to penetrate; or past an opening or barrier which can be overreached by the arm. In the first case, the safe distance would be the length of the longest possible finger, with a safety margin; in the second case, the distance would be determined by the longest arm and finger reach.

For foot safety, the most likely barrier is at the ankle, so that only the toes and foot can penetrate further toward the danger point; or the whole leg may have to be considered, probably restrained in the hip area.

There are many variations of these conditions, such as those in the German Standard DIN 31001 and the British Standard BS 5304. Some of these conditions are shown in Figure 1-9.

In each case, the longest possible body segment should be considered, under the given conditions of barrier and mobility. A predetermined safety margin (of, say, 10 percent) should be applied to those body lengths. Certain conditions, such as holding an object that, if entrapped, might pull the hand to-

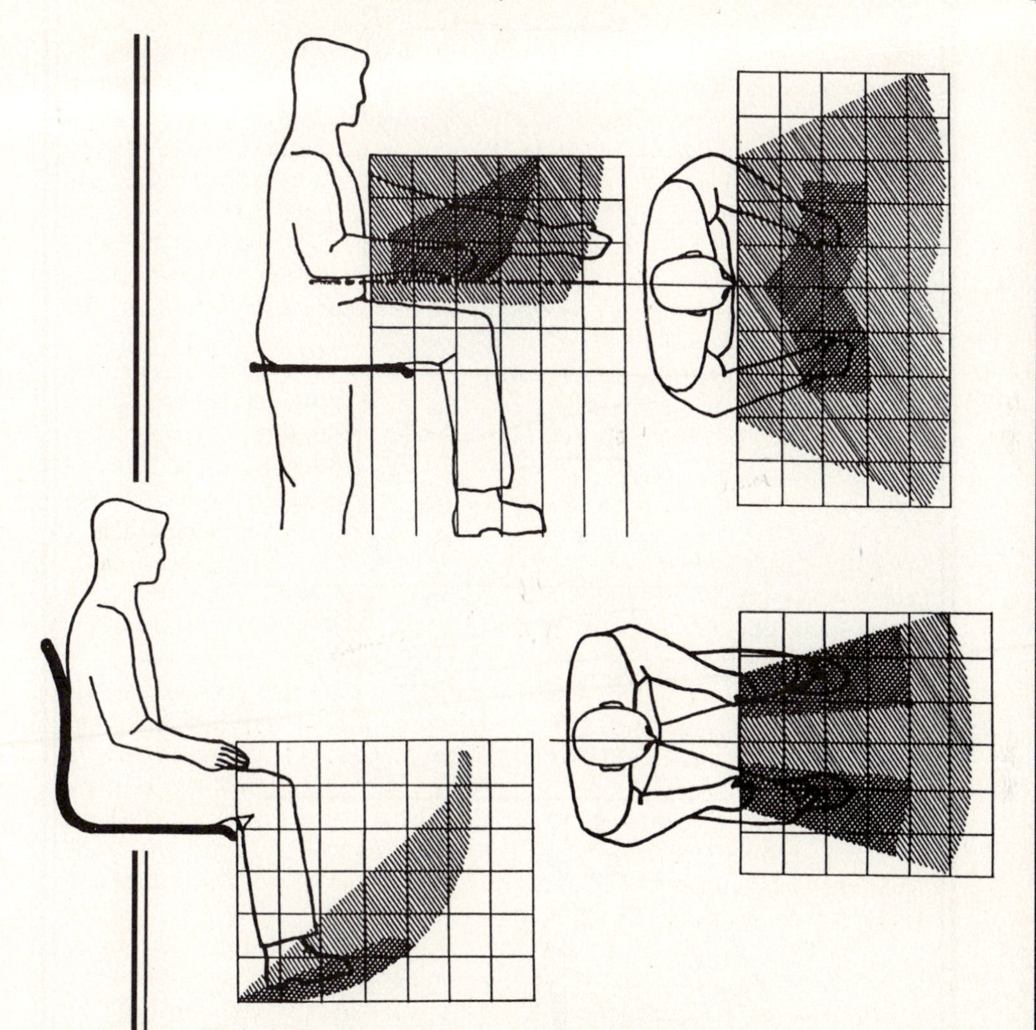

Figure 1-8. The concept of "preferred" working zones of the hands and feet.

ward the hazard point, could be good reason to extend the safety distance further.

A similar strategy should be applied when workstations, tools, and tasks must be designed for small operators. The needs to see, to reach, to apply force are derived from the smallest operators, yet these criteria may not accommodate large persons. Possible solutions would be to have adjustable object dimensions, or to have objects in different sizes (Ayoub and Miller, 1991; Bottoms and Butterworth, 1990; Buckle and David, 1989; Pheasant, 1986; Nicholson, 1991; Thompson, 1989).

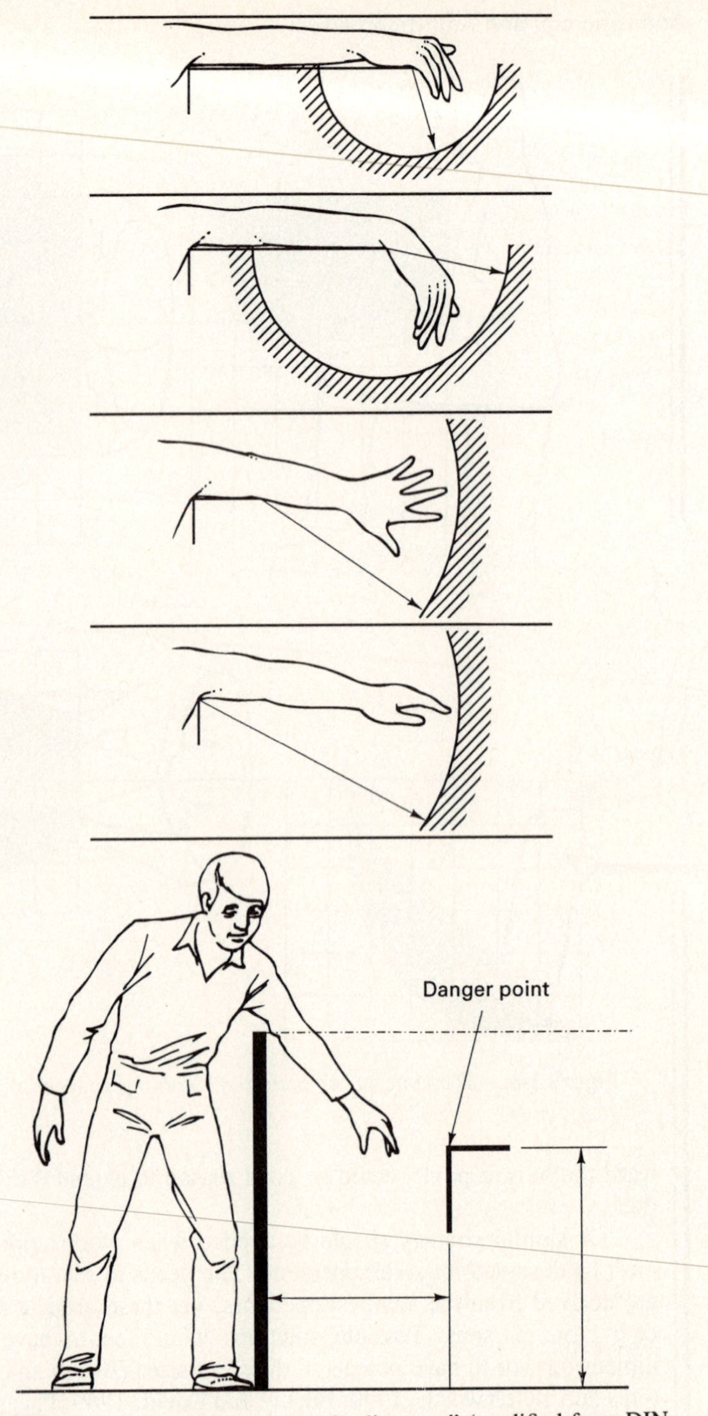

Figure 1-9. Examples for "safe distances" (modified from DIN 31001 and BS 5304).

For a standing operator, some "toe space" should be provided so that one can step close to the workstation. This space should be high enough to accommodate persons wearing thick soles, but shallow enough so that one does not hit the edge of the foot-space cutout with the instep of the foot. Thus, a depth not to exceed 10 centimeters, and a height of not less than about 10 centimeters should be appropriate—see Figure 1-10.

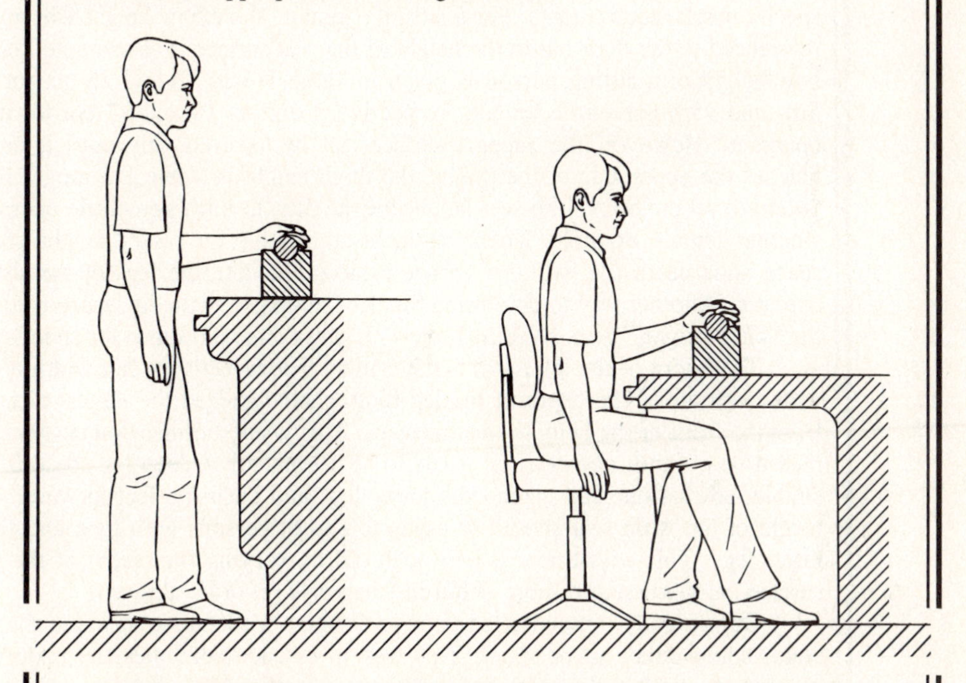

Figure 1-10. Shape of workstations at which the operator stands or sits.

Then, the height of the work surface must be determined. That depends on the physical work to be performed, on the dimensions of the workpiece itself, and on the need to observe the work done. As a general rule, the manipulation itself should be performed at about the height of the elbow of the operator when the upper arm hangs down along side the trunk, or is slightly elevated forward and sideways. For example: Table 1-3 shows an elbow height of about 93 cm for a 5th-percentile standing female operator, and 107 cm for a 95th-percentile standing female. For standing male operators, the respective elbow heights are 100 and 115 cm. One may reduce these heights if the operator does not stand "erect," but that may be offset by the heel height of shoes worn. If the workpiece is large, and the manipulation is performed on its upper part, the support surface (bench height) must be low enough to allow the hands to be at elbow level. However, the work might need close visual observation; this requires an appropriate viewing distance. In that case, particularly if the manipulation requires fairly little force and energy, the work area might be elevated well above elbow height. (But that, in turn, might require support for the elevated hands and forearms.) These conditions are illustrated in Figure 1-10.

To determine actual design values for the workstation, the relevant body dimensions (in particular, elbow height and eye height) of the expected operator population must be selected, and adjusted according to body postures and specific work requirements. For a sitting operator, the elbow height will not be referenced to the floor but to the height of the seat surface. For example: the elbow height of a sitting person is given in Table 1-3 as 18 and 26 cm for the 5th- and 95th-percentile female, respectively; and as 18 and 27 cm for male operators. However, the support surface can be lowered only until it nearly touches the upper side of the thighs: the thigh height in Table 1-3 ranges from 14 cm to 19 cm above the seat height for the 5th- to 95th-percentile operator, whether female or male. These values establish the necessary height of the space underneath the working surface to accommodate the legs of the sitting operator. Another way to determine the needed height of the leg space is to use the "knee height," also given in Table 1-3, plus some allowance for shoe heels.

The width of the leg room is not critical if it exceeds the hip width of the widest operator. The depth of the leg room should exceed the largest distance from the front of the belly to the kneecaps. This is not a dimension customarily measured by anthropometrists; it has to be estimated. A deep leg space is desirable, so that one can extend the lower legs and push the feet forward. The height of the work seat should be adjustable to fit persons with long and short lower legs. This adjustment is best achieved by varying the height of the seat surface (as discussed in more detail in Chapter 9 on office design).

Occasionally, one is called upon to design a workstation at which the operator could either sit or stand. This task in essence combines the major requirements of the stations for either sitting or standing. Specifically, there must be a very tall chair, and a high support surface for the feet. A small board or bar attached to the chair is not recommended, because it reduces the stability of the chair while providing little support surface for the feet, which, accordingly must be kept in place, often by muscle tension instead of being able to move to different positions. The general principles for a combined sit-stand workstation are sketched in Figure 1-11.

Design procedures. Proper procedures are available to develop analogs of the human body (Kroemer, 1989). The "subgroup" method creates models that represent the extreme ends of the body-size range. One identifies critical dimensions and assures that they are fitted. If both the smallest and the largest are taken, one is fairly sure that the intermediate range is accommodated. As Haslegrave (1986) explains, this can be relatively simple, particularly if the problem is one-dimensional or there are no relationships among several relevant dimensions.

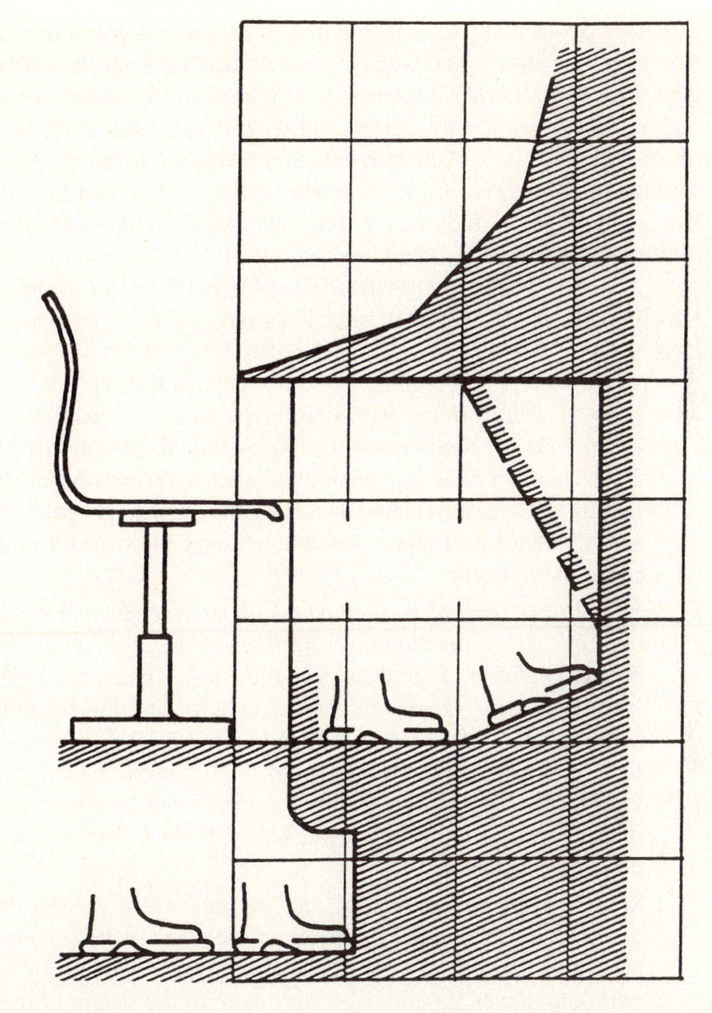

Figure 1-11. Workstation for standing or sitting, or for alternately sitting and standing operation.

☞☞☞ *If one "stacks" values of a given percentile, such as the 5th, one does not end up with a composite figure that, in its sum, is also 5th percentile. For example: 5p hip height plus 5p trunk height plus 5p head height do NOT add up to 5p stature.* ☜☜☜

A combination often used to determine an *n*th-percentile person is that of stature and weight (as in "desirable" weight-height ratios). Yet, the correlation between the two values is low, in the neighborhood of 0.3 for women and 0.4 for men in the general population. Haslegrave concluded that the best way to present a 5p woman (or a 5p man) is to measure a group of women (or men) who have this stature and weight, and to calculate from their measures the median values of all other dimensions of interest in the group. If one then checks each resulting value, this is likely not to be exactly at the 5th (or 95th) percentile, but close to it. An alternative is to use regression equations. The major advantage of the regression-based procedure is that the resulting values are additive, which percentile values are not (McConville and Churchill, 1976; McConville, Robinette, and Churchill, 1981).

A useful and correct general design procedure (Kroemer, Kroemer, and Kroemer-Elbert, 1986, 1991, and Pheasant, 1986) entails four steps, as follows:

Step 1: *Select those anthropometric measures that directly relate to defined design dimensions*. Examples are: hand length related to handle size; shoulder and hip breadth related to escape-hatch diameter; head length and breadth related to helmet size; eye height related to the heights of windows and displays; knee height and hip breadth related to the leg room in a console.

Step 2: *For each of these pairings, determine whether the design must fit only one given percentile of the body dimension, or a range along that body dimension*. Examples are: the escape hatch must be fitted to the largest extreme values of shoulder breadth and hip breadth, considering clothing and equipment worn; handle size of pliers is probably selected to fit a smallish hand; the leg room of a console must accommodate the tallest knee heights; the height of a seat should be adjustable to fit persons with short and with long lower legs. (Table 1-4 shows how to calculate percentile values.)

Step 3: *Combine all selected design values in a careful drawing, mock-up, or computer model to ascertain that they are compatible*. For example: the required leg-room clearance height, needed for sitting persons with long lower legs, may be very close to the height of the working surface, determined from elbow height.

Step 4: *Determine whether one design will fit all users*. If not, several sizes or an adjustment must be provided to fit all users. Examples are: one large bed size fits all sleepers; gloves and shoes must come in different sizes; seat heights are adjustable.

☞☞☞ *The following appeared in the* Washington Post *of May 25, 1984:*

The Navy has adopted new flight training standards that will require its aviators, as a whole, to have longer arms and shorter legs. The standards will exclude 73 percent of all college-age women and 13 percent of the college-age men, according to a military spokesman. [He] said the new standards were devised because some avia-

tion candidates could not reach rudder pedals or see over instrument panels. Some taller pilots were so tightly wedged that their helmets bumped the aircrafts' canopies. "We found out that manufacturers are still building airplanes the way they want, but God is not making people to fit them." Previously, 39 percent of the female applicants and 7 percent of the men were ineligible to become aviation candidates because of their size.

Six years later, the cockpit dimensions of aircraft used throughout the world (Boeing 737-200, 747, 757, and Lockheed TriStar) were evaluated with respect to eight critical body dimensions of pilots, including eye height, hand and leg sizes, and reaches. In many cases, the fit was marginal, at best. For example, based on eye height, 13 percent of the British male and 73 percent of the female pilot candidates would have to be excluded from being crew members (Buckle, David, and Kimber, 1990).

HUMAN BIOMECHANICS

In biomechanics, one attempts to understand characteristics of the human body in mechanical terms. The biomechanical approach is not new. Biomechanics has been applied to the statics and dynamics of the human body, to explain effects of vibrations and impacts, to explore characteristics of the spinal column, and to the use of prosthetic devices, to mention just a few examples.

Leonardo da Vinci (1452–1519) and Giovanni Alfonso Borelli (1608–1679) combined mechanical with anatomical and physiological explanations to describe the functioning of the biological body. Since Borelli, the human body has often been modeled as consisting of long bones (links) that are connected in the articulations (joints), powered by muscles that bridge the articulations. The physical laws developed by Isaac Newton (1642–1727) explained the effects of external impulses applied to the human body.

Treating the human body as a mechanical system entails gross simplifications, such as disregarding mental functions. Still, many components of the body may be well considered in terms of analogies, such as:

bones—lever arms, central axes, structural members

articulations—joints and bearing surfaces

tendons—cables transmitting muscle forces

tendon sheaths—pulleys and sliding surfaces

flesh—volumes, masses

body contours—surfaces of geometric bodies

nerves—control and feedback circuits

muscles—motors, dampers, or locks

organs—generators or consumers of energy

The Skeletal System

The human skeleton is composed of 206 bones, together with associated connective tissue and articulations.

The main function of human skeletal bone is to provide an internal framework for the whole body, see Figure 1-12. The long, more or less cylindrical bones that connect body joints are of particular interest to the biomechanist. They are the lever arms at which muscles pull.

While bone is firm and hard, and thus can resist high strain, it has certain elastic properties. In childhood, when mineralization is relatively low, bone is rather flexible. In contrast, bones of the elderly are highly mineralized and therefore more brittle. Also, they change their geometry, similar to pipes getting wider in diameter but thinner in their walls. Thus, osteoporosis in the elderly means, mechanically speaking, a hollowing of bones, decrease in bone mass, and a brittling of the thinned walls (Ostlere and Gold, 1991). Yet, the moment of inertia, $I = \frac{1}{4}\Pi(R^4_{outer} - R^4_{inner})$, remains about the same.

Bone cells are nourished through blood vessels. Bone material is continuously resorbed and rebuilt throughout one's life. Local strain encourages growth and disuse encourages resorption, as long as a suitable threshold is not exceeded (Wolff's law): overstrain can cause structural damage in terms of micro- or macro-fracture.

Connective tissues are of several types:

Cartilage is a translucent, viscoelastic flexible material capable of rapid growth, located at the ends of the ribs, as disks between the vertebrae, and in general as joint surfaces at the articulations.

Ligaments connect bones and provide capsules around joints.

Tendons are strong yet elastic elongations of muscle, connecting it with bones.

The design of the bones at their joints, the encapsulation by ligaments, the supply of cartilaginous membranes, and the provision of disks or volar plates determine, together with the action of muscles, the mobility of body joints.

Some bony joints have no mobility left, such as the seams in the skull of an adult; some have very limited mobility, such as the connections of the ribs to the sternum. Joints with "one degree of freedom" are simple hinge joints, like the elbow or the distal joints of the fingers. Other joints have two degrees of freedom, such as the ill-defined wrist joint (discussed later), where the hand may be bent in flexion and extension, and laterally pivoted. (The capability to twist is located in the fore-

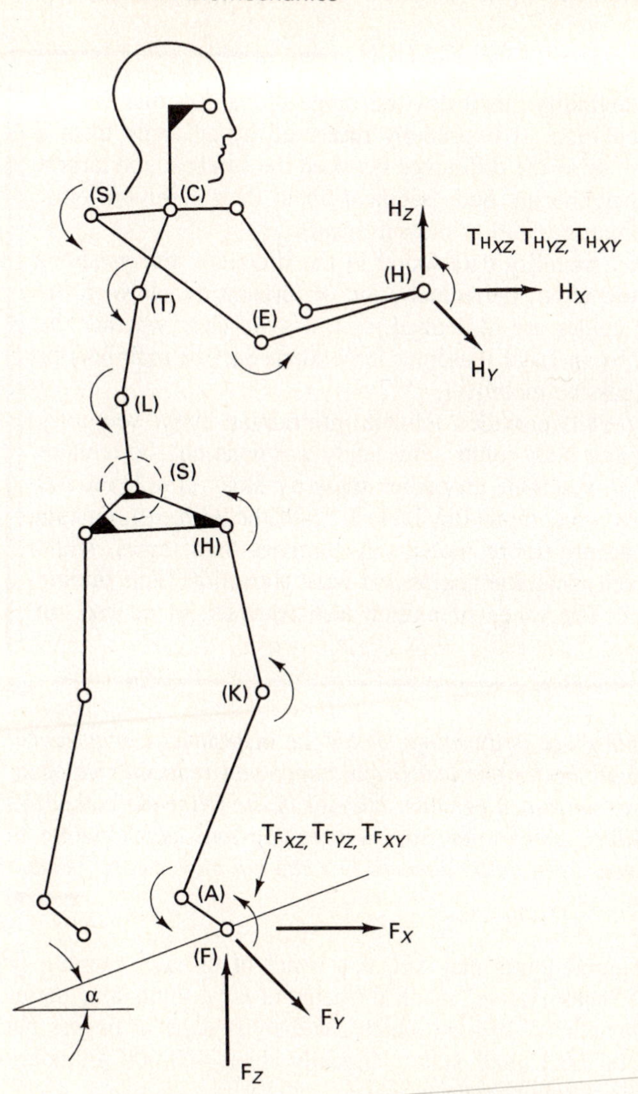

Figure 1-12. Human skeleton simplified as "links connected in joints" (H for hand, E for elbow, S for shoulder, etc.). (From Kroemer, Kroemer, and Kroemer-Elbert, 1990, *Engineering Physiology: Bases of Human Factors/Ergonomics*, 2d ed. With permission by the publisher, Van Nostrand Reinhold. All rights reserved.)

arm, not in the wrist.) Other joints have three degrees of freedom, such as shoulder and hip joints.

Synovial fluid in a joint facilitates movement of the adjoining bones by providing lubrication. For example, while a person is running, the cartilage in the knee joint can show an increase in thickness of about 10 percent, brought about in a short time by synovial fluid seeping into it from the underlying bone marrow cavity. Similarly, fluid seeps into the spinal disks (which are composed of fibrous cartilage) when they are not compressed, for example during sleep. This makes them more pliable directly after getting up than during the day, when they are "squeezed out" by the load of body masses and their accelerations. Thus, immediately after getting up, one stands taller than after a day's effort.

APPLICATION

The term *mobility,* or flexibility, indicates the range of motion that can be achieved in a body articulation. It is properly measured by an angle from a known reference position, or as the difference between the smallest and largest angles enclosed by the neighboring body segment about their common joint. Figure 1-13 indicates common mobility measurements.

Unfortunately, many mobility data found in the literature are questionable or unexplained. Quite often, the actual point of rotation moves with the motion, the arms of the angles are ill defined, and it is not clear whether the positions were achieved by internal muscular force alone (active mobility) or with help from outside (passive mobility).

A study by Staff (1983) provided reliable information about voluntary (unforced) mobility in major body joints. This study was done on 100 females and carefully controlled to resemble an earlier study by Houy (1982) on 100 male subjects. The results are compiled in Table 1-7. Of the 32 measurements taken, 24 showed significantly more mobility by females than males, while men were more flexible only in ankle flexion and wrist abduction. This finding confirmed earlier studies. The range of motion also depends, of course, on training and on age.

Intuitively, "flexibility" in body joints should be of practical importance: more flexibility should indicate better physical performance and reduced risk of injury. Yet, in comparison with untrained persons, athletes in the following categories have been found less flexible: soccer players, runners, persons participating in sports for five years or longer, even ballet dancers in some hip movements (Burton, 1991).

Artificial joints. Natural joints may fail as a result of disease, trauma, or long-term wear and tear. If "conservative" medical treatment fails, joints may be replaced with artificial, manufactured devices, which routinely is done in fingers but mostly in hips and knees. In the United States, annually about 400,000 joints are implanted, predominantly in elderly persons.

The degeneration of cartilage in the major joints (such as the hip or knee) due to trauma or to rheumatoid or osteo-arthritis may lead to the replacement of the articulating surfaces with artificial joints. Although total joint replacement typically restores function and mobility to the patient, its primary and most-appreciated purpose is to relieve pain.

For the patient, joint degeneration is associated with pain and with progressive and severe limitations of motion. If needed, the articulating surfaces are typically replaced in their entirety: in the hip, the head of the femur (thigh bone) is removed and replaced by a spherical metallic ball on a stem, and the acetabular cup is resur-

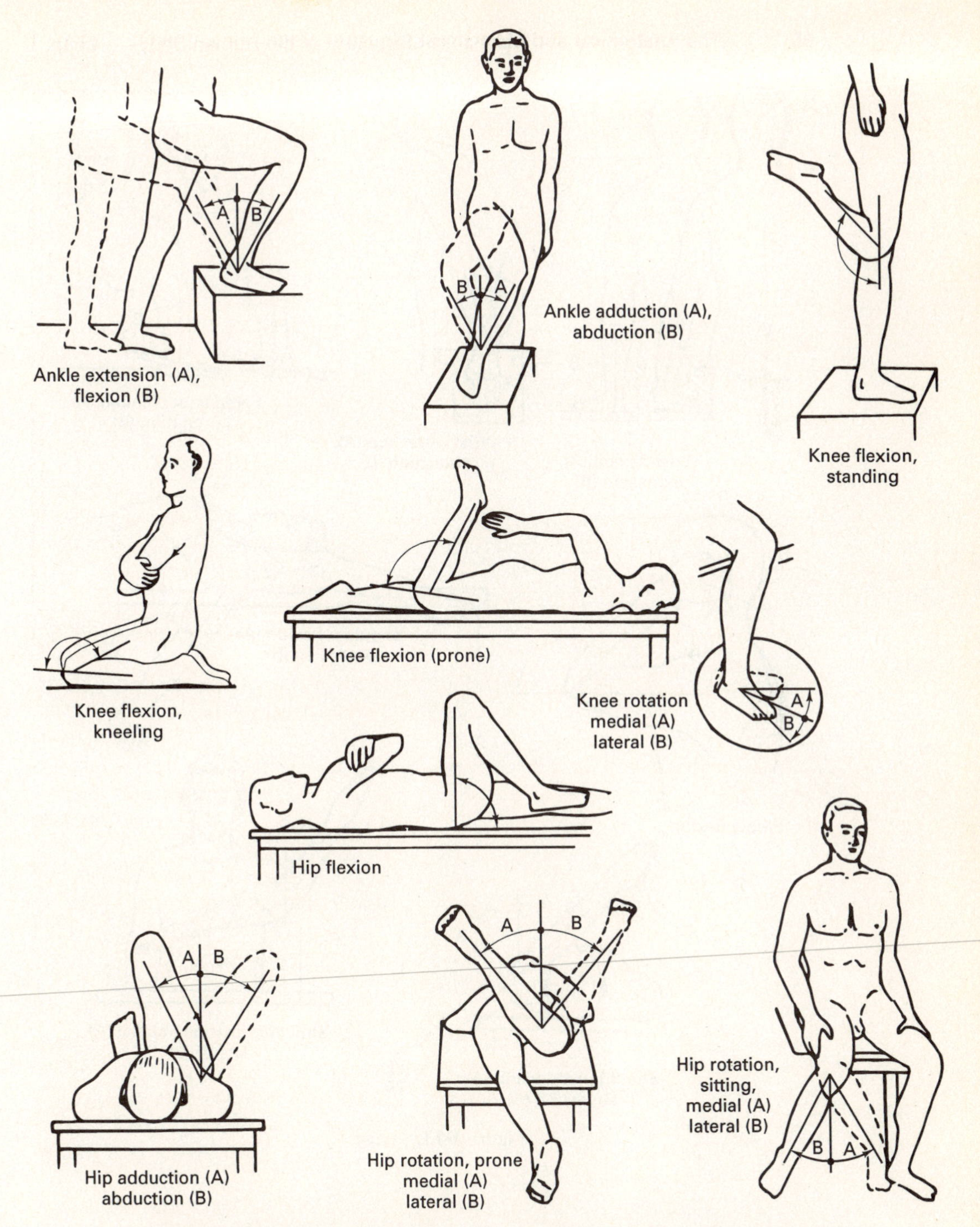

Figure 1-13. Maximal displacements in body joints. (From Kroemer, Kroemer, and Kroemer-Elbert, 1990, *Engineering Physiology: Bases of Human Factors/Ergonomics,* 2d ed. With permission by the publisher, Van Nostrand Reinhold. All rights reserved.)

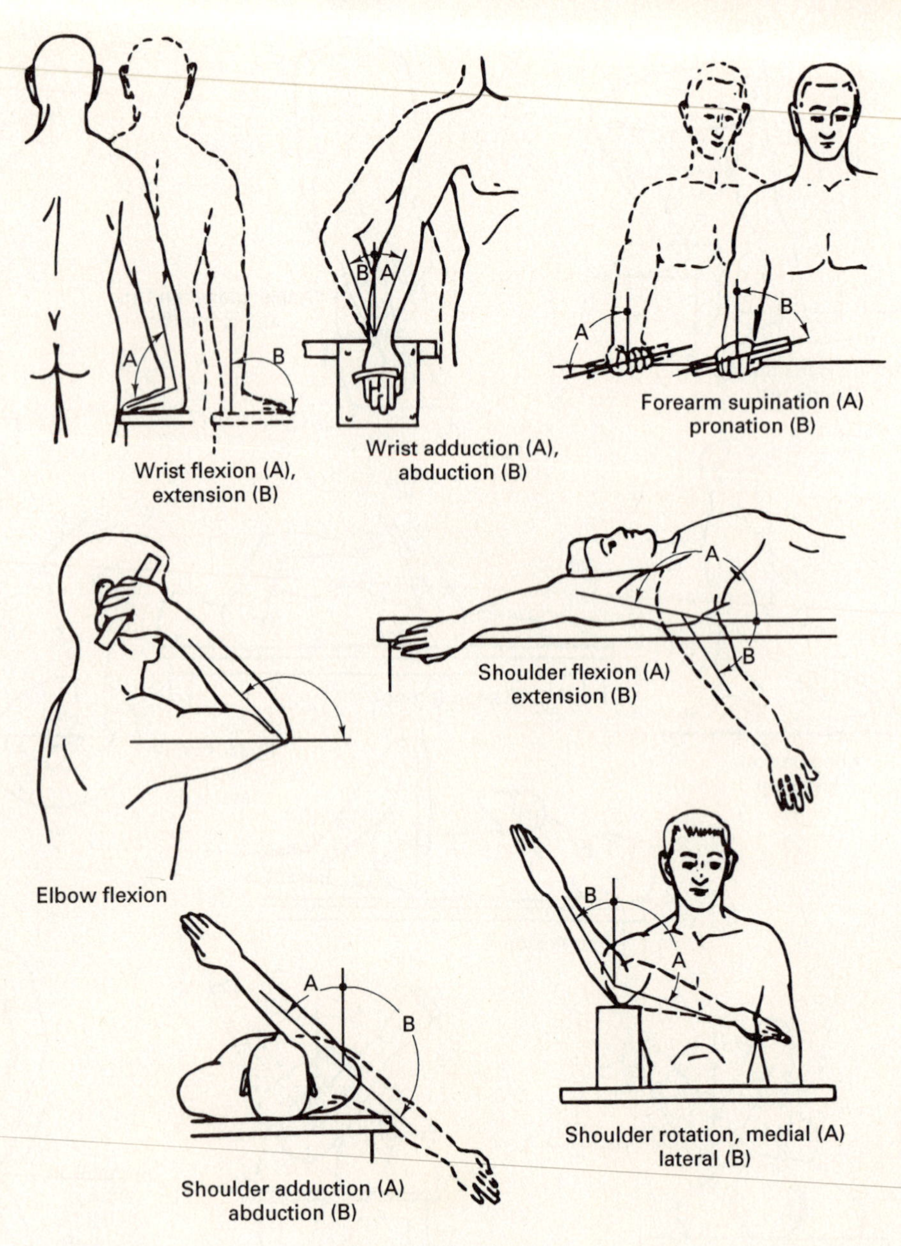

Figure 1-13. (*cont.*)

TABLE 1-7. COMPARISON OF MOBILITY DATA (in degrees) FOR FEMALES AND MALES

		5th Percentile		50th Percentile		95th Percentile		Difference*	
Joint	Movement	Female	Male	Female	Male	Female	Male	Female	Male
Neck	Ventral flexion	34.0	25.0	51.5	43.0	69.0	60.0	+8.5	
	Dorsal flexion	47.5	38.0	70.5	56.5	93.5	74.0	+14.0	
	Right rotation	67.0	56.0	81.0	74.0	95.0	85.0	+7.0	
	Left rotation	64.0	67.5	77.0	77.0	90.0	85.0	NS	
Shoulder	Flexion	169.5	161.0	184.5	178.0	199.5	193.5	+6.5	
	Extension	47.0	41.5	66.0	57.5	85.0	76.0	+8.5	
	Adduction	37.5	36.0	52.5	50.5	67.5	63.0	NS	
	Abduction	106.0	106.0	122.5	123.5	139.0	140.0	NS	
	Medial rotation	94.0	68.5	110.5	95.0	127.0	114.0	+15.5	
	Lateral rotation	19.5	16.0	37.0	31.5	54.5	46.0	+5.5	
Elbow-forearm	Flexion	135.5	122.5	148.0	138.0	160.5	150.0	+10.0	
	Supination	87.0	86.0	108.5	107.5	130.0	135.0	NS	
	Pronation	63.0	42.5	81.0	65.0	99.0	86.5	+16.0	
Wrist	Extension	56.5	47.0	72.0	62.0	87.5	76.0	+10.0	
	Flexion	53.5	50.5	71.5	67.5	89.5	85.0	+4.0	
	Adduction	16.5	14.0	26.5	22.0	36.5	30.0	+4.5	
	Abduction	19.0	22.0	28.0	30.5	37.0	40.0	−2.5	
Hip	Flexion	103.0	95.0	125.0	109.5	147.0	130.0	+15.5	
	Adduction	27.0	15.5	38.5	26.0	50.0	39.0	+12.5	
	Abduction	47.0	38.0	66.0	59.0	85.0	81.0	+7.0	
	Medial rotation (prone)	30.5	30.0	44.5	46.0	58.5	62.5	NS	
	Lateral rotation (prone)	29.0	21.5	45.5	33.0	62.0	46.0	+12.5	
	Medial rotation (sitting)	20.5	18.0	32.0	28.0	43.5	43.0	+4.0	
	Lateral rotation (sitting)	20.5	18.0	33.0	26.5	45.5	37.0	+6.5	
Knee	Flexion (standing)	99.5	87.0	113.5	103.5	127.5	122.0	+10.0	
	Flexion (prone)	116.0	99.5	130.0	117.0	144.0	130.0	+13.0	
	Medial rotation	18.5	14.5	31.5	23.0	44.5	35.0	+8.5	
	Lateral rotation	28.5	21.0	43.5	33.5	58.5	48.0	+10.0	
Ankle	Flexion	13.0	18.0	23.0	29.0	33.0	34.0	−6.0	
	Extension	30.5	21.0	41.0	35.5	51.5	51.5	+5.5	
	Adduction	13.0	15.0	23.5	25.0	34.0	38.0	NS	
	Abduction	11.5	11.0	24.0	19.0	36.5	30.0	+5.0	

*Listed are only differences at the 50th percentile, and if significant ($\alpha < 0.5$).

SOURCE: From Kroemer, Kroemer, and Kroemer-Elbert, 1990, *Engineering Physiology: Bases of Human Factors/Ergonomics*, 2nd ed. with permission by the publisher, Van Nostrand Reinhold.

faced with a plastic liner. In the knee, the articulating surfaces on the bottom of the femur are replaced with metal, and articulating surfaces at the top of the tibia (shin bone) and on the patella (knee cap) are resurfaced with plastic. To date, the metals used have been stainless steel, a cobalt-chromium alloy, and a titanium alloy. These joints all have the same type of design for the major load-bearing components: the metallic component is convex and the plastic component is concave. The plastic now used (after an early, disastrous attempt with Teflon) is an ultra-high molecular-weight polyethylene.

Replacements for the ball-and-socket joint at the hip have been attempted for about a century. Routinely successful total hip replacement started with Charnley's work in England in the 1960s. He pioneered the use of the metal-on-plastic articulations and the use of PMMA (poly methyl methacrylate) as a bone cement. Today, at least 90 percent of patients with hip and knee replacements are pleased with their new joints and function well ten years and longer after surgery. Since surgical technique and device design have improved over the last decade, it is expected that today's joint-replacement patients will enjoy even higher success rates.

A polymeric bone cement is often used to fill the space between the metal implant surfaces and the reamed bone cavity. This cement has no adhesive properties of its own, but serves (as a grout) to link mechanically the prosthesis and the bone. Recently, devices have been designed that attempt to fix the bone directly to the metal implant surfaces, avoiding the need for bone cement. These implant surfaces may be coated with small beads or thin wires to create a pore size of less than 1 mm into which the bone is supposed to grow. To encourage bony ingrowth, an osteoinductive or osteoconductive chemical coating may be sprayed on the "porous" surface of the implant.

A new development in total joint replacement has been the use of ceramics. Although ceramic-on-ceramic hip replacements have been attempted, better success has been obtained with ceramic-on-polyethylene. For these, the acetabular cup liner is made from polyethylene and the head of the femur replacement is ceramic. This ceramic head is placed on a metal stem, which is inserted into the prepared femur. Ceramics have theoretical and practical advantages in reducing wear of the artificial joint, but their high cost, low material toughness, and difficulties in manufacturing quality control continue to be problems. To date, ceramic joint designs for hip and knee replacements have not performed better than metal-on-plastic joint designs, which may have device success rates of 90 to 95 percent at five- to ten-year follow-up.

If necessary, finger joints are usually replaced by a one-component, molded plastic integral hinge. This simple artificial joint is successful for several reasons, including the low loads carried by the joints, and the minimal debris generated by wear.

The design of joint replacements is constrained by biologic and mechanical considerations. Biologically, the device must be compatible with the body, both *in toto* and in particulate form (such as wear debris). The material from which the device is made must be analyzed for toxicity, reactivity, and strength, particularly in such corrosive and warm environment. The interaction between materials and rate of wear with and without lubrication must also be ascertained. The device must be implantable (in terms of complexity and in terms of size) and should yield near-normal range of motion. Finally, the design of the device should consider the possibility of salvage: sufficient bone and soft tissue should remain to allow for replacement of the device or fusion of the joint if needed.

When joint placements fail, the symptom to the patient is usually severe pain. Upon examination, infection is commonly found, and the device is often no longer firmly attached to the surrounding bone. This is mostly a mechanical problem, frequently associated with debris (wear particles) from the metallic component, the plastic component, and/or the cement. The wear particles may trigger a biologic response which leads to resorption of the bone and loss of implant support, as well as inflammation, reduced range of motion, and pain (Dumbleton, 1988; Elbert, 1991; Galante, Lemons, Spector, Wilson, and Wright, 1991).

APPLICATION

Testing of artificial joints is done in physical and numerical experiments. Physical testing may include animal models, especially for final device testing or to evaluate novel designs and/or materials. The testing may also be *ex vivo,* such as on isolated bones and joint segments. Implantation techniques or the range of motion of a device may be evaluated using these isolated segments. Strain gages, brittle stress coatings, and photoelasticity may be used to determine the strains (and infer the stresses) on the surfaces of the devices and implant material (cortical bone), but it is difficult to measure the strain response inside the material.

Numerical testing may rely on elasticity analysis, such as composite beam theory, plate/shell theory, torsion theory, and beam-on-elastic-foundation theory. More recently, linear and nonlinear finite element analyses have been used to calculate the stresses inside and on devices and idealized bone models. The accuracy of the numerical models is limited by the assumptions made in the formulation of the model (such as assuming certain boundary conditions) and by the sophistication of the material models (such as homogeneity, elasticity, continuity).

To design better and longer-lasting artificial joints, one must understand the loads to which the joints are subjected. Indirect and some limited direct methods to determine joint loads have been used. Loading of the joint may be estimated indirectly using the classical technique of correlating limb position, velocity, acceleration, and/or force-plate readings or motion to determine the balance of forces across the joint. Some sophisticated studies incorporate electrical activities of muscles to model the distribution of forces at the joint: recent numerical work has focused on nonlinear optimization of the forces in the tendons and ligaments at a joint along with electromyographic information of muscle activity to predict force distribution during various activities (Elbert, 1991).

Recently, there have been successful attempts to measure directly the joint loads in the human body. For example, a special total hip replacement with a three-axes load cell in the neck of the metal component has been implanted in the thigh bone. The forces occurring during various activities have been telemetrically sent to recorders in the laboratory. The actual *in vivo* loads were found to be somewhat smaller than calculated previously, using less invasive means (Davy, Kotzar, Brown, Heiple, Goldberg, Heiple, Jr., Berilla, and Burstein, 1988; Huiskes, 1985; Ladin and Wu, 1991).

The spinal column. The spine is a complex structure. It consists of 24 movable vertebrae (7 cervical, 12 thoracic, 5 lumbar), and the sacrum with the coccyx, which are a fused group of rudimentary bones. These sections are held together in cartilaginous joints of two different kinds. First, there are fibro-cartilage disks between the main bodies of the vertebrae. Second, each vertebra has two protuberances extending posterior-superiorly, the superior articulation processes, which end in rounded surfaces fitting into cavities on the underside of the next-higher vertebra. These synovial facet joints are covered with sensitive tissue, while the disks between the main bodies of the vertebra have no pain sensors.

The spine transfers forces and both bending and twisting moments from head and shoulder bones to the pelvis. It also protects the spinal cord, which runs through openings at the posterior (spinal canal), carrying signals between the brain and all sections of the body. This complex rod, transversing the trunk and keeping the shoulders separated from the pelvis, is held in delicate balance by ligaments connecting the vertebrae and by muscles that pull along the posterior and the sides of the spinal column. Longitudinal muscles located along the sides and the front of the trunk also both balance and load the spine.

Figures 1-14 through 1-16 illustrate specific aspects of the spinal column. Figure 1-14 shows schematically the stack of the vertebrae, indicating that—in the side view—the column is not straight but has two forward bends (lordoses) in the cervical and lumbar sections, and one backward bend (kyphosis) in the chest area. Only in the frontal view is the spinal column straight; if it is distorted, one speaks of scoliosis. Figure 1-15 is a scheme of the lumbar section of the spinal column, showing particularly the bearing surfaces at the main bodies and at the facets. Figure 1-16 shows a top view of a vertebra, indicating its main body, the structure surrounding the canal for the spinal cord (the vertebral foramen), and the five major protuber-

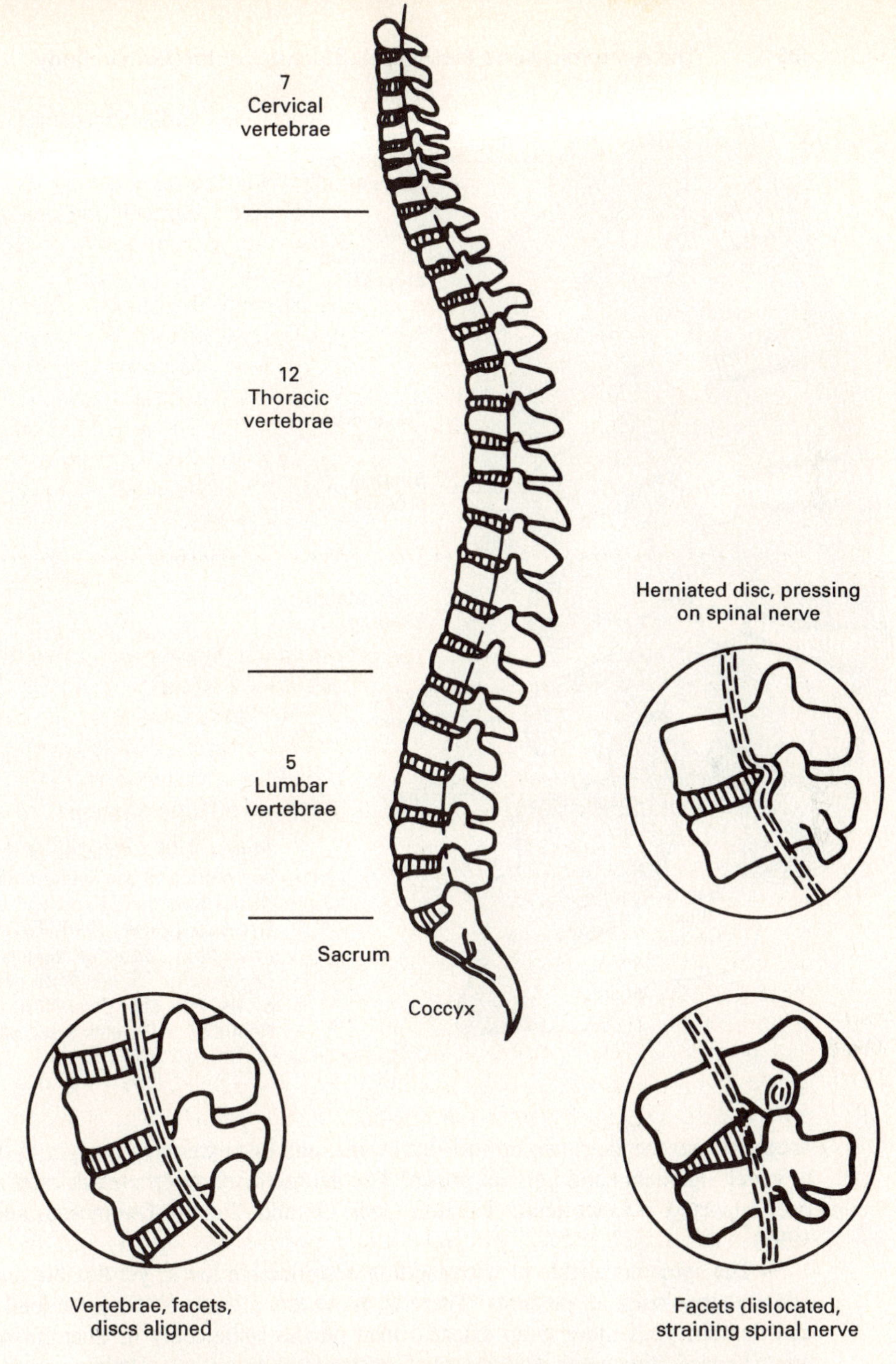

Figure 1-14. Scheme of the human spinal column, seen from the left side. (From Kroemer, Kroemer, and Kroemer-Elbert, 1990, *Engineering Physiology: Bases of Human Factors/Ergonomics,* 2d ed. With permission by the publisher, Van Nostrand Reinhold. All rights reserved.)

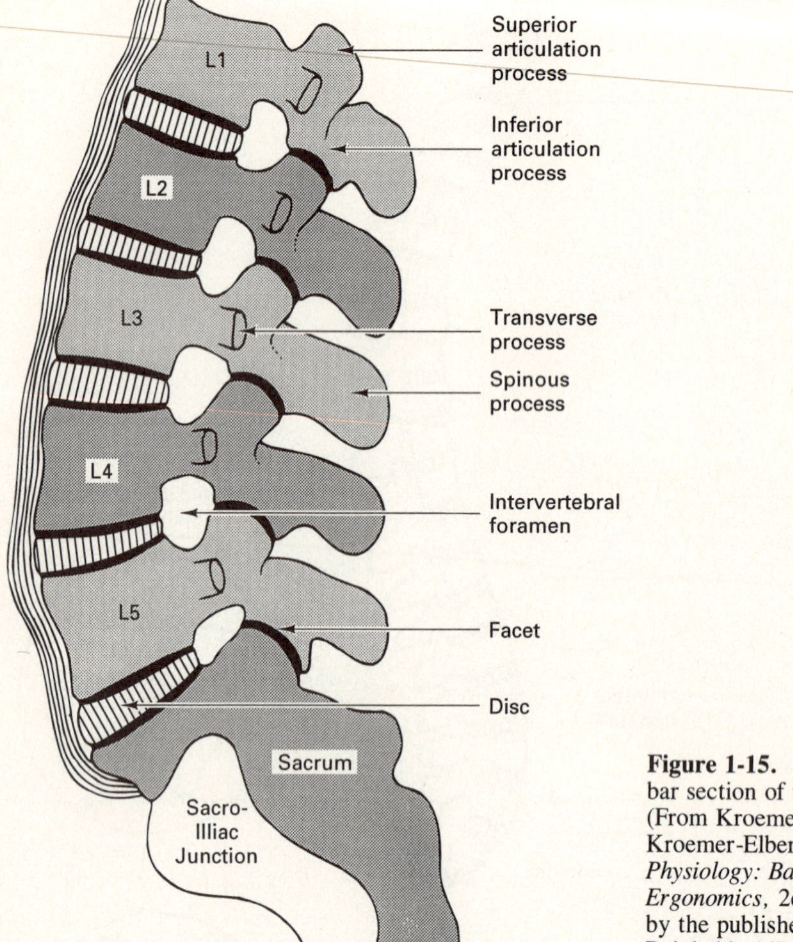

Figure 1-15. Scheme of the lumbar section of the spinal column. (From Kroemer, Kroemer, and Kroemer-Elbert, 1990, *Engineering Physiology: Bases of Human Factors/ Ergonomics,* 2d ed. With permission by the publisher, Van Nostrand Reinhold. All rights reserved.)

ances, two to the side, two upward-backward, and one extending straight to the rear, to which ligaments and muscles attach. The geometry of vertebrae has only recently been reported in exact detail (Panjabi, Goel, Oxland, Takata, Duranceau, and Krag, 1992).

The spine is capable of withstanding considerable loads, yet flexible enough to allow a large range of postures. There is, however, a trade-off between load carried and flexibility. If there is no external load on the spine, only its anatomical structures (joints, ligaments and muscles) restrict its mobility. Applying load to the spinal column reduces its mobility until, under heavy load, the range of possible postures is very limited.

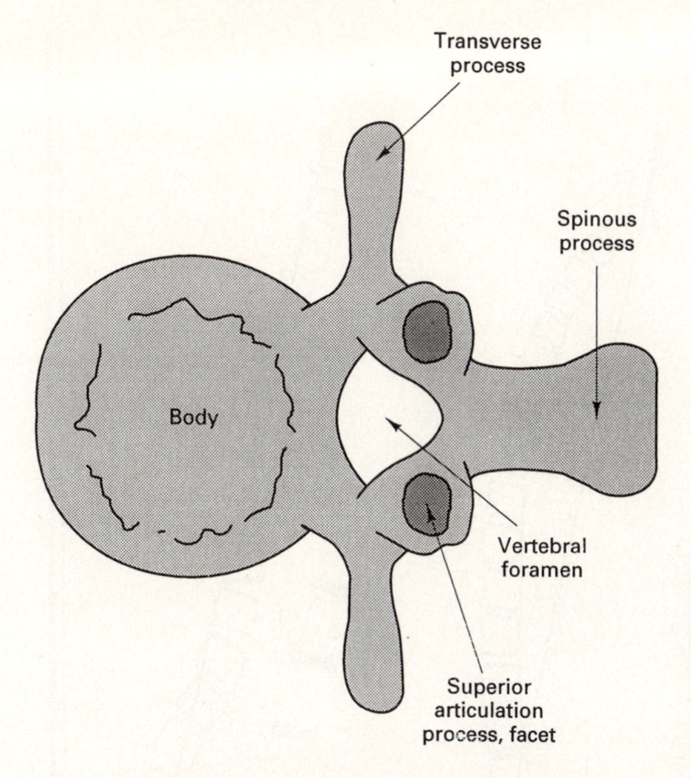

Figure 1-16. Scheme of a vertebra. (From Kroemer, Kroemer, and Kroemer-Elbert, 1990, *Engineering Physiology: Bases of Human Factors/Ergonomics,* 2d ed. With permission by the publisher, Van Nostrand Reinhold. All rights reserved.)

APPLICATION

The traditional model of the spine has been that of a straight column, as depicted in Figure 1-17(a). This simplification allows a unique description of its geometry and strain under the applied load (Aspden, 1988; Yettram and Jackman, 1981). If one considers the spinal column as an arch (Figure 1-17(b)), its load-bearing mechanism depends on its curvature. Since the geometry of spinal arching is not fixed, there is no unique solution that describes its strain. Force along the arch is thought to be transmitted along a straight line, called thrust line. The theorem of plasticity assumes, if the arch is to be stable, that the thrust line must lie within the cross section of the arch components throughout its entire length. If at any point the thrust line lies outside the arch, tensile force must keep the arch within its possible position range, or it buckles—see Figure 1-17(b).

A major load on the spine is compression. Figure 1-18 illustrates that the compressive force (C) results from the pull force (M) of trunk muscles and the weight due to segment masses and external load. Spinal compression is somewhat relieved by the upward-directed force (P) due to intraabdominal pressure (IAP). Yet, owing mostly to the slanted arrangement of load-bearing surfaces

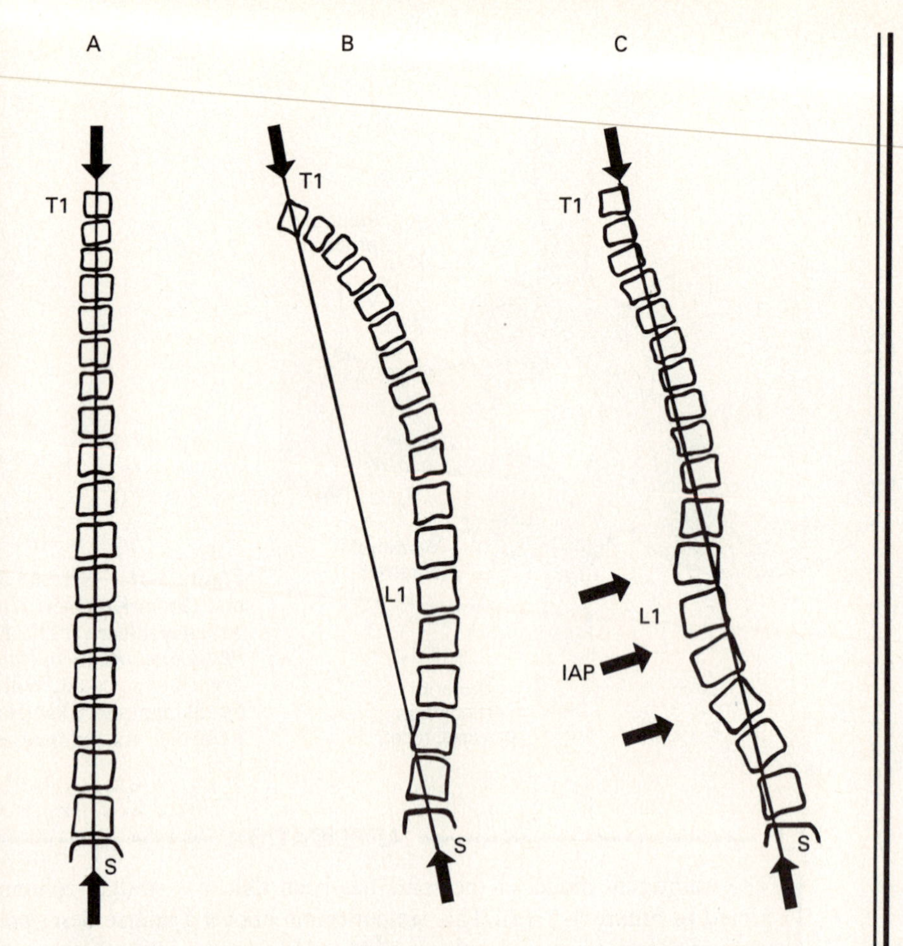

Figure 1-17. Models of the spinal column: (A) as a straight column, (B) arched; (C) supported by intraabdominal pressure. (Modified from Aspden, 1988.)

at disks and facet joints, the spine is also subjected to shear (*S*). Furthermore, the spine must withstand both bending and twisting torques (*T*).

Aspden (1988) calculated spinal strain according to his model and obtained three interesting results:

1. The calculated compression loads in a stable arched spine are considerably lower than those computed using the straight model.
2. These loads depend on the adopted posture, i.e., on the geometry of the spine.
3. Intraabdominal pressure (*IAP*) can stiffen the lumbar spine.

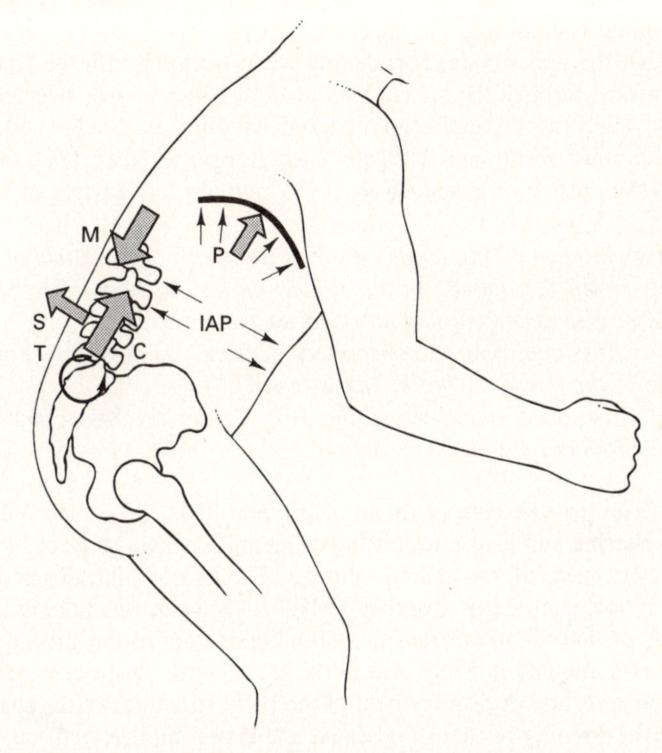

Figure 1-18. Intraabdominal pressure (IAP) and its resulting force vector (*P*) reduce the compressive force (*C*) which is produced by trunk muscle pull (*M*). Shear force (*S*) and torque (*T*) also load the spine.

The effect of the *IAP* in stiffening the lumbar spine is shown in Figure 1-17(c). The thrust line would be outside the spinal column if kyphotic flattening of the lumbar area were maintained. Yet, if lumbar lordosis is introduced, the thrust line can be kept within the spinal components, and the arch is stable. The larger *IAP*, the better lordosis can be maintained even under heavy axial loading (compression) of the spinal column. Such lordotic curvature of the spinal column is, supposedly, used by competitive weight lifters so that they may lift large weights with relatively small compressive force in the spine. In contrast, the straight spinal column (as presumed in traditional spinal modeling) generates large compression forces for the same external load (Aspden, 1988).

The spinal column is often the location of discomfort, pain, and injury because it transmits many internal and external strains. For example, when standing or sitting, impacts and vibrations from the lower body are transmitted primarily through the spinal column into the upper body. Conversely, forces and impacts experienced through the upper body, particularly when working with the hands, are transmitted downward through the spinal column to the floor or seat structures that support the body. (See the discussion of material handling in Chapter 10.) Thus, the spinal column must absorb and dissipate much energy, whether it is transmitted to the body from the outside or generated inside by muscles for exertion of work to the outside.

In engineering terms, strain is the result or the effect of stress: stress is the input, strain the output. In the 1930s, the psychologist Hans Selye introduced the concept of stressors causing stress (or distress if excessive).

It is confusing to use the term "stress" as having two meanings: either the cause or the result. (What is "job stress"?)

To avoid confusion, in this text the engineering terminology will be used: stress produces strain.

The gel-like core of the intervertebral disk, the nucleus pulposus, is the main load-bearing and load-transmitting element. It is kept in place by surrounding layers of elastic material, the annulus fibrosus. The disk and its surrounding ligaments form a "physiological shock absorber" which, when not functioning properly because of injury or deterioration, transmits unsuitable strain to the cartilaginous end plates of the vertebrae and possibly also to the facet joints. Displacement of its cartilaginous components and/or displacement of the bony structures of the spinal column may reduce the opening of the neural canal and cause impingement on the spinal cord.

The nucleus pulposus has no blood supply but is nourished through exchange of tissue fluid which circulates through the disk as a result of osmotic forces, gravitation, and the pumping effects of body movements on the spinal column. Thus, disk nourishment improves with activity and is adversely affected by immobilization. (Smoking also seems to have a negative influence on disk maintenance—Andersson, 1991.) Tissue fluid circulation is needed to provide a proper balance of water, solutes, glycosaminoglycans, protein, and collagen. If this proper balance is not achieved, degeneration and fissures may develop in the annulus through which nuclear material may penetrate and herniate peripheral areas. These are sensitive to mechanical and chemical stimulation—we feel that "something is wrong."

Low back pain (LBP) is the result of disorders that have been with humans since ancient times. It has been diagnosed among Egyptians 5000 years ago, and was discussed in 1690 by Bernadino Ramazzini. The problem is not confined to mankind, since quadrupeds suffer from low back pain as well. Everyone has an "8-in-10 chance" of suffering from back pain some time during one's life (Snook, 1988).

Low back pain is just that: a painful sensation of disorder apparently existing in the low back area. LBP may stem from a large number of possible sources, many believed to be basically associated with time-related changes in the spinal column and its supporting ligaments and muscles, starting in the teen years and usually increasing as one gets older: they result from a combination of repetitive trauma and the normal aging processes. Strong activity demands may trigger the occurrence of various low back symptoms. However, except in cases of acute injuries, the causes or reasons for LBP usually remain unclear. Rowe (1983) found that only 4 percent in a large sample of industrial LBP cases were related to traumatic injuries during industrial work. Classification of low back pain is difficult, and different clinicians frequently diagnose it differently (Nachemson and Andersson, 1982). Furthermore, many persons who do have signs of spinal degeneration (as diagnosed by spinal imaging, such as x-rays, CT scans, MRI scans, myelograms, and diskograms) do in fact not suffer from pain. Among the changes commonly found in the spinal column are damage to the cartilaginous endplates at the main body of the vertebrae; degeneration of the annulus fibrosus and of the nucleus pulposus of the disk; and "drying out" of the disk structure. All make the disk behave less like a hydrostatic device, and change the spine's biomechanical motion characteristics.

A theory popular in recent decades is that many overexertion injuries of the spinal column can be traced to, or explained by, compression of the spinal disks. Such excessive compression is thought to damage, temporarily or permanently, the fibro-cartilage disk, which, in consequence, may lead to the intrusion of disk structures into its surroundings, particularly toward the spinal cord. Yet, experimental measurements of the compression within the disk in the living human body are difficult to perform. Thus, many assessments rely on calculations using biomechanical models (see Chapter 7), which, in turn, often simply assume static strains. The spinal joints (disks and facets) are not subject only to compression but also to shear, bending, and twisting. To consider such combined strains, and to take into account muscles and ligaments, is a major task of biomechanical research and modeling in the near future.

Muscle

☞☞☞ *The Greek physician Galen (129 to about 199) studied human physiology, first at a gladiator school in Pergamon and then in Rome. He identified muscles and showed that they worked in groups. He also showed the importance of the spinal cord by cutting it in various positions in animals and noting the extent of the resulting paralysis (Asimov, 1989).* ☜☜☜

In the human, there are three types of muscle, which together comprise about 40 percent of the weight of the body. *Cardiac* muscle brings about contractions of the heart. *Smooth* muscle works on body organs, such as by constricting blood vessels. *Skeletal* muscle is under voluntary control of the somatic nervous system (see Chapter 3) and serves two purposes. One is to maintain postural balance by generating tensile force. The other is to cause local motion of body segments by pulling on the bones to which the muscles are attached, thereby creating torques or moments around the body joints which serve as pivots. There are several hundred skeletal muscles in the human body, identified by their Latin names. The Greek words for muscle, *mys* or *myos,* are often used as prefixes.

Muscles perform their functions by contracting, that is, by quickly and reversibly developing internal lengthwise tension, often but not always accompanied by a shortening and followed by elongation. The end of the muscle that remains essentially stationary is called the origin; the other end moving with the moving body segment is called the insertion.

Architecture of muscle.

Proteins. The main components of muscle besides water (about 75 percent by weight) are proteins (20 percent). *Collagen,* an abundant protein in the body, constitutes the insoluble fiber of the binding and supportive substance in muscle tissue. The proteins *actin* and *myosin* form rod-shaped polymerized molecules that attach end-to-end with some overlapping, creating thin strands.

Filaments. The actin and myosin *filaments* form the "contracting microstructure" of the muscle. As seen in a cross-cut through the muscle, each myosin molecule is surrounded by six actin molecules. The thin actin strands are wound around the thicker myosin in form of a spiral (double helix). The actin filaments project from the Z-disks (see below) like bristles of a brush. If the filaments are "at rest," *troponin* and *tropomysin* proteins separate actins and myosins. During a contraction, troponins and tropomysin are pulled away, and the actin strands slide along the myosin, temporarily connecting with each other by so-called cross-bridges, looking like miniature golf-club heads. Figure 1-19 illustrates the relative locations of actin and myosin within contracted, relaxed, and stretched muscle.

Fibrils and Striation. Between 10 to 500 muscle filaments are packed tightly into a bundle known as fibril (also termed myofibril) and covered by a membrane called endomysium. Since contraction of fibrils can be easily observed, they are often called "contractile elements." Within each fibril, the endings of actin and myosin rods overlap; this dense region appears in the electron microscope as a band dark. Lighter regions consist mainly of actin molecules and are called isotropic or I-bands. The darker stripes consist mainly of myosin molecules and are called anisotropic or A-bands. The banding or striping has led to the name "striated" muscle. These striations show a repeating pattern about every 250 Å* which makes the fibril appear to have a series of disklike partitions. These are called Z-disks (or Z-lines). They are dense membranes across fibrils containing the transverse part of the "plumbing and fueling" system of the muscle, which provides fluid transport and carries chemical and electrical messages for control of muscular activities. The distance between Z-lines, the *sarcomere,* is about 0.002 mm. Spaces among the fibrils are filled with a network of tubules, sacs, cisterns, and channels which are connected to the larger blood vessels in the Z-disks.

Fibers. Fibrils, in turn, are also packed into bundles, wrapped by a connective tissue called fascia. These bundles are separated from each other and bathed in a

* 1 Å (angstrom) = 10^{-10} mm.

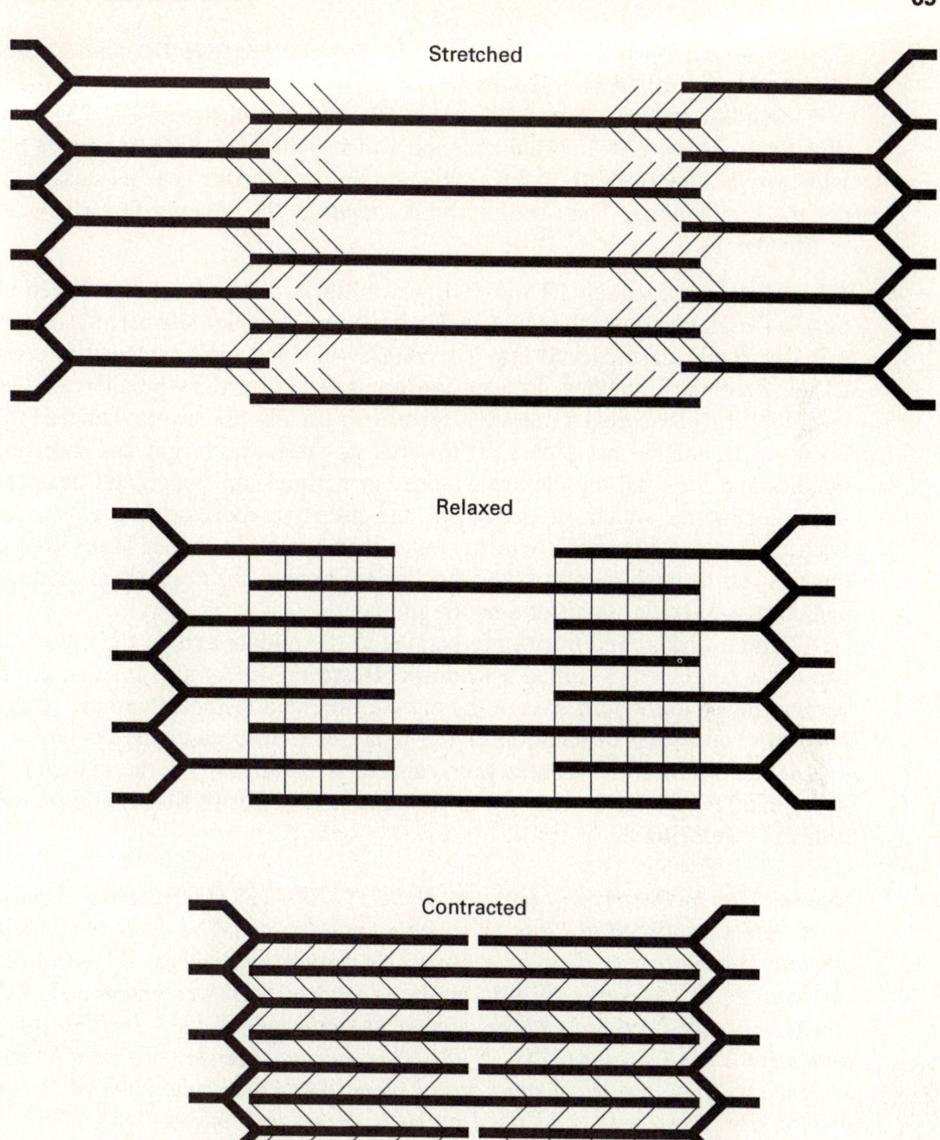

Figure 1-19. Schemes of the location of actin and myosin filaments, and of the cross-bridges, within stretched, relaxed and contracted muscle. (From Kroemer, Kroemer, and Kroemer-Elbert, 1990, *Engineering Physiology: Bases of Human Factors/Ergonomics,* 2d ed. With permission by the publisher, Van Nostrand Reinhold. All rights reserved.)

fluid called sarcoplasm. Bundles of fibrils are grouped together and covered by a thin membrane called sarcolemma.

Bundles of fibrils packed together constitute the muscle *fiber*. This is a cylindrical unit, 10^5 to 10^6 Å in diameter and 1 to 50 mm long. Each is a single large cell with several hundred cell nuclei located at regular intervals near its surface, just under the sarcolemma. They contain *mitrochondria,* the "energy factories" discussed in Chapter 2.

Muscle. Between 10 and 150 muscle fibers constitute the so-called primary bundle *(fasciculus),* which is wrapped in a membrane called the perimysium. Several primary bundles are packed into a secondary bundle, again wrapped in connective tissue. Secondary bundles, in turn, are grouped into tertiary bundles, and so forth until the structure called a muscle is formed. It usually has several hundred thousand fibers in its middle and tapers off towards its ends, the origin and insertion. The bundles and the total muscle are wrapped in perimysium, tough and dense collagenous membranes which, at the origin and insertion, develop into elastic tendons. Each tendon attaches directly to the outer membrane of a bone. Many (but not all) tendons run through tough tissue tubes, called tendon sheaths, which guide tendons and contain synovial fluid for smooth gliding.

Often, fibrils (and fibers) are parallel to the middle axis of the muscle. Such a parallel arrangement is called a fusiform. However, fibers also may be arranged at various angles to the long axis of the muscle, like in a feather: there are several penniforms. The actual orientation of the fibers with respect to the centerline of the muscle determines the contraction capability (strength) of the muscle. Muscle strength also depends on the size of the muscle (number of fibers) and on the types of fibers (see below).

The Swiss physiologist von Haller (1708–1777) conducted experimental work which, published in 1766, showed that muscles could be made to constrict by a stimulus transmitted to them by a nerve. He showed that nerves all led to the brain or spinal cord, indicating them to be the centers of sense perception and of responsive actions. He is considered the founder of modern neurology. In 1780, the Italian anatomist Luigi Galvani (1737–1798) observed that muscles of dissected frog legs twitched when an electrical spark struck them. While his explanation was wrong, the finding that electricity was involved with nerve and muscle action was correct (Asimov, 1989).

The "motor unit." Several muscle fibrils share a common portion of a sarcolemma where the axon end of a motor nerve, called alpha-motoneuron, attaches. This junction between nerve and muscle is called "motor endplate." Each nerve innervates a number, usually hundreds or thousands, of muscle fibers. These fibers under common controls, called a "motor unit," are all stimulated by the same signal. (One alpha-motoneuron may innervate more than one motor unit, however.)

Motor units can be classified by the innervation ratio, describing the number of fibers innervated by one neuron. Muscles used for finely controlled actions (e.g., rotation of the eyeball) have a ratio such as 1 : 7, while muscles for gross activities may have ratios of 1 : 1000 or more.

Another classification of motor units describes their types of muscle fibrils. Type I is usually associated with muscles exerting fine and enduring control, e.g., over the fingers or eyes or those in the back. The Type I fibril is short and appears red because it is penetrated by many capillaries, which provide good blood supply and oxygen storage: hence it resists fatigue. A twitch is produced by a relatively low action potential, but takes relatively a long time, 60 to 120 ms, to peak.

Type II fibrils appear light in comparison because they are not profusely capillarized. They are less resistant to fatigue but perform better under anaerobic conditions. They require high action potentials but produce fast twitches, 15 to 50 ms to peak. (Type II fibrils are subdivided according to their supply with capillaries.)

Note that the times just given for slow or fast twitches have been measured in isolated muscle preparations, usually taken from cats, and with artificial stimulation. In the human muscle, distinct groupings of muscle fibrils groupings are not as prevalent, and the actual behavior of the stimulated muscle depends on many factors, such as fatigue and external resistance. Basmajian and De Luca (1985) warn, therefore, against simplistic use of this classification of intramuscle fibrils, because it may not provide reliable information on human motor unit behavior during a voluntary contraction.

Activation of the Motor Unit. As previously described, the muscle fiber consisting of thousands of fibrils is covered by a semipermeable membrane called the sarcolemma. At rest, sodium (positive) and some potassium (positive) ions accumulate on the outside of the membrane, while (negative) chlorine ions are on the inside. This establishes a polarized sarcolemma, with a transmural electrical potential of nearly 100 mV.

An action impulse arriving from an alpha-motoneuron at the motor endplate must be strong enough to depolarize the membrane potential by at least 40 mV. If this threshold is not achieved, the motor unit does not react; if achieved, the motor unit contracts completely. This *all-or-none principle* governs muscle contraction.

Given sufficient depolarization, the sarcolemma permeability is increased so that (positive) sodium ions can penetrate and neutralize (negative) chlorine ions. This local depolarization, called endplate potential, propagates at a speed of about 5 m s^{-1} along the membrane. The depolarization wave, acting like an electric current, causes hydrolysis of water molecules. This releases hydrogen and hydroxal ions, which, in turn, split off a phosphate group from adenosine triphosphate (ATP), with activated myosin (ATPase) present as a catalyst (see the section on energy release in Chapter 2).

The decomposition of ATP results in the formation of ADP and of phosphoric acid, H_3PO_4, a reaction which liberates about 11 kcal per mole of ATP.* That reaction supplies the primary source of energy for muscular contraction (see Chapter 2).

*One mole is the quantity of a chemical compound whose weight in grams equals its molecular weight.

In death, the ADP complex is disintegrated and firmly bonds the bridges to the actin molecules, leading to "rigor mortis.")

At the same time, the action potential moving along the sarcolemma also propagates into the fibrils through a system of tubules, called the sarcoplasmic reticulum. This "plumbing" network starts at the sarcolemma at each Z-disk and transverses the muscle fibrils. Within its cisterns, calcium ions are trapped, because the membranes of the sarcoplasmic reticulum do not allow them to escape when the muscle is at rest. However, the arriving action potential makes the cistern membranes permeable to the positive calcium ions, which, discharged into the myofibrillar fluid, raise its calcium concentration a thousandfold. This, in turn, allows some calcium ions to combine with myosin molecules to form the activated myosin catalyst (ATPase). In addition, the released calcium binds to the troponin-tropomysin molecules, pulling the protein strands away from the binding sites between the actin and myosin molecules. This allows cross-bridges to be established between actin and myosin and the actual contraction process of the muscle fibril commences.

Twitch. After an action impulse arrives at the motor unit, there is an initial "latency" (of about 10 ms) during which no perceptible muscle tension or change in muscle length takes place. However, during this time the alpha-motoneuron signal generates an endplate potential, releases calcium ions from the cisterns, activates the ATP fuel element, and establishes cross-bridges between actin and myosin. Then follows the period of "contraction" or shortening, which takes up to 40 ms in isolated fast-twitch fibrils and up to 110 ms in isolated slow-twitch fibrils. (In the living muscle, those times are considerably longer.) As mentioned earlier, splitting a phosphate group off the ATP releases heat energy. This causes the cross-bridges between myosin and actin to vibrate, alternately establishing and breaking connections to the actin molecule. Although the details of this process are not yet fully understood (Astrand and Rodahl, 1986), one observes a "ratchet" action which tries to pull the actin along the myosin filaments. This reduces the length of a sarcomere, and hence shortens the length of the entire muscle, depending on the number of motor units involved in series.

Then follow 30 to 50 milliseconds of "relaxation" if no new alpha-motoneuron impulse arrives. During this period, ATP causes dissociation of actin from the positive calcium ions, so that calcium is again transported through the sarcoplasmic reticulum membranes into the cisterns, where it remains confined until the next action impulse. Actin and myosin filaments separate, and the troponin-tropomyosin complexes again occupy the binding sites of the cross-bridges. With disengaged cross-bridges and hence no contraction, the muscle relaxes and elongates to its resting length, possibly helped by the pull of gravity or an antagonistic muscle. During the relaxation period ATP is resynthesized. ADP picks up a phosphate group from the breakdown of phosphocreatine into creatin and phosphoric acid. This is called the Lohman reaction, which "recharges the ATP battery." The breakdown of phosphocreatine provides energy additional to that gained in the earlier breakdown of ATP. (See Chapter 2 for more details.)

The final phase is "recovery." This lasts 30 ms or longer, if no contraction signal arrives. During this time, leftover phosphoric acid combines with glucose to form diophosphate. This may be completely oxidized to carbon dioxide and water in the mitochondria, if sufficient oxygen is available on site. In this case, 675 kcal are yielded per mole of glucose, which is enough to regenerate 38 moles of ATP. However, if the oxygen supply is insufficient, or more energy is needed, "anaerobic glycolysis" is used to reduce the diphosphate to lactic and phosphoric acids. Most of the produced lactic acid is finally converted to glycogen, which then is oxidized to carbon dioxide and water—provided that oxygen is available from myoglobin or arterial blood.

Still looking at a single motor unit, one sees *summation* (or superposition) of twitches when they are initiated frequently after each other, so that a contraction is not yet completely released by the time the next stimulus arrives. In this case, the new contraction builds on a level higher than if the fiber were completely relaxed, and accordingly higher contractile tension is generated in the muscle. Such "staircase effect" takes place when excitation impulses arrive at frequencies of 10 or more per second. When a muscle is stimulated above frequencies of 30 to 40 stimuli per second, successive contractions fuse together, resulting in a maintained contraction called tetanus. In superposition of twitches, the muscle tension generated may be two or three times as large as a single twitch, and a full tetanus may build up to five times the single-twitch tension.

Thus, for a single motor unit, the frequency of contractions and the strength of contraction are controlled by *rate coding* of the exciting nervous signals. Fibers belonging to one motor unit are generally in various locations within the muscle. Therefore, activation of one motor unit brings about "weak contractions" throughout the muscle. If more motor units are recruited at the same time, the contractile strength exerted by the muscle increases. Thus, nervous control of muscle "strength" (see later) follows a complex pattern of rate and recruitment coding.

If not enough time is provided for relaxation and recovery, remaining metabolic byproducts such as lactic acid are not removed sufficiently, potassium accumulates, the muscle experiences an "oxygen debt," and ATP breakdown is hindered—see Chapter 2. This leads to muscular *fatigue,* which prevents the continuation of strong contractions. Muscular fatigue is overcome by rest, during which the accumulated metabolic byproducts are removed.

Fatigue of a single motor unit, or of a whole muscle, is a function of the frequency and intensity of muscular contraction, and of the period of time over which it is maintained. The more strength exertion is required of a given muscle, the shorter the period through which this strength can be maintained. Figure 1-20 schematically shows this relationship between (static, isometric) strength exertion and *endurance*. Maximal muscle strength can be maintained for only a few seconds; 50 percent of strength is available for about one minute; less than 20 percent can be applied continuously for long times.

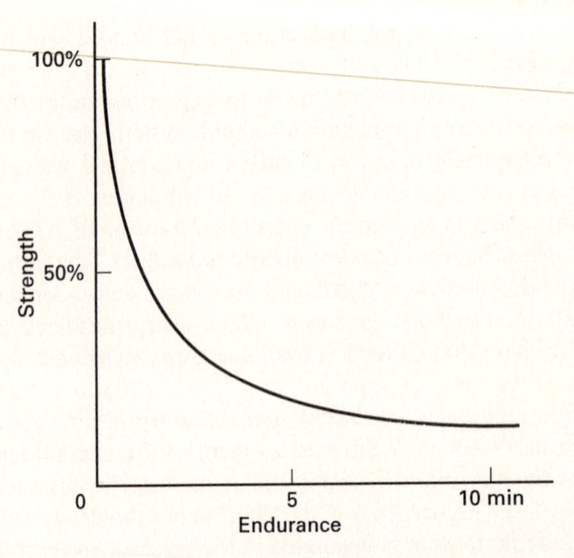

Figure 1-20. Strength and endurance of muscular exertion. (From Kroemer, Kroemer, and Kroemer-Elbert, 1990, *Engineering Physiology: Bases of Human Factors/Ergonomics,* 2d ed. With permission by the publisher, Van Nostrand Reinhold. All rights reserved.)

Length-strength relationships within striated muscle. In engineering terms, muscles exhibit "viscoelastic" behavior. They are "viscous" in that their behavior depends both on the amount by which they are deformed, and on the rate of deformation. They are "elastic" in that, after deformation, they return to their original length and shape. These behaviors, however, are not pure in the muscle, because it is nonhomogeneous, anisotropic, and discontinuous in its mass. Nevertheless, nonlinear elastic theory and viscoelastic descriptors can be used to describe major features of muscular performance (Enoka, 1988; Schneck, 1985, 1990).

Under a "no-load" condition, in which no external force applies and no internal contraction occurs, the muscle is at its *resting length.* Fiber stimulation, with no external load, causes the muscle to contract to its smallest possible length, which is at about 60 percent of resting length. In this condition, the actin proteins are completely curled around the myosin rods, so that the Z-lines are as close as possible. This is the shortest possible length of the sarcomere. In this position, the muscle cannot develop any active contraction force.

Stretching the muscle beyond resting length by application of external forces between insertion and origin lengthens it passively, i.e., increases the length of the sarcomere. The actin and myosin fibrils are slid along each other by the external stretch force. When the muscle is stretched to 120 or 130 percent of resting length, the cross-bridges between the actin and myosin rods are in an optimal position to generate a contractile force. If the fibrils are enlongated further, the cross-bridge overlap between the protein rods is reduced. At about 160 percent of resting length, so little overlap remains that no force can be developed internally. Thus, the curve of active contractile force developed within a muscle is zero at approximately 60 percent resting length, about 0.9 at resting length, at unit value at about 120 to 130 percent of resting length, and then falls back to 0 at about 160 percent resting length. (These values apply to an isometric twitch contraction.)

Passive reaction to external stretch also occurs, as in a rubber band. This passive stretch resistance increases nearly linearly from resting length to the point of muscle or tendon (attachment) breakage. Thus, above resting length, the tension in the muscle is the summation of active and passive strain. This explains why we stretch ("preload") muscles for a strong force exertion, as in bringing the arm behind the shoulder before throwing.

The viscoelastic theory also explains why the tension that can be developed isometrically is the highest possible, while in active shortening, muscle tension is decidedly lower. The higher the velocity of muscle contraction, the faster actin and myosin filaments slide by each other and the less time is available for the cross-bridges to develop and hold. This reduction in force capability of the muscle holds true for both *concentric* and *eccentric activities*. In eccentric activities, however, where the muscle becomes increasingly lengthened beyond resting length, the total force resisting the stretch increases with larger length, owing to the summing of active and passive tension within the muscle.

☞☞☞ *In 1680, the book* De Motu Animalium *(about the motion of animals) of the Italian physiologist Giovanni Alfonso Borelli (1608–1679) was posthumously published. He explained muscular action on a mechanical basis, describing the actions of bones and muscles in terms of a system of levers. This showed that natural laws govern life and nonlife alike (Asimov, 1989). In 1721, Jean Bernoulli wrote his "Physiomechanicae Dissertatio" "De Motu Musculorum" (biomechanics of the motion of muscles), followed in 1799 by Charles-Augustin de Coulomb's "Memoire sur la force des hommes" (about human strength).* ☜☜☜

Biomechanical Description of the Body

Do muscles work optimally? The question "How does the body perform activities if free to chose different ways?" is of basic interest (Bohannon, 1991). It becomes quite important in biomechanical modeling (see Chapter 7), where the human body is considered to be a system of bones, joints, and masses, activated by muscles that span joints. The muscles develop forces which are transmitted as torques across the joints. Tolerance limits in muscles, joints, and joint-enclosing ligaments are of particular biomechanical interest, especially in maximal efforts; but even in submaximal exertions, the body apparently uses some kind of "optimization" in the shared efforts of muscles.

Genaidy and Houshyar (1989) discussed various optimization theories and modeling techniques. The overall goal may be "minimization" of total energy spent, contraction effort in a muscle, number of muscles employed, forces endured in joints and ligaments, muscle fatigue, perceived pain, and bone compression. In other cases, one can presume "maximization" of the overall output, of strength exerted to the outside, etc.

Modeling requires the identification of the elements involved, for example, of all muscles participating in an activity, or of all joint loadings. This can result in a large number of unknown forces and torques, often much in excess of the available number of equations of equilibrium. To solve this indeterminate problem, "optimization" assumptions and techniques may be employed. Many of these use linear or nonlinear techniques to find a unique solution. The mathematical optimization techniques need to be bound by limits and assumptions which realistically describe the human body. They must be based on sound physiological and biomechanical principles, rather than simply be conceptually or mathematically convenient—much research remains to be done in this area.

APPLICATION

Links, joints, and masses. The human skeletal system is often simplified into a relatively small number of straight-line links (representing long bones) and joints (representing major articulations). Figure 1-21 shows a typical link-joint system with identifying numbers. In it, the feet are not subdivided into their components, and the spinal column is represented by only

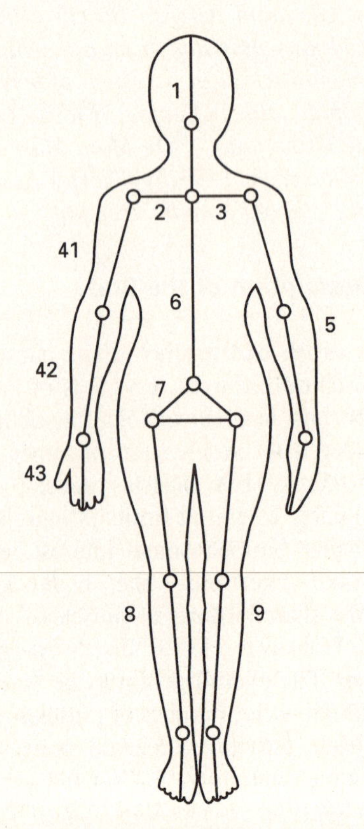

Figure 1-21. Numbering scheme for the body link system.

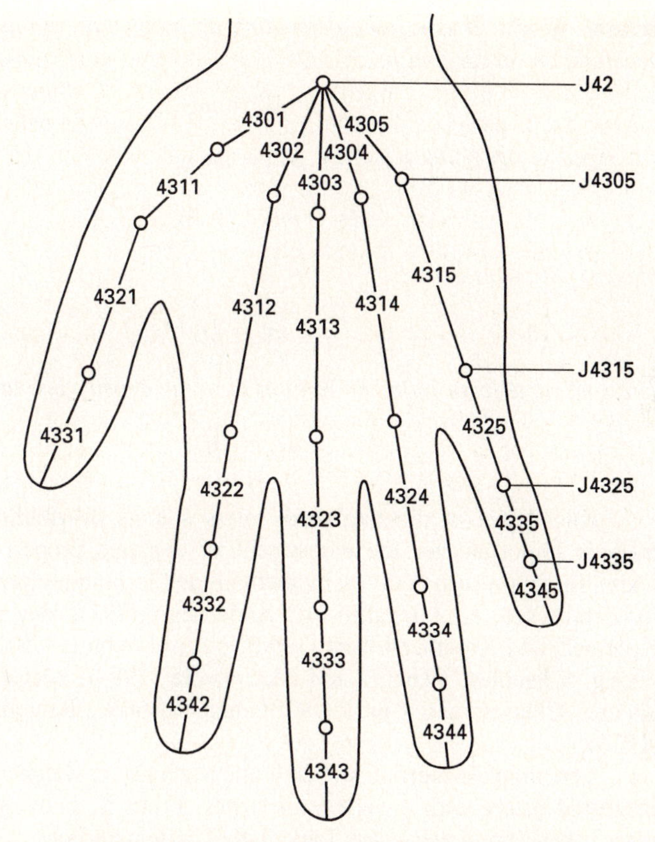

Figure 1-22. Numbering scheme for the right hand.

three links. If more details are needed, they can be generated—as done, for example, for the hand shown in Figure 1-22.

The determination of the location of the joint center-of-rotation is relatively easy for simple articulations, such as the hinge joints in fingers and elbows. However, it is much more difficult for complex joints with several degrees of freedom, such as in the hip or in the shoulder. Once the joints are established, a straight-line link length is defined as the distance between adjacent joint centers. For definitions of the joint centers and of the links between them, see NASA/Webb (1978), Chaffin and Andersson (1984), or Kroemer, Kroemer, and Kroemer-Elbert (1990).

Knowledge of whole body or of segment volume is necessary to calculate inertial properties. Use of the *Archimedes' principle* provides the body volume: the subject is immersed in a container filled with water, and the displaced water yields the volume. Other data can be obtained from cadaver dissection or from model calculations (Kroemer, Snook, Meadows, and Deutsch, 1988).

Weight W *is a force depending on body mass* m *and the gravitational acceleration, according to*

$$W = mg$$

Density D is the mass per unit volume:

$$D = mV^{-1} = Wg^{-1}V^{-1}$$

Mass is, of course,

$$m = DV$$

The specific density D_s is the ratio of D to the density of water, D_w:

$$D_s = DD_w^{-1}$$

Knowledge of the total body mass and its distribution throughout the body is important for the assessment of dynamic properties of the human body. To obtain such data, many methods and techniques have been developed: for details see, e.g., Chaffin and Andersson (1991), Hay (1973), Kroemer, Kroemer, and Kroemer-Elbert (1990), NASA/Webb (1978), Kaleps, Clauser, Young, Chandler, Zehner, and McConville (1984), McConville, Churchill, Kaleps, Clauser, and Cuzzi (1980), and Roebuck, Kroemer, and Thomson (1975).

The simplest inertial property under gravity is weight, *W*, which can be measured easily with a variety of scales. From it, body segment weight or mass can be predicted—see Table 1-8. The human body is not homogeneous throughout; its density varies depending on cavities, water content, fat tissue, bone components, and so on. Still, in many cases it is sufficient to assume that either the body segment in question, or even the whole body, is of constant (average) density.

Lean body mass or *lean body weight* is often used to distinguish between body compositions. Basic structural components such as skin, muscle, and bone are relatively constant in percentage from individual to individual. Fat, however, varies in percentage of total mass or weight throughout the body and for different persons. This allows body weight to be expressed as lean body weight plus fat weight. There are several techniques to determine body fat. Many rely on skinfold measures: in selected areas of the body, the fold thickness of skin with its underlayment of fat tissue is measured with a special caliper.

The body mass is often considered as concentrated in one point in the body, where its physical characteristics respond in the same way as if distributed throughout the body. The location of the "center of mass" (also called center of gravity) shifts with body posture, with muscular contractions, food

TABLE 1-8. PREDICTION EQUATIONS TO ESTIMATE SEGMENT MASS (IN KG) FROM TOTAL BODY WEIGHT *W* (KG)

Segment	Empirical equation	Standard error of estimate
Head	$0.0306W + 2.46$	0.43
Head and neck	$0.0534W + 2.33$	0.60
Neck	$0.0146W + 0.60$	0.21
Head, neck and torso	$0.5940W - 2.20$	2.01
Neck and torso	$0.5582W - 4.26$	1.72
Total arm	$0.0505W + 0.01$	0.35
Upper arm	$0.0274W - 0.01$	0.19
Forearm and hand	$0.0233W - 0.01$	0.20
Forearm	$0.0189W - 0.16$	0.15
Hand	$0.0055W + 0.07$	0.07
Total leg	$0.1582W + 0.05$	1.02
Thigh	$0.1159W - 1.02$	0.71
Shank and foot	$0.0452W + 0.82$	0.41
Shank	$0.0375W + 0.38$	0.33
Foot	$0.0069W + 0.47$	0.11

SOURCE: Adapted from Kroemer, Kroemer, and Kroemer-Elbert, 1990.

and fluid ingestion or excretion, and respiration. Table 1-9 lists relative locations of mass centers. For further information about mass properties of the human body, see Chandler, Clauser, McConville, Reynolds, and Young (1975), NASA (1989), and Roebuck, Kroemer, and Thomson (1975).

☞☞☞ *Newton's first law states that a mass remains at uniform motion (which includes being at rest) until acted upon by unbalanced external forces. The second law, derived from the first, indicates that force is proportional to the acceleration of a mass. The third law states that action is opposed by reaction.*

Newton's second law makes it clear that force is not a basic unit but a derived one. Relatedly, no device exists that measures force directly. All measuring devices for force (or torque) rely on other physical phenomena, which are then transformed and calibrated in units of force (or torque). The events used to assess force are usually either displacement (such as bending of a metal beam) or acceleration experienced by a mass.

The correct unit for force measurement is the newton; one pound-force unit is approximately 4.45 newtons, and 1 kg_f *(also called 1 kilopond, kp) equals 9.81 newtons. The pound (lb), ounce (oz), and gram (g) are usually not force but mass units.*

TABLE 1-9. LOCATIONS OF THE CENTERS OF MASS OF BODY SEGMENTS, MEASURED ON THE STRAIGHT BODY IN PERCENT FROM THEIR PROXIMAL ENDS

	Harless (1860)	Braune and Fischer (1889)	Fischer (1906)	Dempster (1955)	Clauser, McConville and Young (1969)
Sample Size	2	3	1	8	13
Head	36.2%	—	—	43.3%	46.6%
Trunk*	44.8	—	—	—	38.0
Total arm	—	—	44.6%	—	41.3
Upper arm	—	47.0%	45.0	43.6	51.3
Forearm and hand*	—	47.2	46.2	67.7	62.6
Forearm*	42.0	42.1	—	43.0	39.0
Hand*	39.7	—	—	49.4	18.0
Total leg*	—	—	41.2	43.3	38.2
Thigh*	48.9	44.0	43.6	43.3	37.2
Calf and foot	—	52.4	53.7	43.7	47.5
Calf	43.3	42.0	43.3	43.3	37.1
Foot	44.4	44.4	—	42.9	44.9
Total body	58.6**	—	—	—	58.8**

*The values on these lines are not directly comparable, since the different investigators used differing definitions for segment lengths.

**Percent of stature, measured from the floor up.

SOURCE: Adapted from Kroemer, Kroemer, Kroemer-Elbert, 1990.

Torque (also called moment) is the product of force and its lever arm (distance) to the articulation about which it acts; the direction of the force must be at a right angle to its lever arm. In kinesiology, the lever arm is often called the "mechanical advantage." Force (as well as torque) is a vector, which means that it must be described not only in magnitude but also by direction, by its line of application, and even by the point of application.

Body kinetics. The "stick-person" concept, consisting of links and joints, embellished with volumes and masses and driven by muscles, can be used to model human motion and strength capabilities—see Chapter 7. Figure 1-12 shows a model of the human body exerting forces or torques with the hand to an outside object. Within the body, the forces are transmitted along the links. First, the force exerted with the right hand, modified by the existing mechanical advantages, must be transmitted across the right elbow (E). (Also, at the elbow additional force or torque must be generated to support the mass of hand and forearm. However, for the moment, the model will be considered massless.) Similarly, the shoulder S must transmit the same efforts, again modified by existing mechanical conditions. In this manner, all subsequent joints transmit the effort exerted with the hands throughout the trunk, hips, and legs, and finally from the foot to the floor. Here, force and torque vectors can be separated again into their component directions, similar to the vector analysis at the hands. Still assuming a massless body, the same sum of vectors must exist at the feet as was found at the hands (Oezkaya and Nordin, 1991).

Of course, the assumption of no body mass is unrealistic. This can be remedied by incorporating information about mass properties of the human body, as mentioned earlier. Body motions, instead of the static position assumed here, would complicate the model.

Describing human motion (kinematics). In the past, human movement has been described with anatomically derived terms used in the medical profession: the nouns *flexion, tension, duction, rotation,* and *nation* together with the prefixes *ex, hyper, ad, ab, in,* and *out*. Unfortunately, the same terms are applied indiscriminately to rotational movements about joints and to translational displacements of limbs; certain motions said to occur in a given plane may also occur in others. This makes it very difficult to describe relative locations of body parts, and a "zero position" is often not specified.

Based on Roebuck's 1968 attempt to generate a more appropriate terminology (described in detail by Roebuck, Kroemer, and Thomson, 1975), a new taxonomy uses as coordinate system the common frontal, medial, and transverse planes which may be centered at each body articulation considered—see Figure 1-23.

To describe movement verbally, the verb "twist" indicates rotation about a long axis. The familiar "flexion" and "extension" are maintained. "Pivot" indicates rotation about an axis perpendicular to the flexion-extension axis (Kroemer, Marras, McGlothlin, McIntyre, and Nordin, 1989). Roebuck's prefixes are kept, "e-" signifying out or up from the standard position; "in-" the opposite of e- (i.e., in or down), while "posi-" and "negi-" are self-explanatory. Table 1-10 lists the terms. Any motion can be identified by a descriptive term composed of the designations for the plane, the direction, and the type of motion. For example, "sag-in-flexion" indicates that flexion occurs in the sagittal plane and is inward, e.g., toward the origin of the reference system.

Consistent use of reference points, planes, and motion terms should reduce uncertainty that exists about human motion capabilities and allow their measurement and reporting, utilizing computer models and systems.

Human Strength

As previously discussed, muscular contraction occurs by activation, concurrently or subsequently, of many motor units. The cooperative effort of the participating units, controlled by both rate and recruitment coding, determines the contraction of the whole muscle. In general, one cannot voluntarily contract more than about two-

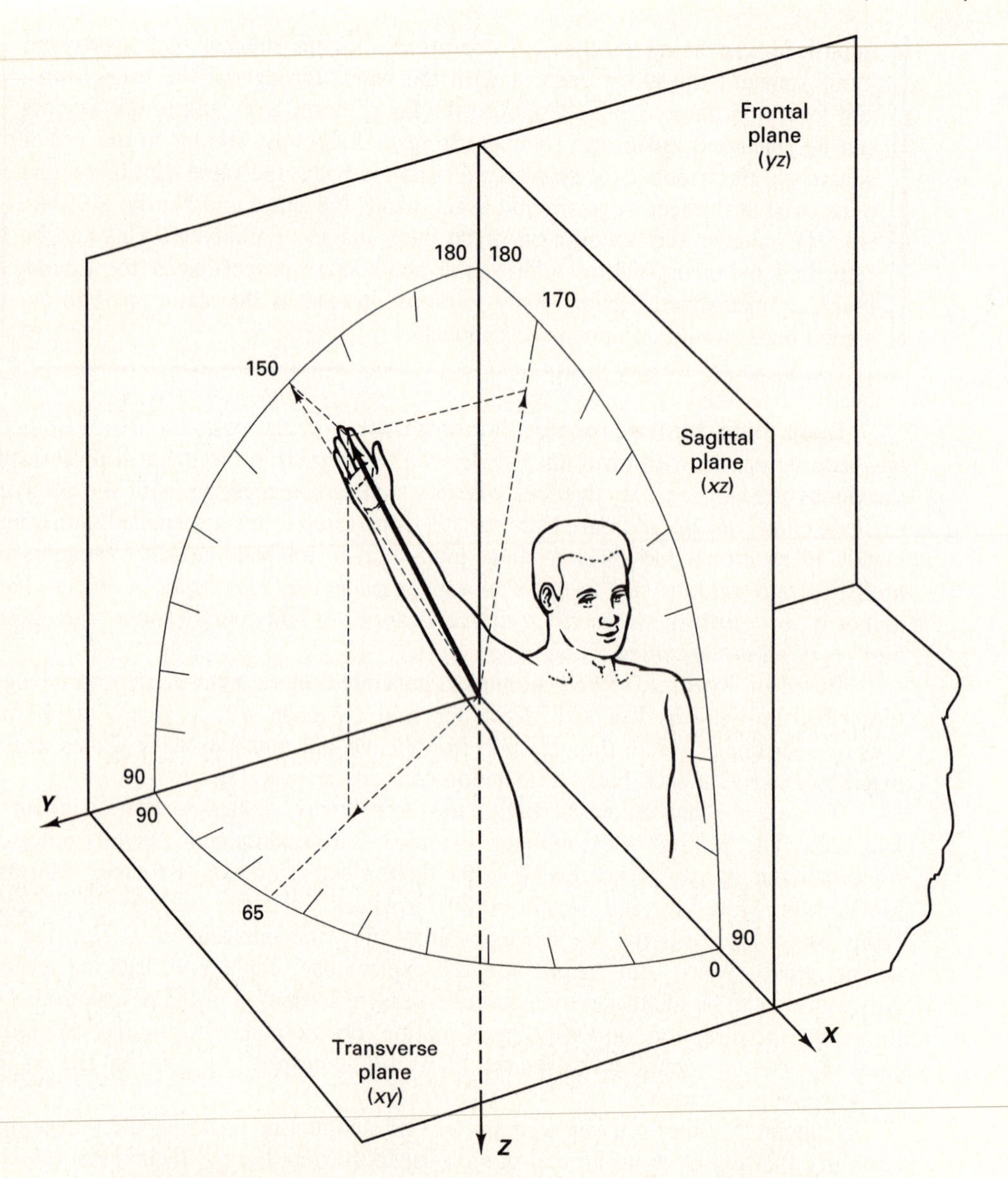

Figure 1-23. Descriptor of the angular position of the right arm in the new notation system. The position of the hand point on the arm vector is $F = 150°$, $S = 80°$, $T = 25°$. Note that the directions for x, y, and z are commonly used in biomechanics to describe accelerations of the "vehicle" that carries a human occupant. (From Kroemer, Kroemer, and Kroemer-Elbert, 1990, *Engineering Physiology: Bases of Human Factors/Ergonomics,* 2d ed. With permission by the publisher, Van Nostrand Reinhold. All rights reserved.)

TABLE 1-10. NEW TERMINOLOGY TO DESCRIBE BODY MOTIONS

New terms	Meaning	Replacing
Pivot and flexion, extension	Rotation of a body segment about its proximal joint	duction vection rotation
Twist	Rotation of a body segment about its internal axis	pronation supination rotation
Ex-	Away (up, out) from zero	ab- e-
In-	Toward (in, down) zero	ad-
Clock(wise) (or none)	In clockwise direction, as seen on own body	supi- or pro- (depending on body segment)
Counter- (clockwise)	In counterclockwise direction, as seen on own body	pro- or supi- (depending on body segment
Front-, trans-, sag-	Reference to the plane in which motion is described	uncertainty and confusion

Examples:
Front-ex-pivot = pivoting movement in the frontal plane, away from zero.
Trans-in-twist = twisting movement in the transverse plane, clockwise.

SOURCE: From Kroemer, Kroemer, and Kroemer-Elbert, 1990, *Engineering Physiology: Bases of Human Factors/Ergonomics,* 2d ed. With permission by the publisher, Van Nostrand Reinhold.

thirds of all fibers of a muscle at once. Apparently, this limitation assures that structural tensile capacity is not exceeded, although this can occur as a result of a reflex, which might damage or even tear a muscle or tendon.

In the human body, muscles are arranged in pairs or groups that act in opposite directions around their common joint. For example, the triceps muscle opens the elbow angle, while the biceps, brachialis and brachioradilis muscles reduce the elbow angle. Simultaneous activation (coactivation) of these *agonistic* (also called *protagonistic*) and *antagonistic* muscles allows the body to control strength and motion.

Voluntary muscle strength is the torque that a given muscle (or group of muscles) can maximally develop voluntarily around a skeletal articulation which is spanned by the muscle(s). (For convenience, this torque may be measured as the force acting perpendicularly on a (known) lever arm around this articulation.) This definition acknowledges the fact that, with current technology, it is impossible to measure the force or tension developed within a muscle of a living human. If it becomes feasible to measure this internal force directly, voluntary muscle strength can be redefined as "the maximal voluntary force that a muscle can exert along its length, under voluntary control."

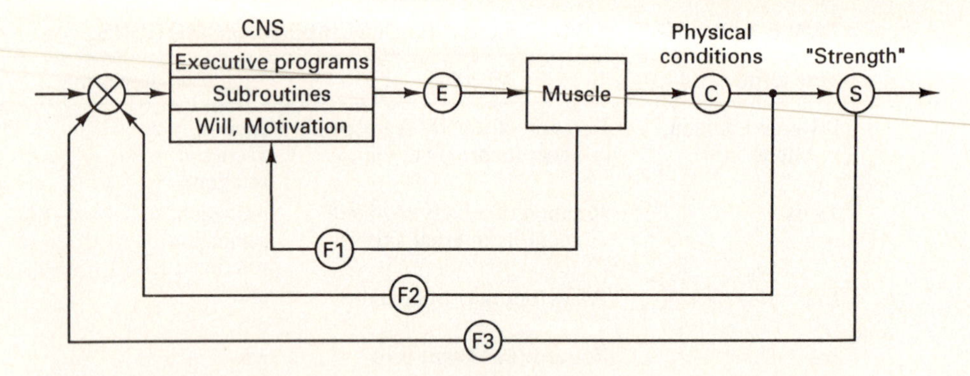

E: Efferent excitation impulses generated in the CNS
F: Afferent feedback loops

Figure 1-24. Model of the regulation of muscle strength exertion. (From Kroemer, Kroemer, and Kroemer-Elbert, 1990, *Engineering Physiology: Bases of Human Factors/Ergonomics,* 2d ed. With permission by the publisher, Van Nostrand Reinhold. All rights reserved.)

The model shown in Figure 1-24 helps to understand the events involved in the exertion of muscular strength. The control initiatives generated in the central nervous system start with calling up an "executive program" (called "engram" by Clark and Horsch, 1986) which exists for all normal muscular activities, such as walking or pushing and pulling objects. The general program is modified by "subroutines" appropriate for the specific case, such as walking quickly upstairs, or pulling hard, or pushing carefully (Schmidt, 1988). These in turn are modified by "motivation," which determines how (and how much of the structurally possible) strength will be exerted under the given conditions. A qualitative listing of circumstances which may increase or decrease one's willingness to exert strength is given in Table 1-11.

These complex interactions result in excitation signals *E* transmitted along the efferent nervous pathways to motor units, which are triggered to contract. The contraction strength developed in the muscle depends on the motor units involved, on the rate and frequency of signals received, on muscle length and thickness, and possibly on fatigue existing in the muscle as a consequence of previous contractions. It also depends on whether the muscle length changes under contraction or remains constant—see the discussion of static and dynamic exertions later in this chapter.

The output of the muscular contraction is modified by the existing mechanical conditions, such as the lever arms at which muscle tendons pull with respect to the bridged articulation and the pull angle with respect to the lever arm. These conditions would change, of course, in dynamic activities, while they are assumed constant in a static effort.

Thus, the output of this complicated chain of controllers, feedforward signals, controlled elements, and modifying conditions is the "strength" measured at the interface between the body segment involved and the measuring device (the object

TABLE 1-11. FACTORS LIKELY TO INCREASE (+) OR DECREASE (−) MAXIMAL MUSCULAR PERFORMANCE

	Likely effect
Feedback of results to subject	+
Instructions on how to exert strength	+
Arousal of ego involvement, aspiration	+
Pharmaceutical agents (drugs)	+
Startling noise, subject's outcry	+
Hypnosis	+
Setting of goals, incentives	+ or −
Competition, contest	+ or −
Verbal encouragement	+ or −
Fear of injuries	−
Spectators	?
Deception	?

SOURCE: From Kroemer, Kroemer, and Kroemer-Elbert, 1990, *Engineering Physiology: Bases of Human Factors/Ergonomics*, 2d ed. With permission by the publisher, Van Nostrand Reinhold.

against which strength is exerted). The assumption is that at any moment at least as much resistance is available at this interface as can be exerted by a person. If this were not the case (i.e., if Newton's third law were violated), no reliable muscle-strength measurement could be performed.

The model also shows a number of feedback loops through which the muscular exertion is monitored for control and modification. The first feedback loop, $F1$, is in fact a reflex-like arc which originates at proprioceptors, such as Ruffini organs in the joints (signaling location), Golgi tendon-end organs (indicating changes in muscle tension), and muscle spindles (indicating length)—see Chapter 3. These interoceptors and their signals are not under voluntary control and directly influence the signal generator in the spinal cord for a quick response. The other two feedback loops originate at exteroceptors and are rooted through a comparator modifying the input signal into the central nervous system. Loop $F2$ originates at kinesthetic receptors sensitive to events related to touch, pressure, and body position in general. For example: as one pulls on a handle, body position is monitored together with the sensations of pressure in the hand and of forces throughout the body, including the sensation of pressure felt in the feet as they press down on the floor. Feedback loop $F3$ originates at exteroceptors and signals such events as sounds and motions related to the effort to the comparator. For example, this feedback may be the sounds or movements generated in the objects on which one pulls; it may be the pointer of an instrument that indicates the strength applied; or it may be the experimenter or coach shouting feedback and exhortation to the subject.

This model indicates measurement opportunities. Considering the feedforward section of the model, it becomes apparent that there is (with current technology) no suitable means to measure the executive programs, the subroutines, or the effects of will or motivation on the signals generated in the central nervous system. (Only very general information can be gleaned from an electroencephalogram, EEG.) Efferent excitation impulses from the motor nerves to the muscles can be recorded through electrodes by an electromyogram, EMG.

Interpretation of an electromyogram, the record of the electrical signals associated with the activation of motor units in muscle in terms of frequency and amplitude, relies on several basic assumptions. One assumption is that the signals stem from the same motor units. A surface electrode can be used for isometric contractions, because in this static case theoretically the muscle does not move under the electrode on the skin; in dynamic muscle use, indwelling (wire or needle) electrodes can follow the moving muscle. Another assumption concerns the relationship between EMG amplitude and muscle strength. In the isometric case, calibration often shows a nearly linear increase in EMG intensity from rest to maximal voluntary muscle exertion; but in dynamic muscle use, the EMG-force relationship is complex and difficult to establish. Another complication comes with exertion time, when rate and recruitment coding of motor unit excitation often vary; this is particularly the case with muscle "fatigue" where the EMG generally shows a shift toward lower frequencies, usually together with an increase in amplitude (Basmajian and DeLuca, 1985; Chaffin and Andersson, 1991; Soderberg, 1992). Thus, electromyography is not an "easy" technique to assess muscle strength.

The contraction activities at the muscle can be observed qualitatively, but not quantitatively, in the living human: at this time no instruments are available to measure directly the tensions within muscle filaments, fibrils, fibers, or groups of muscles *in situ*.

The physical conditions accompanying the strength exertion can be observed and recorded if they are external to the body: location of the coupling between body and measuring device, direction of force or torque, time history of exertion, body position, temperature, humidity, and so on. However, the mechanical advantages internal to the body are much more difficult and often practically impossible to record and control. This applies, for example, to lever arms of the tendon attachments and to pull angles within the muscle with respect to their lever-arm attachments.

Hence, the resulting output labeled "strength" is the only clearly definable and measurable event (given present technology). It is the amount and direction of torque or force, over time, exerted to and detected by the measuring device.

APPLICATION

Assessment of Human Muscle Strength

From the foregoing, two conclusions are obvious for strength measurements on living humans:

Muscular strength is what is measured by an instrument. (A dissatisfying but realistic statement.)

Strength is influenced by motivation and the physical conditions under which it is exerted.

In mechanics, one distinguishes between "statics" and "dynamics." In physiological terms, the static condition is generated in an *isometric* muscle contraction where, presumably, the muscle length remains constant. (The Greek *iso* means unchanged or constant, and *metrein* refers to the measure or length of the muscle.) If there is no change in muscle length during the isometric effort, there is no motion of the involved body segments, hence Newton's first law requires that all forces acting within the system are in equilibrium. Because no displacement results from the muscular contraction, the physiological "isometric" case is equivalent to the "static" condition in physics. This theoretically simple and experimentally well controllable condition has lent itself to rather easy measurement of muscular strength, and most of the information currently available on human strength is limited to the static (isometric) muscular effort (Caldwell, Chaffin, Dukes-Dobos, Kroemer, Laubach, Snook, and Wasserman, 1974).

Dynamic muscular efforts are much more difficult to describe and control than static contractions. In dynamic activities, muscle length changes, and therefore involved body segments move. Thus, displacement is present, and its time derivatives (velocity, acceleration, and jerk) must be considered. This is a much more complex task for the experimenter than that encountered in static testing. Only recently, a systematic breakdown into independent and dependent experimental variables has been presented for dynamic and static efforts (Kroemer, Marras, McGlothlin, McIntyre, and Nordin, 1990). *Independent* variables are those that are purposely manipulated during the experiment in order to assess resulting changes in the *dependent* variables.

For example, if one sets the displacement (muscle-length change) to zero—the *isometric* condition—one may measure the force generated and possibly the number of repetitions that can be performed until force is reduced because of muscular fatigue. This case is described in Table 1-12. Of course, with no displacement, its time derivatives velocity, acceleration, and jerk are also zero. In the isometric technique, one is also likely to control the mass properties, probably by keeping them constant.

One may also choose to control velocity, i.e., the rate at which muscle length changes, as an independent variable. If velocity is set to a constant value, one speaks of *isovelocity* or isokinematic (often falsely called isokinetic) muscle-strength measurement. Time derivatives of constant velocity, namely acceleration and jerk, are zero. Mass properties are usually controlled in isovelocity tests. The variables displacement, force, and repetition can be chosen as either dependent variables or controlled independent variables. Most likely, force and/or repetition are chosen as the dependent variables to assess the result of the testing. Following the scheme laid out in Table 1-12, one also can devise tests in which acceleration or its time derivative, jerk, is kept constant.

If one sets the amount of force (or torque) to a constant value, it is most likely that mass properties and displacement (and its time derivatives) are controlled independent variables, and repetition a dependent variable. This *isoforce* condition, in which muscle tension is kept constant (*isotonic*), is, for practical reasons, often combined with an isometric condition, such as in holding a load motionless (i.e., displacement is zero).

EXAMPLE

The term *isotonic* has often been wrongly applied. Some older textbooks used lifting or lowering of a constant mass (weight) as typical for isotonics. This is physically false for two reasons. The first is that according to Newton's law, acceleration and deceleration of a mass requires application of changing (not constant) forces. The second fault lies in overlooking the changes that occur in the mechanical conditions (pull angles and lever arms) under which the muscle functions during the activity. Finally, it is virtually impossible to generate a constant tension in a muscle when it acts in motion against an external resistance. Hence, practically, a truly isotonic muscle activity occurs only in combination with an isometric one.—It is certainly misleading to label all dynamic activities of muscles isotonic, as is occasionally still done.

In the *isoinertial* condition, the mass properties are controlled by setting mass to a constant value. In this case, repeated movement of such constant mass (as in lifting) may be either a controlled independent or, more likely, a dependent variable. Also, displacement and its derivatives may become dependent outputs. Force (or torque) applied is likely to be a dependent value, according to Newton's second law (*force* equals *mass* times *acceleration*).

Table 1-12 also contains the most general case of muscle-performance measurement, labeled *free dynamic*. In this case, the independent variables (displacement and its time derivatives) as well as force are unregulated, i.e., left to the free choice of the subject. Only mass and repetition are usually controlled, although they may be used as dependent variables. Displacement and its time derivatives may be dependent variables. Force, torque, or some other performance measure is likely to be chosen as a dependent output.

The strength-test protocol. After the type of strength test to be done, and the measurement techniques and the measurement devices, have been chosen, an experimental protocol must be devised. This includes the selection of subjects, their information and protection; the control of the experi-

TABLE 1-12. TECHNIQUES TO MEASURE MOTOR PERFORMANCE BY SELECTING SPECIFIC INDEPENDENT AND DEPENDENT VARIABLES

Names of technique	Isometric (Static)		Isovelocity (Dynamic)		Isoacceleration (Dynamic)		Isojerk (Dynamic)		Isoforce (Static or dynamic)		Isoinertial (Static or dynamic)		Free dynamic	
Variables	Indep.	Dep.	Indep.	Dep.	Indep.	Dep.	Indep.	Dep.	Indep.	Dep.	Indep.	Dep.	Indep.	Dep.
Displacement linear/angular	constant* (zero)		C or X		C or X		C or X		C or X		C or X			X
Velocity, linear/angular	O		constant		C or X		C or X		C or X		C or X			X
Acceleration, linear/angular	O		O		constant		C or X		C or X		C or X			X
Jerk, linear/angular	O		O		O		constant		C or X		C or X			X
Force, torque	C or X		C or X		C or X		C or X		constant		C or X			X
Mass, moment of inertia	C		C		C		C		C		constant		C or X	
Repetition	C or X		C or X		C or X		C or X		C or X		C or X		C or X	

SOURCE: Modified from Kroemer, Marras, McGlothlin, McIntyre, and Nordin, 1990.

Legend

Indep = independent
Dep = dependent
C = variable can be controlled
* = set to zero
O = variable is not present (zero)
X = can be dependent variable

The boxed constant variable provides the descriptive name.

mental conditions; the use, calibration, and maintenance of the measurement devices; and (usually) the avoidance of training and fatigue effects.

Regarding the selection of subjects, care must be taken that they are in fact a representative sample of the population for which data are to be gathered. Regarding the management of the experimental conditions, the control of motivational aspects is particularly difficult. It is widely accepted (outside sports and medical function testing) that the experimenter should *not* give exhortations and encouragements to the subject—see Table 1-11. The so-called "Caldwell regimen" (Caldwell, Chaffin, Dukes-Dobos, et al., 1974) pertains to isometric strength testing but can be adapted for a dynamic test—see Table 1-12. Edited excerpts are given in the definition and list below.

Definition: Static strength is the capacity to produce torque or force by a maximal voluntary isometric muscular exertion. Strength has vector qualities and therefore should be described by magnitude and direction.

1. Measure static strength according to the following conditions:
   (a) Static strength is assessed during a steady exertion sustained for four seconds.
   (b) The transient periods of about one second each, before and after the steady exertion, are disregarded.
   (c) The strength datum is the mean score recorded during the first three seconds of the steady exertion.

2. Treat the subject as follows:
   (a) The subject should be informed about the test purpose and procedures.
   (b) Instructions to the subject should be kept factual and not include emotional appeals.
   (c) The subject should be instructed to "increase to maximal exertion (without jerk) in about one second and then maintain this effort during a four-second count."
   (d) Inform the subject during the test session about his/her general performance in qualitative, noncomparative, positive terms. Do not give instantaneous feedback during the exertion.
   (e) Rewards, goal setting, competition, spectators, fear, noise, etc., can affect the subject's motivation and performance and, therefore, should be avoided.

3. Provide a minimal rest period of two minutes between related efforts; more if symptoms of fatigue are apparent.

4. Describe the conditions existing during strength testing:
   (a) Body parts and muscles chiefly used.
   (b) Body position (or movement).
   (c) Body support/reaction force available.

(d) Coupling of the subject to the measuring device (to describe location of the strength vector).
(e) Strength measuring and recording device.

5. Describe the subject:
(a) Population and sample selection.
(b) Current health and status: medical examination and questionnaire recommended.
(c) Gender.
(d) Age.
(e) Anthropometry (at least height and weight).
(f) Training related to the strength testing.

6. Report the following data:
(a) Mean (median, mode).
(b) Standard deviation.
(c) Skewness.
(d) Minimum and maximum values.
(e) Sample size.

It is much more complex to measure or calculate forces or torques for dynamic than for static conditions. In particular, accelerations (changes in velocity) generate additional forces (e.g., tangential, centrifugal, Coriolis) and torques which may be substantially larger than those experienced in static conditions. Several publications provide further information; see, e.g., Chaffin and Andersson (1991) and Oezkaya and Nordin (1991). Typical applications of strength measurements are in material handling—see Chapter 10.

☞☞☞ If you can't calibrate, you can't measure. ☜☜☜

SUMMARY

Anthropometry and biomechanics of the human body are among the most basic descriptors needed to design "fitting" equipment and work procedures.

Information about body size is available in a rather complete manner only for military populations. Yet, soldiers are a select sample of the general population. They are youngish, healthy, and do not include people of very unusual body dimensions. Furthermore, until recently there were not many female soldiers. Thus, estimates of the civilian population derived from information on soldiers have been rather unreliable in the past.

In North America, a recent survey of the U.S. Army has remedied many of the problems associated with missing anthropometric information. Apparently the U.S. Army is fairly similar to the general population in its composition, including females. Thus, this information is used in this book as indicative of the body dimensions of the general population of the United States. Statistical treatment of the measured data allows us to establish correlations among them and to predict, using this knowledge, data that were not actually measured.

Statistics applied to anthropometric data provide a wealth of information. For example, the standard deviation divided by the mean is a measure of the variability of the recorded data. Large variability could indicate either a truly diverse sample, or problems in data acquisition or treatment. Much variability stems from the fact that most surveys are performed as cross sections of the population, meaning that at one moment in time measurements are taken on persons of different ages. While this reflects interindividual differences, it does not inform us about intraindividual changes occurring with aging, or with changes in health.

In recent decades, body dimensions have been increasing in many parts of the earth. In particular, growth in many length dimensions has been observed, such as in stature and leg length, often accompanied by an increase in body weight. Such information is important for the design of closely fitting equipment, particularly clothing.

It is inadequate and inexcusable to design tools, equipment, or workstations to the phantom of the "average person." Instead, the ranges of body dimensions must be considered, and the ways in which certain body dimensions change with respect to each other. For example, it is indefensible to establish height-weight indices (such as in desired body weight) when in fact the correlation between height and weight is very low. Instead, fairly simple rules apply which consider variations in body dimensions. For ergonomic design of equipment, a four-step procedure has been established that facilitates proper "sizing" of designed products.

The human body is often modeled as a rigid skeleton with joints that allow movement. The body members have mass properties, and are moved by muscles.

The long bones of the segments establish the lever arms at which muscles attach. Articulations are of different kinds, and some wear out and can be replaced by artificial joints. Their design, construction, implantation, and use has developed into a specific ergonomic subdiscipline and a specialized industry, and helped millions of people to mobility which they had lost due to accidents, diseases, or wear and tear.

Muscles convert chemically stored energy into mechanically useful force and work. The arrangement of muscles within the human body is rather complex. Usually, a group of agonistic muscles is opposed by antagonistic muscles, which together regulate motion and energy development. This poses a rather interesting challenge for biomechanical modeling. Furthermore, voluntary and involuntary control of muscular exertions, fatigue, posture, motion characteristics, and motivation strongly influence the muscular output.

Until recently, understanding of the musculo-mechanical conditions was hampered by insufficient definition of dynamic circumstances. Therefore, isometric (or static) muscle efforts are fairly well researched, while little knowledge exists about dynamic capabilities. Newly developed procedures to test muscle strength, both statically and dynamically, should provide both theoretical insights and practical information to be used in the ergonomic design of work tasks and equipment.

CHALLENGES

How different are the body dimensions of soldiers from those of civilians?

If such differences exist, how important would they be for the design of special tools, equipment, workstations, or work procedures?

Are there "recipes" that describe how to derive "functional" body dimensions from those measured in static, standardized postures?

How do secular changes in body dimensions affect the design of technical products?

Is designing for "fit from the 5th to the 95th percentile" appropriate?

Are "women small men," anthropometrically?

Which procedures are appropriate to select fair subsamples from a general population for measurement purposes?

How can correlations among body dimensions reduce the needed number of measurements taken on a population sample?

How exactly must body dimensions be known to suffice for design purposes?

Can the "staircase effect" be generated by external electrical stimulation of muscles? If so, might there be a danger of overexertion?

Is external electrical stimulation of muscle contraction useful for building stronger muscles?

How does cocontraction of agonistic and antagonistic muscles about a joint influence the loading of that joint?

Could wearing of a tight belt around the waist increase intraabdominal pressure? To what, if any, effect?

Which tissues in the trunk determine the posture of the spinal column?

What difference does it make, with respect to forces at the spinal disk, if one gives up the assumption of a straight vertebral column in favor of a column with kyphosis and lordosis?

What are the factors limiting a person's exertion of strength?

What are some practical means to affect (or control) a person's motivation for strength generation?

Is it possible to develop "conversion algorithms" to determine dynamic muscle strength from static strength, and vice versa?

2

How the Body Does Its Work

OVERVIEW

In many respects one may compare the way in which the body generates energy with the functioning of a combustion engine: fuel (food) is combusted, for which oxygen must be present. The combustion yields energy that moves parts mechanically. The fueling and cooling system (blood vessels) moves supplies (oxygen, carbohydrate, and fat derivatives) to the combustion sites (muscle, other organs) and removes combustion byproducts (lactic acid, carbon dioxide, water, heat) for dissipation (at skin and lung surfaces).

☞☞☞ *The Greek physician Alcmaeon (sixth century B.C.) was apparently the first to deliberately and carefully dissect human cadavers. He saw the difference between arteries and veins and noticed that the sense organs were connected to the brain by nerves. Praxagoras (fourth century B.C.) distinguished between arteries and veins, but he thought arteries carried air because they were usually found empty in corpses. (The word artery is from Greek words meaning "carrying air.") About 280 B.C., Herophilus noticed that the arteries pulsed and thought that they carried blood (Asimov, 1989).* ☜☜☜

The processes are controlled by several complex and overlapping control systems (central nervous system, hormonal system, limbic system). Control centers (in the brain and spinal cord) rely on feedback from various body parts and provide feedforward signals according to general (autonomic, innate, learned) principles and according to situation-dependent (voluntary, motivational) rules.

The *respiratory system* provides oxygen for the energy metabolism and dissipates metabolic byproducts. The *circulatory system* carries oxygen from the lungs to the consuming cells, to which it also brings the "fuel," i.e., derivatives of carbohydrates and fats. Furthermore, it removes metabolic byproducts from the combustion sites. The *metabolic system* provides the chemical processes in the body, particularly those that yield energy (Kroemer, Kroemer, Kroemer-Elbert, 1990).

THE RESPIRATORY SYSTEM

There is close interaction between the respiratory system, which absorbs oxygen and dispels carbon dioxide, water, and heat, and the circulatory system, which provides the means of transport. Figure 2-1 indicates that interaction schematically.

The respiratory system moves air to and from the lungs, where part of the oxygen contained in the inhaled air is absorbed into the bloodstream; it also removes carbon dioxide, water, and heat from the blood into the air to be exhaled. Between 200×10^6 and 600×10^6 alveoli provide a grown person with about 70 to 90 m^2 of exchange surfaces in the lungs.

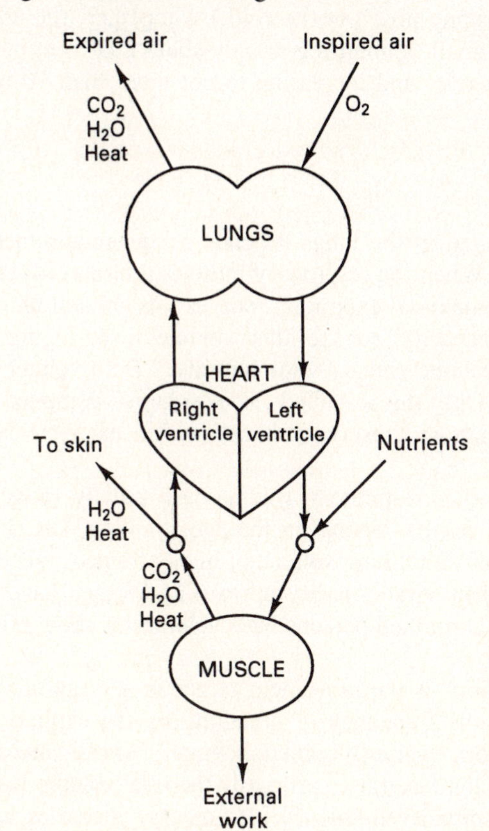

Figure 2-1. The interrelated functions of the respiratory and circulatory systems. (From Kroemer, Kroemer, and Kroemer-Elbert, 1990, *Engineering Physiology: Bases of Human Factors/Ergonomics*, 2nd ed. With permission by the publisher, Van Nostrand Reinhold. All rights reserved.)

Air exchange is brought about by the pumping action of the thorax. The diaphragm separating the chest cavity from the abdomen descends about 10 cm when the abdominal muscles relax. Muscles connecting the ribs contract and raise the ribs. Hence, the dimensions of the rib cage and of its included thoracic cavity increase both toward the outside and in the direction of the abdomen: air is sucked into the lungs. When the inspiratory muscles relax, elastic recoil in lung tissue, thoracic wall, and abdomen restores their resting positions without involvement of expiratory muscles: air is expelled from the lungs. When ventilation needs are very high in heavy physical work, the recoil forces are augmented by activities of expiratory (intercostal) muscles, and contraction of the muscles in the abdominal wall further assists expiration.

At the mucus-covered surfaces in the nose, mouth, and throat, the temperature of the inward-flowing air is adjusted to body temperature, moistens or dries the air, and cleanses it of particles. In a normal climate, about 10 percent of the total heat loss of the body, whether at rest or work, occurs in the respiratory tract. This percentage increases to about 25 percent at outside temperatures of about −30°C. In a cold environment, heating and humidifying the inspired air cools the mucosa; during expiration, some of the heat and water is recovered by condensation from the air to be exhaled. (Hence the "runny nose" in the cold.) Altogether, the energy required for breathing is relatively small, amounting to only about 2 percent of the total oxygen uptake of the body at rest, and increasing to not more than 10 percent during heavy exercise.

Respiratory Volumes

The volume of air exchanged in the lungs depends on the requirements associated with the work performed. When the respiratory muscles are relaxed, there is still air left in the lungs. A forced maximal expiration reduces this amount of air in the lungs to the so-called "residual capacity" (or "residual volume"); see Figure 2-2. A maximal inspiration adds the volume called "vital capacity." Both volumes together are the "total lung capacity." Only the so-called "tidal volume" is moved, leaving both an inspiratory and an expiratory "reserve volume" within the vital capacity during rest or submaximal work.

Vital capacity and other respiratory volumes are usually measured with the help of a *spirometer*. The results depend on the age, training, sex, body size, and body position of the subject. Total lung volume of highly trained, tall young males is between 7 and 8 L and their vital capacity up to 6 L. Women have lung volumes about 10 percent smaller. Untrained persons have volumes of about 60 to 80 percent of their athletic peers.

"Pulmonary ventilation" is the movement of gas in and out of the lungs. It is calculated by multiplying the frequency of breathing by the expired tidal volume. This is called the (respiratory expired) "minute volume." At rest, one breathes 10 to 20 times every minute. In light exercise, primarily the tidal volume is increased, but with heavier work the respiratory frequency also quickly increases up to about 45

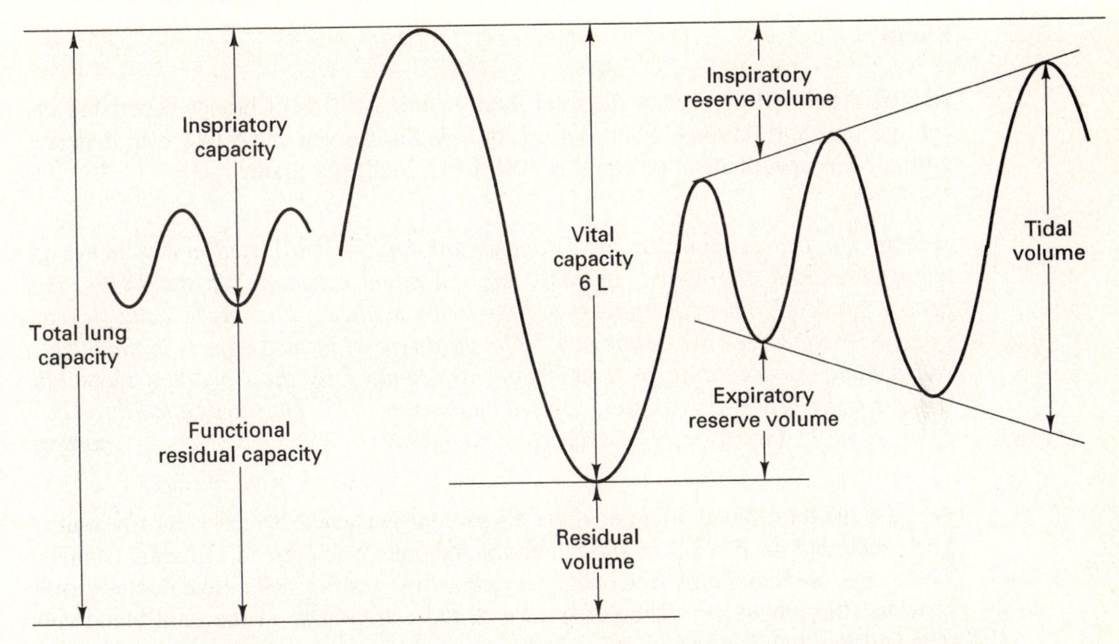

Figure 2-2. Respiratory volumes. (From Kroemer, Kroemer, and Kroemer-Elbert, 1990, *Engineering Physiology: Bases of Human Factors/Ergonomics,* 2nd ed. With permission by the publisher, Van Nostrand Reinhold. All rights reserved.)

breaths/min. This indicates that breathing frequency, which can be measured easily, is not a reliable indicator of the heaviness of work performed.

The respiratory system is able to increase its moved volumes and absorbed oxygen by large multiples. The minute volume can be increased from about 5 L/min to 100 L/min or more; that is an increase in air volume by a factor of 20 or more. Though not exactly linearly related to it, the oxygen consumption shows a similar increase.

THE CIRCULATORY SYSTEM

The circulatory system carries oxygen from the lungs to the cells, where nutritional materials, also brought by circulation from the digestive tract, are metabolized. Metabolic byproducts (CO_2, heat, and water) are dissipated by circulation. The circulatory and respiratory systems are closely interrelated, as shown earlier in Figure 2-1.

Water is the largest weight component of the body: about 60 percent of body weight in men, about 50 percent in women. In slim individuals, the percentage of total water is higher than in obese persons, since adipose tissue contains very little water. The relation between water and lean (fat-free) body mass is rather constant in "normal" adults, about 72 percent.

Blood

Approximately 10 percent of the total fluid volume consists of blood, depending on age, gender, and training. Four to 4.5 L of blood in women and 5 to 6 L in men are normal. The specific heat of blood is 3.85 J (0.92 cal) per gram.

The Dutch naturalist Jan Swammerdam (1637–1680) used newly improved microscopes and discovered, in 1658, the red blood corpuscle. In the 1840s, the British physician Thomas Addison (1798–1866) studied white blood cells, leucocytes, from the Greek for "white cell." The third type of formed objects in the blood, called platelets according to their shape, was studied by the Canadian physician William Osler (1849–1919) who reported on them in 1873. They are called thrombocytes, from the Greek for "clot cells" (Asimov, 1989).

Of the total blood volume, about 55 percent is plasma, which is mostly water. The remaining 45 percent of the blood volume consist of formed elements (solids). These are predominantly red cells (erythrocytes), white cells (leukocytes), and platelets (thrombocytes). The percentage of red-cell volume in the total blood volume is called hematocrit.

Blood groups

In 1930, the Austrian physician Karl Landsteiner (1868–1943) showed that human blood could be divided into four types. This knowledge made blood transfusions safe (Asimov, 1989).

According to the content of certain antigens and antibodies, blood is classified into four groups: O, A, B, AB. The importance of these classifications lies primarily in their incompatibility reactions in blood transfusions. There are other subdivisions, in particular the one according to the rhesus (Rh) factor.

Functions. The blood carries dissolved materials, particularly oxygen and nutritive materials as well as hormones, enzymes, salts, and vitamins. It removes waste products, particularly dissolved carbon dioxide and heat.

The red blood cells perform the oxygen transport. Oxygen attaches to hemoglobin, an iron-containing protein molecule of the red blood cell. Each molecule of hemoglobin contains four atoms of iron, which combine loosely and reversibly with four molecules of oxygen. Hemoglobin molecules can react simultaneously with oxygen and carbon dioxide. There is high affinity of hemoglobin to carbon monoxide, which takes up spaces otherwise taken by oxygen; this explains the toxicity of CO.

☞☞☞ *The body can lose up to 15 percent of its blood volume without dramatic effects. Yet, at 20 percent loss, blood pressure is reduced and pulse and breathing are affected; gravely so at 30 percent loss, where heart rate is increased as well. At 40 percent loss, death is imminent if blood and particularly fluids are not infused.* ☜☜☜

Architecture of the Circulatory System

The circulatory system is nominally divided into two subsystems: the *systemic* and the *pulmonary* circuits, each powered by one half of the heart (which can be considered a double pump). The left side of the heart supplies the systemic section, which branches from the arteries through the arterioles and capillaries to the metabolizing organ (e.g., muscle); from there it combines again from venules to veins to the heart's right side. The pulmonary system starts at the right ventricle, which powers the blood flow through pulmonary artery, lungs, and pulmonary vein to the left side of the heart.

Each half of the heart has an antechamber (atrium) and a chamber (ventricle), the pump proper. The atria receive blood from the veins, which is then brought into the ventricles through a valve. In essence the heart is a hollow muscle which produces, via contraction and with the aid of valves, the desired blood flow.

☞☞☞ *Based on Galen's (129–199) beliefs, it was thought until the early 1600s that blood was manufactured in the liver and carried to the heart, from which it was pumped outward through arteries and veins alike and consumed in the tissues. The Italian physician Girolamo Fabrici (1537–1619) noticed the valves in the veins that prevent the blood from flowing in the direction that Galen had postulated; yet, Fabrici did not dare to contradict Galen's doctrine. Galen had thought that the heart was a single pump and that there were pores in the thick muscular wall separating the two ventricles. In 1242, the Arabic scholar Ibn an-Nafis wrote that the right and left ventricles were totally separated. He explained how blood pumped by the right ventricle was led by arteries to the lung where the blood collected air. The enriched blood then was collected into increasingly larger vessels, which brought it back to the left ventricle, from where blood was pumped to the rest of the body. Unfortunately, an-Nafis' book did not become known outside the Arabic world until 1924. Yet, in 1553 the Spanish physician Miguel Serveto published a book in which he correctly described some features of the circulation. In 1559, the Italian anatomist Realdo Colombo also described major features of the circulatory system. His work was widely used in the medical profession. The English physician William Harvey published in 1628 his book* De Motu Cortis et Sanguinis, *which describes functions of the heart and blood. This book is generally considered the beginning of modern physiology (Asimov, 1989).* ☜☜☜

The Heart as Pump

The ventricle is filled through the valve-controlled opening from the atrium. The heart muscle contracts (*systole*), and when the internal pressure is equal to the pressure in the aorta, the aortic valve opens and the blood is ejected from the heart into the systemic system. Continuing contraction of the heart increases the pressure further, since less volume of blood can escape from the aorta than the heart presses into it. Part of the excess volume is kept in the aorta and its large branches, which act as a "windkessel," an elastic pressure vessel. Then, the aortic valve closes with the beginning of the relaxation (*diastole*) of the heart, while the elastic properties of the aortic walls propel the stored blood into the arterial tree, where elastic blood vessels smooth out the waves of blood volume. At rest, about half the volume in the ventricle is ejected (stroke volume), while the other half remains in the heart (residual volume). Under exercise load, the heart ejects a larger portion of the contained volume and increases its contraction frequency. When much blood is required but cannot be supplied, such as during very strenuous physical work with small muscle groups or during maintained isometric contractions, the heart rate can become very high. At a heart rate of 75 beats/min, the diastole takes less than 0.5 second and the systole just over 0.3 second; at a heart rate of 150 beats/min, the periods are close to 0.2 second each. Hence, an increase in heart rate occurs mainly by shortening the duration of the diastole.

The events in the right heart are similar to those in the left, but the pressure in the pulmonary artery is only about one-fifth of those during systole in the left heart.

Specialized cardiac cells (the sinoatrial nodes) serve as "pacemakers," determining the frequency of contractions by propagating stimuli to other cells of the heart muscle. The heart has its own intrinsic control system, which operates, without external influences, at (individually different) 50 to 70 beats/min. Changes in heart action stem from the central nervous system.

☞☞☞ *Since Luigi Galvani's time (1737–1798) it had been known that muscular contractions were associated with small electric potentials. The demonstration that this applied also to the heart muscle was made possible by the development of a specific galvanometer by the Dutch physiologist Willem Einthoven (1860–1927) in 1903. It allowed the recording of the electrocardiogram, abbreviated often as EKG instead of ECG because, in German, cardio is spelled with a k. Einthoven received the Nobel prize for medicine and physiology in 1924 (Asimov, 1989).* ☜☜☜

Myocardial action potentials are recorded in the electrocardiogram (ECG). The different waves have been given alphabetic identifiers: the P wave is associated with the electrical stimulation of the atrium, while the Q, R, S, and T waves are associated with ventricular events. The ECG is mostly employed for clinical diagnoses; however, with appropriate apparatus it can be used for counting and recording the heart rate. Figure 2-3 shows the electrical, pressure, and sound events during a contraction-relaxation cycle of the heart.

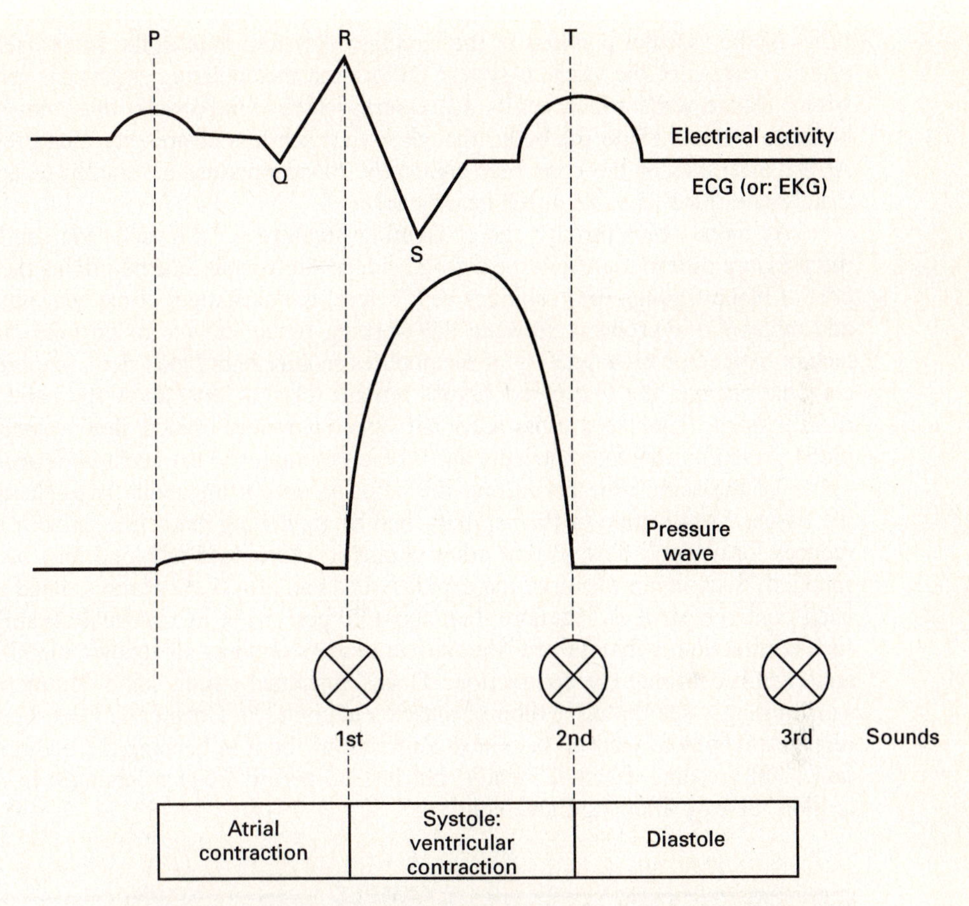

Figure 2-3. Scheme of the electrocardiogram, the pressure fluctuation, and the phonogram of the heart with its three sounds. (From Kroemer, Kroemer, and Kroemer-Elbert, 1990, *Engineering Physiology: Bases of Human Factors/Ergonomics,* 2nd ed. With permission by the publisher, Van Nostrand Reinhold. All rights reserved.)

The Pathways of Blood

Since the available blood volume does not vary, the cardiac output can be affected by two factors: the frequency of contraction (heart rate) and the pressure generated by each contraction in the blood. Both determine the so-called (cardiac) *minute volume*. The cardiac output of an adult at rest is around 5 L/min. When performing strenuous exercise, this level might be raised five times to about 25 L/min, while a well-trained athlete may reach up to 35 L/min.

A healthy heart can pump much more blood through the body than usually needed. Hence, a circulatory limitation is more likely to lie in the transporting capability of the vascular portions of the circulatory system than in the heart itself. The arterial section of the vascular system (before the metabolizing organ) has relatively strong elastic walls which act as a pressure vessel (windkessel), thus transmitting pressure waves far into the body, though with much loss of pressure along the way. At the arterioles of the consumer organ, the blood pressure is reduced to approximately one-third its value at the heart's aorta.

As blood seeps through the consuming organ (e.g., a muscle) via capillaries, the pressure differential from the arterial side to the venous side maintains the transport of blood through the "capillary bed." Here, the exchanges of oxygen, nutrients, and metabolic byproducts between the working tissue and the blood take place. If lack of oxygen or accumulation of metabolites require high blood flow, smooth muscles that encircle the fine blood vessels remain relaxed, thus allow the pathways to remain open. The large cross-sectional opening reduces blood flow velocity and blood pressure, allowing nutrients and oxygen to enter the extracellular space of the tissue, at the same time permitting the blood to accept metabolic byproducts from the tissue. Constriction of the capillary bed by tightening encircling smooth muscle reduces local blood flow so that other organs in more need of blood may be better supplied. Such compression of the capillary bed can also occur if the striated muscle itself contracts strongly, at more than about 20 percent of its maximal capability. If such contraction is maintained, the muscle hinders or shuts off its own blood supply and cannot continue the contraction. Thus, "sustained strong static contraction" is self-limiting—see the discussion of muscle endurance in Chapter 1. One should not require anybody to work in sustained contractions, be it in keeping the body in position or in grasping a handle tightly, but instead permit frequent changes in muscle tension, best by allowing movement.

EXAMPLE

A typical example for muscles cutting off their own blood flow is overhead work where muscles must keep the arms elevated. After a fairly short time, one must let them hang down to allow muscle relaxation and renewed blood flow.

The venous portion of the systemic system has a large cross section and provides low flow resistance; only about one-tenth of the total pressure loss occurs here. Valves are built into the venous system, allowing blood flow only toward the right ventricle.

Pascal's law states that static pressure in a column of fluid depends on the height of that column. However, the hydrostatic pressure in, for example, the feet of a standing person is not as large as expected from physics, because the valves in the veins of the extremities modify the value: in a standing person, the arterial pressure in the feet may be only about 100 mmHg higher than in the head. Nevertheless, blood, water, and other body fluids in the lower extremities are pooled there, leading to a well-known increase in volume of the lower extremities (swollen ankles), particularly when one stands or sits still.

Regulation of Circulation

If metabolite concentration in a muscle increases, this local condition directly causes smooth muscles encircling blood vessels to relax, allowing more blood flow. At the same time, the central nervous system (CNS) signals can trigger constriction of other less important vessels supplying organs. This leads to quick redistribution of the blood supply, which favors skeletal muscles over the digestive system ("muscles-over-digestion" principle). However, even in heavy exercise the systemic blood flow is so controlled that the arterial blood pressure is sufficient for an adequate blood supply to the brain, heart, and other vital organs. To accomplish this, neural vasoconstrictive commands can override local dilatory control. For example, the temperature-regulating center in the hypothalamus can affect vasodilation in the skin if this is needed to maintain a suitable body temperature, even if it means a reduction of blood flow to the working muscles ("skin-over-muscles" principle).

Thus, circulation at the arterial side, at the organ/consumer level, is regulated both by local control and by impulses from the central nervous system, the latter having overriding power. At the same time, the heart increases its output by higher heartbeat frequency; also, the blood pressure increases. At the venous side of circulation, constriction of veins, combined with the pumping action of dynamically working muscles and the forced respiratory movements, facilitate return of blood to the heart. This makes increased cardiac output possible, because the heart cannot pump more blood than it receives.

Heart rate generally follows oxygen consumption and hence energy production of the dynamically working muscle in a linear fashion from moderate to rather heavy work. However, the heart rate at a given oxygen intake is higher when the work is performed with the arms than with the legs. This reflects the use of different muscles and muscle masses with different lever arms to perform the work. Smaller muscles doing the same external work as larger muscles are more strained and require more oxygen. Also, static (isometric) muscle contraction increases the heart rate, apparently because the body tries to bring blood to the tensed muscles. Work in a hot environment causes a higher heart rate than at a moderate temperature, as discussed in Chapter 5. Finally, emotions, nervousness, apprehension, and fear can affect the heart rate at rest and during light work.

☞☞☞ *In 1614, the Italian physician Santorio Sanctorius (1561–1636) reported on experiments in which he sat in an elaborate weighing machine while he ate and drank and eliminated wastes. He found that he lost more weight than the waste alone could account for and attributed this to "insensible perspiration." Sanctorius' experiments were the beginning of the study of metabolism (Asimov, 1989).* ☜☜☜

THE METABOLIC SYSTEM

Over time, the human body maintains a balance (homeostasis) between energy input and output. The input is determined by the nutrients, from which chemically stored energy is liberated during the metabolic processes within the body. The output is mostly heat and work, measured in terms of physically useful energy, i.e., energy transmitted to outside objects. The amount of such external work performed strains individuals differently, depending on their physique and training. There is close interaction between the metabolic, circulatory, and respiratory systems, as sketched in Figure 2-4.

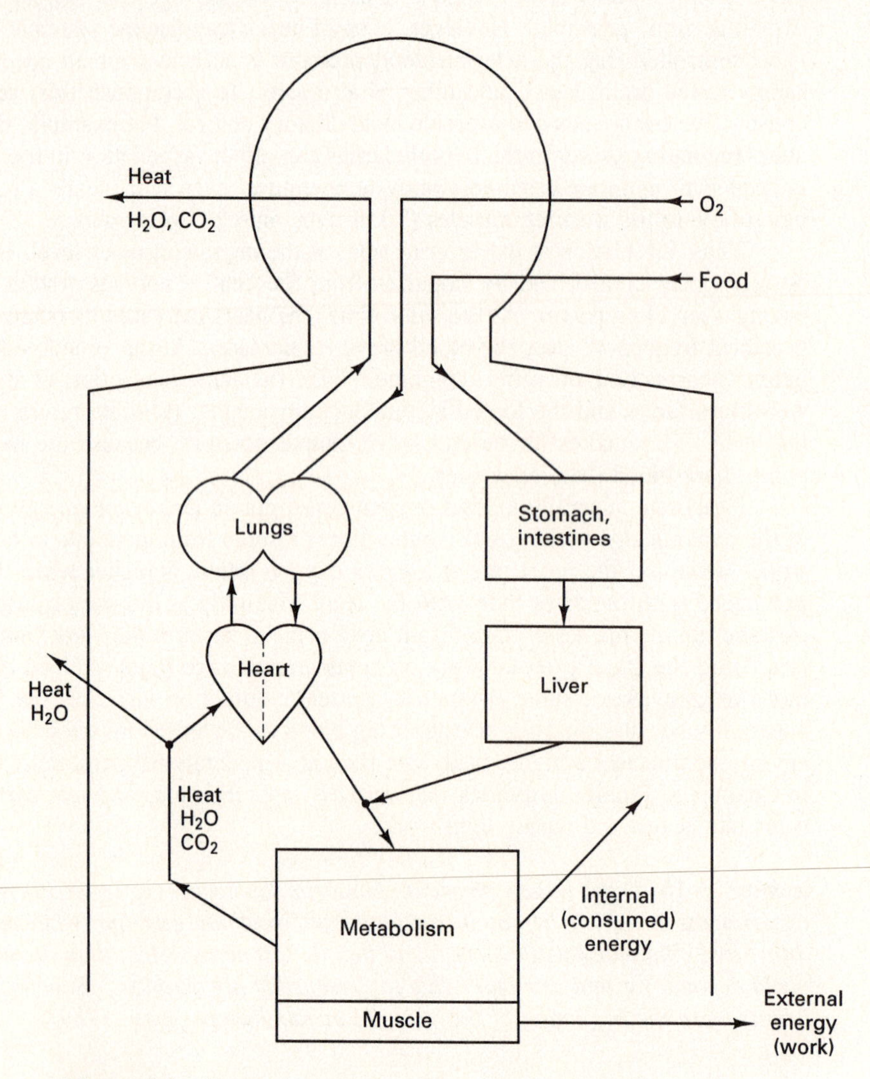

Figure 2-4. Interactions among energy inputs, metabolism, and outputs of the human body. (From Kroemer, Kroemer, and Kroemer-Elbert, 1990, *Engineering Physiology: Bases of Human Factors/Ergonomics*, 2nd ed. With permission by the publisher, Van Nostrand Reinhold. All rights reserved.)

☛☛☛ The "Human Energy Machine." *Astrand and Rodahl (1977) used an analogy between the human body and an automobile. In the cylinder of the engine, an explosive combustion of a fuel-air mixture transforms chemically stored energy into physical kinetic energy and heat. The energy moves the pistons of the engine, and gears transfer their motion to the wheels of the car. The engine needs to be cooled to prevent overheating. Waste products are expelled. This whole process can work only in the presence of oxygen and when there is fuel in the tank. In the "human machine," muscle fibers are both cylinders and pistons: bones and joints are the gears. Heat and metablic byproducts are generated while the muscles work. Nutrients (mostly carbohydrates and fats) are the fuels which must be oxidized to yield energy.* ☚☚☚

Human Metabolism and Work

The term "metabolism" includes all chemical processes in the living body. In a narrower sense, it is here used to describe the (overall) energy-yielding processes.

The balance between energy input I (via nutrients) and outputs can be expressed by an equation:

$$I = M + W + S \tag{2-1}$$

where M is the metabolic energy generated, W the work performed on an outside object, and S the energy storage in the body (negative if energy loss from it).

☛☛☛ *The measuring units for energy (work) are Joules (J) or calories (cal) with 4.2 J = 1 cal. (Exactly: 1 J = 1 Nm = 0.2389 cal = 10^7 ergs = 0.948×10^{-3} BTU = 0.7376 ft lb.) One uses the kilocalorie, kcal or Cal = 1000 cal, to measure the energy content of foodstuffs. The units for power are 1 kcal hr^{-1} = 1.163 W, with 1 W = 1 J s^{-1}* ☚☚☚

Assuming no change in energy storage and also that no heat is gained from or lost to the environment, one can simplify the energy-balance equation to

$$I = M + W \tag{2-2}$$

Human energy efficiency e (work efficiency) is defined as the ratio between work performed and energy input:

$$e = \frac{W}{I} 100 = \frac{W}{M + W} 100 \text{ (in percent)} \tag{2-3}$$

In everyday activities, only about 5 percent or less of the energy input is converted into "work," which is energy usefully transmitted to outside objects; highly trained athletes may attain, under favorable circumstances, perhaps 25 percent. The remainder of the input is finally converted into heat.

Work (in the physical sense) is done by skeletal muscles. They move body segments against external resistances. The muscle is able to convert chemical energy into physical work or energy. From resting, it can increase its energy generation up to fiftyfold. Such enormous variation in metabolic rate not only requires quickly adapting supplies of nutrients and oxygen to the muscle but also generates large amounts of waste products (mostly heat, carbon dioxide, and water), which must be removed. Thus, while performing physical work, the ability to maintain the internal equilibrium of the body is largely dependent on the circulatory and respiratory functions which serve the involved muscles. Among these functions, the control of body temperature is of particular importance. This function interacts with the external environment, particularly with temperature and humidity, as discussed in Chapter 5 in more detail.

Energy Liberation in the Human Body

Energy transformation in living organisms involves chemical reactions that either liberate energy, most often as heat, or require energy. The first kind of reaction is called *exergonic* (or *exothermic*). The other kind of reaction requires energy input; it is called *endergonic* (or *endothermic*). Generally, breakage of molecular bonds is exergonic, while formation of bonds is endergonic. Depending on the molecular combinations, bond breakage yields different amounts of released energies. Often, reactions do not simply go from the most complex to the most broken-down state, but achieve the process in steps with intermediate and temporarily incomplete stages.

The Energy Pathways

Energy is supplied to the body as food or drink. Chewing destroys the structure of the food mechanically, while the saliva starts breaking down the food chemically. Saliva is 99.5 percent water (a solvent) and 0.5 percent salts, enzymes, and other chemicals. One enzyme (lysozyme) destroys bacteria, thus protecting the mucous membranes from infection and the teeth from decay. Another enzyme (salivary amylase) breaks down starch.

While swallowing, breathing stops and the epiglottis closes for a second or two so that a food "bolus" can avoid the windpipe (trachea) and slide down the gullet (esophagus). Liquids need only about one second, but solid food takes up to 8 seconds to slide to the stomach.

The stomach generates gentle waves, 2 to 4 per minute, which mix the food with gastric juice. The gastric juice starts breaking up the proteins contained in the food/water mixture chemically, but does little to break down fats and carbohydrates, while alcohol is mostly absorbed in the stomach. Carbohydrate-rich foods leave the stomach within two hours, while protein-containing foods are slower; fatty foods stay up to 6 hours.

Digestion was long thought to be a physical action, the result of the grinding of the stomach. The French physicist de Reaumur (1683–1757) showed in 1752 that digestion was a chemical process (Asimov, 1989).

The contents of the stomach empties into the top of the small intestine, called duodenum (Latin for 12 fingers, indicating the length of that intestine section). Here, the true chemical energy extraction begins. It takes the food 3 to 5 hours to move through the small intestine, a tube about 3 cm in diameter and 7 m in length. During this time, about 90 percent of all the nutrients are extracted. The inner surface of the small intestine is enlarged by many fingerlike projections (villi). Through the surface, absorption takes place into blood and lymph. In the following large intestine, final processing is completed and disposal of solid waste through feces accomplished.

Does stopping smoking mean gaining weight?

Nicotine prolongs the time during which food stays in the stomach before being sent to the intestines—a full stomach turns off the appetite. Nicotine dulls the desire for sweets and it lessens the activity of the lipoprotein lipase, an enzyme that regulates the storage of fat carried in the blood.

When one stops smoking, the nicotine content of the body is strongly reduced. Consequently, sweet food becomes more attractive, the stomach is emptied faster, and fat is stored in the cells more easily. Being prepared for these effects, not giving in to the desire to eat more, particularly more sweets, and perhaps even introducing nicotine via special medications counteract the possible weight gain.

All together, it takes 5 to 12 hours after eating a meal to extract its nutrients. However, having digested the food does not mean having it "assimilated" yet. The foodstuffs still need to be degraded, either to release their energy content or to use them as raw material for body growth and repair.

Foodstuffs

Our food consists of various mixtures of organic compounds (foodstuffs), of water, salts, minerals, vitamins, etc., and of fibrous material (mostly cellulose). This roughage itself does not release energy; energy is derived from carbohydrates, fats, proteins, and alcohol.

The primary foodstuffs are carbohydrates, fats, and proteins. Their average nutritionally usable energy contents per gram are: 4.2 kcal (18 kJ) for carbohydrate, 9.5 kcal (40 kJ) for fat, and 4.5 kcal (19 kJ) for protein. Alcohol yields about 7 kcal (30 kJ) per gram.

Carbohydrates range from small to rather large molecules, and most are composed of only the three chemical elements: C, O, and H. (The ratio of H to O usually is 2 to 1, as in water, hence the name carbohydrate, meaning watered carbon.) Carbohydrates are digested by breaking the bonds between monosaccharides so that the compounds are reduced to simple sugars, such as glucose, which can be absorbed through the walls of the intestines into the bloodstream.

In the blood, the glucose ($C_6H_{12}O_6$, a monosaccharide) passes through the hepatic portal vein to the liver. From here, glucose is either sent to the central nervous system (brain and spinal cord) or to the muscles for direct use. Glucose is the primary source of energy, which is almost exclusively used by the central nervous system and which provides the quick energy for muscular actions.

In 1856, the French physiologist Claude Bernard (1813–1878) discovered a form of starch in the liver. Because it was easily broken down into glucose, he called it glycogen (Greek for glucose-producer). He showed that glycogen could be built up from glucose to act as an energy store, and that it could be broken down to glucose again when energy was needed (Asimov, 1989).

The liver may alternately change glucose into a long-chain molecule called glycogen, a polysaccharide $(C_6H_{10}O_5)_x$. It is deposited in the liver and near skeletal muscle as energy storage. When these glycogen storage areas are filled, the liver converts glucose to fat and stores it in the adipose tissue. Here, excess energy input becomes felt and seen.

Fat is the other major energy source for the body. Fat is a triglyceride, a molecule formed by joining a glycerol nucleus to three fatty-acid radicals. Unsaturated fat has double bonds between adjacent carbon atoms, hence, the compound is not saturated with all the hydrogen atoms it could accommodate. Most plant fats are polyunsaturated, hence liquid, while most animal fats are saturated and hence solid.

Digestion of fat means the breakage of bonds linking the glycerol with the three fatty acids. The digestion takes place primarily in the small intestine, where the glycerol and fatty-acid molecules can pass through cell membranes. The bloodstream transports only water-soluble materials such as glycerol, while the water-repellent fatty acids are absorbed into the lymph.

Regulated by the liver, fat is usually stored in adipose tissue for energy conversion as needed (see later). Furthermore, fat is the carrier of vitamins A, D, E, and K in the food. It also cushions vital organs (heart, liver, spinal cord, brain, eyeballs) against impact. A layer of fat under the skin insulates the body against energy transfer to and from the environment. For an illustration, note the appearance of swimmers, who usually have more "rounded" bodies than "skinny" long-distance runners.

The third major food component is *protein*, which consists of chains of amino acids joined together by peptide bonds. Many such bonds exist, and hence proteins come in a variety of types and sizes. As they are assimilated, the protein bonds are broken into amino acids, which are absorbed into the bloodstream. The amino acids are transported to the liver, which disperses some to cells throughout the body to be rebuilt into new proteins. However, most amino acids become enzymes, organic catalysts which control the chemical reactions between molecules without being consumed themselves. Still others become hemoglobin, the oxygen carrier in the blood, or antibodies, or hormones, or collagen. Usually the body employs protein for these important functions, but it can use amino acids, derived from protein, for energy.

Energy Storage

Under normal nutritional and exertion conditions, fat accounts for most of the stored energy reserves, while glucose and glycogen are the primary and first-used sources of energy at the cell level, particularly in the central nervous system and at muscles. The liver controls the release of glucose for direct consumption, or the generation of glycogen and of fat for energy storage. (One can become "fat" without eating it.)

In terms of stored energy, fat amounts by far to the most. On the average, about 16 percent of body weight is fat in a young man, increasing to 22 percent by middle age and more if he "is fat." Young women average about 22 percent of body weight in form of fat, rising to 34 percent (or more) by middle age. Athletes generally have lower percentages, about 15 percent in men and 20 percent in women. Assuming 15 percent of a 60-kg person as a low value, body fat amounts to 9 kg. Twenty-five percent of a 100-kg person, or 34 percent of a 70-kg person, means approximately 25 kg of body fat. Given that each gram of fat yields about 9.5 kcal, this means an energy storage in form of fat of about 85,500 Cal for a skinny lightweight person, and nearly 240,000 Cal for a heavy person.

Much less energy storage is provided by glycogen. Most of us have about 400 grams of glycogen stored near the muscles, about 100 grams in the liver, and some in the bloodstream. With an energy value of about 4.2 kcal per gram, we have only some 2,200 Cal as energy available from glycogen.

Energy Use

We have an enormous amount of energy stored in our bodies, most of it in the form of fat (which needs to be converted for use) while some is easily available in form of glycogen, even as glucose.

The use of the energy in the human body is achieved by catabolism (destructive metabolism), in which organic molecules are broken down, releasing their internal bond energies. This is primarily, and in an overall sense, accomplished in the human body by *aerobic metabolism*. For example, glucose can be oxidized according to the formula:

$$C_6H_{12}O_6 + 6\ O_2 = 6\ CO_2 + 6\ H_2O + \text{Energy} \qquad (2\text{-}4)$$

This means that one molecule of glucose combines with six molecules of oxygen and results in six molecules each of carbon dioxide and water, while energy (about 690 kcal/mole) is released.

Another way oxidation occurs is by the breakdown of glucose and glycogen molecules into several fragments, which become oxidized by each other. In this case, energy yield is *anaerobic*, and the processes are called glycolysis and glycogenolysis, respectively.

☞☞☞ *The British physiologist Archibald Vivian Hill (1886–1977) demonstrated in 1913 that heat was developed, and oxygen consumed, not during but after a muscu-*

lar contraction, when the muscle was at rest. The German biochemist Otto Meyerhof (1884–1951) independently demonstrated the same fact in chemical terms. This indicated that during muscular contraction, lactic acid is developed from glycogen. When the muscle is at rest again, lactic acid is oxidized, thus paying off the oxygen debt that was incurred during the preceeding reaction of "anaerobic glycosis" (Greek for sugar-splitting without air). Hill and Meyerhof were awarded shares of the Nobel prize for medicine and physiology in 1922 (Asimov, 1989).

Glucose (and fat) catabolism takes place in steps. The first is anaerobic, a biochemical reaction in which intermediary metabolites are produced, such as pyruvic acid. The second step is aerobic, meaning that oxygen is present. Here, intermediary metabolites, such as pyruvic acid, are completely metabolized, and six CO_2 molecules and six H_2O molecules are produced, as per Equation (2-4).

Energy Release

Living cells store "quick-release" energy in the molecular compound *adenosine triphosphate*, ATP. Its phosphate bonds can be broken down easily by hydrolysis according to the exergonic reaction:

$$\text{ATP} + H_2O \longrightarrow \text{ADP} + \text{energy (output)} \tag{2-5a}$$

The ATP supply needs to be replenished constantly. This is done through *creatine phosphate*, CP, which transfers a phosphate molecule to adenosine diphosphate, ADP. Energy must be supplied for this endergonic reaction:

$$\text{ADP} + \text{CP} + \text{energy (input)} \longrightarrow \text{ATP} + H_2O \tag{2-5b}$$

While ATP can provide energy anaerobically for a few seconds, resynthesis of ATP is necessary for continuous operation. The energy source for this process is the breakdown of complex molecules (provided by the absorbed foodstuffs) to simpler ones, ultimately to CO_2 and H_2O. Hence, glucose, fats, and possibly proteins provide the ultimate source of energy, keeping the ATP/ADP conversion going.

The first few seconds. At the very beginning of muscular effort, breaking the phosphate bond of ATP releases the "quick energy" for muscular contraction. However, the contracting muscle consumes its local supply of ATP in about two seconds.

The first 10 seconds. The next source of immediate energy is creatine phosphate, CP. It transfers a phosphate molecule to the just-created molecule of ADP and thus turns it back into ATP. (This cycle of converting ATP into ADP and back to ATP does not require the presence of oxygen.) Since one has three to five times more ADP than ATP, there is enough energy available for a muscle to perform up to 10 seconds of high activity.

After 10 seconds. After about 10 seconds of ATP-ADP-ATP reactions, energy must be supplied to sustain the reformatting of ATP. Now, the energy absorbed from the foodstuffs comes into play: glucose is broken down, releasing energy for the recreation of ATP.

Given that only a few seconds have elapsed since the activity started, there simply has not been enough time to use oxygen in the energy-conversion process. Thus, while releasing energy, the breakdown of glucose (generating carbon dioxide and water) is not complete, but other metabolic byproducts are generated also, particularly lactic acid. (If this metabolic byproduct is not resynthesized within about a minute in the presence of oxygen, the muscles simply cannot continue to work further.)

This anaerobic energy release primarily relies on the breakdown of glucose, although some glycogen is also involved. This is why glucose is called the primary, most easily accessible, and most metabolized energy carrier for the body.

After minutes and longer. If the physical activity has to continue, it must be performed at a level at which oxygen is sufficiently available to maintain the energy-conversion processes. Hence, the energy generated by a quick burst of maximal effort cannot be maintained at this level for extended periods of time.

In enduring work, the energy demanded from the muscles is so low that the oxygen supply at the cell level (mitochondria) allows "aerobic" energy conversion, meaning that sufficient oxygen is available to maintain the energy processes. Without oxygen, a molecule of glucose yields two molecules of ATP. With oxygen, the glucose energy yield is 36 molecules of ATP. Even richer in energy is the fat molecule palmitate, which yields about 130 molecules of ATP.

Aerobic and Anaerobic Metabolism

Because energy yield is so much more efficient under aerobic conditions, where no metabolic byproducts are generated, one can keep up a fairly high energy expenditure as long as ATP is replaced as fast as it is used up. The aerobic energy yield of glucose under aerobic condition, described earlier [in Equation (2-4)] releases energy in the amount of 690 kcal per mole. (The conversion is complete; no lactic acid is developed.) A more complex conversion takes place in the utilization of fats (glycerol and fatty acids). They are converted to intermediary metabolites and enter the so-called Krebs cycle. Their final energy yield is approximately 2,340 kcal per mole, more than three times that of glucose.

Still, if very heavy expenditure is required over long periods of time, such as in a marathon run, the interacting metabolic system and the oxygen-supplying circulatory system might become overtaxed. The runner who "hits the wall" has most likely used up the body's glycogen supply and has gone into "oxygen debt."

However, in our regular activities the energy output is regulated to agree with the body's abilities to develop energy under a sufficient supply of oxygen. If needed, one simply takes a break. While one is resting, accumulated metabolic byproducts are resynthesized, and the metabolic, circulatory, and respiratory systems are returned to their normal states.

Most of the single intermediate steps in the metabolic reactions are in fact anaerobic, but finally, oxygen must be provided. Thus, overall, sustained energy use is aerobic; i.e., it requires oxygen.

A simple suitable analogy of the human energy machine is that of an energy generator (the ATP-to-ADP energy release) driven by a combustion engine (fueled by carbohydrate and fat derivatives in the presence of oxygen).

Energy Use and Body Weight

Equation (2-1) described the balance between energy input, energy output, and energy storage. If the input exceeds the output, storage (body weight) is increased; conversely, a weight decrease takes place if the input is less than the output. Approximately 7,000 to 8,000 kcal make a difference of 1 kilogram in body weight.

The body tries to maintain a given energy storage. This means that a person's body weight (which in addition to water mostly consists of the weight of bones, of tissues, and of fat as stored energy) is kept at this level unless the "set point" is altered by radical changes in health, in energy expenditures (such as continued vigorous exercising, or reduction of physical activities), and in nutritional habits (such as severe overeating or starvation). As long as the set point is kept by the body, body weight remains rather constant. Even if one reduces food intake, the body tries to extract enough energy from the (reduced) food intake to maintain the old body weight. A continued starvation diet usually lowers storage, hence body weight. Returning to the previous eating habits after having achieved the desired lowered body weight allows the body to reattain its previous weight, unless the set point has been lowered.

APPLICATION

ASSESSMENT OF ENERGY EXPENDITURES

The ability to perform physical work is different from person to person, and depends on gender, age, body size, health, the environment, and motivation—as sketched in Figure 2-5.

To match a person's work capacity with the job requirements, one needs to know the individual's energetic capacity, and how much a given job demands of this capacity. To measure an individual's capacity, one makes the person perform a known amount of work (usually on a bicycle ergometer or a treadmill), and measures the subject's reactions. To measure the energy requirements of a given work task, one lets a "standardized" person perform the job and again measures the person's reactions to the task. (Since "standard persons" are usually not available, one simply measures the reaction of the workers actually doing the job, assuming that they are "normal.")

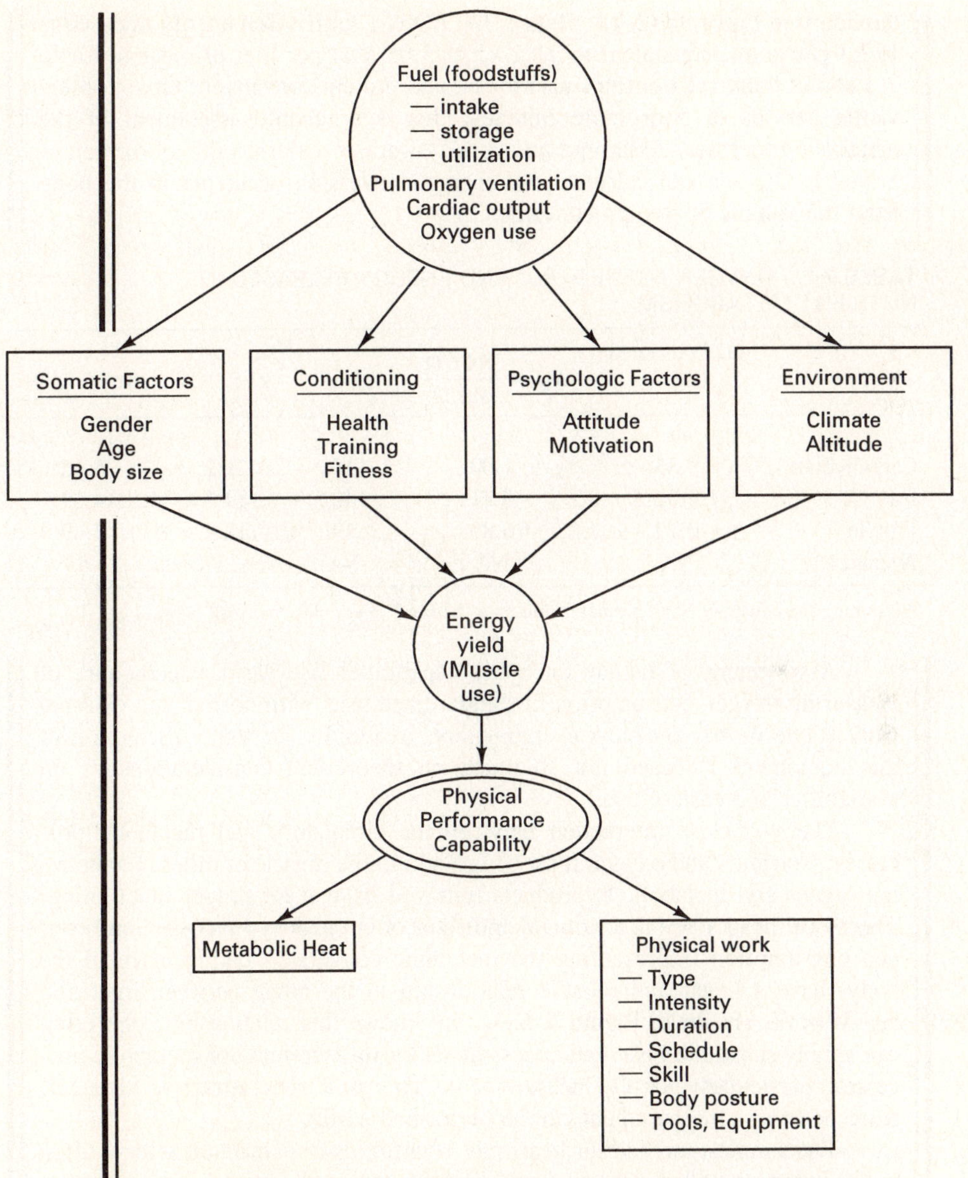

Figure 2-5. Determiners of individual physical work capacity.

While performing work, a person's oxygen consumption (and CO_2 release) is a measure of metabolic energy production. The instruments used for this purpose rely on the principle that the difference in O_2 and CO_2 contents between the exhaled and inhaled air indicates the oxygen absorbed and carbon dioxide released in the lungs. The respiratory exchange quotient RQ compares the carbon dioxide expired to the oxygen consumed. One gram of carbohydrate needs 0.83 L oxygen to be metabolized and releases the same volume of carbon dioxide [see Equation (2-4)]. Hence, the RQ is 1 (unit). The energy released is 18 kJ per gram, equivalent to 21.2 kJ or 5.05 kcal per liter of oxygen. Table 2-1 shows these relationships also for fat and protein conversion. Given observation periods of 5 or more minutes, this is a reliable assessment of the metabolic processes. Assuming an overall "average" caloric value of oxygen of 5 kcal/L O_2, one can calculate the energy conversion occurring in the body from the volume of oxygen consumed.

TABLE 2-1. OXYGEN NEEDED, *RQ,* AND ENERGY RELEASED IN NUTRIENT METABOLISM

	O_2 Consumed (Lg^{-1})	$RQ = \frac{\text{Vol } CO_2}{\text{Vol } O_2}$	$kJ\ g^{-1}$	$kJ\ L^{-1}O_2$	$Kcal\ L^{-1}O_2$
Carbohydrate	0.83	1.00	18	21.2	5.05
Fat	2.02	0.71	40	19.7	4.69
Protein	0.79	0.80	19	18.9	4.49
Average*	NA	NA	NA	21	5

*Assuming the construct of a "normal" adult on a "normal" diet doing "normal" work.

Assessments of human energetic capabilities use various techniques of measuring oxygen consumption in standardized tests with normalized external work, done mostly on bicycle ergometers, treadmills, or steps. (Selection of this equipment is based not so much on theoretical considerations as on availability and ease of use.)

There is close interaction between the circulatory and metabolic processes. Nutrients and oxygen must be brought to the muscle or other metabolizing organs and metabolic byproducts removed from it for proper functioning. Therefore, heart rate (as a primary indicator of circulatory functions) and oxygen consumption (representing the metabolic conversion taking place in the body) have a linear and reliable relationship in the range between light and heavy work, shown in Figure 2-6. If one knows this relationship, one often can simply substitute heart-rate assessments for measurement of metabolic processes, particularly for O_2 measurement. This is a very attractive shortcut, since heart-rate measurement can be performed easily.

The simplest method for heart-rate counting is to palpate an artery, often in the wrist or perhaps in the neck, or to listen to the sound of the beating heart. All the measurer needs to do is count the number of heartbeats over a given period of time (such as 15 seconds) and from this calculate an average heart rate per minute. More refined techniques utilize various plethysmographic methods, which rely on deformations of tissue due to changes in filling of the imbedded blood vessels. These methods range from measuring mechanically the change in volume of tissues, for example in a finger, to using photoelectric techniques that react to changes in transmissibility of light depending on the

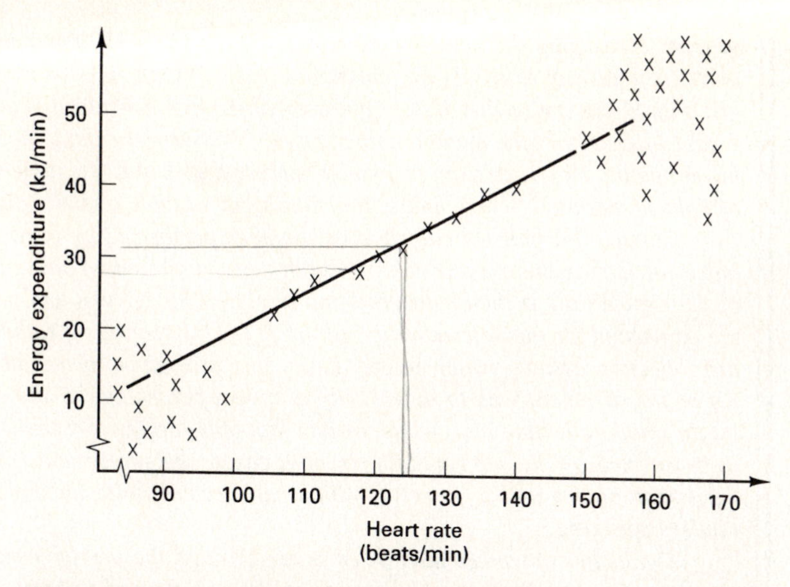

Figure 2-6. Scheme of the relationships between energy expenditure and heart rate. (From Kroemer, Kroemer, and Kroemer-Elbert, 1990, *Engineering Physiology: Bases of Human Factors/Ergonomics,* 2nd ed. With permission by the publisher, Van Nostrand Reinhold. All rights reserved.)

blood filling, such as of the ear lobe. Other common techniques rely on electric signals associated with the pumping actions of the heart (ECG) sensed by electrodes placed on the chest.

These techniques are limited in their reliability of assessing metabolic processes primarily by the intra- and interindividual relationships between circulatory and metabolic functions. Statistically speaking, the regression line (shown in Figure 2-6) relating heart rate to oxygen uptake (aerobic energy production) is different in slope and intersect from person to person and from task to task. In addition, the scatter of the data around the regression line, indicated by the coefficient of correlation, is also variable. The correlation is low at light loads, where the heart rate is barely elevated and circulatory functions can be influenced easily by psychological events (excitement, fear, etc.) which may be completely independent of the task proper. With very heavy work, the O_2-HR relation may also fall apart, for example when cardiovascular capacities may be exhausted before metabolic or muscular limits are reached. Presence of heat load also influences the O_2-HR relationship.

For more information on techniques to assess metabolic processes with commercially available equipment see, for example, Astrand and Rodahl (1986); Eastman Kodak (1983, 1986); Kinney (1980); Kroemer, Kroemer, and Kroemer-Elbert (1990); Mellerowicz and Smodlaka (1981); Stegemann (1984), Webb (1985).

Measuring the oxygen consumed over a sufficiently long period of time is a practical way to assess the metabolic processes. (A physician or physiologist should supervise this test.) One liter of oxygen consumed releases about 5 kcal of energy in the metabolic processes if the subject eats a "normal" diet, has a healthy body oxidizing primarily carbohydrates and fats under conditions of light to moderate work, and is functioning in suitable climatic conditions.

Classically, indirect calorimetry has been performed by collecting all exhaled air during the observation period in airtight (Douglas) bags. The volume of the exhaled air is then measured and analyzed for oxygen and carbon dioxide as needed for the determination of the RQ. This requires a rather elaborate air-collecting system, which mostly limits this procedure to the laboratory. A major improvement was to divert only a known percentage of the exhaled air into a small collection bag, which means that only a relatively small device has to be carried by the subject. Still, in both cases, the subject must wear a face mask with valves and a noseclip, which can become quite uncomfortable and hinders speaking.

Significant advances have been made through the use of instantaneously reacting sensors which can be placed into the air flow of the exhaled air, allowing a breath-by-breath analysis without air collection and without a constraining face mask. The differences in oxygen measured with different equipment are usually rather small; for example, a comparison between the classical Douglas bag method, the partial (Max-Planck) gas meter, and the Oxylog®, a small portable instrument, showed variations in the mean of less than 7 percent; the linear regression coefficients were better than 0.90 (Louhevaara, Ilmarinen, and Oja, 1985). For most field observations, the accuracy of the bagless procedures is quite sufficient.

Use of heart rate has a major advantage over oxygen consumption as an indicator of metabolic processes: it responds faster to work demands, hence indicates more easily quick changes in body functions due to changes in work requirements.

Subjective Rating of Perceived Effort. The human is able to perceive the strain generated in the body by a given work task and to make absolute and relative judgments about this perceived effort: one can assess and judge the relationships between the physical stimulus (the work performed) and its perceived sensation (Pandolf 1983). This correlation between the psychologically perceived intensity of physically described stimuli has been used probably as long as people exist to express one's preference of one type of work over another. Weber, in 1838, and Fechner, in 1860, established formal relationships between a physical stimulus and its perceptual sensation. In the 1970s, Borg developed formal techniques to rate the perceived exertion associated with different kinds of efforts. For example, one can assess the perceived effort using a nominal scale ranging from "light" to "hard." Such a verbally anchored scale can be used to "measure" the strain subjectively perceived while performing standardized work, allowing (similar to the methods previously described) a relative assessment of a person's capability to perform stressful work.

In 1960, Borg developed a "category scale" for the *rating of perceived exertions* (RPE). The scale ranges from 6 to 20 (to match heart rates from 60 to 200 beats/min). Every second number is anchored by verbal expressions:

THE 1960 BORG RPE SCALE (MODIFIED 1985)

6 — no exertion at all
7 — extremely light
8
9 — very light
10
11 — light
12
13 — somewhat hard
14
15 — hard
16
17 — very hard
18
19 — extremely hard
20 — maximal exertion

In 1980, Borg proposed a "category scale with ratio properties" which could yield ratios and levels and allow comparisons but still retain the same correlation (of about 0.88) with heart rate as the RPE scale, particularly if large muscles are involved in the effort.

THE BORG GENERAL SCALE (1980)

0 — nothing at all
0.5 — extremely weak (just noticeable)
1 — very weak
2 — weak
3 — moderate
4 — somewhat strong
5 — strong
6 —

7 — very strong
8 —
9 —
10 — extremely strong (almost maximal)

(*Note*: The terms "weak" and "strong" may be replaced by "light" and "hard," or "heavy," respectively.)

The instructions for use of the scale are as follows (modified from Borg's publications): While the subject looks at the rating scale, the experimenter says:

> I will not ask you to specify the feeling, but do select a number which most accurately corresponds to your perception of (experimenter specifics symptoms).
>
> If you don't feel anything, for example, if there is no (symptom), you answer *zero*—nothing at all.
>
> If you start feeling something, just about noticeable, you answer *0.5*—extremely weak, just noticeable.
>
> If you have an extremely strong feeling of (symptom) you answer *10*—extremely strong, almost maximal. This is the absolute strongest which you have ever experienced.
>
> The more you feel, the stronger the feeling, the higher the number which you choose. Keep in mind that there are no wrong numbers; be honest; do not overestimate or underestimate your ratings. Do not think of any other sensation than the one I ask you about.
>
> Do you have any questions?

Let the subject get well acquainted with the rating scale before the test. During the test, let the subject do the ratings toward the end of every work period, i.e., about 30 seconds before stopping or changing the workload. If the test must be stopped before the scheduled end of the work period, let the subject rate the feeling at the moment of stoppage.

ENERGY REQUIREMENTS AT WORK

Basal Metabolism

A minimal amount of energy is needed to keep the body functioning, even if no activities are done at all. This *basal metabolism* is measured after fasting for 12 hours, accompanied by protein intake restriction for at least two days, with complete physical rest in a neutral ambient temperature. Under these conditions, the basal metabolic values depend primarily on age, gender, height, and weight. Altogether there is relatively little interindividual variation, hence a commonly accepted figure is 1 kcal (4.2 kJ) kg^{-1} hr^{-1} or 4.9 kJ min^{-1} for a person of 70 kg.

Resting Metabolism

The highly controlled conditions under which basal metabolism is measured are rather difficult to accomplish for practical applications. Therefore one usually simply measures the *resting metabolism* before the working day, with the subject as well at rest as possible. Depending on the given conditions, resting metabolism is 10 to 15 percent higher than basal metabolism.

Work Metabolism

The increase in metabolism from resting to working is called *work metabolism*. This increase above resting level represents the amount of energy needed to perform the work.

At the start of physical work, oxygen uptake initially follows the demand sluggishly. As Figure 2-7 shows, after a slow onset, oxygen intake rises rapidly and then slowly approaches the level at which the oxygen requirements of the body are met.

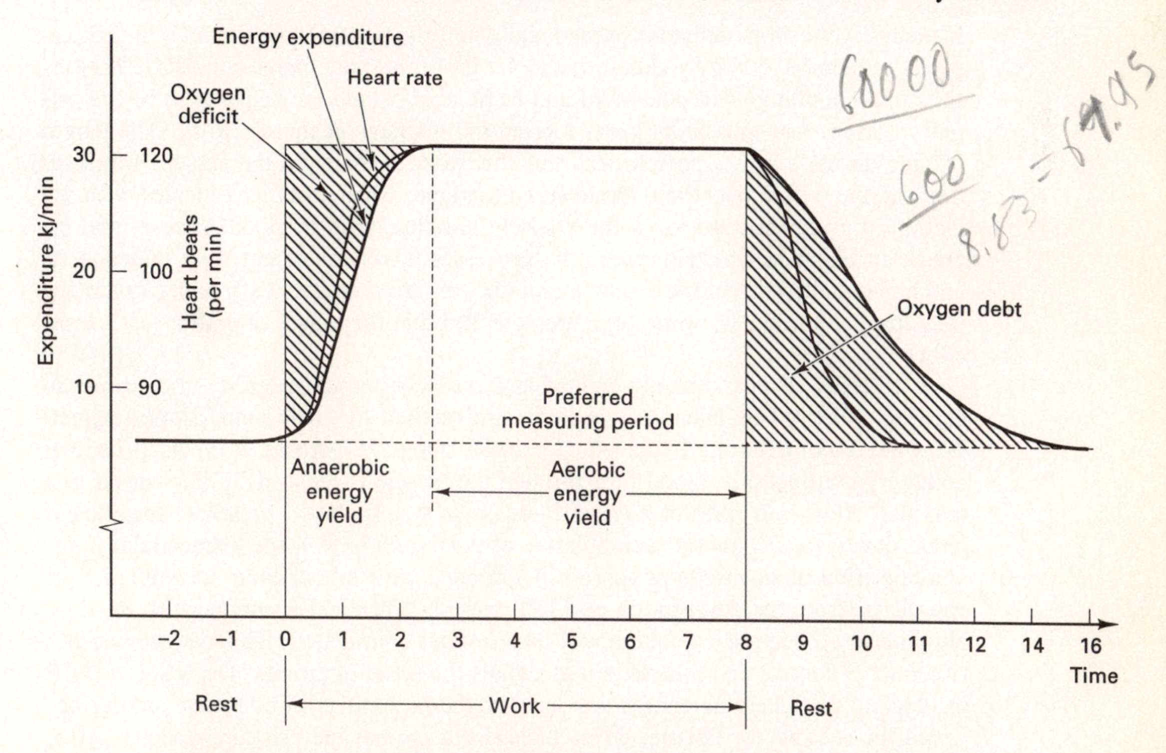

Figure 2-7. Scheme of energy liberation, energy expenditures, and heart rate at steady state work. (From Kroemer, Kroemer, and Kroemer-Elbert, 1990, *Engineering Physiology: Bases of Human Factors/Ergonomics,* 2nd ed. With permission by the publisher, Van Nostrand Reinhold. All rights reserved.)

During the first minutes of physical work, there is a discrepancy between oxygen demand and available oxygen. (During this time, the energy liberation in the body is largely anaerobic.) This *oxygen deficit* must be repaid at some time, usually during rest after work. The amount of this deficit depends on the kind of work performed and on the person, but the *oxygen debt* repaid is approximately twice as large as the oxygen deficit incurred. Of course, given the close interaction between the circulatory and the metabolic systems, heart rate reacts similarly; but as it increases faster at the start of work than oxygen uptake, it also falls back more quickly to the resting level.

If the workload does not exceed about 50 percent of the worker's maximal oxygen uptake, oxygen uptake, heart rate, and cardiac output can finally achieve the required supply level and can stay on this level. This condition of stabilized functions at work is called "steady state." Obviously, a well-trained person can attain this equilibrium between demands and supply even at a relative high workload, while an ill-trained person would be unable to attain a steady state at this high requirement level but could be in equilibrium at lower demand.

Fatigue

If the energetic work demands exceed about half the person's maximal O_2 uptake capacity, anaerobic energy-yielding metabolic processes play increasing roles. This results in accumulations of potassium and lactic acid, which are believed to be the primary reasons for "muscle fatigue," forcing the stoppage of muscle work. The length of time during which a person performs this work depends on the subject's motivation and the will to overcome the feeling of fatigue, which usually coincides with depletion of glycogen deposits in the working muscles, drop in blood glucose, and increase in blood lactate. However, the processes involved are not fully understood, and highly motivated subjects may maintain work that requires very high oxygen uptake for many minutes, while other persons feel that they must stop after just a brief effort.

"Fatigue" is operationally defined as a "reduced muscular ability to continue an existing effort." This phenomenon is best researched for maintained static (isometric) muscle contraction. If the effort exceeds about 15 percent of MVC (maximal voluntary contraction), blood flow through the muscle is reduced, in fact cut off in a maximal effort—in spite of a reflex increase in systolic blood pressure. Insufficient blood flow brings about an accumulation of potassium ions in the extracellular fluid, and depletion of extracellular sodium. Combined with intracellular accumulation of phosphate (from the degradation of ATP), these biochemical events perturb the coupling between nervous excitation and muscle-fiber contraction. This discoupling between CNS control and muscle action signals the onset of fatigue. Depletion of ATP or creatine phosphate as energy carriers, or the accumulation of lactate, so far believed the reasons for fatigue, do occur also but are not the primary reasons. Also, the increase of positive hydrogen ions, resulting from anaerobic metabolism, causes a drop in intramuscular pH, which then inhibits enzymatic reactions, notably those in the ATP breakdown (Kahn and Monod, 1989).

When severe exercise brings about a continuously growing oxygen deficit and an increase in lactate content of the blood because of anaerobic metabolic processes, a balance between demands and supply cannot be achieved; no "steady state" exists, and the work requirements exceed capacity levels—as sketched in Figure 2-8. The resulting fatigue can be counteracted by the insertion of rest periods. Given the same ratio of "total resting time" to "total working time," many short rest periods have more "recovery value" than a few long rest periods.

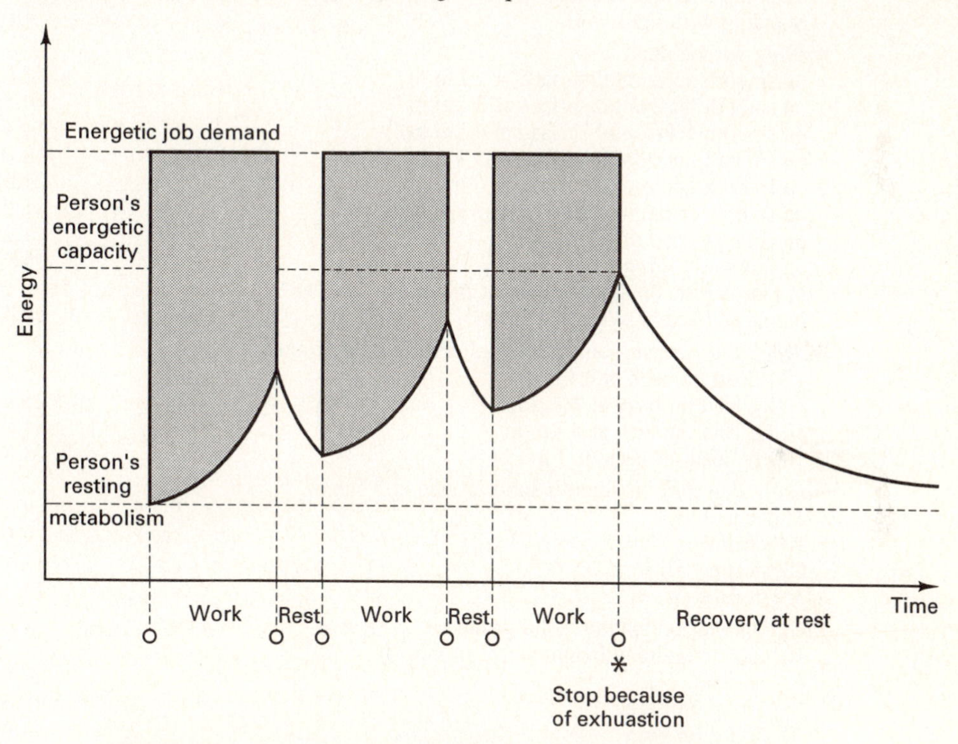

Figure 2-8. Metabolic reactions to the attempt of doing work that exceeds one's capacity even when interspersed rest periods. (From Kroemer, Kroemer, and Kroemer-Elbert, 1990, *Engineering Physiology: Bases of Human Factors/Ergonomics,* 2nd ed. With permission by the publisher, Van Nostrand Reinhold. All rights reserved.)

APPLICATION

Techniques to Estimate Energy Requirements

Tables of energy expenditures have been compiled for body postures and for many professional or athletic activities—see Tables 2-2 and 2-3.

Instead of using such general tables, one can compose the total energetic cost of given work activities by adding together the energetic costs of the work elements that, combined, make up this activity. If one knows the time spent in

TABLE 2-2. ENERGY CONSUMPTION (TO BE ADDED TO BASAL METABOLISM) AT VARIOUS ACTIVITIES

Activity	Energy Consumed, kJ min^{-1}
Lying, Sitting, Standing	
Resting while lying	0.2
Resting while sitting	0.4
Sitting with light work	2.5
Standing still and relaxed	2.0
Standing with light work	4.0
Walking without load	
on smooth horizontal surface at 2 km hr^{-1}	7.6
on smooth horizontal surface at 3 km hr^{-1}	10.8
on smooth horizontal surface at 4 km hr^{-1}	14.1
on smooth horizontal surface at 5 km hr^{-1}	18.0
on smooth horizontal surface at 6 km hr^{-1}	23.9
on smooth horizontal surface at 7 km hr^{-1}	31.9
on country road at 4 km hr^{-1}	14.2
on grass at 4 km hr^{-1}	14.9
in pine forest, on smooth natural ground at 4 km hr^{-1}	18 to 20
on plowed heavy soil at 4 km hr^{-1}	28.4
Walking and carrying on smooth solid horizontal ground	
1-kg load on back at 4 km hr^{-1}	15.1
30-kg load on back at 4 km hr^{-1}	23.4
50-kg load on back at 4 km hr^{-1}	31.0
100-kg load on back at 3 km hr^{-1}	63.0
Walking downhill on smooth solid ground at 5 km hr^{-1}	
5° decline	8.1
10° decline	9.9
20° decline	13.1
30° decline	17.1
Walking uphill on smooth solid ground at 2.5 km hr^{-1}	
10° incline (gaining height at 7.2 m min^{-1})	
no load	20.6
20 kg on back	25.6
50 kg on back	38.6
16° incline (gaining height at 12 m min^{-1})	
no load	34.9
20 kg on back	44.1
50 kg on back	67.2
25° incline (gaining height at 7.2 m min^{-1})	
no load	55.9
20 kg on back	72.2
50 kg on back	113.8
Climbing stairs 30.5° incline, steps 17.2 cm high, 100 steps per minute (gaining 17.2 m min^{-1}), no load	57.5
Climbing ladder 70° incline, rungs 17 cm apart (gaining 11.2 m min^{-1}), no load	33.6

Note: While Rohmert and Rutenfranz (1983) claim that intra- and interindividual differences in energy consumption are within ±10 percent for the same activity, a comparison of data presented in various texts shows a much higher percentage in variation, particularly at activity levels requiring little energy.

SOURCE: Adapted from Astrand and Rodahl 1977, Guyton 1979, Rohmert and Rutenfranz 1983, Stegemann 1984.

TABLE 2-3. TOTAL ENERGY COST PER DAY IN VARIOUS JOBS AND PROFESSIONS

Occupation	Energy expenditure, kcal/day		
	Mean	Minimum	Maximum
MEN			
Laboratory technicians	2840	2240	3820
Elderly industrial workers	2840	2180	3710
University students	2930	2270	4410
Construction workers	3000	2440	3730
Steel workers	3280	2600	3960
Elderly peasants (Swiss)	3530	2210	5000
Farmers	3550	2450	4670
Coal miners	3660	2970	4560
Forestry workers	3670	2860	4600
WOMEN			
Elderly housewives	1990	1490	2410
Middle-aged housewives	2090	1760	2320
Laboratory technicians	2130	1340	2540
University students	2290	2090	2500
Factory workers	2320	1970	2980
Elderly peasants (Swiss)	2890	2200	3860

Note: The physical job demands may be different today from what they were decades ago, when these data were collected.

SOURCE: Adapted from Astrand and Rodahl, 1977.

a given activity element and its metabolic cost per time unit, one can simply calculate the energy requirements of this element by multiplying its unit metabolic cost by its duration time.

EXAMPLE

An example of daily energy expenditure: for a person resting (sleeping) eight hours per day, at an energetic cost of approximately 5.1 kJ/min, the total energy cost is about 2,450 kJ (5.1 kJ $\text{Min}^{-1} \times 60$ min $\text{hr}^{-1} \times$ 8 hr). If the person then does six hours of light work while sitting, at 7.4 kJ/min, this adds another 2,664 kJ to the energy expenditure. With an additional six hours of light work done standing, at 8.9 kJ/min and further with four hours of walking at 11.0 kJ/min, the total expenditure during the full 24-hr day would come to about 10,960 kJ (or approximately 2,610 kcal).

Energy requirements of work allow one to judge whether a job is (energetically) easy or hard. Given the largely linear relationship between heart rate and energy uptake, one can often simply use heart rate to judge work as "light" or "heavy." Of course, such labels reflect judgments that rely very much on the current socioeconomic concept of what is permissible, acceptable, comfortable, easy, or hard. Depending on the circumstances, one finds a diversity of opinions about how "hard" a given job is—but its demands on the body can be objectively measured, such as by heart rate or oxygen uptake.

EXAMPLE

DEFINING THE "HEAVINESS" OF WORK:

CLASSIFICATION	TOTAL ENERGY EXPENDITURES IN KJ MIN^{-1}	TOTAL ENERGY EXPENDITURES IN KCAL MIN^{-1}	HEART RATE IN BEATS PER MINUTE
Light work	10	2.5	90 or less
Medium work	20	5	100
Heavy work	30	7.5	120
Very heavy work	40	10	140
Extremely heavy work	50	12.5	160 or more

"Light" work is associated with rather small energy expenditure (about 10 kJ/min including the basal rate) and accompanied by a heart rate of approximately 90 beats/min. In this type of work, the energy needs of the working muscles are covered by the oxygen available in the blood and by glycogen at the muscle. Lactic acid does not build up. At "medium" work, with about 20 kJ and 100 beats/min, the oxygen requirement at the working muscles is still covered, and initially developed lactic acid is resynthesized to glycogen during the activity. In "heavy" work, with about 30 kJ and 120 beats/min, the oxygen required is still supplied if the person is physically capable to do such work and specifically trained in this job. However, the lactic acid concentration incurred during the initial minutes of the work is not reduced but remains till the end of the work period, to be brought back to normal levels after cessation of the work.

With light, medium, and even heavy work, metabolic and other physiological functions can attain a steady-state condition throughout the work period (provided the person is capable and trained). This is not the case with "very heavy work," where energy expenditures are in the neighborhood of 40 kJ/min, and heart rate is around 140 beats/min. Here, the original oxygen deficit increases throughout the duration of work, making intermittent rest periods necessary or even forcing the person to stop this work completely. At even higher energy expenditures, such as 50 kJ/min, associated with heart rates of 160 beats/min or higher, lactic acid concentration in the blood and oxygen deficit are of such magnitudes that frequent rest periods are needed, and even highly trained and capable persons may be unable to perform this job through a full work shift.

Note that only "dynamic work" can be suitably assessed by energy demands. "Static" efforts, where muscles are contracted and kept so, hinder or completely occlude their blood supply by compression of the capillary bed. Thus, the heart makes an effort to overcome the resistance by increasing beating rate and blood pressure; but because blood flow remains insufficient, relatively little energy is supplied to contracted muscles and consumed there. Hence, such static effort may be exhausting but is not well assessed by energy measures.

The engineer determines the "effort" required and how work is to be done, and also has control over the work environment. To arrange for a suitable match between capabilities and demands, the engineer needs to adjust the work to be performed (and the work environment, discussed in Chapter 5) to the body's energetic capabilities.

A typical example of the use of "effort" measurements is in the selection of either ramps, stairs, or ladders: the usual measurements taken are oxygen consumption, heart rate, and subjective ratings/preferences. Figure 2-9 indicates resulting recommendations, depending on the angle of ascent (MIL-STD 1472). Accordingly, for angles of up to 20 degrees, ramps are preferred; between 20 and 50 degrees, stairs; at steeper angles, stair ladders; and for angles above 75 degrees, ladders. The recommended design features are shown in Figure 2-10 (MIL-STD 1472), but better data for design are needed (Irvine,

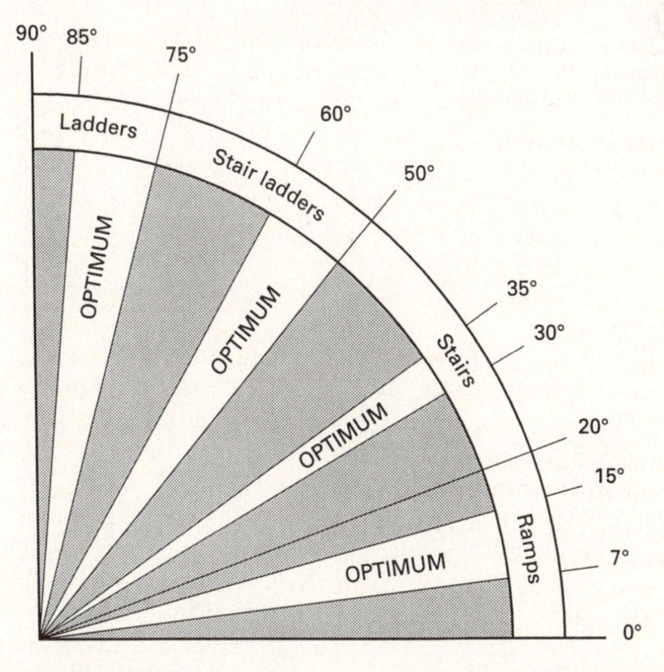

Figure 2-9. Selection of ladders, stairs, or ramps according to the angle of ascent (MIL-STD 1472).

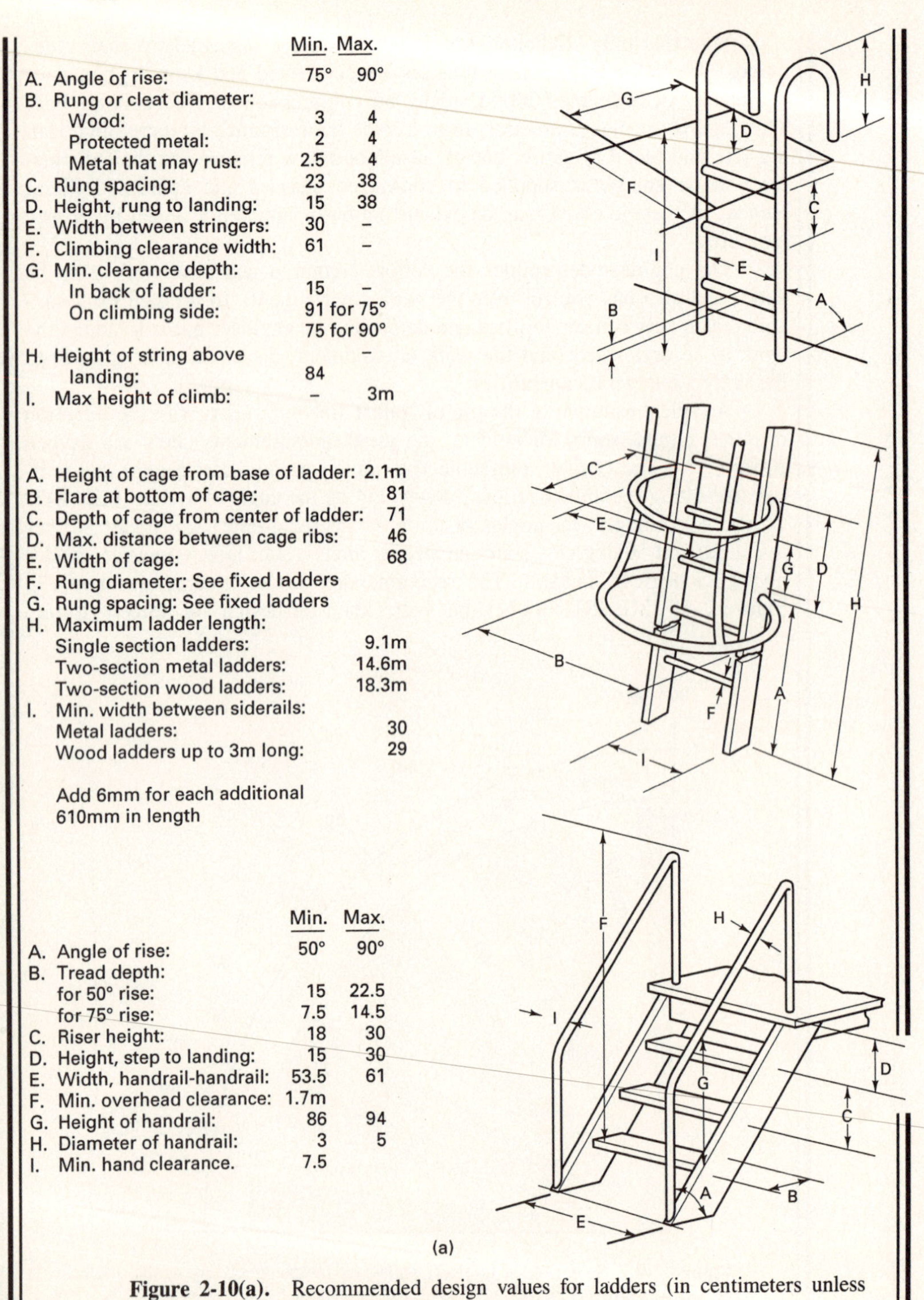

	Min.	Max.
A. Angle of rise:	75°	90°
B. Rung or cleat diameter:		
Wood:	3	4
Protected metal:	2	4
Metal that may rust:	2.5	4
C. Rung spacing:	23	38
D. Height, rung to landing:	15	38
E. Width between stringers:	30	–
F. Climbing clearance width:	61	–
G. Min. clearance depth:		
In back of ladder:	15	–
On climbing side:	91 for 75° 75 for 90°	
H. Height of string above landing:	84	–
I. Max height of climb:	–	3m

A. Height of cage from base of ladder:	2.1m
B. Flare at bottom of cage:	81
C. Depth of cage from center of ladder:	71
D. Max. distance between cage ribs:	46
E. Width of cage:	68
F. Rung diameter: See fixed ladders	
G. Rung spacing: See fixed ladders	
H. Maximum ladder length:	
Single section ladders:	9.1m
Two-section metal ladders:	14.6m
Two-section wood ladders:	18.3m
I. Min. width between siderails:	
Metal ladders:	30
Wood ladders up to 3m long:	29

Add 6mm for each additional 610mm in length

	Min.	Max.
A. Angle of rise:	50°	90°
B. Tread depth:		
for 50° rise:	15	22.5
for 75° rise:	7.5	14.5
C. Riser height:	18	30
D. Height, step to landing:	15	30
E. Width, handrail-handrail:	53.5	61
F. Min. overhead clearance:	1.7m	
G. Height of handrail:	86	94
H. Diameter of handrail:	3	5
I. Min. hand clearance.	7.5	

(a)

Figure 2-10(a). Recommended design values for ladders (in centimeters unless otherwise stated) (MIL-HDBK 759B).

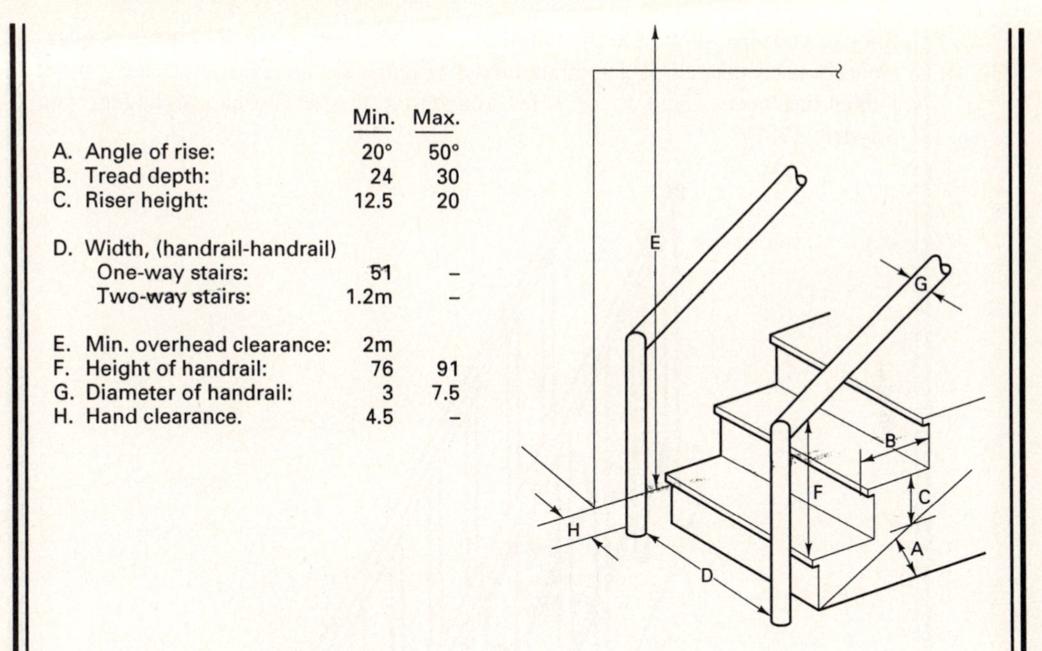

	Min.	Max.
A. Angle of rise:	20°	50°
B. Tread depth:	24	30
C. Riser height:	12.5	20
D. Width, (handrail-handrail)		
One-way stairs:	51	–
Two-way stairs:	1.2m	–
E. Min. overhead clearance:	2m	
F. Height of handrail:	76	91
G. Diameter of handrail:	3	7.5
H. Hand clearance.	4.5	–

	Min.	Max.
A. Angle of rise:	–	20°
B. Height of handrails:	96	110
C. Width: Determined by function and usage; particularly size of rolling stock and loads.		
D. Diameter of handrail:	2.5	7.5
E. Clearance around handrail:	5	–

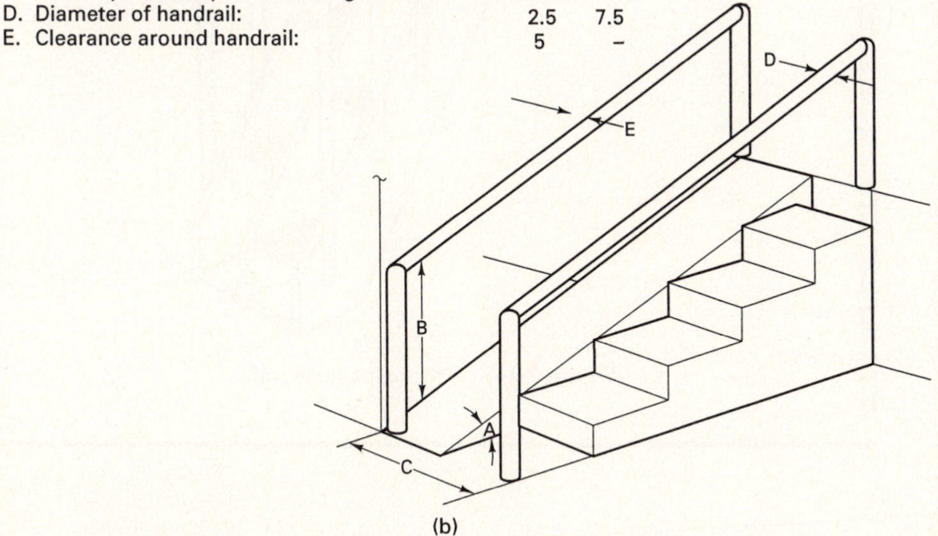

(b)

Figure 2-10(b). Recommended design values for stairs (in centimeters unless otherwise stated) (MIL-HDBK 759B).

Snook, and Sparshatt, 1990). On ships, an "alternating tread" ladder is occasionally used instead of a regular ladder (Figure 2-11). This alternating tread ladder has been found to be safer and easier to use (Jorna, Mohageg, and Snyder, 1989).

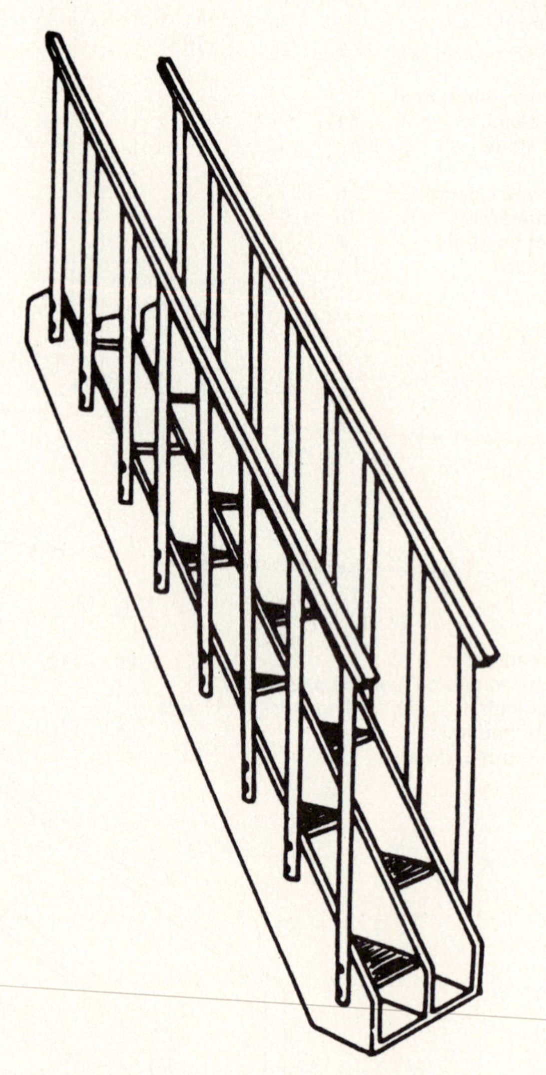

Figure 2-11. Alternating tread stair.

☞☞☞ ***What your body does—how and why?***

Sneezing. *The sneeze is a very fast expiration of air. It is caused by an irritation in the nasal passage. Its purpose is to clear the cause of irritation.*

The sneeze starts when one begins to exhale. The vocal chords come together to plug the windpipe and a sound evolves. This response is involuntary and difficult to control.

Children sneeze through their noses, which often creates a mess. Adults usually try to send the sneeze on a detour through the mouth.

Yawning. *Yawning appears to be a reaction to either the lack of oxygen, or the accumulation of carbon dioxide, in the lungs or breathing pathways. This occurs particularly if one is inactive, such as when sitting and listening to a boring presentation. The breathing frequency is reduced, and the "air quality" in lungs and breathing pathways becomes intolerable. A deep inhalation, facilitated by opening the mouth, flushes all the stale air out, taking carbon dioxide with it and introducing oxygen.*

This does not explain why yawning is "contagious"; perhaps "communal yawning" is a social sign of belonging.

Shivering. *As a muscle contracts to generate energy to do work, heat is produced as a side product. If no work to the outside is done, a muscle contraction generates heat alone. This occurs in shivering, where the jaw makes your teeth chatter, or other parts of the body are moved by quick muscle contractions. All of this generates heat to keep the body warm in a cold environment.*

Muscle cramps. *Cramps usually occur in response to a contraction signal sent to a muscle already stretched. This may occur in a sports event, or while you are asleep or just waking up.*

A cramp may involve edema formation, the accumulation of fluid inside the muscle but outside its blood vessels. This generates pressure, which causes pain. Another explanation may be an inadequate supply of oxygen to muscle tissues of persons who suffer from low blood pressure. A third guess is that local controllers, or in the central nervous system, run amok.

Yet, none of these explains why the muscle suddenly contracts in a strong and painful manner, and maintains that contraction painfully.

Rumbling stomach. *Stomach muscles contract and move gas inside the stomach. This occurs particularly when there is no food in the stomach. The gases come from swallowed air. The bubbles make audible noises.*

Putting food into the stomach takes care of the problem.

Snoring. *The uvula is a small soft structure hanging from the palate above the root of the tongue. This is at the entrance to the pharynx, which carries air from the nasal cavity to the larynx, the voice box. When the uvula vibrates in the air* *stream, a snoring sound occurs because the uvula touches other structures. This is most likely to occur when you are lying on your back.*

Thus, turning over on the side is likely to stop the snoring.

Swelling of ankles. *When one sits without leg motion for a long time, such as in an airplane, swelling (edema) of the lower legs is often experienced, particularly of the feet and ankle regions. This occurs because a serumlike fluid, mostly plasma, seeps from the capillaries into the interstitial spaces of the connective tissue, making it expand. The resulting tissue pressure hinders lymphatic flow as well as blood flow through the capillaries and other small blood vessels. The preventive action is to move the lower legs and feet often while sitting, or even better is getting up and walking around. After swelling has occurred, it is best to raise the feet and let the accumulated fluid dissipate.*

Catching a cold. *Contrary to popular belief, a "cold" is not brought about by a chill, wetness, or draft: they do not even predispose to infection. Any one of about 200 known cold viruses can infect the upper respiratory tract including nose, sinuses, and throat. The increased occurrence of "colds" during the colder weather seasons merely stems from the fact that people spend more time crowded together indoors, making it easy for viruses to spread from person to person.*

SUMMARY

The human body converts foodstuffs, in the presence of oxygen, into energy, which then is used to perform work. The respiratory system absorbs oxygen into the bloodstream, and extracts from the blood carbon dioxide, water, and heat, which are expelled into the exhaled air. Breathing rate and breathing volume are not convenient measures for the physical effort of the body.

The circulatory system transports oxygen and energy carriers (mostly glycogen) to the consuming organs (e.g., muscles) and moves metabolic byproducts (CO_2, water, heat) to the lungs and (water and heat) to the skin for dispersion. The heart pumps the blood through the system. The number of heartbeats per minute provides a convenient and rather accurate assessment for the circulatory efforts, and for the physical workload of the body in general.

The metabolic system breaks the energy carriers of the food into molecules that the body can use. Carbohydrate is broken into polysaccharides, glycogen and glucose. Fat is broken into glycerol and fatty acids. Protein is broken into amino acids. Of these, glucose and glycogen are the most rapidly used energy carriers, while fat serves as energy storage, used mostly in longer-lasting efforts.

The metabolic process of converting polysaccharides and fats into energy used by the muscles occurs in several steps, some of which are anaerobic; yet, overall, oxygen must be provided. Since it is known how much oxygen must be provided for the conversion of each energy carrier, measurement of oxygen consumption provides an accurate means to assess the body's energy needs. Oxygen consumption and heart rate are highly correlated.

Another way to assess the workload of the body is to subjectively rate the perceived effort. This procedure has been well standardized and can be used in combination with, or instead of, the objective measures.

It is one of the tasks of the ergonomist to keep the job requirements in terms of energy demands within reasonable limits. This can be done if the energy demands of the job are measured, and matched with the energy capabilities of the operator.

CHALLENGES

How would special training through exercise, or the development of deficiencies by disease, of any one of the respiratory, circulatory, or metabolic systems affect a person's ability to perform work: either of the physical, strenuous type or of the psychological, mental type?

Which events or functions in the respiratory system are easily measurable by the ergonomist and can be used to assess work-related loading of the body?

What are the specific tasks of the blood in supporting the functions of the working body?

By what means can the heart increase blood flow into the aorta?

How is blood supply to a working organ regulated?

What are the functions of an artificial pacemaker for the heart?

Why do well-trained athletes often have a very low resting heart rate?

What specific features limit the use of heart rate as an indicator of circulatory loading, and particularly of metabolic efforts?

The energy-balance equation does not specify the duration of processes and observation. How does that affect its validity and use?

How would the consideration of energy storage, or loss, affect the use of the energy-balance equation?

What are the limitations of the equation given for human energy efficiency?

How can one "become fat" if one eats no fat?

Which are the important functions of fat deposits in the body?

Which are the two major means used by the body to store energy for use as needed?

Which kind of energy storage in the body is used to support a physical effort?

Why does one have anaerobic processes involved, if indeed the overall process of metabolism is aerobic?

Isn't it reasonable to assume that "average persons" perform everyday activities?

What limitations and assumptions apply to tables of energy consumption for certain activities, found in the literature and also given in this chapter?

Are there differences in "experimental control" that must be applied if either "objective" or "subjective" measures are taken to assess the workload associated with certain activities?

3

How the Mind Works

OVERVIEW

The traditional psychological model of the human information system is sequential: input is sensed, then processed, and output follows. While this model has been criticized, nearly all currently available information is based on it.

The nervous system controls the body. It receives information from body sensors via its afferent peripheral part. Information is processed, decisions are made, and control signals are generated in the central part. These signals are transmitted to body organs, including muscles, in the efferent peripheral part.

Job and environmental conditions can be very stressful, not only in "regular" tasks but especially in confined and dangerous environments, such as during space exploration. The assessment of "workload" is of importance for subsequent design recommendations.

INTRODUCTION

"Of course, the brain is a machine and a computer . . . But our mental processes, which constitute our being and life, are not just abstract and mechanical but personal, as well—and, as such, involve not just classifying and categorizing, but continual judging and feeling also. If this is missing, we become computer-like . . ." (Sacks, 1990, p. 20).

In the "traditional" system concept of engineering psychology, the human is considered a receptor and processor of information or energy, who outputs informa-

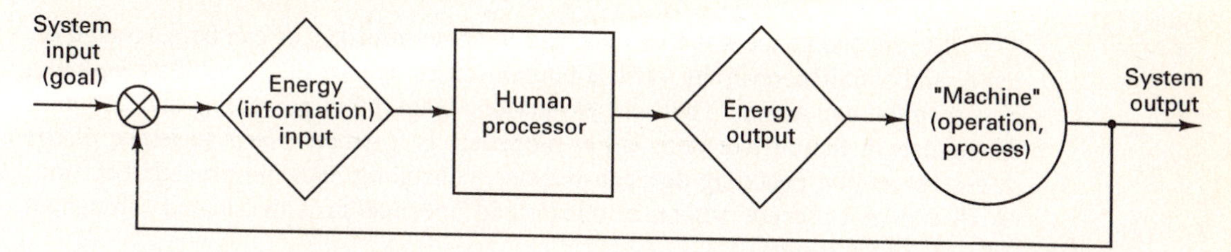

Figure 3-1. The human as energy or information processor.

tion or energy. Input, processing, and output follow each other in sequence. The output can be used to run a "machine," which may be a simple hand tool or a space craft. This basic model is depicted in Figure 3-1.

The actual performance of this human-technology system (in the past often called a man-machine system) is monitored and compared with the desired performance. Hence, one feedback loop connects (or several feedback loops connect) the output side (or one of its elements) with the input side. The difference between output and input is registered in a comparator, and corrective actions are taken to minimize any output/input difference. The human in this system compares, makes decisions, and corrects.

This "human processor" is the object of research, either to understand human basic functions or to observe human actions and reactions within the system. (Especially in the 1970s and 80s, information theory, including signal channelling and processing, were major research topics. These aspects are beyond the scope of this book, but there are many related publications in the field of cognitive psychology.) "Input" and "output" are the sites of application of ergonomics and human-factors engineering. The design of the "machine" is the classical engineering task, although with significant help from the ergonomist.

The "Traditional" and the "Ecological" Concepts

"Traditional" psychologists believe that our activities can be described as a linear sequence of stages, from perception to encoding to decision to response. Research is done separately on each of these stages, on their substages, and on their connections. Such independent, stage-related information is then combined into a linear model to provide information for the engineering psychologist.

This behavioristic model is thought invalid by "ecological" psychologists (Flach, 1989; Vicente and Harwood, 1990). Based on the thoughts of Brunswik (1956), Gibson (1966), and Meister (1989), this approach to the study of human perception and action assumes simultaneous rather than sequential interactions. Two major concepts in the ecological approach are *affordance* and its *perception*.

Affordance is the property of an environment that has certain values to the human. Flach gives the example of a stairway, which affords passage for a person who can walk but not for a person confined to a wheelchair. Thus, passage is a property of the stairway, but its affordance value is specific to the user. Accordingly, ergonomics or human engineering provides affordances.

The second concept is that information about affordances can be perceived directly and simultaneously by various human senses as part of the intimate coupling of perception and action. Thus, the closed-loop coupling of perception and action of the human in the environment is *not* modelled as a simple linear sequence of the stages perception-encoding-decision-response; encoding does not precede decision, which does not precede response, but instead information is distributed throughout the closed-loop system.

If one follows this concept, then indeed research on behavior stages, assumed to follow each other in sequence, does not realistically explain human behavior in a technological system. This leads to "the unfortunate conclusion that not only much of the data produced by traditional academic psychology [are] irrelevant to system design, but may be irrelevant to the science of behavior" (Flach, 1989, p. 3). Instead, the study of behavior must be at the level of an "ecological human-environment system." This approach would require fundamentally new models and research of information, of cognition, and of performance assessment, different from those associated with traditional psychology.

Yet, current knowledge is almost completely based on the "traditional" sequential-system concept.

ORGANIZATION OF THE NERVOUS SYSTEM

Anatomically, one divides the nervous system into three major subdivisions. The *central* nervous system (CNS) includes brain and spinal cord; it has primarily control functions. The *peripheral* nervous system (PNS) includes the cranial and spinal nerves; it transmits signals, but usually does not control. The *autonomic* nervous system consists of the sympathetic and the parasympathetic subsystems. Together they regulate, among other involuntary functions, those of smooth and cardiac muscle, of blood vessels, digestion, and glucose release in the liver. The autonomic system is not under conscious control. It generates the "fright, flight or fight" responses.

Functionally, one divides the nervous system into two major subdivisions: the autonomic system (just discussed), and the *somatic* (Greek: soma, body) nervous system, which controls mental activities, conscious actions, and skeletal muscle.

☞☞☞ *While at the "Museum" in Alexandria, Egypt, about 280 B.C. Herophilus distinguished sensory and motor nerves. He described the liver and the retina of the eye. Erasistratus (about 250 B.C.) distinguished between the cerebrum and the cerebellum of the brain and thought that the many convolutions of the brain, more numerous than in any other animals, were related to superior human intelligence. In 1810, the German physician Franz Joseph Gall (1758–1828) published the first volume of his treatise on the nervous system. He stated that the gray matter on the surface of the brain and in the interior of the spinal cord was the active and essential part, and that the white matter was just connecting material. Gall also believed that the shape of the brain had something to do with mental capacity; he went so far as to relate the shape of the brain, and of the skull, with emotional and temperamental qualities, which started the pseudoscience of phrenology. The French anthropologist Pierre-Paul Broca (1824–1880) demonstrated Gall's belief that the brain controlled different parts and functions of the body (Asimov, 1989).*

The brain is usually divided into *forebrain, midbrain,* and *hindbrain*. Of particular interest for the neuromuscular control system is the forebrain with the *cerebrum,* which consists of the two (left and right) cerebral hemispheres, each divided into four lobes. Control of voluntary movements, sensory experience, abstract thought, memory, learning, and consciousness are located in the cerebrum. The multifolded *cortex* enwraps most of the cerebrum. The cortex controls voluntary movements of the skeletal muscle and interprets sensory inputs. The *basal ganglia* of the midbrain are composed of large pools of neurons, which control semivoluntary complex activities such as walking. Part of the hindbrain is the *cerebellum,* which integrates and distributes impulses from the cerebral association centers to the motor neurons in the spinal cord. Figure 3-2 is a schematical sketch of the major brain parts.

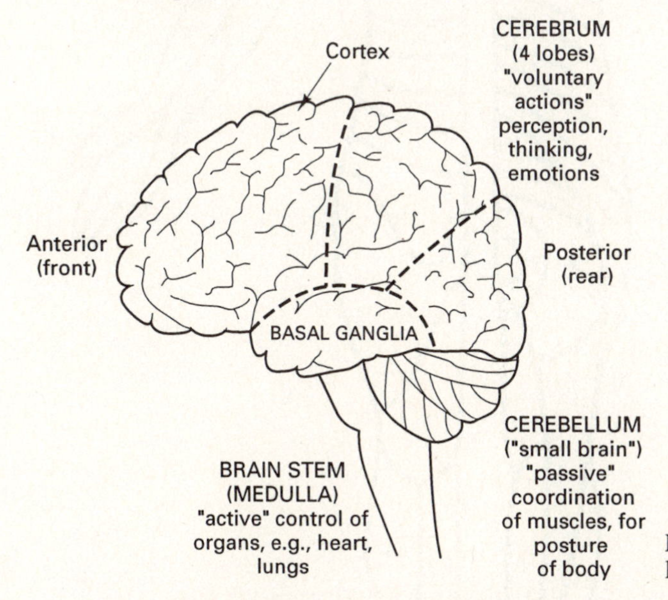

Figure 3-2. Side view (from the left) of the human brain.

The *spinal cord* is an extension of the brain. The uppermost section of the spinal cord contains the twelve pairs of *cranial nerves,* which serve structures in the head and neck, as well as the lungs, heart, pharynx and larynx, and many abdominal organs. They control eye, tongue, facial movements, and the secretion of tears and saliva. Their main inputs are from the eyes, the tastebuds in the mouth, the nasal olfactory receptors, and touch, pain, heat, and cold receptors of the head. Thirty-one pairs of *spinal nerves* pass out between the appropriate vertebrae and serve defined sectors of the rest of the body. Nerves are mixed sensory and motor pathways, carrying both somatic and autonomic signals between the spinal cord and the muscles, articulations, skin, and visceral organs. Figure 3-3 shows how the

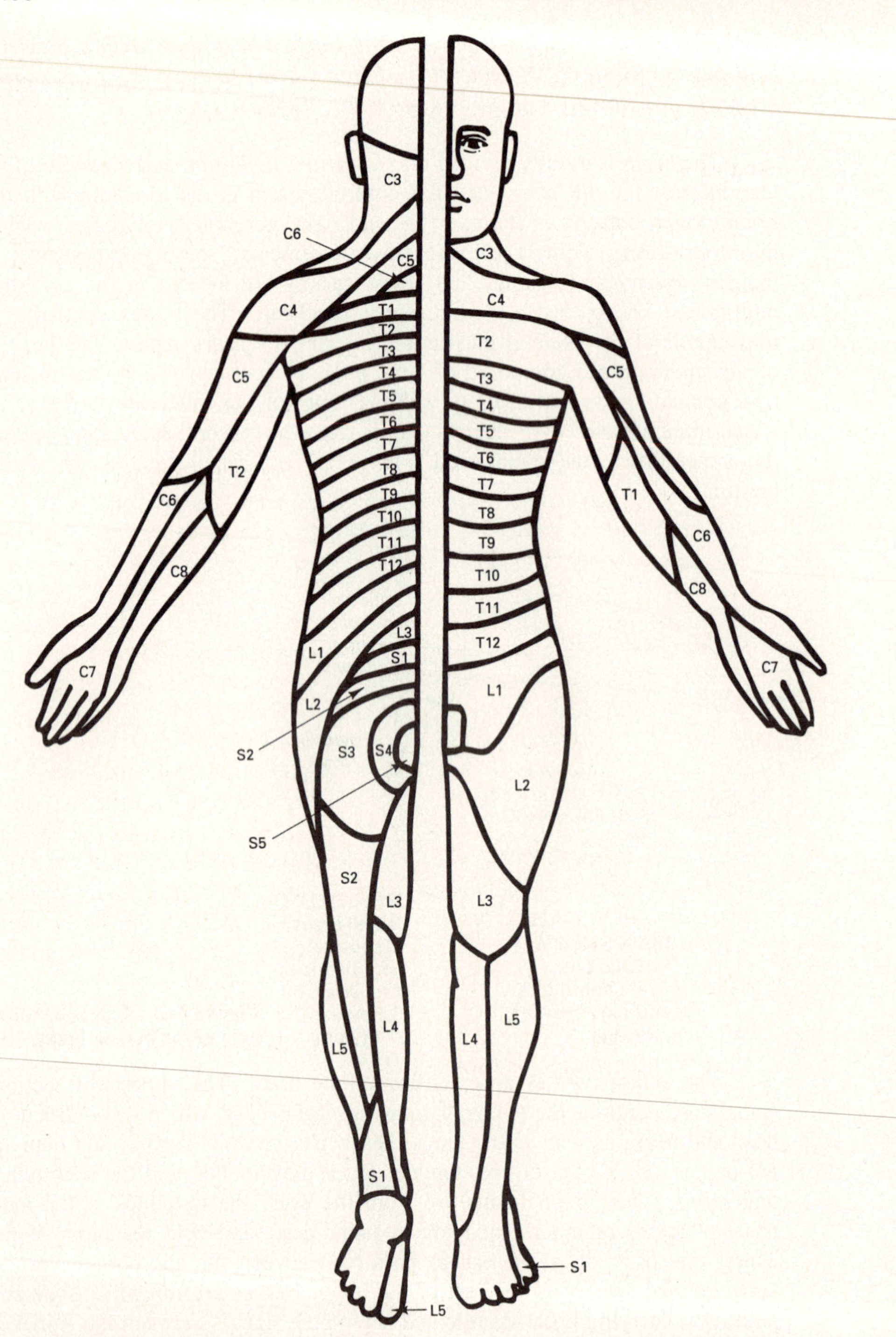

Figure 3-3. Sensory dermatomes with their spinal nerve roots. (Modified from Sinclair, 1973.)

spinal nerves emanating from sections of the spinal column ("spinal nerve roots") innervate defined areas of the skin ("dermatomes"). Figure 3-4 associates major nerves of the body with their respective areas of cutaneous sensitivity.

Sensors

The central nervous system receives information from internal receptors, *interoceptors,* which report on changes within the body; on digestion, circulation, excretion, hunger, thirst, sexual arousal, and feeling well or sick. *Exteroceptors* respond to light, sound, touch, temperature, electricity, and chemicals. Since all of these sensations come from various parts of the body (Greek, *soma*), external and internal receptors together are also called somesthetic sensors.

Internal receptors include the *proprioceptors*. Among these are the muscle spindles, nerve filaments wrapped around small muscle fibers which detect the amount of stretch of the muscle. Golgi organs are associated with muscle tendons and detect their tension, hence report to the central nervous system information about the strength of contraction of the muscle—see Chapter 1. Ruffini organs are kinesthetic receptors located in the capsules of articulations. They respond to the degree of angulation of the joints (joint position), to change in general, and also to the rate of change.

The sensors in the vestibulum are also proprioceptors. They detect and report the position of the head in space and respond to sudden changes in its attitude. This is done by sensors in the semicircular canals, of which there are three, each located in another orthogonal plane. To relate the position of the body to that in the head, proprioceptors in the neck are triggered by displacements between trunk and head.

Another set of interoceptors, called *visceroceptors,* reports on the events within the visceral (internal) structures of the body, such as organs of the abdomen and chest, as well as on events within the head and other deep structures. The usual modalities of visceral sensations are pain, burning sensations, and pressure. Since the same sensations are also provided from external receptors and because the pathways of visceral and external receptors are closely related, information about the body is often integrated with information about the outside.

External receptors provide information about the interaction between the body and the outside: sight (vision), sound (audition), taste (gustation), smell (olfaction), temperature, electricity, and touch (taction)—see Chapter 4 for detailed information. Several of these are of particular importance for the control of muscular activities: the sensations of touch, pressure, and pain can be used as feedback to the body regarding the direction and intensity of muscular activities transmitted to an outside object—see Chapter 2. Free nerve endings, Meissner's and Pacinian corpuscles, and other receptors are located throughout the skin of the body, although in different densities. They transmit the sensations of touch, pressure, and pain. Since the nerve pathways from the free endings interconnect extensively, the sensations reported are not always specific for a modality; for example, very hot or cold sensations can be associated with pain, which may also be caused by hard pressure on the skin. Figure 3-5 sketches receptors in the skin.

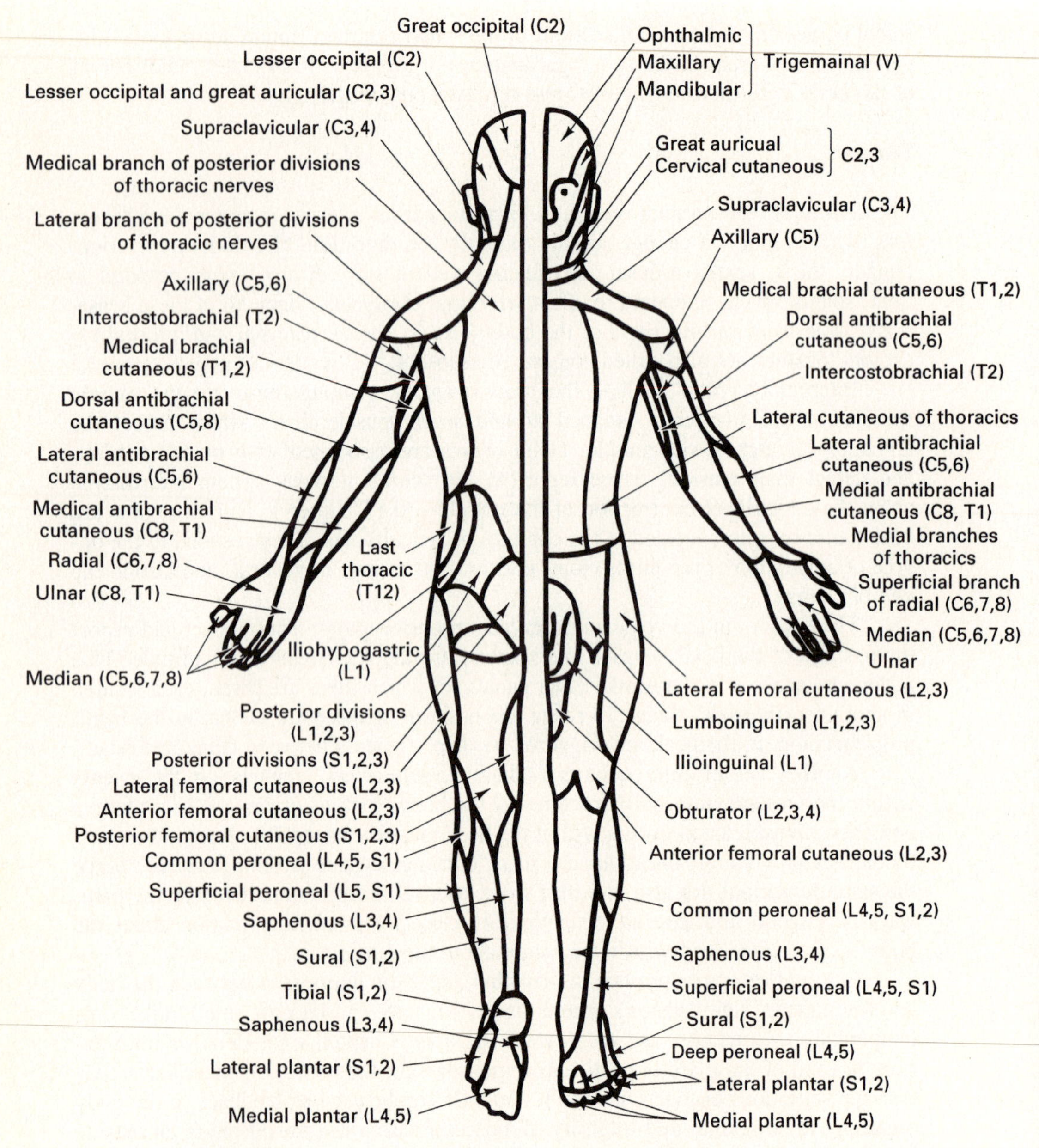

Figure 3-4. Territories of the major nerves. (Modified from House and Pansky, 1967.)

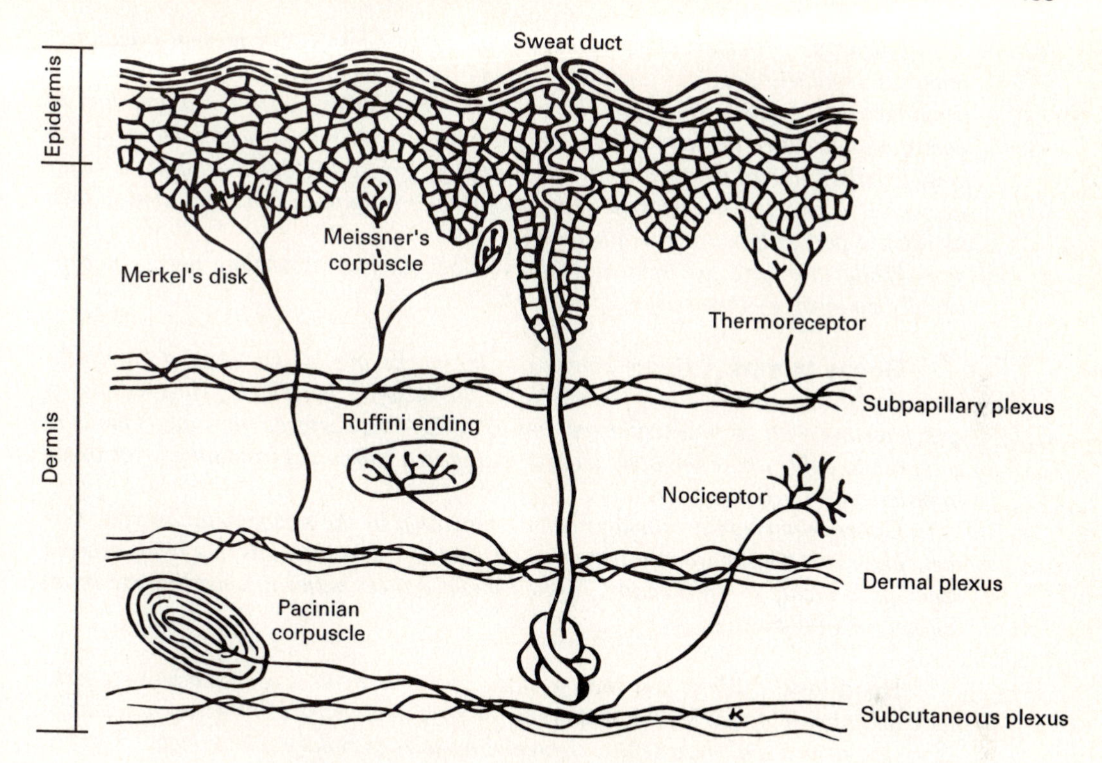

Figure 3-5. Skin receptors. (Modified from Griffin, 1990).

☞☞☞ ***What the nervous system does: how and why?***

Butterflies in the stomach. *The underlying event is not clear. It may be that the heart beats faster when a person is under stress, or the diaphragm that separates chest and abdominal cavities might be fluttering. Yet, the butterfly feeling may in fact not be an event close to the stomach: the ability to localize a sensation in one's own body is often poor. For example, a heart attack is sometimes felt in the arm, and shoulder pain may actually come from the diaphragm.*

Blushing. *This is a response of the body to embarrassment by filling the capillaries near the skin surface with blood.*

Why this reaction occurs is unknown. How it occurs is well known: it is stimulated by the autonomic nervous system, which controls involuntary body functions, such as the heart, glands, and blood vessels.

Lump in throat. *This is another example of the actions of the autonomic nervous system. When threatened, the sympathetic nervous system becomes active and, in this case, makes saliva thick. This is felt as something interfering with swallowing, like a lump in the throat.*

Hiccups. *Hiccups are breathing gone out of control. For normal breathing, impulses regularly progress along the phrenic nerve leading to the diaphragm that separates the abdomen from the chest cavity. The signals cause the diaphragm to contract and reduce the lung volume for expiration. If there is a sudden burst of nervous impulses, the diaphragm contracts abruptly, generating quick expirations. Hiccups can occur with no apparent stimulus but tend to happen at certain times, such as when a person has just eaten heavily.*

While there are folk remedies for hiccups, such as a startling noise, hiccups usually go away by themselves.

Goose bumps. *Goose bumps are the puckering of skin around hair follicles, due to the contraction of a muscle at the base of the follicle. Hair stands up and, together with the now enlarged and irregular skin surface, traps air. This is in response to cold, where the body tries to maintain a warm and insulating layer of air in order not to lose heat.*

Goose bumps are involuntary events generated by the sympathetic nervous system. They can appear when a person is stricken by fear. Perhaps this is an archaic reaction to create a "thicker fur," which might protect better against the bite of an animal, or of cold.

Head jerk. *When you start to doze off, your head drops. Suddenly it jerks back up. Apparently, a sensor in a joint, or in a neck muscle, received a signal of excessive stretch and, in a reflex, tightened muscles to reduce this stretch.*

Why this occurs is not clear.

Bedtime twitch. *As in the head jerk, muscles relax as you begin to fall asleep. In this phase, the brain is still very active. Apparently, there is a phase in falling asleep when suddenly muscle contraction signals come through which make your muscles contract, suddenly and unexpectedly.*

The purpose of this twitch is unknown.

The "Signal Loop"

Following the traditional concept, one can model the human as a processor of signals in some detail, as shown in Figure 3-6. Information (energy) is received by a sensor and impulses sent along the afferent (sensory) pathways of the *peripheral nervous system* (PNS) to the *central nervous system* (CNS). Here, the signals are perceived, usually in comparison with information stored in the short- or long-term memories. The signal is processed in the CNS and an action (including "no action") chosen. Appropriate action (feedforward) impulses are generated and transmitted along the efferent (motor) pathways of the PNS to the *effectors* (voice, hand, etc.). Of course, many feedback loops exist, although only a few are shown in the model.

Both sides of the processor model can be analyzed further. Figure 3-7 shows how *distal stimuli* provide information which may be visual, auditory, or tactile. To be sensed, the stimulus must appear in a form to which human sensors can respond;

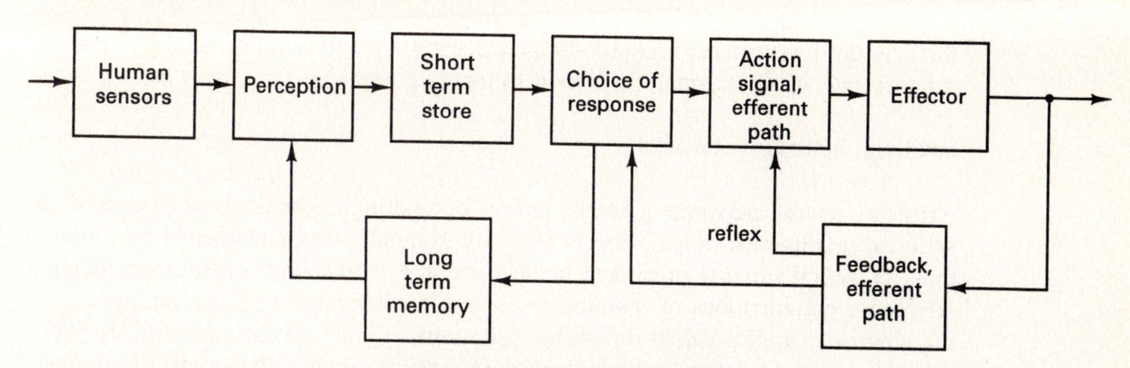

Figure 3-6. The human processor.

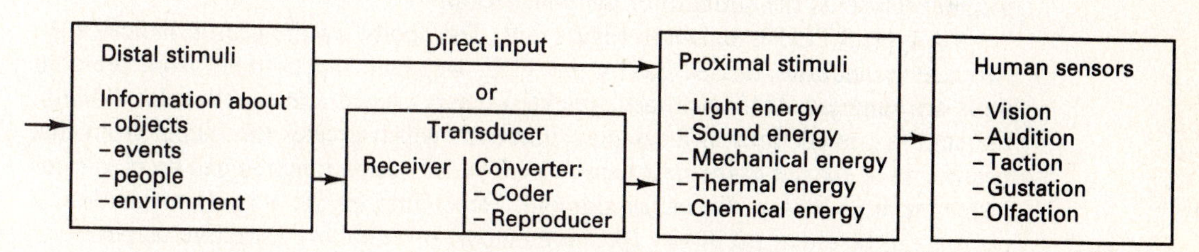

Figure 3-7. Energy input side.

it must have suitable qualities and quantities of electromagnetic, mechanical, electrical, or chemical energy. If the distal events do not generate *proximal stimuli* that can be sensed directly, the distal stimuli must be transformed into energies that can trigger human sensations. For this, *transducers* are designed by the ergonomist. For example, a "display" of some kind, such as a computer screen, dial, or a light can serve as transducer.

On the output side, the actions of the human effector (such as hand or foot) may directly control the "machine," or one may need another transducer. For example, movement of a steering wheel by the human hand may be amplified by auxillary power ("power steering"). Figure 3-8 portrays the model. Of course, recognizing

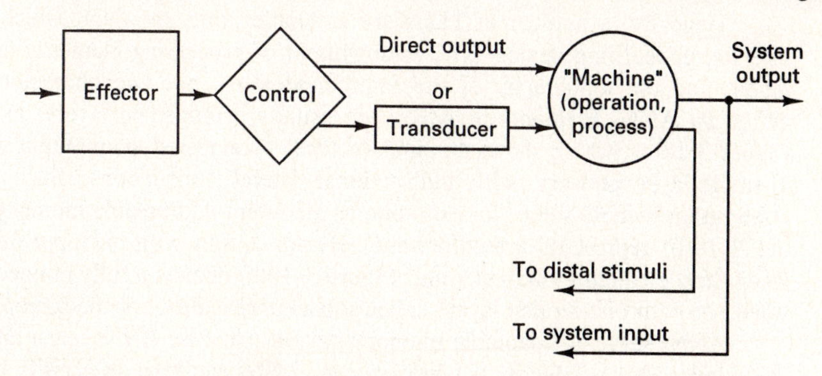

Figure 3-8. Energy output side.

the need for a transducer and providing information for its suitable design is again a primary task of the ergonomist/human factors engineer.

Artificial Neural Networks

Artificial neural networks (ANNs) are a computing technology in the field of artificial intelligence (Spelt, 1991). They are computer-based mathematical simulations of neural circuits presumed to be operating in the brain. More specifically, ANNs are combinations of (nonlinear) computing elements, i.e., neurons, processing elements, nodes, called threshold logic units (TLUs). These elements are typically arranged in layers interconnected in various ways. The connections have weights which multiply the values passed out of one TLU and on to the next; a weight represents the strength of the connection.

A typical TLU is shown in Figure 3-9. The "body" of the neuron indicates the two major functions carried out by the TLU: weighted inputs ($a_i\omega_i$) from previous units are summed. If the summed weighted inputs exceed a threshold value ($\Theta_j\omega_{oj}$), the sum is passed through a ramping function, which creates the output from that neuron, to be passed to the next neuron(s). The activation function can be of various forms, such as binary, linear, or sigmoid. Depending on the activation function, a neuron can be either excitatory (positive output) or inhibitory (negative output).

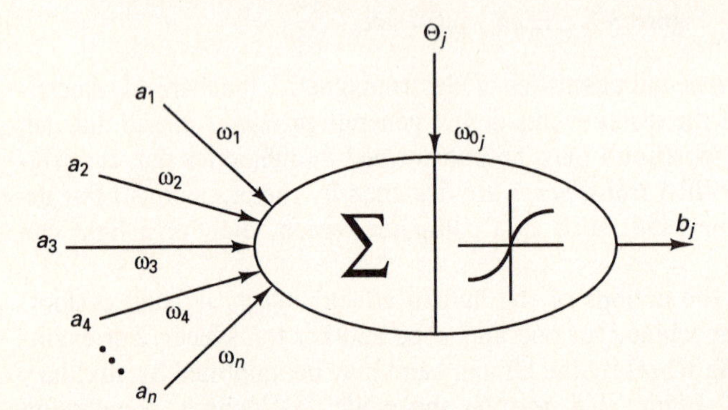

Figure 3-9. Diagram and notation for a typical threshold logic unit of an ANN (Spelt, 1991).

Typically, a number of TLUs are assembled into an ANN, which serves as a filtering or learning device. The arrangement of processing elements into layers or fields, and the topography of the interconnections, are two characteristics which define an ANN. Elements in ANNs are usually arranged in layers, as depicted in Figure 3-10(a), which shows an input vector (x) connected to an output vector (y) by a single-layer network with bidirectional lateral connections. Such a one-layer configuration could serve, for example, as a content-addressable memory ANN. Figure 3-10(b) represents a feedforward network design with an input and an output layer, hence called a two-layer net. Figure 3-10(c) depicts a fully connected feedforward net with one hidden layer, called a backpropagation (or: backprop) network.

ANNs serve as adaptable memory systems; that is, ANNs learn and then store the acquired knowledge in weight matrices. ANNs can learn to classify either spatial

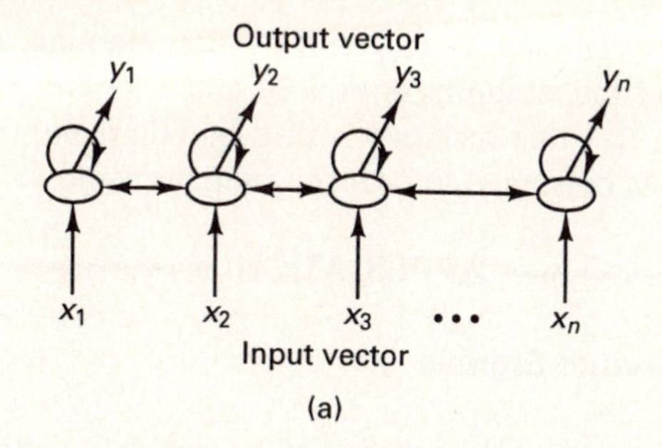

(a)

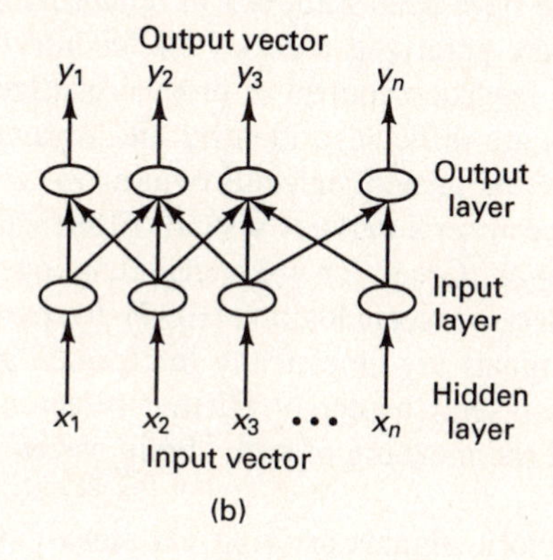

(b)

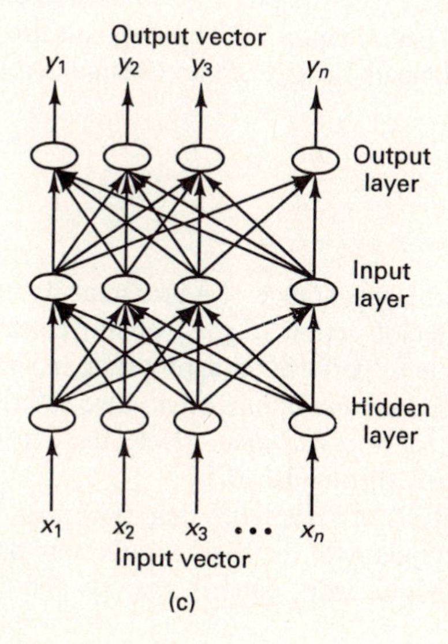

(c)

Figure 3-10. Various types of ANN configurations. See text for description. Reprinted with permission from "Introduction to Artificial Neural Networks for Human Factors" by Phillip F. Spelt, *Human Factors Society Bulletin,* Vol. 34, No.7, 1991. Copyright 1991 by the Human Factors and Ergonomics Society, Inc. All rights reserved.

or spatiotemporal patterns. The ability to perform, after training, with only partial input information is one of the strengths of ANNs, and is referred to as "fault tolerance." It results from the fact that memory is distributed among the weights in the matrices, similar to what is presumed to occur in human memory.

APPLICATION

Ergonomic Uses of Nervous Signals

Given current understanding of the human mind, and with the limited current technologies to pick up and use nervous signals, it is still "futuristic" to expect that devices can be controlled by thinking about it. Yet, some successes based on empirical knowledge have been achieved in rehabilitation engineering: one may electrically stimulate paralyzed muscles for feedforward, or use stimulation of skin sensors for feedback control of prostheses (Szeto and Riso, 1990).

Afferent impulses are difficult to identify and separate from efferent signals, mostly because of the anatomical intertwining between the sensory and motor pathways of the peripheral nervous system. Electrical events occurring in the cortex (the *encephalon,* Greek for wrapping) that covers the central brain, are recorded in the electroencephalogram (EEG) by means of surface electrodes. The recorded signals are empirically interpreted and used particularly to classify sleep stages—see Chapter 6. Further research is hoped to lead to better understanding of the meaning of such "brain waves" and of their quantitative values.

The effects of motoric signals arriving via alpha-motoneurons at muscle fibers are picked up (via indwelling or surface electrodes) as electromyogram (EMG) at skeletal muscle in general (see Chapter 1)—and specifically as electrooculogram at eye muscle (see Chapter 6)—or as electrocardiogram (ECG or EKG) at heart muscle—(see Chapter 2). These signals are well understood and their recording is technically easily done (Basmajian and DeLuca, 1985; Soderberg, 1992).

RESPONDING TO STIMULI

The time passing from the appearance of a proximal stimulus (e.g., a light) to the beginning of an effector action (e.g., foot movement) is called *reaction time*. The additional time to perform an appropriate movement (e.g., stepping on a brake pedal) is called *motion* or *movement time*. Motion time added to reaction time results in the *response time*. (Note that, in everyday use, these terms are often not clearly distinguished.)

Experimental analysis of "reaction" time goes back to the very roots of experimental psychology: many of the basic results were obtained in the 1930s, with additional experimental work done in the 50s and 60s. Innumerable experiments have been performed; hence, many different tables of such times have been published in engineering handbooks. Some of these apparently have been consolidated from various sources; however, the origin of those data, the experimental conditions (e.g., the intensity of the stimulus) under which they were measured, the measuring accuracy, and the subjects participating are no longer known. The following table is typical of generally used but fairly dubious information, often applied without much consideration or confidence:

electric shock	130 ms
touch, sound	140 ms
sight, temperature	180 ms
smell	300 ms
taste	500 ms
pain	700 ms

Accordingly, there appears little practical time difference in reactions to electrical, tactile, and sound stimuli. The slightly longer times for sight and temperature sensations may be well within the measuring accuracy, or within the variability among persons. However, the time following a smell stimulus is distinctly longer, and the time for taste again considerably longer, while it takes by far the longest to react to the infliction of pain.

Another somewhat dubious listing concerns the time "delays" that occur between start and arrival of the signal at different sections of the nervous system:

at receptor	1 to 40 ms
along the afferent path	2 to 100 ms
in CNS processing	70 to 300 ms
along the efferent path	10 to 20 ms
muscle latency and contraction	30 to 70 ms

Simply adding the shortest times leads to the theoretically shortest possible delay. Of course, in reality there is little reason to assume a situation in which all the delays are shortest. Yet, the best chances to reduce delays are obviously in the processing time needed by in the central nervous system. Thus, "clear signals leading to a unambiguous choice of action" would be the best approach to reduce delays—that is, to react most quickly.

Reaction Time

If a person knows that a particular stimulus will occur, is prepared for it, and knows how to react to it, the resulting reaction time (RT) is called "simple reaction time." Its duration depends on the stimulus modality and the intensity of the stimulus.

If one of several possible stimuli occurs, or if the person has to choose among several possible reactions, one speaks of "choice reaction time." Choice reaction time is a logarithmic function of the number of alternative stimuli and responses:

$$RT = a + b \log_2 N$$

where a and b are empiric constants and N is the number of choices.

N may be replaced by the probability of any particular alternative $p = 1/N$:

$$RT = a + b \log p^{-1} \qquad \text{(the "Hick-Hyman" equation)}$$

(To be quite exact, $RT = a + bH$, where H is the transmitted information.)

Under optimal conditions, simple auditory, visual, and tactile reaction times are about 0.2 second. If conditions deteriorate, such as uncertainty about the appearance of the signal, reaction slows. For example, simple reaction time to tones near the lower auditory threshold (30 to 40 dB) may increase to 0.4 second. Similarly, visual reaction time is dependent on intensity, flash duration, and size of the stimulus. Figure 3-11 indicates these relations, as mea-

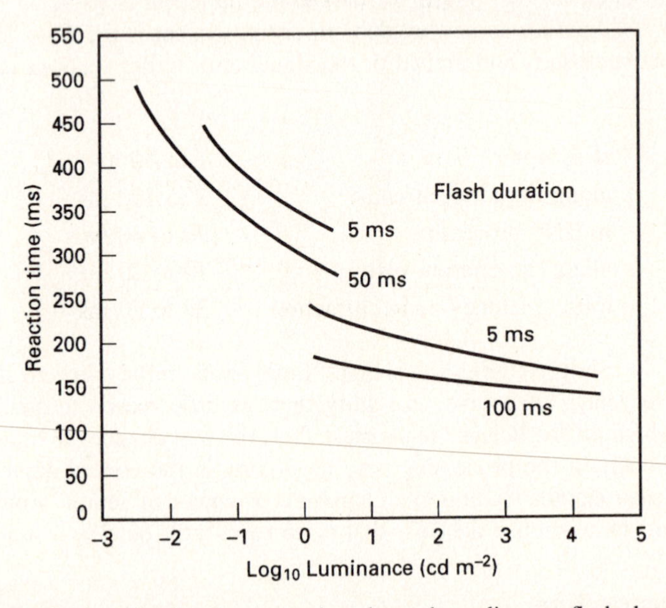

Figure 3-11. Simple visual reaction times depending on flash duration and flash intensity (illumination). SOURCE: Modified from "Motor Control" by Keele, S. W. in K. R. Boff, L. Kaufman, and J. P. Thomas (eds.), *Handbook of Human Perception and Human Performance*, copyright © 1986 by John Wiley & Sons, Inc. Reprinted by permission of John Wiley & Sons, Inc.

sured by various researchers: for reasonable sizes of the light source (between 0.5 and 1.7 degrees), reaction time is shortened by increased luminance and by increased flash duration. The reaction time is at minimum with a weak light source (of about 3 cd m^{-2}, which is near cone threshold), regardless of exposure duration or size. Reactions to visual stimuli in the periphery of the visual field (such as 45 degrees from the fovea) are about 15 to 30 ms slower than to centrally located stimuli (Boff and Lincoln, 1988). (These signal characteristics are further explained in Chapter 4.)

Reaction times of different body parts to tactual stimuli vary only slightly, within about 10 percent, for finger, forearm, and upper arm. When these times are divided into pre-motor time (the time from stimulus to onset of electromyographic activity in muscles) and motor time (the time from the onset of EMG activity to the beginning of movement), there are no differences in pre-motor time (Anson, 1982). This indicates that it takes longer to move a more massive limb than a lighter one—as one would expect by applying simple mechanical considerations.

Simple reaction time changes little with age from about 15 to 60 years, but is substantially slower at younger ages and slows moderately as one grows old.

Reaction time slows if it is difficult to distinguish between several stimuli which are quite similar, but only one of which is the one that should trigger the response—as to be expected from the Hick-Hyman formula. Measured reaction times are shown in Table 3-1 and Figure 3-12.

TABLE 3-1. MERKEL'S 1885 DATA ON REACTION TIMES FOR VISUALLY PRESENTED NUMERALS. SUBJECTS HAD TO PRESS THE APPROPRIATE BUTTON OF 10 BUTTONS

Number of Alternatives	Reaction Time (ms)
1	187
2	316
3	364
4	434
5	487
6	532
7	570
8	603
9	619
10	622

SOURCE: Adapted from Keele, 1986. Reprinted by permission of John Wiley & Sons, Inc.

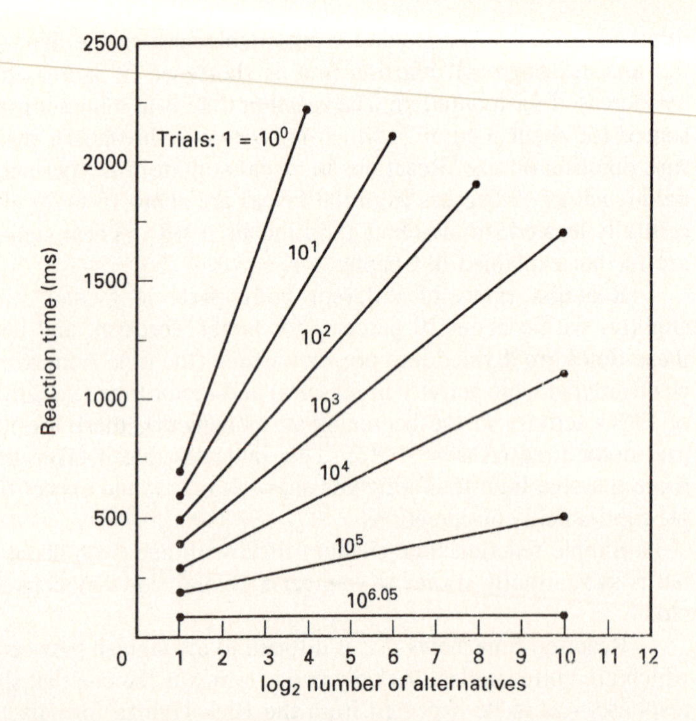

Figure 3-12. Schematic relation between reaction time, practice, and number of alternatives. (Modified from Keele, 1986, who used data from Teichner and Kreb, 1974.) Reprinted by permission of John Wiley & Sons, Inc.

Motion Time

Motion time follows reaction time. Movements may be simple, such as lifting a finger in response to a stimulus, or quite complex, such as swinging a tennis racket. Swinging the racket contains not only more complex movement elements, but also larger body and object masses that must be moved, which takes time. Movement time also depends on the distance covered and on the precision required. Related data are contained in many systems of time and motion analyses, often used in industrial engineering.

Well-designed, well-controlled studies of movement times were performed by Paul Fitts in the early 1950s, which have become classics. Fitts found that when target precision was fixed, movement time increased with the logarithm of distance. If distance was fixed, movement time increased with the logarithm of the reciprocal of target width. Distance and width almost exactly compensated for each other.

These relations have been expressed in a motion-time (MT) equation called "Fitts' law":

$$MT = a + b \log_2 \frac{2D}{W}$$

where D is the distance covered by the movement, and W is the width of the target. The expression $\log_2 (2D/W)$ is often called index of difficulty. (The factor 2 is used simply to help avoid negative logarithms.) The constants a and b depend on the situation (such as body parts involved, masses moved, tools or equipment used), on the number of repetitive movements, and on training (read Keele, 1986, for more details). Fitts' law has been found to apply to many movement-related tasks, even to the "capture time" of a moving target (Hoffmann, 1991). Apparently, all reaction-time models can be considered subsets of one general model (Kvalseth, 1991).

Mental Workload

The assessment of workload, whether psychological or physical, commonly relies on the "resource construct," meaning that there is a given (measurable) quantity of capability and attitude available of which a certain percentage is demanded by the job. If less is required than available, a reserve exists—see Figure 3-13. Accordingly, workload often is defined as the portion of resource (i.e., of the maximal performance capacity) expended in performing a given task.

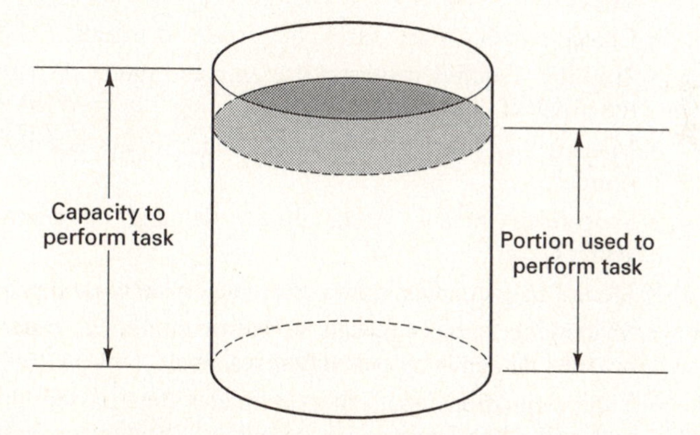

Figure 3-13. Traditional "resource" model.

Following this concept, one should obviously avoid any condition in which more is demanded from the operator than can be given, because the task performance will not be optimal and the operator is likely to suffer, physically or psychologically, from the overload. However, a task demand below capacity would leave a residual capacity. Its measurement provides an assessment of the actual workload. Refined models have been proposed, for example in terms of a multiple-resource model in which separate reservoirs are postulated, such as for stages of information processing (afferent, central, efferent), codes of processing (e.g., verbal or spatial), and input/output modalities (visual, auditory/verbal, manual): see the overviews by Hancock and Meshkati (1988); Moray

(1988); O'Donnell and Eggemeyer (1986); Wickens and Kramer (1985); Wierwille and Casali (1983) for more details.

Workload is empirically assessed using four different approaches: objective measures of primary task performance, of secondary task performance, and of physiological events, and subjective assessment. Measures of task performance, as well as subjective assessment, presume that both "zero" and "full" capacities are known, since they assess the portion of its loading (Derrick, 1988; Nygren, 1991). Measuring performance on a secondary concurrent task is intended to assess the spare capacity that remains after allocating capacity resources to the primary task. If the subject allocates some of the resources truly needed for primary task performance to the secondary task, the secondary task is *intrusive* (or *invasive*) on the primary task. An intrusive secondary task modifies the condition meant to be assessed.

Examples of secondary tasks used in workload measurement are as follows (Boff and Lincoln, 1988; Paas and Adam, 1991; Yeh and Wickens, 1988):

- Simple reaction time: draws on perceptual and response execution resources.
- Choice reaction time: same, but greater demands.
- Tracking: requires central processing and motor resources, depending on the order of control dynamics.
- Monitoring of the occurrence of stimuli: draws heavily on perceptual resources.
- Short-term memory tasks: heavy demand on central processing resources.
- Mental mathematics: draws most heavily on central processing resources.
- Shadowing (subject repeats verbal or numerical material as presented): heaviest demands on perceptual resources.
- Time estimations: (a) subject estimates time passed: draws upon perceptual and central processing resources; (b) subject indicates sequence of regular time intervals by motor activity: large demands on motor output resources.

Physiological measures, such as heart rate, eye movements, pupil diameter, or muscle tension, can often be done without intruding on the primary task (Kramer, 1991). However, these measures may be insensitive to the task requirements, or may be difficult to interpret (Wierwille and Casali, 1983; Wierwille, Casali, Connor, Rahimi, 1985).

Subjective assessments of the perceived workload rely on internal integration of the demands, but they may be unreliable, invalid, or inconsistent with other performance measures (Nygren, 1991; Wickens and Yeh, 1983). If

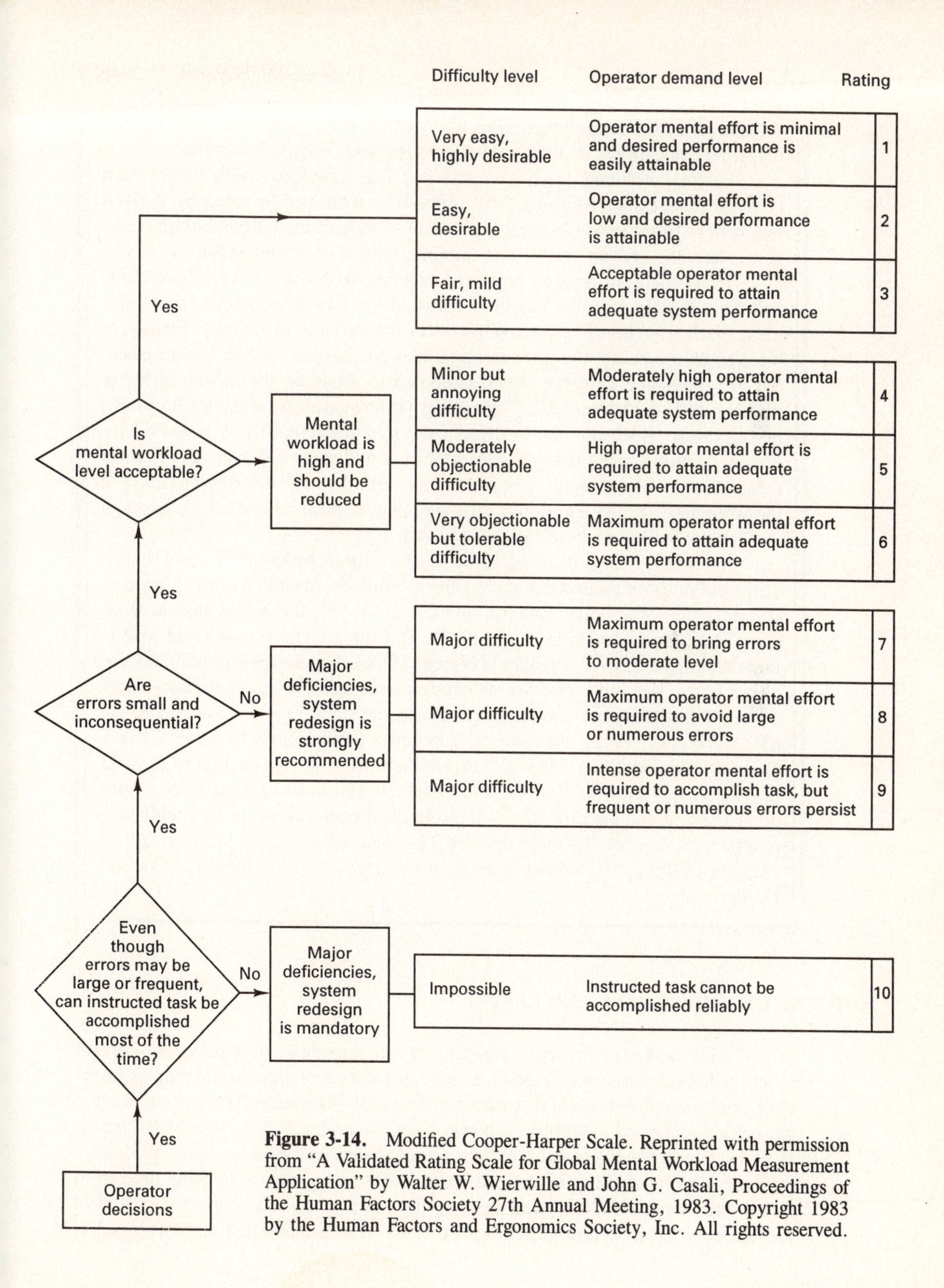

Figure 3-14. Modified Cooper-Harper Scale. Reprinted with permission from "A Validated Rating Scale for Global Mental Workload Measurement Application" by Walter W. Wierwille and John G. Casali, Proceedings of the Human Factors Society 27th Annual Meeting, 1983. Copyright 1983 by the Human Factors and Ergonomics Society, Inc. All rights reserved.

subjective measures are taken after the task has been completed, they are not real-time evaluations; if performed during the task, they may intrude.

Further problems arise from the fact that individuals differ from each other in their capacities to perform tasks. The workload imposed by a given task differs from person to person; also, the workload may depend on the temporal state of a person, such as on training, fatigue, and motivation.

Thus, there are serious problems, starting with the concept of workload and what exactly is to be measured. Next, it must be assured that the assessment method chosen is indeed sensitive to the variable of interest. Finally, it must be assumed that the measurement does not intrude on task performance. Seven (1989) and Doherty (1991) proposed to focus on the measurement of primary task performance, by observing performance on components of the primary task which are "non-critical": the hypothesis is that as workload increases, performance changes measurably. Candidates for such unobtrusive measures may be the status of speech, depletion of stock, disorder or clutter at the workplace, or the length of a customer queue. Such embedded measures of workload would not add to the task at hand.

Pragmatic measures of workload that have been widely used, even though criticized on both theoretical and technical grounds, are the modified Cooper-Harper scale (Wierwille, Rahimi, and Casali, 1985), the NASA task load index (TLX; Heart and Staveland, 1988), and the subjective workload assessment technique (SWAT; Reid and Nygren, 1988). The first is a unidimensional rating scale, the other two are subjective techniques using multidimensional scales. While there are good theoretical and statistical reasons to use TLX and SWAT (Nygren, 1991), they are more complex to administer than the Cooper-Harper scale, which is fairly self-explanatory. It is shown in Figure 3-14 as modified by Wierwille, Rahimi, and Casali (1985), to be applicable to systems other than piloted aircraft. The ratings obtained from this scale are highly correlated with those in the more detailed TLX and SWAT scales, which might be employed after initial ratings have been obtained from the modified Cooper-Harper scale.

"STRESS" ON INDIVIDUALS AND CREWS

"Stress" and "anxiety" are core concepts of psychopathology: a prevalent general model (called diathesis-stress) assumes that most disorders arise from complex interactions between environmental stressors and (usually biological) predispositions that can make a given individual break down. There is evidence that "stress" and physical illness are related; for example, the social readjustment rating scale of Holmes and Rahe (1967) lists life events (ranging from death of spouse over being fired from a job to vacation and holiday) and subjects are asked to indicate those that they have experienced recently. High scores on that scale have been related to heart attacks, other health problems, even colds. These relations are striking, though caution needs to be applied, since they indicate only correlations, not cause-effect relations. A moderating factor is "hardiness" of a person, a constellation of behavior and attitude associated with resistance to the effects of stress.

In his classic studies in the 1930s, Selye saw "stress" primarily as a physical trauma to which the human responded. Now it is generally accepted that it is not only simply connected with physical impacts, but also with the appraisal of that situation by the individual; hence it is also a cognitive phenomenon.

"Mind and body are intimately interconnected, intertwined, and interdependent—of this there is no longer room for doubt. If, once upon a time, it was possible to believe that physical and mental events belong to essentially different spheres, in contact only in the sense of a spiritual mind inhabiting the wonderous mechanisms of the body, that is no longer so. And whilst we may know less in detail than we would like, we do know a great deal about the physiology of emotion and of stress" (Blinkhorn, 1988, p. 29). Yet, in psychology, there is lack of consensus on even a definition of stress. Thus, the "term stress is enshrouded by a thick veil of conceptual confusion and divergence of opinion" (Motowidlo, Packard, and Manning, 1986, p. 618).

APPLICATION

The concept of "stress on the job" is both common and elusive. We all have had the experience of being driven to the margin of physical and psychological capability by strenuous physical exertion, hot climate, schedule pressure, unreasonable behavior of bosses or colleagues, oncoming illness, or the feeling of useless efforts. Some of these stressors are physical, others psychological; self-imposed or external; short-term or continual (Cox, 1990).

Yet, the concept is elusive because what may be stimulating under one condition may become excessive under other circumstances. The simple stress-produces-strain sequence, which engineers use, may dissolve into the complex relations familiar to psychologists: a stressor may generate a positive "stress" which spurs more activity, or it may result in "dis-stress" that overloads the person and generates ineffectiveness, evasive behavior, anxiety, even illness.

The confusing situation may be clarified by the model shown in Figure 3-15. It shows three major aspects of ergonomic concern.

Job demands depend on type, quantity, and schedule of tasks; the task environment (in physical or technical terms); and the task conditions, i.e., the psycho-social relations existing on the job. These (and possibly other related) work attributes are the job "stressors" that are imposed on the human.

A person's capability to perform the job demands, and the person's attitude (influenced both by physical and psychological well-being) must be "matched" with the job demands. If the job demands require only a portion of the person's abilities and attention, she/he is likely to feel underloaded, under-

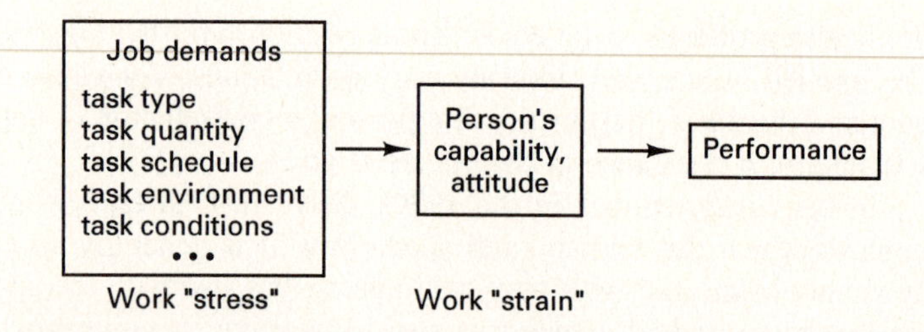

Figure 3-15. Simple model of the relations between job demands ("stress"), human responses ("strain"), and procedures.

estimated, and might become bored, inattentive, and underachieving, or, on the positive side, might seek more challenge. If the job demands exceed the person's capability, he/she is overloaded and would seek either to reduce the workload, or to increase the capability. Of course, other sources of stress besides those at work might exist that require a portion of the person's capabilities or attitude, influencing the physical and psychological well-being. In this case, the experienced strain depends on the sum total of job and other demands in relation to the person's capability and attitude.

While we usually assume that most "job stresses" are due to a person's overloading, not demanding enough of the individual's capacity is not infrequent either. A good match between job demands and a person's capability and attitude is, obviously, a desirable condition. The construct of a "U-shaped function" (often postulated but seldom proven), shown in Figure 3-16, relates the stress imposed by the work to the resulting strain experienced by the person. According to the "U theory," both too little and too much stress produce undue strain ("distress").

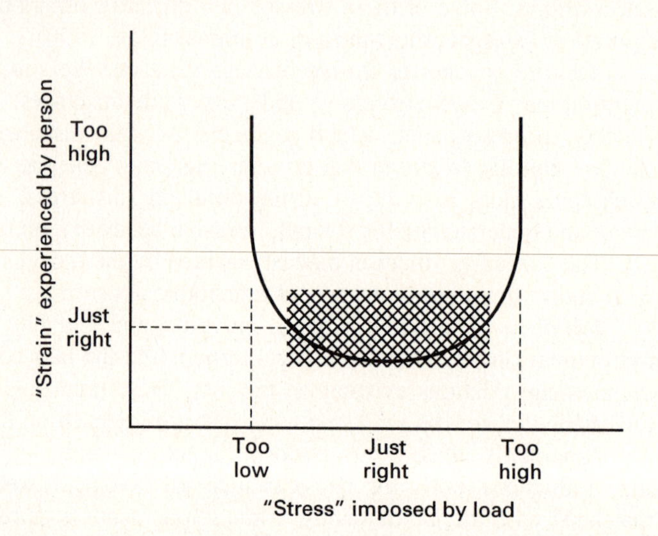

Figure 3-16. Postulated "U-function" relating stress and strain.

Monotony is the opposite of variety, which are both characteristics of the environment perceived by an individual. Monotony is produced by an environment where either there is no change or else changes occur in a repetitive and highly predictable fashion over which the individual has little control. A varied environment provokes interest and sometimes even human emotion; in contrast, an unvaried environment produces boredom, which also can be considered an emotion. Thus, boredom is an individual's emotional response to an environment that is perceived as monotonous. A "bored" person often complains of feeling "tired" or "fatigued" (terms used in this book as related to a physiological rather than psychological status).

If the job demands (work stress) exactly match the person's capabilities and attitude, proper "strain" exists and the on-the-job performance is satisfying, both objectively and subjectively. If the job demands are far below the person's abilities, an "underload" condition exist and the on-the-job performance is most likely (but not necessarily) reduced. On the other hand, if the work requires more than a person is able and willing to give, an "overload" condition exists, and work performance is likely to suffer.

Understanding people's capabilities, and developing job demands and conditions that are matched to these, is the main focus of the ergonomist. The following brief discussion of attributes and conditions of the human may help to understand proper stress-strain relationships, and to avoid in particular occupational overload conditions. (Many of these aspects fall into the domain of psychopathology, where many books and journals provide much more information than can be provided here.)

Strain Experienced by an Individual

Overworked nurses exhibit these behavioral consequences: fear, dread, anxiety, irritation, annoyance, anger, sadness, grief, and depression (Motowidlo, Packard, and Manning, 1986). The extent of the experienced strain depends, among other factors, on job experience, age, attitude, self-esteem, and coping ability (Heilman, Rivero, Brett, 1991; McEvoy and Cascio, 1989; Frone and McFarlan, 1989; Hansen, 1989; Motowidlo, et al., 1986; Parkes, 1990). Behavior patterns are highly correlated with the experience of strain. The "type A" behavior pattern is exhibited by persons who are engaged in an (often chronic) struggle to obtain (often poorly defined) things from their environment in the shortest possible period of time, if necessary against the opposing effects of other things or persons. Type A behavior is characterized by aggressiveness, competitiveness, impatience, and urgency in overcoming obstacles. Type A persons are likely to act in ways which make job events more stressful for themselves and find the resulting strain particularly intense. Many researchers, but not all, have found these persons to be more susceptible to emotional and cardiovascular disorders than persons showing the opposite type of behavior: type B individuals are relaxed, unhurried, less aggressive, and not so competitive and usually have fewer disorders (Hackett, Rosenbaum, and Tesar, 1988; Kamarck and Jennings, 1991).

"Stress management," a popular term seldom defined, has been used by psychologists in a variety of ways, but some general statements can be made. The term management implies that stress is felt to be an unavoidable component of living with which individuals can learn to cope. Selye (1978) stated that to talk about being "under stress" is as pointless as talking about "running a temperature." What is of concern is an excess of stress, a distress. Lazarus and Cohen (1977) identified various categories of stressors, including cataclysmic stressors that happen to several people at once (e.g., natural disasters, massive corporate layoffs) and "personal" stressors.

Various life events have been used to scale personal stressors. The Social Readjustment Rating Scale, mentioned above, is a 43-unit checklist of differently stressful events, ranging from death of a spouse to holiday, that have been correlated with symptoms of physical illness and psychological distress. Background stressors, so-called "daily hassles," have been ordered into a Hassles Scale and also related to health and well-being (Kanner, Coyne, Schaefer, and Lazarus, 1981). Examples of daily hassles in the workplace are inconsiderate co-workers and noisy work environments. On an intuitive level it is easy to recognize the cumulative impact that such daily annoyances may have on the physical and emotional well-being of an individual, especially if they are perceived as being uncontrollable. Additionally, individuals are often not adequately aware of the frequency or intensity of background stressors, particularly if they increase gradually over time. An analogy might be found in the animal world: in 1899, (when such studies were not subject to panel review), Scripture found that frogs would submit to being boiled if the water temperature was increased gradually.

Individual attitudinal variables (such as beliefs and character predispositions) are mitigators of the degree of distress elicited in an individual by a given environmental stressor. Tache and Selye (1986) likened stress reactions to allergy attacks. In an allergic reaction, the overmobilization of the immune system—not the allergen itself—causes individuals to feel ill. Individuals have different coping styles. Folkman and Lazarus (1988) define coping as the cognitive and behavioral efforts to manage specific external and/or internal demands that are appraised as taxing or exceeding the resources of the person. Resources of an individual may be material, physical, intrapersonal, interpersonal, information/education, and cultural. Cognitive components of effective and adaptive coping strategies include rationality, flexibility, and farsightedness (Antonovsky, 1979). Effective coping strategies are either problem oriented or emotion focused. Problem-oriented coping is directed at controlling an environmental stressor to reduce its impact and is effective if a stressor is objectively controllable. Examples are time management, environmental control or environmental adaptations, assertive communication and limit setting. Emotion-focused coping strategies include structured relaxation exercises, the use of humor, exercise or hobbies, and guided cognitive reevaluation ("reframing") of a stressor (Lazarus, 1984).

The workplace has become a common setting for "stress management" programs. They often emphasize developing healthier lifestyles (smoking cessation, weight management, exercise programs, treatment for drug and alcohol abuse). Such programs are obviously beneficial to the individual employee who has needs for such interventions, but they also affect corporate economy if they reduce the number of sick days and the expense of medical insurance coverage.

Strain Experienced by Confined Groups

A special kind of "on-the-job stress" is experienced by crews in confined spaces, such as in space ships. Even during the relatively short missions done by U.S. astronauts, severe intraindividual and interindividual problems have appeared, ranging from physiological deficiencies (see Chapter 5), to anxiety, to difficulties in interaction with other crew members and disagreements with ground control. Such situations have also occurred under other isolated and confining working conditions, such as in the Antarctic (Harrison, Clearwater, and McKay, 1991).

☞☞☞ *During one of the U.S. Skylab missions, the commander Gerald Carr and his fellow astronauts not only "plotted" to hide items from ground control, but also went on a "strike" until some disagreements could be worked out. Similar problems with ground controllers in the USSR, and interpersonal conflicts among cosmonauts, developed aboard the Salyut and Mir space stations (Holland, 1991).* ☜☜☜

APPLICATION

Changes in environment and physiological functions affect psychological and psycho-social well-feeling.

Among the physiological effects of space conditions (further discussed in Chapter 5) on astronauts are

- Dimming of vision, peripheral light loss, eventually blackout under increasing plus-Z-direction acceleration.
- Diminished vision, red-out, increase in accommodation time, blurring or doubling of vision in minus-Z-direction acceleration.
- Lengthening of visual reaction time by increased g-level.
- Degradation of estimates of size and distance of objects in space because objects appear sharper and in higher contrast while foreground and background distance cues are absent.
- More time required for the eyes to adapt to the existing wide ranges of illumination and luminance levels when shifting the gaze.
- Reduction of visual performance by vibration such as during liftoff and landing, when vision is particularly important.

- Shift in perceived colors, reduction in contrast sensitivity, reduction in near acuity but improved visual acuity for distant objects (according to anecdotal and contradictory reports).
- Sensitivity to "noise" appears to increase, while auditory functions do not change significantly in space.
- Reduced cabin pressure requires crew members to talk louder to be heard, owing to the reduction in sound transmission.
- The sensitivity of smell is decreased (probably due to the ever-present nasal congestion in microgravity).
- Unpleasant odors (for instance associated with medical symptoms, food, or body waste) can become very annoying; yet, pleasant odors are also met with increased responsiveness.
- The sense of taste is degraded, food judged to be well seasoned on earth tastes bland in space. (This may be associated with nasal congestion and the upward shift of body fluids.)
- Vestibular effects are mostly evident in space sickness and spatial disorientation.
- Larger forces must be applied to generate the same kinesthetic sense stimuli as in a one-*g* environment.
- Motor skills are impaired upon entering microgravity, but the decrement is reduced or eliminated after a short period of adaptation.
- Perception of body language and facial expression is changed because crew members appear in different postures and their faces are puffed.
- Anxiety, often present in the initial and final phases of a space mission, or during emergency situations, can greatly reduce the effectiveness of individual crew members, particularly of the commander, and of a team in general.
- Rigid hierarchical structures (such as prevalent in the U.S. armed forces) usually dissolve somewhat during long missions and develop into a more collegial, democratic sharing of decisions and activities.

While it is known that personal, interpersonal, and psychosocial factors play major roles regarding the success of failure of long missions, information on these factors is still sparse. The USSR made extensive experiences with long-duration space flights, but the United States has had only space flights of less than two weeks duration since 1974. Crew-member compatibility and cohesion, and leadership style, are very important during the long periods of confinement, isolation, monotony, and danger. Sells (1966) reviewed the literature relevant to long-duration space missions with respect to social systems previously experienced and found that submarine missions and isolated exploration parties had the highest similarity to space flight.

Three stages in the reactions to isolation have been experienced. In the first stage, there is heightened anxiety, apparently dependent on the degree of present danger that a person perceives. Heavy workloads appear to reduce the anxiety level during this time. The second stage is associated with depression-like feelings, usually occurring during that long segment of time of settling down to routine duties. This may result from the absence of familiar social roles as spouse, father, mother, or club member. The final stage is at the end of the mission, often with increased expression affects, and anticipatory behavior, frequently coupled with aggressiveness. In this stage, work performance is commonly degraded, and serious errors of judgement or omission occur (Holland, 1991). Christensen and Talbot (1986) reviewed the existing literature on psychological and sociological aspects of space flight. Table 3-2 summarizes the main issues that affect crew behavior and performance. It also lists measures that can be taken to influence behavior and performance positively.

Until the early 1990s, all long space missions were performed by crews of males. Mixed-gender crews will generate new facets of intragroup behavior. One idea to stabilize relations is to use married couples. Yet, even then the conditions of confined spaces, working together closely, and social and sexual desires are likely to generate difficult relations among the crew members, with possibly large changes over time (Frazer, 1991). A related problem is that of pregnancy in space: it is at present unknown whether microgravity, radiation, and other space conditions might affect the development of the fetus.

ENHANCING PERFORMANCE

There is generally good reason to achieve and perform at one's best. Learning, knowledge, communicating, creativity, concentration, skill, and "performance under stress" are important on the job or in sports; the military is particularly interested in attaining fearlessness, cunning, courage, one-shot effectiveness, fatigue reversal, or nighttime fighting capabilities.

Training should help to attain such performance ideals. Thus, techniques for enhancement of human performance have received much attention, particularly in the popular press.

Many entrepreneurs, most of them outside academia, advocate techniques to concentrate on specific targets, accelerate learning, improve motor skills, alter mental states, reduce stress, increase social influence, foster group cohesion, or perform parapsychological "remote viewing." Examples are "biofeedback," by which to attain information about internal processes and control them; "hemispheric synchronization" or "split-brain learning," based on assumptions about right- and left-brain activities; "neurolinguistic programming," such as procedures for influencing another person; "mind reading," and even nontactile psychokinetic control of devices.

Some of these techniques are promoted as easy to do, such as by just watching a videotape or listening to subliminal information during sleep. But the advertised

TABLE 3-2. FACTORS THAT INFLUENCE BEHAVIOR AND PERFORMANCE

Psychological, Psychosocial, and Psychophysiological	Environmental	Space System	Support Measures
Limits of performance (perceptual/motor)	Spacecraft habitability	Mission duration and complexity	Inflight psychosocial support
Cognitive abilities	— confinement	Organization for command and control	Recreation
Decision making	— physical isolation	Division of work between human and machine	Exercise
Motivation	— social isolation	Crew performance requirements	Work-rest/avoiding excess workloads
Adaptability	— lack of privacy	Information load	Job rotation
Leadership	— noise	Task load/speed	Job enrichment
Productivity	Weightlessness	Crew composition	Training
Emotions/moods	Artifical life support	Spacecrew autonomy	— preflight environmental adaptation
Attitudes	Work-rest cycles	Physical comfort/quality of life	— social sensitivity
Fatigue (physical/mental)	Shift change	Communications (intracrew and space/earth)	— for team effort
Crew composition	Desynchronization of body rhythms	Competency requirements	— self-control
Crew compatibility	Hazards	Time compression	Inflight maintenance of proficiency
Psychological stability	Boredom		Earth contacts
Personality variables	Stresses		
Social skills	— single		
Human reliability (error rate)	— multiple		
Space adaptation	— sequential		
Spatial illusions	— simultaneous		
Time compression			

SOURCE: Adapted from Christensen and Talbot, 1986.

successes, some of them quite surprising, usually are supported only by personal experiences and testimonial statements rather than by scientifically acceptable proof. Thus, the National Research Council appointed a committee which investigated some of these techniques and their possibilities of success. The results of that committee's work have been reported in detail by Druckman and Swets (1988), Swets and Bjork (1990), and Druckman and Bjork (1991). The following text relies on their statements and findings.

General Findings

- Some *theoretical bases* quoted by promoters are unproven or questionable and, in many cases, the advertised results were not substantiated by provable findings. In other instances, the topic still has not been thoroughly researched, or may be difficult to research, because some of the underlying assumptions do not conform with our currently accepted scientific procedures and understandings—for example, in transcendental meditation, see below.

APPLICATION

- *Successful learning* results from the quality of instruction, from practice and study time, motivation of the learner, and matching of training procedures to the demands of the task. Some of the so-called super-learning programs may be effective if they combine these factors suitably, but there is little or no evidence that they include effective instructional techniques from outside the main realm of accepted research and practice.
- Many *measures of the effectiveness of training procedures* rely on measurements of performance during training or at the end of the training period, but these assessments often are not indicative of the actual performance weeks, months, or years in the future. Little research has been done to generate measures of performance at a time when the performance is required.
- *Mental practice* is, in theory and by some limited experience, effective in enhancing the performance of motor skills. Yet, simply listening to experts, or observing the performance of experts on videotapes, has not been proven to be effective.
- Positive effects of *biofeedback* on skilled performance remain to be determined.
- The literature refutes claims that link differential use of *brain hemispheres* to performance.
- There is no scientific justification from research conducted over a period of 130 years for the existence of *parapsychological phenomena*.

Specific Findings

- "Subliminal self-improvement by audio tape" is promoted as contributing to improvement of a person's attitude, confidence, performance; to reduction of anxiety; to help in dieting or stopping smoking. There is some scientific evidence that subliminal learning is possible, such as during sleep or otherwise without conscious awareness, because stimuli can be provided that are not consciously registered (for example, because they are very short) but are perceived nevertheless. Yet, there is apparently no mysterious essence that invades the unconscious mind and generates knowledge or confidence during waking hours.

  On tapes investigated, the researchers could not find (by objective methods or by subjective actions of subjects) any contained subliminal messages, although the manufacturers claimed that such messages were embedded. Some users claimed effects in the direction of the missing messages, but these listeners probably simply convinced themselves that messages had the desired effect, and hence generated a kind of self-fulfilling expectancy. In this case, the "subliminal tapes" were successful, self-administered placebos.
- *Meditation* is part of several kinds of eastern mysticism, such as Buddhism. Mystical traditions, including Christian and Jewish ones, involve dancing and chanting for long periods of time in order to achieve an altered state of mind. Yet, the effects of self-induced hypnotic states and of focussed attention are difficult to assess. In the late 1950s, a yogi from India introduced "transcendental meditation" to the West. It is a set of techniques to influence a person's conscientiousness through the regulation of attention. While there is no complete agreement on that concept, most meditation techniques include the need to sit or lie quietly in a particular position, to attend to one's breathing, to adapt a passive attitude, to be at ease, and, in some techniques, to repeat a word or phrase (mantra).

  Thus, meditation may lead to physiological and physical benefits by distracting a person from stressors, either environmental or self-generated. However, it is difficult to evaluate the effects, because the "control group" subjects involved in testing procedures must be experienced in meditation: disciples of meditation, particularly followers of yoga techniques, claim that the benefits of meditation are available only to those who practice it. Very few Westerners of scientific background have mastered yoga-related meditation, and there are no scientific publications by such researchers.

  Comparing the effects associated to meditation in persons who only rest, one may find their reductions in somatic arousal (such as heart rate, respiration rate, blood pressure, skin responses) smaller than those found in experienced meditators. But since relaxation and meditation go together, it is difficult to attribute any positive effects to meditation by itself, particularly since to follow meditation usually includes a change in lifestyle.
- *Performing under pressure* has been researched by psychologists in the natural laboratory of sports. A common-sense and appealing concept is that of mental

health related to athletic performance, which can most easily be stated in a negative form: anxious, depressed, hysteric, neurotic, introverted, withdrawn, confused, or fatigued athletes do not perform well, while athletes with more positive attributes do. Some performance gains may also be achieved by "mental practices" such as the imagined rehearsal of activities, by goal setting, by rewards for performance, and by relaxation and biofeedback. The enhancing techniques appear most successful when they include multiple approaches, directly administered by a mentor rather than via tape, and often repeated. Apparently, they are most effective for persons who do have problems with precompetitive anxiety or concentration. Yet, the gains from these techniques are seemingly small, and they have not been investigated with elite athletes.

Preperformance routines are also commonly used: for example, after a played point, a tennis player relaxes for a few seconds, then walks back confidently to the serving line, there relaxes for a moment, then concentrates on adjusting the strings of the racquet and finally purposely prepares to either serve or receive the next ball. Such patterns of thoughts, actions, and images, consistently carried out before performance of a skill, are meant to divert attention from negative and irrelevant information and to establish the appropriate physical and mental state for the action to be done. Performed at critical time, such cognitive-behavioral techniques appear indicative of better attentional focus and result in improved overall performance.

Apparently, the "ideal mental performance state" includes the following: clear focus on task requirements; absorption in the task to be done; high intrinsic motivation; concentration of the consciousness on the task without distraction; feeling of exercising control by using well-set and rehearsed cognitive and motor routines.

APPLICATION

- *Mens Sana in Corpore Sano?* In the original Latin text, this axiom is followed by the verb "sit," indicating the hope that there be a sane mind in a sane body. Recently, the relationship between aerobic fitness and psychological well-being has received considerable attention, triggered by the recognition that the lifestyles of many people are disrupted by anxiety and depression. Physicians routinely prescribe exercise to counteract emotional disorders. However, reviews have failed to show that exercise is psychologically more beneficial than more conventional psychological techniques applied by themselves, but carefully performed exercise can have additional health benefits (such as weight loss and otherwise improved physiological conditions), which may in turn help to promote overall well-being and adherence to a psychological intervention program involving exercise. Thus, though in an indirect way, physical exercise can help people to cope better with psychosocial stressors.

Enhancing Team Work

Enhancing team performance is of interest at work, in sports, or in the military, where one does not perform alone but rather together with other people who do similar, parallel, or complementary tasks. Teams may be highly coordinated (such as pilot and co-pilot in an airplane) or little organized (such as academic research teams).

A team should provide greater resources than an individual, but teamwork also involves often difficult interpersonal coordination and management problems. The performance of such an aggregate of individuals is not simply the sum of the individual efforts. It is common experience (although positive exceptions exist) that team performance often falls short of reasonable expectations. Proper selection of members according to their physical, mental, and interpersonal capabilities may enhance team performance. Members should be motivated to participate, otherwise some do not contribute their share to the team effort but do "social loafing." Yet, not all suboptimal group performance is due to lowered input of individual members, but it may be the result of faulty interpersonal processes for combining individual capabilities. "Brainstorming," for example, long believed to be an effective and idea-stimulating team technique, is usually not more successful than when performed by isolated individuals, possibly due to "blocking," i.e., the inability of a team member to produce ideas while others are talking.

Another disappointing result has recently been found for "consensus decision making": group decisions tend to be more "risky" than individual decisions. Popular are Delphi and the Nominal Group techniques. In the Delphi technique, individual judgments or opinions are privately elicited, then summarized, and the results are circulated to all team members for further modification until individual positions stabilize. Thus, anonymity of the individual members' inputs is retained, while in the Nominal Group technique, after the initial stages, group members meet face-to-face to exchange information. In spite of the popularity of such techniques, little reliable research has been carried out to determine their effectiveness. However, the evidence on team performance available today is not encouraging: neither technique appears to improve on the performance of freely interacting groups.

☞☞☞ *"The review of what we know about group performance is more striking for what is missing than for what is known" (Druckman and Bjork, 1991, p. 257).* ☜☜☜

DETECTING DECEPTION

Deception is an important survival skill for animals; it is also common in humans, for example, in giving the impression that one is more knowledgeable than is actually true, in bending the truth in a "white lie," or in intentionally proposing something known to be false. Folk wisdom has it that a liar can be detected by blushing, facial expressions, gaze shifts, and involuntary body movements ("body language") as well as by changes in voice loudness and speaking patterns. Yet, a person may not exhibit any or all of these cues, because an experienced liar can suppress them, or because these reactions depend on attitude and upbringing. Thus, these cues are in general not reliable indicators of either lying or speaking the truth. Still, most observers believe they can detect liars. Studies have shown, however, that professional lie detectors such as customs inspectors, police detectives, CIA and FBI agents, and judges do no better than chance in actually detecting deception. Altogether, there is a large discrepancy between scientific evidence and anecdotal subjective assessments (Druckman and Bjork, 1991).

☞☞☞ *Folk wisdom about how liars are acting (e.g., they don't look you in the eye, they don't smile, they fidget, they take long to respond) is not a reliable diagnostic of deception.* ☜☜☜

❑ APPLICATION

Interest in "scientific lie detection" originated in Italy, Germany, Austria, and Switzerland from the late 1880s until World War II. Psychophysiological measurement techniques were introduced at that time and have been developed ever since (Ben-Shakhar and Furedy, 1990; Gale, 1988). *Polygraphic lie-detector testing* is based on the concept that specific physiological responses are provoked by lying, and that the associated emotions of the liar can be qualitatively detected and qualitatively interpreted. The physiological events mostly recorded are associated with breathing, blood pressure, heartbeat, and galvanic skin responses, particularly in the palms of the hands.

Polygraphy is in wide use in the United States in three contexts: criminal investigation, security vetting, and personnel selection. In several states some of these tests are admissible in court, in others they are not. Polygraphy is also widely used in Canada, Japan, Australia, and Israel. It is commonly used in the United Kingdom, but is of little interest, in fact frowned upon, in Norway, Sweden, Holland, and Germany (where, however, handwriting analysis is popular).

Lie-detector procedures rely on an assumed intimate interconnection of mind and body. Thus, they purportedly discover facts about the mind by observation of physical responses to questions (Blinkhorn, 1988). The polygraph is simply the instrument that records physiological events in connection with questions asked by the tester. These reactions are then interpreted and conclusions are drawn regarding the truthfulness of the examinee.

"[U]se of the polygraph as a measure of good character in general, and of truthfulness and honesty in particular, rests on several dubious [socio-psychological and psychophysical] assumptions. It is assumed that good character can be specified in terms of measurable personality traits. . . . It is assumed that if a person is dishonest on the polygraph test, then he or she will behave in other undesirable ways on other occasions. It is assumed that truthfulness and honesty are inherent qualities of behavior. . . . The polygraph may or may not be a lie detector, but it could never be a personality detector" (Hampson, 1988, p. 64).

"The polygraph used as a lie detector falls very far short of acceptable standards for psychological tests. It is essentially unstandardized; it is internally inconsistent; re-scoring of charts is unreliable; no retest reliability information is available for examinees; it produces a disproportionate number of false positive results. . . . There are no good reasons for placing credence in the results it produces" (Blinkhorn, 1988, p. 39).

Great controversy has been raging regarding several aspects of polygraph lie detection: the theoretical foundations, the procedures used, the interpretation of the results, the conclusions drawn and the future implications for the tested person.

With respect to the *theoretical foundations,* much discussion has concentrated on whether there is indeed a "guilt reaction and emotion" that exists reliably and in a consistent manner. Apparently, such guilt depends on a person's character, upbringing, cultural background, and life experience. Social habits play a major role, for example in concealing the truth about a friend's illness, being defensive of a child's behavior, or bad-mouthing a disliked co-worker. Some civilizations have in fact made lying an art, even built community life on a pattern of lying and cheating (as described, for example, by the anthropologist Margaret Mead, 1901–1978). Growing up in a rough neighborhood certainly instills other life criteria than does living in a pampered and protected environment. Physiological responses to questions can be controlled; there are "manuals" on "how to beat the polygraph" (Gudjonsson, 1988).

Regarding the *procedures*: instrumentation has received relatively little attention in comparison to "experimental design and control." The best-controlled situation is in the laboratory, where persons are randomly assigned to either the guilty or not guilty categories and their reactions to questions tested. Yet, in real life, the status of the examinee is usually not known, which introduces a much more difficult situation. There are, unfortunately, very few field studies that have avoided serious methodological errors (Raskin, 1988). The most common procedure is to determine whether a person is "guilty" or "not guilty," with the label "inconclusive" applied when neither of these conclusions can be drawn. For this, primarily two techniques are used. The first is the "control question" procedure in which "irrelevant" questions ("Is today Thursday?") are mixed with relevant ones which ask, in essence, "Did you do it"? The second is the "guilty knowledge test" which does not attempt to determine whether the respondent is lying or not, but rather whether he or she possesses guilty knowledge: "Which knife did you use?" (Lykken, 1988). The discussion about the relative merits and problems associated with these tests, and whether or not they should be replaced by others, has been going on undecidedly for decades.

That discussion spills over to *interpretation of the results*. Several issues are of great importance. One is the relationship of the measured responses to "guilt," as discussed above, and another the sensitivity of the observed events. Still another problem is that of accuracy: it may be measured as either "false-positive errors" (an innocent individual is wrongly classified as guilty) or "false-negative errors" (a guilty person is classified as innocent). Of course, the "inconclusive" finding also may be false, in one direction or the other.

Various measures of accuracy have been made using these criteria, most of them in laboratory settings but just a few (for obvious reasons) in the real world. The accuracy results obtained were often disturbingly low. In the control-question procedure, guilty subjects were correctly classified in 80 to 84 percent of the studies both in simulated and field tests, but the correct classifications of innocent subjects were only 63 to 72 percent. In guilty-knowledge tests, which were performed only in the laboratory, the average accuracy was about 84 percent for guilty subjects and 94 percent for innocent persons; thus, this method can better protect the innocent.

Reliability is another concern, with respect to both intra- and intertester repeatability. Very few studies have been conducted, almost all in laboratory simulations in which it was found that the false-positive errors were underestimated; it is suspected that false negatives are underestimated in the real world (Ben-Shakhar and Furedy, 1990).

There are other problems beyond sensitivity, accuracy, and reliability. For example, some polygraph techniques score certain results higher than others. Various, nonstandardized procedures are used; some allow the introduction of the tester's personal biases and implications based on impressions gained of the subject that are not derived from the polygraph results.

The final and most important aspect is that of the *conclusions* drawn regarding the person's character, legal judgments, and implications for the future. These conclusions and their consequences can be, obviously, truly critical and of far-reaching importance.

SUMMARY

In a "person-machine-environment" system, the human perceives information simultaneously by various senses, and at the same time plans and executes actions. Thus, the traditional concept of a linear sequence "stimulus to sensory input to perception to processing to effector output" is probably overly simplistic, even unrealistic. Yet, the existing information relies on that stage-follows-stage concept.

Accordingly, the central nervous system, as processor, receives information concerning the outside (and the inside of the body) from receptors which respond to light, sound, touch, temperature, electricity, chemicals, and acceleration vectors. The received stimuli are transmitted in a converging manner along the afferent paths of the peripheral nervous system to the central nervous system. There, the information is processed and action signals are generated. These are sent along the efferent paths of the peripheral nervous system to the effectors, such as hand or voice box. The selection of appropriate external stimuli, and their transformation so that they can be reliably sensed, are among the major tasks of the ergonomic engineer. On the output side, the design task is to select the proper output channels to control a "machine."

Job demands and environmental conditions can be very stressful for the human, both physically and psychologically. Dealing with the "stress" problem is made difficult by widespread confusion in terminology, concepts, and consequences. Different personalities react differently to the same stressors. In spite of the complexity of the problem, procedures for assessing the job demands and for reorganizing and designing the system accordingly are in use. These can help in everyday situations but are of particular importance for unusual situations, such as in space, where tasks, environment, and the crew relations are novel.

Improving performance of individuals is the goal of learning and training. To measure the effectiveness of certain practices is often a challenge for two reasons: performance itself may be difficult to measure, or the specific "treatment variables" of the training may be difficult to define or separate. It is not clear why certain training practices are successful, others not. Some "unconventional" techniques using, e.g., hemispheric training, or parapsychological postulates, have no value; others are dubious or unproven, such as subliminal messages, visualization, observing experts, or use of biofeedback. Much theoretical methodology still needs to be tested before sound techniques for successful training of skillful performance can be developed. How the human "mind" works is still not well understood.

CHALLENGES

What difference would it make if we discussed mental functions simply in terms of "brain" instead of "mind"?

How would our understanding of human mental processing change if we went from the "traditional" concept to the "ecological" approach? In this context: would the concept of "affordance" be modified?

Would the consideration of the various types and sheer number of internal and external sensory receptors make the "traditional" or "ecological" concept more plausible?

What transducers are imaginable to make distal stimuli into signals that trigger human sensors?

Consider driving fast in dense traffic. When is a driver more likely to have an accident: if one can either brake or switch to another lane, or if braking is the only possible action?

Why should the presentation of an emergency signal by several modalities (such as light and sound) together be advantageous over presenting it only to one type of sensor?

What might be a better design of an automobile "braking system" than the one currently used?

How could Fitts' law be applied to designing a workstation for manually assembling many small parts?

Why is it difficult to maintain a conversation while looking for somebody's telephone number in a phone book?

Consider various means to measure the "workload" of a driver during rush-hour traffic.

Under what conditions, and to what extent, should "stress management" programs be carried out by an employer?

What might be suitable training programs for crews that will go into long, confined missions, such as space exploration?

How could one reduce the "end-of-mission" stressors?

Consider the advantages and problems associated with mixed-gender crews during long isolated missions, such as in space travel to Mars.

Which might be means to make subliminal learning feasible?

How could one measure the effectiveness of training in terms of how it affects performance?

Consider carefully planned and executed measures to assure the highest-possible team performance.

4

Human Senses

OVERVIEW

Seeing, hearing, smelling, tasting, and touching comprise the "classical" human senses. But the human is also sensitive to electricity and to pain and has a posture and motion (vestibular) sense. Of all senses, seeing and hearing are thoroughly understood, and related information is well suited for ergonomic application. However, information on the touch sense, including the feel of temperature, and sensitivity to electricity and pain are much less researched, and engineering applications are rather haphazard. The senses of smell and taste are usually not used in engineering.

Proper ergonomic procedures include measures to correct sensory deficiencies, such as in seeing, and to protect sensor functions from damage, such as hearing. Little is commonly done in enhancement or protection of the other sensory capabilities.

INTRODUCTION

The human is able to receive signals through several senses, originally classified by Aristotle (384–322 B.C.) into five different groups: *seeing* (vision), *hearing* (audition), *smelling* (olfaction), *tasting* (gustation), and *touching* (taction). (Taction might include more than one sense, because one feels several mechanical stimuli, such as contact and pressure, often together with pain, electricity, and temperature.) After more than two millennia, these groupings are still commonly used, although they include neither pain nor the vestibular sense by which we primarily balance the body.

Other classifications are by anatomy; by the sensor organs that perceive different stimuli; or by stimuli, i.e., the external objects or energies that trigger a sensation.

☞☞☞ *"Consider all the varieties of pain, irritation, abrasion; all the textures of lick, pat, wipe, fondle, knead; all the prickling, bruising, tingling, brushing, scratching, banging, fumbling, kissing, nudging. Chalking your hands before you climb onto uneven parallel bars. A plunge into an icy farm pond on a summer day when the air temperature and body temperature are the same. The feel of a sweat bee delicately licking moist beads from your ankle. Reaching blindfolded into a bowl of Jell-O as part of initiation. Pulling a foot out of the mud. The squish of wet sand between the toes. Pressing on an angel food cake. The near-orgasmic caravan of pleasure, shiver, pain and relief that we call a back scratch"* (Ackerman, 1990, pp. 80–81). ☜☜☜

BODY SENSORS

Every human sensor can be modeled as a transducer that consists of two components. The first part is the *receptor,* which is stimulated by an appropriate proximal stimulus to produce some reaction. The second part is the *converter,* which codes or reproduces the reaction to generate a potential signal to be sent on its nervous pathway.

The receptor must be stimulated by the right kind and quantity of the proximal stimulus (see Chapter 3) so that it can react. How this functions in reality is not fully understood, nor is it known how the sensor acts as the converter of the signal received by the receptor or how it generates the potential that is sent as sensory signal to the next neuron on its way to the central nervous system.

The problem is quite complex, because numerous sensors are distributed over the *receptive field;* consider, for example, the many tactile sensors in the tip of the finger. The primary receptor cells (transducers) are linked directly, or in branching ways, to a first-order afferent neuron. This is the first collecting and switching point, which receives the input from all sensors within its receptive field. The first-order afferent neuron may or may not be triggered sufficiently by the incoming signals to send an action potential to the next higher, second-order afferent neuron. This receives inputs also from other first-order neurons. Thus, numerous first-order afferent neurons are linked to second-order neurons, which report to the next hierarchy. The signal flow from several low-order neurons to a few higher neurons in the CNS, repeated possibly several times, follows the "principle of convergence," shown in Figure 4-1.

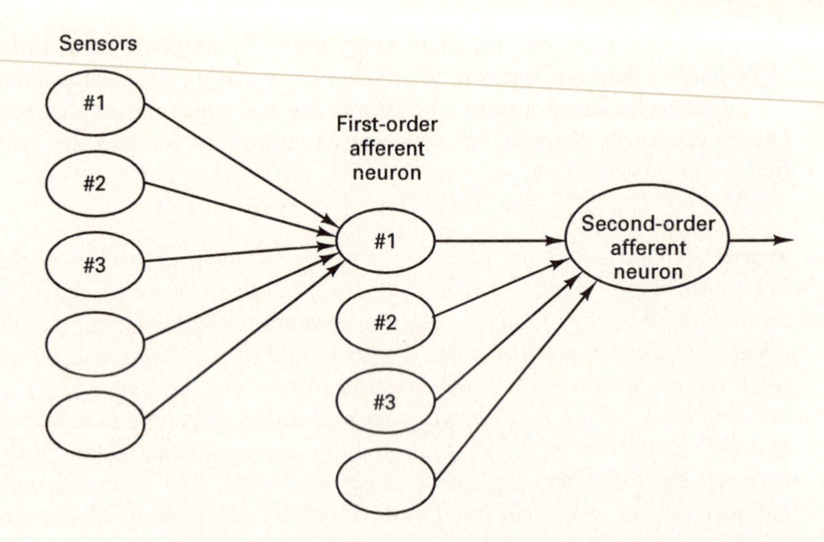

Figure 4-1. Model of converging receptive fields.

Stimuli

Some stimuli trigger only one type of sensor, but there are also many nonspecific stimulations: electromagnetic and mechanical energies can trigger sensations of vision, hearing, touch, and pain and of the vestibulum. Thermal stimuli trigger cold and warm as well as pain sensations. The auditory sense responds to pressure changes at frequencies between 20 and 20,000 Hz. Chemical stimuli affect particularly the senses of taste and smell.

Adaptation and Inhibition

We are bombarded every minute by a variety of stimuli. They would overwhelm us if we could not filter them out according to their importance. Since each of the afferent neurons may or may not transmit an arriving signal further up the chain toward the CNS, this converging network functions as a filtering or coding system, letting only those signals through that are sufficiently strong, of sufficient importance. The sensory system "adapts" by reducing the nervous discharge even while the strength of the proximal stimulus remains the same. A typical example is the adaptation of taction to the feeling of clothes worn. The system may be "inhibited" also by reduction of nervous discharge, such as by a second stimulus appearing in the sensory field. A typical example is vigorous rubbing of a stubbed toe: rubbing triggers new taction sensations which mask the pain. Except for the auditory and vestibular systems, very little is presently known about the excitation or inhibition in the CNS above the first-order neuron.

NEED

There are several possible "sensory" adaptation processes: it may be the reduction in the firing rate or trigger intensity at the sensor, or the reduction in transmission along the afferent nervous path, or the masking of signals. Different researchers have attributed various meanings to the term "adaptation," and diverse research approaches have been used, which, altogether, do not yet provide coherent results. The field still needs much work.

Getting used to a background stimulus level so that we distinguish only important signals is probably a mixture of adaptation and inhibition, which results from the presence of several stimuli (Sherrick and Cholewiak, 1986). However, some receptors adapt slowly, keeping their nervous discharges at the same level for a second or more. Others adapt quickly, their discharge rate falling within milliseconds to low or zero value.

Sensory Thresholds

A stimulus may be of such weak intensity that one cannot sense it: it is below the lower or *absolute* or *minimal threshold*. The upper or *maximal threshold* is the limit above which the sensor does not respond any more (such as to sound frequencies above 20,000 Hz), but where the stimulus might still annoy or even damage the sensors (such as high sound intensities above 20 kHz).

The *proportionality of sensation* to a change in the stimulus is often of interest. For example: how do we react to the numerically same change of 10 dB in sound pressure at low pressure levels as opposed to high ones? The *difference threshold* is the smallest physical difference in the amount of stimulation that produces a *just noticeable difference* in the intensity of our sensation. This was investigated in detail by Fechner and Weber in the nineteenth century. Usually, difference thresholds increase with increasing intensity: "Weber's law" states that the difference threshold is a constant proportion of the stimulus intensity (which may not be true for extremely small or large intensities of stimuli).

EXAMPLE

There is interplay between sensory adaptation and threshold. This is strikingly demonstrated in the classical experiment in which a person immerses one hand in warm water, and the other in cold water. The initial sensations of warm and cold, respectively, subside slowly as one adapts to the water temperatures. Next, both hands are immersed in lukewarm water: now the warm hand signals the sensation of cold, the cold hand the feeling of warmth. It is apparently the difference in water temperatures that makes each hand report the change from the previous state. What would happen if the temperature of the third water container were close to one of the others? If the two temperatures were close

TABLE 4-1. CHARACTERISTICS OF THE HUMAN SENSES

	Vision	Audition	Touch	Taste, smell	Vestibular
Stimulus	Light-radiated electromagnetic energy in the visible spectrum	Sound-vibratory energy, usually airborne	Tissue displacement by physical means	Particles of matter in solution (liquid or aerosol)	Accelerative forces
Spectral range	Wavelengths from ~400 to ~700 nm (violet to red) 10^{-6} mL to 10^4 L	20–20,000 Hz 20 μPa to 200 Pa	Temperature: 3 seconds exposure of 200 cm^2 of skin 0 to 400 pulses per second	Taste: salt, sweet, sour, bitter Smell: fragrant, acid, burnt, and caprylic	Linear and rational accelerations
Spectral resolution	120–160 steps in wavelength (hue) varying from 1–20 nm	~3 Hz for 20–1,000 Hz; 0.3% above 1,000 Hz	~10 percent change in number of pulses per second		
Dynamic range	~90 dB (useful range); for rods = 0.000032–0.0127 cd/m^2; for cones = 0.127–31830 cd/m^2	~140 dB	~30 dB (0.01–10 mm displacement)	Taste: ~50 dB (0.00003% to 3% concentration of quinine sulphate) Smell: 100 dB	Absolute threshold is ~0.2 deg s^{-2}
Amplitude resolution ($\Delta I/I$)*	Contrast = 0.015	0.5 dB	~0.15	Taste: ~0.20 Smell: 0.10–50 dB	~0.10 change in acceleration
Acuity	1 min of visual angle	Temporal acuity (clicks) ~0.001 sec	Two-point acuity ranges from 0.1 mm (tongue) to 50 mm	?	?
Response rate for successive stimuli	~0.1 sec	~0.01 sec (tone bursts)	Touches sensed as discrete to 20/sec	Taste: ~30 s Smell: ~20–60 s	~1–2 s; nystagmus may persist to 2 min after rapid changes in rotation
Best operating range	500–600 nm (green-yellow) at 34.26–68.52 cd/m^2	300–6000 Hz at 40–80 dB		Taste: 0.1% to 10% concentration	~1 g acceleration directed to foot

*I = intensity level; ΔI = smallest detectable change in intensity from I.

SOURCE: Adapted from VanCott and Kinkade, 1972; Boff and Lincoln, 1988.

enough, the hand that moves between them would not signal a noticeable difference, while the other hand would indicate a strong difference in felt temperatures. How would this change if the overall temperature ranges were decreased, or increased? Further, what would happen if one got used to higher or lower temperatures through repeated and continued immersing with stepwise increased temperatures? People who work in high or low temperatures can "tolerate" these more easily. Is this a form of adaptation? How do the just-noticeable differences change in this case?

Table 4-1 shows smallest detectable energies, and largest amounts that are tolerable or practical. Stimuli near either of these limits may lead to unreliable sensing.

APPLICATION

THE VISION SENSE

The characteristics of human vision are well researched and described in the literature: Boff, Kaufman, and Thomas (1986) devote seven chapters to the description of the details of human vision. Fortunately, the model of the "normal young adult eye," underlying most descriptions of the human visual sense, is indeed quite representative, since the functions of individual eyes are very similar (Snyder, 1985; Westheimer, 1986). Therefore, one can formulate general statements about the functions of the eye.

Architecture of the Eye

As sketched in Figure 4-2, the eyeball is a roughly spherical organ of about 2.5 cm diameter, surrounded by a layer of fibrous sclera.

When parallel beams of light from a distant target reach the eye, they first encounter the *cornea,* a translucent bulging round-dome section of the sclera at the front of the eyeball, kept moist and nourished by tears. The cornea provides all the refraction needed to focus on an object more than about 6 meters (20 feet) away (if the eye is young and healthy).

Behind the cornea is the *iris,* tissue surrounding a round opening, the *pupil*. The dilator muscle opens and the sphincter muscle closes the pupil like the aperture diaphragm of a camera, regulating the amount of light entering the eye.

Having passed through the pupil, light beams enter the *lens*. If a distant object needs to be seen, suspensory ligaments keep the lens thin and flat, so that the light rays are not bent. For close objects (which cannot be focused by the cornea), the ciliary muscle around the lens makes it thicker and rounder, so that the light beams are refracted and focused on the retina.

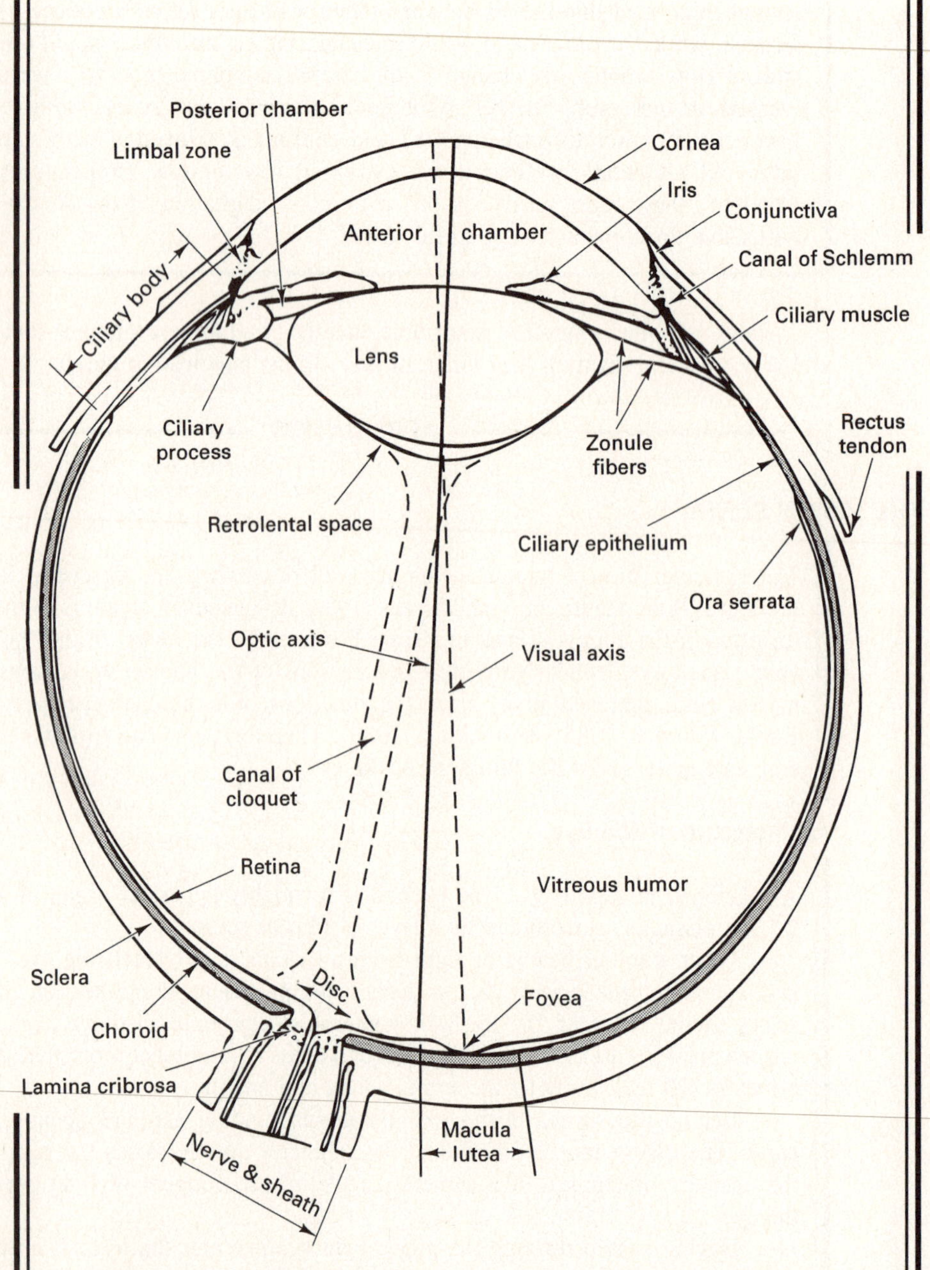

Figure 4-2. Horizontal section through the human eye. (Reprinted with permission from Snyder, 1985.)

The space behind the lens, the interior of the eyeball, is filled with the *vitreous humor,* a gel-like fluid, which has refractory properties similar to water.

Light focused by cornea and lens finally reaches the *retina,* a thin tissue that lines about three-quarters of the inner surface of the eyeball, opposite the pupil. It is supplied by many arteries and veins, and contains about 130 million light detectors.

There are two kinds of light sensors on the retina, named for their shape. The majority, about 120 million, consists of *rods,* which respond to even low-intensity light and provide black-gray-white vision. There are also about 10 million *cones,* which respond to colored bright light. Rods and cones are special types of nerve cells that convert light into electrical signals.

Pigments in rods and cones serve to convert light into electrical signals. Each cone contains a pigment that is most sensitive to either blue, green, or red wavelengths. An arriving light beam, if intense enough, triggers chemical reactions in one of the three types of pigmented cones, creating electrical signals that the brain uses to distinguish among about 150 color hues. Rods contain only one pigment, *rhodopsin,* which is bleached by light into two colorless molecules. This bleaching process sets off electrical impulses that are passed to the brain for the perception of white, black, and shades of gray.

Cones are concentrated at the *fovea,* in the center of the retina directly behind the pupil, where only few rods are found. The fovea, along with its yellowish surrounding area, the *macula,* is the area mostly needed for reading fine print.

The electrical signals coming from rods and cones are transmitted through the *optic nerve* to the brain. The nerve exits the eye at its rear, about 15 degrees off-center toward the inside. Since there are no light sensors at this area, an image at this "blind spot" cannot be seen. However, since the blind spots of both eyes are medially located, they do not overlap in our field of vision and therefore we are unaware of their existence.

☛☛☛ *The Arab physicist Alhazen (965–1039) was the first to discover that vision was made possible by rays of light falling on the eye, and was not the result of rays of light sent out by the eyes. Alhazen studied lenses and attributed their magnifying effect to the curvature of their surfaces. This represents the beginning of the scientific study of optics. At the middle of the thirteenth century, convex lenses for eyeglasses used by the farsighted—that is, mostly by the elderly—were well known both in China and Europe. In 1451, the German scholar Nicholas of Cusa used concave lenses for the nearsighted (Asimov, 1989).* ☚☚☚

Mobility of the Eyes

Each eye is theoretically capable of movements in six degrees of freedom, illustrated in Figure 4-3. Mostly used are rotational movements in *pitch* (around

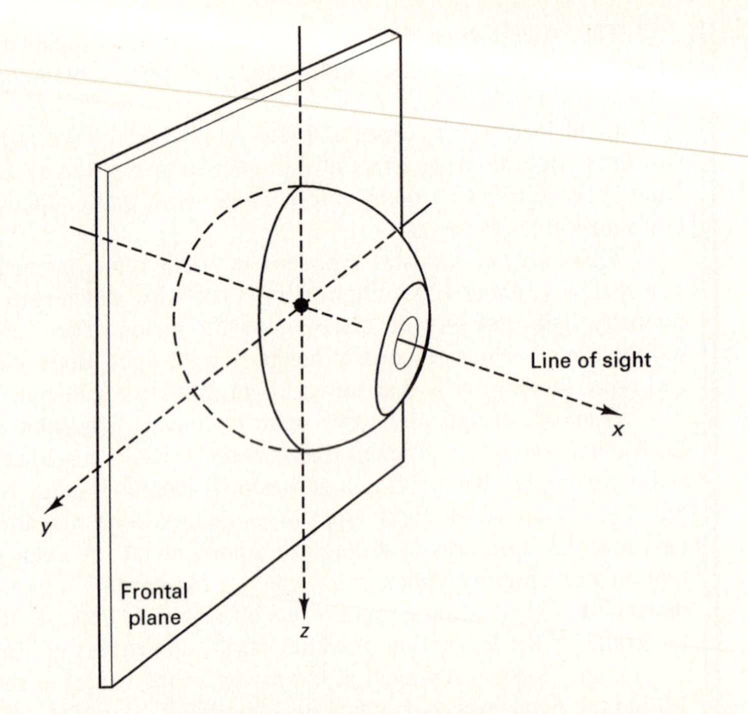

Figure 4-3. Mobility axes of the eye.

y), *roll* (around x) and *yaw* (around z), but the entire eye also moves linearly, most appreciably forward and backward (along x) in "infraduction." Thus, the center of rotation of the eye is not fixed, but the center displacements are fairly small. Therefore, one usually assumes it to remain in place, with the center of rotation of the eye located approximately 13.5 mm behind the cornea.

Six striated muscles are attached to the outside of the eye controlling its movements. They are the medial (internal) and lateral (external) recti, superior and inferior recti, and the superior and inferior oblique muscles—see Figure 4-4. Since the muscles do not attach at orthogonal direction to each other, they interact. The superior and inferior recti muscles are primarily responsible for pitch, i.e., up and down eye rotation. The medial rectus and the lateral rectus muscles provide yaw, i.e., left and right movement. The oblique muscles predominantly provide roll movement.

The six muscles work in concert to produce eye movements. When one looks straight ahead, the eyes are said to be in "primary position." The so-called primary movements of the eye are in the two degrees of freedom associated with up and down gaze motions (pitch) and left and right movements (yaw). Given the complex attachments of the muscles moving the eye, these primary movements are accompanied by some roll. The further the eye pitches and yaws, the more the eye rolls also. The amount of eyeball roll at various pitch and yaw angles away from the primary position is about 1 degree of roll

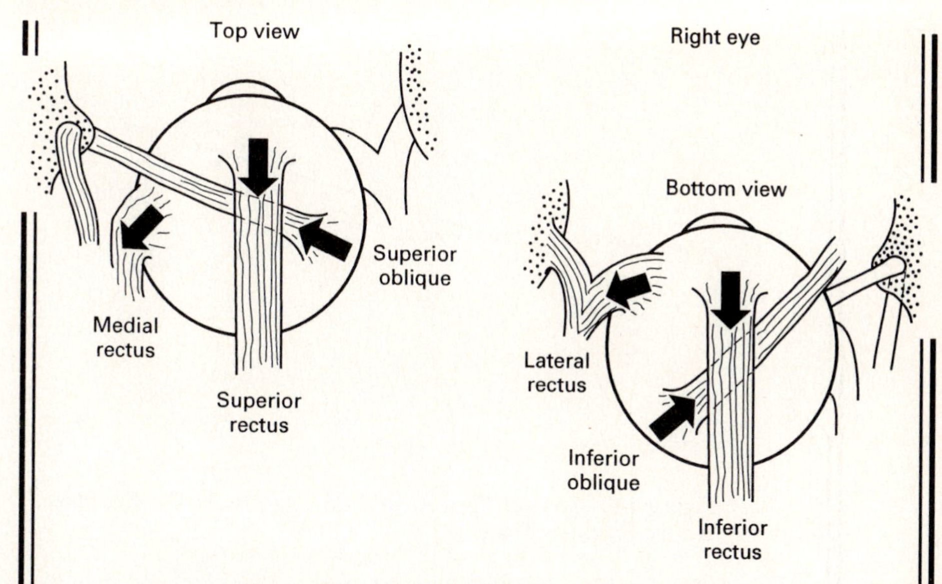

Figure 4-4. Muscles moving the right eyeball.

with 10 degrees each of pitch or yaw; 3 degrees of roll with 20 degrees of pitch or yaw; 8 degrees of roll with 30 degrees; and 15 degrees of roll with 40 degrees each of pitch and yaw, respectively. We are not aware of this "tilting" of the visual image as we move the eyes, because the brain adjusts to it.

The eye can track continuously left and right (yaw) a visual target which is moving less than 30 degrees per second or cycling at less than 2 hertz. Above these rates, the eye is no longer able to track continuously but lags behind and must saccade (move in jumps) to catch up to the visual target.

Line of Sight

If the eye is fixated on a point target, the *line of sight* (LOS) runs from the object through the lens (pupil) to the receptive area on the retina, most likely the fovea. Thus, the LOS is clearly established within the eyeball. To describe the LOS direction external to the eye, a suitable reference is needed. Often, the horizon has been used as reference for the line-of-sight angle (LOSA) in the medial (*xz*) or sagittal plane (see Chapter 1). Yet, it is important to know each of the two portions of LOSA: *eye* pitch angle with regard to the skull of the head, and *head* pitch angle with respect to neck or trunk. The two portions cannot be separated if only the horizon is used as reference.

Various approaches have been taken to define the line-of-sight angle LOSA with respect to the body. One is to define LOSA against the Frankfurt plane (see Glossary), which is attached to the skull. It is easier, however, to use the Ear-Eye plane (see Glossary) because its attachments to the head (the ear canal and the external junction of the eyelids) can be seen and need not be

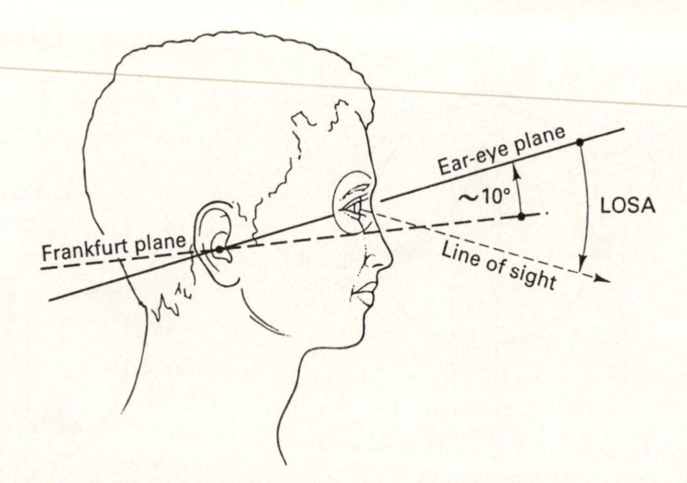

Figure 4-5. Line-of-Sight Angle LOSA against the Ear-Eye Plane, and the Frankfurt Plane.

palpated. The Ear-Eye plane is about 10 degrees more inclined than the Frankfurt plane. For yaw and roll, perpendicular planes, meeting at the eye, can be defined. Figure 4-5 illustrates the definition of the angle of the line of sight as eye pitch angle.

Of course, the angular locations of the Ear-Eye or Frankfurt planes can be described in relation to a stationary reference, such as the horizon or the gravity vector, or to some other reference in the environment. This angle describes the position of the head, and through it LOSA.

It is difficult to establish the position of the head with respect to the trunk. The links between skull and trunk are mechanically ill defined, consisting of at least the cervical vertebrae on top of the thoracic column which have various mobilities in their intervertebral joints. The skull also rotates in three degrees of freedom at the atlas of the first cervical vertebra. A further complication is that there is no easily established reference system within the trunk in regard to which one can describe the relative displacements of the neck and head.

NEED

The topic of trunk, neck, and head positions needs further research and practically useful definitions.

If a visual target is not a point, but can be expressed as the length of a line perpendicular to the line of sight, then the target size is usually expressed as the *subtended visual angle:* this is the angle formed at the pupil. Its size depends on the distance D of the object, and on its size L. The subtended visual angle α is described in Figure 4-6 and is usually given in degrees of arc

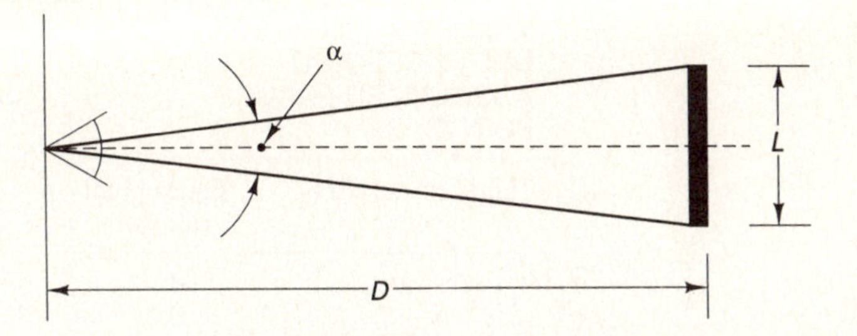

Figure 4-6. The subtended angle.

(1 degree = 60 minutes = 60 × 60 seconds of arc). The equation is:

$$\alpha = 2 \times \text{arc tan } (0.5 \times L \times D^{-1}) \qquad \text{(in degrees)}$$

For visual angles not larger than 10 degrees, this can be approximated by

$$\alpha = 57.3 \times L \times D^{-1} \qquad \text{(in degrees)}$$

or

$$\alpha = 60 \times 57.3 \times L \times D^{-1} = 3438 \times L \times D^{-1} \qquad \text{(in minutes of arc)}$$

Note that the equations do not take into account the distance between the pupil and the lens. This distance of approximately 7 mm has no perceptible effect under most conditions.

The human eye can perceive at a minimum a visual angle of approximately 1 minute of arc, or 1 second of vernier acuity. Table 4-2 presents the visual angles of familiar objects. For ease of use, technical products should be so designed that the angle subtends at least 15 minutes of arc, increasing to at least 21 min of arc at low light levels.

TABLE 4-2. VISUAL ANGLES OF FAMILIAR OBJECTS

Object	Distance	Visual angle (arc)
5-cm diameter gage	0.5 m	5.7 degrees
Sun		about 30 minutes
Moon		about 30 minutes
Character on CRT screen	0.5 m	17 minutes
Pica letter at reading distance	0.4 m	13 minutes
U.S. quarter coin at arm's length	0.7 m	2 degrees
U.S. quarter coin	82 m (90 yards)	1 minute
U.S. quarter coin	3 miles	1 second

The target distance D is often used in its reciprocal value. If D is measured in meters, the unit of the reciprocal, $1/D$, is called the diopter, which indicates the optical refraction needed for best focus. Thus, a target at infinity has the diopter value zero, while a target at 1 meter distance has the diopter value one. Table 4-3 shows values for some typical target distances.

TABLE 4-3. TARGET DISTANCES AND ASSOCIATED FOCUS POINTS

Target distance D (m)	Focus point (diopter)
Infinity	0
4	0.25
2	0.5
1	1
0.67	1.5
0.50	2
0.33	3
0.25	4
0.2	5

The Visual Field

The *visual field* is the area, measured in degrees, which can be seen by both fixated eyes. In its center, the visual field of each eye is occluded by the nose. To the sides, each eye can see a bit over 90 degrees, but only within the inner about 65 degrees can color be perceived, while farther to the periphery shades of gray dominate. (This is because of the location of cones and rods on the retina, discussed earlier.) Upward, the visual field extends through about 55 degrees, where it is occluded by the orbital ridges and eyebrows. However, color can be seen only to about 30 degrees upward. The downward vision is limited by the cheek at about 70 degrees; the area in which color can be seen extends down to about 40 degrees. The "blind spot" on the retina is at approximately 15 degrees to the outside, for each eye.

The references for these angles change as the eyeball within the head and/or the head is moved. Eyeball rotation adds about 70 degrees of visual field to the outside, but nothing in the upward, downward, and inside directions because the orbital ridges, cheeks, and nose stay in place. However, if the head moves in addition to the eyeballs, nearly everything in the environment can be seen, if not occluded by the body or other structures, when head and eyes are turned in that direction.

Accommodation

Accommodation is the action of focusing on targets at various distances. (However, there is some confusion in the meaning and use of optical and visual terms: see Miller, 1990, for a thorough discussion.) The normal young eye can accommodate from infinity to very close distances, meaning that a diopter range from 0 to about 10 can be achieved.

Normally, with visual objects at distances of 6 m or more, the relaxed lens refracts the incoming parallel rays on the retina. To accommodate closer objects, the lens is made thicker by pull of the enclosing ciliary muscle: the radius of the frontal lens is reduced from 10 to 6 mm, and the radius of the rear surface from 6 to 5 mm. Thus, the "optical refracting power of the eye" is changed (by adjusting the curvature of the lens) so that by refraction of the incoming light rays the image of the target falls focused upon the retina at every distance; the retina is made "conjugated to the object" at all accommodations. If one "looks at nothing", the lens accommodates at a distance of about 1 meter.

The point that can be focused at the closest distance is called the "near-point." The farthest point that can be focused without conscious accommodation is called the "far-point." The difference between far- and near-points is called the amplitude of accommodation. Most young people can focus at a near-point of about 10 cm, but the minimal distance changes to about 20 cm at age 40, and to about 100 cm at age 60, on average.

Convergence

If one aims both eyes at the same point, an angle exists between the two lines of sight connecting each eye with the target. This angle is very small, in fact negligible, at objects more than 6 meters away, but becomes fairly large at close objects. While it takes about 200 ms to fixate on one point at reading distance, this time is reduced to about 160 ms when focusing at a point at 6 m or farther. If the eyes are not made to converge, an "error of vergence" exists, also called "fixation disparity." This occurs commonly when the quality of the binocular stimulus is low, when the observer is greatly fatigued or under the influence of alcohol or barbiturates (Heuer and Owens, 1989). A "phoria" exists if the images of one target are not focused on the same spots on the retinas of both eyes, resulting in double images.

Visual Fatigue

People doing "close visual work," such as looking at the screen of a computer display, often complain of eye discomfort, visual fatigue, or eye strain (asthenopia), all of which are vaguely described as "subjective visual symptoms or distress resulting from the use of one's eyes" (National Research Council Committee on Vision, 1983, p. 153). While the occurrence of such eye strain, and its intensity, varies much among individuals, it seems often related to the effort of focusing at a distance that is different from the minimal refractive state. That minimal distance is often called "resting distance of accommodation" or "dark vergence" or "dark focus" position (Miller, 1990). For most people whose eyes look ahead but do not focus on a target, that resting position of binocular vergence is about 1 meter away from the pupils. (This finding is in contrast to the earlier assumption that the automatic resting position is at "optical infinity," i.e., with parallel visual axes of the eyes.) The actual point of vergence is characteristic of the individual's oculomotor resting adjustment, therefore also called a "resting tonus posture" (Owens and Leibowitz, 1983; Heuer and Owens, 1989; Jaschinski-Kruza, 1991).

The individual distance (of about 1 meter) for resting accommodation and vergence may be used as a reference for calibrating the relationship between eye movements and space perception. As one lowers the angle of gaze, or tilts the head, the dark vergence point gets closer to the eyes, to about 80 centimeters for a LOSA of 60 degrees downward direction; the dark vergence distance increases as one elevates the direction of sight, on average to about 140 cm at a 15-degree upward direction (Heuer and Owens, 1989; Ripple, 1952; Tyrrel and Leibowitz, 1990).

If one puts visual objects at points significantly different from the resting distance, ocular mechanisms are forced to exert efforts to focus on these targets. The automatic selection of the dark vergence (resting) distance apparently is not only due to the biomechanics of the extraocular muscles, but is also dependent on (currently unidentified) "central processes" that integrate the information about head and eye positions (Heuer, Bruewer, Roemer, Kroeger, and Knapp, 1991). Elderly people find it more difficult to focus on an elevated target than on a lower one (Rabbitt, 1991). If the visual target is away from the resting point, the effort to focus may lead to "visual fatigue."

APPLICATION

Several aspects are of particular interest for the design of work systems that require exact binocular eye fixation. The first is that the binocular dark vergence position, and the accommodation distance, to which the eyes return at rest, are quite different from individual to individual, but constant for each person. Thus, one should allow and encourage each person to adjust a target that is visually fixated to her or his "personal distance." The second practical important finding is that a similar effect on the natural vergence distance is brought about by tilting either the eyes or the head. This explains why people find it more comfortable to lean back in a chair while tilting the head forward/down to look at a close object (at about eye height) rather than to sit upright: this reduces the natural vergence distance. The third finding is that targets at or near "reading distance" should be distinctly below eye level, particularly if the viewer is elderly. The fourth finding of practical importance is well known from experience: it is difficult to exactly see a visual target of low optical quality when the observer is fatigued or under the influence of drugs.

Paraphrasing Tyrrell and Leibowitz (1990, p. 342): Scientists and grandmothers can now agree that visual fatigue is related to near work. It is easier to look down on it than to look up to it.

Visual Problems

With increasing age, the accommodation capability of the eye decreases, because the lens becomes stiffer by losing water content. This condition is known as *presby-*

opia. The result is difficulty in making light rays converge exactly on the retina. If the conversion is in front of the retina, the condition is called *myopic;* if the focus point is behind the retina, one speaks of a *hyperopic* eye.

A nearsighted (myopic) person finds it difficult to focus on far objects but has little trouble seeing close objects. This condition often improves with age, when the lens remains flattened. In fact, even then far objects are not exactly in focus, but the rays from far objects still strike the retina (although not exactly in focus) so that these distant objects can be sufficiently identified. In contrast, farsightedness becomes more pronounced with age, meaning that it becomes more difficult to focus on near objects. Both problems, myopia and hyperopia, can be fairly easily corrected by either contact lenses or spectacles.

In many people, the pupil shrinks with age. This means that less light strikes the retina, and therefore many older people need to have increased illumination on visual objects for sufficient visual acuity.

Another problem often encountered with increasing age is *yellowing of the vitreous humor*. This absorbs some light energy and requires high illumination of the visual target for good acuity. Also, light rays are refracted within the vitreous humor, bringing about the perception of a light veil (like mist) in the visual field. If bright lights are in the visual field, resulting "veiling glare" can strongly reduce one's seeing ability. Obviously, the yellowing problem cannot be corrected with artificial lenses.

Astigmatism occurs if the cornea is not uniformly curved, so that an object is not sharply focused on the retina, depending on its position within the visual field. Often, the astigmatism is a "spherical aberration," meaning that light rays from an object located at the side are more strongly refracted than those from an object at the center of the field of view, or vice versa. Another fairly common case is "chromatic aberration," where an eye may be hyperopic for long waves (red) and myopic for short waves (violet, blue). Astigmatic problems usually can be solved with an artificial lens in front of the eye.

Another visual impairment is *night blindness,* a condition of a person having less than normal vision in dim light, that is, with low illumination of the visual object.

Color weakness exists if a person can see all colors, but tends to confuse them, particularly in low illumination. Defective color vision is rather common: about 8 percent of men are color-defective, but less than 1 percent of women. Some people are *color-blind,* meaning that they confuse, for example, red, green, and gray. Only very few people can see no color at all, or only one color.

Isaac Newton (1642–1727) performed experiments in which he made a beam of light pass through a glass prism. What he saw emerging was a band of colors, with red being the least bent portion of the light, followed by orange, yellow, green, blue, and violet. Making these colorful lights pass through another prism, he showed that they merged to become white light again. In this way, Newton showed that white light was not "pure" but a mixture of different colors (Asimov, 1989).

APPLICATION

Vision Stimuli

Humans are sensitive to light, meaning that they detect changes in the visual stimulation depending on wavelength, intensity, location, or duration. Usually, the discussion of light sensitivity is limited to wavelength and intensity.

The human eye is sensitive and can adapt to increases and decreases in illumination over a wavelength range of about 380 nm to 720 nm, that is from violet to red. The minimal intensity to trigger the sense of light perception is 10 photons, or an illuminance* of the eye of about 0.01 lux. At such low intensity, shorter wavelengths, i.e., blue-green, are more easily perceived than longer wavelengths; the main perception is of "light," not of color (Hood and Finkelstein, 1986).

Viewing Conditions

Cones perceive colors if the object illumination* is "bright," above about 0.1 lux. This is called the *photopic* condition. If the illuminance falls between 0.1 and 0.01 lux, both cones and rods respond. This is called the *mesopic* condition, present at twilight and dawn, when we can see color in the (brighter) sky, but (dimmer) objects appear only in shades of gray. In dim light, below about 0.01 lux, only rods respond: in this *scotopic* condition, only black and white, and gray shades in between, can be perceived.

Figure 4-7 describes this condition schematically. It shows the sensibility of rods and cones to visible wavelengths, the so-called spectral sensitivity. Rods respond mostly to shorter wavelengths, while cones cover the whole spectrum. However, this schematic is drawn by making the maximum of each curve equal to 100 percent; in reality, the three different pigments of cones have their maximal sensitivities at around 440, 540, and 570 nm, while rods have their single maximal sensitivity at about 510 nm. In terms of luminous intensity (which has been set to 100 percent for the rod and the cone curves in Figure 4-7) rods are more sensitive because they respond to lower intensities than cones. This means that as the ambient light level increases, visual detection shifts from domination by the rod system to predominant use of the cones. "In the dark," the threshold for detecting light of almost any wavelength is determined by rod sensitivity. Under "bright" conditions, the cones provide most of the information, independent of the wavelengths of the stimuli. Between these two extremes of illumination, the spectral distribution of light determines (and other parameters codetermine) which system is more sensitive. The transition from using rods to using cones is largely due to a desensitization of the rods by relatively weak light that leaves the cone sensitivity intact. These con-

*See the section on measurement of light (below) for terms and units of measurement.

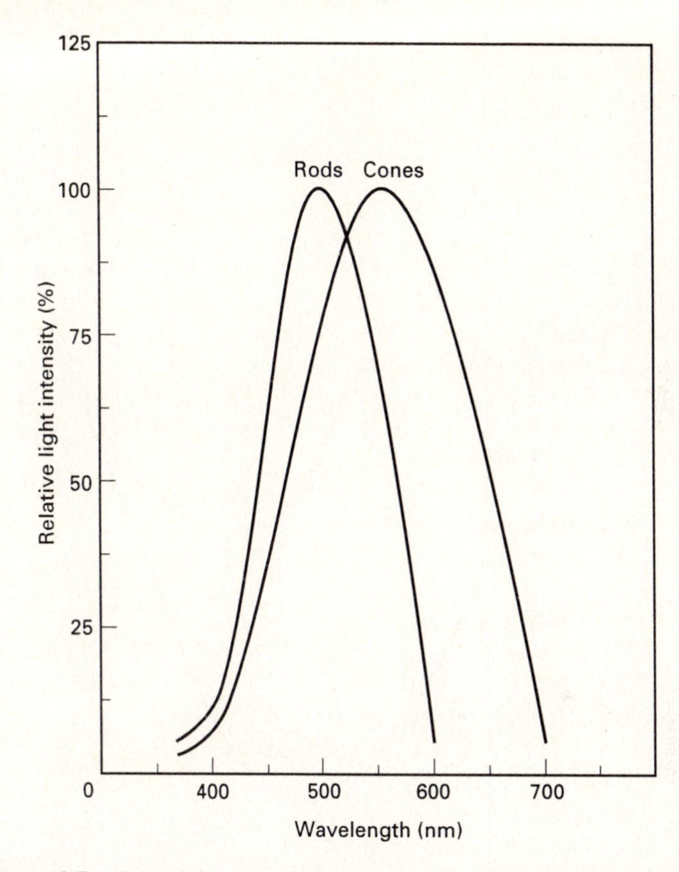

Figure 4-7. Sensitivity of rods and cones to wavelength. Shows conditions at illuminations below 0.01 lx (scotopic) and above 0.1 lx (photopic). but not in the intermediate (mesopic) range.

ditions are shown schematically in Figure 4-8. Cones need, throughout the wavelength spectrum, higher intensities of light in order to be functional.

There are some interesting phenomena associated with vision at night:

At dark, the color-sensitive cones at the retina are inactive, and what is "seen" is perceived through the rods.

Directly behind the lens, there are only few rods at the back of the retina. This constitutes a "blind spot" where objects may not be detected.

If one stares at a single light source on a dark background, the light seems to move. This is called the "autokinetic phenomenon."

If the horizon is void of visual cues, the lens relaxes and focuses at a distance of about 1 to 2 meters, making it difficult to notice far objects. This is known as "night myopia."

Night vision capabilities deteriorate with increasing lack of oxygen. Thus, at 1,300 meters (4000 ft) of altitude, vision is reduced by about 5 per-

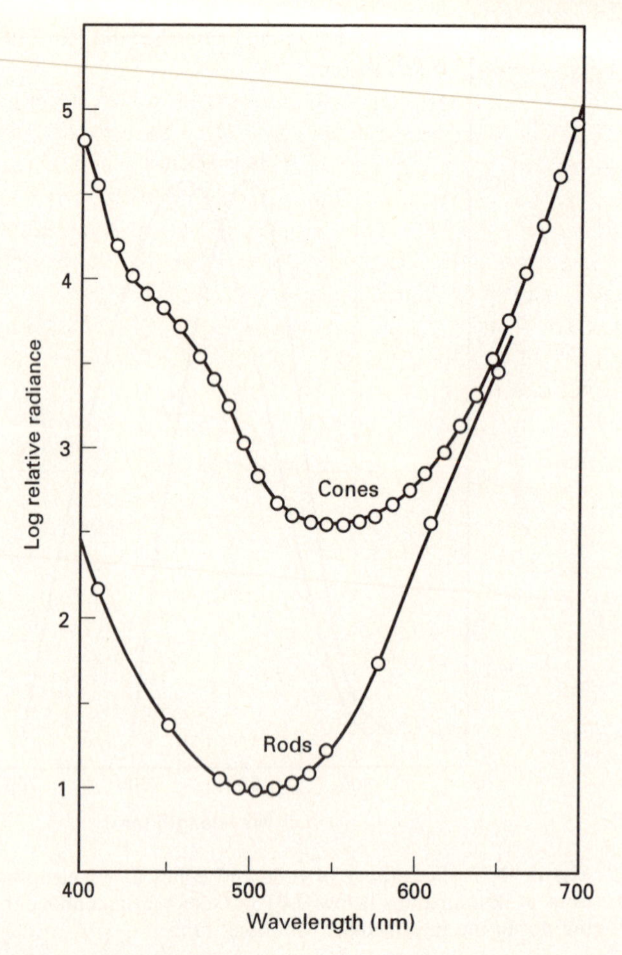

Figure 4-8. Thresholds of cones and rods to light intensity. (Modified with permission from Snyder, 1985.)

cent; at 2,000 m, the reduction is about 20 percent, up to 40 percent in smokers whose blood has lost some capability to carry oxygen.

Ocular Adaptation

The eye can change its sensitivity through a large range of illumination conditions. This is called *visual adaptation,* and one commonly distinguishes between adaptation to light and to dark conditions. The main functions participating in adaptation are pupil adjustments, spatial summation of stimuli, and photochemical functions, including the stimulation of rods and cones. The actual change in response thresholds of the eye during dark or light adaptation depends on the luminance and duration of the previous condition to which the eye had adapted, on the wavelength of the illumination, and on the location of the light stimulus on the retina.

Adaptation from light to darkness takes about 30 minutes; during this period, initially the cones are most sensitive, and the rods thereafter. Even after adaptation, the sensitivity at the fovea (cones) is only about one-thousandth of that at the periphery of the retina (rods). Therefore, weak lights can be noticed in the periphery of the field of view, but not if one directs the gaze at them, i.e., when they are refracted onto the fovea. As Figure 4-9 shows, dark adaptation is mostly governed by the change in threshold of the cones. Maximal sensitivity is at about 510 nm, a shift from light adaptation, when the highest sensitivity is at about 560 nm. People who suffer from "dark blindness" have nonfunctional rods and can adapt only via cones. Persons who are "color-

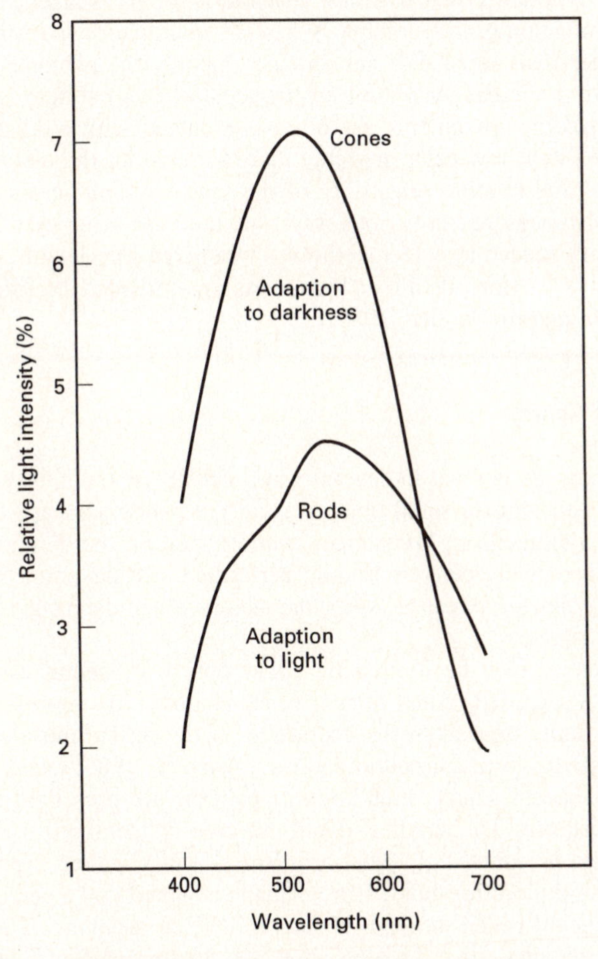

Figure 4-9. Adaptation of cones and rods.

blind" have nonfunctional cones (or only one or two of the three types of cones) and must adapt via rods only.

Adaptation to light condition is, quite in contrast to adaptation to dark, very fast: full adaptation is achieved within a few minutes. In adaptation to light, about 560 nm (yellow) are most easily perceived at the fovea, with a shift to about 500 nm (yellow-green) in dark adaptation. Above 650 nm, the spectral sensitivity of the fovea is not very different during either dark or light adaptation, where there are large differences in response to the different wavelengths at the shorter waves: with increasing adaptation to dark, short-wave stimuli are perceived as having higher luminance. This can be an important consideration for the selection of colors in dark rooms. Also, persons who must adapt to dark conditions after having been in a lighted environment adapt faster to long-wave lights (red, yellow) than to short-wave (blue).

Human visual response characteristics are utilized, for example, in night illumination of instruments. When the illumination is restricted to wavelengths longer than about 600 nm (orange), mainly the photopic (cone) system is excited, while the adaptation of the scotopic (rod) system is largely maintained. Therefore, instruments are often illuminated with reddish or yellowish light, with the yellow color probably more suitable for the older eye.

The relative sensitivity of the visual system across the visual spectrum can be measured in various ways, but the most often used assessment is the intensity needed to detect a stimulus of a given wavelength, called "threshold intensity." (More detailed discussions are provided by Hood and Finkelstein, 1986; and by Snyder, 1985.)

Visual Acuity

Acuity can be defined in several ways, usually as the ability to detect small details and to discriminate small objects. Acuity depends on shape of the object, on wavelength, illumination, luminance, contrast, and on the duration of the stimulus. The measurement of acuity is usually performed at a far point (6 m; 20 ft) and a near (0.4 m) viewing distance, since the factors which determine resolution can differ in the two cases.

Acuity may be limited by either optical or neural causes. Optical problems lead to a degraded retinal image, which can often be improved by corrective lenses. Neural limits stem from the coarseness of the retinal mosaic (receptor distribution) or sensitivity limits of neural pathways (Snyder, 1985; Olzak and Thomas, 1986).

To assess acuity, high-contrast patterns are presented to the observer from a fixed distance. The smallest detail detected or identified is taken as the threshold, expressed in minutes of visual arc. Visual acuity is then expressed as the reciprocal of the resolution threshold. It is normally assumed that a person should be able to resolve a detail which subtends about 1 minute of visual arc. The most common testing procedures use either Landolt rings, or Snellen or Sloan letters; all standardized black stimuli against a highly contrasting white background. A measurement of 20/20 on the Snellen chart is regarded as "perfect vision." A person is defined as "partially sighted" if the vision (after correction) is worse than 20/70, but still better than 20/200 (meaning that this person can see an object at a 20 feet distance that others can see at 70 or 200 ft, respectively). By definition, a person is *legally blind* if the vision in the better eye is, after correction, 20/200 or poorer.

Since the results of acuity measures are dependent on the testing conditions, such as the test pattern and its distance from the eyes, the National Academy of Sciences (1980) recommended standards for acuity assessment. Among the conditions which affect acuity are:

Luminance level: for dark targets on light background, acuity improves as background luminance increases (up to about 150 cd/m^2; for measurement units, see below).

Locus of stimulation on the retina: highest resolution is obtained under photopic illumination with the target viewed foveally. At scotopic illumination, acuity is highest when the target is offset approximately 4 degrees from the fixation point.

Pupil size: highest acuity is observed when the pupil is at an intermediate diameter.

Viewing distance: for a perfect lens, there should be no effect of the distance if the visual angle is used to describe the conditions. However, acuity does change with viewing distance, because the lens of the eye changes its shape to fixate at different distances. If accommodation errors exist, the image may be blurred. Such accommodation errors are more pronounced at low luminance levels.

Age: spatial vision capabilities change considerably with age, generally leading to a decline in acuity starting in the 40s.

Visual acuity depends primarily on the ability to see edge differences between black and white stimuli, measured at rather high illuminance levels. Such measurement of static edge acuity is simple, but it is neither the only nor the best measure of visual resolution capabilities. For example, people with perfect Snellen acuity may not do so well in other measures of contrast sensitivity, such as the ability to detect targets from a busy background or to "see" highway signs at given distances. As one's gaze sweeps, the visual details in the view field generate an image of ever-changing spatial frequencies and contrasts on the retina. Thus, one can consider the visual world as a constantly changing array of textures composed of varying contrasts and spatial frequencies. Thus, contrast sensitivity changes with differing viewing fields: it is said to be "field dependent" (Fine and Kobrick, 1987). This field dependency is operationally defined by the tests that measure it. Many of these use a certain geometric shape, for example a triangle, which is embedded in a more complex geometric figure (Dismukes, 1980).

APPLICATION

Contrast. Complete measures of "contrast sensitivity" assess visual resolution through ranges of spatial frequency and contrast. These tests determine the thresholds for the detection of contrast fluctuations at various frequencies of alternations (Snyder, 1985). Contrast is defined as $(L_{max} - L_{min})(L_{max})^{-1}$ where L_{max} is the highest luminance and L_{min} the lowest luminance of the alternation pattern. It is customary to plot the reciprocal of the contrast threshold function, measured as modulation $M = (L_{max} - L_{min})(L_{max} + L_{min})^{-1}$, and to call this "contrast sensitivity."

Measurement of Light (Photometry)

The measurement of light energies and the perception of that energy by the human observer have been an area of much confusion. The confusion stems from several sources, particularly from the differences between physically described and humanly perceived units, and also from various competing terminologies. Boyd (1982) and Pokorny and Smith (1986) clarified the situation.

Radiometry. Light can be defined as any radiation capable of causing a visual sensation. The natural source of such electromagnetic radiation is the sun. Lamps are common artificial sources. Measurement of the quantity of radiant energy may be done by determining the rise in temperature of a blackened surface which absorbs radiation.

As shown in Table 4-4, there are four fundamental types of energy measurement:

- the total radiant energy emitted from a source, per unit of time (radiant flux);
- the energy emitted from a point in a given direction (radiant intensity);
- the energy arriving at (incident on) a surface at some distance from a source (irradiance);
- the energy emitted from or reflected by a unit area of a surface in a specified direction (radiance).

TABLE 4-4. TERMS, SYMBOLS, AND UNITS OF LIGHT ENERGY

Term	Symbol	Units
Radiant flux	P_e	watts (W)
Radiant intensity	I_e	$W \times sr^{-1}$
Irradiance	E_e	$W\ m^{-2}$
Radiance	L_e	$W\ m^{-2}\ sr^{-1}$

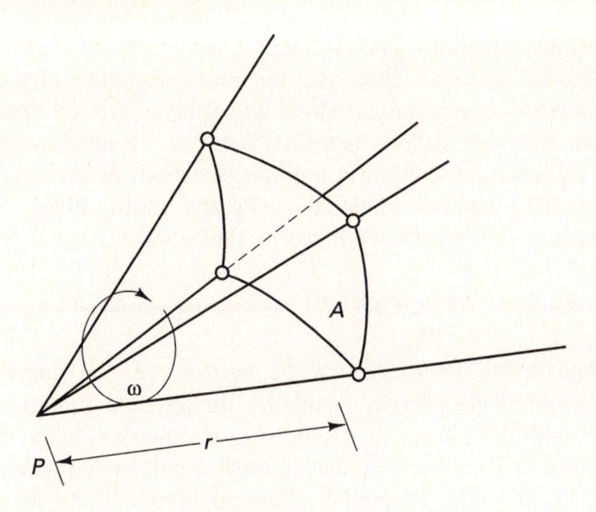

$A = \omega \cdot r^2$
$\omega = 1$ sr for $A = r^2$
ω for a full sphere:
$\omega(\text{sr}) = 4\pi r^2 \cdot r^{-2}$ (sr) $=$ $r\pi$ (sr)

Figure 4-10. Definition of the steradian (sr).

Lines radiating from a point p define a cone with a solid angle ω, with a spherical surface A at a radius r from p—see Figure 4-10. One steradian (sr) as the unit of ω is defined for the condition that $A = r^2$.

The energy emitted from point p is the radiant flux P_e in watts (W). With ω expressed in steradians, the *radiant intensity* per solid angle is $I_e = P_e\omega^{-1}$, in units of W sr^{-1}. Most artificial light sources do not radiate uniformly in all directions; hence, the radiant intensity may be different for different directions. *Irradiance* E_e (radiant flux incident on a surface) of a sphere is, per unit area of surface,

$$E_e = P_e\omega^{-1}r^{-2} \qquad \text{in W m}^{-2}$$

Irradiance can be expressed in terms of the radiant intensity:

$$E_e = I_e r^{-2} \qquad \text{in W m}^{-2}$$

The radiant flux emitted by a point source falls on successively greater areas as the distances from the source increases. Since irradiance E_e changes inversely with the square of the distance r from the source p, this relation is called the "inverse square law of energy flux."

Note that if the surface is flat (instead of being a section of a sphere around p) not all surface elements are at the same distance from the point source p. Thus, areas of the surface are irradiated at varying angles, rather than perpendicularly. If a flat surface section is at an angle from the normal, the irradiance at any point on the surface section is

$$E_e = I_e r^2 \cos \alpha$$

The preceding definitions, equations, and the inverse-square law are valid only for a true point-source light such as a star. However, for most calculations, the error made with artificial light sources is negligible. Most light sources are not "points" but have finite dimensions: they are called "extended sources." Radiance L_e describes the actual density of emitted radiation in a given direction, or arriving at a surface, taking into account the projection angle (Pokorny and Smith, 1986).

Symbols and units used in radiometry are listed in Table 4-4.

APPLICATION

Photometry. The optical conditions of the human eye, the sensory perception of the stimuli, and CNS processing modify the physical conditions described so far. To consider this, the Commission Internationale de l'Eclairage (CIE) developed in 1924 the "standard luminous efficiency function for photometry." In 1951, the CIE adopted a standard luminous efficiency function for dark-adapted (scotopic) vision. Both standards are still valid.

Accordingly, photometric energy is radiant energy modified by the luminous efficiency function of the "standard observer." Thus, a set of units parallel to those specifying radiant energy defines photometric energy. The photometric terms (and their corresponding radiometric terms) are presented in Table 4-5. There are several other terms still in use, several quite unnecessarily, such as "candle power" instead of luminous intensity.

TABLE 4-5. PHOTOMETRIC AND CORRESPONDING RADIOMETRIC TERMS

Radiometry	Photometry
Radiant flux P_e in W	Luminous flux F_v in lumens (lm)
Radiant intensity I_e in W sr^{-1}	Luminous intensity I_v in candela (cd) = lm sr^{-1}
Irradiance E_e in W m^{-2}	Illuminance (or illumination) E_v in lux (lx) = lm m^{-2}
Radiance L_e in W sr^{-1} m^{-2}	Luminance L_v in cd m^{-2} = lm sr^{-2} m^{-2}

SOURCE: Adapted from Pokorny and Smith, 1986.

Manufacturers of lamps often use the expression "lumens per watt" as a measure of lamp efficiency, since the measurement in watts describes only the electrical power intake of the light source. Typical values are: 1700 lm for 100-W incandescent light, and 3200 lm for 40-W fluorescent light.

☞☞☞ *Throughout the ages, every nation, sometimes every region within a nation, developed its own system of measurements. The differences between these were not important as long as trade and communications were slow and infrequent. In France, a commission was appointed in 1790 to develop a system of reasonable measurements. Members of the commission were such great scientists as Laplace, Lagrange, and Lavoisier. The commission generated a system based on basic natural units, such as the meter (from the Greek, to measure), which was equal to 10^{-8} the distance from the North Pole to the Equator.*

As far as possible, other units were developed which interconnected with the meter. Larger and smaller units were derived by multiplying or dividing by 10. This metric system, so far the most useful and logical system of measurements, was slowly accepted throughout the world, with the United States one of the last holdouts against it (Asimov, 1989).

Among the nonmetric units still in use are the following:

Illuminance, *the amount of light falling (incident) on a surface, is occasionally measured in*

$$\text{footcandle (fc)} = 1 \text{ lm ft}^{-2}; \quad 10.76 \text{ fc} = 1 \text{ lx}$$

$$\text{phot} = 1 \text{ lm cm}^{-2}; \quad 10^5 \text{ phot} = 1 \text{ lx}$$

Troland, *the illuminance on the retina from a source with 1 = cd* m^{-2} *luminance, viewed through an artificial pupil of 1* cm^{-2} *size.*

Luminance, *the amount of light energy reflected (or emitted) from a surface, is occasionally measured in*

$$\text{lambert (L)} = \text{cd cm}^{-2}\ \pi^{-1}; \quad 3.183 \text{ L} = \text{cd m}^{-2}$$

$$\text{footlambert (fL)} = \text{cd ft}^{-2}\ \pi^{-1}; \quad 3.426 \text{ L} = 1 \text{ cd m}^{-2}$$

$$\text{stilb} = 10^{-5} \text{ cd cm}^{-2}; \quad 10^5 \text{ stilb} = 1 \text{ cd m}^{-2}$$

$$\text{apostilb} = 1 \text{ cd m}^{-2}\ \pi^{-1}; \quad 0.3183 \text{ apostilb} = 1 \text{ cd m}^{-2}$$

$$\text{nit} = 1 \text{ cd m}^{-2}\ (= 10^5 \text{ stilb})$$

Color Perception

Sunlight contains all visible wavelengths, but objects onto which the sun shines absorb some radiation. Thus, the light that objects transmit or reflect has an energy distribution different from that received. A human looking at the (transmitting or reflecting) object does not analyze the spectral composition of the light reaching the eyes; in fact, what appears to be of identical color may have different spectral contents. The brain simply "classifies" incoming signals from different groups of wavelengths and labels them "colors" by experience. Human color perception, then, is a psychological experience, not a single specific property of the electromagnetic energy which we see as light.

"[W]e judge colors by the company they keep. We compare them to one another, and revise according to the time of day, light source memory." (Ackerman, 1990, p. 152) "Not all languages name all colors. Japanese only recently included a word for 'blue.' . . . Primitive languages first develop words for black and white, then add red, then yellow and green; many lump blue and green together, and some don't bother distinguishing between other colors of the spectrum. . . . The Maori of New Zealand . . . have many words for red—all the reds that surge and pale as fruits and flowers develop, as blood flows and dries." (Ackerman, 1990, p. 253)

Depending largely on the distribution of cones (and rods) over the retina, not all its areas are equally sensitive to all colors. We see all colors while looking fairly straight ahead, but cannot perceive any colors at the very periphery of our visual field. Green, red, and yellow all can be perceived within an angle of about 50 degrees to the side from straight ahead, while blue can be seen to about 65 degrees sideways and white even at 90 degrees.

Colorimetry provides a system of color measurement and specification based on the concept of equivalent-appearing stimuli (Pokorny and Smith, 1986). Colorimetry is an experimental technique in which one simultaneously views two spectral fields and tries to adjust the spectral content of one to make the two fields appear identical.

Color-matching experiments have shown that the human perceives the same color if one variously mixes three independent adjustable primary colors, i.e., red, green, and blue. (This "additive" combination of spectral radiations is not the same as mixing pigments, discussed below.) Lights that contain dissimilar spectral radiations but are nevertheless perceived as the same color by the observer are called metameric. Since it is possible to find a metamer for any color by varying only the three primary colors, human color vision is called trichromatic. Two colors that together appear white are called complementary. For example, when yellow and blue lights in proper wavelength proportions are together projected on a screen, the mixture appears white.

When an artist mixes yellow and blue pigments to generate green, this is done by "double subtraction." The yellow pigment absorbs all blue, but reflects yellow as well as red and green. The blue pigment absorbs yellow and red, but reflects blue and green. Both pigments mixed together reflect only green, since the blue pigment absorbs the yellow and the yellow pigment the blue. However, if light of a different wavelength were to fall on the pigments, they would each absorb different "colors" and, combined, reflect a different wavelength which might not appear green.

We see an object as white when it reflects light about equally throughout all wavelengths; but so do gray and black pigments. The difference between black and white is not one of color but of how much light they reflect. Freshly fallen snow reflects only about 75 percent of the sunlight falling on it, but it appears brightly white to us. Black velvet appears very dark black because it reflects only very little of the light that falls on it. Figure 4-11 shows, schematically, how pigments reflect light of different wavelengths, i.e., colors.

Returning to trichromaticity: it is characterized by the two facts that

- any color can be produced by combining (adding together) a suitable mixture of three arbitrarily selected spectral radiations (primaries), and
- color specifications (wavelength components) can be stated precisely, making it possible to compare colors. Thus, the data of a color-matching experiment can be expressed in terms of vectors in a three-dimensional space.

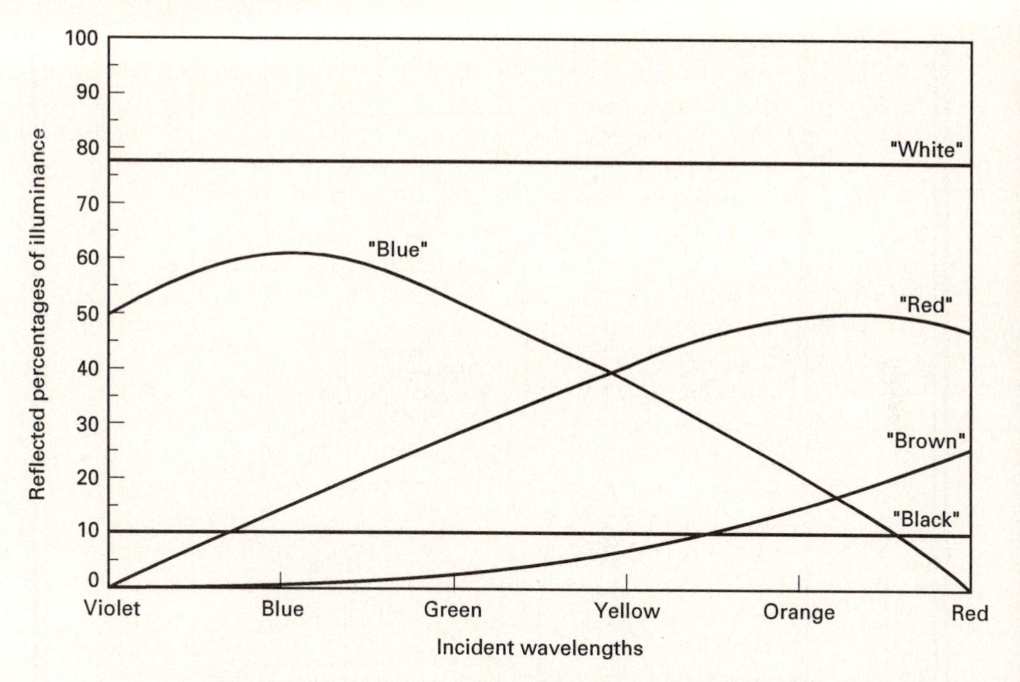

Figure 4-11. Different colors reflected by different pigments (schematic).

The results can be shown in a color plane, or *chromaticity diagram,* which is a two-dimensional representation of the results of color mixtures. This is accomplished by converting the values of the three stimuli into a form where the sum of the three equals unity. By convention X corresponds to the red region of the spectrum, Y to the green region, Z to the blue region: $X + Y + Z = 1$. If the proportions of red and green are plotted along the abscissa and ordinate, respectively, the results of the combined wavelength, the third trichromatic coefficient, fall within a horseshoe-shaped curve called the spectrum locus. This is shown in Figure 4-12.

The chromaticity diagram plots colors by the amounts of standard red, blue, and green primaries that match them. This chart, standardized in 1931 by the CIE, uses red, green, and blue lights which are mathematically specified and called the three primaries. In this diagram, the sum of red, blue, and green required to match a given color is expressed as unity. This means that one standard color component can be determined by subtracting the other two from unity. It is common practice to express the percentages of red and green on the abscissa and ordinate, respectively, and to subtract their sum from unity to determine the amount of blue.

Since the spectrum colors are the most saturated, they are located on the horseshoe curve. Other perceived colors fall inside this boundary. White is located in the center of the diagram, composed of equal amounts of red, green, and blue. This white, consisting of one-third red, one-third green, and one-

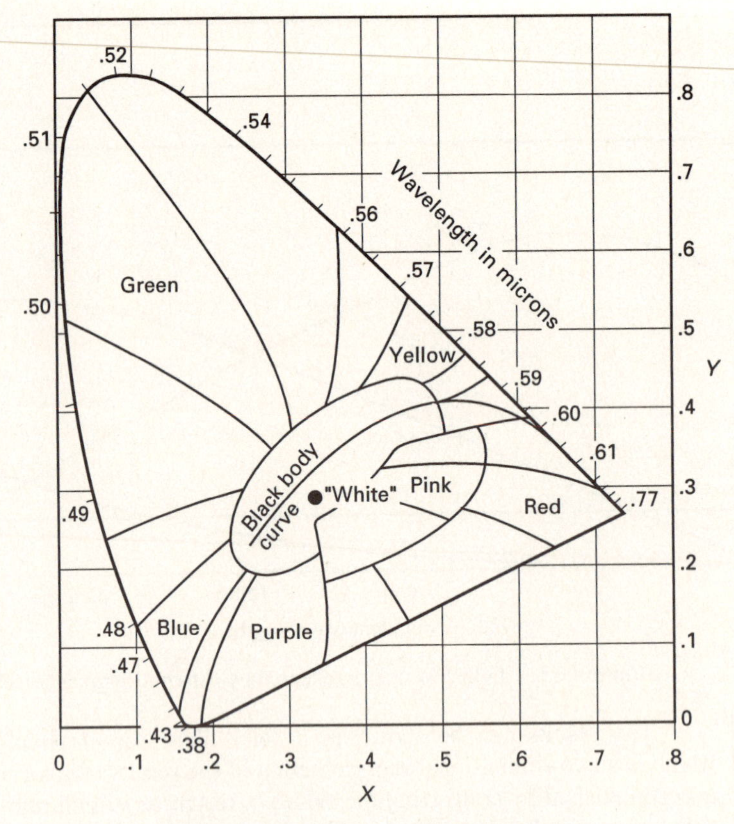

Figure 4-12. Trichromaticity diagram.

third blue, is designated as "illuminant C" by the CIE: it is about the color of light from a northern summer noon sun on a clear day.

The numbers along the curve are wavelengths. A color appears greenish in the upper corner, reddish in the lower right corner, and bluish in the lower left corner. The "purity" of each color is determined by its proximity to the center of the diagram, where it would appear white.

Although the CIE diagram is six decades old, it is still in common use (Sayer, Sebok, and Snyder, 1990).

Theories of Color Vision

The above described (Young-Helmholtz) trichromatic theory has been opposed by Hering's *opponent colors* theory and by Judd's *zone* or *stage* theories. (Other theories are still being developed and tested, as discussed by Pokorny and Smith, 1986.) The complex nature of color appearance is not yet fully understood, and the mathematical models available are not entirely satisfactory. A human's judgment of the perceived color appearance of a visual stimulus depends on subjective impressions experienced when viewing the stimulus, and the judgment varies with the viewing conditions and the kind of stimuli.

Color Ordering Systems

Many systems to order colors according to various variables and criteria are possible; several have been well developed and are often used. Wyszecki (1986) provides a complete overview, from which only a few main points are repeated here.

There are three major groups of color systems. The first group is based on principles of additive mixtures of color. Examples are the Maxwell disk and the Ostwald Color System. The second group uses the principles of color subtraction, such as in pigment mixing. This is widely used in printing. The third group is based on the appearance and perception of colors. Material standards are selected to present scales of, for example, hue, chroma, and likeness. Examples are the Munsel Color System and the OSA system (Optical Society of America).

Terminology

The following terms are often used, although neither universally, uniformly, nor in a standardized manner (Wyszecki, 1986).

BRIGHTNESS and DIMNESS are individual perceptions of the intensity of a visual stimulus.

CHROMA is the attribute of color perception attained by judging to what degree a chromatic color differs from an achromatic color of the same lightness. (The 1942 Munsell Book of Color provides many good examples.) See: saturation.

CHROMATIC or ACHROMATIC INDUCTION is a visual process that occurs when two color stimuli are viewed side by side, when each stimulus alters the color perception of the other. The effect of chromatic or achromatic induction is usually called simultaneous contrast, or spatial contrast.

HUE is an attribute of color perception that uses color names and combinations thereof, such as yellow or yellowish green. The four *unique* (or *unitary*) hues are red, green, yellow, and blue, neither of which is judged to contain any of the others.

ILLUMINANT COLOR is that color perceived as belonging to an area that emits light as primary source.

LIGHTNESS is the individual perception of how much more or less light a stimulus emits in comparison to a "white" stimulus also contained in the field of view.

OBJECT COLOR is that color perceived as belonging to an object.

RELATED COLOR is a color seen in direct relation to other colors in the field of view.

SATURATION is the attribute of color judgment regarding the degree to which a chromatic color differs from an achromatic color regardless of lightness. See: chroma.

SURFACE COLOR is that color perceived as belonging to a surface for which the light appears to be reflected or radiated.

UNRELATED COLOR is a color perceived to belong to an area seen in isolation from other colors.

All of these terms rely on perception and judgment. Occasionally, some terms are inappropriately used instead of physically determined concepts: for example, (physical) luminance is not the same as (psychological) brightness. Thus, confusion is rampant.

APPLICATION

Esthetics and Psychology of Color

While the physics of color stimuli arriving at the eye can be well described (although often with considerable effort), perception, interpretation, and reaction to colors are highly individual, nonstandardized, and variable. Thus, people find it very difficult to describe colors verbally, given the many possible combinations of individually perceived hue, lightness, and saturation.

People believe in and describe nonvisual reactions to color stimuli. For example, reds, oranges, and yellows are usually considered "warm" and stimulating. Violets, blues, and greens are often felt to be "cool" and to generate sensations of cleanliness and restfulness. Pale colors often seem cooler than dark colors, cold colors more distant than warm colors, weak colors more distant than intense colors, soft edges of color patches more distant than hard edges, etc. Although experimental evidence on these effects is controversial (Kwallek and Lewis, 1990), color schemes are often applied to work and living areas to achieve these stereotypical responses.

ILLUMINATION CONCEPTS IN ENGINEERING AND DESIGN

APPLICATION

The characteristics of human vision discussed above provide the bases for engineering procedures in the design of the environment for proper vision. The most important concepts are:

1. Proper vision requires sufficient quantity and quality of illumination.

2. Special requirements on visibility (and particularly the decreased seeing abilities of the elderly) require care in the arrangement of proper illumination.
3. Illuminance of an object is inversely proportional to the distance from the light source.
4. Use of colors, if selected properly, can be helpful; but color vision requires sufficient light.
5. What mostly counts is the luminance of an object, that is, the energy reflected or emitted from it, which meets the eye.
6. Luminance of an object is determined by its incident illuminance, and by its reflectance:

$$\text{luminance} = \text{illuminance} \times \text{reflectance} \times \pi^{-1}$$

Reflectance is the ratio of reflected light to received light, in percent. (The numerical value for illuminance is in lux and for luminance in cd m^{-2}—see Table 4.5. The factor π^{-1} is omitted when the following nonmetric units are used: luminance in footlambert fL, illuminance in footcandle fc.) Figure 4-13 shows luminance levels experienced by humans.
7. The ability to see an object is much influenced by the luminance contrast between the object and its background, including shadows. Contrast is usually defined as the difference in luminances of adjacent surfaces, divided by the larger luminance, in percent:

$$\text{Contrast (in percent)} = (L_{max} - L_{min})(L_{max})^{-1} \times 100$$

8. Avoid unwanted or excessive glare. There are two types of glare. *Direct glare* meets the eye directly from a light source (such as the headlights of an oncoming car); *indirect glare* is reflected from a surface into the eyes (such as headlights of the following car in the rearview mirror). Often, kinds of reflected glare are described as either "specular" (coming from a smooth polished surface such as a mirror), "spread" (coming from a brushed, etched, or pebbled surface), "diffuse" (coming from a mat, flatly painted surface), or "compound" (a mixture of the glare types).

Direct glare can be avoided by:

(a) Placing high-intensity light sources outside the cone of 60 degrees around the line of sight.
(b) Using several low-intensity light sources instead of one intense source, placed away from the line of sight.
(c) Using indirect lighting, where all light is reflected at a suitable surface (within the luminaire, or at the ceiling or walls of a room) before it reaches the work area. This generates an even illuminance without shadows. (But shadows may be desirable to see objects better.)
(d) Using shields or hoods over reflecting surfaces, or visors over a person's eyes, to keep out the rays from light sources.

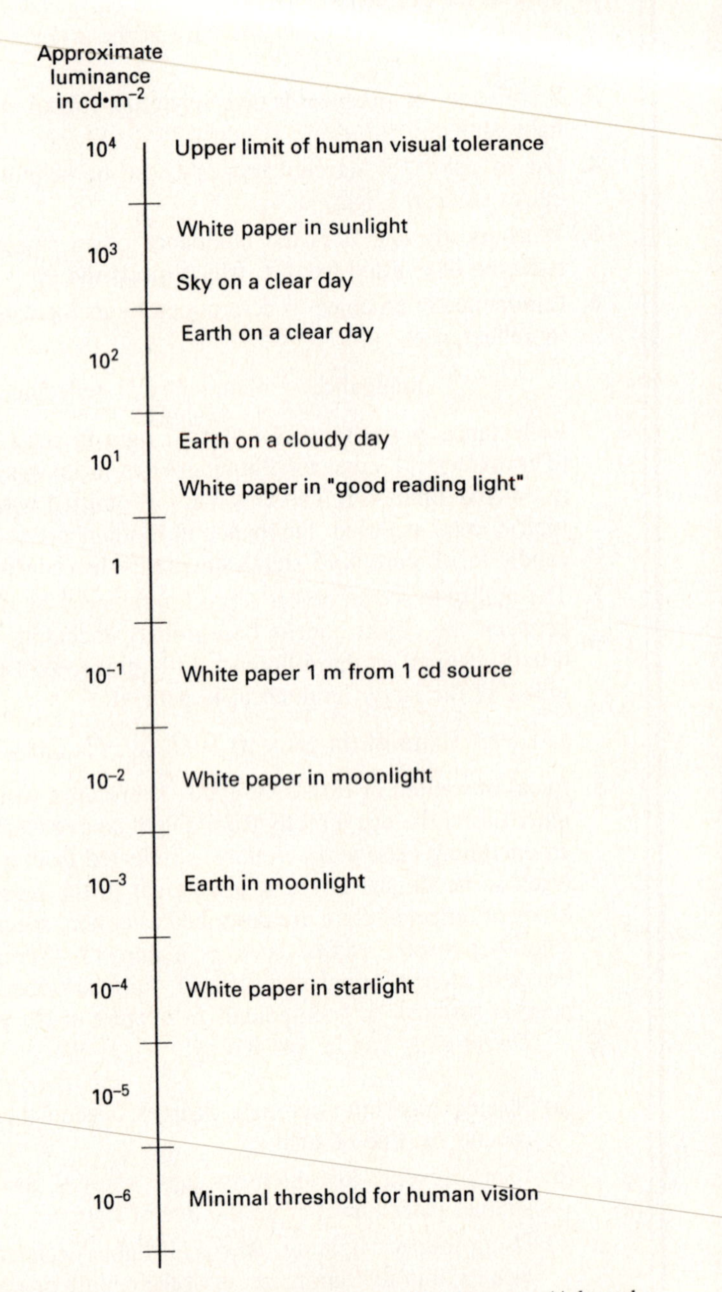

Figure 4-13. Luminance levels experienced by humans. (Adapted from Van Cott and Kinkade, 1972.)

Indirect glare can be reduced by:

(e) Diffuse, indirect lighting.
(f) Dull, matte, or other nonpolished surfaces.
(g) Properly distributed light over the work area.

9. Use of *direct lighting* (when rays from the source fall directly on the work area) is most efficient in terms of illuminance gain per unit of electrical power; but it can produce high glare, poor contrast, and deep shadows. The other way is to use *indirect lighting,* where the rays from the light sources are reflected and diffused at some suitable surface before they reach the work area. This helps to provide an even illumination without shadows or glare, but is less efficient in terms of use of electrical power. A third way also uses diffuse lighting, but the light source is enclosed by a large translucent bowl so that the room lighting is emitted from a large surface. This can cause some glare and shadows, but is usually more efficient in use of electrical power than indirect lighting.

What is best suited depends on the given conditions, such as the task to be done, objects to be seen, and eyes to be accommodated. Thus, general recommendations are difficult to compile. Nevertheless, the *IES Lighting Handbook* (Kaufman and Haynes, 1981) contains recommended illuminance values for certain application categories. An excerpt from these recommendations is provided in Table 4-6. Before they are applied, it should be remembered that given reflectances might require illumination with either the low values (with high reflectance) or the high values (with low reflectance).

THE HEARING SENSE

Acoustics is the science and technology of sound, including its production, transmission, and effects. The acoustical design goal is to establish an environment that

- transmits desired sounds reliably and pleasantly to the hearer.
- is satisfactory to the human with respect to noise.
- minimizes sound-related annoyance and stress.
- minimizes disruption of speech communications.
- prevents hearing loss.

Sound is any vibration (passage of zones of compression and rarefaction, through the air or any other physical medium) which stimulates an auditory sensa-

TABLE 4-6. RECOMMENDED ILLUMINANCE VALUES (EXAMPLES)

Illuminance categories and illuminance values for generic types of activities in interiors

Type of activity	Illuminance category	Ranges of illuminances: Lux	Ranges of illuminances: Footcandles	Reference work-plane
Public spaces with dark surroundings	A	20 to 50	2 to 5	General lighting throughout space
Simple orientation for short temporary visits	B	50 to 100	5 to 10	
Working spaces where visual tasks are occasionally performed	C	100 to 200	10 to 20	
Performance of visual tasks of high contrast or large size	D	200 to 500	20 to 50	Illuminance on task
Performance of visual tasks of low contrast or very small size	E	500 to 1,000	50 to 100	
Performance of visual tasks of low contrast and very small size over a prolonged period	F	1,000 to 2,000	100 to 200	
Performance of very prolonged and exacting visual tasks	H	5,000 to 10,000	500 to 1,000	Illuminance on task, obtained by a combination of general and local (supplementary lighting)
Performance of very special visual tasks of extremely low contrast and small size	I	10,000 to 20,000	1,000 to 2,000	

TABLE 4-6. (continued)

Examples

Area/Activity	Illuminance category	Area/Activity	Illuminance category
Auditorium		Barber shop, beauty parlor	E
Assembly	C	Club and lodge room	
Social activity	B	Lounge and reading	D
Bank		Conference room	
Lobby		Conferring	D
General	C	Critical seeing (refer to individual task)	
Writing area	D	Court room	
Teller station	E	Seating area	C
		Court activity area	E
		Dance hall, discotheque	B
		Booking binding	
Assembly Work		Folding, assembling, pasting	D
Simple	D	Cutting, punching, stitching	E
Moderately difficult	E	Embossing and inspection	F
Difficult	F	Brewery	
Very difficult	G	Brew house	D
Exacting	H	Boiling and key washing	D
Bakery		Filling (bottles, cans, kegs)	D
Mixing room	D	Candy making	
Make-up room	D	Box department	D
Oven room	D	Chocolate department	
Decorating and icing		Husking, winnowing, fat extraction, crushing and refining, feeding	D
Mechanical	D		
Hand	E	Bean cleaning, sorting, dipping, packaging, wrapping	D
Scales and thermometers	D		
Wrapping	D	Milling	E

SOURCE: Adapted from Kaufman and Haynes, 1981.

tion. Commonly, one is concerned with sound that arrives at the ear by air. A sketch of the ear is provided in Figure 4-14.

Ear Anatomy and Hearing

Airborne sound waves arriving from the outside are collected by the outer ear (auricle and pinna) and channeled along the auditory canal to the eardrum (tympanic membrane), which vibrates according to the frequency and intensity of the arriving sound wave. Resonance effects of the pinna and ear canal have amplified the sound intensity (by 10 to 15 dB; see below for an explanation of the units) when it reaches the eardrum.

In the *middle ear,* the sound arriving through the eardrum is mechanically transmitted by the ear bones (ossicles)—hammer (malleus), anvil (incus), and stir-

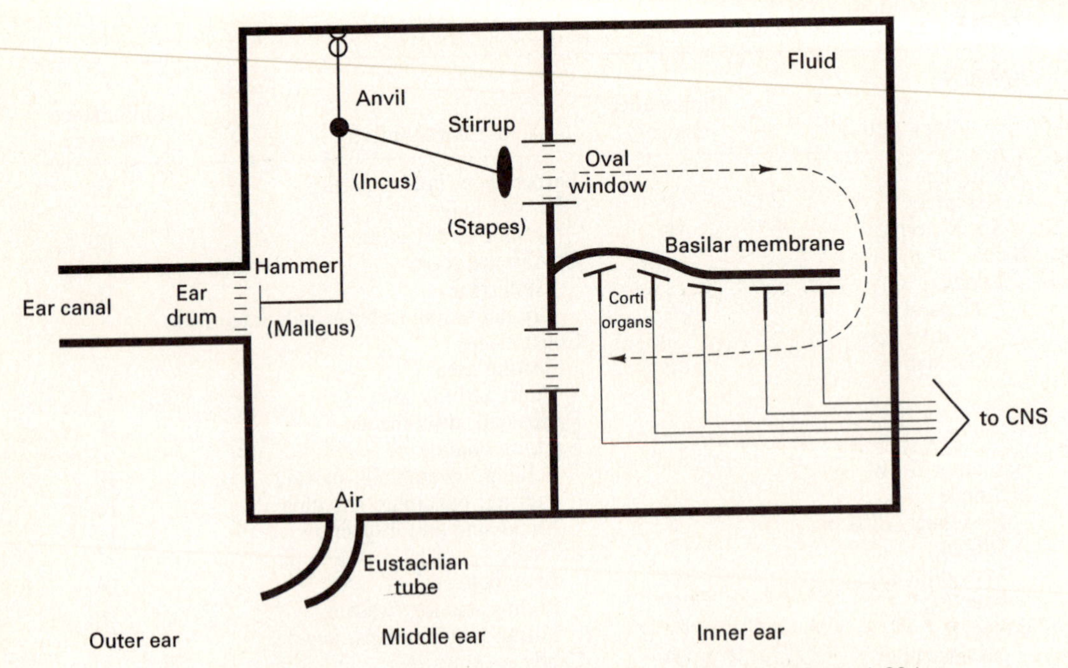

Figure 4-14. Schematic of the human ear. (See also Figure 4-23.)

rup (stapes)—to the oval window. Via lever action, this mechanical structure increases the intensity of the sound from the eardrum to the oval window; also, the area of the eardrum is about 17 times larger than the surface of the oval window, which increases the effective sound pressure (the total gain is about 25 dB). The middle ear is filled with air, and the Eustachian tube connects it with the pharynx. This tube allows the air pressure in the middle ear to become equal with external pressure.

The *inner ear* is filled with fluid (endolymph). In it, sound waves are propagated as fluid shifts from the oval window through the cochlea (which makes about two and one-half turns) along the basilar membrane to the round window. The motion of the fluid deflects the basilar membrane, which stimulates sensory hair cells (cilia) in the *organs of Corti* on the basilar membrane. Depending on their structure and location, the Corti organs respond to specific frequencies. The transformed impulses are transmitted along the auditory (cochlear) nerve to the brain.

A *tone* is single-frequency oscillation, while a sound usually contains a mixture of frequencies. Their frequencies (frequency distributions) are measured in hertz (Hz), their intensities (amplitude, sound pressure levels) in logarithmic units known as decibels (dB; see below). One reason for use of a logarithmic scale is that the human perceives sound pressure amplitudes in a roughly logarithmic manner.

APPLICATION

The Human Hearing Range

Infants can hear tones of about 16 to 20,000 Hz; old people can rarely hear frequencies above 12 kHz. The minimal pressure threshold of hearing is about 20×10^{-6} N m^{-2} (or 20 micropascal) in the frequency range of 1000 to 5000 Hz. The ear experiences pain when the sound pressure reaches about 200 Pa—see Figure 4-15.

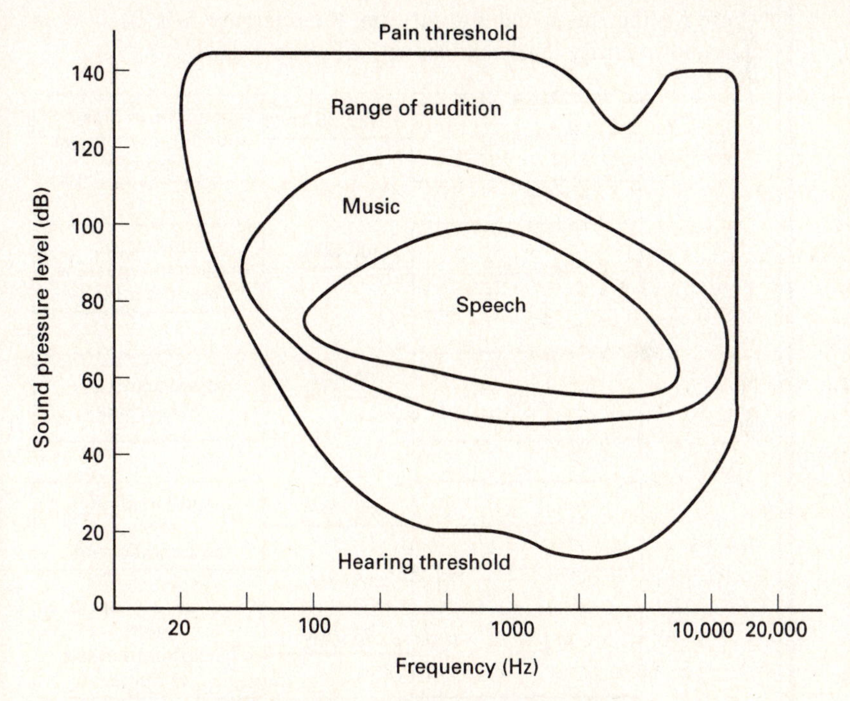

Figure 4-15. Ranges of human hearing.

The sound pressure level is defined as a ratio between two sound *pressures*; P_0, the threshold of hearing, is used as reference. Thus, the definition of sound pressure level, *SPL*, is

$$SPL = 10 \log(P^2 P_0^{-2})$$

or

$$SPL = 20 \log_{10}(P P_0^{-1}) \qquad \text{in dB (decibel)}$$

where P is the root-mean-square (rms) sound pressure for the existing sound.

With these values, the "dynamic range" of human hearing from 20×10^{-6} to 200 Pa is

$$20 \log_{10}\left(\frac{200}{20 \times 10^{-6}}\right) = 140 \text{ dB}$$

In terms of sound *intensity* ("power"), the sound intensity level *SIL* is similarly defined as

$$SIL = 10 \log_{10}(I^2 I_0^{-2})$$

where I is the rms sound intensity and the reference I_0 is 10^{-12} W m^{-2}. Ranges of sound intensity levels are shown in Figure 4-16.

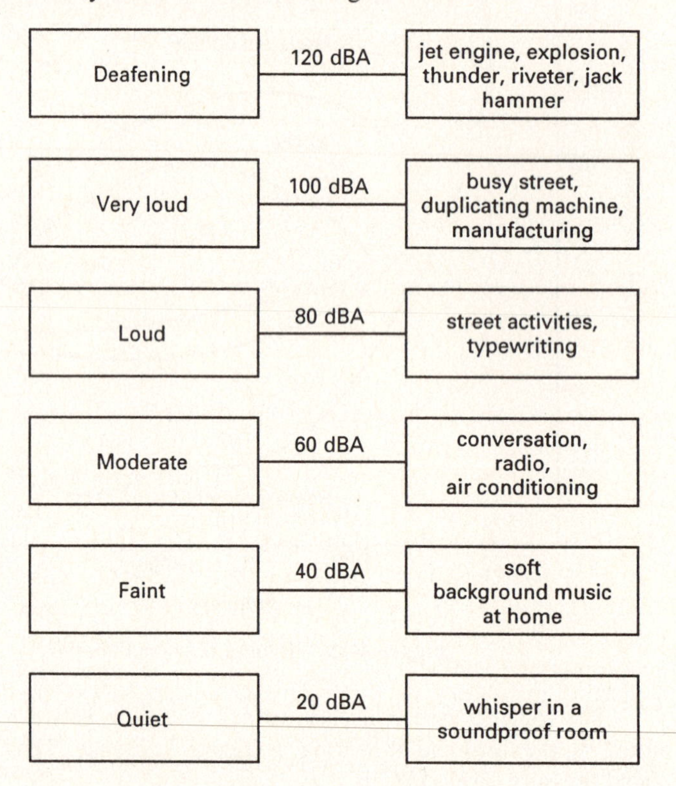

Figure 4-16. Sound intensity levels.

Doubling the *SPL* causes an increase of 6 dB; a 1.41-fold increase in sound pressure causes an increase of 3 dB, which means doubling the sound energy. If two sounds (of frequencies and temporal characteristics that are both random) occur at the same time, their combined *SPL* can be calculated from $SPL_{\text{combined}} = 10^{SPL_1/10} + 10^{SPL_2/10} + \cdots + 10^{SPL_n/10}$. Accordingly, one can approximate the combined *SPL* from the difference d in the intensities as follows: if d is 0 dB, add 3 dB to the louder sound; for d of 1 dB, add 2.5 dB; for

2, add 2; for 4, add 1.5; for 6, add 1; for 8, add 0.5; for d of 10 or more, take the louder sound by itself.

Sound can reach the ear on two different paths. Airborne sound travels through the ear canal and excites the eardrum and the structures behind it, as described above. Sound may also be transmitted through bony structures to the head and ear, but this requires 30 to 50 dB higher intensities (depending on the existing frequencies) to be similarly effective. The propagation through air or structures depends on the intensity and frequency of the sounds, and on the transmission characteristics of the medium. Moving from the common terrestrial gravity and atmosphere to microgravity and no-atmosphere conditions in space, or to immersion in water instead of air, changes the sound transmission characteristics. The velocity V of sound in solids, liquids, and gases is determined by the modulus of elasticity of the medium E and its density D according to $V = (E/D)^{-1/2}$ with V in m s^{-1}, E in N m^{-2}, D in kg m^{-3}.

Human Responses to Sound

Music has long accompanied activities: singing during field work, marching to rhythmical music. Effects of music on industrial work were observed early in the twentieth century. Yet, even today little is known systematically about the psychophysical consequences of different kinds of music and rhythm and their effects on "productivity" over extended periods of time.

Fox (1983) distinguished between two different kinds of music: background and industrial. Background music is like "acoustical wallpaper" in shops, hotels, waiting rooms, etc. This is meant to create a welcoming atmosphere, to relax customers, to reduce boredom, and to cover other disturbing sounds. Its character is subdued, its tempo intermediate, and vocals are avoided. It may produce a monotonous environment for those continuously exposed to it, while it may appear pleasant to the transient customer.

"Music while you work" is in many respects the opposite of background music. It is not continuous but programmed to appear at certain times, it has varying rhythms with vocals, and it may contain popular fashionable "hits." It is meant to break up monotony, to generate mild excitement and an emotional impetus to demanding physical effort or drudgery in a dreary impersonal environment. While improved morale and productivity have been reported in many circumstances, a clear linking of the underlying arousal theory and the specific components of the music is difficult (Fox, 1983). Thus, the musical content, the rhythm, the loudness, the temporal presentation, and their selection for certain activities, environments, and listener populations are still a matter of art rather than science.

Noise

☞☞☞ Noise is any acoustic phenomenon that annoys the listener. Thus, noise is psychological and subjective. Single, short tones of low intensity may be considered noise under certain conditions, just as loud, lasting, complex sounds may be under other circumstances.

Noise is defined as unwanted, objectionable, or unacceptable sound. Any sound may be annoying and thus be conceived as noise. The threshold for noise annoyance varies depending on the conditions, including sensitivity and mental state, of the individual. Noise can create feelings of negative emotions, of surprise, frustration, anger, and fear. Noise can delay the onset of sleep, awaken a person from sleep, disturb rest, and interfere with the hearing of wanted sounds. Noise can interfere with some human sensory and perceptual capabilities and thereby degrade task performance. It may also produce temporary or permanent alterations in body chemistry. Noise can temporarily or permanently change hearing capability.

Physiological Effects of Sound

Exposure to intense sounds may result in a temporary threshold shift (TTS) from which the hearing eventually returns to normal with time away from the source; or it can cause a permanent threshold shift (PTS), which is an irrecoverable loss of hearing. A PTS may be the result of damage to the cochlear cilia, the organs of Corti at the basilar membrane in the inner ear, or the nerves leading to the CNS. Which of these are damaged depends on the frequency and, of course, intensity of the incident noise. The damage is probably due to an overstimulation of cell metabolism, leading to oxygen depletion and destruction, or to exceeding the elastic capacities of the physical structures. The metabolic activity existing at the moment of noise arrival is important: heat, heavy work, infectious disease, or other causes of heightened metabolism increase the vulnerability of the sensory organs. Above about 130 dB, the induced turbulence in the ear may also do mechanical damage (Stekelenburg, 1982). Timing and severity of the loss are dependent upon the duration of exposure, the physical characteristics of the sound (intensity, frequency) and the nature of the exposure (continuous or intermittent). Damage may be immediate, such as by an explosion, or may occur over some time, such as with continuous exposure to noise. The effect of continuous exposure is usually insidious and cumulative. Table 4-7 lists physiological effects of sounds.

Other physiological effects of intense sound have been observed (NASA, 1989; Stekelenburg, 1982). Among these are:

- High-intensity sounds induce muscles of the middle ear to contract, mostly affecting the stapes, which reduces force transmission to the cochlea. However, it takes about 30 milliseconds after the onset of the sound to activate this protective reflex, and about 200 ms for complete contraction to occur; and it lasts for less than a second. Thus, the contraction response may be too late or too short to provide sufficient protection to the inner ear.

TABLE 4-7. PHYSIOLOGICAL EFFECTS OF NOISE

Condition of exposure			
SPL (dB)	Frequency spectrum	Duration	Reported disturbances
175	Low frequency	Blast	Tympanic membrane rupture
167	2,000 Hz	5 min	Human lethality
161	2,000 Hz	45 min	Human lethality
160	3 Hz		Pain in the ears
155	2,000 Hz	Continuous	Tympanic membrane rupture
150	1–100 Hz	2 min	Reduced visual acuity; chest wall vibrations; gagging sensations; respiratory rhythm changes.
120–150	OASPL		Mechanical vibrations of body felt; disturbing sensations
120–150	1.6 to 4.4 Hz	Continuous	Vertigo and occasionally disorientation, nausea, and vomiting
135	20–2,000 Hz		Pain in the ears
120	OASPL		Irritability and fatigue
120	300–9,600 Hz	2 sec	Discomfort in the ear
110	20 to 31.5 kHz		TTS occurs
106	4,000 Hz	4 min	TTS of 10 dB
100	4,000 Hz	7 min	TTS of 10 dB
100		Sudden onset	Reflex response of tensing, grimacing, covering the ears, and urge to avoid or escape
94	4,000 Hz	15 min	TTS of 10 dB
75	8 to 16 kHz		TTS occurs
65	Broadband	60 days	TTS occurs

SPL—Sound Pressure Level re 20 μPa

TTS—Temporary Threshold Shift

OASPL—Overall Sound Pressure Level

SOURCE: Adapted from NASA, 1989.

- Noise exposure causes an increase in the concentration of corticosteroids in the blood and brain and affects the size of the adrenal cortex. Continued exposure is also correlated with changes in the liver and kidneys and with the production of gastrointestinal ulcers.
- Electrolytic imbalances and changes in blood glucose levels have been associated with noise exposure. It is possible that sex hormone secretion and thyroid activity are also affected by noise. Changes in cardiac muscle, fluctuation in blood pressure, and vasoconstriction have been reported with 70 dB SPL and above, becoming progressively worse with higher exposure.
- Abnormal heart rhythms have been associated with noise exposure.

☞☞☞ *High-intensity sound, whether one considers it "pleasant but loud" or "noise," can permanently damage the hearing ability.* ☜☜☜

APPLICATION

Effects of Noise on Human Performance

Observed performance effects of noise are (NASA, 1989):

- As noise intensity increases, increased arousal can cause an improvement in task performance; beyond a certain level of intensity, however, task performance degrades.
- Sudden unexpected noise can produce a "startle" response which interrupts concentration and physical task performance.
- Continuous periodic or aperiodic noise reduces performance on complex tasks such as visual tracking; the decrement increases with increasing noise levels.
- Psychological effects of noise may include anxiety, helplessness, narrowed attention, and other adverse effects that degrade task performance.

Other effects are listed in Table 4-8.

TABLE 4-8. EFFECTS OF NOISE ON HUMAN PERFORMANCE

Conditions of exposure			
SPL (dB)	Spectrum	Duration	Performance effects
155		8 hr; 100 impulses	Hearing TTS 2 minutes after exposure
120	Broadband		Reduced ability to balance on a thin rail
110	Machinery noise	8 hrs	Chronic fatigue
105	Aircraft engine noise		Visual acuity, stereoscopic acuity, near-point accommodation, all reduced
100	Speech		Overloading of hearing due to loud speech
90	Broadband	Continuous	Vigilance decrement; altered thought processes; interference with mental work
90	Broadband		Performance degradation in multiple-choice, serial-reaction tasks
85	1/3-Octave at 16 kHz	Continuous	Fatigue, nausea, headache
75	Background noise in spacecraft	10–30 days	Degraded astronaut performance
70	4,000 Hz		Hearing TTS 2 minutes after exposure

See Table 4-11 for effects on person-to-person voice communication.

SOURCE: Adapted from NASA, 1989.

Avoiding Noise-Induced Hearing Loss

Sounds of sufficient intensity and duration result in temporary or permanent hearing loss. Permanent hearing loss may be from mild to profound and, if related to neural structure, is with current knowledge not treatable.

There are two types of injuries. Short-duration sound of high intensity, such as produced by a cannon or an explosion, can damage any or all of the structures of the ear, in particular the hair cells in the organs of Corti, which may be torn apart. This results in immediate, severe, and permanent hearing loss.

More moderate exposure may initially cause only short-term hearing loss, measured as temporary threshold shift (TTS). During quiet periods, hearing returns to its normal level. Yet, TTS may include subtle mechanical intracellular changes in the sensory hair cells and swelling of the auditory nerve endings. Other potentially irreversible effects include vascular changes, metabolic exhaustion, and chemical changes within the hair cells.

Repeated exposures to sounds that cause TTS may gradually bring about a permanent threshold shift (PTS), i.e., a noise-induced hearing loss (NIHL). Experiments with animals have shown that with each exposure, cochlear blood flow may be impaired, and some hair cells may be damaged. Often, the damage is confined to a special area on the cilia bed on the cochlea, related to the frequency of the sound. With continued noise exposure more hair cells are damaged, which the body cannot replace; also, nerve fibers to that region in the ear are degenerated, accompanied by corresponding impairment within the central nervous system.

Impairment of hearing ability at special frequency ranges indicates noise exposure at these frequencies. In western countries, NIHL usually occurs initially in the range of 3,000 to 6,000 Hz, particularly at about 4,000 Hz, then through high frequencies, culminating around 8,000 Hz. Yet, reduced hearing near 8000 Hz is also an effect characteristic of aging, which makes it often difficult to distinguish between environmental and age-related causes. With continued noise exposure, NIHL increases in magnitude and extends to lower and higher frequencies. NIHL increases most rapidly in the first years of exposure; after many years, it levels off in the high frequencies, but continues to worsen in the low frequencies (National Institutes of Health, 1990).

Measurement of hearing loss usually includes measures of the auditory thresholds (sensitivity) at various frequencies. Such pure-tone audiometry is often combined with measures of speech understanding. Difficulty in understanding speech with NIHL is mostly associated with difficulties of differentiating speech sounds in the high frequency ranges, especially of high-frequency, low-intensity consonants. Thus, with NIHL, important information content of speech is often unclear, unusable, or inaudible. Also, other sounds such as background noise, competing voices, or reverberation may interfere with the listener's ability to receive information and to communicate.

Sounds That Can Damage Hearing

Some sounds are so weak physically that they are not heard. Other sounds are audible but do not have temporary or permanent aftereffects. Some sounds are strong enough to produce a temporary hearing loss. Sounds that are sufficiently strong, and/or long lasting, and involve certain frequencies, can damage one's hearing.

The exact distinction among these sounds cannot be stated simply, because not all persons respond to sound in the same manner. Yet, in general, about 85 dBA of sound level is potentially hazardous. That hazard depends on the actual frequency spectrum and on its duration. Most environmental sounds include a wide band of frequencies above and below the range from 20 Hz to 20 kHz that humans can hear.

Sound level, frequency, and duration of exposure (singly or repeatedly) are critical for determining sounds that can damage hearing.

It appears that sound levels below 75 dBA do not produce permanent hearing loss, even at about 4,000 Hz, where people are particularly sensitive. At higher intensities, however, the amount of hearing loss is directly related to sound level (for comparable durations). In the United States, current OSHA regulations allow 16 hours of exposure to 85 dBA, 8 hours to 90 dBA, 4 hours to 95 dBA, etc. In Europe, also 8 hours at 90 dBA are allowed, but 4 hours at 93 dBA, or 16 hours at 87 dBA—the energy trade-off is 3 dBA for four hours. If the sound level is about 140 dB, damage does not follow the simple energy concept; apparently, impulse noise above that level generates an acoustic trauma from which the ear cannot recover.

☞☞☞ *Simple subjective experiences indicate the existence of hazardous sound exposure. Examples: a sound that is appreciably louder than conversational level; a sound that makes it difficult to communicate; ringing in the ear (tinnitus) after having been in the sound environment; or the experience that sounds seem muffled after leaving the noisy area.* ☜☜☜

Individual Susceptibility to NIHL

In young children, there is little difference in hearing thresholds between girls and boys. Yet, between ages 10 and 20, males begin to show reduced high-frequency auditory sensitivity, and women continue to have better hearing than men into advanced age. (These differences may be due to greater exposure of males to noise, not to their inherent susceptibility.) There is a broad range of individual differences and sensitivities to a given noise exposure. The biological reasons are unknown that would explain why, for example, TTS and PTS in response to a given noise may differ as much as 50 dB among individuals. It is suspected that the anatomy and mechanical characteristics of the individual ears play a role, such as may the use of ototoxic drugs, or previous noise exposure (National Institutes of Health, 1990).

Means to Prevent NIHL

Both the level of noise and its duration, i.e., overall the exposure, contribute to the risk of NIHL. Common sources of noise are guns, power tools, chain saws, airplanes, farm vehicles, firecrackers, automobile and motorcycle races, music (in concerts, or heard through loudspeakers or headphones), and many occupational environments. NIHL can occur whether one likes the sound, or not.

There are three major strategies to counter the damaging effects of noise.

The first strategy is to reduce or avoid the *generation of sound* by properly designing machine parts such as gears or bearings, reducing rotational velocities, changing the flow of air, or replacing the apparatus with one of a quieter kind. "Active countermeasures" are being developed in which sounds in the same frequency and amplitude but of the opposite direction (180 degrees off-phase) of the noise source physically erase the source noise. Currently, this works best at frequencies below 1 kHz.

The second strategy is to *impede the transmission of noise* from the source to the listener. In occupational environments, one might try to put mufflers on the exhaust side of a machine, encapsulate the noise source, put sound-absorbing surfaces in the path of the sound, or physically increase the distance between source and ear.

The third strategy is to *remove humans from noisy places* altogether, at least for parts of the work shift.

Hearing protection devices. The last resource of protection (in fact a subdivision of the second strategy) is wearing of a hearing protection device (HPD) that reduces the harmful auditory and/or annoying subjective effects of sounds. HPDs are either worn externally (sound-isolating helmets or muffs, "caps") or inserted into the ear canal ("plugs"). These hearing protectors are variously effective, partly depending on the intensity and the frequency spectrum of the sound arriving at the human, partly depending on the "fit" of the protector to the wearer's head or ear. Inappropriate initial fit, loosening during activities, and, of course, failure to wear this equipment reduces its effectiveness (Casali and Park, 1990).

Unfortunately, commercially available HPDs cannot differentiate and selectively pass speech versus noise energy. Therefore, the devices do not directly improve the signal-to-noise relation. However, they can occasionally afford intelligibility improvements in intense noise by lowering the total energy of both speech and noise which are incident on the ear, which reduces overload distortion in the cochlea. HPDs have little or no degrading effect on intelligibility in noise above about 80 dBA, though they can cause considerable misunderstanding at lower levels (at which protection usually is not needed anyway). Some of the negative effects may be due to the tendency to lower one's own voice because the bone-conductive voice feedback inside the head is amplified by the presence of the protector, mostly at low frequencies. Therefore, one's own voice is perceived as louder in relation to the noise than is actually true, often resulting in a compensatory lowering of the voice by 2 to 4 dB. Thus, one should make a conscientious effort to "speak louder" when one wears HPDs (Berger and Casali, 1992).

Nonverbal signals, such as warning sounds, or sounds of machinery, are also affected by wearing of HPDs. Signals above 2000 Hz are most likely to be missed, due to the high frequency properties of conventional HPDs, particularly if the person wearing them has impaired hearing at these or higher frequencies. This indicates that warning signals should be specifically designed to penetrate, such as by using low frequencies (below 500 Hz). Such frequencies diffract easily around barriers, which is a positive side effect. Yet, for a person with normal hearing, wearing an HPD does not usually compromise detection of signals. Altogether, wearing of HPDs is highly advisable if the ambient noise arriving at the human ear cannot be lowered otherwise.

Audible noise with constant sound levels of 85 dB(A) or greater is hazardous. (The notation "dB(A)" is explained in the next section.) If humans must be subjected to such noise, hearing protection devices are necessary. People should not be exposed to continuous noise levels exceeding 115 dB(A) rms overall sound pressure level under any circumstances. (Note that this is an OSHA requirement currently existing in the United States. Other countries have different requirements.)

Infrasound and ultrasound. Below and above the regular hearing limits, sound levels may occur which, although inaudible, still have vibration effects on the human body.

To protect the human, the following interventions can be taken:

- *Infrasound* pressure level shall be less than 120 dB in the frequency range of 1 to 16 Hz, for 24-hour exposure (NASA, 1989). To achieve this goal, well-fitted ear plugs provide attenuation at frequencies below 20 Hz similar to that in the 125-Hz band; in contrast, ear muffs are not effective (Berger and Casali, 1992).
- Hearing conservation measures shall be initiated when the *ultrasonic* criteria listed in Table 4-9 are exceeded (NASA). Ear muffs and plugs provide protection with attenuation of at least 30 dB at frequencies between 10 and 30 kHz (Berger and Casali, 1992).

Psychophysics of Hearing

While physical measurements can explain many acoustical events, persons may interpret them and react to them in very subjective manners—for example, finding certain sounds either attractive or noisy. The experienced sensation of a tone or complex sound depends not only on its intensity and frequency, but also how we feel about it.

TABLE 4-9. AIRBORNE HIGH-FREQUENCY AND ULTRASONIC LIMITS

One-third octave band center frequency, kHz	One-third octave band level in dB
10	80
12.5	80
16	80
20	105
25	110
31.5	115
40	115

SOURCE: NASA, 1989.

"Loudness" is the subjective experience of frequency and intensity. Compared to the intensity at 1,000 Hz, at lower frequencies one has to *increase* the sound pressure level to generate the feeling of "equal loudness." For example, the intensity of a 50-Hz tone must be nearly 100 dB to sound as loud as a 1,000-Hz tone with about 60 dB. However, at frequencies in the range of about 2,000 to 6,000 Hz, the intensity can be lowered and still sound as loud as at 1,000 Hz. Yet, above about 8,000 Hz, the intensity must be increased again above the level at 1,000 Hz to sound equally loud. The "equal-loudness contours" (called "phon" curves) were originally developed by Fletcher and Munson in 1933 and revised by Robinson and Dadson in 1957—see Figure 4-17. The actual shapes of the curves are slightly different if the stimulus sounds are presented either by loudspeaker or by earphone.

These perceptions of "equal loudness" indicate that there are nonlinear relationships between "pitch" (the perception of frequency) and "loudness" (the perception of intensity): simply said, it is more difficult to hear a low-frequency sound than a high sound. The differences in human sensitivity to tones of different frequencies are imitated by filters that are applied to sound-measuring equipment. These filters "correct" the physical readings to what the human perceives. Different filters have been used, identified by the first letters of the alphabet. The "A" filter is most often used because it corresponds best to the human's response at 40 dB. A-corrected SPL values are identified by the notation dBA or dB(A).

There are some other interesting acoustic phenomena:

☞☞☞

Directional hearing. *Humans are able to tell where a sound is coming from by using the difference in arrival times (phase difference) or intensities (as a result of the inverse-square law of energy flux) to determine the direction of the sound. Yet, the ability to use stereophonic cues varies among individuals and is much reduced when earmuffs are worn.*

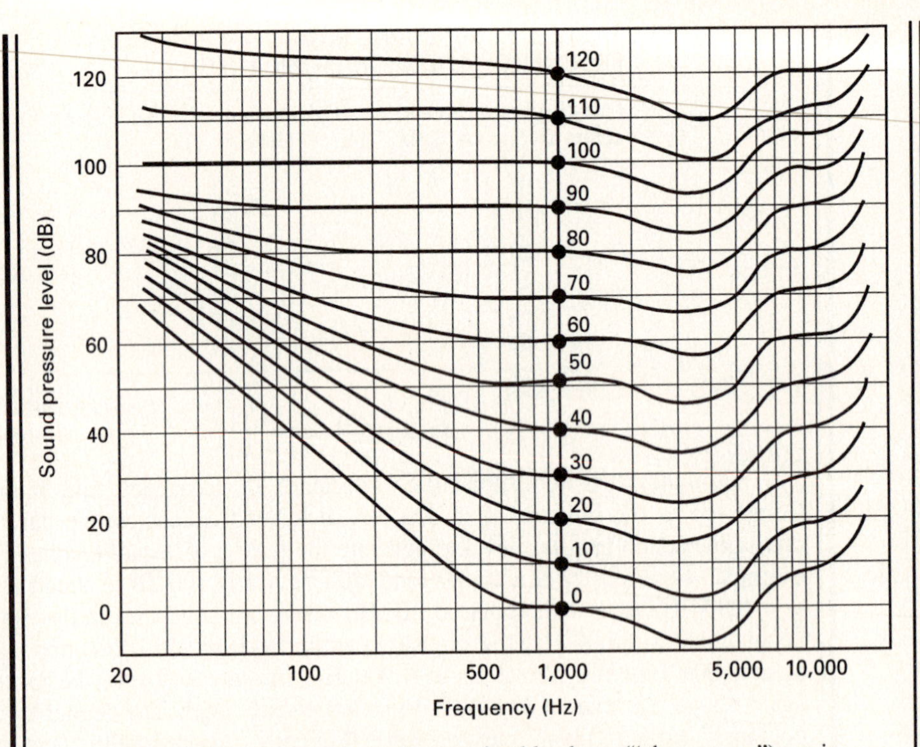

Figure 4-17. Curves of equally perceived loudness ("phon curves") are in congruence with the sound pressure level at 1000 Hz.

Distance hearing. *The ability to determine the distance of a sound source is related to the fact that sound energy diminishes with the square of the traveled distance, but the human perception of energy depends also on the frequency, as just discussed. Thus, a sound source appears more distant when it is low in intensity and frequency, but appears to be closer if it is high in frequency and intensity.*

Difference and summation tones. *Two tones that are sufficiently separated in frequency (so that they excite separate areas on the basilar membrane) are perceived as two distant tones. When two such tones are very loud, one may hear two supplementary tones. The more distinct tone is at the frequency difference between the two tones; the quieter tone is the summation of the original frequencies. For example, two original tones, at 400 and 600 Hz, generate a difference tone at 200 Hz, and a summation tone at 1,000 Hz.*

Common-difference tone. *When several tones appear which are separated by a common frequency interval (of 100 Hz or more), one hears an additional frequency based on the common difference. This explains how one may be able to hear a deep bass tone from a sound system that is physically incapable of emitting it.*

Aural harmonics. *One may perceive a pure sound as a complex one because the ear can generate harmonics within itself. These "subjective overtones" are more pronounced with low- than with high-frequency tones, especially if these are about 50 dB above threshold.*

Intertone beat. *If two tones differ only slightly in their frequencies, the ear hears only one frequency, called the "intertone," at a frequency that is halfway between the frequencies of the two original tones. The two tones are in phase at one moment and out of phase at the next, causing the intensity to wax and wane; thus, one hears a beat. Beating occurs at a frequency equal to the difference in frequencies of the two tones. If the beat frequency is below six per second (that is, if the original tone frequencies are close together), the beat is very distinct, appearing as a variation in loudness; only the intertone is heard. When the beat is above eight per second (i.e., with enlarged frequency interval between the two original tones), the intertone appears to be pulsating or throbbing, and one may hear the original tones as well. When the beats occur more often than about 20 times per second (with even larger frequency separation between the original tones) the intertone becomes faint, and the two original tones are predominant.*

Doppler effect. *As the distance between the sound source and the ear decreases, one hears an increasingly higher frequency; as the distance increases, the sound appears lower. The larger the relative velocity, the more pronounced the shift in frequency. The Doppler effect can be used to measure the velocity at which source and receiver move against each other.*

Concurrent tones. *When two tones of the same frequency are played in phase, they are heard as a single tone, its loudness being the sum of the two tones. Two identical tones exactly opposite in phase cancel each other completely and cannot be heard. This physical phenomenon (called destructive interference or phase cancellation) can be used to suppress the propagation of acoustical or mechanical vibrations and is particularly effective, with current technology, at frequencies below 1,000 Hz.*

Voice Communications

"Intelligibility" is the psychological process of understanding meaningful words, phrases, and sentences. For satisfactory communication of most voice messages in noise, 75 percent intelligibility is at least required. Direct, face-to-face communication provides visual cues that enhance speech intelligibility, even in the presence of background noise. Indirect voice communications lack the visual cues. The distance from speaker to listener, background noise level, and voice level are important considerations. Ambient air pressure and gaseous composition of the air affect voice efficiency and frequency.

The intensity level of a speech signal relative to the level of ambient noise is a fundamental determiner of the intelligibility of speech. The commonly used "speech-to-noise ratio" (*S/N*) is really not a fraction but a signed difference; for a speech of 80 dBA in noise of 70 dBA, the *S/N* is simply +10 dB. With an *S/N* of +10 dB or higher, people with normal hearing should understand at least 80 percent of spoken words in a typical broadband noise. As the *S/N* falls, intelligibility drops to about 70 percent at 5 dB, to 50 percent at 0 dB, and to 25 percent at −5 dB. People with noise-induced hearing loss may experience even larger reductions in intelligibility, while persons used to talking in noise do better.

People have a tendency to raise the voice in noise, and to return to normal when the noise subsidies. (This is called the Lombard reflex, although it is probably not a reflex but a conditioned response.) In a quiet environment, males produce normally about 58 dBA, in a loud voice 76 dBA, and when shouting, 89 dBA. Women have normally a voice intensity that is 2 or 3 dBA less at lower efforts and 5 to 7 dBA less at higher efforts. Thus, people can increase the *S/N* by raising their voices fairly easily at low noise levels, but the ability to compensate lessens as the noise increases. Above about 70 dBA, raising one's voice becomes inefficient, and it is insufficient at 85 dBA or higher. Furthermore, this forced effort of a shouting voice often decreases intelligibility, because articulation becomes distorted at voice output extremes (Casali, 1989).

The *S/N* is a rough estimate for predicting communication effectiveness, but the loss of intelligibility by "masking" through noise depends not only on the intensity but also on the frequency, both of the signal and the noise. In general, at small *S/N* differences, low-frequency noise causes more speech degradation than high-frequency noise.

Most human voices encompass a frequency bandwidth of about 200 to 8,000 Hz in speech, with men using more of the low-frequency energy than women. Intelligibility is little affected by either filtering or masking frequencies below 600 or above 3,000 Hz. But interfering with voice frequencies between 1,000 and 3,000 Hz drastically reduces intelligibility.

High-frequency consonant sounds are more critical for understanding words than are vowels. Unfortunately, consonants generally have less speech energy than vowels and are more readily masked by ambient noise. Noises of predominantly low-frequency nature, particularly those with single-frequency components, have masking effects that spread upward in frequency. They intrude upon speech bandwidths if the noise is between 60 to 100 dBA. Thus, (unamplified) speech communication is very specifically dependent on the frequencies and sound energies of the interfering noise. Therefore, using a broadband dosimeter to measure the exiting noise is usually not sufficient for predicting intelligibility. Several techniques exist which predict speech intelligibility based on narrowband measurements. For this, either a real-time sound analyzer or a sound-level meter with filter sets is needed (Casali, 1989).

The Articulation Index (*AI*) is an often used metric to assess intelligibility. It requires that noise levels in 15 one-third octave bands with centers from 200 to 5,000 Hz be measured in the work environment and compared to speech peaks in the same bands. Then, the noise level in each band is subtracted from the speech level in that band. The differences are weighted, with the highest weight given to the most critical voice frequencies. The weighted differences are then summed to produce a single *AI* value. This technique is described in Table 4-10. Excellent to very good intelligibility can be expected with $1 < AI < 0.7$, good for $0.7 < AI < 0.5$, acceptable with $0.5 < AI < 0.3$, marginal or unacceptable with $AI < 0.3$ (NASA, 1989). Figure 4-18 facilitates comparison between *AI* and other measures of intelligibility.

TABLE 4-10. CALCULATION OF THE ARTICULATION INDEX AI (EXAMPLE)

Band and centers	Observed speech peaks minus noise, dB	Weight factor	Result
200	30	0.0004	0.0120
250	26	0.0010	0.0260
315	27	0.0010	0.0270
400	28	0.0014	0.0392
500	26	0.0014	0.0364
630	22	0.0020	0.0440
800	16	0.0020	0.0320
1,000	8	0.0024	0.0192
1,250	3	0.0030	0.0090
1,600	0	0.0037	0.0000
2,000	0	0.0038	0.0000
2,500	12	0.0034	0.0408
3,150	22	0.0034	0.0758
4,000	26	0.0024	0.0624
5,000	25	0.0020	0.0500
			AI = 0.4738

SOURCE: Example from Sanders and McCormick, 1987.

The *AI* is a fairly complex but rather exact means to predict intelligibility. A more convenient alternative is the Speech Interference Level (*SIL*), which requires less complex instrumentation and fewer data points. The *SIL* is simply the average of the decibel levels in the octave bands 600 to 1,200, 1,200 to 2,400, and 2,400 to 4,800 Hz. It is now often used in a slightly modified form, the Preferred Speech Interference Level (*PSIL*). The *PSIL* is computed as the arithmetic average of three noise measurements taken in octave bands centered at 500, 1,000, and 2,000 Hz. The higher the PSIL, the poorer the communication; for a typical distance between speaker and listener of 1 meter, a PSIL of 80 or above would indicate difficulties in communication. The PSIL should not be used if the noise has powerful low- or high-frequency components. Similar, but even simpler, are the Preferred Noise Criteria (*PNC*) curves, which also

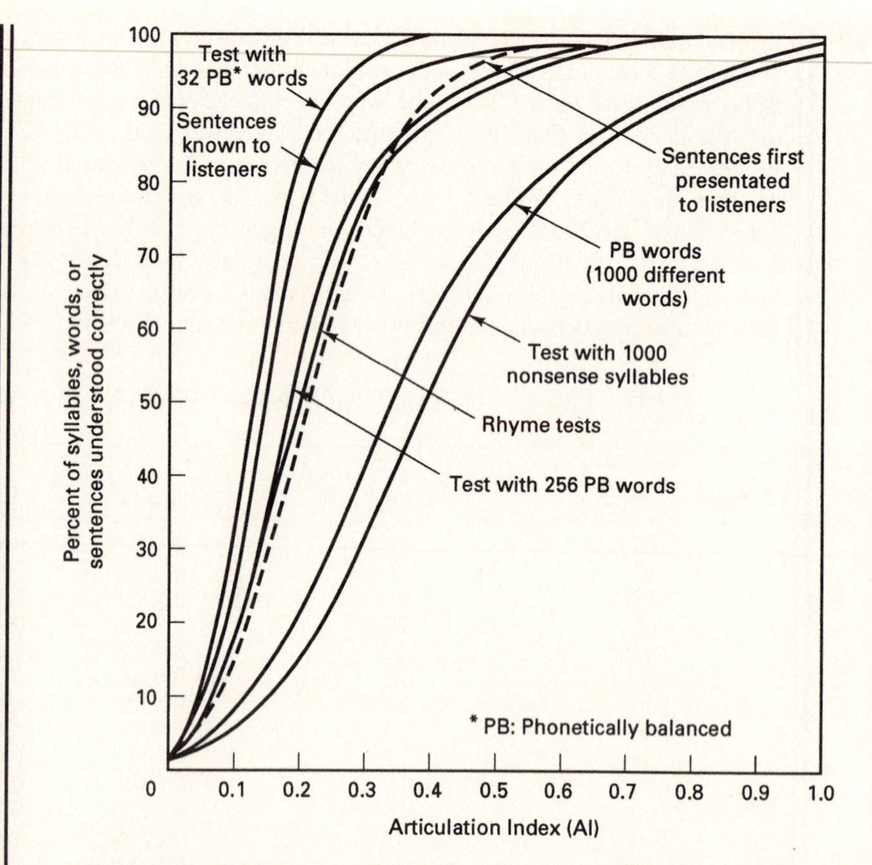

Figure 4-18. Articulation index and speech intelligibility (NASA, 1989). These relations are approximate: they depend upon type of spoken material and skill of talkers and listeners.

use an octave-band analysis of the noise. The highest *PNC* curve penetrated by the noise spectrum is the value indicative of the situation. Rooms in power plants, for example, have recommended *PNC*s of 50 to 60, while offices should be at 30 to 40 (Beranek, Blazier, and Figwer, 1971).

Masking of speech. The frequencies in voice communications range from about 200 to 8,000 Hz. Masking of speech occurs when the presence of environmental sound inhibits the perception of speech sounds. A given frequency sound can mask signals at neighboring frequencies, possibly rendering them inaudible.

Crew-member efficiency is impaired when noise interferes with voice communication. If masking occurs, the time required to accomplish communication is increased through slower, more deliberate verbal exchanges. Not only is this annoying, but it can result in increased human error due to misunderstandings. Table 4-11 and Figure 4-19 indicate speech-interference-level criteria for voice communications.

TABLE 4-11. SPEECH INTERFERENCE LEVEL (NOISE) CRITERIA FOR VOICE COMMUNICATIONS

Speech interference level SIL (dB)	Person-to-person communication
30–40	Communication in normal voice satisfactory.
40–50	Communication satisfactory in normal voice at 1 to 2 m; need to raise voice at 2 to 4 m; telephone use satisfactory to slightly difficult.
50–60	Communication satisfactory in normal voice at 30 to 60 cm; need to raise voice at 1 to 2 m; telephone use slightly difficult.
60–70	Communication with raised voice satisfactory at 30 to 60 cm; slightly difficult at 1 to 2 m. Telephone use difficult. Ear plugs and/or ear muffs can be worn with no adverse effects on communications.
70–80	Communication slightly difficult with raised voice at 30 to 60 cm; slightly difficult with shouting at 1 to 2 m. Telephone use very difficult. Ear plugs and/or ear muffs can be worn with no adverse effects on communications.
80–85	Communication slightly difficult with shouting at 30 to 60 cm. Telephone use unsatisfactory. Ear plugs and/or ear muffs can be worn with no adverse effects on communications.

SOURCE: Adapted from NASA, 1989.

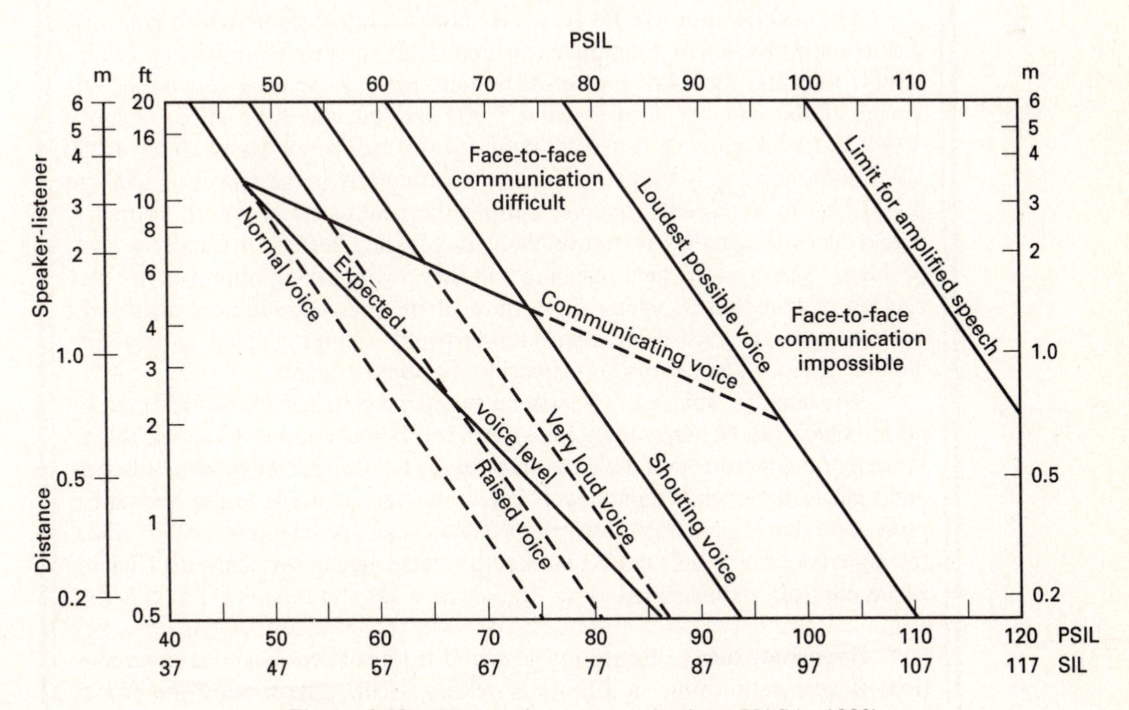

Figure 4-19. Face-to-face communications (NASA, 1989).

Components of speech communication. Speech communication has five major components: the message itself, the speaker, the transmission means, the environment, and the listener.

The *message* itself becomes clearest if its context is expected, its wording is clear and to the point, and the ensuing actions are familiar to the listener.

The *speaker* should use common and simple vocabulary with only a limited number of terms. Redundancy can be helpful (for example: "Boeing 747 jet"). The speed should be slow rather than fast. Phonetically discriminable words should be used. The International Spelling Alphabet is shown in Table 4-12.

TABLE 4-12. INTERNATIONAL SPELLING ALPHABET

A : Alpha	J : Juliet	S : Sierra
B : Bravo	K : Kilo	T : Tango
C : Charlie	L : Lima	U : Uniform
D : Delta	M : Mike	V : Victor
E : Echo	N : November	W : Whiskey
F : Foxtrot	O : Oscar	X : X-ray
G : Golf	P : Papa	Y : Yankee
H : Hotel	Q : Quebec	Z : Zulu
I : India	R : Romeo	

The *transmission* should be by a "high-fidelity system" which has little distortion in frequency, amplitude, or time. If there is either frequency or amplitude filtering, the following considerations apply. *Frequency clipping* affects vowels if the clipping occurs below 1,000 Hz, but has little effect if above 2,000 Hz. Peak clipping is usually not critical if below 600 Hz or above 4,000 Hz. Center clipping is highly detrimental, particularly in the ranges of 1,000 to 3,000 Hz. In *amplitude clipping*, cutting the peaks affects vowels primarily and reduces the quality of transmission in general, but is not usually a great problem. Surprisingly, peak clipping and then reamplifying improves the perception of consonants, which carry most of the message (one can read most written messages even if all vowels are missing). Center clipping, in contrast, garbles the message because it primarily affects consonants.

The hearer's ability to understand the message is, of course, affected by noise, which can be assessed by the speech interference level, discussed above. Wearing of ear-protection devices (plug, muff, helmet) produces some filtering and reduces the overall sound level. Yet, it may be advisable to use special ear protection that is penetrable by specific frequencies, or one may have to adjust the signal so that it will not be masked by noise (see above and also Chapter 11 on controls and displays).

Reverberation. Reflection of sound from surfaces is called reverberation. Reverberation time in a room is defined as the time it takes the *SPL* to decrease by 60 dB after shutting off the sound source. A certain amount of reverberation is desirable, because it makes speech sound alive and natural. But too much reverberation is undesirable, if reflections arrive at the same time a new word is uttered and hence interfere with its perception. A room with little or no reverberation is called "dead": there is little interference between words, and intelligibility is near 100 percent, but because the sound of the word decays before it can propagate through the room, verbal communication may be difficult. If the delay between reflected and original sounds is long, separate sounds (original and echo) are heard. Long room reverberation times can produce bouncing or booming sound. As reverberation time increases, intelligibility decreases in a nearly linear fashion. If the reverberation time increases beyond 6 seconds, intelligibility is cut in half. A highly reverberant room is called "live" or "hard."

Communication at altitude and under water. Communications at high altitude, and under water, require specific technical means. At altitude, where ambient pressure is low, human voice and earphones as well as loudspeakers become less efficient generators of sound, and microphones become less sensitive at certain frequencies. This requires that specific amplification be incorporated in either the source, the transmitter, or the receiver of signals.

Under water, hearing is limited by reverberation of sound and noises made by movement, and by an increased minimal threshold of hearing, which may be raised about 40 dB by the impedance mismatch between water and air at the ear. The difficulties can be overcome by amplification of the transmitted sound and by using a directional receiver to discriminate against sounds coming from other directions. One can talk under water, but the noise that accompanies emitted bubbles masks speech so that the listener hears mainly vowel sounds. If a face mask with built-in microphone is not available, the diver should wait for bubbles to die away before saying the next word, and use simple vocabulary mostly using vowels.

THE OLFACTORY SENSE

Smells are dear to us, but we have no names for them. Instead, we describe how they make us feel, such as intoxicating, sickening, pleasurable, delightful, or revolting. Some smells are fabulous when they are diluted, but truly repulsive when they are not (Ackerman, 1990).

In the upper part of each human nostril, several millions of smell-reacting sensors are located in a patch of 4 to 6 cm^2, called the olfactory epithelium. The nasal airways are bent, so little air flow passes along the olfactory cells, but it can be increased by sniffing. Certain molecules trigger (in still unknown ways) the sensors, which then send signals directly to the olfactory bulbs of the brain. However, distributed throughout the nasal cavity are also free endings of the fifth (trigeminal) cranial nerve which are connected to a different region of the brain. The trigeminal receptors provide the so-called "common chemical sense." Among other odorants, they are triggered by substances that generate irritating, tickling, and burning sensations, which initiates protective reflexes, such as sneezing and interruption of inhalation (National Academy of Sciences, 1979). Most, if not all, odorants in high-enough concentration stimulate both the olfactory and trigeminal sensors.

Not all odorants are external: the body also generates "smells." Body odor arises from the apocrine glands, which are small in infants but develop during puberty. Most of them are at the armpits, face, chest, genitals, and anus.

APPLICATION

Odorants are chemical substances and can be analyzed by chemical methods. Odors are sensations and must be assessed by measuring human responses to them. If the physical and chemical determinants of odor were fully understood, it would be possible to predict the sensory properties of odorous materials from their chemical analysis—in practical terms, one could construct an "odor meter" analogous to a decibel meter for sound. Such understanding is not yet at hand, nor is any such device available. Nonetheless, various instrumental and sensory methods of measurement have been developed and have been applied to sources of odor and to the ambient atmosphere. However, many of the available techniques are costly and time consuming, and not all have been validated (National Research Council, 1979).

Many atmospheric contaminants are odorless, or very nearly so; carbon monoxide is a notorious example. But many other substances are readily detectable even in minute concentrations. For example, an organic sulfur compound at a concentration of one molecule per billion molecules of air is likely to be smelled.

Most existing odorants are mixtures of the basic components, thus generating complex odors to which different people, in different environments, at different lengths of exposure react quite differently. Accordingly, various theories have been proposed to explain smell sensitivity, among them Amoore's stereo-chemical model, which assumes that the dimensions and shapes of certain odor molecules must fit certain receptor sites to generate the sensation of an odor (Amoore, 1970).

While we use the olfactory sense daily, little is known systematically about it, partly because a smell is not easily quantified. The usual test procedure is to present a smelling substance in varying concentrations; the threshold concentration is found when it provokes a sensation in 50 percent of the cases. The sensation (the "smell") is described in relation to other smells; it may be considered pleasant or unpleasant. Assessment of "odor annoyance" has been attempted similarly to noise classification (Hangartner, 1987).

Several systems to describe and distinguish odor qualities have been used in the past. The classic Linacus-Zwaardemarker 9-category system has been largely displaced by either of two other procedures:

- The Crocker-Henderson approach ranks every smell in four categories along a 9-point scale. The categories are fragrant, acid, burnt, and caprylic ("soapy").
- Henning's 6-category system "smell prism" (see Figure 4-20) employs the terms flowery (or fragrant), spicy, fruity, resinous, burnt, and foul (or putrid, rotten) arranged in a prism geometry. Stimuli that typically evoke these olfactory response categories are violets, cloves, oranges, balsam, tar, and rotten eggs.

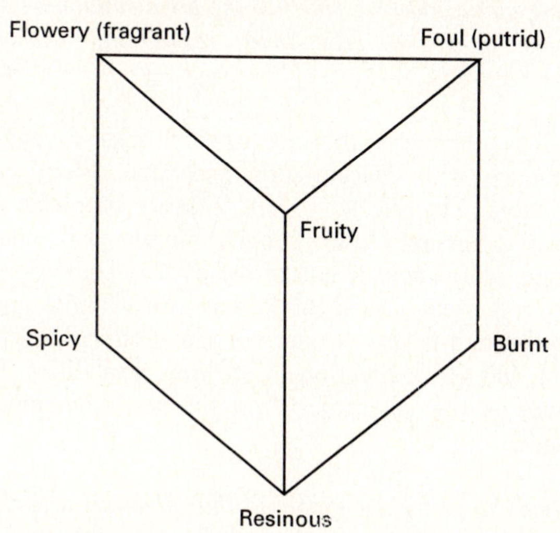

Figure 4-20. Henning's smell prism.

Odors may affect well-being by eliciting unpleasant sensations, by triggering possibly harmful reflexes and other physiologic reactions, and by modifying olfactory function. Unfavorable responses include nausea, vomiting, and headache; induction of shallow breathing and coughing; upsetting of sleep, stomach, and appetite; irritation of eyes, nose, and throat; destruction of the sense of well-being and the enjoyment of food, home, and external environment; disturbance; annoyance; and depression. Exposure to some odorous substances may also lead to a decrease in heart rate, constriction of blood vessels of the skin and muscles, release of epinephrine, and even alterations in the size and condition of cells in the olfactory bulbs of the brain. However, the relationships between the intensity or duration of the exposure to odor and the magnitudes of these symptoms have not been established (National Research Council, 1979).

APPLICATION

It helps the designer's interest in using smell as an engineering means that only minute quantities of some stimulating agents are required to bring about a smell sensation. For example, it suffices to add methyl mercaptan in the amount of only 25×10^{-6} gram to each liter of natural gas so that it can be detected by smell, or to add 3 to 10 ppm pyridine to argon, another inert gas (Cain, Leaderer, Cannon, Tosun and Ismail, 1987). Increasing the concentration to 10 or 15 times the threshold evokes the maximal intensity of smell. Another application is spraying sweetener in the ambient air: if the wearer of a respiratory mask or hood smells it, the seal is proven inadequate.

☞☞☞ *Only 20 percent of the perfume industry's income comes from making perfumes to wear; the other 80 percent comes from perfuming the objects in our lives. Used-car dealers have a "new-car" spray, real estate dealers sometimes spray cake-baking aromas around the kitchen of the house before showing it to a client. Shopping malls add food smell to the air-conditioning system to put shoppers in the mood to visit their restaurants (Ackerman, 1990).* ☜☜☜

There are some serious problems with using the smell sense for engineering purposes. Smells change with concentration and time. Many can be masked easily by other odors. The olfactory sense is easily adaptable and is quite different from person to person. (After spending months in Bombay, India, a city in which strong odors abound, one of the authors could no longer smell the "wild" animals in a circus.) Most children and men cannot smell exaltolide, while many women find it very strong; but sensation varies with the menstrual cycle. Smoking and aging affect one's smelling capabilities. Blocking the nasal passages, such as when one suffers from the cold, can temporarily eliminate the ability to smell.

☞☞☞ *Since ancient times, it has been supposed that pleasant aromas preserve health and that unpleasant odors are injurious. These suspicions formed the basis for the use of aromatic Eau de Cologne and of pomanders stuffed with balsams and for the attribution of diseases to atmospheric "miasmas." Thus, the word "malaria" is derived from the Italian expression for "bad air," mala aria (National Research Council, 1979, p. 3).* ☜☜☜

The magnitude of the human sensory responses to odor (the perceived odor intensity) decreases as the concentration of odorant gets smaller. This effect is used to control indoor odors by ventilation or outdoor odors by the use of tall stacks. However, the relationship between odor intensity and odorant concentration is not directly proportional. When odorous air is diluted with odor-free air, the perceived odor decreases less sharply than the concentration; for example, a tenfold reduction in the concentration of amyl butyrate in air is needed to reduce its perceived odor intensity by half. Nor do all odorants respond by the same ratios; some, like amyl butyrate, show sluggish changes in odor with changes in concentration, and others change more sharply (National Research Council, 1979).

How we delight our senses varies greatly from culture to culture. Massay women, who use cattle dung as a hair dressing, would find American women scenting their breath with peppermint equally bizarre (Ackerman, 1990).

THE GUSTATION SENSE

The sense of taste is only poorly understood. The human can distinguish primary qualities, i.e., categories of taste, but not easily quantitative differences. This is partly due to the fact that the taste sense interacts strongly with sensations of smell, of temperature, and of texture, all of which are present in the mouth. Pepper, for example, tastes as it does because it stimulates several types of receptors. Food becomes almost tasteless if a cold causes temporary loss of the smell sense.

"How strange that we acquire taste as we grow. Babies don't like olives, mustard, hot pepper, beer, fruits that make one pucker, or coffee." — "No two of us taste the same plum. Heredity allows some people to eat asparagus and pee fragrantly afterward (as Proust describes in "Remembrance of Things Past"), or eat artichokes and then taste any drink, even water, as sweet. Some people are more sensitive to bitter tastes than others and find saccharin appalling, while others guzzle diet sodas. Salt cravers have saltier saliva. Their mouths are accustomed to a higher sodium level, and foods must be saltier before they register as salty" (Ackerman, 1990, p. 141).

The receptors for taste qualities are primarily located on the tongue (mostly on its tip, the sides, and the rear of the upper surface) but are also found at the palate, pharynx, and tonsils. The taste buds (really a budlike collection of cells) are arranged in clusters. Humans have about ten thousand taste buds at the tip of the tongue which are continuously replaced every two weeks. It appears that some taste buds react only to one quality, while others respond to several or all of the qualities. The tip of the tongue is particularly sensitive to sweet, the sides to sour, the back to bitter, and all of them to salty stimuli. The number of taste buds seems to diminish with aging after the middle forties, when remaining buds may also atrophy. Taste sensitivity differs from person to person, and decreases with age. Some people cannot taste certain substances; for example, 3 of 10 persons cannot perceive phenylthiocarbamide.

To taste a substance, it must be soluble in the saliva. Taste sensitivity depends on many variables, most of which are interactive: these include the substance of the stimulus, its concentration, its location of activity, its time of application. Furthermore, the sensitivity depends on the prior state of adaptation, on the chemical condition of the saliva, on temperature, and on other variables.

Common taste qualities are salty, sour, bitter, and sweet. They are evoked by the stimuli NaCl, acid hydrogen ions, nitrogen alkaloids, and inorganic carbon, respectively; their perception is strongly affected by temperature. Of the four qualities, bitter and sweet are even less understood than salty and sour. Temperature also af-

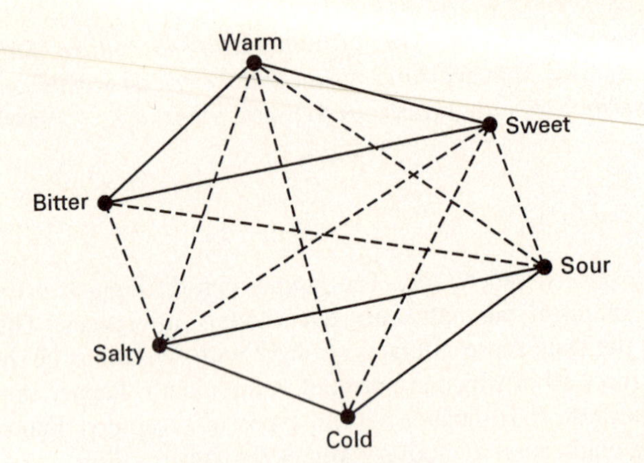

Figure 4-21. Békécy's taste model.

fects taste sensations. One attempt to describe the relations between taste qualities is shown in Figure 4-21, where the solid lines represent interactions, and the dotted lines indicate no interactions.

☞☞☞ *Reading an article on wine: "'Number one was a youngster: fresh, light, passionfruity, straightforward, not oakey, fruit-sweet, fine and understated, well-balanced and clean—a modern style with low-key oak and showing cool-fermentation high-tech winemaking, but not very complex, and needing age. Number two was much more complex, showing some maturity: very smoky, lightly leesy, toasted nut aromas; strongly constituted, tightly structured, powerful, spicy with plenty of wood-derived complexity, very long in flavour and excellent drinking now, but in no danger of falling apart. Number three was, for me, even bigger and richer, fatter and more complex than number two. Strong aromas of butter, grilled nuts, toasted bread, smokiness, very full-bodied, ripe, and alcholic, quite fat and reminding everybody of a good 1978 white Burgundy.' Let's just take this slowly. By my count, and excluding what might be repetitions, there are 28 different qualifiers in that quotation, implying 28 scales of assessment. This writer is asking me to believe that it is possible to discriminate 28 taste scales, let alone assess points along each one? And to do this reliably?" (Applied Ergonomics, 1988, 19(4), p. 324).* ☜☜☜

CUTANEOUS SENSES

❑ **APPLICATION**

The sensory capabilities located in the skin are called cutaneous (from *cutis,* Latin for skin) or *somesthetic* (from *soma,* Greek for body).

They are generally divided into four groups:

- Mechano-receptors, which sense taction, i.e., contact ("touch"), tickle, pressure, and related commonly understood but theoretically ill-defined stimuli.

- Thermo-receptors, which sense warmth or cold, relative to each other and to the body's neutral temperature.
- Electro-receptors, which respond to electrical stimulation of the skin. (It is disputed whether such specific sensors exist.)
- Noci-receptors (Latin *nocere,* to damage), which sense pain. We feel what is commonly but imprecisely called sharp or piercing pain, usually associated with events on the surface; and dull or numbing pain, usually felt deeper in the body. (There is controversy because some researchers state that there are no specific pain sensors, but that pain is transmitted from other sense organs.)

The study and engineering use of the cutaneous senses is hampered by many uncertainties: stimuli are often not well defined, particularly in older research. Sensors are located in different densities over the body. The functioning of sensors is not exactly understood. Many sensors react to two or more distinct stimulations simultaneously, and in similar outputs; it is often not clear whether or which specific sensors respond to a given stimulus. The pathways of signal conduction to the central nervous system are complex (as discussed earlier), and may be joined by other afferent paths from different regions of the body. At the CNS, arriving signals are interpreted in unknown manners. Given all these uncertainties, it is obvious that many of our engineering applications are based mostly on everyday experience and are deduced from very limited experimental data.

NEED

Much more research needs to be performed to provide the ergonomist with complete, reliable, and relevant information on cutaneous senses for the design of systems.

Sensing Taction

The taction sense reacts to "touch" at the skin. The term *tactile* is often used if the stimulus is received solely through the skin, while the term *haptic* is applied when information is obtained simultaneously through cutaneous and kinesthetic senses—that is, through the skin and through proprioceptors in muscles, tendons, and joints. Most of our everyday tactile perception is actually haptic perception (Kubovi, 1986).

Weber demonstrated in 1826 that skin sensors react to the stimulus location, and to the stimuli of force/pressure, warmth, and cold. (Pain and other more diffuse feelings were originally not considered part of the tactile sense, but are nowadays included insofar as they respond to stimuli on the skin.) In spite of its everyday use, and of much research done, this sense is not fully understood. Taction stimuli often are not well defined: what is the relation between pressure, force, and touch? Which sensors respond to each stimulus in what way? Do the sensors respond singly or in groups? If in groups, in what patterns? Does one sensor respond to only one stimulus, or to several stimuli?

Questions exist regarding the association of specific nervous functions with the stimulation of specific sensors, and how these interact to provide a signal to the brain. One often postulates the presence of specialized nerve endings, one for each sensation, and that nerves are connected to specific centers in the brain. However, pattern theory assumes that the particular experience of a sensation depends on the co-action of several separate and elementary nervous events; but not all need to be present in each pattern. (An overview of the current unsettled state of theories and knowledge is provided by Sherrick and Cholewiak, 1986.)

Architecture of the taction system. All tactile sensors are "triggered" by a stimulus, if appropriate in type and intensity, to discharge signals, varying in frequency and amplitude, towards the CNS. Several types of sensors have been identified. The most common is a "free nerve ending," a nervous proliferation that distally dwindles in size and disappears. Thousands of such tiny fibers extend through the skin layers. They respond particularly to mechanical displacements and are very sensitive near hair follicles.

In glabrous (smooth and hairless) skin, encapsulated receptors are also common. Among these are Meissner corpuscles, which are particularly numerous in the ridges of the fingertips. They transmit transient electrochemical surges to the nerves in response to ambient pressures. A single Meissner corpuscle may connect with up to nine separate nerves, which may also branch to other corpuscles. This is an example of simultaneous convergence (discussed above) and diversion of the neural pathways. (How this arrangement can reliably code neural signals is not yet understood.)

While Merkle receptors are similar to Meissner corpuscles, the Pacinian corpuscle is an encapsulated nerve ending of a single dedicated nerve fiber. These highly responsive tactile receptors are located densely in the palmar sides of hand and fingers and in distal joints. They are also prevalent near blood vessels, lymph nodes, and in joint capsules.

Nerves from the various receptors are colligated in peripheral cutaneous nerve bundles which proceed with their neighbors to the dorsal roots of the spinal cord. Fibers originating at the same receptor may proceed to separate dorsal roots, i.e., dermatomal segments. (It is not yet understood how such a complex nervous signal pattern is interpreted by the brain.)

Nerve fibers have been differentiated according to their conduction velocities. Conduction speeds in the human peripheral nerves range from one meter per second to 120 m/s. In general, conduction velocity is larger with higher fiber threshold, thicker fiber, and presence of sheathing (myelination).

Tactile sensor stimulation. Classic experiments concerned a subject's ability to perceive the presence of a stimulus on the skin; to locate one stimulus or two simultaneous stimuli; and to distinguish between one stimulus or two stimuli applied at the same time. While earlier research procedures were highly individual, modern research uses mostly three classes of stimulation: *step functions,* in which a displacement is produced quickly and held for a period of a second or so; *impulse functions,* in which a transient of some given wave form is produced for a few milliseconds; and *periodic functions,* which displace the skin at constant or variable frequencies for several milliseconds. These forms of mechanical energy can be imparted to the skin by different transducers. Most work has used skin displacement, but other experiments rely on units of force, or on measures of transmitted energy. (Again, much work remains to be done.) A compilation of "classic" tactile sensitivity of body parts was done by Boff and Lincoln (1988).

APPLICATION

Sensing Temperature

While there is no question that temperature can be sensed, there is at present no agreement on what its sensors are. Research is hampered by the common experimental procedure of applying pointed metallic cylinders of different temperatures ("thermodes") to the skin, thus generating touch and temperature sensations simultaneously. Cooling or warming by air convection or radiation may be better ways to stimulate. An additional complication arises from the fact that temperature sensations are relative and adaptive. Objects at skin temperature are judged as neutral or indifferent; this value of temperature is called "physiological zero." A temperature below this is called cool; a temperature above, warm. Slowly warming or cooling the skin near physiological zero may not elicit a change sensation. This area is called the "zone of neutrality." For the forearm, this zone of indifference ranges approximately from 31° to 36°C. Neutrality zones are different for different body parts.

There seem to be nerve sensors that respond specifically to cold and falling temperature, while others appear to specialize on heat and increasing temperatures. These scales may overlap, which can lead to paradoxical or contradictory information. For example, spots on the skin which consistently report "cold" when stimulated at less than physiological zero, may also report "cold" when they are stimulated by a warm thermode of about 45°C. (An opposite paradoxical "warmth" sensation has also been reported.) The sensation of warmth can also be aroused, in some instances, by applying a pattern of alternating warm and cold stimulation. This occasionally generates the sensation of heat even if a cold probe is applied.

Changes towards warm temperature from physiological neutral are more easily sensed than changes towards cold, with a ratio of about 1.6 to 1. The longer the stimulus is applied, and the larger the area of the skin to which it is transmitted, the smaller the temperature change that can be discerned. Warm sensations adapt within a short period of time, except at rather high temperature levels. For cold, adaptation is slower and does not seem to occur completely. This may be explained by the fact that in the heat, both vasodilation and sweating respond to a rise in skin temperature, while vasoconstriction is the only countermeasure to cold. Rapid cooling often causes an "overshoot" phenomenon; that is, one feels colder for a short time than one physically is.

Chapter 5 discusses, in detail, interactions of the body with the environments.

Sensing Cutaneous Pain

Touch, pressure, electricity, warmth, and cold can arouse unpleasant, burning, itching, or painful sensations. At present, it is not clear whether or not modality-specific pain sensors per se exist; it is also questionable if there are distinct "pain centers" in the CNS; in fact, there is some discussion on whether pain is a separate modality independent from other sensory receptions.

"The full array of devices and bodily loci employed in the study of pain would bring a smile to the lips of the Marquis De Sade and a shudder of anticipation to the Graf von Sacher-Masoch" (Sherrick and Cholewiak, 1986, pp. 12–39).

Many research results are difficult to interpret because of the various levels and categories that are used under the label "pain." It can range from something barely felt to unbearable. The threshold for pain is a highly variable quantity, probably because it is so difficult to separate from other sensory and from emotional components. Besides cutaneous pain, discussed here, there is visceral, tooth, head, or nerve trauma pain. One can adapt to pain, at least under certain circumstances, and to certain stimuli. In some cases, so-called "second pain" has been experienced, which is a new and different pain wave following a primary pain after about two seconds. "Referred pain" indicates the displacement of the pain location, usually from its visceral origin to a more cutaneous location; an example is cardiac anginal pain, which may be felt in the left arm.

Sensing Electrical Stimulation

While indeed one feels electrical currents in the skin, the presence of specific sensors for electricity has been much discussed. Currently, there is no receptor known which is specialized to sense electrical energy, or is particularly receptive to it. In fact, electricity apparently can arouse almost any sensory channel of the peripheral nervous system. Figure 4-22 shows schematically the sensitivity to two-point electrical stimulation at various body sites.

Research regarding electrical stimulation has shown that the threshold for electrical stimulation depends heavily on electrode configuration and location, on waveform, repetition rate, and on the individual subject. Generally, the threshold is at about 0.5 to 2 mA with 1-millisecond pulse duration. During shorter pulse durations, temporal summation of single pulses occurs (Boff and Lincoln, 1988).

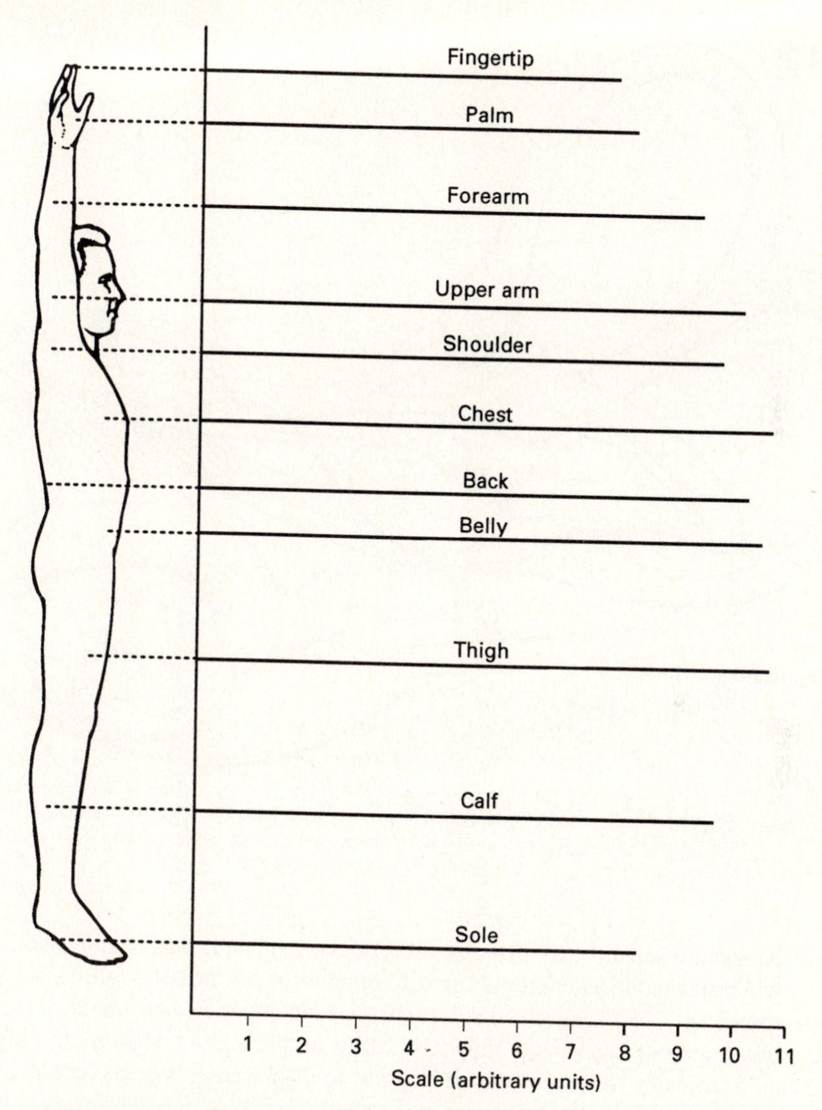

Figure 4-22. Relative sensitivity of body sites to two-joint electrical stimulation. (Modified from Boff and Lincoln, 1988.)

THE VESTIBULAR SENSE

Located next to the cochlea of the inner ear, on each side of the head, are three semicircular canals with two sacklike otolith organs, the utricle and the saccule—shown schematically in Figure 4-23. These nonauditory organs are called the *vestibular system*.

The arches of the three vestibular canals are at about right angles to each other, with one canal horizontal and two vertical when the head is erect. Thus, the three

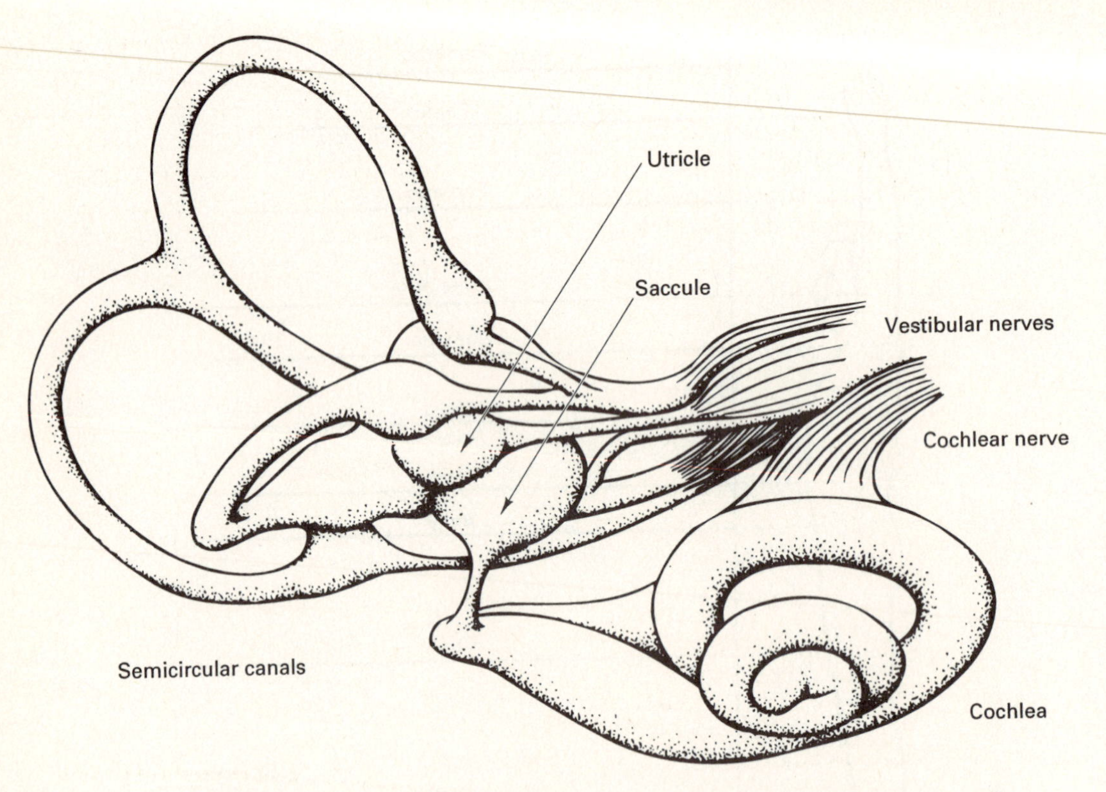

Figure 4-23. The three semicircular canals and the otolith organs of the vestibulum, and the cochlea, with their nervous connections to the 8th cranial nerve. (Adapted from a 1934 drawing by Max Broedel.)

canals are sensitive to different rotations of the head. Each canal functions as a complete and independent fluid circuit, in spite of the fact that all canals share a common cavity in the utricle. Each canal, near its junction with the utricle, has a widening (ampulla) which contains the crista ampullaris, a protruding ridge. It carries cilia, sensory hair cells, which respond to displacements of the endolymph. Cilia are also located in both the utricle and saccule.

The vestibular system responds to magnitude and direction of accelerations, including gravity. The sack-shaped utricle and saccule are sensitive to gravity and other linear accelerations of the head. The response latency to linear accelerations is fairly long, about 3 seconds to 0.1 *G*, which asymptotically reduces to about 0.4 second at accelerations of 1 *G* or higher. The response time to rotational acceleration is about 3 seconds at accelerations slower than 1 deg S^{-2}, reducing to about 0.3 second with accelerations exceeding 5 deg s^{-2}. However, the system adapts to constant acceleration, so that small changes in acceleration may not be perceived. There is, at present, no information about acceleration thresholds in different planes of rotation (Howard, 1986).

A number of peculiar features are associated with the pea-sized vestibular system. Bringing the head into various postures requires that the brain compares signals not only with a new spatial reference system, but also with new reference inputs from the sensors, because the endolymph now loads the cilia in a different manner. Sideways rotation (roll, yaw) initiates different base signals from the two vestibular systems in the head, which must be consolidated in the brain. Given these complex signals, and the other sensory inputs simultaneously arriving at the cerebellum and cerebral cortex, several "vestibular illusions" can occur (Boff and Lincoln, 1988), including

- "Illusionary tilt": interpretation of linear acceleration as body tilt;
- "Unperceived tilt": when the body is aligned with the gravitation vector, a person in a banking airplane does not feel this roll;
- "Inversion illusion": when a person is in the prone position, or in zero gravity, one may feel upside-down;
- "Elevator illusion": change in gravitational force produces apparent rise or lowering of seen objects;
- "Coriolis cross-coupling effect": the feeling of falling sideways when the head is tilted forward while the body rotates about a vertical axis;
- "Motion or space sickness": probably due to conflicting inputs from the vestibular and other sensors—see Chapter 5.

❏ **APPLICATION**

ENGINEERING USE OF SENSORY CAPABILITIES

☞☞☞ *In spite of the thousands of years of everyday experience, fairly little has been done in terms of systematic and conscientious applications of the human sensory capabilities. True, we all feel with our fingertips whether a surface is smooth or not; a welder brings his hand cautiously close to an object to find out whether it is still hot; and a blind person uses Braille to "read" text which cannot be seen. But surprisingly little of the known information has been used by engineers. Round door knobs give no indication, by feel, in which direction they must be turned to open the door. Emergency bars, which one must press to open a hinged door, usually provide no cues (neither for the touch sensors nor the eyes) as to which side of the door will open.* ☜☜☜

While our vision and audition are overstrained in many instances, other information input senses are underused, such as taction, olfaction, and gustation. The cutaneous senses of hands are commonly utilized, but other body segments that have the same or similar sensitivities are hardly ever employed.

Of course, the CNS receives signals from the various human senses simultaneously. This allows us to get a "general picture" of the events taking place within and outside the body. For example: the sensation of exerting a force toward an outside object to move it (such as in lifting) is submitted by muscle spindles which report on muscle stretch (length, and change in length), by Golgi tendon organs which report on muscle tension (force development), by Ruffini joint organs which report limb location and joint angles; together with cutaneous senses, because the bending of any joint stretches some regions of skin around the joint and relaxes others. Furthermore, the sense of vision provides information about the movement of object and body segments; also, sounds associated with the movement provide additional information. Hence, the engineer often provides redundant information, using several sensory modalities at the same time: in addition to instrument displays, a recorded voice tells the pilot to "pull up" if the aircraft noses down.

The most widespread use of human tactual capabilities has been in the coding of controls by shape. In the decade after World War II, much research on controls and displays was performed. Jokingly, that time has been called the "knobs and dials era" by human-factors engineers. Nearly all the shape-coding knowledge which we use today was derived at this time. Knobs on controls were formed like wheels, or like airplane wings. Shape and size coding was used to indicate what will happen after activating these controls, with the information conveyed both by vision (if one looks at the control handle) and by feeling (as one touches the control).—For specific design recommendations, see Chapter 11 on controls and displays.

Incompleteness of research findings is one explanation for the fact that so little systematic engineering use has been made of human sensory skin capacities. To provide the needed information, research must be based on solid theories regarding the various receptors, their stimulation, and their responses; regarding the screening and propagation of the signals to the central nervous system; and regarding the interpretation of the signals there. Theories of somesthetic sensitivity require models, and their testing, for the absorption and propagation of various energy forms in the path from the skin surface to the receptors. In 1986, Sherrick and Cholewiak found the condition of research, regarding both the theoretical underpinnings and the methodological procedures, to still to be in a "primitive state."

Enhancing Human Perception

Just as hair in the human skin amplifies mechanical surface distortion so that the sensor may be more easily activated, a number of engineering means exist to make perception more intense. For taste and smell, as examples, the concentration of the active substance may be enlarged to ensure and hasten perception. Furthermore, signals about the same events simultaneously provided in different modalities, can enhance the speed and accuracy of reception, recognition, and processing of the information, as detailed by Boff and Lincoln (1988).

It is difficult to submit sensory information that is habitually conveyed by one sense through another sense, such as transmitting visual information through taction, as often done for blind persons. The "coding" of the signal from one sensory system to the other is difficult, particularly if the human has never experienced the first kind of sensory input. Kantowitz and Sorkin (1983) found that it took 9 days of extensive practice for a blind person to triple reading performance from about 10 words per minute using an optical-to-tactile converter. However, two persons who could see became very proficient on the same converter after about 20 hours of experience, when they were able to read 70 to 100 words per minute. It might be better, at least in certain circumstances, not to use a natural code which preserves the actual spatial and temporal relationships between the original character (signal) and the one displayed in the other sensor mode; perhaps an artificial code should be employed that allows better and more exact perception. To follow the example above: instead of using Braille to convey a letter shape to the blind, it would be advantageous to "speak" the text.

Given the lack of reliable experimental information, current guidelines for design applications of the taction sense still rely much on extrapolations of previous knowledge, common-sense experiences, and guesses. On this basis, the following engineering recommendations are made, though with much caution.

Using the Taction Sense

Touch information, as transmitted through mechano-receptors, can be differentiated by the human according to

- the magnitude of mechanical deformation,
- the temporal rate of change,
- the size and location of the stimulated skin area (i.e., the number of receptors stimulated).

For static *two-needle point stimulation,* two-point resolution starts at about 2 mm separation at the tip of the finger, at about 4 mm separation at the lips, and at about 40 mm separation at the forearm, but requires about 70 mm separation on the back.

Vibration at the fingertips shows the following minimal thresholds:

200 Hz with a displacement of about 2×10^{-4} mm; 800 Hz with 10^{-3} mm deformation.

Below 10 Hz and above 1,000 Hz only general pressure, not vibration, is sensed.

The highest sensitivity for vibrations appears to be at about 250 Hz, but flying by the "seat of the pants" may be rather dangerous, since sensitivity at the buttocks is very low.

Sensitivity to taction stimulation is highest in the facial area and at the fingertips, and fair at the forearm and lower leg.

One may prefer certain body parts for tactile input sites. Another technique is to amplify the tactual signal, for example by holding a thin piece of paper between the finger and a surface on which unevenness should be detected. Also, the use of gloves can enhance tactile perception, if these have suitable thickness and deformation characteristics (Kantowitz and Sorkin, 1983).

Measured by step-function inputs, the minimal threshold for force varies from 5×10^{-5} N on the face to 35×10^{-5} N on the big toe. Such stimuli, if they last but 1 ms each, must be separated by at least 5.5 ms to be perceived as two stimuli at the fingertip. If the break between stimuli is too short, they fuse into the sensation of a single stimulus: this is called temporal fusion.

Even if succinct information about the characteristics of cutaneous inputs and their nervous responses is often missing, it is worthwhile at least to compare the human response capabilities qualitatively.

In standard vigilance conditions (in which subjects are required to report the appearance of a transient stimulus, or of a change in the stimulus intensity), auditory signals are better detected than weak mechanical vibratory signals and electrocutaneous signals. Longest reaction latencies, most misses, and most false alarms occur with the electrocutaneous signals. However, in complex vigilance environments, it is effective to add electrocutaneous signals to provide redundancy. These and other experiments (discussed by Sherrick and Cholewiak, 1986) indicate that the cutaneous system has utility as either an independent or an additional channel for information input.

Using the Temperature Sense

The temperature sense is difficult to use for communicative purposes. This is due to its relatively slow response and its poor location identification, and to the capabilities for temporal adaptation and for spatial summation of inputs. There are interactions between mechanical and temperature sensations: for example, a colder weight feels heavier on the skin. Thermal sensations can be stimulated chemically, for instance by application of menthol, alcohol, and pepper, but not mechanically.

The strength of the temperature sensation is made stronger by increasing

- the absolute temperature of the stimulus, and its difference from physiological zero.
- the speed of change in temperature.
- the exposed surface (e.g., immersion of the whole body in a bath as compared to partial immersion).

The strength of the thermal sensation depends on the location and size of the sensing body surface.

Assuming a "neutral" skin temperature at about 33°C, the following "rules of thumb" apply for naked human skin:

- 10°C of skin temperature appear "painfully cold"; 18°C feel "cold"; 30°C still feel "cool." Highest sensitivity to changes in coolness exists between 18° and 30°C.
- Heat sensors respond well throughout the range of about 20° to 47°C.
- Thermal adaptation—that is "physiological zero"—can be attained in the range of approximately 18° to 42°C, meaning that changes are not felt when the temperature difference is less than 2°C.
- The ability to distinguish between different temperatures is, for cold sensations, best in the 18° to 30°C range, and for warmth best between 20° and 47°C. Distinctly cold temperatures (near or below freezing) and hot temperatures above 50°C also provoke pain sensations. At about 45°C both cold and heat fibers are stimulated, which may result in the paradoxical sensation of cold when indeed the stimulus is hot. Another interaction may occur between the sensation of pressure and temperature: a force applied in a cold condition appears to be larger than when applied in the heat.
- Cold sensations, as compared to sensations of warmth, are quicker than and have little interference from warm signals, particularly in the face, chest, and the abdominal areas. The body's ability to feel warmth is less distinct, but best in hairy parts of the skin, around the kneecaps, and at fingers and elbows.

Using the Smell Sense

Olfactory information is seldom used by engineers because few research results are available, because people react quite differently to smell stimuli, because smell can be easily masked, and because smell stimuli are difficult to arrange. Among the few industrial applications is adding smelling stuff (methylmercaptan to natural gas, pyridin to argon) to allow people to smell leaking gas.

Using the Taste Sense

Gustation information is not being used in engineering applications at present, but it is an art of much importance to the food and beverage industry.

Using the Electrical Sense

Electricity as an information carrier is only seldom used, although it has great potential for transmitting signals to the human. Electrode attachment is convenient, and the transmitted energies are low, requiring only about 30 microwatts at the electrode-skin junction, up to a tolerable limit of about 300 milliwatts.

Electrical stimulation can provide a clear attention-demanding signal that is resistant to masking. Its major drawback, which it shares with mechanical stimulation, is the problem of pain: stinging, burning, or deep-tissue aching can appear.

Using the Pain Sense

Pain is apparently not a phenomenon that lends itself to engineering applications. In fact, pain may be one of the phenomena that are distinctly "nonapplicable," because one should not cause pain to the human, and because the sensation of pain follows the damage already done too slowly to prevent more damage.

"Fighting the enemy, boredom, Romans staged all-night dinner parties and vied with one another in the creation of unusual and ingenious dishes. At one dinner a host served progressively smaller members of the food chain stuffed inside each other: inside a calf, there was pig, inside the pig a lamb, inside the lamb a chicken, inside the chicken a rabbit, inside the rabbit a dormouse, and so on. Another host served a variety of dishes that looked different but were all made from the same ingredient. . . . Slaves brought garlands of flowers to drape over the diners, and rubbed their bodies with perfumed ungents to relax them. The floor might be knee-deep in rose petals. Course after course would appear, some with peppery sauces to spark the taste buds, others in velvety sauces to soothe them. Slaves blew exotic scents through pipes into the room, and sprinkled the diners with heavy, musky animal perfumes like civet or ambergris. Sometimes the food itself squirted saffron or rose water or some other delicacy into the diner's face . . ." (Ackerman, 1990, p. 144).

SUMMARY

The eye plays a major role to collect information. Thus, the provision of proper visual signals through selection of proper illumination and contrast, and the avoidance of unbecoming circumstances such as glare, are important ergonomic tasks. The eye is seldom used as an output means, with "eye tracking" one of the exceptions.

The hearing sense provides inputs only. There are complex but well-researched relations between frequency and intensity of sounds, hence solidly founded recommendations for the design of communication and sound-signal systems are available. Protection of the ear from damaging noises is also well understood, and related technical means are documented and ready to be applied.

Surprisingly and disappointingly, the human touch sense is not well researched, and much of the existing information is dubious, or difficult to apply. Nevertheless, the sense of touch (for pressure, vibration, electricity, and pain) is often employed in human-operated systems. Temperature sensing is rather well understood, although not very reliable. The vestibular sense provides information about head and body posture, but is largely dependent on the existence of a well-defined gravity vector (which is missing in space). Furthermore, the human processor is easily confused by conflicting information from the vestibular and other senses—leading, for example, to nausea. The senses of smell and taste are not currently used for many engineering applications, but are widely used in industry.

CHALLENGES

What might be reasonable approches to classify the various human senses?

How could one avoid interactions between different senses during experiments? Or would it be reasonable to present signals that simultaneously cover two or more modalities?

What might be a practical and easily used, but still scientifically exact, reference plane or system to identify the direction of the line-of-sight?

Why do many older people desire higher illumination to see objects clearly?

Does one person see a given color, say green, exactly like another person? If not, what could be practically done to compare the two perceptions?

Does the definition of the border between dark and light objects have any effect on the perception of "contrast"?

Do different physical explanations underly the "mixing of three independent primary colors" and the artist's mixing of pigments in paint to generate a given color?

How could one determine whether indeed colors have specific effects on mood, and on task performance, such as "red stimulates"?

How could one assess the effects of different kinds of "music" on mood and productivity?

How could the generation of specific smells (such as in perfumes) be made a systematic science, or engineering approach, instead of remaining an individual art?

What is needed to turn "making things taste good" into a scientific or technological systematic process instead of remaining an art?

Does the replacement of metal objects at different temperatures (thermodes) by the flow of air at different temperatures solve the problem of simultaneous stimulation of temperature and tactile sensors?

Which sensory modalities should be combined, under certain conditions and for certain signals, to increase the likelihood of perception, to increase the speed of recognition, and to enhance the central nervous processing?

5

How the Body Interacts with the Environment

OVERVIEW

Most of us work in a "normal" environment: in a moderate climate and on solid ground not too much above sea level. Yet, experiencing changes in the climate from summer to winter, or working at a high-altitude location, or being subjected to vibrations and impacts as a truck driver, or diving in the deep sea, or flying as a pilot or astronaut—all these challenge the body in many ways. Various technical and behavioral means are at hand to avoid or minimize negative effects on human comfort, health, or performance.

In industry, conditions of dust, of toxic fumes, of chemicals may exist that can be harmful to the human. Recognizing and avoiding such conditions is the domain of "industrial hygiene." Several textbooks exist in this discipline, e.g., by Cralley and Cralley (1985), Fraser (1989), and Plog (1988). For the United States, recommendations and guidelines have been published by the American Conference of Government Industrial Hygienists (ACGIH and ASHRAE, see the Addresses section of the Introduction to this book).

THERMO-REGULATION OF THE HUMAN BODY

The human body generates energy, and at the same time exchanges (gains or loses) energy with the environment. Since a rather constant core temperature must be maintained, suitable heat flow from the body outward must be achieved in a hot climate, and excessive heat loss be prevented in a cold environment.

There is some controversy about the best modelling of the energy production in the body and its exchange with the environment, given conditions of work and clothing worn. Specific topics of concern are whether one may assume a "core" that must be kept at nearly constant temperature (as done here; see also Kroemer, Kroemer, and Kroemer-Elbert, 1990; or Youle, 1990), or whether it is reasonable to assume an average skin temperature (as done by ASHRAE, 1985).

The human body has a complex control system to maintain the deep body "core" temperature very close to 37°C (about 99°F), as measured in the intestines, or rectum, or ear, or (most often) estimated by the temperature in the mouth. While there is minor temperature fluctuation throughout the day due to diurnal changes in body functions (see Chapter 6), the main task is to regulate the energy exchange between (metabolic) heat generated within the body and external energy, which may be gained in hot surroundings or lost in a cool environment.

Keeping the core temperature close to 37°C is the primary task of the human thermo-regulatory system, in cold or hot environments.

Changes in core temperature of ±2° from 37°C affect body functions and task performance severely, while deviations of ±6°C are usually lethal. At the skin, the human temperature-regulation system must keep temperatures well above freezing and below the 40s in its outer layers, but there are major differences from region to region. For example, the toes may be at 25°C, legs and upper arms at 31°, and the forehead at 34°C while one feels, overall, "comfortable" (Youle, 1990).

The Energy Balance

While discussing the metabolic system, the energy equation concerning inputs and outputs was given in Chapter 2 as

$$I = W + M + S \tag{5-1}$$

where I is the energy input via nutrition, W the external work done, M the metabolic heat generated, and S the energy storage in the body.

Assuming for convenience that the quantities I and W remain unchanged, one can concentrate on the energy exchange with the thermal environment via

$$I - W = \text{constant} = M + S \tag{5-2}$$

The system is in balance with the environment if all metabolic energy M is dissipated to the environment without a change in the quantity S.

Energy Exchanges with the Environment

Energy is exchanged with the environment through *radiation, R; convection, C; conduction, K;* and *evaporation, E.*

Heat exchange through the flow of electromagnetic energy by *radiation, R,* depends primarily on the temperature difference between two opposing surfaces, for example between a window pane and a person's skin. Heat is always radiated from the warmer to the colder surface. Hence, the body can either lose or gain heat through radiation. This radiative heat exchange does not depend on the temperature of the air between the two opposing surfaces.

The amount of radiating energy Q_R gained or lost by the human body through radiation depends essentially on the participating body surface s and on the difference between the quadrupled temperatures T (in degrees Kelvin) of the exchanging surfaces:

$$Q_R = f(s, \Delta T^4) \tag{5-3}$$

approximated by

$$Q_R \approx h_p s \times \Delta t \tag{5-4}$$

where h_p is the heat-transfer coefficient. (For more detail, read Kroemer, Kroemer, and Kroemer-Elbert, 1990, or Youle, 1990.)

Energy is also exchanged through *convection, C,* and *conduction, K.* In both cases, the heat transferred is also proportional to the areas of human skin participating in the process, and to the temperature difference between skin and the adjacent layer of the external medium. Hence, in general terms, heat exchange by convection or conduction is

$$Q_{C,K} = f(s, \Delta t) \tag{5-5}$$

approximated by

$$Q_{C,K} \approx k \times s\,\Delta t \tag{5-6}$$

with k the conduction or convection coefficient.

EXAMPLE

Cork and wood "feel warm" because their heat-conduction coefficients are below that of human tissue. Metal of the same temperature accepts body heat easily and conducts it away; therefore, it feels colder than cork or wood.

Exchange of heat through convection, C, takes place when the human skin is in contact with air and fluids, e.g., water. Heat energy is transferred from the skin to a colder gas or fluid next to the skin surface, or transferred to the skin if the surrounding medium is warmer. Convective heat exchange is facilitated if the medium moves quickly along the skin surface, thus maintaining a temperature differential. As long as such temperature gradient exists, there is always some natural movement of air or fluid: this is called "free convection." More movement can be produced by forced action (such as by an air fan, or while swimming in water rather than floating motionless): this is called "induced convection."

Heat exchange by *evaporation* is in only one direction: the human loses heat by evaporation. There is no condensation of water on the skin, which would add heat. Evaporation of water requires an energy of about 580 cal per cm^3 of evaporated water, which reduces the heat content of the body by that amount. Some water is evaporated in the respiratory passages but most (as sweat) on the skin.

The heat lost by evaporation Q_E from the human body depends on the participating wet body surfaces, s, and on air humidity, h:

$$Q_E = f(s, h) \tag{5-7}$$

As with convection, movement of the air layer at the skin increases the actual heat loss through evaporation, since it replaces humid air by drier air. Some evaporative heat loss occurs in hot as well as cold environments, because there is always evaporation of water in the lungs, which increases with enlarged ventilation at work, and there is also continuous diffusion of sweat ("insensible water loss") onto the skin surface.

The efficiency of radiation, convection (and conduction), and evaporation depends on the difference between the temperature of the skin and the environment. Figure 5-1 shows this schematically.

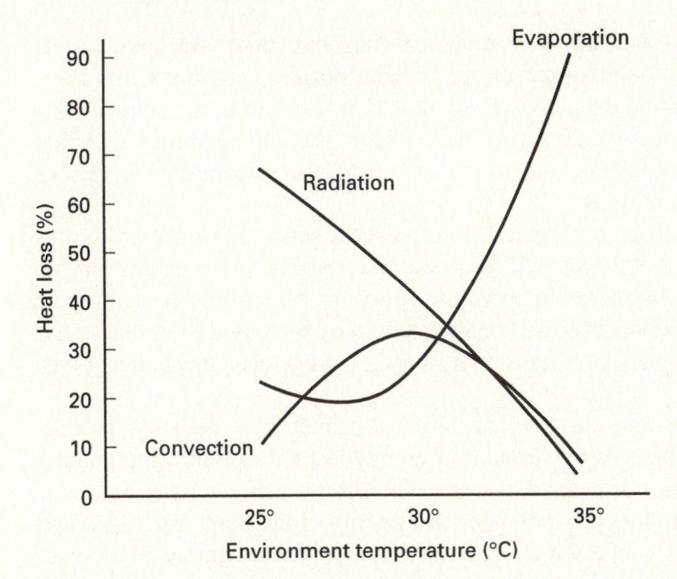

Figure 5-1. Contributions of the different kinds of heat transfer in cool or warm environments.

Storage (loss or gain) of heat energy is determined by the "specific heat" of the human body, which is about 3.5 kJ kg^{-1}. A person of 65 kg who increases the "average body temperature" by 1°C in one hour stores 230 kJ h^{-1}, i.e., 64 W.

Given these variables, heat balance exists when metabolic energy M developed in the body, heat storage S in the body [see Equations (5-1) and (5-2)], and heat exchanges with the environment by radiation R, convection C, conduction K, and evaporation E are in equilibrium. This can be expressed as

$$M + S + R + C + K + E = 0 \tag{5-8}$$

The quantities R, C, K, and E are negative if the body loses energy to the environment, and positive if the body gains energy from the environment; E is always negative.

Temperature Regulation and Sensation

Heat is produced in the body's "metabolically active" tissues—primarily at skeletal muscles, but also in internal organs, fat, bone, and connective and nerve tissue. Heat energy is circulated throughout the body by the blood. The blood flow is modulated by the actions of constriction, dilation, and shunting of blood vessels. Heat is exchanged with the environment at the body's respiratory surfaces and, of course, through the skin.

In a cold environment, heat must be conserved, which is done primarily by reducing blood flow to the skin and by increasing insulation. In a hot environment body heat must be dissipated and gain from the environment prevented. This is done primarily by increases in blood flow to the skin, in sweat production, and in evaporation.

Temperatures in the body are not uniform throughout; there are large differences between "core" and "shell" temperatures. Under normal conditions, the average gradient between skin and deep body is about 4°C at rest, but in the cold the difference in temperature may be 20°C or more. Thus the temperature-regulation system located in the hypothalamus has to maintain various temperatures at various locations under various conditions.

If the body is about to be overheated, internal heat generation must be diminished. Therefore, muscular activities will be reduced, possibly to the extent that no work is being performed anymore. In the opposite case, when more heat must be generated, the work or exercise level will be augmented by increased muscular activities. (Given the low efficiency of muscular work, it generates much heat—see Chapter 2.)

Muscles can generate more heat or less heat but cannot cool the body. In contrast, sweat production influences the amount of energy lost but cannot bring about a heat gain. Vascular activities affect the heat distribution through the body and control heat loss or gain, but they do not generate energy. Muscular, vascular, and sweat-production functions cooperatively regulate the body heat content in interaction with the external climate.

Various temperature sensors are located in the core and the shell of the body (Chapter 4). Hot sensors generate signals (sent to the hypothalamus) particularly in the range of approximately 38° to 43°C. Cold sensors are most sensitive from about 35° to 15°C. There is some overlap in the sensations of "cool" and "warm" in the intermediate range. Between about 15° and 45°C, perception of either "cold" or "hot" condition is highly adaptable. Below 15°C and above 45°C the human temperature sensors are less discriminating but also less adapting. A "paradoxical" effect is that, around 45°C, sensors again signal "cold" while in fact the temperature is rather hot.

Achieving Thermal Homeostasis

The human regulatory system must achieve two suitable temperature gradients: from the *core to the skin* and from the *skin to the surroundings*. The gradient from the core to the skin is internally the most important, because overheating or undercooling of the key tissues in the brain and the trunk must be avoided, even at the cost of overheating or undercooling the shell.

Thermal homeostasis is *primarily* achieved by regulation of the blood flow from deep tissues and muscles to lungs and skin. In the lungs, 10 to 25 percent of the total dissipated heat is transmitted to the environment; most heat is exchanged at the skin.

Secondary activities to establish thermal homeostasis take place at the muscles. They generate heat either by voluntary activities (changes in external work and exercise) or by involuntary shivering. Different actions are taken, depending on the goal of the regulatory system. If heat gain is to be achieved, skeletal muscle contractions are initiated; but if heat gain must be avoided, muscular activities are abolished.

Changes in clothing and shelter are *tertiary* actions to achieve thermal homeostasis. They achieve (together with blood-flow regulation and muscle activities) the appropriate temperature gradient between the skin and the environment. They affect radiation, convection, conduction, and evaporation. "Light" or "heavy" clothes have different permeability and ability to establish stationary insulating layers. Clothes affect conductance, i.e., energy transmitted per surface unit, time, and temperature gradient. Also, their color determines how much external radiation energy is absorbed or reflected.

APPLICATION

Assessing the Thermal Environment

The thermal environment is determined by four physical factors: air (or water) temperature, humidity, air (or water) movement, and temperature of surfaces that exchange energy by radiation. The combination of these four factors determines the physical conditions of the climate and our perception of the climate.

Measurement of the *air temperature* is performed with thermometers, thermistors, or thermocouples. Whichever technique is used, it must be ensured that the ambient temperature is not affected by the other three climate factors, particularly humidity, but also air movement and surface temperatures. To measure the so-called "dry temperature" of ambient air, one keeps the sensor dry and shields it with a surrounding bulb that reflects radiated energy. Hence, air temperature is often measured with a so-called "dry-bulb" thermometer.

Air humidity may be measured with a psychrometer, hygrometer, or other electronic devices. These usually rely on the fact that the cooling effect of evaporation is proportional to the humidity of the air, with higher vapor pressure making evaporative cooling less efficient. Therefore, one can measure humidity using two thermometers, one dry, one wetted. The highest absolute content of water vapor in the air is reached when any further increase would lead to the development of water droplets. The amount of possible vapor depends on air pressure and on the temperature of the air, with lower pressure and higher temperature allowing more water vapor to be retained than lower temperatures. One usually speaks of "relative humidity," which indicates the actual vapor content in relation to the possible maximal content ("absolute humidity") at the given air temperature and air pressure.

Air movement is measured with various types of anemometers using mechanical or electrical principles. One may also measure air movement with two thermometers—one dry and one wet (similar to what can be done to assess humidity), relying on the fact that the wet thermometer shows more increased evaporative cooling with higher air movement than the dry thermometer.

Radiant heat exchange depends primarily on the difference in temperatures between the surfaces of the individual and the surroundings, on the emission properties of the radiating surface, and on the absorption characteristics of the receiving surface. One easy way to assess the amount of energy acquired through radiation is to place a thermometer inside a black globe that absorbs practically all arriving radiated energy.

Personal comfort is not determined by "thermal balance" [Equation (5-8)] alone: in a warm environment, skin wettedness plays a major role; in a cold environment, skin temperature. Two scales, the Bedford and the ASHRAE scales, are widely used to assess individual thermal comfort: as shown in Table 5-1, they yield similar results (Youle, 1990).

In the past, various techniques were used to assess the combined effects of some or all four environmental factors and to express these in one model, chart, or index. They resulted in several "empirical" thermal indices, which are based on data collected from subjects who were exposed to various climates. Most establish a "reference" or "effective" climate that "feels the same"

TABLE 5-1. SCALES TO ASSESS SUBJECTIVE THERMAL COMFORT

Bedford			ASHRAE
Much too warm	7	+3	Hot
Too warm	6	+2	Warm
Comfortably warm	5	+1	Slightly warm
Comfortable	4	0	Neutral
Comfortably cool	3	−1	Slightly cool
Too cool	2	−2	Cool
Much too cool	1	−3	Cold

SOURCE: Adapted from Youle, 1990.

as various combinations of the several climate components. A well-known example is the *effective temperature, ET:* it reflects combinations of dry temperature, wet temperature, and air movement, with various levels of activities and clothing—see Figure 5-2.

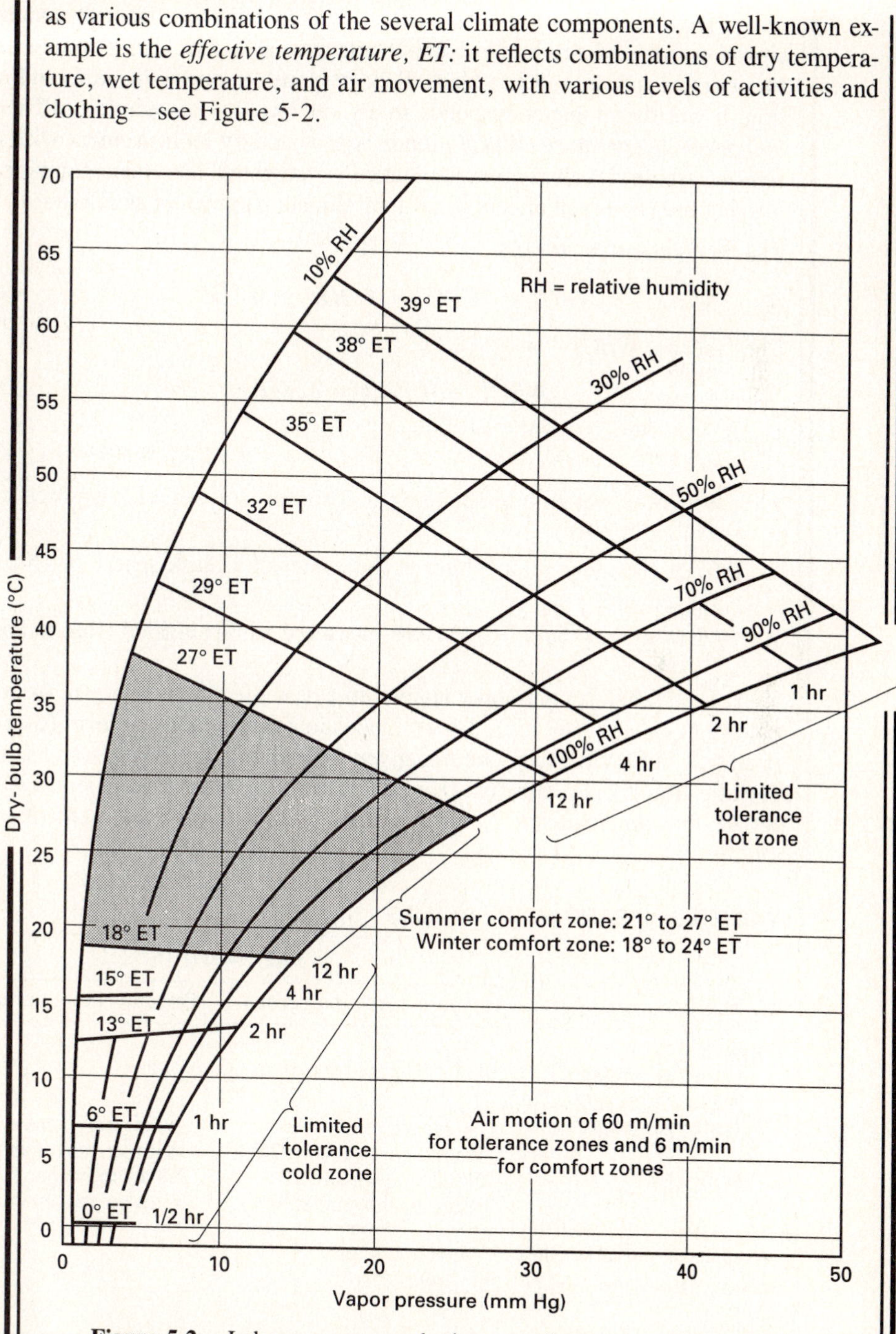

Figure 5-2. Indoors summer and winter comfort zones and thermal tolerance for appropriately dressed, sitting persons doing light work (MIL-HDBK 759).

"Direct" indices are measurements taken with special instruments designed to respond to climate components like a human. For example, a dry thermometer placed inside a black globe responds to air temperature and radiation; a wet thermometer responds to air velocity and air humidity. The *wet-bulb globe temperature (WBGT)* index is provided by an instrument with three sensors whose readings are automatically weighted and then combined. It weights the combined effects of all four climate parameters as follows:

The "outdoors *WBGT*" is

$$WBGT = 0.7WB + 0.2GT + 0.1DB \tag{5-9}$$

The "indoors *WBGT*" is

$$WBGT = 0.7WB + 0.3GT \tag{5-10}$$

where

WB is the wet-bulb temperature of a sensor in a wet wick exposed to natural air current;
GT is the globe temperature at the center of a black sphere of 15 cm diameter; and
DB is the dry-bulb temperature measured while shielded from radiation.

The *WBGT* is commonly applied to assess the effects of warm or hot climates (ACGIH, 1989; ISO, 1989). Depending on the activity level (expressed in watts), the WGBT temperatures given in Table 5-2 are considered "safe" for most healthy people, although there is some concern about the adequacy of the *WGBT* for combinations of high humidity with little air movement (Ramsey, 1987).

TABLE 5-2. "SAFE" WGBT VALUES

	"Safe" WBGT (°C)	
M, Metabolic rate (W)*	Person is heat acclimatized	Person is not acclimatized
$M \leq 117$	33	32
$117 < M \leq 234$	30	29
$234 < M \leq 360$	28	26
$360 < M \leq 468$	No air movement: 25 With air movement: 26	No air movement: 22 With air movement: 23
$M > 468$	No air movement: 23 With air movement: 25	No air movement: 18 With air movement: 20

*Assuming a skin surface area of 1.8 m²

SOURCE: Abbreviated from ISO 7243, 1982.

Reactions of the Body to Hot Environments

In hot environments, the body produces heat and must dissipate it. To achieve this, the skin temperature should be above the immediate environment in order to facilitate energy loss through convection, conduction, and radiation.

If heat transfer is still not sufficient, sweat glands are activated, and the evaporation of the produced sweat cools the skin. Recruitment of sweat glands from different areas of the body varies among individuals. The overall amount of sweat developed and evaporated depends very much on clothing, environment, work requirements, and the individual's acclimatization.

If heat transfer by blood distribution and sweat evaporation is insufficient, muscular activities must be reduced to lower the amount of energy generated through metabolic processes. In fact, this is the final and necessary action of the body, if otherwise the core temperature would exceed a tolerable limit.

EXAMPLE

If the body has to choose between unacceptable overheating and continuing to perform physical work, the choice will be in favor of core-temperature maintenance, which means reduction or cessation of work activities.

APPLICATION

There are several signs of heat strain on the body. The first is sweat rate. In strenuous exercises and hot climates, several liters of sweat may be produced in an hour. Sweat begins to drip off the skin when the sweat generation has reached about one-third of the maximal evaporative capacity. Of course, sweat running down the skin contributes very little to heat transfer.

Heat strain increases the circulatory activities. Cardiac output must be enlarged, which is mostly brought about by a higher heart rate. This may be associated with a reduction in systolic blood pressure. Another sign of heat strain is a rise in core temperature, which must be counteracted before the temperature exceeds the sustainable limit.

The water balance within the body provides another sign of heat strain. Dehydration, indicated by the loss of only 1 or 2 percent of body weight, can critically affect the ability of the body to control its functions. Therefore, the fluid level must be maintained, best by frequent drinking of small amounts of water. Normally, it is not necessary to add salt to drinking water, since in western diets the salt in the food is more than sufficient to resupply the salt lost with the sweat.

Among the first reactions to heavy exercise in excessive heat are sensations of discomfort and perhaps skin eruptions, "*prickly* heat," associated with sweating. As a result of sweating, so-called *heat cramps* may develop, which

TABLE 5-3. HEAT DISORDERS

Disorder	Symptoms	Causes	Treatments
Transient heat fatigue	Decrease in productivity, alertness, coordination and vigilance.	Not acclimatized to hot environment.	Graduate adjustment to hot environment.
Heat rash ("prickly heat")	Rash in area of heavy perspiration; discomfort, or temporary disability.	Perspiration not removed from skin; sweat glands inflamed.	Periodic rests in a cool area; showering/bathing; drying skin.
Fainting	Blackout, collapse.	Shortage of oxygen in the brain.	Lay down.
Heat cramps	Painful spasms of used skeletal muscles.	Loss of salt; large quantities of water consumed quickly.	Adequate salt with meals; salted liquids (unless advised differently by a physician).
Heat exhaustion	Extreme weakness or fatigue; giddiness; nausea; headache; pale or flushed complexion; body temperature normal or slightly higher; moist skin; in extreme cases vomiting and/or loss of consciousness.	Loss of water and/or salt; loss of blood plasma; strain on the circulatory system.	Rest in cool area; salted liquids (unless advised differently by a physician).
Heat stroke	Skin is hot, dry, and often red or spotted; core temperature is 40°C (105°F) or higher and rising; mental confusion; deliriousness; convulsions; possible unconsciousness. Death or permanent brain damage may result unless treated immediately.	Thermo-regulatory system breaks down under stress and sweating stops. The body's ability to remove excess heat is almost eliminated.	Remove to cool area; soak clothing with cold water; fan body; call physician/ambulance immediately

are muscle spasms related to local lack of salt. They may occur after quickly drinking large amounts of fluid, which dilutes the body fluids.

Heat exhaustion is a combined function of dehydration and overloading of the circulatory system. Associated effects are fatigue, headache, nausea, and dizziness, often accompanied by giddy behavior. *Heat syncope* indicates a failure of the circulatory system, demonstrated by fainting. *Heat stroke* indicates an overloading of both the circulatory and sweating systems and is associated with hot dry skin, increased core temperature, and mental confusion. Table 5-3 lists symptoms, causes, and treatment of heat-related disorders.

When human skin touches hot surfaces, the risk of a burn exists. The actual critical contact temperature depends on the duration of the contact and on the material of the contacted object or fluid. The shorter the contact, the higher temperatures can be tolerated, as listed in Table 5-4.

TABLE 5-4. MAXIMAL SURFACE TEMPERATURE THAT CAN BE TOLERATED BY HUMAN SKIN WITHOUT BURN RISK

	Maximal surface temperature (in °C) for contact times of				
Material	1 s	4 s	1 min	10 min	8 hr
Metals					
Uncoated, smooth surface	65	60			
Uncoated, rough surface	70	65	50		
Coated with varnish, 50 μm thick	75	65			
Concrete, ceramics	80	70		all	all
Glazed ceramics (tiles)	80	75	55	48	43
Glass, porcelain	85	75			
Plastics					
Polyamid with glass fibers	85	75			
Duroplast with fibers	95	85	60		
Teflon, Plexiglas	NA	85			
Wood	115*	95	60		
Water	65	60	50		

*Up to 25°C higher for very dry and very light woods.

SOURCE: Siekmann, 1990.

Reactions of the Body to Cold Environments

The human body has few natural defenses against a cold environment. Most of the actions taken are behavioral in nature, such as putting on suitably heavy clothing, covering the skin, seeking shelter, or using external sources of warmth.

In a cold climate, the body must conserve heat while producing it. To conserve heat, the temperature of the skin is lowered, reducing the temperature difference against the outside. This is done by displacing the circulating blood toward the core, away from the skin; for example, the blood flow in the fingers may be reduced to 1 percent of that in a moderate climate. Thus, cold fingers and toes may result, with possible damage to the tissue if the temperatures get close to freezing.

The development of "goose bumps" on the skin helps to retain a layer of stationary air close to the skin, which is relatively warm. The stationary layer has the effect of an insulating envelope, reducing energy loss at the skin.

The other major reaction of the body to a cold environment is the increase of metabolic heat generation. This may occur involuntarily, i.e., by shivering (thermogenesis). Shivering usually begins in the neck, apparently because warmth is important to supply blood to the brain. The onset of shivering is normally preceded by an increase in overall muscle tone in response to body cooling. With increased firing rates of motor units (see Chapter 1), but no actual movements generated, a feeling of stiffness is generally experienced. Then suddenly shivering begins, caused by muscle units firing at different frequencies of repetition (rate coding) and out of phase with each other (recruitment coding). Since no mechanical work is done to the outside, the total activity is transformed into heat production, allowing an increase in the metabolic rate to up to 4 times the resting rate. If the body does not become warm, shivering may become rather violent; its motor-unit innervations become synchronized so that large muscle units are contracted. While such shivering can generate heat that is five or more times the resting metabolic rate, it can be maintained only for a short period. There may be another mechanism to produce heat, called nonshivering thermogenesis: body organs, particularly in the liver and the viscera, increase their metabolism. The existence of this response in humans is debated. Of course, muscular activities also can be done voluntarily, such as by increasing the dynamic muscular work performed, or by moving body segments, contracting muscles, flexing the fingers, etc. Such dynamic muscular work may easily increase the metabolic rate to ten or more times the resting rate.

Activation of cutaneous vasoconstriction is apparently under the control of the sympathetic nervous system in addition to local reflex reactions to direct cold stimuli. An interesting phenomenon associated with cutaneous vasoconstriction is the "hunting reflex" (cold-induced vasodilation): after initial vasoconstriction has taken place, there is a sudden dilation of blood vessels which allows warm blood to return to the skin, such as that of the hands, which rewarms that body section. Then, vasoconstriction returns again, and this sequence may be repeated several times. If vasoconstriction and metabolic rate regulation cannot prevent serious energy loss through the body surfaces, the body will suffer some effects of cold strain.

EXAMPLE

To reduce the temperature difference to the outside, the body lowers skin temperature in a cold environment. Thus, the skin is first exposed to cold damage, while the body core is protected as long as possible.

APPLICATION

As the skin temperature is lowered to about 15° to 20°C, manual dexterity begins to be reduced. Tactile sensitivity is severely diminished as the skin temperature falls below 8°C. If the temperature approaches freezing, ice crystals develop in the cells and destroy them, a result known as "frostbite." Reduction of core temperature is more serious, because vigilance begins to drop at temperatures below 36°C. At core temperatures of 35°C, one may not be able to perform even simple activities. When the core temperature drops even lower, the mind becomes confused, with loss of consciousness occurring around 32°C. At core temperatures of about 26°C, heart failure may occur. At very low core temperatures, such as 20°C, vital signs disappear, but the oxygen supply to the brain may still be sufficient to allow revival of the body from hypothermia.

Severe reductions in skin temperatures are usually accompanied by a fall in core temperature. At local temperatures of 8° to 10°C, peripheral motor nerve velocity is decreased to near zero; this generates a "nervous block." Hence, severe cooling of skin and central body goes along with increasing inability to perform activities, even if they could save the person ("cannot light a match"), leading to apathy ("let me sleep") and final hypothermia.

Hypothermia can occur very quickly if a person is exposed to cold water. While one can endure up to two hours in water at 15°C, one is helpless in water of 5°C after 20 to 30 minutes. The survival time in cold water can be increased by wearing clothing that provides insulation; also, obese persons with much insulating adipose tissue are at an advantage over skinny ones. Floating motionless results in less metabolic energy being generated and spent than when swimming vigorously.

How cold does it feel? In a cold environment, a subject's decision to stay in the cold or to seek shelter depends on the subjective assessment of how cold the body surface or the body core actually is. It is dangerous if a person fails to perceive and to react to the body's signals that it is becoming dangerously cold, or if the body temperature becomes so low that further cooling is below the threshold of perception.

The perception of the body's getting cold depends upon signals received from surface thermal receptors, from sensors in the body core, and from some combination of these signals. As skin temperatures decrease below 35.5°C, the intensity of the cold sensation increases; cold sensation is strongest near 20°C, but at lower temperatures the intensity of perception decreases. It is often difficult to separate feelings of cold from pain and discomfort.

The conditions of cold exposure may greatly influence the perceived coldness. It can make quite a difference whether one is exposed to cold air (with or without movement) or to cold water, whether or not one wears protective clothing, and what one is actually doing. When the temperature plunges, each downward step can generate an "overshoot" sensation of cold sensor receptors, which react very quickly not only to the difference in temperature, but also to the rate of change. Yet, if the temperature stabilizes, the cold sensations become smaller as one adapts to the condition. Exposure to very cold water accentuates the overshoot phenomenon observed in cold air. The reason may be that the thermal conductivity of water is about a thousand times greater than that of cold air at the same temperature. Thus, cold water causes a convective heat loss that may be twenty-five times that of cold air (Hoffman and Pozos, 1989).

EXAMPLE

In experiments, subjects (wearing a flotation suit) were immersed into cold water at 10°C. Their temperatures at groin, back, and rectum were continuously recorded, and the subjects rated how cold they perceived these areas to be. The results of the experiment showed that the subjects were unable to reliably assess how cold they actually were. Neither their core nor surface temperatures correlated with their cold sensations (Hoffman and Pozos, 1989).

Altogether, the results of many experiments and experiences indicate that the subjective sensation of cold is a poor, possibly dangerous indicator of core and surface temperature of the body. Measuring ambient temperature, humidity, air movement, and exposure time and reacting to these physical measures is probably a better strategy than relying on subjective sensations.

Acclimatization

Continuous or repeated exposure to hot or cold conditions brings about a gradual adjustment of body functions, resulting in a better tolerance of the climatic stress. Acclimatization to heat is more pronounced than to cold. Improvement in heat tolerance is demonstrated by an increased sweat production, lowered skin and core temperature, and a reduced heart rate, compared with the first reactions of the unacclimatized person to heat exposure. The process (called acclimation) is very pronounced within about a week, and full acclimatization is achieved within about two weeks. Interruption of heat exposure of just a few days reduces the effects of acclimatization, which, upon return to a moderate climate, are entirely lost after about two weeks.

While a healthy and well-trained person acclimates more easily than somebody in poor physical condition, training cannot replace acclimatization. However, if physical work must be performed in a hot climate, then such work should also be included in the acclimatization phase.

Adjustment to heat takes place whether the climate is hot and dry, or hot and humid. Heat acclimatization seems to be unaffected by the type of work performed, whether heavy and short or moderate but continuous. It is important, during acclimatization and throughout heat exposure, that fluid and salt losses be replaced.

Acclimatization to cold is much less pronounced; in fact, there is doubt that true physiological adjustment to moderate cold takes place when appropriate clothing is worn. The first reaction of the body exposed to cold temperature is shivering — the generation of metabolic heat to counteract heat loss. Also, some changes in local blood flow are apparent. There are so-called "local" acclimatizations, particularly in blood flow in the hands and face. However, normally the adjustment to cold conditions is more one of proper clothing and work behavior than of pronounced changes in physiological and regulatory functions. Thus, the body has little or no need to change its rate of heat production or, relatedly, of food intake, in "normally cold" temperatures.

There are no great differences between females and males with respect to their ability to adapt to either hot or cold climates, with women possibly at a slightly higher risk for heat exhaustion and collapse and for cold injuries to extremities. However, these slight statistical tendencies can be easily counteracted by ergonomic means and may not be obvious at all when observing only a few persons of either gender.

Working Strenuously in Heat and Cold

Hot and cold climatic conditions (as well as air pollution and high altitude—discussed later) affect in various ways the human abilities to perform short or long, moderate or heavy work. The following text is a synopsis of the known effects for use by engineers and managers (Kroemer, 1991).

Effects of heat. When exposed to whole-body heating, the human body must maintain its "core" temperature near 37° Celsius. It does so by raising its skin temperature, increasing blood flow to the skin, accelerating heart rate, and enlarging cardiac output. The change in blood routing reduces the blood that can be supplied to muscles and internal organs. Yet, if muscles must work, their raised metabolism poses increased demands on the cardiovascular system.

Cardiovascular Effects. The pumping capacity of the heart is between about 25 ("average" adults) and 40 (elite athletes) liters per minute. The blood vessels in skin and internal organs can accept up to 10 liters, and all muscles together up to 70 liters per minute. Since the available cardiac output is half or less of these 80 liters, the ability of the heart to pump blood is the limiting factor for muscular work in a hot climate.

Effects on Muscles. An increase of muscle temperature above normal does not affect the maximal isometric contraction capability of muscle tissue, but power output of muscles is reduced at higher (and lower) temperatures. Muscle overheating accelerates the metabolic rate, which can make the muscle ineffective if it must work over some period of time. The loss of power and endurance owing to excessive muscle temperature can be counteracted by lowering muscle temperature before exercise. This reduces the cardiovascular strain and blood lactic acid concentration, and depletes muscle glycogen at a lower rate.

Dehydration. When working in a hot environment, the body loses water, i.e., it gets dehydrated. Acute water loss, incurred in a short time (in a few hours or less), called hypohydration, does not reduce isometric muscle strength (or reaction times) if the water loss is less than 5 percent body weight. However, fast and large water loss (such as introduced by diuretics) generates the risk of heat exhaustion, which is primarily the result of fluid volume depletion. Dehydration reduces the body's capacity to perform work of the aerobic or endurance type.

To counteract water loss, one must drink fluid. Plain water is best. If strenuous activities last longer than one or two hours, diluted sugar additives may help to postpone the development of fatigue by reducing muscle glycogen utilization and improving fluid-electrolyte absorption in the small intestine. Regular liberally salted food during meals (as customary in the United States) is normally sufficient to counteract salt loss. In fact, salt tablets have been shown to generate stomach upset, nausea, or vomiting in up to 20 percent of all athletes who took them.

Acclimatization. The body can adapt to heat, but not to dehydration. Most heat acclimatization takes place during the first week of exposure. First, the cardiovascular system adjusts, with blood plasma volume enlarged and heart rate decreased from its initial reaction to heat. Liberal drinking of water during the acclimatization period is helpful. Body core temperature is back to normal after 5 to 8 days in the heat. The chloride concentration in the sweat takes up to 10 days to adapt, as does the production of sweat volume and its controlled evaporation on the skin. Thus, within two weeks of staying in a hot climate, the body has acclimated completely. (Most of the heat acclimatization is lost after two to three weeks of return to normal conditions.)

Effects on Mental Performance. It is difficult to evaluate the effects of heat (or cold) on mental or intellectual performance because of large subjective variations and a lack of practical, yet objective, testing methods. However, as a rule, mental performance deteriorates with rising room temperatures, starting at about 25°C for the unacclimatized person. That threshold increases to 30 or even 35 degrees if the individual is acclimatized to heat. Brain functions are particularly vulnerable to heat; keeping the head cool improves the tolerance to elevated deep body temperature. A high level of motivation may also counteract some of the detrimental effects of heat. Thus, in laboratory tests, mental performance is usually not significantly affected by heat as high as 40°C (104°F) WBGT.

APPLICATION

Working in the Heat: Summary. Short-term maximal muscle-strength exertion is not affected by heat or water loss. The ability to perform high-intensity endurance-type physical work is severely reduced during acclimatization to heat, which takes normally up to two weeks. Even after having achieved acclimatization, the demands on the cardiovascular system for heat dissipation and for blood supply to the muscles continue to compete. The body prefers heat dissipation, with a proportional reduction of performance capability. Dehydration further reduces the ability of the body to work; hypohydration poses acute health risks. Mental performance is usually not affected by heat as high as 40°C WBGT.

Effects of cold. As in a hot climate, the body must maintain its "core" temperature near 37° Celsius in a cold environment. When exposed to cold, the human body first responds by peripheral vasoconstriction, which lowers skin temperature, in order to decrease heat loss through the skin. Such reduction in blood flow occurs in all exposed areas of the body with the exception of the head, where up to 25 percent of the total heat loss can take place. If control of blood flow away from the periphery is insufficient to prevent heat loss, shivering sets in, which is a regular muscular contraction mechanism, but without generating external work, since all energy is converted to heat. Muscular activities of shivering and of physical work require increased oxygen uptake, which is associated with increased cardiac output.

Cardiovascular Effects. The necessary increase in cardiac output is brought about mostly by increasing the stroke volume, while heart rate remains at low levels. (An explanation for the increased stroke volume has been found in higher catecholamine levels and in the shifting of blood volume from the periphery into more central circulation, associated with heightened blood pressure.) Yet, keeping the heart rate low as a reaction to cold exposure opposes the response associated with physical exercise—that is, to increase the heart rate to help enlarge cardiac output.

Effects on Body Temperature. The two opposing cardiac responses to cold and exercise affect body temperature. At light work in the cold, core temperature tends to fall after about one hour of activity. Cold sensations in the skin regularly initiate reactions leading to lowered skin temperature, yet areas over active muscles can remain warmer due to the heat generated by muscle metabolism. Thus, in the cold, relatively much heat is lost through convection (and evaporation). Which of the opposing physiological cold responses predominates depends on the special conditions, i.e., on ambient temperature, type of body activity, and the clothing insulation.

While one can feel the coldness of air in the upper respiratory tract, the warming efficiency of the upper respiratory passages is sufficient to preclude cold injuries to lung tissues under normal conditions. Discomfort and constriction of airways may be felt when inspiring very cold air through the mouth. Yet, air temperature is seldom too cold for exercise and physical work.

Effects on Energy Cost. For submaximal work in the cold, oxygen consumption is increased as compared to working at normal temperatures. (Some of this increased oxygen cost at low work levels is due to shivering.) At higher exercise intensities, oxygen cost in the cold is about the same as at normal temperatures.

However, an extra effort is required to "work against" heavy clothing worn to insulate against heat loss.

Regarding maximal exercise levels, fairly little experimental work has been performed. The limited available information indicates that at maximal work levels, a cold climate does not affect the ability for maximal exercise, as long as the exposure does not exceed about five hours. In this case, the physiological stimuli provoked by exercise appear to override those of cold. However, if core temperature gets lower, maximal work capacity is reduced, apparently mostly by suppressing heart rate and thus reducing the transport of oxygen to the working muscles in the bloodstream.

Little is known about the effects of cold exposure on endurance. However, a decrease in muscle temperature affects muscle contraction capability negatively, inducing an early onset of fatigue.

Dehydration. Dehydration occurs surprisingly easily in the cold, partly because sweating is increased in response to the increased energy demands of working in the cold, and owing to a suppressed thirst sensation. Also, urine production is increased in the cold, which can trigger water loss through more frequent urination. While dryness of cold air may cause respiratory irritation and discomfort, severe dehydration through the lungs does not occur, since exhaled air is cooled on its way out to nearly the temperature of the inhaled air, returning water vapor by condensation onto the surface of the airways. (This explains the common experience of a "running nose" in the cold.)

Acclimatization. Acclimation of the human to a prolonged stay in the cold is less effective than body adjustment to heat; in fact, it is questionable whether humans moving into a cold environment truly acclimate to cold at all, although decreases in shivering have been observed. It is uncertain whether fitness training facilitates the adaptation to cold. Most of the counteractions to cold exposure are taken to improve "insulation," such as by wearing proper clothing and staying within sheltered areas. Thus, the human usually carries a fairly normal "microclimate" while being in the cold, and hence the body has no need to acclimate.

Effects on Mental Performance and Dexterity. If the core temperature of the body drops, vigilance is reduced at about 36°C. Central nervous system coordination suffers at about 35°, apathy sets in, and loss of consciousness occurs near 32°C. Manual dexterity is reduced if finger skin temperatures fall below 20°, tactile sensitivity is reduced at about 8°C. At about 5°C, skin receptors for pressure and touch cease to function; the skin feels numb. ("Frostbite" is the result of ice crystals destroying tissue cells.) While muscle spindles are initially more active as muscle temperature drops, at about 27°C their activity is reduced 50 percent and at about 15°C is completely abolished. This explains the difficulty of performing finely controlled movements in the cold.

APPLICATION

Working in the Cold: Summary. Strong isometric muscle exertions are impaired only if the muscles are cold. The ability to do light work is reduced in the cold. Endurance activities are impaired only if core or muscle temperatures are lowered and if dehydration occurs. Clothing worn for insulation may hinder work. Dexterity and mental performance suffer in extreme cold.

Designing the Thermal Environment

There are many ways to generate a thermal environment that both is suitable to the physiological functions for the (acclimatized or nonacclimatized) person and brings about thermal comfort. One can adjust each of the physical conditions of the climate (humidity, air movement, temperatures) to influence the heating or cooling of the body via radiation, convection, conduction, and evaporation. These interactions are listed in Table 5-5, and they must be carefully considered when designing and controlling the environment.

TABLE 5-5. DESIGNING THE THERMAL ENVIRONMENT TO INCREASE (+) OR DECREASE (−) BODY HEAT CONTENT

					Temperatures (as compared to skin) of					
	Air humidity		Air movement		Air, Water		Solids		Opposing surface	
Heat transfer	Dry	Moist	Fast	Calm	Hotter	Colder	Hotter	Colder	Hotter	Colder
Radiative	No direct effect		No direct effect						+	−
Convective	No direct effect		−	(−)	+	−				
Conductive							+	−		
Evaporative	−	(−)	−	(−)	−	(−)				

The negative sign in parentheses indicates relatively small heat loss.

APPLICATION

What is of importance to the individual is not the climate in general, the so-called macroclimate, but the climatic conditions with which one interacts directly. Every person prefers an individual *microclimate* that feels "comfortable" under given conditions of adaptation, clothing, and work. The suitable microclimate is not only highly individual but also variable. It depends on gender and on age: with increasing years the muscle tonus is reduced; older persons tend to be less active and to have weaker muscles, to have a reduced caloric intake, and to start sweating at higher skin temperatures. It depends on the surface-to-volume ratio, which for example in children is much larger than in adults, and on the fat-to-lean body-mass ratio.

Thermal comfort depends largely on the type and intensity of work performed. Physical work in the cold may lead to increased heat production and hence to decreased sensitivity to the cold environment, while in the heat, hard physical work could be highly detrimental to the achievement of an energy balance. The effects of the microclimate on mental work are rather unclear, with the only sure thing being the common-sense statement that extreme climates (particularly very cold temperatures) hinder it.

Of course, clothing also affects the microclimate. Air bubbles contained in the clothing material or between clothing layers provide insulation, both against hot and cold environments; permeability to fluid (sweat) and air also plays a major role (Nielsen, Gavhed, and Nillson, 1989; Parsons 1988). Colors of the clothes are important in a heat-radiating environment, such as in sunshine, with darker colors absorbing heat radiation and light ones reflecting incident energy. An "Index of Required Clothing Insulation" is being discussed by the International Organization for Standardization (Aptel, 1988; Haslam and Parsons, 1988).

EXAMPLE

The insulating value of clothing is measured in clo units, with 1 clo = $0.16°C^{-1}\ W^{-1}\ m^{-2}$, the value of the "normal" clothing worn by a sitting subject at rest in a room at about 21°C and 50 percent relative humidity.

Clothing also determines the surface area of exposed skin. More exposed surface areas allow better dissipation of heat in a hot environment but can lead to undercooling in the cold. Fingers and toes need special protection in the cold. Head and neck have warm surfaces which release much heat, often desirable in a hot environment but not in the cold.

Convection heat loss is increased if the air moves swiftly along exposed surfaces. Therefore, with increased air velocity, body cooling becomes more pronounced. During World War II, experiments were performed on the effects of ambient temperature and air movement on the cooling of water. These physical effects were also assessed psychophysically in terms of the *wind-chill* sensation at exposed human skin. Table 5-6 shows how exposed human skin reacts to energy losses brought about by air velocity at various air temperatures. It lists the *wind-chill equivalents* in degrees Celsius which reflect the effects of air velocities at various temperatures. Note that these wind-chill temperatures are based on the cooling of exposed body surfaces, not on the cooling of a clothed person. Also, these numbers do not take into account air humidity. Under hu-

TABLE 5-6. "WIND CHILL TEMPERATURE" DEPENDING ON AIR TEMPERATURE AND AIR MOVEMENT

Wind speed ($m\ s^{-1}$)	Actual air temperature (°C)															
CALM	+10	5	2	−1	−4	−7	−9	−12	−15	−18	−23	−29	−34	−40	−46	−50
2.2	+9	+5	−1	−3	−7	−9	−12	−15	−18	−21	−26	−32	−38	−44	−50	−56
4.5	+5	+3	−7	−9	−12	−16	−18	−23	−26	−30	−36	−43	−50	−57	−64	−71
6.7	+2	−2	−9	−13	−18	−21	−23	−28	−32	−38	−43	−50	−58	−65	−73	−80
8.9	0	−8	−12	−16	−18	−23	−26	−32	−34	−40	−47	−55	−63	−71	−79	−87
11.2	−1	−9	−12	−18	−21	−26	−29	−34	−37	−42	−51	−59	−67	−76	−83	−92
13.4	−2	−11	−15	−19	−23	−28	−32	−36	−40	−44	−53	−62	−70	−78	−87	−96
15.6	−3	−12	−15	−20	−23	−29	−34	−37	−40	−45	−55	−63	−72	−81	−90	−98
17.9	−3	−12	−18	−21	−26	−30	−34	−38	−43	−47	−56	−65	−73	−82	−91	−100
Winds above 18 $m\ s^{-1}$ have little *additional* effect							Danger: flesh may freeze within 1 minute					Great Danger: flesh may freeze within 30 seconds				

SOURCE: Adapted from U.S. Army, 1981; ACGIH, 1989.

mid conditions, freezing of flesh may occur at wind-chill values as low as 800 kcal m^{-2} hr^{-1}.

Thermocomfort, obviously, is also affected by acclimatization, i.e., the status of the body (and mind) of having adjusted to changed environmental conditions. A climate that was rather uncomfortable and restricted one's ability to perform physical work during the first day of exposure may be quite agreeable after two weeks. Relatedly, seasonal changes in climate, usual work, clothing, and attitude play a major role in what appears to be acceptable or not. In the summer, most people are willing to find warmer, windier, and more humid conditions more comfortable than they would in the winter.

Various combinations of climate factors (temperature, humidity, air movement) can subjectively appear as similar. The WBGT discussed earlier is most often used to assess the effects of warm or hot climates on the human; for a cold climate, various similar approaches have been proposed but are not universally accepted yet—see Youle (1990) for a critical overview and for details. ❑

Given the many climate variables, it is not surprising that the same temperature is considered by some people to be too warm and by others too cold. However, at truly cold temperatures much agreement on "cold" is expected, while under a very hot condition most people will consent to "hot."

The effects of climate: summary. With appropriate clothing and light work, comfortable temperature ranges are about 21° to 27°C ET in a warm climate or during the summer, but lower at 18° to 24°C ET in a cool climate or during the winter. In terms of body measurements, skin temperatures in the range of 32° to 36°C are considered comfortable, associated with core temperatures between 36.7° and 37.1°C. Preferred ranges of relative humidity are between 30 and 70 percent. Deviations from these zones are uncomfortable or even intolerable. Air temperatures at floor level and at head level should differ by less than about 6°C. Differences in temperatures between body surfaces and side walls should not exceed approximately 10°C. Indoors, air velocity should not exceed 0.5 m/sec, preferably remaining below 0.1 m/s. Further information for the built environment, such as in offices, is contained in the ANSI-ASHRAE Standard 55, latest edition. In many cases, attention must be paid to proper clothing and to management of work-rest ratios (White, Hodous, and Vercruyssen, 1991). For outdoors activities, recommendations in military standards or ISO standards are applicable.

WORKING IN POLLUTED AIR

Natural events such as forest fires, dust storms, and volcanic eruptions can fill the air with contaminants, mostly smoke, soot, and dust. Pollution of air is often a man-made problem, well known from the smog conditions in Los Angeles, London, Beijing, and other areas. Primary pollutants in the air are carbon monoxide, oxides of sulphur and nitrogen, and particulates. They affect directly the respiratory system of the body, but indirectly also circulation and metabolism—see Chapter 2.

Effects of CO. Carbon monoxide (CO) is the most important primary pollutant with respect to physical work performance. Hemoglobin in the human blood has an affinity 230 times greater for carbon monoxide than for oxygen. Hence, CO easily attaches to hemoglobin and takes the place of oxygen, thus reducing the ability of blood to provide cells with oxygen. Furthermore, CO attached to hemoglobin causes the remaining binding sites on the hemoglobin molecule to develop a high affinity for oxygen, thus making it more difficult to release oxygen to the cells that need it.

Effects of SO_2 and NO_2. Sulphur oxide (SO_2) increases the flow resistance in the upper respiratory tract. This is bothersome for asthmatics but does not appear to decrease the submaximal exercise capability of healthy individuals. Its effect on maximal exercise capabilities has not been studied. Likewise, nitrogen dioxide (NO_2, which is potentially harmful to humans) does not seem to affect submaximal exercise capabilities, although it can be an irritant in the upper respiratory tract. Inhaling particulates from soot of cigarette smoke or of dust can also irritate the respiratory tract. Their effects on maximal exercise capabilities have not been studied.

Effects of SO_2 and NO_2. Sulphur dioxide (SO_2) increases the flow resistance in the upper respiratory tract. This is bothersome for asthmatics but does not appear ondary pollutants include ozone as well as peroxyacetyl nitride and other aerosols. Ozone is formed by the interaction of oxygen, nitrogen dioxide, hydrocarbons and ultraviolet light; thus ozone formation is closely tied to sunlight. Ozone is a potent irritant of airways, but no clear physiological impairment of submaximal or maximal performance capability has yet been demonstrated, though it is suspected.

Effects of Exhausts and Aerosols. Automobile exhausts are the primary source of atmospheric peroxyacetyl nitride. While blurred vision and eye irritation are known symptoms of exposure, its effect on submaximal or maximal work efforts has not been studied sufficiently. Aerosols, formed by the interactions of various acids and salts, can cause discomfort, but have not been linked to decrements in work-performance capabilities.

APPLICATION

Working in Polluted Air: Summary. Only carbon monoxide shows a clear detrimental effect on maximal aerobic performance capabilities. Other compounds can cause irritations, but currently there is no evidence that they decrease work capabilities. However, the lack of definitive studies is a serious problem.

WORKING STRENUOUSLY AT HIGH ALTITUDE

The ability to perform strenuous work depends much on the supply of oxygen to the working muscles—see Chapter 2. While the oxygen content in the ambient air remains constant (20.93 percent) to an altitude of at least 100 km, the barometric pressure is reduced with increasing height. Multiplying the percentage of oxygen by the barometric pressure yields the partial pressure of oxygen. At sea level, where the barometric pressure is 760 torr, the partial pressure of oxygen in the ambient air is 159 torr. At 3,000 m height (nearly 10,000 feet), the barometric pressure is about 252 torr; hence the partial pressure of oxygen is about 110 torr, a reduction of nearly 30 percent from sea level.

Effects on Oxygen Transfer. Oxygen (like any other gas) moves from higher to lower concentrations. According to this general rule, a reduction of the partial pressure of oxygen in the inspired air at altitude must reduce the ability of the body to supply its cells with oxygen. This is of critical importance for the mitochondria in muscle, where most of the energy for physical work is generated.

A series of processes determines the ability of the body to bring oxygen to the mitochondria. Breathing moves air in and out of the lungs. In the lungs, oxygen is transferred from the air across lung tissue into the bloodstream, where it combines with hemoglobin. Oxygen-carrying hemoglobin is then transported in the bloodstream to the muscle cells. Here, oxygen diffuses out of capillaries into the cell and finally to the mitochondrion.

Effects on Breathing. The first process, ventilation, is facilitated (for physical and physiological reasons) and automatically increased in altitude above 3,000 meters. Physical work also increases ventilation. Thus, there are no effects on the ability of the human to breathe that would limit work capacity at altitude.

Effects on Blood Oxygenation. The second process is the diffusion of oxygen from the lungs to the blood. The reduction of partial oxygen pressure in the lungs at altitude generates a smaller difference between lung air pressure and blood pressure. This slows diffusion. Since velocity of blood flow is increased during physical work (owing to increased cardiac output), the time available for oxygen diffusion in the lungs to each passing hemoglobin cell is reduced. Hence, blood oxidation falls with increasing altitude.

The next step in the process is to make oxygen-rich blood available in the arterial (systemic) branch of circulation. The oxygen content of arterial blood depends on the hemoglobin concentration in the blood, and on the ability of the hemoglobin to attract oxygen. Within the first few hours of exposure to altitude, the ability to carry oxygen is not altered from sea level, but the actual oxygen content in the blood is reduced, owing to reduced diffusion in the lungs, as just discussed. After a few hours of altitude exposure above 3,000 m there is a shift in fluid distribution in the body: blood plasma volume in the circulatory blood vessels decreases, because up to 30 percent of volume moves into cells. But with acclimatization, blood volume may again slightly increase and red-cell production is stimulated, resulting in an increase in hemoglobin concentration in the blood. Thus, in spite of the volume shift and because of the increase of hemoglobin concentration in the flowing blood, the ability of arterial blood to carry oxygen remains at approximately sea-level values.

Effects on Oxygen Supply to Muscles. The next step in the oxygen-transport process is the provision of oxidated blood to the working muscles, specifically the mitrochondria. During the first few days of exposure to altitude, oxygen supply is diminished according to the reduced oxygen content in the blood (which is the result of reduced diffusion in the lungs). As the exposure to altitude continues and altitude acclimation is achieved (see below), the oxygen content in arterial blood returns to sea-level values.

The final process in the oxygen transport chain is the oxidation of tissue. At altitude, hemoglobin releases oxygen more easily to the tissues than at sea level. With acclimatization to altitude, the capillaries in muscle tissues are enlarged, and hence the diffusion of oxygen from blood to the cell becomes facilitated.

Cardiovascular Effects. A person's ability to perform stressful aerobic work depends much on the heart's ability to move blood through the body, so that oxygen can be provided to metabolizing muscles and that metabolic byproducts (e.g., lactic acid, carbon dioxide, heat, water) be removed from them. The cardiac output, or minute volume, is essentially the product of stroke volume and heart rate. Cardiac output is not much affected at lower altitudes but shows marked changes in heights above about 1,500 m. The pumped blood volume actually increases at rest and during submaximal efforts, but after about two days exposure to altitude the volume becomes progressively reduced. After about two weeks of staying at altitude, cardiac output is lowered—at all levels of effort—to a minute volume below that at sea level, and stays at that low volume for the duration of the stay at altitude. This reduction is primarily due to reduced stroke volume, which, in turn, mostly follows from the reduced blood-plasma volume.

Physiological Adjustments to Altitude. The initial response of the body to high altitude is to increase ventilation, i.e., the number of breaths taken per minute, and the depth of each breath. The increased ventilation enlarges the pressure of oxygen within the lungs, and facilitates the release of carbon dioxide to the air.

Another adjustment is the redistribution of fluid in the body, as just discussed. The reduction in blood-plasma volume occurs within hours of arriving at altitudes of more than 3,000 meters. With long altitude stays, some of the blood is redistributed, but it does not return to sea-level conditions.

During the first few hours of exposure to altitude, reduced oxygen content in the arterial blood is brought about by the difficulties of diffusing sufficient oxygen from the air in the lungs into the blood. Adaptation to height begins after a few hours of exposure, mostly via relative and absolute increase of hemoglobin in the circulating blood, and by a slight increase in blood volume. This restores the capability of the blood to bring oxygen to the working cells. However, cardiac output capability remains suppressed (see above) throughout the stay at higher altitudes. Thus, human capability for physically highly demanding activities remains reduced at altitudes above 1500 m, even after height adaptation.

"Altitude Sickness." Rapid change in fluid distribution is associated with several well-known altitude-related medical problems. *Acute mountain sickness* (AMS) occurs commonly at heights above 3,000 meters. It brings about headaches, lassitude, nausea, insomnia, irritability, and depression. Its appearance is directly related to the rate of ascent and to the final altitude. Symptoms become apparent after several hours of exposure and reach their peak severities within one or two days, after which they recede over the next days. AMS can be reduced or eliminated by gradual ascent, and by medication. While AMS is often debilitating, it is self-limiting. More seldom, excessive accumulation of fluid in spaces between cells or in the cells can occur in the brain or lungs of persons who rapidly gain altitude. Both edema conditions are potentially life-threatening but can be counteracted by immediate evacuation to lower elevations, and by medical aid.

APPLICATION

Working at Altitude: Summary. Short-time high-intensity activities ("explosive efforts") do not suffer with increasing altitude because they are anaerobic and hence do not depend on oxygen transport. Likewise, short exertion of muscle strength is not reduced during acute exposure to high altitude.

The ability for (submaximal and maximal) aerobic work remains at sea-level ability up to about 1,500 m altitude. Submaximal work capacity is not affected up to about 3,000 m, but any given task requires a larger percentage of the available (reduced) maximal capacity than at sea level. Hence, the ability to endure such submaximal efforts is also reduced at altitude: the longer the effort, the greater the decrement.

Above 1,500 meters, maximal work capacity decreases at a rate of approximately 10 percent per 1000 m (with much variability among individuals). This reduction persists for the entire stay at height.

THE EFFECTS OF VIBRATION ON THE HUMAN BODY

Vibration is defined as oscillatory motion about a fixed point. The motion is called periodic if it repeats itself.

Most research has been done on two types of vibration. In the first case, the body continues to vibrate, at the same frequency, over a considerable period of time. The simplest way of describing this is by a sinusoidal equation. Even complex vibrations can be decomposed through a Fourier analysis into a superposition of sinusoidal events. The other case is that of non-periodic vibrations, particularly of shocks and impacts.

The human body reacts to the different kinds of vibration in various manners. If the body were rigid, all its parts would undergo the same motion, but only if the driving movement were translational (linear). When the body is rotated, even if it is rigid, not all its parts have the same motion. Of course, the human body is not rigid, and different body parts vibrate differently even if under the influence of the same source of linear vibration.

Of particular interest have been the responses of the spinal column, the head, and the hands. Measurements at the vertebrae are difficult to perform. Previously, x-rays had been used, but they have now been largely abandoned because of radiation danger. Also, chemically inert nails have been driven through the skin of the back into the posterior processes so that vertebra motion could be observed on the protruding shafts: understandably, it is not easy to find volunteers for this procedure. Measurement of motion of the head is much simpler, because one can firmly hold between the teeth a dental mold from which a rigid bar extends outward between the lips; this allows observation of the skull motions. The vibration responses at the hand are easily observed: the transmitted vibrations should be measured at the interface between the tool and the hand of the operator. This is often impossible to do, because available instruments affect the handling of the tool. Therefore, one has to attach measurement devices such as accelerometers to either the object or the hand or forearm, i.e., away from the body-object interface. This is likely to distort the actual conditions. (An example for this problem is the assessment of impacts between finger and key in keyboard operation.) In spite of this difficulty, extensive information is available from experimental measurements and epidemiological investigations—see, for example, Griffin (1990), Guignard (1985), Putz-Anderson (1988), or Wasserman (1987).

The best-known effect of vibrations was originally described in 1862 by the French physician Maurice Raynaud: it is a vibration-induced condition where fingers become pale and cold, often experienced by jackhammer operators. Thus, the condition was called "white" or "dead" finger or hand; this is now named "Raynaud's disease" (see the discussion of overuse disorders in Chapter 8). A second major area of research and application interest is in the effects of impacts and vibration on the spinal column, particularly prodded by complaints of truck and bus drivers, and of operators of earthmoving equipment, about back (and stomach) problems. Another topic of interest is that of head movements, both with respect to the ability to see visual targets and to avoid motion sickness.

Measurements of Vibration

The coordinate system used in describing mechanical vibration of the human is, unfortunately, different from systems used in other applications, such as in astronautics (see the section on gravity effects on the human). In vibration research, the x axis is also in forward direction in reference to the vibrating body, but the y axis goes to the left and the z axis upward—see Figure 5-3. (Note that in aerospace work, different direction conventions are used—see later in this chapter.)

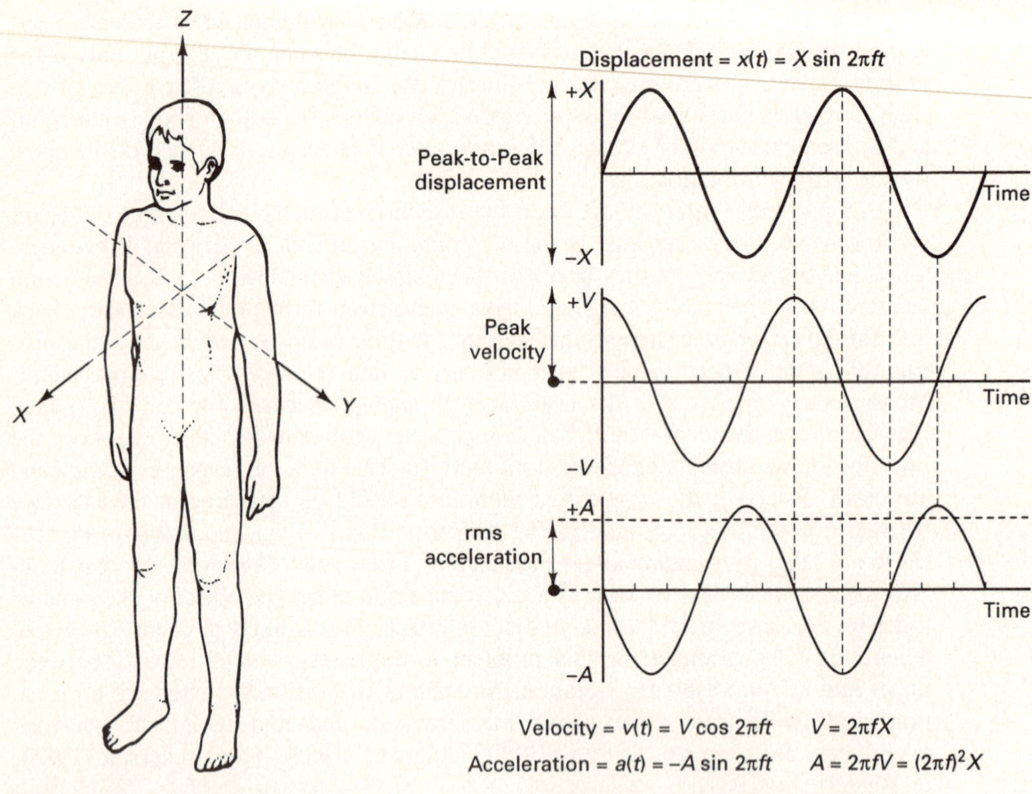

Figure 5-3 Convention on directions of vibrations and impacts—compare with Figure 5-14.

Figure 5-4. Sinusoidal vibration described by displacement, velocity, and acceleration over time.

The "magnitude" of vibration is described by displacement and its time derivatives, velocity and acceleration, over time—see Figure 5-4.

The *displacement* of a mass can be described as the maximal amplitude above (or below) the stationary location; this is called the "peak amplitude."

Another way of describing the magnitude of oscillation is by *velocity*. While this is an appropriate way to describe vibration, it is seldom used, because the instrumentation for measuring *acceleration* is, presently, more convenient. A common measure of acceleration, besides the mean (or average), is the "RMS" value: this is the square root of the mean of the squared values. Sometimes it is described in units of g, the acceleration due to gravity (9.80665 m s^{-2}). Occasionally, a logarithmic scale is used for the vibration magnitude, expressed in decibels (dB), such as used in sound assessment—see Tables 5-7 and 5-8. Griffin (1990) states that this is neither advantageous nor necessary because the range between perception threshold and pain threshold is in a range of about 1,000 : 1—i.e., much smaller than for sound.

If the motion is of a simple harmonic type with a peak magnitude of A, the peak-to-peak magnitude is $2A$, while the RMS magnitude is $0.707A$. Thus, it is nec-

TABLE 5-7. REFERENCE QUANTITIES DEFINED IN ISO 1683 (1983) ($\lg = \log_{10}$)

Description	Definition (dB)	Reference quantity
Sound pressure level in air	$20 \lg (p/p_0)$	2×10^{-5} Pa
Other-than-air sound pressure level	$20 \lg (p/p_0)$	10^{-6} Pa
Vibration acceleration level	$20 \lg (a/a_0)$	10^{-6} m s^{-2}
Vibration velocity level	$20 \lg (v/v_0)$	10^{-9} m s^{-1}
Vibration force level	$20 \lg (F/F_0)$	10^{-6} N
Power level	$10 \lg (P/P_0)$	10^{-12} W
Intensity level	$10 \lg (I/I_0)$	10^{-2} W m^{-2}
Energy density level	$10 \lg (w/w_0)$	10^{-12} J m^{-3}
Energy level	$10 \lg (E/E_0)$	10^{-12} J

TABLE 5-8. CONVERSIONS BETWEEN DECIBELS AND OTHER UNITS OF ACCELERATION AND VELOCITY (REFERENCE LEVELS DEFINED IN TABLE 5-7)

Decibel (dB)	Acceleration (m s^{-2})	Velocity (m s^{-1})
−20	10^{-7}	10^{-10}
0	10^{-6}	10^{-9}
20	10^{-5}	10^{-8}
40	10^{-4}	10^{-7}
60	10^{-3}	10^{-6}
80	10^{-2}	10^{-5}
100	10^{-1}	10^{-4}
120	1	10^{-3}
140	10	10^{-2}
160	10^{2}	10^{-1}
180	10^{3}	1
200	10^{4}	10

essary to be careful about the descriptors peak, peak-to-peak, or RMS. None of these measures reflects the effects associated with the duration of the motion. For this, a "dose" value is often used that indicates the effects of the time of exposure.

In a simple *harmonic* motion, there is sinusoidal oscillation at a single frequency, but most "real" motions contain vibration of several frequencies. In some cases these frequencies are harmonics, i.e., integer multiples of the lowest frequency, also called the *fundamental frequency*. A "spectrum" describes how the vibration magnitude varies over a range of frequencies. The most often used way is to determine the magnitude in either octave or third-octave bands. An octave is the interval between two frequencies when one frequency is twice the other. If f_1 and f_2 are the lower and upper frequencies of the bands, then

$$1/1 \text{ octave:} \quad f_2 = 2f_1$$

$$1/2 \text{ octave:} \quad f_2 = 2^{1/2}f_1$$

$$1/3 \text{ octave:} \quad f_2 = 2^{1/3} f_1$$

$$1/6 \text{ octave:} \quad f_2 = 2^{1/6} f_1$$

For example: when centered on 1 Hz, the octave band is from 0.707 to 1.414 Hz, the third-octave band from 0.891 to 1.122 Hz, respectively. In these cases, the bandwidth increases in proportion to frequency. Another method is to determine the frequency content using a constant bandwidth (such as 0.1 or 1.1 Hz) at all frequencies.

In the following, it is generally assumed that the waveform of the vibratory motion is sinusoidal. The reciprocal of the period T is the *frequency* f, i.e., the number of cycles of motion per second, expressed in hertz. One often uses the *angular frequency*, ω, expressed in radians per second: since a complete cycle (360 degrees) corresponds to 2π radians, $\omega = 2\pi f$, in rad s^{-2}. At the maximum displacement, the velocity is zero and the acceleration is at a minimum; when the displacement is zero, the velocity is maximal and the acceleration is zero, as shown in Figure 5-4.

At a time t, the *instantaneous displacement* x is described by

$$x(t) = X \sin(2\pi f t + \varphi)$$

where X is the peak displacement and φ is the phase angle (time delay).

The *instantaneous velocity* v of the motion is the first time derivative of displacement

$$v(t) = 2\pi f X \cos 2\pi f t = V \cos 2\pi f t$$

where $V = 2\pi f X$ is the *peak velocity*.

The *instantaneous acceleration* a of this motion is the time derivative of velocity:

$$a(t) = -(2\pi f)^2 X \sin 2\pi f t = -A \sin 2\pi f t$$

where $A = (2\pi f)^2 X = 2\pi f V$ is the *peak acceleration*.

Jerk is the time derivative of acceleration: The *peak jerk* equals $(2\pi f)^3 X$, or $(2\pi f)^2 V$, or $2\pi f A$.

For sinusoidal vibration, the measures of displacement, velocity, acceleration, and jerk can be converted; also, one can convert between peak, peak-to-peak and rms values: see Table 5-9.

All of the discussed methods of measuring vibration incorporate covertly the duration of the vibration. For example, there are many possible rms values on the same vibration, depending on the time period over which the rms value is determined. Thus, the "time window" used must be defined.

Effects of Vibration

Vibration can produce a wide variety of effects, depending on the intensity and direction of vibration, and on the body parts to which it is transmitted. Commonly, one groups the observed or suspected effects of vibration into interference with comfort, with activities, and with health.

TABLE 5-9. CONVERSIONS OF VIBRATION PARAMETERS (FOR SINUSODIAL MOTIONS)

	Displacement X	Velocity V	Acceleration A
Displacement X	X	$X = \frac{V}{2\pi f}$	$X = \frac{A}{(2\pi f)^2}$
Velocity V	$V = 2\pi f X$	V	$V = \frac{A}{2\pi f}$
Acceleration A	$A = (2\pi f)^2 X$	$A = 2\pi f V$	A

	Peak	Peak-to-peak	RMS
Peak	Peak	Peak = 0.5 peak-to-peak	Peak = 0.707 rms
Peak-to-peak	Peak-to-peak = twice peak	Peak-to-peak	Peak-to-peak = 1.414 rms
rms	rms = 1.414 peak	rms = 0.707 peak-to-peak	rms

While vibration is most often unwelcome, there are also "good vibrations," such as associated with the pleasant feeling of shaking hands, sitting on a rocking chair or a swing, or laughing. It can be a source of excitement, for instance on the fairground, on sailboats or skis, or in motor cross racing. It has been advocated for improving joint mobility of athletes or of patients suffering from arthritis (Griffin, 1990).

The cause-effect relationships between vibrations and human responses are complex and often difficult to research. Even in the well-controlled laboratory environment, equipment to generate vibrations is often not able to generate "pure" sinusoidal motions, or it may not be capable of generating, at the same time, high accelerations and large displacements together with considerable forces. The reactions of the human body are quite different from person to person, and may depend on muscle tension and posture of the subject, as well as on restraining devices used. In the "real world," vibration effects on the human are often not the sole stressors; for example, it may be quite difficult to determine how vibrations or impacts experienced throughout the day by a truck driver may affect that person's performance and well-being.

Whole-body vibration. "Whole-body" vibration occurs when one stands, lies, or sits on a vibrating surface. Vibration is then transmitted in some way throughout the whole body; yet, for example when sitting on a vibrating seat, the feet may not experience much motion. Thus, the distinction between "whole-body" and "local" vibration is not always clear. Furthermore, seated persons are often simultaneously exposed to local vibrations, e.g., the head from a head rest, the back from a backrest, the hands from a steering wheel, and the feet from the floor.

APPLICATION

Responses to Vertical Vibration. Most parts of the human body move together under vertical oscillation at frequencies below 2 Hz; the associated sensation is that of alternately being pushed up and then floating down. The eyes are able to follow objects that either move with the body or are stationary. Yet, free movements of the hand may be disturbed, which can cause problems in activities that require exact positioning of the hands. If the vibration has a frequency below 0.5 Hz, it may cause symptoms of motion sickness (Griffin, 1990).

Oscillation at frequencies above 2 Hz causes amplification of the vibration within the body. Yet, the frequencies with greatest amplification, i.e., the resonance frequencies, are different for different parts of the body, for different individuals, and for different body postures. At frequencies between 4 and 5 Hz, resonances occur, for example, in the head and hands, and, altogether, discomfort is strongly felt. At frequencies above 5 Hz, the force required to generate a given vertical acceleration falls rapidly with increasing frequency; thus, the vibration reaching the head, and its associated discomfort, decrease. But at frequencies between 10 and 20 Hz, the voice may warble, and vision may be affected—particularly at frequencies between 15 and 60 Hz because of resonances of the eyeballs within the head.

When standing, keeping the knees straight or bent can greatly influence the effects of frequencies above 2 Hz. When sitting, the design of the seat has fairly little influence at frequencies below 2 Hz, but "soft" seats, such as in many automobiles, can greatly amplify vertical vibrations, such as doubling the experienced frequency. Thus, the design of seats to be used in vibrating environments can be of great importance for the vibration effects experienced by the seated person.

Responses to Horizontal Vibration. Sideways or fore-and-aft vibration of the seated body below 1 Hz sways the body, even if resisted by muscle action. Between 1 and 3 Hz, it is difficult to stabilize the upper parts of the body, a situation which is associated with great discomfort. With increasing frequency, horizontal vibration is less readily transmitted to the upper body, so that at frequencies above 10 Hz the vibration is mostly felt at the seat surface. A back rest can greatly influence the effect of horizontal vibration: at low frequencies it can help to stabilize the upper body, but at high frequencies it strongly transmits vibration to the upper body, of course primarily in anterior-posterior direction (Griffin, 1990).

Subjective Assessment of Vibration Effects

Stevens' power law can be used to establish subjective ratings of the sensation associated with vibrations. It relates the perceived sensation P to the magnitude of the stimulus I by $P = KI^n$, where K is a constant. The values for n have been found mostly to be in the range between 0.9 and 1.2, with some values outside this range for special conditions (Griffin, 1990). Semantic scales have been developed by several authors, notably for the assessment of vehicle ride (SAE, 1973). These and similar approaches have been used to establish "comfort contours," similar to those developed to assess noise.

In general, persons of different body size, age, or gender report the same sensations associated with vibration, although larger subjects tend to be a bit less sensitive to frequency below 6 Hz.

ISO Standard 2631 (1985) offers guidance for the evaluation of whole-body vibration with regard to comfort boundaries. It is shown in Figure 5-5. Although the standard needs improvement (Griffin, 1990), it indicates the time-dependent comfort boundaries for combinations of frequencies and accelerations in whole-body vibrations. For example, vertical vibrations at a frequency of 4 Hz are given as comfortable for a period of eight hours if the rms acceleration is about 0.1 m s^{-2}, but the time is reduced to about one hour if at about 0.4 m s^{-2}.

Combined vibration and noise. In many experiments, subjects are subjected to both noise and vibration, particularly because the vibratory equipment also

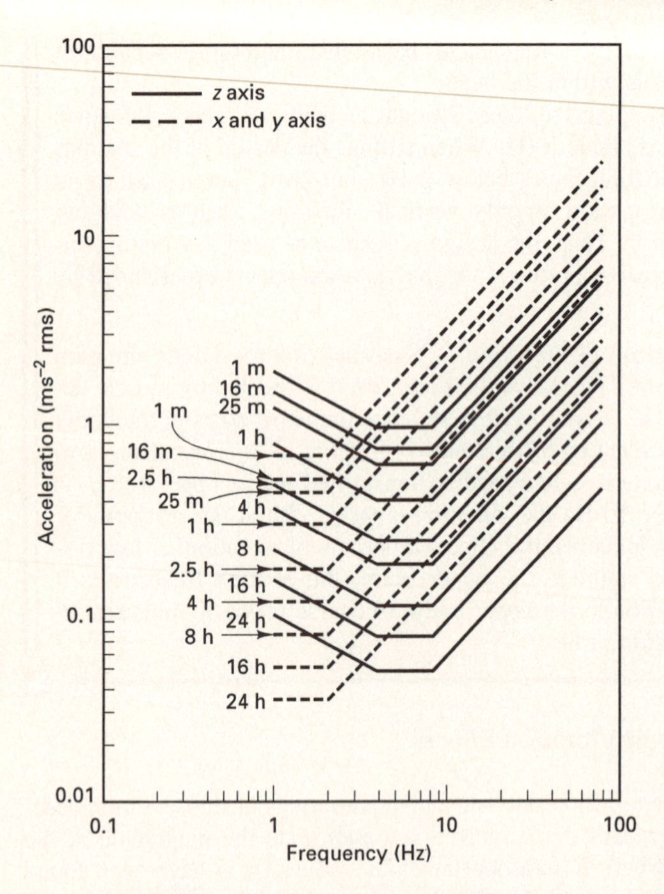

Figure 5-5. Comfort/discomfort boundaries for vertical and horizontal whole-body vibrations (ISO 2631, 1985).

generates noise. There appears to be similarity between the sensations of noise and vibration, as expressed by Stevens' power law, which allows the formulation of conditions of equivalence (Griffin,1990). Thus, in conditions where both mechanical and acoustical vibrations are present, the reduction in one stimulus may not only improve the overall perception of the vibratory environment, but might in fact reduce the perception of the other, unchanged stimulus.

APPLICATION

Effects of Vibration on Performance

The human may be modelled as a system that receives information through sensors, makes decisions on the information, and then performs activities based on these decisions (see Chapter 3). Little research has been done on the effects of mechanical vibrations on mental activities, state of arousal, ability to make decisions, or attitudes.

Vibration of the body mostly affects the principal input ports, the eyes, and the principal output means, hands and mouth. Either the external world (visual or motoric target) may vibrate, or the body may do so while the outside is stationary.

Control operations may be discrete, for instance when pressing a button, or continuous, such as when driving an automobile. Automobile driving is also an example of a "pursuit tracking task," which is in contrast to "compensatory tracking," where only the difference between existing and desired location of the controlled vehicle is known.

The effects of vibration are rather similar for tracking tasks, whether of the pursuit or compensatory kind. Performance reductions are mostly dependent on vibrations induced in the upper body (between 6 and 10 Hz) and in the shoulder region (between 2 and 10 Hz). The hands have resonances at 4 to 5 Hz (which, however, depend on the vibration direction, seating condition, posture, etc.). There is also a strong effect of acceleration, with larger accelerations effecting control operations more than small accelerations. The effects of (induced or resulting) vibration that occurs in several axes simultaneously are complex and difficult to model.

An operation error related to vibration (called breakthrough or feedthrough) is dependent on combined, often complex transfer functions of the biodynamic system, of the control, and of the controlled machinery (Griffin, 1990). Fortunately, the characteristics of many controls and controlled machines considerably attenuate vibration errors at frequencies above 1 Hz, but highly sensitive systems, or tasks that require small or precise movement, may be strongly affected by vibratory environments: examples are the activation of pushbuttons, or handwriting.

If a visual target oscillates slowly, the eyes pursue the movement and maintain a stable image on the retina—see Chapter 4. This is called a reflex response. The human eyes are able to perform pursuit reflexes for display oscillations of up to about 1 Hz. At higher frequencies, the saccadic movements are too slow and the image becomes blurred. When the observer is vibrating, the head and eyes experience both translational and rotational movements. The complex motoric compensatory activities become increasingly insufficient as the frequencies exceed 8 Hz; vision problems occur when the apparent displacement of the visual object gets larger than one minute of arc (Griffin, 1990; Oborne, 1983). In general, the effects of translational vibration decrease with large viewing distance, while the effects of rotational rotation are independent of distance.

The airflow through the larynx as well as breathing irregularities, and related changes in general body tension, may change the pitch of the voice, particularly in vertical oscillations at frequencies between 5 and 20 Hz.

Vibrations Causing Injuries and Disorders

Whole-body vibration with a peak magnitude below 0.01 m s^{-2} is hardly felt, while accelerations of 10 m s^{-2} rms or higher may be assumed hazardous. The effects of intermediate accelerations depend on the actual frequency, direction, and duration of the vibration. Unfortunately, the available information is still insufficient to establish causal relationships; only a small number of physiological responses to whole-body vibration are well documented (Griffin, 1990). Exposure to vibration often results in short-lived changes in various physiological parameters such as heart rate: vertical vibration in the range of 2 to 20 Hz can produce a cardiovascular response which is similar to that during moderate exercise (Guignard, 1985). At the onset of vibration exposure, increased muscle tension and initial hyperventilation have been observed (Dupuis and Zerlett, 1986). The musculo-skeletal system is, by mechanical reasoning, strongly suspected to be affected by the motions and energies that it must resist or counteract. Reflex responses can be inhibited (Martin, Roll, and Gauthier, 1986).

While motions between skin and underlying structures make the interpretation of electromyographic (EMG) recordings difficult (Robertson and Griffin, 1989), they have been used widely to record the activity of back muscles during whole-body vibrations (see Chapters 1 and 11). The observed muscular contractions do not necessarily protect the body but may in fact enhance the strain beyond that of a passive system (Seidel, 1988), partly because of untimely contraction due to phase lags.

Among the responses of the sensory system, vision is easily disturbed by motions of the body, as discussed before. Since vibration and noise often occur together, negative effects on hearing have been reported, in some cases supposedly even in the absence of noise. Tactile perception is affected in many circumstances (Griffin, 1990).

☞☞☞ *"The respectable scientist may wish to avoid the danger of compromise upon the uneven, unproven, and often undefined, ground on which the democratic production of national, international and other standards take place. Imperfect guidance can be beneficial, but attempts to 'whitewash' the imperfections in a standard are more likely to create a prestigious white elephant than to preserve the standard in pristine condition. Environmental standards can be immensely valuable, but they should be evolved today in the recognition that they must assist the evolution of, and eventually make room for, improved guidance tomorrow. Consequently, users of any standard would be wise to assess the areas of expertise, the interests and the knowledge available to those who formulated the standard" (Griffin, 1990, pg. 633).* ☜☜☜

In spite of many uncertainities, standards have been published in the attempt to describe acceptable vibration exposures. Figure 5-6 describes vertical acceleration conditions.

While chronic degenerative effects on bone or cartilage have been reported after prolonged vibration exposure, vibration may be beneficial for the skeletal system of persons who experience only weak stimuli for tissue maintenance or growth (Wolff's law), such as paraplegics and astronauts. On the other hand, prolonged vibration may lead to degeneration of the spinal column, for instance in tractor and truck drivers, or to the "vibration white-finger disease" in workers using vibrating

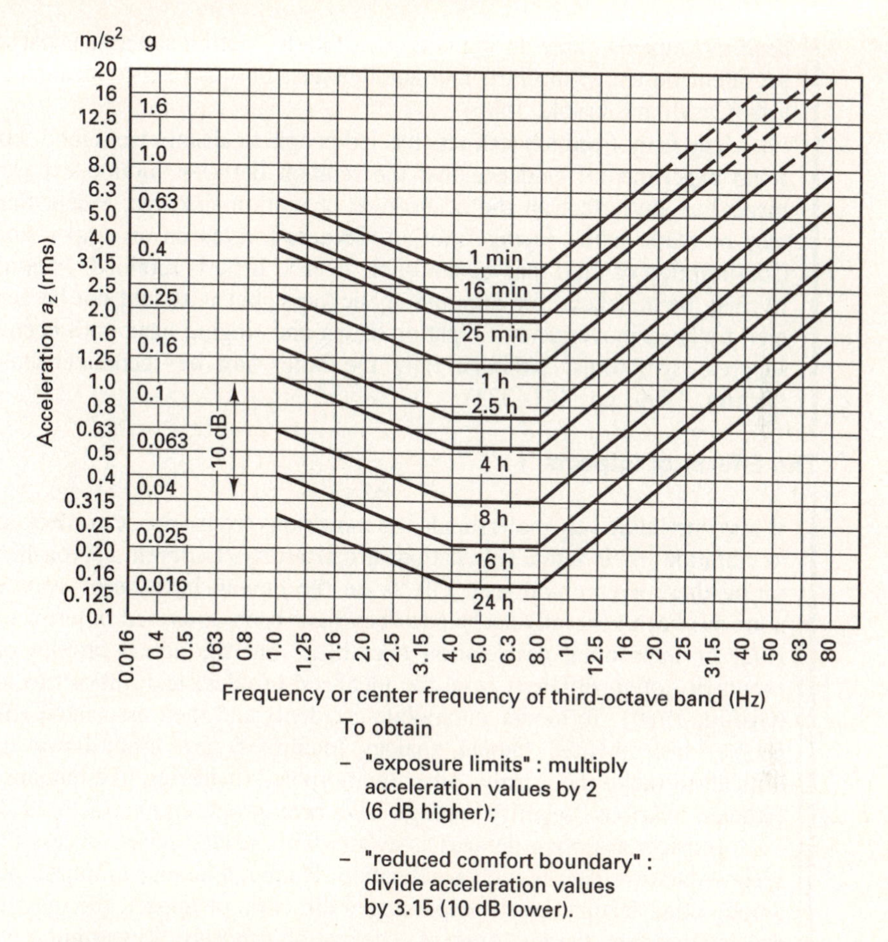

Figure 5-6. Vertical accelerations limits in terms of frequency and exposure time (ISO 2631, 1978).

tools such as jack hammers (Dupuis and Zerlett, 1986; Oborne, 1983; Wasserman, 1982; Wasserman, Phillips, and Petrofsky, 1986).

APPLICATION

Motion sickness. The vestibular system is of major importance with respect to the appearance of vibration-induced "motion sickness," often accompanied by nausea. Previously, it was theorized that the problem arises from a conflict among the sets of information received from two or more sensory systems, particularly the vestibular and visual sensations. While the conflict theory needs further refinement, the involvement of the vestibular system is critical, such as indicated by the strong tendency toward motion sickness caused by head movements alone during body oscillation. Motion sickness is particularly prevalent on ships, aircraft, and automobiles, but also exists in flight simulators (Casali and Frank, 1986).

The fear of motion sickness has led people to abstain from food, but there is no experimental evidence that the relation between motion exposure and meals has any effect on the occurrence of motion sickness, except that there may be truth in the saying "motion sickness thrives on an empty stomach." Consumption of fluid may be advisable even if little is retained. Mental activity may be beneficial for minimizing sickness, but it should not be accompanied by head movements. "Fighting at sea and singing have both been said to suppress symptoms, although only the latter can be recommended here!" (Griffin, 1990, pp. 327–328).

The Effects of Impacts

While one intuitively understands the transitions from impact to shock to bump to vibration, their actual delineations are arbitrary. Different approaches to describe these events and their effects on the human have been proposed, but none has been generally accepted. They have been described in terms of triangular or trapezoidal acceleration over time, but the actual profiles of these events are often different from the idealized profiles, and difficult to measure (Griffin, 1990). To model automobile accidents and their associated effects on people, instrumented human analogs (dummies) are used; however, it is difficult to make them truly "anthropomorphic" (behaving like humans). Nevertheless, various "severity" indices have been proposed (Versace, 1971).

Impacts are often defined as events with sudden onset, of less than one second duration, and of high acceleration. Human tolerance to impact depends, among other factors, on the experienced direction of impact, the magnitude of deceleration, and the total time of deceleration exposure. Linear impacts occurring at right angles to the spinal axis are better tolerated than those parallel to the spine. Among life-threatening skeletal fractures, damage to the vertebrae is most common, while at high impacts head injury is most frequent and severe. Figure 5-7 summarizes the conditions under which humans have survived impacts. Of course, a wide variety of conditions affect impact survival. For example, a 2-meter head-first fall of a child onto a flat solid surface can result in skull fracture; tolerance limits appear to be about 150 to 200 G for three ms^{-2} average acceleration and 200 to 250 G for peak accelerations (R. G. Snyder, personal communication, 18 February, 1991).

Human tolerance to multiple-G forces sustained for a few seconds depends on the directions of these forces relative to the body. Experience and experimentation have shown that they are best tolerated in the plus or minus x-direction, i.e., when the body is supported either on its belly (prone) or on its back (supine), perpendicular to the direction of action. Fairly little is known about sideways actions, i.e., in the $\pm y$ direction. Force along the z axis is

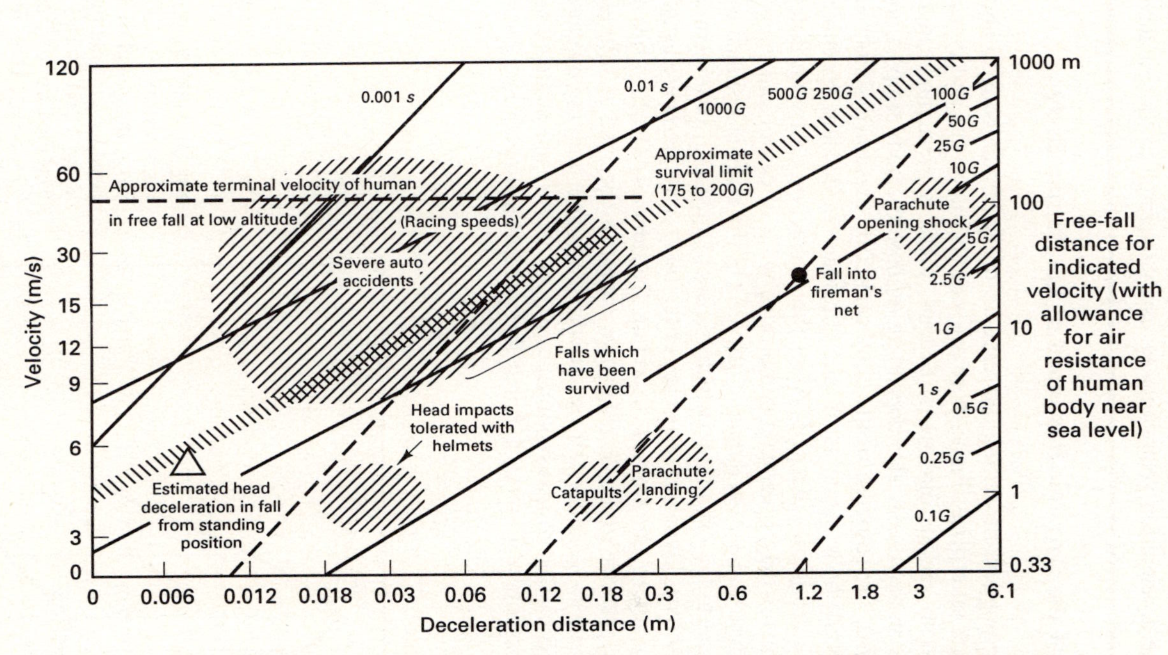

Figure 5-7. Approximate conditions under which humans have survived impacts (modified from Snyder, 1973).

difficult to tolerate, particularly when downward, because it hinders the blood supply to the brain and, therefore, may lead to vision disturbances, greyout or blackout. This has caused many problems and accidents for pilots of aircraft. Astronauts have been put onto "precast" couches during blast-off from earth so that the acceleration vector may be easily tolerated in the $-x$ direction.

Models of the Dynamic Response of the Human Body

Understanding and modelling the response of the human body to impact or vibration is a difficult enterprise. First, there are various intensities and types of impacts and vibrations, in various directions. Second, there is great inter- and intraindividual variability in responses, also depending on posture, body support, and restraint systems. Third, the response of the body cannot be explained simply in terms of resonances. While body elements, if taken in isolation, do show specific natural frequencies, these are highly damped. Also, the interactions between differently vibrating body segments may generate a complex network of causal and temporal sequences.

In most studies, the dynamic response of the body is assumed to be *linear*, i.e., proportional to the excitation (which is either sinusoidal, or a recorded or simulated vehicle motion). Dynamic responses of the body or its parts are usually described by *transfer functions* determined at certain frequencies. Griffin (1990) ordered these into two types: one group describes two measures obtained at different points; this is called *transfer impedance*. *Transmissibility* compares only the magnitudes, for example of head motion to seat motion. "Comfort" curves are assumed to be the inverse of transmissibility. Figures 5-8 and 5-9 present examples for observed transmissibilities.

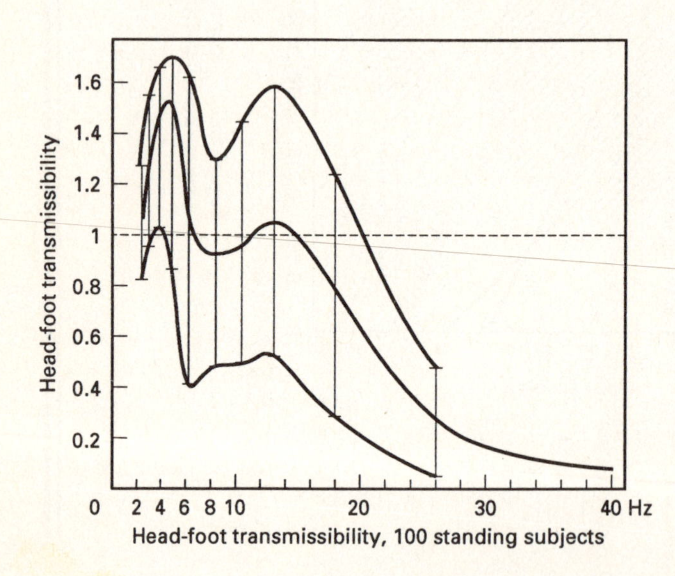

Figure 5-8. Typical head-foot transmissibility: 100 standing subjects, mean and 67 percent boundaries. Note the resonances at about 4 and 14 Hz. SOURCE: Modified from "Vibration of Work" by Oborne, D. J. in D. J. Oborne and M. M. Gruneberg (eds.), *The Physical Environment at Work*. Copyright © 1983 by John Wiley & Sons, Ltd. Reprinted by permission of John Wiley & Sons, Ltd.

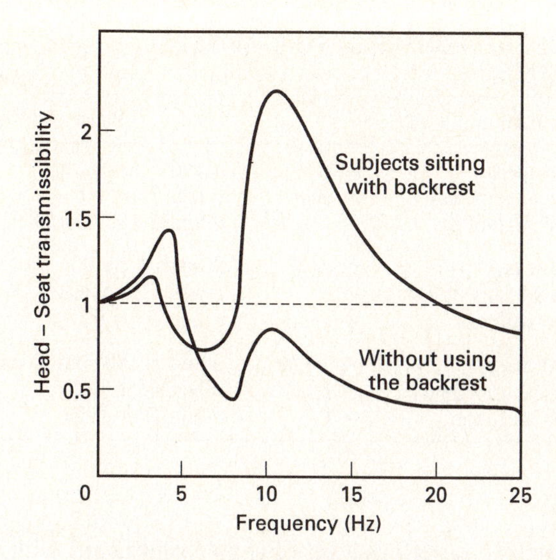

Figure 5-9. Example of head-seat transmissibility of a sitting subject, either leaning against a backrest or without using the backrest (modified from Griffin, 1990).

The other group describes the ratio of two different measures obtained at the same point, called *mechanical impedance*. It describes all relations between the force that drives a system at a particular frequency and the resulting movements, displacement, velocity, and acceleration.

When a force is applied to a mass, it produces an acceleration which is proportional to and in phase with the force. According to Newton's second law, the constant of proportionality is called mass (m). If force is applied to a (massless, ideal) damper, it produces a velocity which is also proportional to and in phase with the force. The constant of proportionality is called "damping" (d). Application of a force to a (massless, ideal) spring produces a displacement. The constant of proportionality is called "stiffness" (k) of the spring.

For a rigid body, force and acceleration are always in phase and thus, at any frequency, the ratio of their rms magnitudes indicates the mass of the object. At high frequencies, however, the human body does not behave as if rigid, and force and acceleration are out of phase in a manner which depends on the stiffness and the damping at each frequency. Of course, one can still calculate the ratio of force to acceleration, but it no longer equals the static mass of the object. Therefore, the term "apparent mass" or "effective mass" is used. Similarly, the properties of dampers and springs have different effects on movement of masses at changing frequencies, thus the terms "impedance" or "dynamic stiffness" are used instead of "damping" or "stiffness."

The term force/acceleration is called the apparent mass (or *effective mass*); the expression force/displacement is called *dynamic stiffness* (or dynamic modulus). Acceleration/force is called *accelerance* (or *inertance*), velocity/force *mobility*, and displacement/force *dynamic compliance* (Griffin, 1990).

For a simple system, there are fixed relations between acceleration, velocity and displacement. Changing from velocity to acceleration adds a 90-degree phase lag, and the values change by $2\pi f$. Table 5-10 displays these relationships.

TABLE 5-10. DYNAMIC RESPONSES OF PURE MASSES, DAMPERS, AND SPRINGS

Element	Modulus		Phase
Mass, m	Apparent mass	$= m$	a and F in phase
	Mechanical impedance	$= i\omega m$	v lags F by 90°
	Dynamic stiffness	$= -\omega^2 m$	d and F 180° out of phase
Damper, c	Apparent mass	$= c/i\omega$	a leads F by 90°
	Mechanical impedance	$= c$	v and F in phase
	Dynamic stiffness	$= i\omega c$	d lags F by 90°
Spring, k	Apparent mass	$= -k/\omega^2$	a and F 180° out of phase
	Mechanical impedance	$= k/i\omega$	v leads F by 90°
	Dynamic stiffness	$= k$	d and F in phase

SOURCE: Griffin, 1990.

One may model a complex system as a set of simple subsystems with discrete components (of mass, damping and stiffness) which, together, have the same mechanical impedance as the complex body. Of course, this model is "true" only within boundaries that must be clearly defined. A typical model of the sitting body involves two masses: m_1, the mass of the body that moves relative to the platform supporting the seated body; and m_2, the mass of the body and legs that does not move relative to the platform—see Figure 5-10. If the feet and legs (m_3) do not move in phase with the seat, the model must be extended.

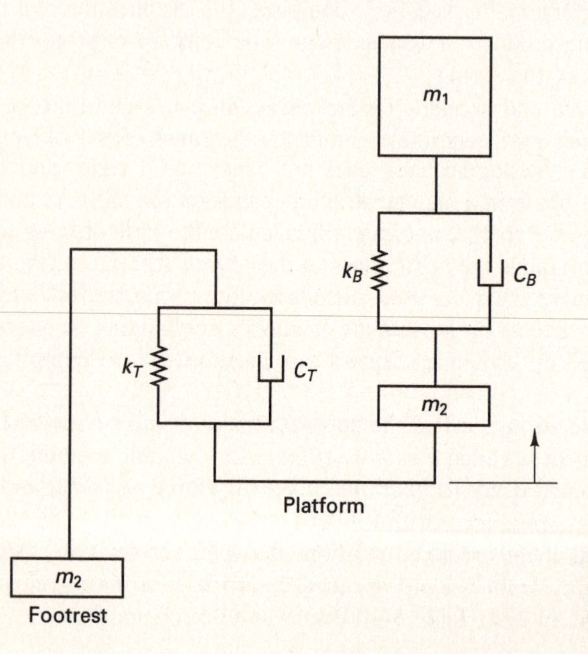

Figure 5-10. Model of a subject sitting on a vibrator (modified from Griffin, 1990).

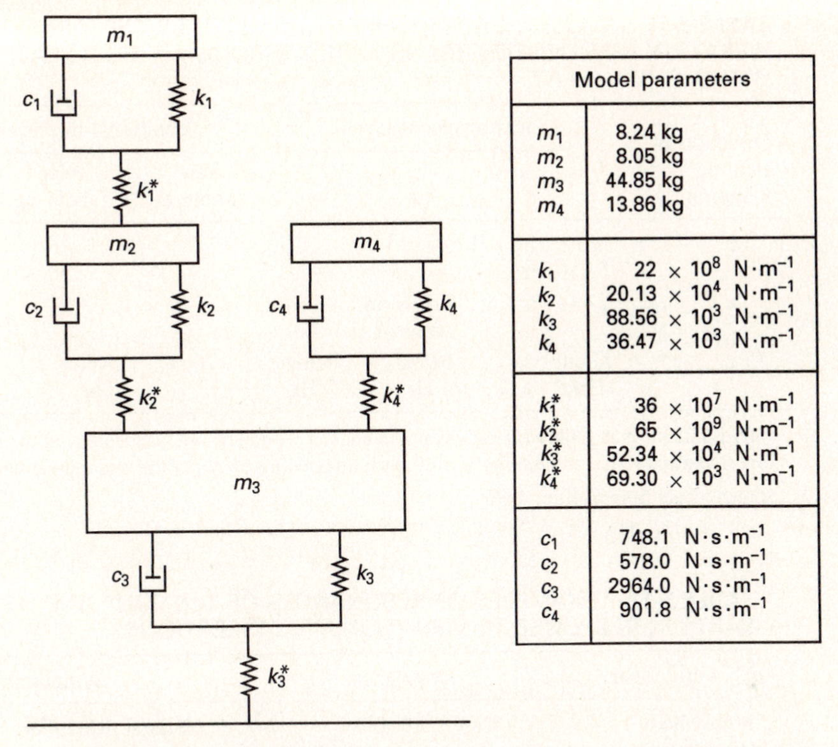

Model parameters	
m_1	8.24 kg
m_2	8.05 kg
m_3	44.85 kg
m_4	13.86 kg
k_1	22×10^8 N·m^{-1}
k_2	20.13×10^4 N·m^{-1}
k_3	88.56×10^3 N·m^{-1}
k_4	36.47×10^3 N·m^{-1}
k_1^*	36×10^7 N·m^{-1}
k_2^*	65×10^9 N·m^{-1}
k_3^*	52.34×10^4 N·m^{-1}
k_4^*	69.30×10^3 N·m^{-1}
c_1	748.1 N·s·m^{-1}
c_2	578.0 N·s·m^{-1}
c_3	2964.0 N·s·m^{-1}
c_4	901.8 N·s·m^{-1}

Figure 5-11. Four-degree-of-freedom model for calculating the vertical transmissibility of the human either sitting or standing (ISO 7962, 1987).

For a person standing on a vibrator, the model that attempts to describe the reactions of the different body parts is more complex. Figure 5-11 shows it according to ISO Standard 7962 (ISO, 1987), which is meant to be applicable up to a frequency of 31.5 Hz.

Such models are only estimates of the actual vibration responses. There are large interindividual variations, and variations exist in the same subject due to body-posture changes; horizontal and vertical impulses may be present at the same time. Body parts may move in several planes even if stimulated only in one direction: for example, the head performs pitch motions in the medial plane even if the vibration applied to the body is strictly vertical. The main head movements induced by vibrations in x, y, and z directions, transmitted through a rigid seat without backrest, are listed in Table 5-11. Resonances in response to vibrations in the z direction are compiled in Table 5-12.

Griffin (1990) stresses the difficulty of correctly modelling the vibration responses of the human body. The underlying data are too meager to allow us to predict with certainty forces on and movements of the body, and at present insufficient to predict the effects of vibration on comfort, health, or performance. Thus, currently available "standards" must be used with great caution.

TABLE 5-11. MAJOR HEAD MOVEMENTS CAUSED BY CERTAIN VIBRATION FREQUENCIES TRANSMITTED THROUGH A RIGID SEAT WITHOUT BACKREST

Direction of exciting vibration	Linear displacement in			Rotational displacement in		
	z	*x*	*y*	Yaw (about *z*)	Roll (about *x*)	Pitch (about *y*)
z	most with 5–10 Hz*	5–12 Hz*				5–12Hz 5 Hz
x	2–12 Hz*	below and above 4 Hz				1 Hz and above
y	about 2 Hz	below 4 Hz	below 5 Hz		below 6hz	

* Transmissibility is reduced if back is not in contact with backrest, particularly at the higher excitation frequencies. Transmissibility is also much affected by body posture and muscle tension.

SOURCE: Data from Griffin, 1990.

TABLE 5-12. EXAMPLES OF RESONANCES OF THE BODY AND ITS PARTS IN RESPONSE TO VIBRATIONS IN *Z*-DIRECTION*

Body part	Resonances (Hz)	Symptoms
Whole body	4 to 5, 10 to 14	General discomfort
Upper body	6 to 10	
Head	5 to 20	
Eyeballs	1 to 100, mostly above 8 strongly 20 to 70	Difficulty seeing
Skull, jaw	100 to 200	
Larynx	5 to 20	Change in pitch of voice
Shoulders	2 to 10	
Lower arms	16 to 30	
Hands	4 to 5	
Trunk	3 to 7	Chest pain at 5 to 7
Heart	4 to 6	
Chest wall	60	
Stomach	3 to 6	
Abdomen	4 to 8	Abdominal pain
Bladder	10 to 18	Urge to urinate
Cardiovascular and respiratory systems	2 to 20	Reactions similar to those in response to moderate work
Brain	below 0.5 1 to 2	Motion sickness Sleepiness

* Note that displacement and acceleration interact with the vibratory frequency to generate specific effects on sensation, performance, and health. Excitation in one direction (e.g., *z*) may produce body responses in other directions (such as *x* or *y*). Body posture, muscle tension, or restraint system may strongly affect responses.

☞☞☞ *"Many models have achieved complexity without representing the known behavior of the body" (Griffin, 1990, p. 181).* ☜☜☜

Vibration: summary. Much empirical and anecdoctal experience exists regarding the reactions of the human (with respect to comfort, performance or health) to experienced vibrations and impacts. In experimental research, vibrations are usually modeled as single or combined sinusoidal displacements of the body, while impacts are modeled in terms of short-term, high-onset single events. Responses of the body are often expressed in terms of resonances at certain frequencies. Yet, the dynamic responses of the body are complex mechanical functions of mass properties, dampers, and springs, which do not comprise all physiological and psychological reactions. Actual vibration responses are subject to large interindividual variations and depend on body posture, restraint systems, and the actual directions and kinds of impulses that exist in the real world, for example in a vehicle driven over rough terrain. Thus, currently available standards and design recommendations must be used with great caution. Nevertheless, the existing information provides guidelines with respect to those vibrations and impacts that have major effects on human comfort, health, and performance, and it provides information on technical means to avoid those conditions that do affect the human.

ASTRONAUTS IN "WEIGHTLESSNESS"

During the 1960s and 70s, the decision was made to send humans into space, instead of just shipping earth- or self-controlled machines. There were two main reasons, one the ability of the human to respond to unexpected situations, the other probably sheer excitement and curiosity.

A human in space must, in order of priority, survive, perform within the spacecraft, and do extravehicular activities (EVAs) for construction and maintenance of the space station and for exploration. Major problems to be overcome by the human astronauts are medical, physiological, psychological, and psycho-sociological in nature (Christensen and Talbot, 1986; Frazer, 1991; Guidi, 1989; Holland, 1991; Hunt, 1987; Jenkins, 1991; Nicogossian, Huntoon, and Pool, 1989).

The technical challenges (beyond that of constructing a shell) are to keep the astronauts well and functioning, by providing suitable atmosphere, temperature, humidity, etc.—see Table 5-13, extracted from a 1991 NASA Committee Report (Jenkins, 1991). Specific requirements are to protect the astronauts from radiation and possibly to generate artificial gravity, both discussed in more detail below. Altogether, this is a typical example of a systems engineering task (Blanchard and Fabrycky, 1990).

"Habitability" is a major design aspect: the space available to each individual must be of sufficient volume, provide privacy and room for storing personal items, allow personalization of decor, and give shelter from uncomfortable temperature and air flow, intrusive noises, lights, and odors.

TABLE 5-13. TECHNOLOGY NEEDS FOR LONG-DURATION SPACE FLIGHTS

Function	Technology needs
Atmospheric pressure and composition control	• Order-of-magnitude improvement in the reliability of the hardware to be utilized in long-duration missions for: total and partial pressure control and monitoring; fire detection and suppression; etc.
Temperature and humidity control	• Improvement in current thermal control technology including: nontoxic heat-transfer fluids, metal hydride heat pumps, and rotating bubble membrane radiator
Atmospheric revitalization (CO_2 control/removal/reduction, O_2 and H_2 makeup, trace gas monitoring and control)	• Minimization of the weight, volume and power demands in closing the CO_2 loop by electrical, chemical, absorption/desorption, or molecular sieve processes • Development of long-duration quality monitoring, including sensors and control technology for trace contamination, toxic compounds, and pathogens
Food supply (storage, processing/preparation, growth chambers)	• Development of closed-loop bioregenerative food production systems • Development of automated systems for harvesting and processing edible biomass or space crops
Water management (wastewater collection/processing, water quality monitoring, storage and distribution of recovered water)	• Development of closed-loop portable water recycling systems • Investigation of alternative ways to minimize the weight/volume/power demands of waste-processing equipment: e.g., vapor compression/distillation, vapor-phase catalytic ammonia reduction, supercritical water oxidation • Development of techniques for long-duration quality monitoring
Waste management	• Development of waste-management systems (physical/chemical, or bioregenerative) to collect/process urine, collect/store fecal matter, and recycle waste
Portable life-support systems	• Development of automated control concepts, EVA heat-storage concepts, and atmospheric control concepts to minimize weight, pressure, volume, and power demands
Health maintenance	• Development of concepts of occupational medicine to include personal hygiene, exercise, diagnostics/therapeutics, and surgery/medical aid capabilities for utilization in reduced-gravity environments

SOURCE: Adapted from Jenkins, 1991.

In addition to the psycho-social aspects of living under "space stress" (discussed in Chapter 3), special problems can arise if crews of male and female astronauts are put together for long-duration space activities (Frazer, 1991) and if medical problems arise (Nelson, Gardner, Ostler, Schultz, and Logan, 1990). So far, only rather limited experiences have been gained in actual spaceflights, most of which have been conducted by Soviet "cosmonauts" and by American "astronauts," but Americans have not done any long-term space missions since the last Skylab activity in February 1974. Therefore, one tries to extrapolate from the experiences gained in other long and isolated missions, such as those involving submarines, exploration (particularly in Antarctica), navy ships, bomber crews, prisoners of war, professional athletic teams, mental hospital wards, prison stays, industrial work groups, shipwrecks, and disasters (Holland, 1991).

Many technical aspects of the construction of space vehicles are fairly well mastered (Griffin and French, 1991). Yet, protection of the astronaut from sudden radiation, from medical emergencies, and from the slow effects of the lack of gravity are still largely unsolved problems. Psychological stress associated with long-duration space exposure is discussed elsewhere in this book.

Microgravity

Astronauts in space experience under almost all conceivable conditions some weak "gravity pull," either because adjacent celestial masses exert attraction forces, or because the spacecraft is being accelerated, linearly or angularly. Since there exists no true "zero gravity" (except in the center of a spacecraft in a stable orbit), one calls any acceleration level below $10^{-4}G$ "microgravity," or "weightlessness." (The unit value of G is the acceleration of gravity on earth, 980.665 cm s^{-2}.)

Microgravity has detrimental physiological effects. However, on trips into space, many crewmembers become accustomed to working and living there within just a few hours, and are able to perform well in about three days. Complete adjustment so that one feels "at ease" takes about three weeks. (Personal communication by U.S. astronauts Carr and Lousma, 11 January 1990.) It takes about as long to readjust again to earth gravity. However, many astronauts suffer from nausea initially, and a few throughout a space mission. This can hinder, possibly severely, the execution of tasks.

For long-duration spaceflight—that is, lasting 60 days or longer—physiological problems associated with microgravity become serious and need to be counteracted by technical and behavioral means (Guidi, 1989; Holland, 1991; Jenkins, 1991). Most problems can be grouped under "radiation," "musculo-skeletal," "fluids and blood," and "nervous control" categories.

Radiation

Radiation can be life threatening, is unpredictable, and increases as the astronaut moves further away from earth. There are three primary natural sources of radiation: the earth's magnetosphere, solar flares and wind, and cosmic radiation. The magnetospheric radiation is strongest between about 2,400 and 19,000 km above ground. Standard spacecraft shielding can protect astronauts from exposure, particularly since they usually pass quickly through the radiation zone. More shielding is necessary to protect astronauts against solar winds, which are composed of high-energy particles.

The most serious threats are solar flares and cosmic radiation. Solar flares run in about 11-year cycles, corresponding to sunspot activity. During each cycle, between 20 and 30 solar flares expose the astronaut to up to 100 rem of radiation, while up to 5 flares may generate up to 1000 rem, and 2 flares may be even more powerful than 5000 rem. (NASA has set a radiation exposure limit of 400 rem not to be exceeded during an astronaut's career. A standard x-ray exposes a patient to about 0.01 rem; Denver, Colorado, residents are exposed to approximately 0.2 rem per year. Exposure to 100 rem causes acute radiation sickness, 300 rem can be lethal.) Another major problem is that solar flares are difficult to predict; particles that precede a flare do so by only about one hour. This gives little time for space crews to seek shelter.

For protection, space ships, and space stations such as on the moon, must be provided with radiation shielding to protect against even the most intense bursts of solar radiation, which usually last for less than a day. Solar storms must be monitored at all times, and astronauts on the moon be kept within one hour of travel time to a radiation shelter.

It is very difficult to protect astronauts against cosmic radiation, which has very high energy particles. Cosmic radiation not only damages or destroys cells, but can also damage genetic material, which affects the children of astronauts. Very dense shielding is required. This may be fairly easily accomplished on a moon station, because one can use natural topographic features, such as valleys, which provide some shelter, or burrow into the moon rock, or heap lunar material on the roof of a built shelter.

While in the spaceship, the astronauts must be protected not only from solar and cosmic high-energy particles, but also from radiation generated by an on-board nuclear rocket engine, which is the currently most promising technique for generating power for space exploration. This adds additional mass to the space ship; using current technology, the ship's mass would be at least 500 tons.

Musculo-skeletal System

A major problem associated with long duration spaceflight is the *deconditioning of the musculo-skeletal system*. Owing to reduced use in low gravity, muscles lose some of their volume. Muscle atrophy, if untreated, creates a condition similar to that experienced by paraplegics, whose muscles diminish from lack of use. Such loss of muscle mass can cause significant problems. For example, after a 211-day mission, Soviet cosmonauts could not maintain a standing position on earth for more than a few minutes. Such inability could cause severe problems for astronauts who, for example, have just landed on Mars. To counter the effects of muscle disuse in low gravity, the muscles must be exercised extensively.

Exercising muscles also helps against another skeletal problem, the *demineralization of bones* in space. Calcium is lost through urine and fecal excretions. This increases the possibility of a bone fracture. Bones that have been found to be most susceptible are those in the legs, the hips, and the lower lumbar vertebrae. These bones support much of the human body weight when in a gravity environment. In long-duration spaceflights, cosmonauts have lost up to 11 percent of the mass of certain bones.

This demineralization process raises other concerns. For example, increased calcium flow through the kidneys may augment the hazard of formation of kidney stones.

Both loss of muscle mass and of bone minerals can be counteracted by a daily extensive regimen of exercise, such as against resistance machines and against a motion-restricting suit. The effectiveness was demonstrated by cosmonauts who spent a full year in orbit and, upon landing on earth, were able to walk several yards without aid.

Blood and Fluid Distribution

Another major problem that occurs in low gravity is the *redistribution of body fluids*. The body's circulatory system operates well in a 1*G* environment. For example, valves in the veins prevent blood from pooling in the feet and legs, but instead force it toward the right ventricle of the heart. The antipooling system still operates in space, although there is little gravity to work against. This results in a reduction of fluid volume in the legs and an accumulation in head, neck, and chest.

This fluid redistribution has important ramifications. Internal sensors indicate that there is too much blood in the upper body, for which the body compensates by reducing blood flow. This aids the upper body, but worsens the problem in the lower body, where the fluid level is already low: up to 20 percent reductions in leg fluid volume after more than 200 days of spaceflight have been encountered. As a side effect, the astronaut's thirst is decreased. Drinking insufficient fluid reduces the overall fluid level in the body. A decrease in fluid volume adds to the demineralization of bones, and increases the retention of sodium.

Because the brain "thinks" that there is an excess of blood, the production of new red blood cells drops. This may result in anemia in the astronauts, but probably only after about 100 days in microgravity.

Changes in fluid distribution within the body also have a negative effect on the heart. The increased volume of fluid in the upper body increases blood pressure in the head and neck veins. Sensing this heightened pressure, the heart actually reduces its output by lowering its contracting rate and shortening the length of the diastole. In cosmonauts who stayed in space for months, actual decreases in the heart size were noted.

After some time of weightlessness, the body establishes a new circulatory pattern, which seems to have no ill effects on the body while in space. However, upon returning to earth, astronauts experienced frequent episodes of dizziness and even fainting. This was probably due to the reduced heart rate and the reduced total blood volume. Back in gravity, the antipooling circulatory mechanisms of the body are again needed and the available blood is redistributed, which involves moving of blood from the upper to the lower body. The brain then is insufficiently supplied by blood, resulting in dizziness and fainting. The problem is usually cured within a week or so on earth, since more blood and red blood cells are produced, thus supplying enough blood throughout, but the recovery may be much slower in a low-gravity environment such as on the moon or Mars. During this period of recovery, astronauts are quite sensitive to rapid changes in posture, such as in standing up rapidly.

There are methods available to counter the redistribution of body fluids. One is the use of a reduced-pressure pressure suit around the legs, which has been successful in aiding blood flow to the legs.

Together with the changes in the cardiovascular systems, *changes in blood composition* also occur. These include decreases in plasma volume and in the mass of erythrocytes, and increases in the cholesterol level together with decreases in the concentration of phosphor. The red blood cell content also decreases, reducing the amount of oxygen in the astronaut's blood. Finally, decreases in the level of ATP and in the intensity of glycolysis (Chapter 2) have been found in microgravity. This is of little concern in microgravity, where less energy is needed to move the body than on earth. However, the reduction in the ability to generate energy within the body through metabolic processes (for which ATP glycolysis and available oxygen are important) can reduce an astronaut's overall work capacity level, which is needed upon return to a gravity environment. Such decrease in capacity could be quite critical after a long trip to Mars, where there is only one-third earth gravity.

Nervous Control

During the first few days of spaceflight, *disturbances in the vestibular system* (Chapter 4) have been experienced, possibly mostly related to the otolith receptors. In low gravity, the disturbed vestibular sense, particularly if combined with conflicting stimuli from the visual and proprioceptive systems, can cause disorientation, vertigo, dizziness, and postural and movement illusions. This generates some deficiency of sensory-motor coordination during the first few days, often accompanied by nausea. About every second space crew member suffers profound discomfort and motion sickness over some time, which varies by individual from hours to days, until the body adapts and reorientates itself (Nicogossian, Huntoon, and Pool, 1989).

Similar symptoms occur when astronauts reenter a gravity environment, to which body (and mind) must adjust. This usually causes reduced work capacity and generates errors in interpreting the visual environment, such as movement illusions, and brings about internal discomfort (see Chapter 3). Such symptoms can be expected for the first few days after landing on extraterrestrial bodies.

Depending on the work-rest cycle and external time clues, an *astronaut's circadian rhythm* (see Chapter 6) *may be altered*. So far, astronauts near the earth have been kept on a 24-hour cycle, with 8 to 10 hours of work and 8 hours allocated to sleep. This has avoided problems with conflicting internal rhythms. It might be advisable to maintain this 24-hour cycle even in space ships traveling to Mars.

Microgravity affects sleeping and eating. Sleep disturbances have been common during spaceflight, but conditions other than low gravity may play a role. Helped by improved food technology, eating and drinking practices have been developed that are not substantially different from those on earth.

A problem that must have more to do with psychological than with physiological events is an *increased sensitivity to vibrational and acoustic stimuli*. After about 30 days in low gravity, the tolerance level to sounds and vibrations has been found greatly reduced, which may affect the sleep patterns and communications of astronauts, and which can generate general annoyance.

Another problem is the *suppression of the body's immune system* in space. Reasons for the suppression are not well understood, but cosmonauts returning to earth after more than 200 days in space displayed severe allergic reactions which they had not had before leaving earth. This problem needs to be studied, because currently no countermeasures are known. The problem might be quite severe if astronauts were sent to Mars, for example. Even if there was nothing on Mars that could harm humans, upon return to earth after two or more years, the astronauts could be susceptible to many diseases here.

The *endocrine system* also suffers ill effects from long exposure to microgravity. Primary effects are changes in plasma composition and decreases in hormones. Together with other changes, these effects are likely to stem from a lack of both physiological and psychological stress during orbital flights. Whether these changes are even more pronounced on long-duration spaceflights, and what the consequences thereof would be, is not known at this time.

Training. Long-term space flight without countermeasures will produce major changes in the cardiovascular, respiratory, muscular-skeletal, and neuromuscular systems of the astronaut. Physical exercises are key countermeasures to many of these unwanted adaptations.

Since endurance and strength seem to be most important for performance of space tasks, relatively high aerobic fitness and muscular strength, especially of the upper body musculature, are expected to be a criterion for selection and training of astronauts, particularly of those who are involved in extravehicular activities. Yet, astronauts who have built unusually large (hypertrophied) capacities, such as marathoners and weight lifters, will suffer significant losses in these.

Actually doing extravehicular activities at regular intervals will probably be sufficient to maintain the endurance and strength required to perform such work effectively. Yet, if insufficient EVAs occur, a minimum of one maximal aerobic exercise every 7 to 10 days during spaceflight may be all that is necessary for maintenance of normal cardiovascular functions and replacement of body fluids.

Means for such exercise may be in electro-myo-stimulation, though conventional exercises appear generally appropriate. These are likely to include workouts on a bicycle ergometer, rowing machine, or adaptations/combinations thereof. Soviet cosmonauts wore an elasticized suit, called a pigeon suit, for 12 to 16 hours a day. This suit resisted motions of its wearer, who therefore had to develop extra energy to perform actions; thus the wearer was forced to "exercise" continually.

Effects on Performance

Absence of gravity is in general a bonus for locomotion in space. Once one is accustomed to microgravity, motion is accomplished with minimal effort and "acrobatic maneuvers" are done with ease. However, for force exertions, the body must be braced so that counterforces can be developed for the active exertion of force vectors.

A problem encountered usually only in the early stages of spaceflight is a *decrease in motor performance*. Astronauts find it difficult to accurately estimate the amount of physical work required to perform certain tasks, and also it takes them longer than on earth to perform activities, owing to the lack of gravity-created references and resistances. After a few days in low gravity, the movements and activities of astronauts become more precise, and the perceived level of difficulty decreases.

☞☞☞ *Upon return to earth after the Skylab 3 flight in 1973, the astronaut, Jerry Carr, said "I didn't faint, but I felt pretty clumsy. My head felt like a big watermelon and I had to work hard to support it. I'd been a butterfly for 84 days and suddenly weighed something again"* (Final Frontier, *1989, 2(3), p. 29).* ☜☜☜

Upon returning to earth gravity after a prolonged stay in space, astronauts' *vertical stability* was found to have declined, together with the reduction of muscle strength already discussed. These problems disappeared in about a week upon return to full gravity. It is unknown, however, how long it would take to recover in a $\frac{1}{3}G$ environment, as on Mars.

It is somewhat amusing to note that astronauts, having returned to earth, tend to drop things for the first few days. They seem to be carrying on the habit of simply letting an item go in low gravity, where it floats.

In space, there is also a *change in body posture, and in some body dimensions*. If relaxed, the body assumes a semicrouched position—that is, knee and hip angles are about 130 degrees; flattening of the curvature of the lumbar and thoracic spine section also occurs, and the pelvic angle changes (resulting in an extension of the body of up to 10 cm). Head and cervical column become bent forward, and the upper arms "float up" against the trunk to about 45 degrees. This posture, reported by NASA in 1978, shown in Figure 5-12, is quite similar to one found in relaxed underwater postures, reported by Lehmann in 1962, shown in Figure 5-13. This posture seems to result mostly from a new balance of muscular and other tissue forces acting about the various body joints.

While the relaxed space posture has by itself no ill effects, it needs to be considered in the design of workstations. For example, it is difficult to work at waist level, since the astronaut must force the arms continually down to this level. Bending forward requires effort by abdominal muscles. Also, it is difficult to sit or stand

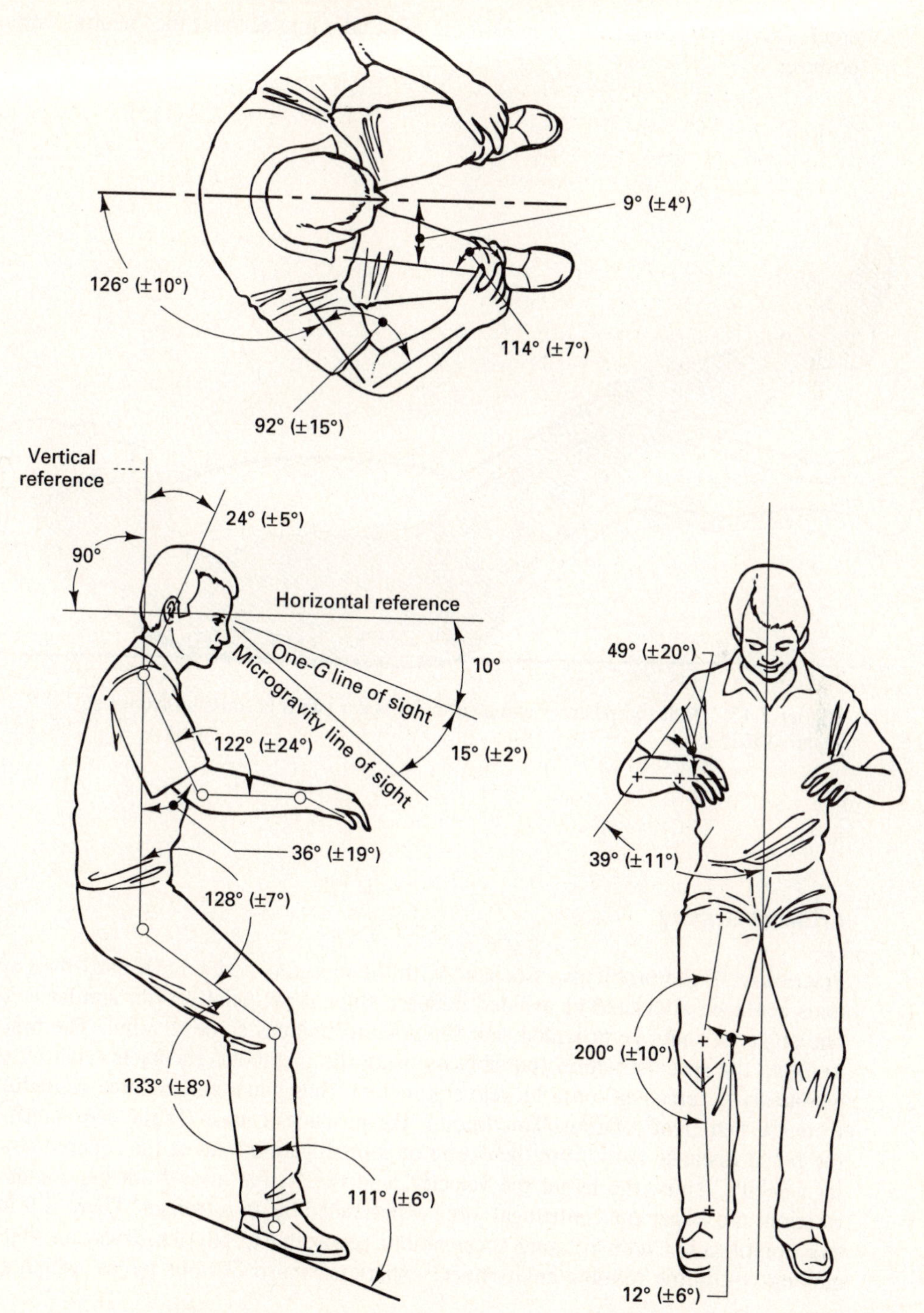

Figure 5-12. Relaxed posture assumed in space (NASA, 1989).

erect. Obviously, design of space clothing must take into account the "neutral" space posture.

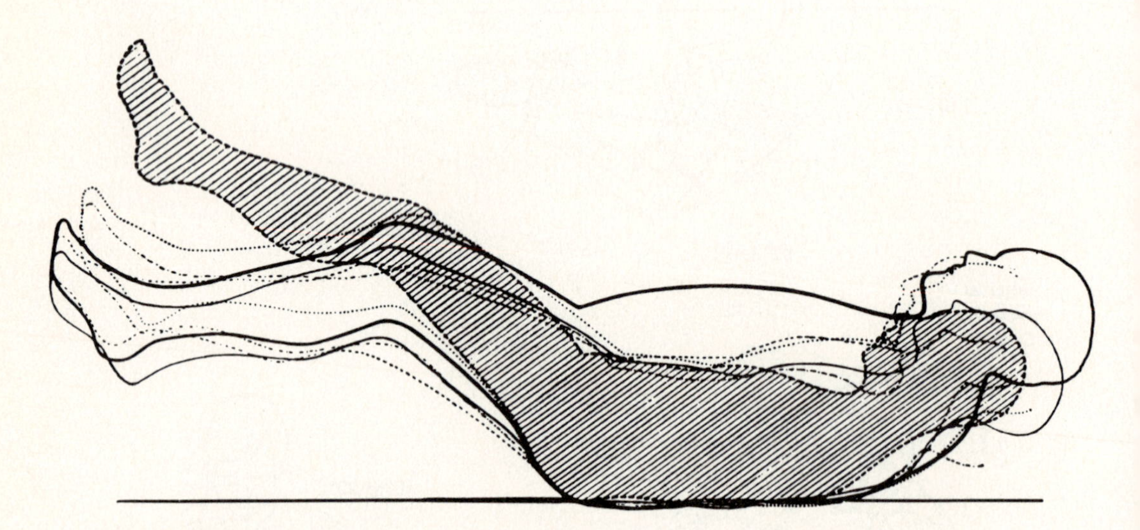

Figure 5-13. Relaxed postures assumed under water (with permission from Lehmann, 1962).

Artificial Gravity

Practically all the problems associated with the musculo-skeletal and body-fluid systems could be alleviated or avoided in space ships if artificial gravity similar to the one on earth could be provided. For this, two techniques come to mind. The first is to rotate the space structure (possibly by using the metabolic energy exerted by the astronauts "exercising" on a bicycle ergometer), thus generating a force centrifugal from the center of rotation. This force is the product of mass (of the astronaut), of the radial distance away from the center of centrifugation, and of the squared angular velocity. Thus, the larger the velocity, and the farther away from the center of rotation, the larger the centrifugal force experienced by an astronaut. There is a further complication with rotating acceleration: according to physical laws, an object moving within the rotating environment experiences also Coriolis forces, which are perpendicular to both the vector of rotation and the vector of individual movement. This "confuses" the vestibular system and is likely to cause severe motion-sickness problems which, unfortunately, probably will last throughout the whole duration of being in this environment (Kennedy, 1991).

Another, better solution is to generate a linear acceleration environment. For example, on a trip to Mars, the spaceship could be linearly accelerated in the desired direction for half the distance. Then, it would be turned around and, through use of the same accelerating engines, be decelerated until it arrived at Mars with zero speed. The condition of constant linear acceleration would avoid all problems associated with a rotating environment. Unfortunately, present technology provides neither engines nor fuel to generate the required linear acceleration over the necessary long period. Technological progress may alleviate this problem.

For these reasons, current countermeasures are limited to the use of exercise equipment and restrictive space suits, the generation of an artificial vacuum around the lower body, and the application of training and drugs to overcome the negative effects of low gravity on the body and mind.

ACCELERATIONS IN AEROSPACE

The pilot of a high-performance aircraft, or an astronaut departing from earth, or arriving at earth or elsewhere, may be subjected to linear or rotational accelerations. These may be of the impact type—that is, of quick onset and cessation—or they may be sustained for some time.

In aerospace work, the direction of acceleration is usually described relative to the direction in which the eye (or another body organ) is displaced by the acceleration. This system has a positive acceleration in x-direction back-to-chest ("eyeballs in"), a positive y-direction left-to-right ("eyeballs left"), and a positive z-direction head-to-toe ("eyeballs up"). This system is depicted in Figure 5-14 and described in Table 5-14. There is a special convention of opposite signs for acting and resulting accelerations in the z axis.

Similarly, a descriptive system exists for angular motions. A roll ("cartwheel") is about x, a pitch ("somersault") about y, and a yaw ("pirouette") about z.

On earth, there is (for all practical purposes) a constant $1G$ level, pointing toward the center of earth. With current spacecraft technology, during launch, entry, and abort operations, $+6G_x$ may be experienced, and up to $+2G_x$ during stage separation. In transorbital flight, very low sustained accelerations are experienced, such as $10^{-6}G$, omnidirectional. During orbital maneuvers, angular accelerations up to ± 1.5 deg s^{-2} are possible in all directions.

Impact accelerations may be encountered as a result of thruster firing, ejection, seat/capsule firing, escape-device deployment, flight instability, air turbulence, and

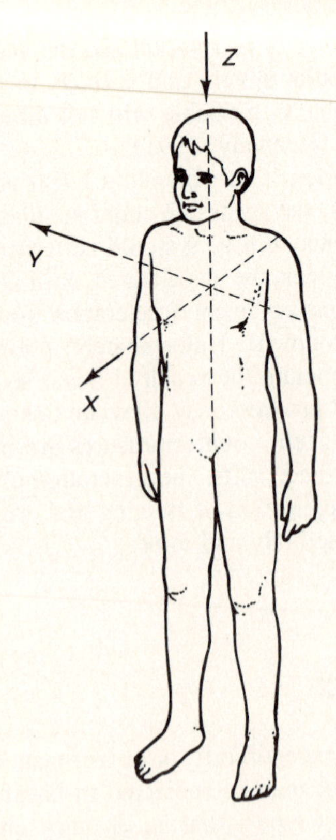

Figure 5-14. Aerospace convention of acceleration directions (NASA, 1989)—compare with Figure 5-3.

TABLE 5-14. AEROSPACE TAXONOMY

	Direction of acting acceleration		Inertial resultant of the acceleration of the body	
Linear motion	Action	Acceleration description	Reaction	Vernacular description
Forward	$+a_x$	Forward	$+G_x$	Eye balls in
Backward	$-a_x$	Backward	$-G_x$	Eye balls out
To right	$+a_y$	R. lateral	$+G_y$	Eye balls left
To left	$-a_y$	L. lateral	$-G_y$	Eye balls right
Upward	$-a_z$	Headward	$+G_z$	Eye balls down
Downward	$+a_z$	Footward	$-G_z$	Eye balls up

SOURCE: Adapted from NASA, 1989.

crash landings. Expected accelerations are:

- Violent maneuvers: up to $6G$, omnidirectional.
- Parachute opening shock: approximately $10G_z$.
- Aircraft ejection firings: up to $17G_z$.
- Crash landings: from 10 to more than $100G$, omnidirectional.

Depending on the magnitude and direction of acceleration, the effects may be imperceptible and normal, they may generate discomfort and impairment, or they may be dangerous or fatal. During upward acceleration, while seated, one experiences increased weight and drooping of skin and body tissue at $+2G_z$; impossibility to raise oneself and dimming of vision at $+3G_z$; blackout and loss of consciousness at 5 to $6G_z$. Downward acceleration effects include facial congestion, reddening of vision at $-3G_z$; few subjects can tolerate $-5G_z$ over more than 5 seconds. Further information about linear and rotational accelerations is given in the preceeding section on vibration and is contained in NASA Standard 3000 (NASA, 1989), which also establishes guidelines for designing aerospace equipment, shown in Table 5-15.

TABLE 5-15. IMPACT DESIGN LIMITS FOR ACCELERATIONS LASTING UP TO 1 SECOND

Direction of impact acceleration	Impact limit	Rate of impact
$\pm G_x$	$20G$	$10{,}000G$/sec
$\pm G_y$	$20G$	$1{,}000G$/sec
$\pm G_z$	$15G$	$500G$/sec
45 deg off-axis (any axis)	$20G$	$1{,}000G$/sec

SOURCE: NASA, 1989.

EXAMPLE

Some high-performance U.S. fighter aircraft were equipped in the mid-70s with "reclining seats," which during high-acceleration maneuvers (such as in tight turns) would support the pilot's body in a half-supine position. Yet, experience has shown that declining the body 35 or 40 degrees backward is often not effective, and may in reality slightly decrease the tolerance to high-G forces with sharp onsets. While this is in contrast to theoretical considerations, this experience had already been reported as the result of experiments performed during World War II in Germany and Canada. Since a completely supine position (which would counteract greyout and blackout) cannot be assumed in these aircraft, the newest recommendations are to sit upright, even slightly crouched forward, in preparation for and during high-G maneuvers (Wood, Code, and Baldes, 1990).

Aerospace conditions: summary. The technology to construct and launch space vehicles exists, but protection of the astronaut from cosmic radiation is still an unsolved problem. There are also slow, lingering and dangerous effects of the lack of gravity in space: the body's musculo-skeletal, vascular, cardiac, and sensory systems suffer in substance and performance. This, in turn, negatively affects health and the ability to perform tasks both in space and on the ground, either upon return to earth or upon landing on another celestial body. Generation of artificial gravity in the space vehicle would alleviate these problems.

SUMMARY

The human body functions in interaction with its environment.

The climate is characterized by temperature, humidity, and air movement. The processes of interacting between the environment and the body are described by the physical events of radiation, evaporation, convection, and conduction. Their effectiveness depends on several conditions: they include the clothing worn, the energetic content of the work conducted, and the exposure time. The temperature difference between the skin and the environment is very important, but humidity and air flow also play major roles. Within reasonable limits, and given suitable job demands and clothing, the human can function in both hot and cold environments.

Air pollution is a detriment not only to health, but also to the ability to perform physically demanding work, particularly if carbon monoxide is present. Further studies may also show negative effects of other pollutants.

Arrival at altitude, after having lived near sea level, influences the ability to perform physical work in various manners. Up to about 1500 meters, the ability for aerobic work is not much affected. At higher altitudes, both the short-term maximal capacity and the long-term sub-maximal ability are diminished, and they remain reduced for the entire stay at altitude.

The effects of vibration on the human depend very much on the amplitude, frequency, direction, and point of application. Vibration in one direction, for instance foot-to-head, can bring about response movements in other directions, such as in head nodding. Whole-body vibration can have consequences quite different from vibration of only body parts, such as the hands. Body posture, and body restraint systems, can greatly affect the results of vibrations. Some information is available about suitable ergonomic measures, depending on circumstances. However, more research and better modeling are needed.

Space is an environment new to humans. So far, its exposure effects have been experienced by just a few persons, mostly over periods of days, in some cases over months. Microgravity influences the function of the body in many ways, particularly the musculo-skeletal and the circulatory systems. Countermeasures to avoid health problems, and to protect against radiation, still need to be developed. The design of work tasks and work environments suitable for long-term space missions is a challenge.

CHALLENGES

Consider the effects, in theory and practice, of the concept of either "constant core temperature" or "average skin temperature."

Why is it difficult to control heat transfer through the head?

The energy-balance equation, given in this chapter, does not consider the time domain. What are the consequences?

While it is true that the temperature of air between a radiating surface and the body does not affect the energy transferred by radiation, it should have an affect on energy transfer by convection. How?

Why is there no energy transfer between the human body and the environment through condensation?

Why is it more important to avoid overheating and undercooling of the body's core than of the shell?

Is it conceivable that procedures other than the current subjective ones be used to establish indices of "effective climate"?

Which engineering means exist to control the environmental climate in (a) offices, (b) workshops, such as a machine shop or a foundry, and (c) outdoors?

Are sensations of feeling hot, or cold, reliable indicators of climate strain?

Is exercising a practical means of acclimatizing oneself to working in heat, in cold, or at altitude?

Through which body parts is vibration most likely transmitted?

Is it reasonable to assume that vibration is transmitted to the body either only horizontally, or only vertically, as done in most research?

How would the descriptors of vibration change if the acting vibration were not sinusoidal?

What professionals other than truck drivers are particularly exposed to vibrations, or impacts?

What might be the effects of several sources of vibration arriving simultaneously at different body parts?

Could one imagine that vibrations on the job might, under certain circumstances, be helpful in job performance?

What explanations other than "conflicting CNS information" might be applied to the problem of motion sickness?

What factors are likely to affect the likelihood of withstanding impacts?

Which body postures, and physical behaviors, might be suitable to combat the effects of sustained strong accelerations?

Why is it difficult to make "crash dummies" anthropomorphic?

What are reasons for, or against, sending humans into space?

Discuss the aspects of "systems engineering" associated with space engineering.

Consider details that make a confined living (or working) space "habitable."

What means exist to physically "exercise" astronauts in space?

6

Body Rhythms, Work Schedules, and Alcohol Effects*

OVERVIEW

The human body changes its physiological functions throughout the 24-hour day. During waking hours, the body is prepared for physical work, while during the night sleep is normal. Attitudes and behavior also change rhythmically during the day. Work should be arranged to least disturb physiological, psychological, and behavioral rhythms in order to avoid negative health and social effects and to avoid reductions in work performance. Alcohol usually reduces work performance.

INTRODUCTION

The human body follows a set of daily fluctuations, called *circadian rhythms* (from the Latin *circa,* about, and *dies,* the day; also called diurnal from the Latin *diurnus,* of the day). They are regular physiological occurrences, observable, for example, in body temperature, heart rate, blood pressure, and hormone excretion. These rhythms are systems of temporal programs within the human organism. Each is thought to be controlled within the body by a self-sustained pacemaker or "internal clock," which runs on a daily cycle. Several rhythmic programs, such as core temperature, blood pressure, and "sleepiness," are coupled with each other. Usually, their synchronization or "entrainment" is enforced by time markers, often called

*Sections of this chapter use material presented in our 1990 book, *Engineering Physiology*. We thank the publisher, Van Nostrand Reinhold, for permission to use that material.

zeitgeber (from the German *Zeit*, time, and *geber*, giver). Among the zeitgebers are daily light and darkness, true clocks, and temporally established activities, such as work tasks, office hours, meal times, etc. Human social behavior (the inclinations to do certain activities as well as to rest and sleep) both follows and reinforces biological rhythms.

The female menstrual cycle is a well-documented "chrono-biological" rhythm. Seasonal mood changes have been noted for hundreds of years. Other cycles are postulated or mythical, such as dependency on the moon phases. So-called bio-rhythms were a fad a few decades ago: they were said to be regular waves of physiological and psychological events, starting at birth but running in different phases and phase lengths. Whenever "positive" phases of any of these rhythmic phenomena coincided, a person was believed to be under positive conditions, able to perform exceptionally well. In contrast, if "negative" phases concurred, the person was supposedly doing badly. Research has shown, conclusively, that no such accumulations of positive or negative phases were demonstrable; also, several or all of the supposedly existing rhythms were either myths or artifacts (Hunter and Shane, 1979; Persinger, Cooke, and Janes, 1978).

FEMALE MENSTRUAL CYCLE

The female menstrual cycle is regulated through synchronization of the activities of the hypothalamus, pituitary, and ovary. The 28-day time period is usually divided into five phases: (1) preovulatory or follicular, (2) ovulatory, (3) postovulatory or luteal, (4) premenstrual, and (5) menstrual. Main hormonal changes occur in the release of estrogen and progesterone around the twenty-first day of the menstrual cycle; estrogen shows a second peak at ovulation. Hormonal release is low during the premenstrual phase.

It has been commonly assumed that hormonal changes during the menstrual cycle have profound effects on a woman's psychological and physiological states. However, after reviewing the existing scientific work, Patkai (1985) states that the information available almost exclusively tries to correlate certain behavioral or physiological events with the menstrual phase. But the observable events are fairly weak in their occurrence and, of course, an existing correlation does not necessarily indicate a causal relationship.

The bulk of existing research relies on self-reported changes in mood and physical complaints in the course of the menstrual cycle. Fairly little information is available on changes in arousal and on objective measures of performance. While there is neurophysiological evidence that estrogen and progesterone affect brain function, it must be considered that these two hormones have antagonistic effects on the central nervous system, with estrogen stimulating and progesterone inhibiting. Varying hormone production during the menstrual cycle may affect the capacity to perform certain tasks, but the extent to which hormones actually determine performance depends on how a decrease in total capacity may be offset by increased effort.

Patkai cites a study in which secretaries showed the highest typing speed before the onset of menstruation and during the first three menstrual days. The idea of a higher effort on these days was rejected by the secretaries, since they considered themselves to be working at full capacity all the time.

The occurrence of negative moods and physical complaints in the majority of women before and during menstruation is fairly well established, but the precise nature of the so-called premenstrual syndrome is not yet determined. According to Patkai, there is evidence that menstruation can bring about negative behaviors, which, however, are usually mediated by social and psychological factors. Medication can often counteract discomfort and negative mood.

CIRCADIAN RHYTHMS

The prerequisite for human health is the maintenance of physiological variables in spite of external disturbances. This state of balanced control is called homeostasis. However, a close look at this supposedly "steady state" of the body reveals that many physiological functions are in fact not constant but show rhythmic variations. Rhythms with a cycle length of 24 hours are called circadian (or diurnal) rhythms, those which oscillate faster than once every 24 hours are called ultradian; those which repeat less frequently, infradian.

Among the circadian rhythms, the best-known physiological variables are body temperature, heart rate, blood pressure, and the excretion of potassium—see Figure 6-1. Most of these variables show a high value during the day and lower values during the night, although hormones in the blood tend to be more concentrated during the night, particularly in the early morning hours. The amount by which the variables change during the diurnal variation, and the temporal locations of rhythm extremes during the day, are quite different among individuals and can change even within one person (Minors and Waterhouse, 1981; Folkard and Monk, 1985).

One way to observe the diurnal rhythms, and to assess their effects on performance, is simply to observe the activities of a person. During daytime, a person is normally expected to be awake, active, and eating, while at night sleeping and fasting. Physiological events do not exactly follow that general pattern. For example, body core temperature falls even after the person has been sleeping for several hours; it is usually lowest between 3 and 5 o'clock in the morning. Temperature then rises, quickly so when one gets up. It continues to increase, with some variations, until late in the afternoon. Thus, body temperature is not a passive response to our regular daily behavior, such as getting up, eating meals, performing work, and doing other social activities, but is self-governed.

In the human, the circadian pacemaker is located in the suprachiasmatic nucleus of the hypothalamus. Under constant living conditions, the underlying *physiological rhythms of the body are solid, self-regulated, and remain in existence even if daily activities change*. Yet, variations in observed rhythmic events (due to exogenous influences) may "mask" the internal regular fluctuations. For example, skin temperature (particularly at the extremities) increases with the onset of sleep, re-

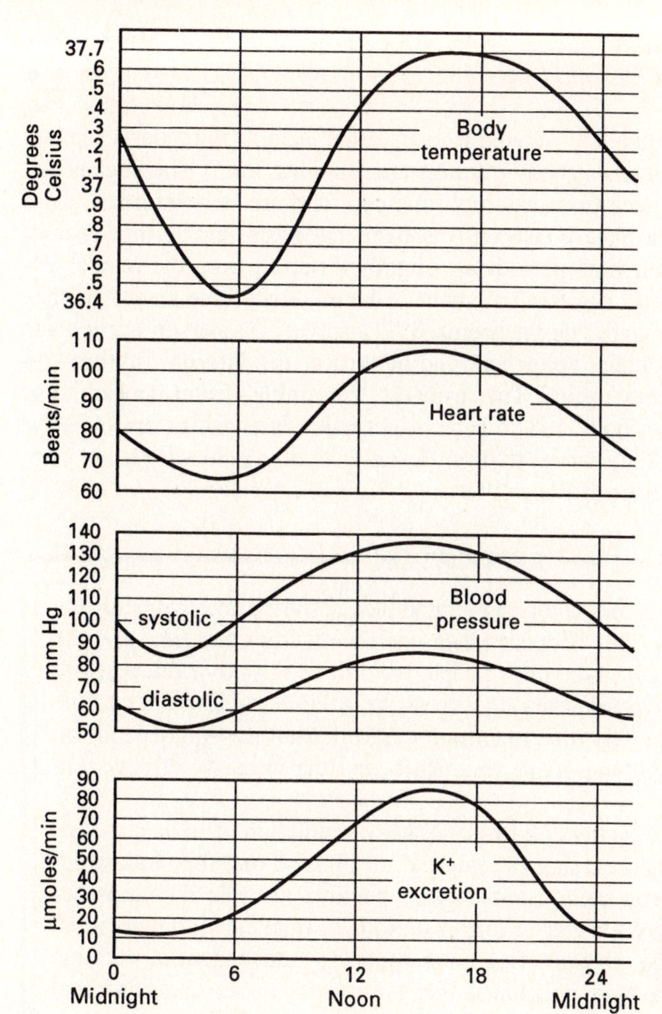

Figure 6-1. Typical variations in body functions over the day (adapted from Colligan and Tepas, 1986).

gardless of when this occurs. Turning the lights on increases the activity level of caged birds, at any time. Thus, observed skin temperature or activity level do not necessarily indicate the internal rhythm but may in fact mask it.

Under regular circumstances there is a well-established phase coincidence between the external activity signs and the internal events. During the night, the low values of physiological functions, for example of core temperature and heart rate, are primarily due to the diurnal rhythm of the body; however, they are further helped by nighttime inactivity and fasting. During the day, peak activity usually coincides with high values of the internal functions. Thus, normally, the observed diurnal rhythm is the result of internal (endogenous) and external (exogenous) events that concur. If that balance of concurrent events is disturbed, consequences in health or performance may appear.

The number of deaths due to heart disease is highest during the morning hours, as is the number of human births.

When a person is completely isolated from external factors (zeitgebers), including regular activities, the internal body rhythms are "running free." This means that the circadian rhythms are free from external time cues and are only internally controlled. Many experiments have consistently shown that circadian rhythms persist when running free, but their time periods are slightly different from the regular 24-hour duration: most rhythms "run free" at about 25 hours, some take longer. If they run at different phases, they are "desynchronized." However, if a person is put again under regular (such as 24-hour) zeitgebers and activities, the internal rhythms resume their coordinated steady cycles. The strongest "entraining" factor appears to be the light-dark sequence (Aschoff, 1981): very bright lights or sunshine are effective in advancing or retarding the diurnal rhythm (Czeisler et al., 1989; 1990a, b). Activity is also a strong enforcer (Turek, 1989).

APPLICATION

Providing new zeitgebers and enforcing related new activity times can shift the body's internal rhythm. This happens when one travels from one time zone to another and stays there. It takes most people from three days to a full week to adjust their circadian rhythms when they cross the Atlantic by airplane, a time shift of five or six hours. The time to entrain a new diurnal rhythm depends on the individual, the magnitude of the time shift, and the intensity of new zeitgebers.

Manipulation of zeitgebers allows laboratory simulation of jet lag or shift work. Entraining or synchronizing the internal rhythms so that they follow periodic time cues has been demonstrated to be possible at cycle durations between 23 to 27 hours. (At shorter or longer periods of time cues, the circadian rhythms are free running, though often not completely independent of the time cues.) Most researchers have concluded that it is easier to set one's internal clock "forward," such as occurs in the spring when "daylight saving time" is introduced, or during a West-to-East flight such as from North America to Europe. However, a recent study on shift work did not indicate a significant beneficial effect of forward as compared to backward rotation (Duchon, Wagner, and Keran, 1989).

Individual differences. Experimentation has shown that some people have consistently shorter (or longer) free-running periods than others. Females have, on the average, a free-running period that is about 30 minutes shorter than that of males. Wever (1985) suggests that this pattern may make females more prone to rhythm disorders than males. Those who have short periods are likely to be "morning types," while those with longer internal rhythms are probably "evening types." There are (at least) three scales—based on questionnaires—that purport to assess one's diurnal type, especially "morningness" or "eveningness" (Folkard, Monk, and Lobban, 1979; Horne and Oestberg, 1976; Torsvall and Akerstedt, 1980). However, there is evidence that these properties may be suituation-or habit-dependent, that they may vary with culture, age, or work schedule, and that the scales may have different reliability and validity (Greenwood, 1991).

There are changes in circadian rhythms with aging. Rhythm amplitudes usually are reduced with increasing age. This is particularly obvious for body temperature. In the elderly, there is a shift toward morningness. Also, the oscillatory controls seem to lose some of their power with aging, which means a greater susceptibility to rhythm disturbances (Kerkhof, 1985).

Daily performance rhythms. Given the systematic changes in physiological functions during the day, one expects corresponding changes in mood and performance. Of course, attitudes and work habits are also, and often strongly, affected by the daily organization of getting up, working, eating, relaxing, and going to bed. Experimentally, one can separate the effects of internal circadian rhythms and of external daily organization. For practical purposes, one wants to look at the results (e.g., as they affect performance) of the internal and external factors combined.

APPLICATION

Early in this century it was thought that the morning hours would be best for mental activities, with the afternoon more suitable for motoric work. "Fatigue" arising from work already performed was believed to reduce performance over the course of the day. Performance of simple mental work, such as recording numbers, was observed to show a pronounced reduction early in the afternoon (and a sharp decrease about two to five o'clock in the night). This was labeled the "postlunch dip." However, this reduction in performance was not paralleled by similar changes in physiological functions; for example, body temperature remains fairly constant during that period of the day. Hence, it was postulated that performance was affected by the interruption of activities by the noon meal and that the subsequent digestion would lower the psychological "arousal" level, bringing about an increased lassitude, a status of deactivation, together with increased blood glucose and changed blood distribution in the body as results of food ingestion. Accordingly, the postlunch dip appears to be caused by the exogenous "masking" effect of the "food break" intake rather than by endogenous circadian events. In activities with medium to heavy physical work no such dip has been found (except when the food and beverage ingestion was very heavy and when true physiological fatigue had been built up during the prelunch activities).

Many different activities can be performed during the day. They may follow a circadian rhythm or may differ because of external requirements. They may be physical or mental in nature. Personal interest, fatigue or boredom, rewards or urgency usually have stronger effects on performance than diurnal rhythms during day/night hours. For example, information processing in the brain (including immediate or short-term memory demands), mental arithmetic activities, or visual searches, are strongly affected by personality, or by the length of the activity, and by motivation. Thus, it appears that one cannot make "normative" statements about diurnal performance variations during regular working hours.

SLEEP

Two millennia ago, Aristotle thought that during wakefulness some substance ("warm vapors") in the brain built up which needed to be dissipated during sleep. In the nineteenth century, there were two opposing schools of thought: one that sleep was caused by some "congestion of the brain by blood," the other that blood was "drawn away from the brain."

"Behavioral" theories were common in the nineteenth century, such that sleep was the result of an absence of external stimulation or that sleep was not a passive response but an activity to avoid fatigue from occurring. Early in the twentieth century, it was thought that various sleep-inducing substances accumulated in the brain, an idea taken up again in the 1960s. In the 1930s and 1940s, various "neural inhibition" theories were discussed, including sleep-inducing "centers," such as for arousal in the reticular formation of the brain (Horne 1988).

"Restorative" theories about the function of sleep focus on various types of "recovery" from the wear and tear of wakefulness. Alternative theories claim that sleep is not restorative but simply a form of instinct or "nonbehavior" to occupy the unproductive hours of darkness; yet, relative immobility of the body during sleep is a means to conserve energy (Horne, 1988). It appears that the three aspects of sleep function—i.e., restoration, occupying time, and energy conservation—all explain certain characteristics of sleep but none of them completely or sufficiently.

According to Horne, it is convenient to model the regulation of alertness, wakefulness, sleepiness, sleep, and many physiological functions as under the control of two "central clocks" of the body. One controls sleep and wakefulness, the other physiological functions, such as body temperature. Under normal conditions the internal clocks are linked together, so that body temperature and other physiological activities increase during wakefulness and decline during sleep. However, this congruence of the two rhythms may be disturbed, for instance by night-shift work, where one must be active during nighttime and sleep during the day. As such patterns continue, the physiological clocks adjust to the external requirements of the new sleep/wake regimen. This means that the formerly well-established physiological rhythm flattens out and, within a period of about two weeks, reestablishes itself according to the new sleep/wake schedule.

Sleep phases. Sleep is not homogeneous but has several stages which repeat, more or less regularly, during the night. After falling asleep, the sleep stages become progressively "deeper," indicated by more synchronous and less frequent brain activities. In deep sleep, heart rate and respiration are slow, and muscles retain their tonus. Yet, in the deepest-sleep phase, the brain becomes nearly as active as in wakefulness, heart rate and breathing vary, dreams are frequent, and the eyes move rapidly under closed lids. Accordingly, this is called "rapid-eye-movement" (REM) sleep. The REM phase becomes longer and the light-sleep phases get shorter as the sleep-stage sequence repeats itself, about every 90 to 100 minutes throughout the night.

The brain and muscles show large changes from wakefulness to sleep, which can be observed fairly easily by electrical means (polysomnography). Electrodes attached to the surface of the scalp pick up electrical activities of the cortex, which is also called "encephalon" because it wraps around the inner brain. Thus, the measuring technique is named electro-encephalo-graphy, EEG. The EEG signals provide information about the activities of the brain.

EEG signals can be described in terms of amplitude and frequency. The amplitude is measured in microvolts, and it rises as consciousness falls from alert wakefulness through drowsiness to deep sleep. EEG frequency is measured in hertz; the frequencies observed in human EEG range from 0.5 to 25 Hz. Sleep researchers call frequencies above 15 Hz "fast waves," those under 3.5 Hz "slow waves." Frequency falls as sleep deepens; "slow-wave sleep" (SWS) is of particular interest to sleep researchers.

Certain frequency bands have been given Greek letters. The main divisions are:

Beta, above 15 Hz. These fast waves of low amplitude (under 10 microvolts) occur when the cerebrum is alert or even anxious.

Alpha, between 8 and 11 Hz. These frequencies occur during relaxed wakefulness, when there is little information input to the eyes, particularly when they are closed.

Theta, between 3.5 and 7.5 Hz. These frequencies are associated with drowsiness and light sleep.

Delta, under 3.5 Hz. These are slow waves of large amplitude, often over 100 microvolts, and occur more often as sleep becomes deeper.

Also, certain occurrences in the EEG waves have been labelled *vertices, spindles,* and *complexes,* which appear regularly associated with sleep characteristics.

Muscular activities are recorded in their electrical activities via electromyography, EMG (see the muscle section in this book). In sleep observation, the muscles that move the eyes and those in the chin and neck regions are of particular interest.

The importance placed upon EEG and EMG events by sleep researchers has been changing over decades. Currently, EMG outputs of the eye muscles are most often used as the main determiner: sleep is divided into periods associated with rapid eye movements, REM, and those without, non-REM. Non-REM conditions are further subdivided in four stages according to their associated EEG characteristics. Table 6-1 lists these.

Within the human body, only the brain assumes a physiological state during sleep that is unique to sleep and cannot be attained during wakefulness. While, for example, muscles can rest during relaxed wakefulness, the cerebrum remains in a condition of "quiet readiness," prepared to act on sensory input, without diminution in responsiveness (Horne, 1985, 1988). Only during sleep do cerebral functions show marked increases in thresholds of responsiveness to sensory input. In the deep-sleep stages associated with slow-wave non-REM sleep, the cerebrum apparently is functionally disconnected from subcortical mechanisms. The brain needs sleep to restitute, a process that cannot take place sufficiently during waking relaxation (Horne, 1985, 1988). It seems that the first 5 to 6 hours of regular sleep (which happen to contain most of the slow-wave non-REM sleep and at least half the REM sleep) are obligatory to retain psychological performance at normal level, but that more sleep may be called "facultative" or "optional" because it mostly serves to "occupy unproductive hours of darkness" with dreams the "cinema of the mind" (Horne, 1988, pp. 54 and 313).

TABLE 6-1. SLEEP STAGES

Condition	Muscle EMG	Brain EEG	Sleep stage	Average percent of total sleep time
Awake	Active	Active, alpha and beta	0	—
Drowsy, transitional "light sleep"	Eyelids open and close, eyes roll	Theta, loss of alpha, vertex sharp waves	1, non-REM	5
"True" sleep		Theta, few delta, sleep spindles K-complexes	2, non-REM	45
Transitional "true" sleep		More delta, SWS (<3.5 Hz)	3, non-REM	7
Deep "true" sleep		Predominant delta SWS (<3.5 Hz)	4, non-REM	13
Sleeping	Rapid eye movements, other muscles relaxed	Alert, much dreaming, alpha and beta	REM	30

SOURCE: Adapted from Horne, 1988.

Sleep loss and tiredness. If a person does not get the usual amount of sleep, the apparent result is tiredness, and the obvious cure is to get more sleep. Figure 6-2 shows the effects of sleep loss on body temperature—the temperature is raised but keeps its phase.

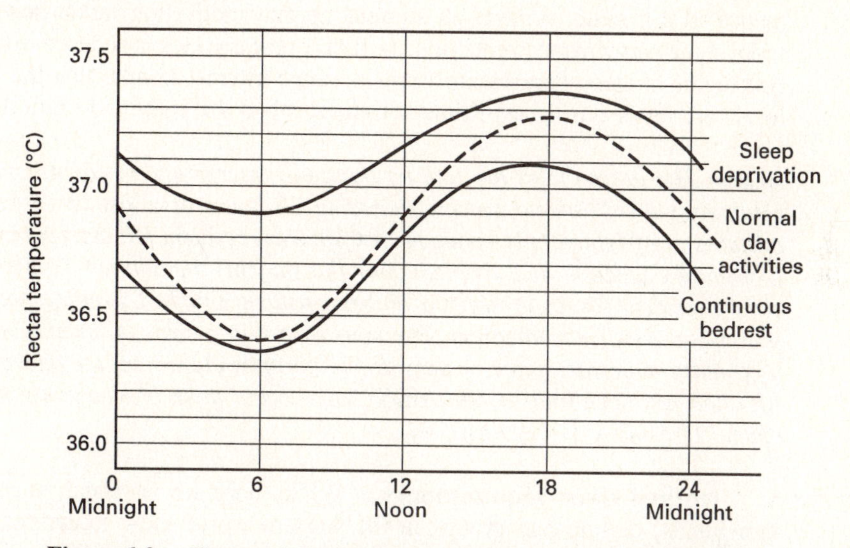

Figure 6-2. Changes in body temperature associated with normal activities, bedrest, and sleep deprivation (schematically from Colligan and Tepas, 1986).

As discussed above, it is not entirely clear why humans (or animals) need sleep. There is the general opinion that sleep has recuperative benefits, allowing some sort of restitution or repair of tissue or brain following the "wear and tear" of wakefulness. However, what is meant by restitution or repair is usually not clearly expressed, nor fully understood. Certainly, sleep is accompanied by rest and thus by energy conservation. But a human can attain similar relaxation during wakefulness, when not forced to be active.

Many experiments have failed to show restitutive physiological effects of sleep; in fact, even moderate sleep deprivation has little physiological effect, as discussed by Horne (1985, 1988). For example, sleep deprivation does not impair muscle restitution or the physiological ability to perform physical work. Apparent reductions in physical exercise capability, owing to sleep deprivation (such as reported by Froeberg 1985), may be mostly due to reduced psychological motivation rather than to a decrease in physiological capabilities. The effects of moderate sleep deprivation on body functions are not clear but may be less consequential than often believed.

In contrast, the restorative benefits of sleep to the brain are fairly well researched. Two or more nights of sleep deprivation bring about psychological performance detriments, particularly reduced motivation to perform (but apparently not a reduction of the inherent cognitive capacity), behavioral irritability, suspiciousness, speech slurring, and other performance reductions. However, while these changes indicate some impairment of the central nervous system owing to sleep deprivation (apparently, the brain needs to sleep), Horne states that they are not as extensive as one might expect if a person needed eight hours of sleep per day for brain restitution. After up to two days of sleep deprivation, even though one feels tired, mental performance is still rather normal on stimulating and motivating tasks; however, boring tasks show performance reduction. (All task performance is reduced after more than two nights of sleep deprivation.) It is of some interest to note that the performance levels are lower during nighttime activities, when the body and brain usually rest.

Harrowing tales are told by [hospital] interns and residents, many of whom routinely work 120-hour weeks, including 36 hours at a stretch. Some admit that mistakes are frighteningly common. A California resident fell asleep while sewing up a woman's uterus—and toppled onto the patient. In another California case, a sleepy resident forgot to order a diabetic patient's nightly insulin shot and instead prescribed another medication. The man went into a coma. Compassion can also be a casualty. One young doctor admitted to abruptly cutting off the questions of a man who had just been told he had AIDS: "All I could think of was going home to bed" (Time, December 17, 1990, p. 80).

Normal sleep requirements. While there are, as usual, variations among individuals, certain age groups in the western world show rather regular sleeping habits. Newborns sleep 16 to 18 hours a day, mostly in sets of a few hours duration. Young adults sleep, on the average, 7.5 hours (with a standard deviation of about one hour). Some adults are well rested after 6.5 hours of sleep, or less, while others take habitually 8.5 hours and more. The amount of slow-wave sleep in both short and long sleepers is about the same, but the amounts of REM and non-REM sleep periods differ considerably.

APPLICATION

If people can sleep for just a few hours per day, many are able to keep up their performance levels even if the attained total sleep time is shorter than normal. The limit seems to be around 5 hours of sleep per day, with even shorter periods still being somewhat beneficial.

PROLONGED HOURS OF WORK AND SLEEP DEPRIVATION

There are conditions in which persons must work continuously for long periods, such as 24 hours or longer. This not only means working without interruption but also encompasses deprivation of sleep. Hence, negative effects of such long working spells on performance are partly a function of the long work itself and partly of "sleepiness." Appearance of negative effects depends on the types of tasks performed, on motivation of the worker, and on timing, because wakefulness and sleepiness appear in cycles during the day.

As a general rule, task performance is influenced by three factors: the internal diurnal rhythm of the body, the external daily organization of work activities, and personal motivation and interest in the work. Each factor can govern, influence, or mask the effects of the others on task performance.

Performance of different types of work is affected differently by long periods of work with concurrent sleep loss. Execution of a task that must be performed uninterruptedly for half an hour or longer is more affected by sleep loss than a short work task. Also, if such a task must be replicated, performance is likely to become worse with each successive repetition. Performance of monotonous tasks is highly reduced by sleep deprivation, but execution of a task new to the operator is less affected. On the other hand, doing a complex task can be more affected than doing a simple one. Work that is paced by the work itself deteriorates more with sleepiness than does an operator-paced task. Accuracy in performing a job may be still quite good even after losing sleep, but it takes longer to perform the job. Froeberg (1985), from whose work the statements above are taken, also found that a task which is interesting and appealing, even if it does include complex decision making, can be performed rather well even over long periods of time. But if the task is disliked and unappealing, decision making is prolonged. Long-term and short-term memory degrades.

APPLICATION

In general, sleep loss (particularly if associated with long periods of work such as a full day) deteriorates performance in terms of reaction time, of failure to respond or false response, of slowed cognition and diminished memory. Most deterioration occurs when the circadian rhythms are set for a night's sleep. Performance becomes more reduced after two or three nights of sleep deprivation. After missing four nights, very few people are able to stay awake and to perform, even if their motivation is very high (Froeberg 1985).

Thus, performance of all mental tasks (except short and interesting ones) diminishes with long hours of work associated with sleep loss. The longer the work period, and the more monotonous, repetitive, uninteresting, and disliked the task, the more performance degrades.

The coffee (tea) pickup

Caffeine is a stimulant (of the chemical family methylated xanthines) that is quickly absorbed into the bloodstream. For about half an hour after drinking a strong cup of coffee, most people feel more awake and better able to pay attention; heart rate is increased by 2 to 10 beats per minute over a period of 5 to 15 minutes. Drinking five to ten cups of strong coffee is likely to have an "overdosing" effect, generating a condition called caffeinism, whose symptoms include lightheadedness, tremor, headache, palpitation, and difficulty in falling asleep.

Caffeine is contained in coffee, tea, cocoa, and in most chocolate products. It is added to many cola-type drinks and is a component of some medications for headache, cold, and allergy. Cocoa and chocolate products often contain theobromine, which has effects similar to caffeine on behavior and body functions. Theophylline, with similar effects, is contained in tea.

The amounts of caffeine and related substances are:

- *In coffee: 60 to 150 mg per 250 cm^3 (cup), with instant coffee usually in the low range.*
- *In tea: 8 to 50 mg per 250 cm^3 (cup), with instant tea usually in the middle range.*
- *In cocoa: about 15 mg per 250 cm^3 (cup).*
- *Soft drinks contain usually between 40 and 70 mg per 355 cm^3 (12 fl. oz.).*
- *Chocolate has about 12 mg caffeine and 120 mg theobromine per 50 g (2 oz. av.), while most baked chocolate goodies have about half these amounts.*

Incurring sleep deprivation and recovering from it. The following discussion assumes sleep deprivation of at least one night.

During long working times coupled with lack of sleep, periods of "no performance" occur, also known as "lapses" or "gaps": these are short periods of reduced arousal or even of light sleep (Wedderburn, 1987). With increasing time at work, so-called "microsleeps" occur more frequently: the subject falls asleep for a few seconds, but these short periods (even if frequent) do not have much recuperative value, because the subject still feels sleepy and performance still degrades.

Naps lasting one to two hours improve subsequent performance. However, if a person is awakened from napping during a deep-sleep phase, "sleep inertia" with low performance can appear which may last up to 30 minutes. Temporal placing of a nap may have differing effects: for example, the common short early-afternoon nap has little effect on performance of subsequent work (in laboratory tests), yet many people say they simply need that nap after lunch. On the other hand, naps of at least two hours taken in the late evening or during nighttime, when the diurnal rhythm is low, have positive effects on subsequent performance lasting several hours, provided that the amount of sleep loss incurred until this moment is moderate, such as one night without sleep (Gillberg, 1985; Minors and Waterhouse, 1987; Rogers, Spencer, Stone, and Nicholson, 1989).

APPLICATION

If long periods of mental work are unavoidable, one may try to interspace physical activities, e.g., exercises. Also, "white noise" (see Chapter 4) may improve performance slightly. Hot snacks are often welcome, as are hot and cold (usually caffeinated) beverages. Occasional "stirring" music may help. "Bright" illumination, such as of 2,000 lux or more, helps to suppress production of the hormone melatonine, which causes drowsiness. Drugs, particularly amphetamines, can restore performance to nearly normal level, even when given after three nights without sleep (Froeberg, 1985).

Recovery from sleep deprivation is quite fast. A full night's sleep, undisturbed, probably lasting several hours longer than usual, restores performance efficiency almost fully.

SHIFT WORK

One speaks of shift work if two or more persons, or teams of persons, work in sequence at the same workplace. Often, each worker's shift is repeated, in the same pattern, over a number of days. For the individual, shift work means attending the same workplace either regularly at the same time ("continuous" shift work) or at varying times (discontinuous, including rotating, shift work).

☛☛☛ *Shiftwork is not new. In Ancient Rome, deliveries were to be done, by decree, at night to relieve street congestion. Bakers habitually have worked through the late night hours. Soldiers and firefighters always have been accustomed to night shifts.* ☚☚☚

With the advent of industrialization, long working days became common, with teams of workers relaying each other to maintain blast furnaces, rolling mills, glass works, and other workplaces where continuous operation was desired. The 24-hour period was covered with either two 12-hour workshifts or three 8-hour workshifts (Kogi, 1985, 1991). Since the industrial revolution, when 12-hour shifts were common, drastic changes in work systems have occurred. In the first part of the twentieth century, the then-common 6-day workweeks with 10-hour shifts were shortened. Today, many work systems use the 8 hours per day/5 workdays per week arrangement, which was introduced in many countries in the 1960s. In general, the number of days worked per week was reduced, usually to allow two weekend days to be free; also, the number of hours worked per day was reduced. Yet, working "overtime" has become a fairly regular feature in many employment situations.

Shift systems. Shift work is different from "normal" day work, in that such work is performed regularly during times other than morning and afternoon, and/or that, at a given workplace, more than one shift is worked during the 24-hour day. A shift may last for an 8-hour work period, or may be shorter or longer.

Though estimates vary and depend on the definition of "shift work," probably about one of four workers in developed countries is on some kind of shift schedule other than regular day work (Kogi, 1991; Monk and Tepas, 1985; Tepas and Mahan, 1989).

❑ **APPLICATION**

There are many diverse shift systems. For convenience they can be classified into several basic patterns, but any given shift system may comprise aspects of several patterns. Kogi (1985) lists four particularly important features of shift systems: Does a shift extend into hours that would normally be spent asleep? Is it worked throughout the entire 7-day week or does it include days of rest, such as a free weekend? Into how many shifts are the daily work hours divided; i.e., are there two, three, or more shifts per day? Do the shift crews rotate or do they work the same shifts permanently? These features are shown in Figure 6-3.

Other identifiers of shifts and shift patterns are the starting and ending time of a shift; the number of workdays in each week; the hours of work in each week; the number of shift teams; the number of holidays per week or per rotation cycle; the number of consecutive days on the same shift, which may be a fixed or variable number; and the schedule by which an individual either works or has a free day, or free days (Kogi, 1985). All of these aspects may

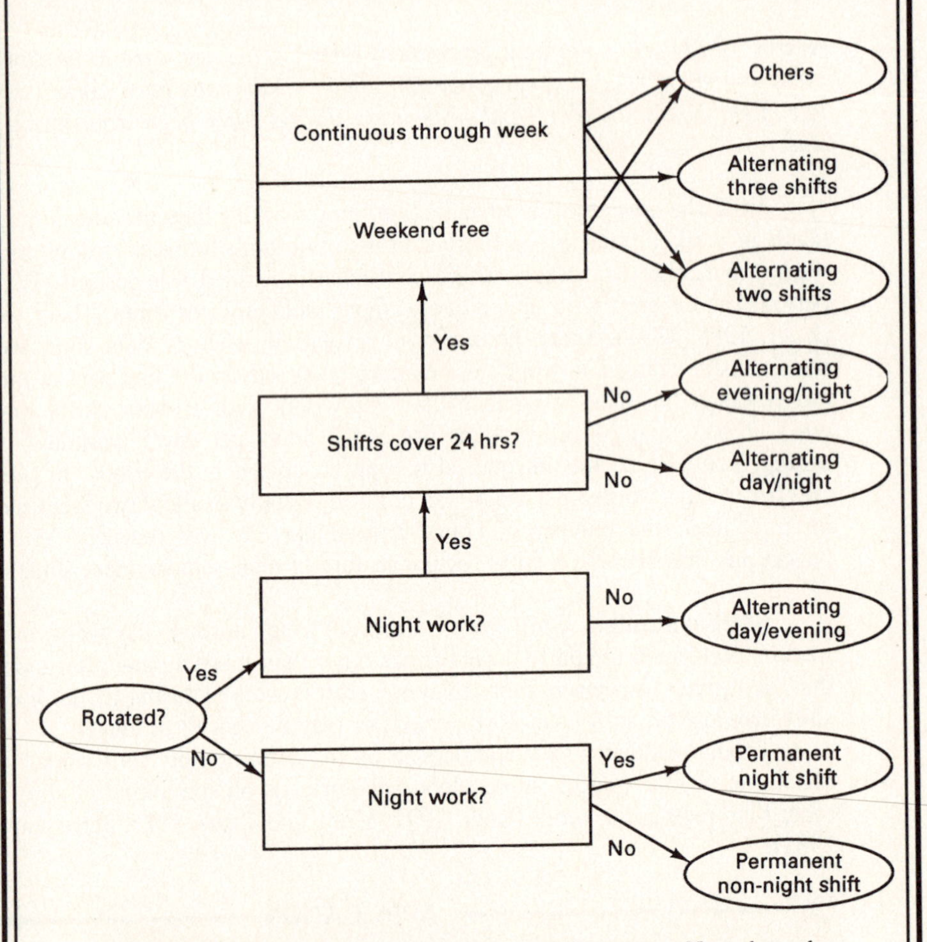

Figure 6-3. Flowchart of key features of shift systems. Note that other shift attributes are possible. SOURCE: Adapted from "Introduction to the Problems of Shiftwork" by K. Kogi (1985) in S. Folkard and T. H. Monk (eds.), *Hours of Work*. Reproduced by permission of John Wiley and Sons Limited.

TABLE 6-2. EXAMPLES OF 5-WORKDAYS-PER-WEEK SHIFT SYSTEMS

System	Workdays/free days	Shift sequence
Permanent day shift	5/2	1-1-1-1-1-0-0, 1-1-1-1-1-0-0, . . .
Permanent evening shift	5/2	2-2-2-2-2-0-0, 2-2-2-2-2-0-0, . . .
Permanent night shift	5/2	3-3-3-3-3-0-0, 3-3-3-3-3-0-0, . . .
Rotations		
Alternating day-evening	10/4	1-1-1-1-1-0-0, 2-2-2-2-2-0-0, . . .
Alternating day-night	10/4	1-1-1-1-1-0-0, 3-3-3-3-3-0-0, . . .
Alternating day-evening-night	15/6	1-1-1-1-1-0-0, 2-2-2-2-2-0-0, . . . 3-3-3-3-3-0-0 (forward rotation) . . . or 1-1-1-1-1-0-0, 3-3-3-3-3-0-0, . . . 2-2-2-2-2-0-0 (backward rotation) . . .

Legend
1 represents day shift, 2 evening shift, 3 night shift, 0 free day, i.e., without scheduled shift.

SOURCE: Adapted from Kogi, 1985. Reproduced by permission of John Wiley & Sons Limited.

affect the welfare of the shift worker, the work performance, and the organizational scheduling.

In terms of organizing the schedule, it is easiest to set up a permanent or weekly rotation schedule. Several such solutions are shown in Table 6-2.

Many examples of work schedules are presented and discussed in the literature. (See, e.g., Colligan and Tepas, 1986; Colquhoun, 1985; Eastman Kodak Company, 1986; Folkard and Monk, 1985; Johnson, Tepas, Colquhoun, and Colligan 1981; Kogi, 1991; Tepas and Monk, 1986.) In most systems used today, the same shift is worked for five days, usually followed by two free days during the weekend. This regimen does not, however, evenly cover all 21 8-hour shift periods of the week; additional crews are needed to work on weekends or other "odd" periods. If one uses three shifts a day with four teams, the shift system (for one team) is 1-1-2-2-3-3-0-0 with a 6/2 work/free day ratio and a cycle length of eight days; this is known as the "metropolitan rotation." The "continental rotation," which also assumes three shifts per day and four crews, has the sequence 1-1-2-2-3-3-3, 0-0-1-1-2-2-2, 3-3-0-0-1-1-1, 2-2-3-3-0-0-0; its work/free day ratio is 21/7, its cycle length exactly four weeks.

The ratio of work days versus free days in a complete cycle is an important characteristic of any shift system. Table 6-3 presents a number of other features that describe different shift systems.

Compressed workweek/extended workday. One speaks of a *compressed workweek* when the required hours of work per week (such as 40 hours) are performed in only 4 or even 3 days per week, allowing the worker to have 3 or 4 free days each week. Apparently this is an attractive idea for many persons: it reduces the number of trips to and from work, and there are fewer "setups" and "closedowns" at work. However, there are concerns about in-

TABLE 6-3. CHARACTERISTICS OF SHIFT ARRANGEMENTS

Cycle length	$C = W + F$	
Free days per year	$D = 365F(W + F)^{-1}$	
Number of days worked before the same set of shifts re-occurs on the same days of the week	$R = C = W + F$	if $(W + F)$ is multiple of 7
	$R = 7(W + F)$	if $(W + F)$ is not a multiple of 7

Legend
W = number of work days, F = number of free days

SOURCE: Kogi, 1985. Reproduced by permission of John Wiley & Sons Limited.

creased fatigue due to long workdays and about reduced performance and safety (Kogi, 1991).

Compressing the weekly working hours into a few days means extending the duration of each workshift. The type of work to be performed is a major determiner of whether this can and should be done. Long working days have been mostly used in cases where one waits or remains on "standby" much of the shift, such as firefighters do. Also, activities have been done in long shifts that require only few or small physical efforts which are diverse and interesting, yet fall into routines. Examples are nursing, clerical work, administrative work, technical maintenance, computer supply operations, and supervision of automated processes. Little experience exists from long shifts that include manufacturing, assembly, machine operations, and other physically intensive jobs. Information has been mostly gathered from subjective statements of employees, some limited psychological test batteries, and scrutiny of performance and safety records in industry.

The results are contradictory, spotty, and apparently dependent on the given conditions. In some cases, production and performance are high shortly after introduction of a compressed workweek but fall off after prolonged periods on such a schedule (Ong and Kogi, 1990). However, other observations have not shown this trend. The people involved often indicate significantly increased satisfaction, which may be due more to easily and better arranged leisure time rather than to improvements of the work (Dunham, Pierce, and Castaneda, 1987; Lateck and Foster, 1985).

Working very long shifts, such as 12 hours, is likely to introduce drowsiness (Wedderburn, 1987) and some reduction in cognitive abilities, motor skills, and generally in performance during the course of each shift as the workweek progresses. It appears that there is a potential for "careless shortcuts to completion of a job" by a fatigued worker, and that work practices may be less safe in tasks "that are tedious because of high cognitive or information-processing demands, or those with extensive repetition" (Rosa and Colligan, 1988, p. 315).

Table 6-4 lists potential advantages and disadvantages of compressed workweeks/extended workdays.

TABLE 6-4. POTENTIAL ADVANTAGES AND DISADVANTAGES OF COMPRESSED WORKWEEKS/EXTENDED WORKDAYS

Potential advantages
Generally appealing
Increases possibilities for multiday off-the-job activities
Reduces commuting problems and costs
More time per day for scheduling meetings or training sessions
Fewer start-up and/or warm-up expenses
Increases production rates
Improvement in the quantity or quality of services to the public
Better opportunities to hire skilled workers in tight labor markets
Potential disadvantages
Decrements in job performance due to long work hours, or to "moonlighting" on "free" days
Overtime pay required
More fatigued workers
Increases tardiness and "leaving work early"
Increases absenteeism
Increased employee turnover
Increases on-the-job and off-the-job accidents
Decreases production rates
Scheduling problems if the organization operations are longer than the workweek
Difficulty in scheduling child care and family life during the workweek
Contrary to traditional objectives of labor unions
Increases energy and maintenance costs
Longer exposure (per day) to physical and chemical hazards (acceptable doses are usually for 8-hour exposure)

SOURCE: Adapted from Kogi, 1991; Tepas, 1985 in S. Folkard and T. H. Monk (eds.), *Hours of Work* (Reproduced by permission of John Wiley and Sons Limited).

Flextime. A recent trend has been toward *flextime,* meaning a somewhat "flexible" arrangement of work hours during the day. This arrangement allows the employee to distribute the prescribed number of working hours per shift (for example, 8 hours) over a longer block of time (such as 10 hours) or to distribute the required weekly work time more or less freely over the work days. Often it is required that a "core" time (of, say, 6 hours) is covered every workday, during which all workers must be present. Thus, one can "slide" or "float" the working time across the core time so the start of work is at any time before the core, and the end of work anytime after the core. Table 6-5 lists potential advantages and disadvantages of flextime.

There are other more flexible work arrangements than the regular "8 hours a day, 5 days a week": they include hour-averaging among days, staggered work hours, seasonally adapted hours, part-time work (often combined with job-sharing), even "telecommuting," where a person works at home on a computer connected to the office. All of these decouple the individual from

TABLE 6-5. POTENTIAL ADVANTAGES AND DISADVANTAGES OF FLEXTIME

Potential advantages
Generally appealing
Flexible work times with no loss in base pay
Increased day-to-day flexibility for free time
Reduces commuting problems and costs
Workforce size can adjust to short-term fluctuations in demand
Less fatigued workers
Reduces job dissatisfaction/increases job satisfaction
Increases democracy in the workforce
Recognition and utilization of employee's individual differences
Reduces tardiness
Reduces absenteeism
Reduces employee turnover
Increases production rates
Better opportunities to hire skilled workers in tight labor markets
Potential disadvantages
Irregularity in workhours produced by short-term changes in demand
Difficulty covering some jobs at all required times
Difficulty in scheduling meetings or training sessions
Poorer communication within the organization
Poorer communication with other organizations
Increases energy and maintenance costs
Increases buffer stock for assembly-line operations
Requires more sophisticated planning, organization, and control
Reduces quantity or quality of services to the public
Requires special time-recording
Requires additional supervisory personnel
Extension of health and food service hours

SOURCE: Adapted from Kogi, 1991; Tepas, 1985 (Reproduced by permission of John Wiley and Sons Limited).

strict job requirements, timewise and/or locationwise. This requires that the employer changes from traditional organizational patterns regarding supervision, communication, job requirements, work equipment and scheduling; the employee must increase self-reliance, independence, responsibility, and skills. Kogi (1991) recommends that such major changes in work organization, timing, and content be preceded by careful assessment of the issues before (even a trial) implementation, a process that requires cooperation between management and employees. ❑

Which are suitable shift systems? The human is used to daylight activity, and to sleeping at night. This inherent feature is governed by the internal clocks of diurnal rhythms. Night work, then, is "unnatural," since it forces to work when one should sleep. This may just be "difficult" to do, or it may offset the circadian rhythm. Furthermore, loss of sleep or disturbed sleep experienced by nighttime workers may reduce well-being and performance. Also, social and domestic interactions may be impeded, since our societal structure is generally based on "work at day, play in the evening, sleep during the night." This does not necessarily mean that night work is harmful, but it may be stressful, more or less, depending on the circumstances and the person.

Organizational criteria by which to judge the suitability of shift systems include the number of shifts per day, the length of every shift, or the times of the day during which there is no work done; the coverage of the week by shifts; shift work on holidays; etc. These "independent variables" (to use the terminology of experimental design) have been discussed earlier.

An important "dependent variable" is the health of the shift worker. Do certain shift regimes affect physiological or psychological well-being? For example: does the inability to sleep at night, when a night shift needs to be worked, affect the worker's health? What is the effect of shift work on "social" interactions with family, friends, and the society in general?

Another "dependent variable" is the performance of workers on shift schedules. Is the same output to be expected regardless of the time of work during the 24-hour day? Are specific activities better performed at certain shifts? Do changes in shift-work schedule affect the shift worker's output? Are certain shifts higher in accidents?

EXAMPLE

Monk and Wagner (1989) examined mining accidents that occurred during a full decade of regular seven-day shifts. While one should have expected a steadily decreasing number of accidents as the body adapted to the schedule, accidents were highest on Sunday nights, the fourth night shift. Apparently, the workers did not get enough sleep during Sunday, which would indicate that social obligations overrode the body's need for rest.

Thus, there are several aspects of shift work: health and well-being, performance and accidents, psychological and social adjustments. All interact with each other, but not always in the same direction. Thus, conclusions for selecting a suitable work regimen depend on the given conditions.

Apart from organizational ease, the aspects of work performance, well-being, and social interaction are important in the selection of suitable shift regimes.

Health and Well-being. A strong circadian system exists within the body which is remarkably resistant to sudden large changes in routine. This internal system being so stable, theoretical findings, common sense, and personal experiences suggest that the normal synchrony of behavior in terms of rest and activity sequences should be maintained as well as possible. Thus, work schedules should be arranged to accord with the internal system or, if this is impossible (for example, if night work is necessary), to disturb the internal cycles as little as possible. One logical conclusion is that work activities which contradict the internal rhythm should be kept as short as feasible, so that one can return to the "normal" cyclicity as quickly as possible. For example, one should schedule single night shifts, interspaced by normal workdays, instead of requiring a worker to do a series of night shifts, as shown in Table 6-2. Such a series of night shifts upsets the internal clock, while a single night shift would not severely disturb the entrained cyclicity.

The other solution, also theoretically sound and supported by experience, is to entrain new diurnal rhythms. It takes regular and strong zeitgebers to "overpower" the regular signals, such as light and darkness. For shift work this means that the same setup (such as working the night shift) should be maintained for long periods (several weeks, even months) and not be interrupted by different arrangements (in theory, not even by "free" days, such as on weekends). It appears that certain people are more willing and able to conform to such regular "non-day" shift regimes than others.

Health complaints of shift workers are often voiced and suspected, but actual negative effects are difficult to prove. Night-shift workers have, on the average, about half an hour shorter sleep time than persons who are permanently on day shift. However, Carvalhais, Tepas, and Mahan (1988) found that persons who permanently work the evening shift sleep about half an hour longer than persons on the day shift. Also, many persons on night shift complain about the reduced quality of sleep that they get during the day, with noise often mentioned as particularly disturbing.

Some authors have found statistically significant health complaints as a function of shift work, particularly digestive disorders and gastrointestinal complaints, while other researchers have failed to prove significance (Alfredsson, Akerstedt, Mattsson, and Wilborg, 1991; Cooper and Smith, 1985; Folkard and Monk, 1985). Altogether, no differences have been found in the mortality of night-shift workers as compared to workers in other shifts. However, it appears fairly clear that persons who suffer from health disturbances are more negatively affected by night shifts than by other shift arrangements. It also appears that older workers, possibly due to deteriorating health and difficulties to get sufficient restful sleep (both phenomena apparently increasing with age), may be more negatively affected by shift work than younger workers.

Performance. The reduced quantity and quality of sleep experienced by night workers leads to the conclusion that many suffer from a chronic state of partial sleep deprivation. Negative effects of a sleep deficit on behavioral aspects and work performance have been well demonstrated, as discussed earlier. For some tasks, the interaction of circadian discrepancies (between work demand and body state) and sleep deprivation may result in significant detriments to night-work performance, including safety (Monk, 1989; Monk and Wagner, 1989). (Yet, accident statistics are usually confounded by many variables in addition to the shift factor, such as the work task, worker age and skill, shift schedule, etc.)

During the first night shift or shifts, performance is likely to be impaired between midnight and the morning hours, with the lowest performance around 4:00 A.M. Such impairment, which may be absent for physical work or minimal for interesting cognitive tasks, varies in level but is similar to that induced by "legal doses" of alcohol (Monk, 1989). However, as the worker continues to do night shifts, given suitable conditions, the internal clocks realign with the new activity rhythms, and a suitable daily routine of social interactions, sleeping, and going to work is established.

Tolerance of shift work is different from person to person and varies over time. Three out of ten shift workers have been reported to leave shift work within the first three years due to problems encountered. Tolerance of those remaining on shift work depends on personal factors (age, personality, troubles, and diseases; the ability to be flexible in sleeping habits, to overcome drowsiness), on social-environmental conditions (family composition, housing conditions, social status), and of course on the work itself (workload, shift schedules, income and other compensation). These factors interact, and their importance differs widely from person to person and changes over one's work life (Costa, Lievore, Casaletti, Gaffuri, and Folkard, 1989; Moog, 1987; Rosa and Colligan, 1988; Volger, Ernst, Nachreiner, and Haenecke, 1988). Evening-shift workers suffer particularly in their social and domestic relations, while night-shift workers are more affected by the conflicts between the requirement to work while physiologically in a resting stage and by insufficient sleep during the day. However, physiological and health effects are not abundant in shift workers who have been on such assignments for years, possibly because persons who cannot tolerate these conditions abandon shift work soon after trying it (Bohle and Tilley, 1989; Tepas and Mahan, 1989).

Social Interaction. People's needs are individually different. For example, parents of small children usually need to be home for family interaction and are unlikely to accept unusual work assignments, while older persons who do not need to interact with their children so intensely may be more inclined to work "nonnormal" hours.

A major problem associated with shift work is the difficulty of maintaining normal social interactions when the schedule forces one to work or sleep during times when social relations usually occur. This makes it difficult to share in family life, to bond with friends, and to participate in "public events" such as sports. Common daily activities may not be easily executed, such as shopping or watching television.

It is noteworthy that, in different countries and cultures, certain events or conditions may be present or not, may be regularly scheduled at different times, and may be of different value to individuals and the society at large. The southern siesta time is not commonly known in northern regions. Shops that are open continuously in the United States close in the late afternoon and stay closed on weekends in Europe. Family ties are much more important in some cultures than in others and may vary among individuals. Thus, statements regarding the effects of shift work on social interactions may apply in one case but not in another. However, it is generally true that shift work and its consequences to the individual worker often interfere with social relations. Whether this has a demonstrable effects on well-being and performance depends on the individual case.

APPLICATION

How to select a suitable work system. The foregoing discussions should have made it clear that, if at all possible, the working hours should be during daylight. However, in many cases this "normal" arrangement is replaced by other shift work, covering either the late afternoon and evening hours, or the night.

There seems to be an inevitable fall in performance during overnight work, related to the circadian rhythm. This is of particular concern with long periods of duty, and when the overnight work period follows poor sleep. Reports of reduced performance during the night are numerous, as summarized by Rogers, Spencer, Stone, and Nicholson (1989), and by Tepas and Mahan (1989).

The argument for permanent assignment to either an evening or a night shift is well founded in theory, on the grounds that such a permanent arrangement allows the internal rhythms to become entrained according to this rest/work pattern. However, that reasoning is not as convincing as it might appear, as most shift arrangements are not truly consistent or permanent, because the weekend interrupts the cycle. Furthermore, strong zeitgebers during the 24-hour day (such as light and dark) remain intact even for the person on regular evening or night shifts, thus hindering a complete entrainment of the internal functions. This leads to the opposite conclusion, also well founded in theory. Accordingly, it is better to work only occasionally outside the morning/afternoon period and to work only one such evening or night shift. In this case, most people are able to perform their unusual work without much detriment during this one work period, while they remain entrained on the usual 24-hour cycle. Of course, some individuals are able and willing to adjust fairly easily to different work patterns. Thus, unusual work patterns may be acceptable to "volunteers."

EXAMPLE

For crews of airplanes who must cross time zones in their long-distance flights and catch some sleep at their destination before returning, several problems exist. The first is that the quality and length of sleep at the stopover location is often much worse than at home. The resulting tiredness may be masked or counteracted by use of caffeine, tobacco, and alcohol. The second problem is the extended time of duty, which includes preflight preparations, the flight period itself, and the wrap-up after arrival at the stopover. Negative effects are substantially larger after an eastward flight than a westward one; also, crew members over 50 years old are more affected than their younger colleagues (Graeber, 1988).

Recommendations for the "shift" arrangement for flight crews are fairly well established (Eastman Kodak, 1986). In general, but particularly when flying eastward, flight crews should adhere to well-planned timing. This should duplicate, as far as possible, the sleep-wake activities at home, meaning that the crew should try to go to bed at the regular home time and get up at regular home time. Thus, they maintain their regular diurnal rhythm. Of course, the next flight duty should be during their regular time of wakefulness.

With respect to the length of a work shift, one general recommendation is that physically demanding work should not be expected over periods longer than 8 hours unless frequent rest pauses are available; but an 8-hour shift may be too long for very strenuous work. The same applies to work that is mentally very demanding, requiring complex cognitive processes or high attention. For other "everyday" work, durations of 9, 10, even 12 hours per day can be quite acceptable. Flextime arrangements often are welcomed by employees, possibly in combination with "compressed" workweeks, particularly if they allow extended free weekends.

Recommendations for shift arrangements. Shift work is often desired for organizational and economic reasons. Individual acceptance of shift work depends on a complicated balance of professional and personal concerns, including physiological, psychological, and social aspects. Of ten persons assigned to shift work, seven or eight are likely to stay on this schedule while the others drop out.

Preexisting health problems may be exacerbated by shift work. Gastrointestinal/digestive problems are fairly frequent among shift workers. Workers on permanent night shifts also often complain about insufficient sleep and general fatigue, but cardiovascular or nervous diseases are apparently not more prevalent among shift workers than in the general population.

Regarding work performance of shift workers, or accidents, little reliable information is available, because work performed on evening and night shifts often differs from that performed during the day. But work (other than cognitive) on the first night shift is likely to be impaired in the early morning hours, when fatigue affects the worker like light doses of alcohol (see below).

In deciding which one of the many possible shift plans to select, criteria must be established for a justifiable system. For example, one might establish the following requirements:

- Daily work duration should not be more than 8 hours.
- The number of consecutive night shifts should be as small as possible;

preferably, only a single night shift should intervene among the other work shifts.

- Each night shift should be followed by at least 24 hours of free time.
- Each shift plan should contain free weekends—at least two consecutive work-free days.
- The number of free days per year should be at least as large as for the continual day worker.

Using these criteria, Knauth, Rohmert, and Rutenfranz (1979) discussed a large number of shift plans in order to select those that comply with the requirements. Following similar methods, one can carefully determine suitable work schedules which fit the given requirements and conditions (Eastman Kodak, 1986).

For evening or night shifts, high illumination levels should be maintained at the workplace, such as 2,000 lux or more. Furthermore, environmental stimuli should be employed to keep the worker alert and awake, such as occasional "stirring" music and provision of hot snacks and of (caffeinated) hot and (non-alcoholic) cold beverages. The work should be kept interesting and demanding, since boring and routine tasks are difficult to perform efficiently and safely during the night hours.

The shift worker should use "coping strategies" for setting the biological clock, obtaining restful sleep, and maintaining satisfying social and domestic interactions. Unless the shift worker is on a very rapidly rotating schedule, the aim is to reset the biological clock appropriately to the shift-work regimen. For example, sleep should be taken directly after a night shift, not in the afternoon. Sleep time should be regular and kept free from interruptions. Shift workers should seek to gain their family's and friends' understanding of their rest needs. Certain times of the day should be set aside specifically and regularly to be spent with family and friends.

Body Rhythms and Shift Work: Summary

Human body functions and human social behavior follow internal rhythms. Aside from the female menstrual cycle, the best-known rhythms are a set of daily fluctuations called circadian or diurnal rhythms. They appear in such functions as body temperature, heart rate, blood pressure, and hormonal excretions. Under regular living conditions, these temporal programs are well established and persistent.

The well-synchronized rhythms and the associated behavior of sleep (usually during the night) and of activities (usually during the day) can be desynchronized and put out of order if the time markers (zeitgeber) during the 24-hour day are changed and if activities are required from the human at unusual times. Resulting sleep loss and tiredness influence human performance in various ways. Mental performance, attention, and alertness usually are reduced, but execution of most physical activities is not. Furthermore, concerns exist that disturbing the internal rhythm, such as by certain types of shift work, might have negative health effects. However, only gastrointestinal problems are proven to be more frequent with workers on night shift than with persons on day work. Certainly, being excluded by shift work from participating in family and social activities is difficult for many persons.

From a review of physiological, psychological, social and performance behaviors, the following recommendations for acceptable regimes of working hours and shift work can be drawn:

- Job activities should follow entrained body rhythms.
- It is preferable to work during the daylight hours.
- Evening shifts are preferred to night shifts.
- If shifts are necessary, two opposing rules apply: (1) either work only one evening or night shift per cycle, then return to day work, and keep weekends free, or (2) stay permanently on the same shift (whichever it is).
- A shift duration of eight hours of daily work is usually adequate, but shorter times for highly (mentally or physically) demanding jobs may be advantageous; and longer times (such as 9, 10, or even 12 hours) may be acceptable for some types of routine work.
- Compressed workweeks often are acceptable for routine jobs—for example, 4 days with 10 hours.

EFFECTS OF ALCOHOL ON PERFORMANCE

Alcoholic beverages contain many chemical substances; approximately 200 congeners have been identified in wine. It is still not clear which of these (in addition to ethanol) are responsible for the undesired physiological and psychological effects, or for the feelings of temporal euphoria and freedom from inhibition. While there is some evidence that alcoholic beverages, if taken in limited doses, may be beneficial for the cardiovascular system (Avogaro, 1990), alcohol taken excessively affects the human negatively.

Alcohol in the human bloodstream has neurological effects. First it impairs the functioning of the cerebral cortex, which houses "intelligence." Then the limbic system is affected, which among other functions controls mood. The next impaired area is the brainstem, where the "fight-or-flight" response is generated, and where such functions as heart rate, blood pressure, and respiration are controlled. Thus, the central nervous system is impaired, more with increasing blood alcohol content. Impairment occurs in judgment; in language; insight, memory, ability to understand and plan; in motor control and body posture. Also, reception and perception of sensory inputs and appropriate responses are diminished. Large mood and emotion swings may occur, typically from laughter and giddiness ("the soul of the party") to sadness and tearfulness ("crybaby").

Long-term excessive users of alcohol (alcoholics) are likely to show pathological effects. Toxic changes occur in organs such as the brain and muscle. In the intestinal system, "metabolic derangement" is common, since alcohol interferes with absorption, digestion, and metabolism and utilization of nutrients, especially vitamins (Miller and Gold, 1991).

Blood Alcohol Content

The effects of alcohol are usually stated in relation to blood alcohol content (BAC, in percent). BAC is best measured in a blood sample, or often approximated from a sampling of exhaled air ("breathalizer"). The "Widmark formulae" can be used to calculate BAC:

$$BAC(\%) = 0.0318d + 0.1652d^2 - 0.0998d^3 \quad \text{for males}$$

$$BAC(\%) = 0.1660d + 0.0803d^2 + 0.0348d^3 \quad \text{for females}$$

where d, in mL, is the dosage of ethanol per kilogram of body weight.

Absorption

Absorption of orally taken alcohol takes place by simple diffusion through cell boundaries in the gastrointestinal tract. There are large intra- and interindividual differences in the development of BAC.

☛☛☛ *Distilled alcohol (booze) is absorbed faster than beer. The more alcohol is drunk (over a given time), the more slowly (!) it is absorbed, owing to reduced gastric emptying, which itself is caused by the presence of alcohol. (This may also explain the higher tolerance, i.e., the slower absorption rate, of drinkers than of abstainers.) Absorption is also slowed by a "full stomach," again due to delayed gastric emptying, which slows the arrival of alcohol at the lining of the intestinal tract. Full absorption may take up to six hours with a heavy meal.*

Absorption is

- *slowest after a meal high in carbohydrates*
- *medium after a meal high in fats*
- *fastest after a meal high in proteins*

Alcohol does not dissolve in body fat but is freely diffused in the body. Thus, people who have less body water have higher BAC. Therefore, women, obese people, and the elderly generally have higher BAC than men, slim people, and younger people who have drunk the same dosage of alcohol. ☚☚☚

Elimination of Alcohol

Elimination begins at the moment of ingestion and proceeds at uniform rate until concentration is very low. The rate per hour is about 0.015 percent for men and 0.019 percent for women, little affected by physical work. Alcohol is oxidized, and the byproducts, carbon dioxide and water, excreted by breathing; developed water increases the need to urinate. Elimination is not sped up by drinking coffee, since oxidation is not affected by caffeine.

Too much alcoholic good cheer can result in an un-jolly hangover. Headache, nausea, and stomach irritation are caused by undigested byproducts of alcohol, particularly acetaldehyde and lactic acid, that build up in the blood stream as the liver is falling behind in digesting alcohol. Since the speed of alcohol conversion cannot be changed, either by drinking coffee or by breathing fresh air, the simple cure is to wait long enough to give the liver the time needed to eliminate the alcohol in the body. Symptoms of a hangover may be relieved by over-the-counter medications, including antacids, aspirin, and other pain relievers, yet they can cause further stomach irritation.

Women taking oral contraceptives metabolize ethanol more slowly than other women, or males, meaning that they could stay intoxicated longer. Also, highest BACs occur directly before the menstrual flow date.

Specific Effects of Alcohol

Once absorbed, alcohol is distributed by the blood. Hence, alcohol content of organs with a good blood supply (such as brain, lungs, liver, and kidneys) becomes quickly the same as that of blood. The highest alcohol content in the blood occurs about half an hour after ingestion. Since ethanol freely mixes with water, it quickly reaches all cells bathed in water.

The strength of alcohol effects depends on the time of day, i.e., on the circadian rhythm (discussed earlier in this text). For example, its effects are stronger in the early afternoon than in the evening (Horne and Gibbons, 1991).

Effects of alcohol on the nervous system. Alcohol has several effects on the *peripheral nervous system:*

- nerve excitation is increased by a low BAC but inhibited by a high one.
- transmission of neural impulses at the synaptic junction may be reduced according to the depressant effects of alcohol.

The effects on the *brain* (central nervous system) are:

- at BAC levels below 0.05 percent, inhibition is reduced and judgment is impaired (Reduction of inhibition is responsible for the illusion of stimulation by alcohol.)
- at 0.1 percent BAC, depression of sensori-motor functions occurs.
- at 0.2 percent BAC, control of emotion is lost.
- at 0.3 percent BAC, lack of comprehension and stupor occur.
- at 0.4 to 0.5 percent BAC, coma occurs.
- at 0.6 percent BAC, breathing and heartbeat are depressed and death can occur.

Effects of alcohol on the senses. *Visual* acuity is relatively insensitive to BAC, as are light and dark adaptation. But alcohol increases the sensitivity to dim lights, and decreases the ability to discriminate between bright lights. Resistance to glare is reduced. Color sensitivity is affected: "light" red, green, or yellow are less easily discerned from white, but blue and violet hues are more easily discriminated.

Alcohol causes the eyes to converge at long viewing distances and diverge at short ones. Depth perception is impaired at rather high BAC, such as 0.1 percent. The ability to judge distances is reduced. Visual accommodation is impaired, and eye-movement latency increased. Critical fusion frequency (the highest rate at which light is perceived as flashing) is decreased by large dosages of alcohol (BAC about 0.1 percent) but not by smaller doses. Peripheral vision is somewhat reduced by alcohol, but only under heavy general information load. Together, these findings lead to the conclusion that alcohol impairs the ability to see rapidly changing events.

Auditory acuity seems to be rather insensitive to alcohol. However, the ability to glean information from auditory stimuli is impaired.

Both *smell* (olfactory) and *taste* (gustatory) sensitivity are diminished with rather small amounts of alcohol (as little as 0.01 percent BAC).

Sensitivity of *touch* is reduced, particularly with respect to two-point discrimination. Sensitivity to *pain* is diminished by alcohol.

Effects of alcohol on motor control. *Motor control* is greatly reduced by alcohol. For example, standing sway, touching of index fingers, and other measures of hand steadiness and gait control show much decrease at 0.1 percent BAC or less.

Simple *reaction time* is increased by alcohol, but only at BAC levels of 0.07 percent or more; BAC of 0.08 to 0.1 percent increases reaction time by about 10 percent. Choice reaction time is even more affected by alcohol, and the incidence of wrong choices is increased as well.

Response to an auditory signal seems to be more prolonged than to a visual signal.

Effects of alcohol in the cognitive domain. Regarding *verbal performance,* alcohol increases superficial, egocentric, and inappropriate associations. Alcohol reduces verbal fluency and verbal mastery (Pisoni, Johnson, and Bernacki, 1991). Arithmetical *calculations* are impaired. *Time* seems to pass more quickly for a subject under alcohol.

Alcohol leads to *memory* deficiencies regarding events that took place when the person was intoxicated. Alcohol reduces the ability to mentally store information, particularly if much information must be stored (Salame, 1991), but retrieval from memory does not seem to be affected by alcohol.

Simple *auditory* or *visual vigilance* tasks are not affected by alcohol, but complex vigilance tasks are impaired. Impairment of judgment under alcohol is well known, both as it concerns judging one's own performance, and somebody else's.

Willingness to take *risks* is increased under alcohol.

APPLICATION

Effects of alcohol on industrial task performance. Trends indicating performance decrements at various alcohol dosages have been demonstrated by Price (1985). They show psychomotor performance least impaired and perceptual-sensory performance most impaired. Reduction of cognitive performance is intermediate.

Alcohol reduces the ability to perform industrial work tasks. Errors are increased, output is decreased with increasing BAC (experiments with subjects were not carried out, however, beyond BACs of 0.09 percent.) In assembly tasks, productivity was reduced up to 50 percent at 0.09 BAC. Negative effects on quantity and quality of work results were also demonstrated for machine-tool operation (punch press, drill press) and for welding. While different operators react differently to different alcohol dosages, over different application times, and with different work tasks, the performance loss with increasing BAC is a clear trend.

Effects of alcohol on automobile driving. The driver under the influence of alcohol appears to have a "shrunk visual field"; particularly, information is collected at shorter viewing distances and less frequently. Response latency and response errors are increased. Exact steering (staying in lane, or parking) is impaired. Driving speed may be increased or decreased, but the ability to judge driving speed is impaired. The willingness to take risks is increased.

☞☞☞*Signs of an alcohol-impaired driver:*

- *Approaches red traffic light fast, then makes sudden stop*
- *Changes lanes often ("weaving")*
- *Straddles center line*
- *Changes speed often*
- *Drives very fast*
- *Drives with no light in the dark*
- *Does not dim bright lights for oncoming traffic*
- *Drives very carefully and slow*
- *Has difficulty parking*

☜☜☜

Effects of alcohol on pilot performance. Drinking alcohol reduces performance for considerable time. In an experiment using a flight simulator, both younger (mean age 25 years) and older (mean age 42 years) pilots showed reduced flying performance, including communication, for at least two hours after having reached 0.10 percent BAC. The overall performance remained impaired for up to eight hours (Morrow and Jerome, 1990).

The Federal Aviation Agency adopted in 1985 a rule that no person with a BAC of 0.04 percent or higher may act as a crew member of a civil aircraft. Twelve male pilots, all with relatively few flying hours (50 to 315) and without instrument rating, performed simulated flight activities either under placebo conditions, or with alcohol dosages that brought about a BAC of about 0.04 percent. In many of the task segments, performance was reduced with alcohol, but the main flying tasks were relatively little affected. Pilots under alcohol were often inattentive to important secondary tasks and violated "safe" procedures. The researchers (Ross and Mundt, 1988) concluded that pilot performance even under such low alcohol levels would reduce the margin of safety for routine flying conditions, and that in circumstances of increased demands on the pilot, the probability of an accident would be increased significantly.

Effects of Alcohol: Summary

In most people, even at relatively small blood alcohol content, motor performance is diminished. Cognitive performance is even more reduced, while sensory perception and the making of correct decisions and of fast responses are most severely impaired. Unfortunately, the affected person usually is not aware of these impairments, because alcohol also reduces the ability to make judgments about one's own performance.

SUMMARY

The human body functions in patterns that, in essence, reflect resting at night and being active during the day. The circadian rhythms of the body should not be interrupted by work requirements; yet, on occasion it is necessary to work over long periods, or during the night. If this is the case, detriments in certain kinds of performance are likely, and health consequences may exist.

For shift work, it is advisable to keep the body on the same natural rhythm and just interspace one evening or night shift. Another solution is to try to adjust the internal rhythms to continual working at these times.

Alcohol, even in small doses, affects negatively the functioning of the nervous system, the senses, motor control, and cognitive processes. Thus, performance of work tasks is hampered by alcohol content in the blood. Various legal requirements apply, such as for driving automobiles or piloting aircraft.

CHALLENGES

Discuss the interactions between activities of the body according to internal rhythms, and the effects of time markers.

What may explain the large individual differences in daily rhythms, sleep, and activities?

Should one try to design different work schedules for "morning" or "evening types"?

What would you do if you were forced (not) to take a noon break?

Consider the theory that it is only the brain that needs sleep, not the body.

There are two extreme theories: one considers dreams as expressions of mental states (Freud, Jung, and others), the other merely as "cinema of the mind" (Horne).

Consider the interactions between having to work extremely long periods, and missing sleep, as they affect performance.

How would your work be affected if you were forced to get along with, say, five hours of sleep per night?

Given certain types of jobs, such as mental, physical, and combinations thereof: which means might be appropriate to help one perform during long work periods?

Consider the possibilities of dividing long time periods, such as a full year, into divisions other than seven-day weeks, 24-hour days, or weeks with "weekends."

Is it conceivable that people might consent to work/free divisions different than the five/two-day arrangement now common?

Consider how shift arrangements quite different from the common ones might affect social interactions.

Is "absorption" of alcohol the only factor that explains the higher tolerance of habitual drinkers as opposed to that of abstainers?

Does drinking beverages with caffeine affect a person's behavior positively, even if caffeine does not influence the oxidation of alcohol?

Consider the importance of "impaired judgment" after alcohol intake.

What can be done in practice to counter the effects of alcohol on "work" performance and behavior?

7

Models of the Human Operating Equipment

OVERVIEW

We derive understanding of our roles in cooperation with other people, or with tasks and equipment, from "models." They characterize us as teacher and student. They describe us while doing our job, or while driving our car along a grid of roads to the post office. The designer of human-operated machinery (such as spacecraft, airplanes, or automobiles) utilizes computerized models of the human (as pilot, driver, or passenger) to design proper shells and interfaces so that the human is safe, comfortable, and competent in using the equipment. Similar models of the human, though probably less complex, are also important for design and evaluation of shop workplaces where equipment is assembled or repaired. Current models of the human have come a long way from simplistic assumptions, such as represented by static two-dimensional templates. Yet, realistic behavior of the human body, either passively reacting to external forces or actively performing tasks, is still incompletely understood and modelled.

INTRODUCTION

EXAMPLE

Ergonomists/human-factors engineers often use models, i.e., paradigms (patterns) which describe or imitate, in a systematic and well-organized matter, the appearance and the behavior of the human, often in some stressful situation.

We all have ideas, concepts, images, and patterns that help us to understand our roles, or other people's roles, in the private or work environment. If these are well organized and describable, they are normally called "models."

Regarding the human's role in the modern work world, we often follow a distinction between what people can and should do, and what "machines" do better. A general distinction is that people can think and feel, and are vulnerable; machines are strong and without personality, and they may be discarded when they have

TABLE 7.1 PEOPLE OR MACHINES?

Capability	Machines	People
Speed	Much superior.	Lag about 1 second.
Power	Consistent and as large as designed.	1.5kW for about 10 seconds, 0.4 kW for a few minutes, 0.1 kW for continuous work over a day.
Manipulative abilities	Specific.	Great versatility.
Consistency	Ideal for routine, repetition, precision.	Not reliable.
Complex activities	Multichannel, as designed.	Single (or few) channel(s).
Memory	Best for literal reproduction and short-term storage.	Large store, long-term, multiple access. Best for principles and strategies.
Reasoning	Good deductive.	Good inductive.
Computation	Fast, accurate, but poor at error correction.	Slow, subject to error, but good at error correction.
Input sensitivity	Can be outside human senses; depends on design.	Wide range and variety of stimuli perceived by one unit, e.g., eye deals with relative location, movement, and color.
	Insensitive to extraneous stimuli.	Affected by heat, cold, noise, and vibration.
	Poor for pattern detection.	Good at pattern detection. Can detect signals in high noise levels.
Overload reliability	Sudden breakdown.	Can function selectively, may degrade.
Intelligence	None(?)	Can deal with expected and unpredicted events. Can anticipate. Can learn.
Decision making	Dependent on program and sufficient inputs.	Can decide even on incomplete and unreliable information.
Flexibility, improvision,	None.	Have.
Creativity, emotion	None.	Have.
Expendable	Yes.	No.

SOURCE: Modified from W. E. Woodson and D. W. Conover (1964), *Human Engineering Guide for Equipment Designers*. Berkeley, CA: University of California Press, pp. 1–23.

served their purpose. A more detailed description of the respective roles is presented in Table 7-1.

MODEL DEFINITION

In ergonomics and human-factors engineering, the term "model" is usually used according to the following definition: ***A model is a mathematical/physical system, obeying specific rules and conditions, whose behavior is used to understand a real (physical, biological, human-technical, etc.) system to which it is analogous in certain respects.***

Several aspects of that definition deserve attention.

- First, the model is "obeying specific rules and conditions." This means that the model is itself restricted: for example, the model may be a simple design template that is only two-dimensional or of only one single-percentile size, or it may be restricted to static conditions.
- Second, the model is "analogous" to the real system "in certain respects." This means that the model is restricted in its validity (or fidelity), with its boundaries often so tight that they barely overlap the actual conditions. Thus, the model's internal limitations and its limits of applicability need to be kept in mind.

The most important type of model is the "concept model," which is like a hypothesis. It incorporates the current and deduced state of knowledge and can be verified by consulting available data, or by conducting new experiments. The first stage in the formulation of a concept model is the identification of the relevant variables, independent and dependent. The next step, the modeling stage, is the formulation of the relations among the variables. The final stage is that of validation. Williges (1987) provides a thorough discussion of modeling the human interfacing with the computer.

Commonly, models are constructed of the equipment and its functioning, and of the human operator working with the equipment; these two models are usually linked together to show the interfacing between them. Thus, an overall model of the human/equipment behavior is generated. Frequently, one also needs to model the user of the operator-machine model. In this case, there are three submodels: the machine, the operator, and the model user (McMillan, Beevis, Salas, Strub, Sutton, and Van Breda, 1989).

EXAMPLE

This book contains a large number of models. In the first chapter, for example, the human body is "modelled," according to Borelli's seventeenth-century concept, as a skeleton of "links," articulated in the "joints," and powered by muscles as "engines" that move the links about the joints. The use of muscles to produce a strength ouput is modelled in form of a flow system with feedforward and feedback. In the second chapter, the body's energy production is compared to the processes in a combustion engine, and models are given of the interactions among the circulatory, respiratory, and metabolic subsystems. In the third chapter, control of the human body is described in terms of "neural networks" contained in the nervous system. Inputs of information to the body are thought to occur in the various sensors. The information is transmitted along the efferent pathways to the brain, where the information content is processed and decisions made regarding actions, which are initiated by signals sent along the efferent pathways to "output effectors" such as mouth or hands. Within this complex model, subsystems can be modelled, such as the eye or the ear. Many other models are used throughout this book.

While there are several ways of defining "simulation," in this book it is defined as "exercising" a model.

TYPES OF MODELS

One distinguishes between "open" and "closed" models. An open model is affected by outside circumstances, such as by climate conditions, vibrations, impacts, or changes in workload. A closed model excludes these external effects; it functions within its own cocoon.

"Open-loop" models do not consider the results of the activities of the model on itself. An example is a person firing a gun in the dark: when the bullet has left, the person does not know whether it hit the target or not, and hence the firing of the next shot is probably not done differently, because no feedback about the success is obtained. A "closed-loop" system, in contrast, utilizes feedback about previous actions to modify the next activities.

Another major distinction is between "normative" and "descriptive" models. A normative model assumes there is necessarily some sort of a "normal" appearance or behavior, which is perfect, ideal, in a standardized and nonvarying way—in the extreme case, that there is a singular appearance or behavior which the model represents. Thus, a normative model is often deterministically constructed, presuming that the effects of variables within the system, or acting upon the system from the outside, can be clearly predicted and hence modelled. A normative model is used to predict "normal" behavior, to which actually observed behavior is compared.

A descriptive model is one that reflects actual, variable behavior. Such changes in behavior are due to variations (often assumed to be stochastic) in internal or external variables. Descriptive models are often used for "simulation," i.e., exercising a model to imitate actual conditions and behaviors.

Judgment of the "value" of a model is usually by a set of criteria: *validity* is the agreement of outputs of the model with the performance of the actual system that it purportedly represents (this is also called fidelity, or realism). *Utility* is the model's ability to accomplish the objectives for which it was developed. *Reliability* is the repeatability of the model, in the sense that the same or similar results are obtained when exercising it repeatedly. Reliability may also be considered the ability to apply the model to similar, but not exactly the same, systems. This aspect leads to *comprehensiveness,*—the applicability of the model to various kinds of systems (Meister, 1990). *Ease of use* is, obviously, a very important criterion. If highly trained and skillful capabilities are required from the user, a model is not likely to be used often and by many. On the other hand, oversimplification of a model to achieve ease of use is not desirable either. For example, a CADAM (computer-aided design and manufacturing) model was described in 1991 (*CSERIAC Gateway* 2(3), 11-12) in which the body contours of human models, called ADAM and EVE, are based on hypothetical 95th-percentile homunculi, which then are multiplied by 0.93 to supposedly represent "average persons," or multiplied by 0.8725 to depict 5th-percentile phantoms. Thus, ease was achieved by sacrificing validity.

While in the past most models were physical (such as templates or manikins), they are now often mathematical and computerized. A *mathematical model* has the advantage of being precise, formal, and often general: the variables can be manipulated easily, and parameters in the equations assumed freely. Disadvantages of mathematical models may be in their rigid mathematical structure, often including equations the nature of which needs to be presumed and cannot be changed without changing the model (see below). Thus, some mathematical models "fit" reality poorly, often being either too general or too specific. Furthermore, if the boundary conditions are not explicitly and carefully stated, computerized mathematical models can be extrapolated inappropriately.

Often, models are classified by the academic disciplines in which they are primarily used. Thus, a number of anatomical models exist, usually relying on specific anthropometric and biomechanical formulae. A large subset consists of physiological models, primarily as they reflect metabolic or circulatory events within the body, usually related to the environment or to conditions of work. Anatomical/biomechanical/physiological models comprise the majority of "engineering body models," used, for example, in the design of spacecraft, aircraft, or automobiles.

Inadequate Models

Some models seem to have been developed by persons who know a lot about how to manipulate the computer system, but too little about the human and how the human functions with and within the system. Their models of the human are likely to be inaccurate, unrealistic, and overly simplified (as in the case just discussed)—but they may "work well" in terms of the model mathematics and computerization.

For example, one may incorporate motions of the human based on observations of how such movements were performed under certain conditions. Re-creation of these motions is called "animation." Too often, these are simplified or exaggerated, as in cartoons, or the modeller may unthinkingly transfer observations made under certain circumstances to other conditions. A typical case is depiction of the dynamic motion involved in lifting an object, from a sequence of static positions observed in strength testing. Smooth as many of these animations appear to be, they are superficially "true" only for that situation which was previously observed, but likely to be false and misleading in other situations.

A related fallacy is the assumption, born from the desire to keep the model simple, that the behavior will be "linear," meaning that if one variable (say, workload) increases, the associated dependent variable (say, speed of human activities) will increase linearly as well. But many human behavior traits do not respond linearly to changes in the task. If, therefore, a system is based on linear algorithms, then the system behavior will fail to be truly descriptive (or predictive), the more extreme (nonlinear) the conditions are.

Misuse of Models

Here is an example of the usefulness, and the possible misuse, of models. In 1986, a biomechanical model of the human body was developed to explain the stresses on the spinal column while performing lifting tasks. It was more advanced than its predecessors, because it included more details and it attempted to explain dynamic activities and their effects on the body, while previous models were static in nature. However, many assumptions were made in the development of the model, including the following:

- dimensions of the 50th-percentile male;
- movement at constant velocity (after initial acceleration and before final deceleration);
- body segments treated as cylinders;
- curvature of the lumbar section considered constant under all conditions;
- joint locations taken from erect standing posture;
- constant lever arm of posterior back muscles about the spinal column assumed, in particular a constant lever arm of the abdominal muscles at 10 percent of stature;
- all involved muscle forces summed to minimal total effort (no coactivation of muscles).

Obviously, these assumptions are overly simplistic, in fact unrealistic, and severely limit the model's validity. Yet, the temptation is strong to overlook or disregard some of the basic model assumptions, and the limitations which they impose, in order to expand the application of the model to wider boundaries. Thus, in 1988, the model was described (not by the original author) in a shortened text with the titillating title, "A Knowledge-Based Model of Human . . . Capability." In this publication, the application of the model to a variety of actual working conditions was proposed, some of which were clearly outside the stated boundaries of the original model.

Another misuse is feeding of incorrect data to the model. Not able to evaluate the correctness of the input data, the model spits out results anyhow. A related problem is hidden under the euphemism "fitting input data to the model." This may simply mean that the data format needs to fit input format requirements—in this case, there is no problem. However, if fitting data really means a modification of their content, their meaning, their "behavior," then such fitting is really data falsification.

Finally, one may misinterpret and misapply the model outputs, such as by transferring static strength calculations to dynamic activities.

There are three main misuses of model outputs:

- *Model itself is inappropriate: "Whatever in, garbage out."*
- *Inputs to the model are false: "Garbage in, garbage out."*
- *Outputs are misapplied: "Garbage use."*

To avoid these problems, the model user must be knowledgeable—able to judge the appropriateness of the model for the situation, and to assess the validity of the input data,—and must refrain from applying model outputs to conditions outside the model constraints. *Validation* is one way to check whether the model (re)presents reality. Validation of the model means, in the simplest sense, feeding "true" data into the model and comparing the model output (prediction) with the behavior of the "true" system. If the model does not describe reality correctly, then internal algorithms and/or the basic structure of the model are insufficient.

EXAMPLE

Neglecting to assess the validity of a model is like buying an airplane or car without trying it out.

REVIEWING THE PAST AND GUIDING THE FUTURE OF MODELLING THE "HUMAN AT WORK"

The feasibility of developing an integrated ergonomic model of the "human at work" was the topic of a two-day workshop in 1985, convened by the Committee on Human Factors (COHF) of the National Research Council (Kroemer, Snook, Meadows, and Deutsch, 1988; Kroemer, 1989a, b). In 1988, the NATO Research Group 9 organized a Workshop on Applications of Human Performance Models to System Design (McMillian, Beevis, Salas, Strub, Sutton, and Van Brenda, 1989). In 1989, a review of computer models of the human body was conducted for the German Department of Defense (Aune and Juergens, 1989) and in 1990 for the U.S. Army (Paquette, 1990).

Specific objectives were to

- assess the usefulness of current anthropometric, biomechanical, and interface models;
- identify critical points of compatibility and disparity among existing models;
- review the feasibility of using these models for the development of an integrated ergonomic model;
- recommend research approaches for the development of such an integrated model.

One way to make such a task feasible is to consider three major classes of models:

1. anthropometric models, i.e., representations of static body geometry;
2. biomechanical models, i.e., representing physical activities of the body in motion, for which anthropometric data are the primary inputs; and
3. interface models, i.e., specific combinations of anthropometric and biomechanical models with regard to their interfacing with the technological system (the "machine"), representing human-technology interactions.

The general findings of the reviews are presented below.

Anthropometric Models

In the past, human body models have been mostly physical in the form of templates, manikins, and dummies, but future development is likely to concentrate on computer analogs of the human body. Such models need exact anthropometric information (see Chapter 1) in order to be accurate representations of body size, shape, and proportions. In the United States, anthropometric information is most often drawn from the anthropometric data bank at the U.S. Air Force's Armstrong Aerospace Medical Research Laboratory (CSERIAC, 1990). This repository contains many survey results of military samples, but the information on civilian populations is weak, because no comprehensive anthropometry study of the civilian population has ever been undertaken in any western country.

Most existing anthropometric information does not contain three-dimensional body data, but only univariate descriptors. Furthermore, many of these univariate dimensions lack a common reference system to which the individual measurements are related. This fact causes much conceptual and practical difficulty in the development of computer models of the human body size. Hence, various techniques for three-dimensional anthropometric data acquisition have been proposed, including stereophoto techniques, and the use of the laser as a distance-measuring device. For this purpose, mathematical-statistical techniques need to be developed that collect, organize, and summarize as well as display the huge number of collected data. Surface definition has been much improved by "facet algorithms," which allow a complete topographic description of the body surface.

Of course, the current use of landmarks and reference points on the body, now often palpated below the skin, needs to be modified for the use of photographic or laser measuring techniques. This poses the question of whether traditionally measured dimensions can be compared with body dimensions gathered by newer technologies—see the discussion in Chapter 1 and the following quote:

EXAMPLE

The viewpoint adopted for the aforementioned discussion of anthropometric models is that *external* shape dimensions are the dominant concern. Indeed, such has been the traditional main concern of the field of physical anthropology and most anthropometric literature. However, all of the physical templates, models and dummies which incorporate any types of joint simulations must also be concerned with questions of *internal* joint locations and the distances between them (links) of the human body. In the past such concerns were seldom mentioned, though they were real enough to the designer of drafting templates, models and three-dimensional dummies regardless of whether they were designing "anthropometric" models or "biomechanical" models. . . . With the advent of modern concerns for mathematical computer modeling there has been and will be new emphases on how to measure and specify such joint center locations and link lengths as a normal course of doing anthropometry and biomechanics. . . . The somewhat artificial distinction between anthropometry and biomechanics will become even more blurred and indistinct than it was in the past. (Reprinted with permission from a letter of John A. Roebuck, Jr., June 16, 1991, to the first author of this book.)

Biomechanical Models

Most models simulate the body as a series of rigid links, in two or three dimensions, reacting to external impulses, forces, and torques. Many of the early models were built to describe body displacements as a result of externally applied vibrations and impacts; to study body-segment positions at work or in motion, such as gait; or to predict static forces and torques that can be applied to outside objects. A separate set of models describes the stresses in human bones resulting from external loads. These are often combined with models of body articulations (knee, hip, and intervertebral joints)—a difficult task, particularly because of the involvement of many muscles and ligaments (see the discussion in Chapter 1) and the consideration of elastic or plastic properties of human tissues.

Very little has yet been achieved with respect to the true internal activation of muscles and the resulting loading of joints, bones, and connective tissue. For example, the simultaneous use of agonistic and antagonistic muscles (coactivation) is neither well understood nor modelled. Hence, the loads on joints are calculated simply from the resultant force, and therefore may render the internal loading incorrectly, i.e., as too small. Consideration of muscle dynamics is only beginning (Marras, 1989; Kroemer, Marras, McGlothlin, McIntyre and Nordin, 1990; Schneck, 1992).

Nevertheless, a large variety of models exist which differ in inputs, outputs, model structure, optimization, etc. (Aune and Juergens, 1989). An extensive table, prepared by Marras and King and contained in the *Proceedings* of the 1985 COHF Workshop, lists the model types, their input and output variables, and particularly their underlying assumptions. This list not only shows the successes achieved in modelling, but also indicates the often severe restrictions in model coverage, usually limiting the applicability and validity of models to a few given cases and conditions.

Incorporation of cognitive characteristics remains untouched by biomechanical modelers. People are information processors (see Chapter 3) who can modify the activation of their musculo-skeletal system: for example, in life-threatening danger, one can short-circuit internal protective mechanisms and can perform usually "impossible" actions. Such cognitive control processes are virtually nonexistent in biomechanical models at present.

It appears that progress in biomechanical modelling is currently more hindered by our limited basic understanding of the human body and mind than by computational abilities.

Human-Machine Interface Models

Anthropometric and biomechanical models combine to attain the next higher level in the hierarchical structure, i.e., that of interface models. Interface models describe the interactions between the modelled person and with the equipment in a human-machine system.

While the origin of such models is difficult to determine, the first published models in today's sense of the term appeared in the late 1960s and in the 70s. Subsequent developments were usually known by their acronyms, such as ATB, BOEMAN, CAPE, CAR, PLAID-TEMPUS, SAMMIE (Porter, Case, and Bonney, 1990), ADAM (Bartol, Hazen, Kowalski, Murphy, and White, 1990), COMBIMAN and CREW CHIEF (McDaniel, 1991), and MAN3D (Verriest, Trasbot, and Rebiffe, 1991). In the 1990s decade, they represented the state of the art as follows:

- The models are specific to given designs, purposes, and characteristics.
- Their usefulness is basically limited by their anthropometric and biomechanical components. Predictive models of the effects of the dynamics of either their workstations, their tasks, or the modelled human, are not yet available.
- Effects of stress, motivation, fatigue, or injuries are not adequately quantified, hence not modelled. Furthermore, the effects of environmental conditions on human performance usually are not included.
- Validity of the models is largely unknown.

Future Model Developments

For future development, three major guiding tenets are apparent:

1. There is a need for an integrated model of the human, of its performance char-

acteristics and limitations, and of its interactions with technological systems while doing "real" tasks.

2. The development of such an integrated model of the human body is feasible now, and is becoming easier with increasing knowledge about the human, and with increased usefulness of computer systems.
3. An integrated ergonomic model can guide future research and improve engineering applications.

There are two major approaches to the development of an integrated model. The first relies on the development of one *supermodel* that integrates the best qualities of all other models. Interface models such as COMBIMAN, CREW CHIEF, PLAID-TEMPUS, SAMMIE, and JACK (to mention a few) follow largely the supermodel approach. But these models are not compatible with each other, owing to different data formats, different model complexity, different model theories and techniques, and the use of different computer systems.

An alternative to the "integrated" model is the continued use of specific limited submodels, or *modules*. Yet, if one attempts to link together modules, for instance those describing the body's size and the dimensions of its surroundings, or the human's behavior and the movement of the structures encapsulating it, one finds that such building-block process of joining compatible modules requires "translators" between them. This would be much facilitated by a standard structure, i.e., by some supermodel.

Whether the future approach is that of a supermodel or of the modular type, general research needs must be fulfilled. Some research recommendations expressed in the 1985 COHF Workshop are:

- Establish the objectives, procedures, and outline for the development of an integrated ergonomic model, including a supermodel strategy.
- Review and integrate existing anthropometric and biomechanical databases.
- Develop submodels and modular groups.
- Develop generic interfaces between human models and workstation models.
- Develop methods and criteria for the validation of the ergonomic model.

As of today (1992), the 1985 recommendations are still as valid as they were when first expressed. Yet, given the advances made in defining and measuring human biodynamics and in the use of the computer as a tool, and considering the now widespread use of computer-aided design, modelling the "human working with equipment," such as the driver in an automobile, is important and useful. Designers use models of the human now, and will use them increasingly in the future. But such models must reflect solid knowledge, based on results of research that largely still needs to be done. For example, it is known that human motion does not follow optimization with respect to a single criterion, such as energy (Lee, Wei, Zhao, and

Badler, 1990). Thus, fairly complex "objective functions" and algorithms must be developed to make models realistic. An "animated" model will not do, it must be anthropomorphous.

EXAMPLE

Design of vehicles, and of workstations and work procedures, has come a long way from the simplistic two-dimensional design templates (which still serve some purposes for simple tasks). Certainly, a wealth of information has been collected, and has been expressed on paper, for example in various standards used in the automotive industry. Yet, many rely on underlying assumptions which overly simplify reality, or tend to proliferate existing conditions. For example, a drawing template is rather inflexible with respect to postures, clothing, or simply changes in body dimensions. Or, a measured reach envelope or field of vision depends on certain presumed sitting postures and reflects current locations of structures and controls, which are not necessarily those of the future. Another example is the presumption of static postures used at work, such as when assembling automobiles or maintaining airplanes. Such static models of the operator are inadequate or even misleading if rapid changes in postures and body motions occur.

Oversimplification and overreliance on existing conditions can be overcome by basing body models on knowledge about changing body contours, varying biomechanical properties, and predicted behavior including the results of cognitive and emotional processes. In the physiological and biomechanical domains, much progress has been made toward such predictive (instead of normative) modeling, and the presumption of a static condition has given way to the recognition that people are dynamic, not static.

For the automotive industry, as an example, such *biodynamic* (in contrast to *static*) modelling yields many gains and advantages, both in the design of new vehicles and in the design of workstations and work procedures. Better restraining systems have been designed both on the basis of experimental results and of biodynamic modelling. Useful arrangement of controls and of control panels is better understood now than just a few years ago. Assembly work has been made much safer and less injurious, to a large extent owing to the development of biomechanical models of the body. Yet, some design tasks are still a matter of experience and trial-and-error, instead of being solved by computer-based modelling. This includes the old problems of ingress and egress, or of maintenance and repair jobs. Regarding maintenance, much effort is spent by the U.S. armed forces on modelling of related activities and the appropriate design of vehicles and aircraft to facilitate such work. The results of these modelling efforts doubtlessly will be directly applicable to the automotive industry.

NEED

Progress in advanced modelling requires systematic knowledge that does not yet exist. For example, little is known about cognitive processes, decision making, or "instinctive actions" in the case of an emergency. Even the rules for modelling different curvature, flesh compression, or the movement of head and neck are not well defined. Thus, much "basic research" still needs to be done until proper modelling is possible.

Certainly, the better, more complete, and more realistic the model of the human interfacing with the equipment is, the better the "fit" of equipment and task to the human. Given the complexities of the human body (and mind), one would expect that paper-based information, and physical models (such as templates and dummies), will be replaced soon by computer-based analogs which incorporate complex and variable information about the human. This will facilitate the design of equipment (vehicle, workstation or hand tool) for ease of use, and allow for safer operating and working procedures.

SUMMARY

The more becomes known systematically about the human body and mind, and its functioning within systems, the better one is able to express that knowledge in formal "models." Simple models are of the descriptive kind, while more complex models are based on proven theories and make allowances for the effects of the environment and for adaptability and learning through feedback. Models are useful if they are valid (realistic) and reliable. It is also necessary that they can be used without requiring excessive specific knowledge and experience from the user.

To achieve simplicity of model and use, too often the criteria of validity and reliability are neglected, such as by using false ideas of proportions of the human body, or unrealistic animation for body movements. While these problems can be fairly easily recognized and corrected, there are more serious ones that incorporate complex though unproven hypotheses: this is not uncommon in behavioral models, but can often be spotted by such simplistic details as assumed linear relations between variables.

Realistic modelling has enabled many improvements in the design of simple and complex human-machine systems. For example, accessibility of parts that need maintenance in machines can be designed into the product; restraint systems for people in automobiles have become very effective; workplaces can be designed to avoid bending and twisting body movements of the worker.

The intent to develop a model also guides research. To fill the knowledge need of a model concept, information is gathered in a systematic manner to meet exactly that objective. This focuses the research approach—but the model concept must be valid and comprehensive. Significant steps have been made in that direction, but many more need to follow.

CHALLENGES

Consider the exceptions and boundary conditions made in physical models versus behavioral models of the human.

What are the consequences of either limiting or enlarging the numbers of "specific rules and conditions" which a model obeys?

How can one test a "concept model"?

What are the practical advantages of using a "closed" versus an "open" model?

Consider the models employed in Chapters 4 and 5: are they "normative" or "descriptive"? Also, check these models in terms of validity, reliability, and comprehensiveness.

What is more important to the modeller: mastery of the computer aspects, or knowledge of the functioning of the human?

Why does modelling give guidance to future research?

8

Designing to Fit Body Posture

OVERVIEW

Standing and sitting are the most frequent working postures, but others such as lying are used, for instance, in repair work. Furthermore, in many nonwestern countries, sitting and kneeling on the ground is very common during work.

The sitting posture is particularly useful when a relatively small workspace must be covered with the hands, and finely controlled activities performed. For this, the workstation and work object must be suitably designed. Fewer restrictions apply for the ergonomic design of the standing workstation.

The workspace of the hands depends on body posture and work requirements. Hence, various suitable workspace envelopes can be described. Yet, vision requirements at work determine also the suitable working volume.

Operation of controls is done usually either with the hands or the feet. Foot operation is stronger but slower, and should be required from seated operators only. Hand operation of controls is faster, weaker, but more versatile.

The "hand side" of tools and equipment should be designed for proper fit to the hand. This requires not only proper sizing of a handle, but also its arrangement so that the wrist or arm is not brought into straining positions.

Improper posture, and repeated and forceful operations, may lead to "overuse disorders," often associated with the repetitive use of hand tools, particularly if they vibrate. Another common source of overuse disorders is the frequent operation of keyboards.

Ergonomic recommendations for proper design of workstations are at hand.

EXAMPLE

An excerpt from ISO International Standard 6385:

1. *The work area shall be adapted to the operator:*
   - Height of work surface shall be adapted to body dimensions and work performed.
   - Seating arrangements shall be adjusted to the individual.
   - Sufficient space shall be provided for body movements.
   - Controls shall be within reach.
   - Grips and handles shall fit the hand.
2. *The work shall be adapted to the operator:*
   - Unnecessary strain shall be avoided.
   - Strength requirements shall be within desirable limits.
   - Body movements should follow natural rhythms.
   - Posture, strength and movement should be harmonized.
3. *Particular attention shall be paid to:*
   - Alternating in and between sitting and standing postures.
   - Sitting is preferable to standing (if one must be chosen).
   - Keeping chain of force vectors through body short and simple.
   - Allowing suitable body posture, providing appropriate support.
   - Providing auxiliary energy if strength demands are excessive.
   - Avoiding immobility, preferring motions.

INTRODUCTION

One usually distinguishes three major body positions: lying, sitting, and standing. Yet, there are many other postures, not just transient ones between the three major positions, but postures that are independently important—for example, kneeling on one or both knees, or squatting, or stooping, often employed during work in confined spaces such as loading cargo into aircraft, in agriculture, and in many daily activities. Reaching, bending, and twisting of body members are usually short-term activities.

In "western civilization," work is seldom done when lying supine or prone, but it does occur, for example, in repair jobs, or in low-seam underground mining. Prone or supine positions have been used in high-performance aircraft to better tolerate the acceleration forces experienced in aerial maneuvers. For example, during World War II, experiments were performed to use a lying pilot, which reduces the profile of the aircraft. In some fighter airplanes and tanks of the 1980s, the pilot or driver is in a semireclining posture.

Sitting and standing are usually presumed to be associated with more or less "erect" posture of the trunk, and of the legs while standing; sitting at work was thought to be "properly" done when the lower legs were in essence vertical, the thighs horizontal, and the trunk upright. This convenient model of all major body joints at zero or 90 or 180 degrees is suitable for standardization of body measurements, but the "0-90-180 posture" is not one found commonly employed, subjectively preferred, or even "healthy"—see the discussion in the section on computer workstation design in Chapter 9.

Evaluation of "Suitable" Positions at Work

By the rules of "experimental design," body postures can be considered "independent variables." If all other conditions and variables are controlled (such as work task, environment, etc., which often requires a laboratory setting), "dependent variables" can be observed, measured, and evaluated to assess the effects of body postures. In various disciplines, different dependent variables have been recorded, such as

- in physiology: oxygen consumption, heart rate, blood pressure, electromyogram, fluid collected in lower extremities, etc.
- in medicine: acute or chronic disorders, including cumulative trauma injuries.
- in anatomy/biomechanics: x-rays, CAT scans, changes in stature, disc and intra-abdominal pressure, model calculations.
- in engineering: observations and recordings of posture; forces/pressures on seat, backrest, or floor; amplitudes of body displacements; "productivity."
- in psychophysics: interviews (structured or unstructured) and subjective ratings by either the experimental subject or the experimenter.

These techniques have become standard procedures, but their appropriateness, and the interpretation of their results, have recently been questioned. During the last decade, the electromyographic procedure has been used extensively to assess the strain generated in the trunk muscles of seated persons. Most EMG techniques assume static (isometric) tension of the observed muscles, with the maximal voluntary contraction used as reference amplitude (Basmajian and DeLuca, 1985; Soderberg, 1992). Sitting is not necessarily static, however. Furthermore, the importance of small changes in electromyographic events is subject to question. In fact Wiker, Chaffin, and Langolf (1989) recently warned against the use of EMG information alone for design and evaluation of working postures if exertion of less than 15 percent of maximal voluntary muscle strength is required.

Results of measurements and model calculations of the pressure within the spinal discs of sitting and standing persons have been put in doubt by Jaeger (1987); Adams and Hutton (1985) pointed out that the contributions of the facet joints (apophyseal joints) to load bearing of the spinal column have been largely neglected; Aspden (1988) calculated much less pressure within a bent spine than in a straight column. Boudrifa and Davies (1984) investigated the relationships between intraabdominal pressure and back support, and McGill and Norman (1987) as well as Marras and Reilly (1988) researched the relationship between intraabdominal pressure and spinal compression when lifting: they found that abdominal pressure may not relieve the spinal column, as had been assumed.

Observations of body motions (voluntary and unconscious) while sitting are difficult to interpret. Much movement may be the result of discomfort *or* because a chair facilitates changes; sitting still may be enforced by a confining chair design *or* indicate that a comfortable posture is maintained (Green, Briggs, and Wrigley, 1991; Graf, Guggenbuehl, and Krueger, 1991; Occhipinti, Colombini, and Grieco, 1991).

Table 8-1 lists observations and recording techniques of body postures and contains K. H. E. Kroemer's "subjective assessment" of their status and usefulness. It appears that in many cases the threshold values, separating suitable from unsuitable conditions, are unknown or unclear. Thus, the interpretation of the results obtained by many of the listed techniques is difficult, to say the least: most currently useful techniques are based on subjective assessments by the seated persons. Their judgments presumably encompass all phenomena addressed in physiological, biomechanical, and engineering measurements, and they appear to be most easily scaled and interpreted.

Many studies use subjective ratings (by the subject or the experimenter) to assess the suitability of an existing sitting situation. While the procedures vary widely among researchers, most rely on the initial work by Shackel, Chidsey, and Shipley (1969) and by Corlett and Bishop (1976). "Discomfort" or "pain" questionnaires have been developed, for example by Occhipinti, Colombini, Frigo, Pedotti, and Grieco (1985) in Italy; in Scandinavia by Kuorinka, Jonsson, Kilbom, Vinterberg, Biering-Sorensen, Andersson, and Jorgensen (1987), in the United States by Chaffin and Andersson (1991).

Administration of a questionnaire may generate heightened awareness of problems; different manners of administration may result in different outcomes (Andersson, Karlehagen, and Jonsson, 1987). An often-used, well-standardized inquiry tool is the "Nordic Questionnaire" (Dickinson, Campion, Foster, et al., 1992; Kuorinka, Jonsson, Kilbom, et al., 1987). It is well structured and requires forced, binary or multiple-choice answers. It consists of two parts, one asking for general information, the other specifically focusing on low back, neck, and shoulder regions. The general section uses a sketch of the human body, seen from behind, divided into nine regions. The interviewee indicates if there are any musculo-skeletal symptoms in these areas. Figure 8-1 shows the body area sketch. If needed, further body sketches can be used, showing the body in side or frontal views, or giving further details (van der Grinten, 1991).

TABLE 8-1. METHODS FOR POSTURE ASSESSMENT

Observation/ techniques	Measurement procedures	Assessment criteria	Threshold values	Relevant
Oxygen consumption	E	E	V	probably
Heart rate	E	E	V	yes
Blood pressure	E	E	V	yes
Blood flow	E-D	E-D	V	yes
Innervation	D-V	D-V	U	yes
Leg/foot volume	E	V	U	perhaps
Temperature, skin or internal	E-D	U	?	yes
Muscle tension	E-D	V	U	yes
Electromyography	D	D	V	yes
Joint diseases	E-D	D	V	yes
Musculo-skeletal disorders	D	D	V	yes
Cumulative trauma disorders	D	D	V	perhaps
Spinal phenomena:				
—disk pressure	E-V	D	V	yes
—disk disorders	D-V	D	V	yes
—disk shrinkage	D	D	U	yes
—facet disorders	D-V	V	U	yes
—alignment of vertebrae	D-V	V	V	yes
—spine curvature	D-V	V	U	yes
—mechanical stresses, including model calculations	D-V	V-U	V	yes
Intra-abdominal pressure	E-D	V	U	perhaps
Surface (skin) pressure at				
—buttocks	E-D	V	U	yes
—back	E-D	V	U	yes
—thighs	E-D	V	U	yes
Upper extremity posture	E-D	V	U	yes
Posture				
—of head/neck, trunk, legs	E-D	V	?	yes
—changes in posture	E	?	?	yes
—change in stature	E	E	V	perhaps
Sensations (ratings) of				
—ailments	E-D	V	D	yes
—pain	E-D	V	D	yes
—discomfort	E-D	V	D	yes
—comfort, pleasure	E-D	D	D	yes

Legend

Well established	E
Being developed	D
Variable or unknown	V, U
Questionable	?

SOURCE: Modified from Kroemer, 1991.

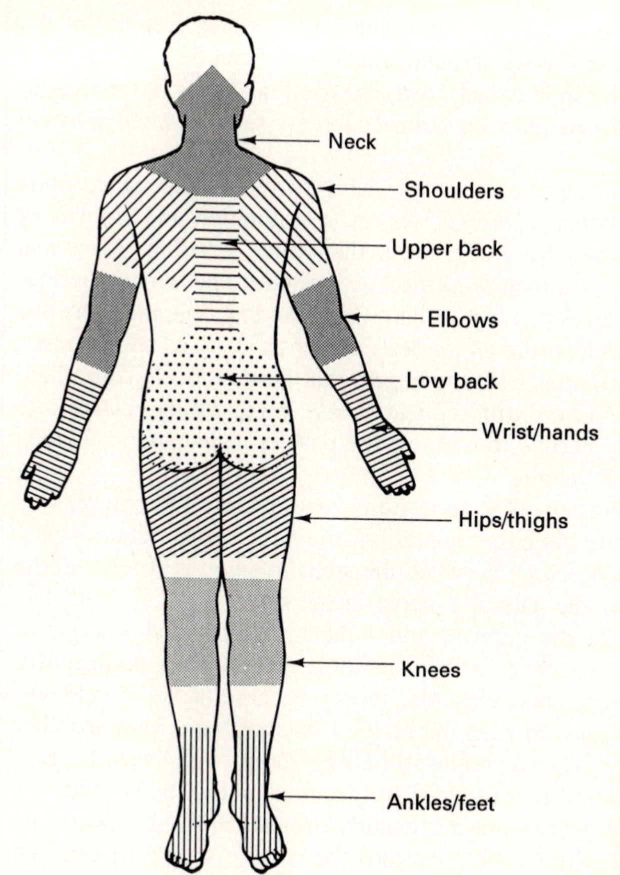

Figure 8-1. Body sketches used in the "Nordic Questionnaire" (Kuorinka, Jonssen, Kilbom, et al., 1987).

A specific section of the Nordic Questionnaire concentrates on body areas in which musculo-skeletal symptoms are most common, such as the neck and low back areas. The questions probe deeply with respect to the nature of complaints, their duration, and their prevalence. A modification of the test for use in the United States has been developed by Chaffin and Andersson (1991); further changes may be advantageous to check on particular conditions. Yet, adherence to the "Nordic Questionnaire" provides internationally standardized information.

APPLICATION

Body Postures at Work

In terms of physical effort, such as measured by oxygen consumption or heart rate, *lying* is the least strenuous posture. Yet, it is not well suited to performing physical work with the arms and hands, because for most activities they must be elevated, which is strainful by itself.

Standing is much more energy-consuming, but it allows free use of the arms and hands, and, if one walks around, much space can be covered. Furthermore, it facilitates dynamic use of arms and trunk and thus is suitable for the development of large energies and impact forces, such as when splitting wood with an axe.

Sitting is, in most aspects, between these two postures. Since body weight is partially supported by a seat, energy consumption and circulatory strain are higher than when lying, but lower than when standing. Arms and hands can be used freely, but their workspace is limited if one remains seated. The energy that can be developed is less than when standing, but given the better stability of the trunk supported on the seat, and possibly by using arm rests, it is easier to perform activities with the fingers that must be finely controlled. Operation of pedals or controls with the feet is easy in the sitting posture, because the feet are hardly needed to stabilize the posture and support the body weight and thus are fairly mobile.

The two most important working postures are standing and sitting. It is "common experience" that, in either condition, the most easily sustained posture of the trunk and neck is one in which the spinal column is straight in the frontal view, but follows the natural S-curve in the side view, i.e., with lordoses (forward bends) in the cervical and lumbar regions, and a kyphosis (backward bend) in the thoracic area. Yet, maintaining that trunk posture over long periods becomes very uncomfortable, mostly because of the muscle tension that must be maintained to keep the body in this position. Also, inability to move the legs and feet when standing still is very disadvantageous, because the feet and lower legs swell as a result of the accumulation of body fluids—a problem to which many women are particularly prone. Thus, either standing still or sitting still is "unphysiologic"; instead, the posture should be changed often. This includes interludes of walking by the standing operator and at least occasionally by the seated person as well, and motions of head, trunk, arms, and legs.

The posture and movements of the spinal column have been of great concern to physiologists and orthopedists. This is due to the fact that so many persons suffer from discomfort, pain, and disorders in the spinal column, particularly in the low back and in the neck areas. Explanations have been sought in the human body's "not being built for long sitting or standing," or not being "fit" because of lack of exercise, or having undergone degeneration, particularly of the intervertebral discs; the latter causes could be counteracted by physical activities and special exercises. To change the posture of sitting persons, various devices have been proposed, such as cushions of the seat or backrest that "pulsate," or frequent readjustments of the chair configuration, particularly of the seat pan and the backrest. Many suggestions have been made regarding the angle of the seat pan, the angle of the backrest, or the use or nonuse of the backrest—see the section in Chapter 9 on computer workstations. In fact, it has been proposed not to provide a backrest at all, so that trunk muscles must be employed to stabilize the body: yet, this contradicts the experience that use of a suitable backrest takes some of the load away from the spinal column. The muscles that stabilize the spinal column run in essence between the pelvic and shoulder areas. Since they can only contract, not push, their intensive use increases the compression force on the spinal column, as sketched in Figure 8-2.

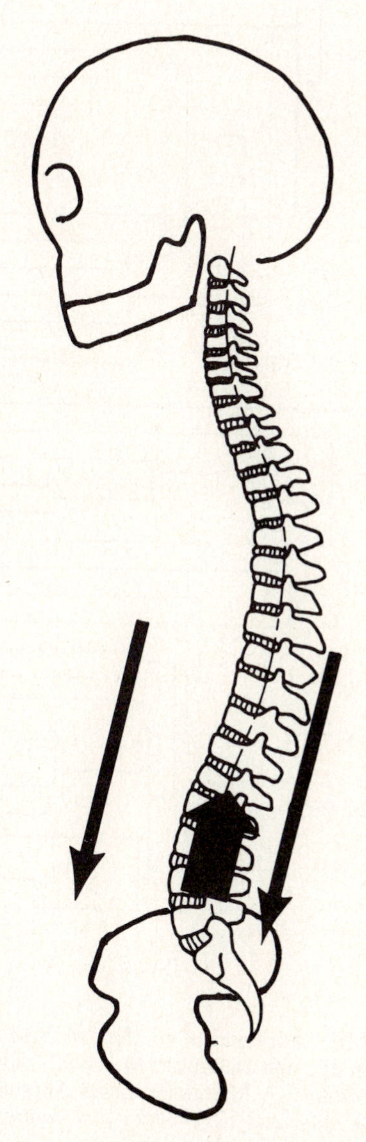

Figure 8-2. Activation of longitudinal trunk muscles generates spine compression.

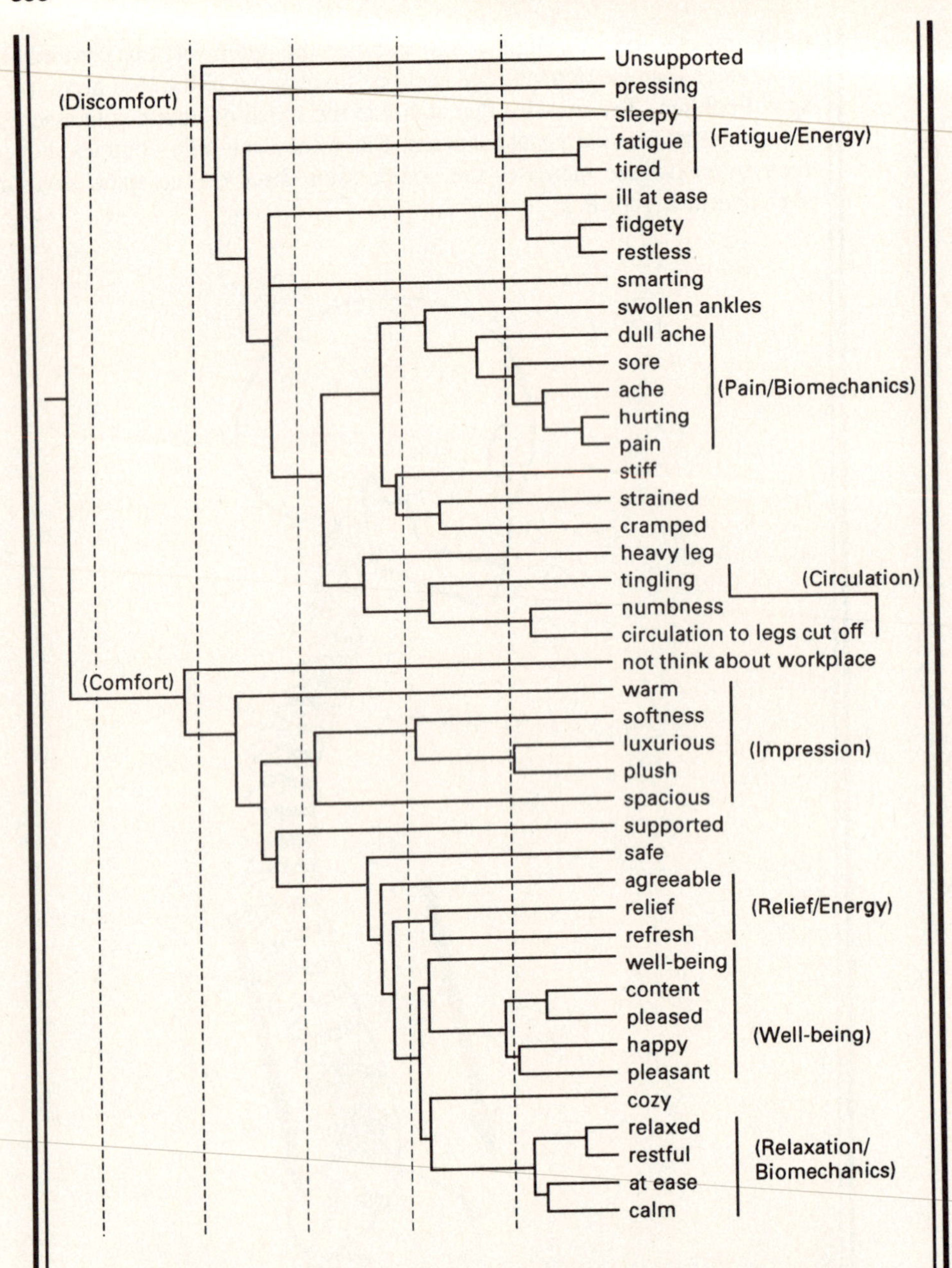

Figure 8-3. Attributes of comfort and discomfort while sitting (courtesy of Zhang, L., and Helander, M. (1992), Identifying Factors of Comfort and Discomfort: A Multidimensional Approach, in Kumar, S. (Ed.), *Advances in Industrial Ergonomics and Safety,* IV, pp. 395–402. London: Taylor & Francis).

Grandjean (1973) summarized these considerations:

- Changes in posture help to avoid continued compression of the spinal column as well as muscular fatigue.
- The seat should be so designed that the sitting posture can be changed frequently, including a reclining position supported by a tall backrest.

Recent experiences with "rest-less" seats support Grandjean's statement (for more information, see the discussion of office design in Chapter 9).

Comfort. The concept of "comfort," as related to sitting, is elusive. Apparently, we perceive a condition as comfortable if we are under the "impression" of warmth, softness, plushness, spaciousness: in terms of "well-being," if we are supported, safe, pleased, content; and in terms of "biomechanics," if we are relaxed and restful. These are major attributes of comfort found by Helander (personal communication, October 23, 1991) in a study with 41 subjects—see Figure 8-3.

Comfort has often been defined, simplistically but conveniently, as the absence of discomfort. Helander found that discomfort is expressed in such terms as stiff, strained, cramped, tingling, numbness; or as unsupported, fatiguing, restless; or even associated with feelings of soreness and pain. Some of these attributes can be explained in terms of circulatory, metabolic, or mechanical events in the body; others go beyond such physiologic and biomechanical phenomena. One can rather easily describe design features that will result in feelings of discomfort (such as chairs in wrong sizes, too high or too low, hard surfaces or edges), but avoiding these mistakes does not make a chair comfortable per se (Stevenson, Maher, McPhee, Long, and Lusted, 1991). Upholstery, for example, that is neither too soft nor too stiff, that "breathes" by letting heat and humidity escape, contributes towards the feeling of comfort. However, exactly what is comfortable for a given condition, a given period of time, depends very much on the individual, on habits, and on the task (read, e.g., Blair, 1991; Corlett, 1990; Lueder, 1992; Nag, Chintharia, Saiyed, and Nag, 1986; van der Grinten, 1991—and Chapter 9 of this book).

Recording and Evaluating Postures at Work

There are two approaches to record posture. One is to postulate given postures and to observe how often they actually occur. For example, so-called anterior, middle, and posterior sitting postures (where the subject either leans forward, sits centered on the seat, or leans backward) have been defined (Schoberth, 1962). Yet, these "pure" postures are hardly ever seen at work. The other technique is to describe, in detail, the actual positions of body members, and to record these. This procedure is facilitated if one concentrates on particularly important body parts and records their positions, using standardized terminology either in descriptive terms (Occhipinti, Columbini, and Grieco, 1991), or in angles of deviation against a reference (Priel, 1974; Corlett, Madely, and Manenica, 1979; Gil and Tunes, 1989). Other techniques provide a set of predrawn body-segment positions from which the observer selects those that best represent the actual conditions (Karhu, Harkonen, Sorvali, and Vepsalainen, 1981; Graf, Guggenbuehl, and Krueger, 1991; Kroemer, 1991; Malone, 1991; Pheasant, 1986; Tracy and Corlett, 1991). The observations may be recorded at the workplace, or taken from a movie or videotape at a later time (Keyserling, 1986). These methods and techniques have been employed with some success, though with various degrees of fidelity, repeatability, and time consumption (Fisher and Tarburtt, 1988; Malone, 1991; Ziobro, 1991). Yet, a truly satisfactory technique still needs to be developed that combines fidelity, accuracy, repeatability, and usability.

Some of these techniques include "judgments" about the suitability of the observed postures. Unfortunately, these are generally ill defined and ill supported; not surprisingly so, given the large number of possible criteria and circumstances that might contribute to the judgment—see the discussion above. While some postures and body movements are clearly undesirable (such as twisting the trunk), the suitability of others depends very much on the given circumstances. Thus, incorporating an evaluation of observed postures in schemes of posture recordings is a rather difficult and so far unresolved task.

EXAMPLE

Brennan (1987) described the ergonomic design challenges associated with a trencher. The machine moves forward, and the operator sits in a position to look in that direction, but the trenching tool is attached to the rear of the machine—see Figure 8-4(a). To observe the trenching operation, the operator must rotate trunk and neck nearly 180 degrees—see Figure 8-4(b). While all the regular controls to move the vehicle are located, as is common, in front of the operator, the controls for operating the trenching attachment are located to the side—see Figure 8-4(c).

This arrangement is ergonomically faulty in several aspects: it enforces a much twisted posture on the operator during the trenching operation and it is likely to result in mistakes in operation of the machine. Given these unfavorable conditions, the trenching operation is likely to be executed in a faulty manner. Unfortunately, similar situations are often found in underground mining equipment, earthmoving machinery, and motorized lift trucks: contorted body postures imposed on the operator; controls improperly located; vision blocked; noise, jolts, and impacts from the ground transmitted to the operator.

Figure 8-4(a). Trencher (with permission from Brennan, 1987).

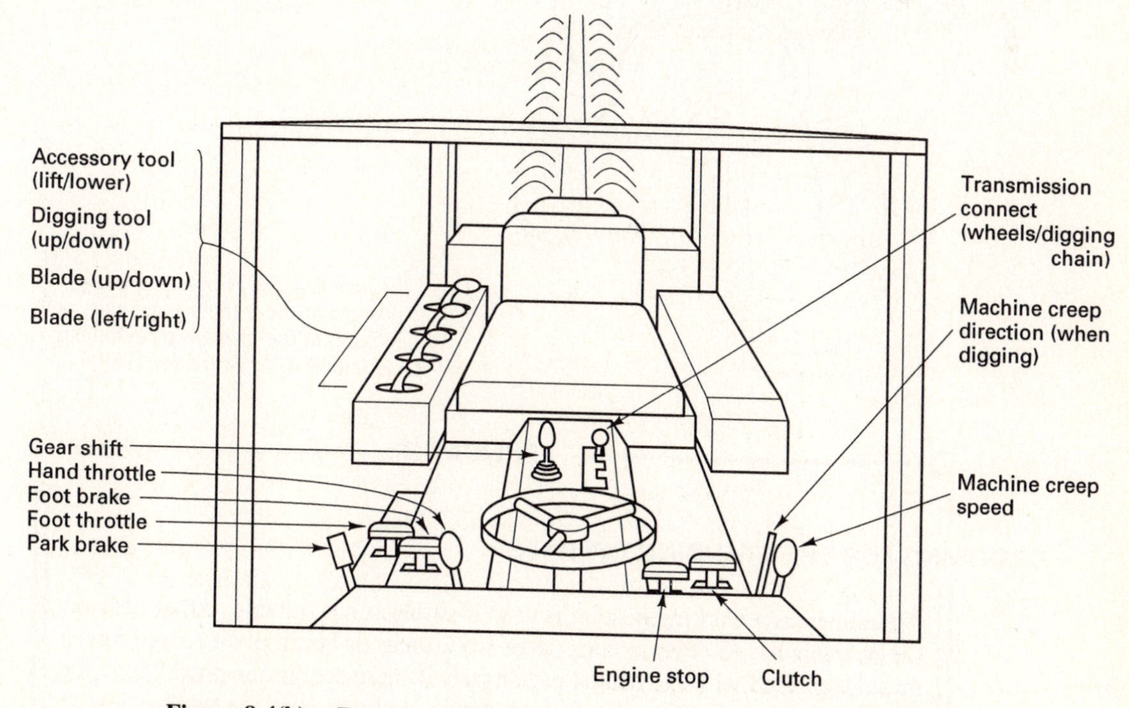

Figure 8-4(b). Frontal view of the trencher cab (with permission from Brennan, 1987).

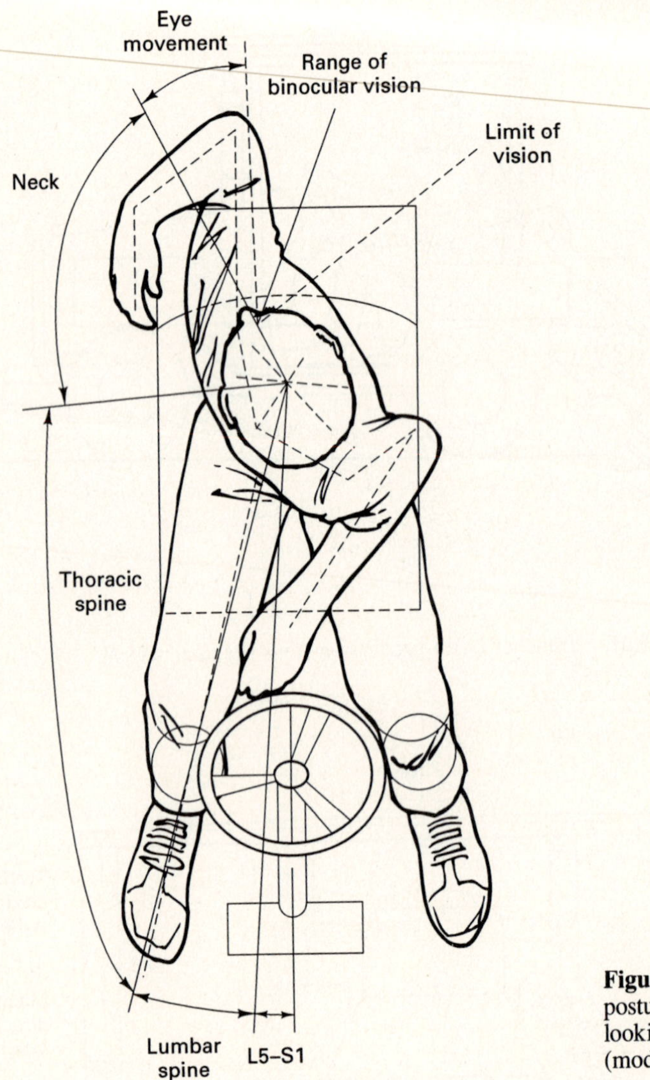

Figure 8-4(c). Contorted body posture of the trencher operator looking at the trenching equipment (modified from Brennan, 1987).

APPLICATION

DESIGNING FOR THE STANDING OPERATOR

"Standing" as a working posture is used if sitting is not suitable, either because the operator has to cover a fairly large work area, or because very large forces must be exerted with the hands, particularly if these conditions prevail only for a limited period. Forcing a person to stand simply because the work object is

Figure 8-5. Work spaces designed for a standing operator: large forces required over a large area, forceful exertions with the hand, working on large objects.

customarily put high above the floor is seldom a sufficient justification. For example, in automobile assembly, car bodies have been turned or tilted, and parts have been redesigned, so that the worker did not have to "stand and bend" in order to reach the work object. Examples for workstations designed for standing operators are shown in Figure 8-5, according to the need to exert large forces over large spaces, to make strong exertions with visual control, or to work with large objects.

The height of the workstation depends largely on the activities to be performed with the hands, and the size of the object. Thus, the main reference point is the elbow height of the worker (who, however, often is not standing upright, but bent or reaching). As a general rule, the strongest hand forces and largest hand mobility are between elbow and hip heights. Thus, the support surface (for example, a bench or table) is determined by the working height of the hands and the size of the object on which a person works. Sufficient room for the feet of the operator must be provided, which includes toe and knee space to move up close to the working surface. Of course, the floor should be flat and free of obstacles. Use of platforms should be avoided, if possible, because one may stumble over an edge. While the movements of the body associated with standing work are, basically, a desirable physiological feature, they should not involve excessive bends and reaches, and especially should not include twisting motions of the trunk.

People should never be forced to stand still at a workstation just because the equipment was originally ill designed, or badly placed, such as unfortunately too often is the case with drill presses used in continuous work. Also, many other machine tools such as lathes have been so constructed that the operator must stand and lean forward to observe the cutting action, and at the same time extend the arms to reach controls on the machine.

So-called stand-seats may allow the operator to assume a somewhat supported posture halfway between sitting and standing. Examples of these are shown in Figure 8-6. Occasionally, high stools have been employed to allow

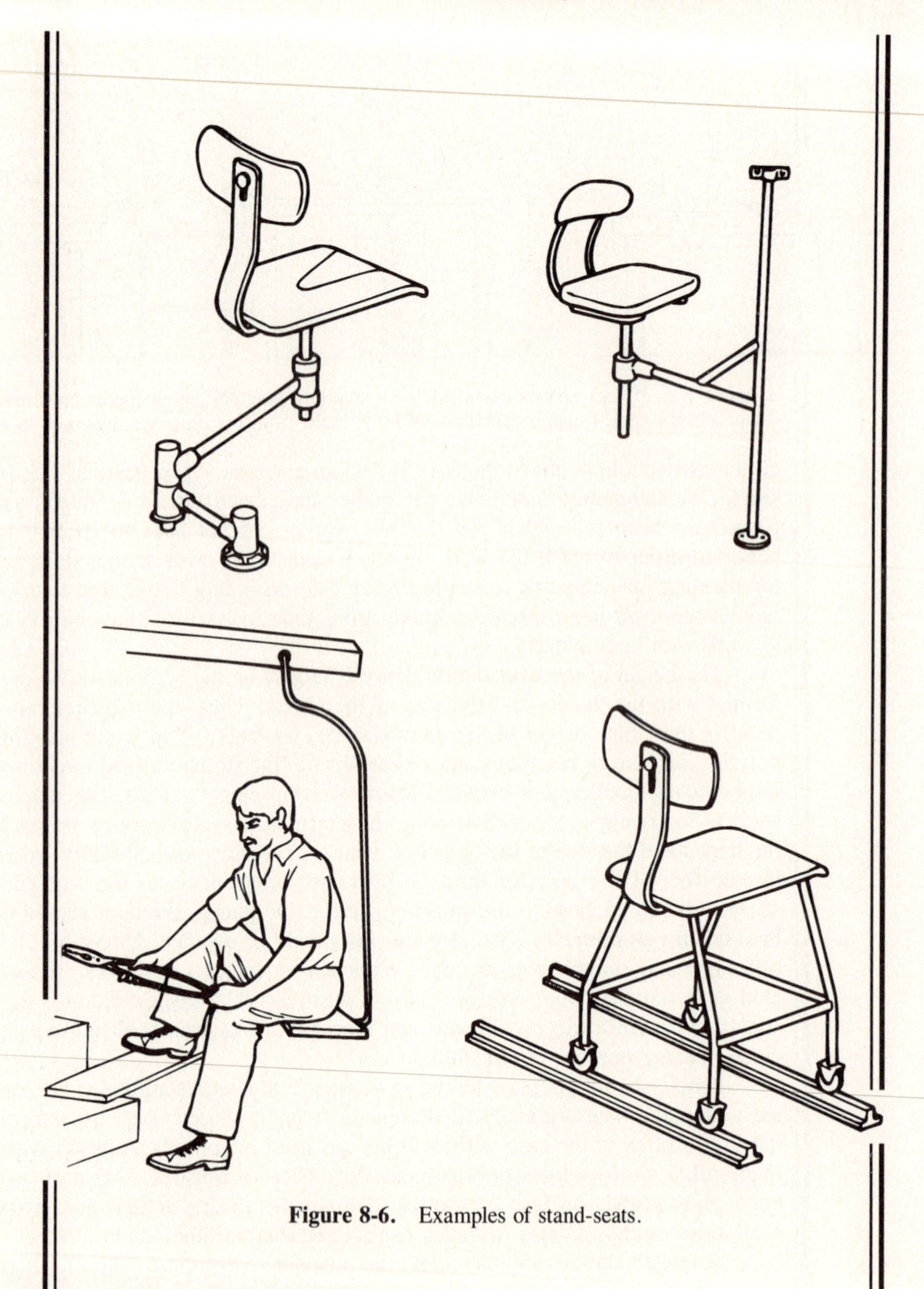

Figure 8-6. Examples of stand-seats.

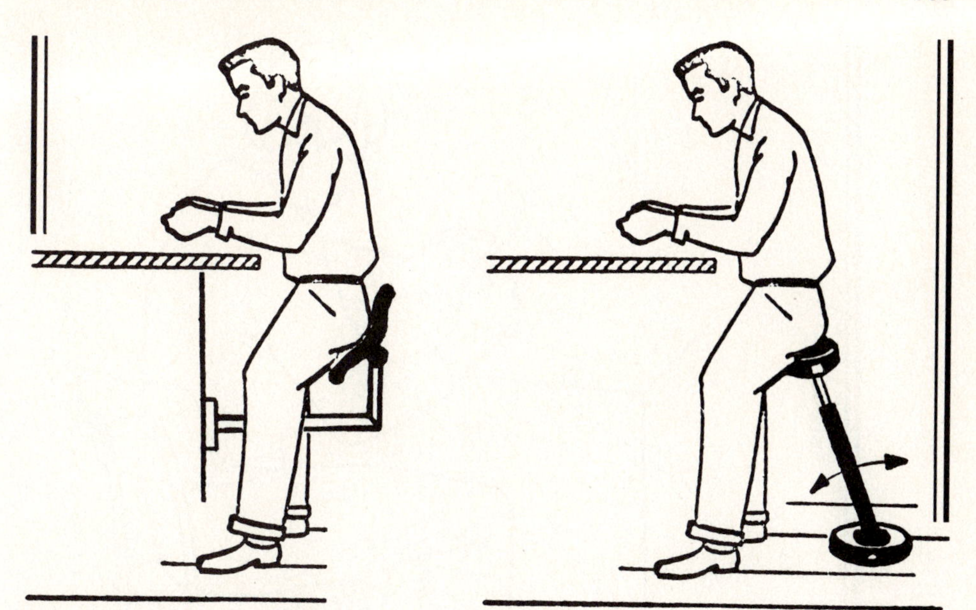

Figure 8-7. Stools used to temporarily allow (rather uncomfortable) sitting at workstations designed for a standing operator.

(rather uncomfortable) sitting at workstations otherwise requiring the operator to stand. Examples are shown in Figure 8-7. Unfortunately, these stools usually do not have backrests, for stability reasons, and it is difficult to support the feet. Thus, neither of these solutions is truly satisfactory.

DESIGNING FOR THE SITTING OPERATOR

Sitting is a much less strainful posture than standing. It allows better-controlled hand movements, but coverage of only a smaller area and exertion of smaller forces with the hands. A sitting person can operate controls with the feet and do so, if suitably seated, with much force. When designing a workstation for a seated operator, one must particularly consider the required free space for the legs. If this space is severely limited, very uncomfortable and fatiguing body postures result, as shown in Figure 8-8.

The height of the working area for the hands is again mostly determined with respect to the elbows. At about elbow height, with the upper arms hanging, the preferred working area is in front of the body so that manipulations with the fingers are facilitated. Many activities performed by seated operators require close visual observation, which codetermines the proper height of the manipulation area, depending on the operator's preferred visual distance and the preferred direction of gaze.

Figure 8-8. Leg space for the seated operator must be provided.

In western civilization it has become customary to provide seats that are at about popliteal height of the sitting person; thus seat heights range from about 35 to 50 centimeters (see the anthropometric tables in Chapter 1).

"Sitting" in nonwestern countries. The tourist visiting India and other Southeastern Asian countries notices sitting behavior different from that in North America or Europe. Sitting cross-legged on the ground, somewhat similar to the "lotus position," is quite widespread. For this, feet and legs are crossed in front of the body. The body weight is transmitted mostly through the buttocks, while the legs and feet serve to stabilize the posture, but some people even sit on their feet. Another common position is the "squat" posture, in which the soles of the feet are on the ground, the knees bent severely, the thighs close to the trunk, and the person nearly sits on the heels—see Figure 8-9. Often, one leg is extended away from the body while the other is kept

Figure 8-9. Working postures often encountered in Asia.

close. Also, quite frequently the kneeling position is used, occasionally with the feet rotated outward. This posture is often assumed if work requires that the upper body be bent forward.

These "low" postures reduce the heights of eyes and elbows with respect to the ground. For workplace and equipment design, this means that "sitting westerners'" working-height dimensions may not suit Asian operators.

Chinese cultural relics from the Shang through the Han dynasties (1600 B.C. to 220 A.D.) show that people sat and slept on mats in small rooms with low ceilings. Kneeling or sitting with crossed legs were common postures, considered proper according to current etiquette and ritual—see Figure 8-10. Sitting with extended legs and with feet extended forward, and squatting, were considered impolite or immoral. (It is interesting to note that "pointing" with the feet toward a person, or showing the sole of the foot, today is still considered rude in Thailand.)

The opening of the Silk Road allowed envoys of the Chinese Han dynasty to visit western Asia. There they saw stools and other chairlike furniture, which they then introduced to China. But it took until the third century B.C. before folding stools appeared in the Chinese imperial court. Yet, by the fourth century, traditional

Figure 8-10. For thousands of years, Chinese lived on mats (Xing, 1988).

rituals and formalities had been changed. Houses had higher ceilings and rooms were more spacious. New items of furniture, including stools with about the same height as in the West, were gradually used, as stone carvings and paintings indicate. Persons until then depicted as sitting cross-legged, with robes covering their feet, now were shown seated on hourglass-shaped stools, on four-legged stools, or on beds raised on legs—see Figure 8-11. During the seventh to the tenth centuries, the traditional life on mats gradually disappeared, and the folding stool became popular.

Around the year 1200, complete sets of raised furniture, including stools, chairs, tables, screens, dressing tables, racks, etc., existed in China. Drawings of the Ming dynasty (1368–1644) show a variety of styles, often elevated on legs, with forms of classical simplicity. However, in noble households it was still regarded improper for women to sit on chairs (Xing, 1988).

☞☞☞ *While in India, one of the authors was invited to see a modernized small manufacturing plant. In it were many hand-operated machines, mostly drill presses and punch presses, elevated on pedestals; and the operators sat on stools. Yet, the overall impression was that of a "staged" situation. The visitor inquired steadfastly and was finally told that all of these machines originally had been placed directly on the floor, without pedestals; and the operators sat, squatted, and kneeled on the floor as well, as they were accustomed to doing. Then, management decided to "improve"*

Figure 8-11. By the Tang Dynasty (618–907 A.C.), stools and chairs had appeared in China and were gaining in popularity (Xing, 1988).

the working conditions according to western images, and put the machines on pedestals and the operators on stools. When left alone, the operators would assume their regular traditional postures on these stools, with their feet at seat height. For visitors, however, the operators were exhorted to put their feet down.

APPLICATION

DESIGNING FOR WORKING POSITIONS OTHER THAN SITTING OR STANDING

Semisitting is one example of a working posture that is neither sitting nor standing. In many cases other postures must be assumed at work, although often only briefly, such as when reaching to a barely accessible object, or straining to perform work inside a narrow opening. Little can be done in terms of systematically designing body supports related to such unusual and awkward postures, except not to design equipment that requires them.

☞☞☞ *A visitor to a high-tech manufacturing facility was impressed by the manager's explanations of how highly automated the production was, and how few people were needed to run it. "Why," he asked, "are there so many cars parked outside?" The answer was that these belonged to the repair people who did the oldfashioned bloody-knuckles dirty repair work on the automated manufacturing machinery.* ☜☜☜

Some jobs, though, habitually require stooped, bent, and twisting working—for example, when loading and unloading luggage of aircraft passengers, both behind the check-in counter and in aircraft cargo holds. Repair, maintenance, and cleaning jobs often require awkward body postures. Low-seam mining is notorious for requiring bent, stooped, kneeling, even crawling and lying working postures from the miners. In the building trades, unusual body postures are frequent.

These and other examples indicate a need for a systematic ergonomic design approach. First, it must be established whether or not such postures are indeed necessary. If not, better solutions for the work can be found that no longer include them. If they cannot be avoided, special body supports must be designed: for example, military standards and specifications describe the body supports of tank crews, or of pilots in fighter aircraft. For the construction industry, Helander (1981) has compiled recommendations. A semireclining "seat" for overhead tasks has been proposed (Lee, Hosni, Guthrie, Barth, and Hill, 1991), and space requirements of operator compartments in low-seam mining equipment have been described (Conway and Unger, 1991)—see Figure 8-12.

Work in Restricted Spaces

In some cases work must be performed in restricted spaces, such as in cargo holds of airplanes or in underground mining. The primary restriction usually lies in the lowered ceiling of the workspace. Work becomes more difficult and stressful as the ceiling height forces workers to bend neck and back, or requires squatting, or even lying. Thus, if work must be done in low-ceilinged spaces, equipment and mechanical aids should be developed that alleviate the human's task. For example, in aircraft baggage handling, it is advantageous to first collect the luggage in containers and then put these in place within the cargo hold, rather than "handling" individual pieces.

Other space-restricted spaces are, for example, passageways, walkways, hallways, and corridors. Minimal dimensions for these are given in Figure 8-13. For tight places, where one may have to squat, kneel, or lie on the back or belly, dimensions are given in Figure 8-14 and Table 8-2. Dimensions for escape hatches, shown in Figure 8-15, need to accommodate even the largest

Body Position	Height H (cm)		Depth D (cm)	
	Minimum	Preferred	Minimum	Preferred
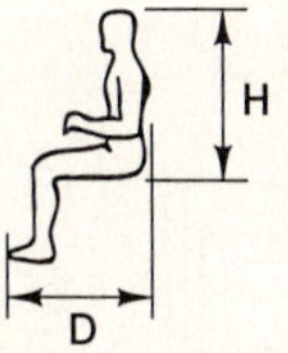 H / D	103	110	94	100
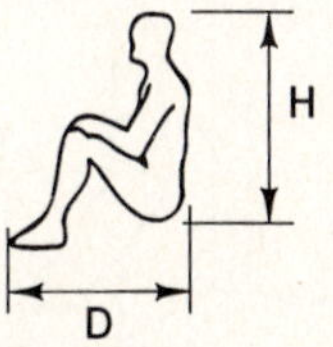 H / D	98	110	94	100
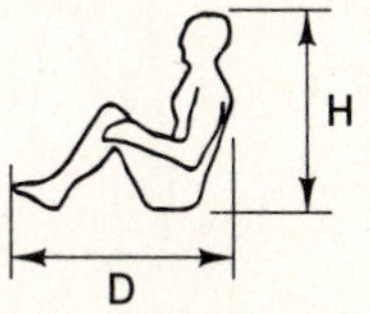 H / D	98	110	88	90
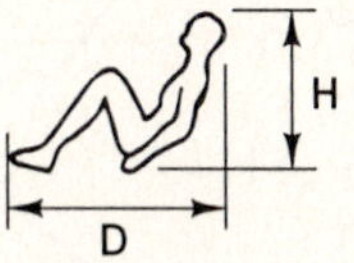 H / D	64	65	125	140
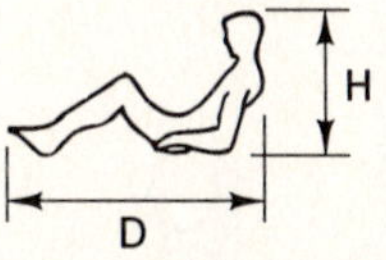 H / D	60	62	150	160
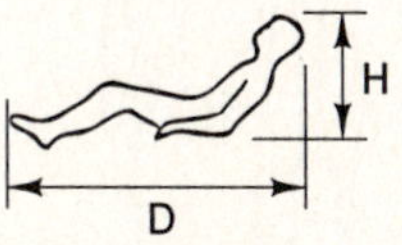 H / D	50	54	170	180
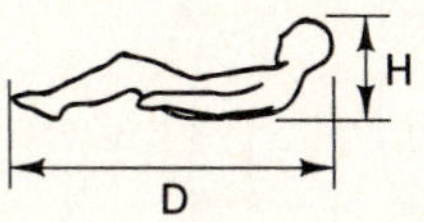 H / D	48	52	190	195
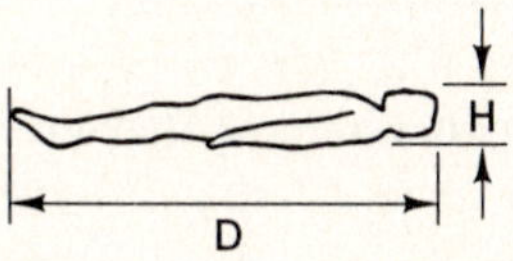 H / D	38	46	200	210

Figure 8-12. Spaces required to accommodate U.S. coal miners (adapted from Conway and Unger, 1991).

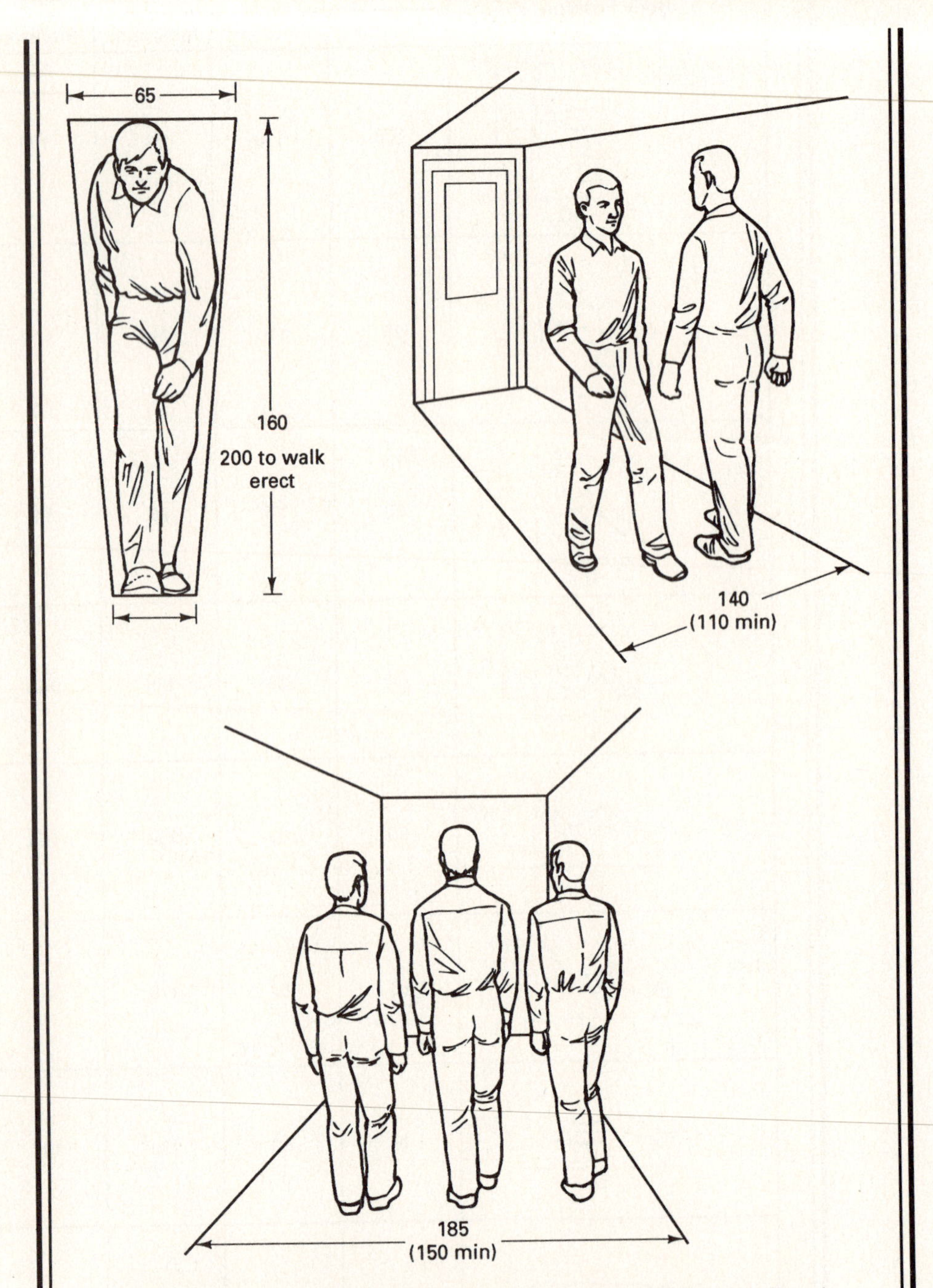

Figure 8-13. Minimal dimensions for passageways and hallways (adapted from VanCott and Kinkade, 1972).

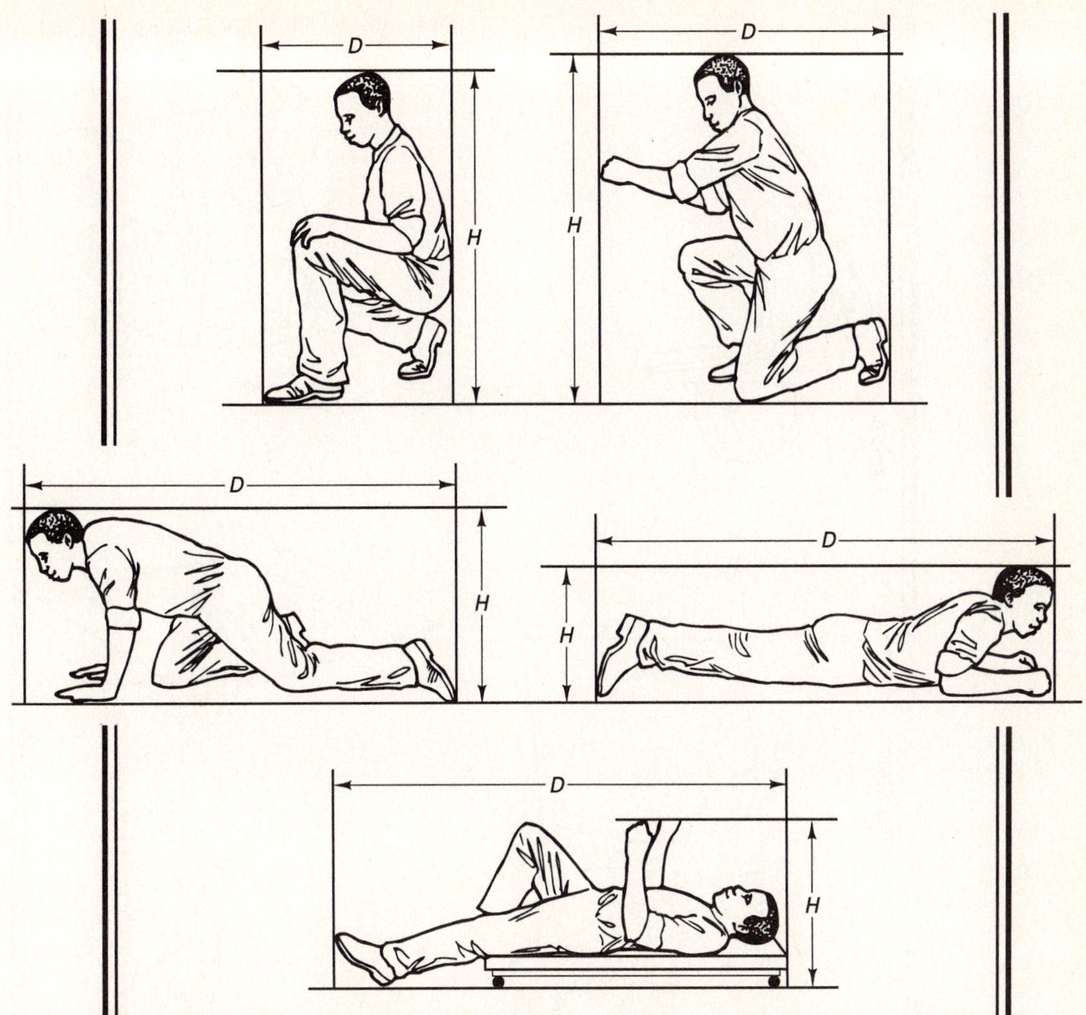

Figure 8-14. Minimum height and depth dimensions for "tight" work spaces (adapted from MIL-STD 759).

TABLE 8-2. DIMENSIONS (IN CM) FOR "TIGHT" WORKSPACES

	Height *H*			Depth *D*		
	Minimal	Preferred	Arctic clothing	Minimal	Preferred	Arctic clothing
Stooped or squatting	66	—	130	61	90	—
Kneeling	140	—	150	106	122	127
Crawling	79	91	97	150	—	176
Prone	43	51	61	285	—	—
Supine	51	61	66	186	191	198

SOURCE: Adapted from MIL-STD 759.

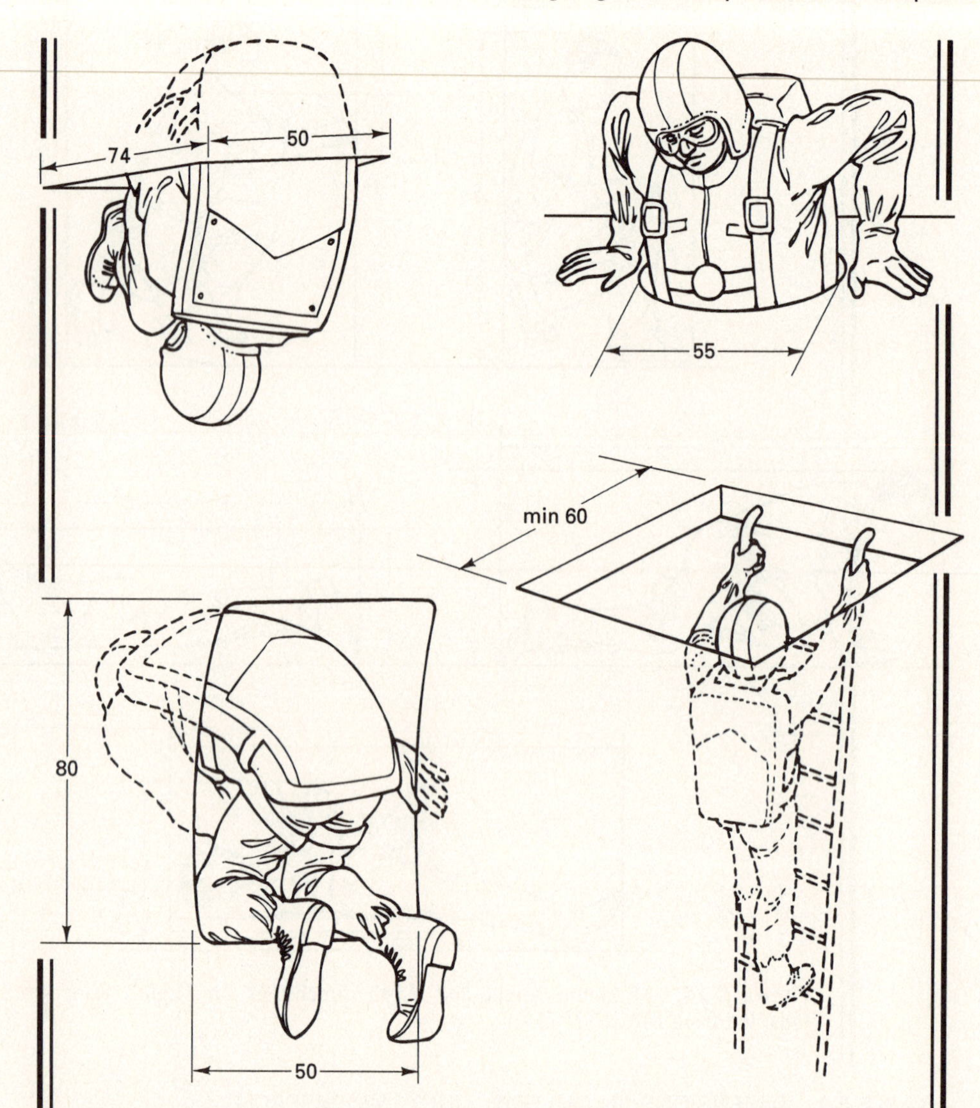

Figure 8-15. Minimal openings for escape hatches (adapted from Van Cott and Kinkade, 1963).

persons wearing their work clothes and possibly equipment. These openings can be made somewhat smaller for maintenance workers who need to get through access openings in enclosures of machinery; recommended dimensions are shown in Figure 8-16. The size for openings through which one hand must pass, holding and operating a tool, depends on the given circumstance, yet some recommended dimensions are shown in Figure 8-17. These dimensions

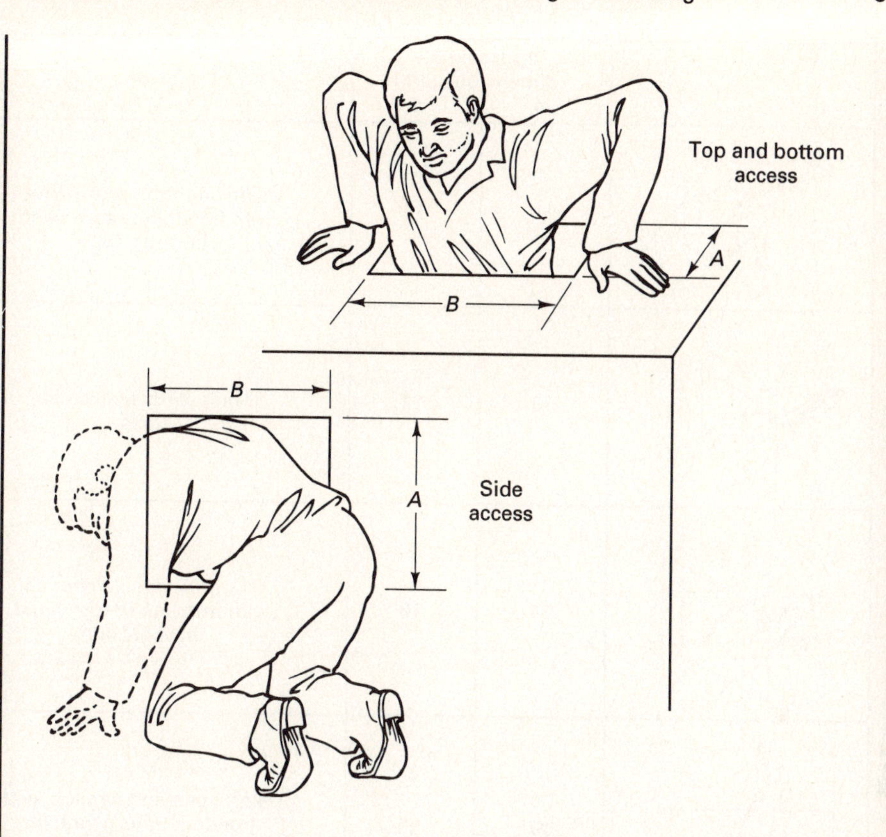

Dimensions	A, depth		B, width	
Clothing	Light	Bulky	Light	Bulky
Top and bottom access	33 cm	41 cm	58 cm	69 cm
Side access	66 cm	74 cm	76 cm	86 cm

Figure 8-16. Access openings for enclosures (adapted from MIL-HDBK 759).

need to be modified if the operator also must see the object through the opening, and if special tools must be used and movements performed with one hand. In some cases, both hands and arms must fit through the opening, which then needs to be about shoulder-wide . For further information, see the standards issued by ISO, NASA, U.S. Military, and various design handbooks, e.g., by Eastman-Kodak (1986); Mital and Karwowski (1991); Sanders and McCormick (1987); Salvendy (1987); Woodson, Tillman, and Tillman (1991).

	Approximate dimensions (cm) A	B	Task
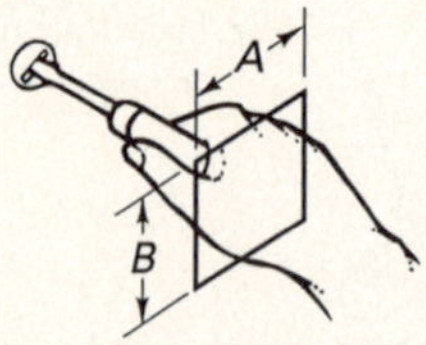	11	12	Using common screwdriver, with freedom to turn hand through 180°
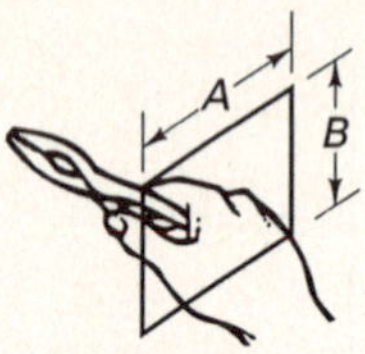	13	12	Using pliers and similar tools
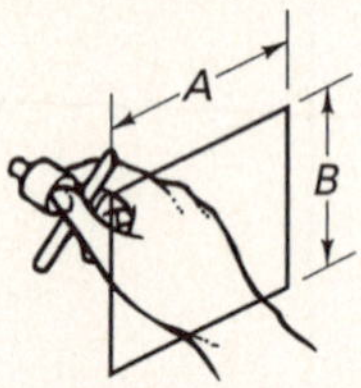	14	16	Using "T" handle wrench, with freedom to turn hand through 180°
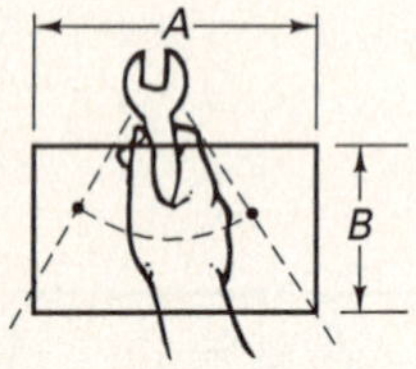	27	20	Using open-end wrench, with freedom to turn wrench through 60°
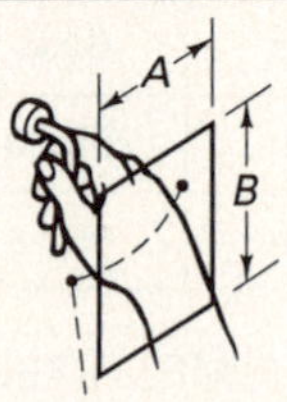	12	16	Using Allen-type wrench with freedom to turn wrench through 60°
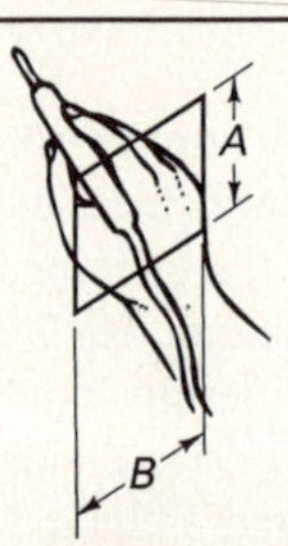	9	9	Using test probe, etc.

Figure 8-17. Minimal opening sizes (in cm) to allow one hand holding a tool to pass (adapted from MIL-HDBK 759).

DESIGNING FOR FOOT OPERATION

In comparison to hand movements over the same distance, foot motions consume more energy, are less accurate and slower, but are more powerful, as one would expect from biomechanical considerations (Astrand and Rodahl, 1986; Hoffmann, 1991).

If a person stands at work, fairly little force and fairly infrequent operations of foot controls should be required, because, during these exertions, the operator has to stand on the other leg alone. For a seated operator, however, operation of foot controls is much easier, because the body is largely supported by the seat. Thus, the feet can move more freely and, given suitable conditions, can exert large forces and energies.

A typical example for such an exertion is pedaling a bicycle: all energy is transmitted from the leg muscles through the feet to the pedals. These should be located directly underneath the body, so that the body weight above them provides the reactive force to the force transmitted to the pedal. The crank radius should be about 15 centimeters for short-legged persons, and up to 20 centimeters for persons with long legs. Suitable pedal rotation is usually between 0.5 and 1 Hz, but depends on such factors as the gear ratio of the bicycle (often variable), the ground surface, and the purpose of bicycling, either for leisure, exercise, or competition. In some cases it is desired to lower the center of mass of the combined person–bicycle system. In this case, the cranks may be moved forward and upward, to nearly the height of the seat. Placing the pedals forward makes body weight less effective for generation of reaction force to the pedal effort, hence some sort of backrest should be provided, against which the buttocks and back press while the feet push forward on the pedal. Instead of using the bicycle principle to propel the body, one can use this approach to generate energy, for instance when "pedaling" an electricity generator. (This may also be done with the hands, but the arms are less powerful than the legs.)

The traditional arrangement of foot controls in the automobile is, by all human factors rules, atrocious: the gas pedal requires that the foot be kept in the same position over long periods of time; the brake pedal must be reached by a complex and time-consuming motion of the foot from the gas toward the body, to the left, and then again forward. It makes no sense that pushing forward on the gas accelerates the vehicle, but also pushing forward on the brake decelerates. The current arrangement encourages use of the right foot alone, while the left foot is usually idle.

Small forces, such as for the operation of switches, can be generated in nearly all directions with the feet, with the downward or down-and-forward directions preferred. The largest forces can be generated with extended or nearly extended legs, in the downward direction limited by body inertia, in the more forward direction both by inertia and the provision of buttock and back support surfaces. These principles are typically applied in automobiles. For example, operation of a clutch or brake pedal can normally be performed easily with the knee at about a right angle. But if the power-assist system fails, very large forces must be exerted with the feet: in this case, one must thrust one's back against a strong backrest and extend the legs to generate the needed pedal force.

Figures 8-18 through 8-23 provide information about the forces that can be generated with the legs via the feet, depending on body support and hip and

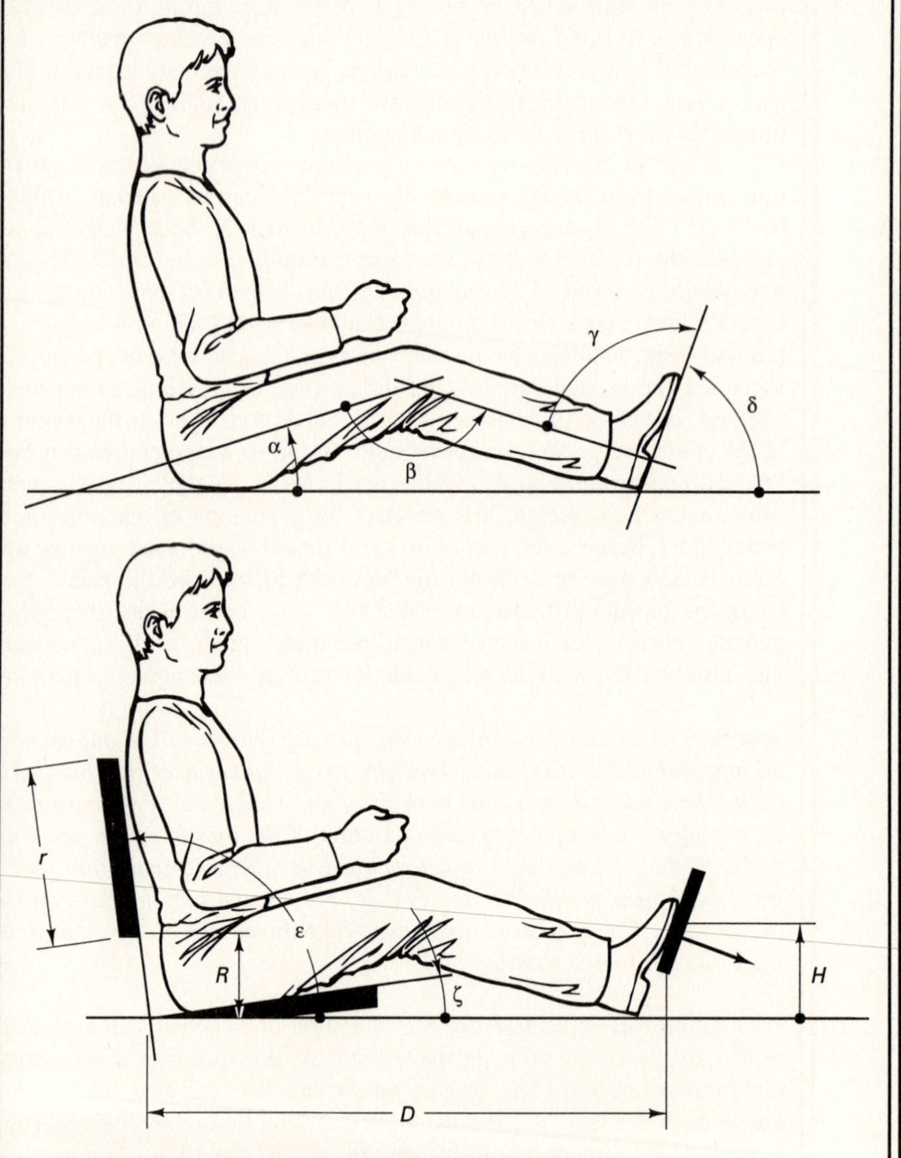

Figure 8-18. Conditions affecting the force that can be exerted on a pedal: body angles (upper part), workspace dimensions (lower part).

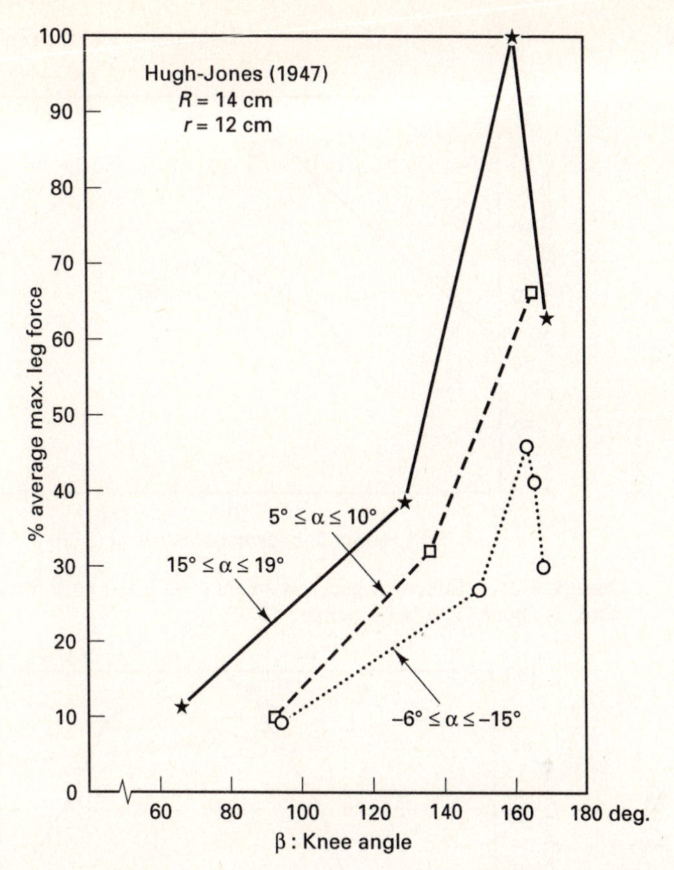

Figure 8-19. Effects of thigh angle α and knee angle β on pedal push foce—maximum force is at least 2100 N (three studies reported by Kroemer, 1971).

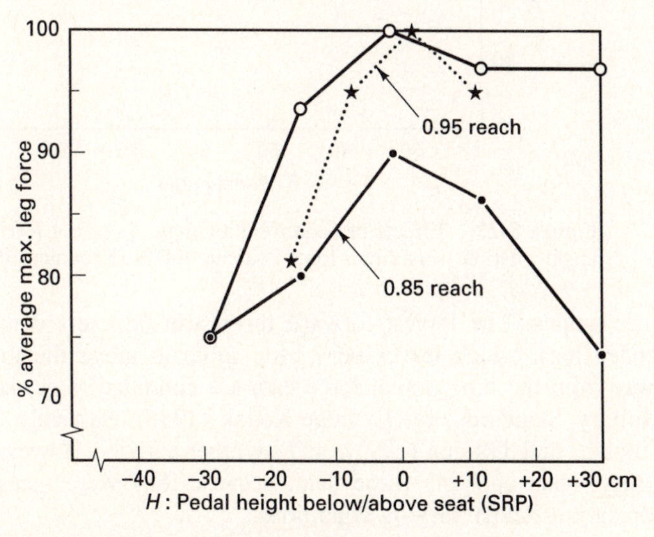

Figure 8-20. Effects of pedal height H on pedal push force—maximal force is about 2600 N (two studies reported by Kroemer, 1972).

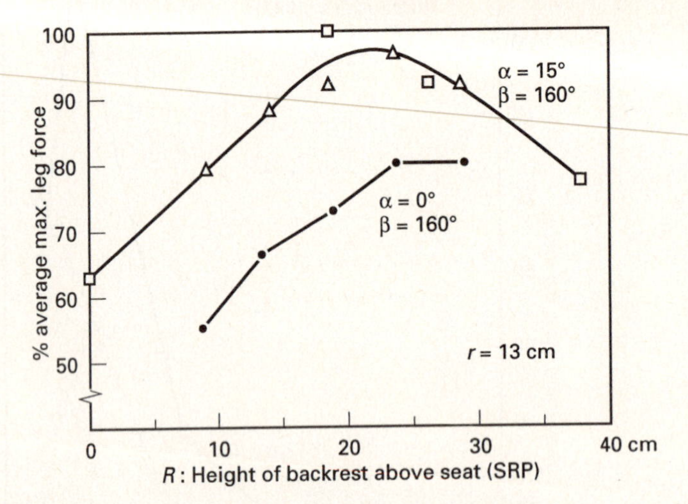

Figure 8-21. Effects of backrest height R on pedal push force—maximal force is about 1700 N (Kroemer, 1972).

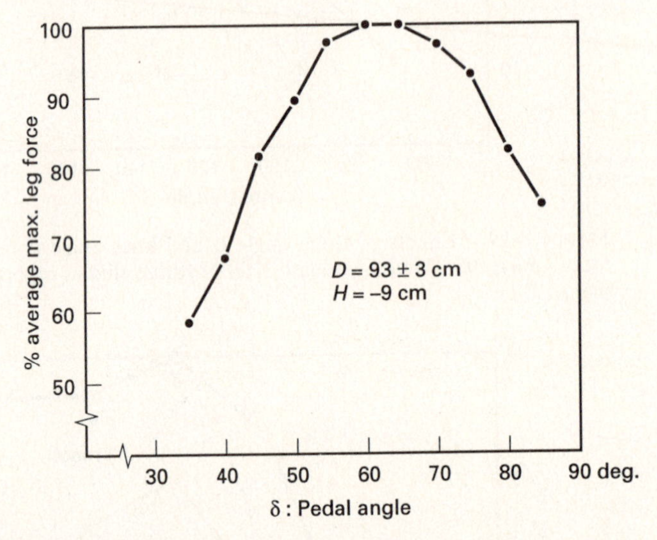

Figure 8-22. Effects of pedal (ankle) angle δ on foot force generated by angle rotation—maximal force is about 600 N (Kroemer, 1971).

knee angles. The largest forward thrust can be exerted with the nearly extended legs, which leaves very little room to move the foot control further away from the hip. Actual force data are compiled, e.g., in NASA and U.S. Military Standards; by Eastman-Kodak (1986); Salvendy (1987); Woodson, Tillman, and Tillman (1991), and by other authors. However, caution is necessary when applying these data, because they were measured on different populations under varying conditions.

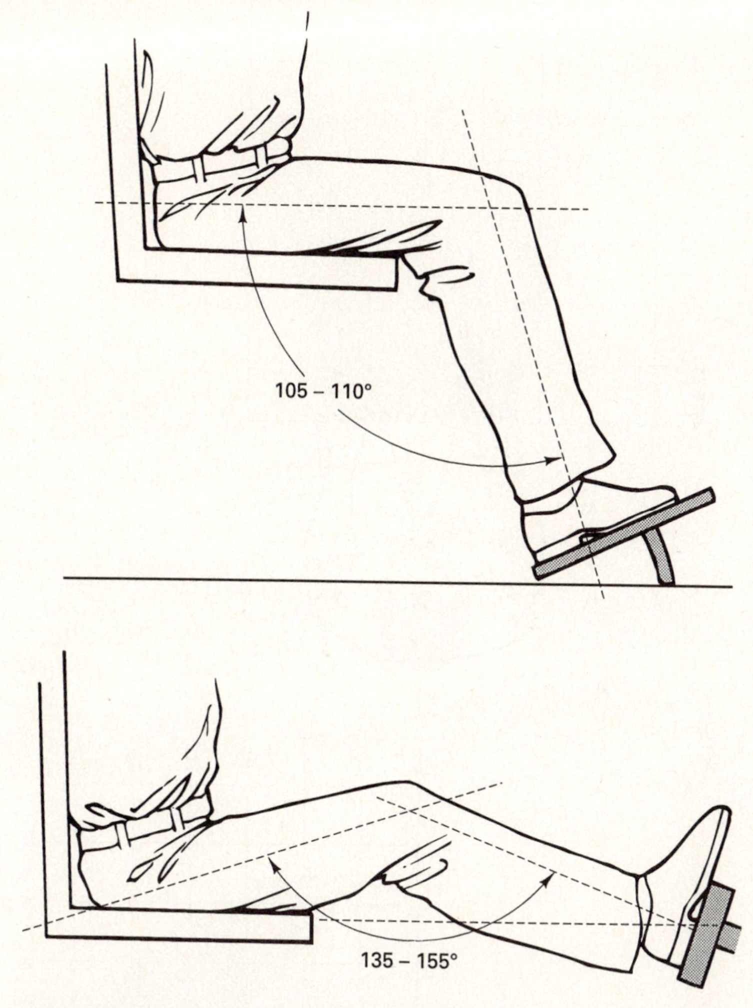

Figure 8-23. Light downward forces can be exerted at knee angles of about 105 to 110 degrees, while strong forward forces require knee extension at 135 to 155 degrees (adapted from VanCott and Kinkade, 1972).

Similar to the hand workspace discussed earlier, a preferred workspace for the feet results from the foregoing discussions. It is shown in Figure 8-24.

☞☞☞ *In automobiles, power-assisted brakes and steering systems generate a difficult design problem. As long as auxiliary power is available, brakes can be operated easily, in almost any conceivable leg posture. Yet, if the auxiliary system fails, suddenly forces are required from the operator which are three to ten times as high for braking (or steering). Not only must the operator recognize*

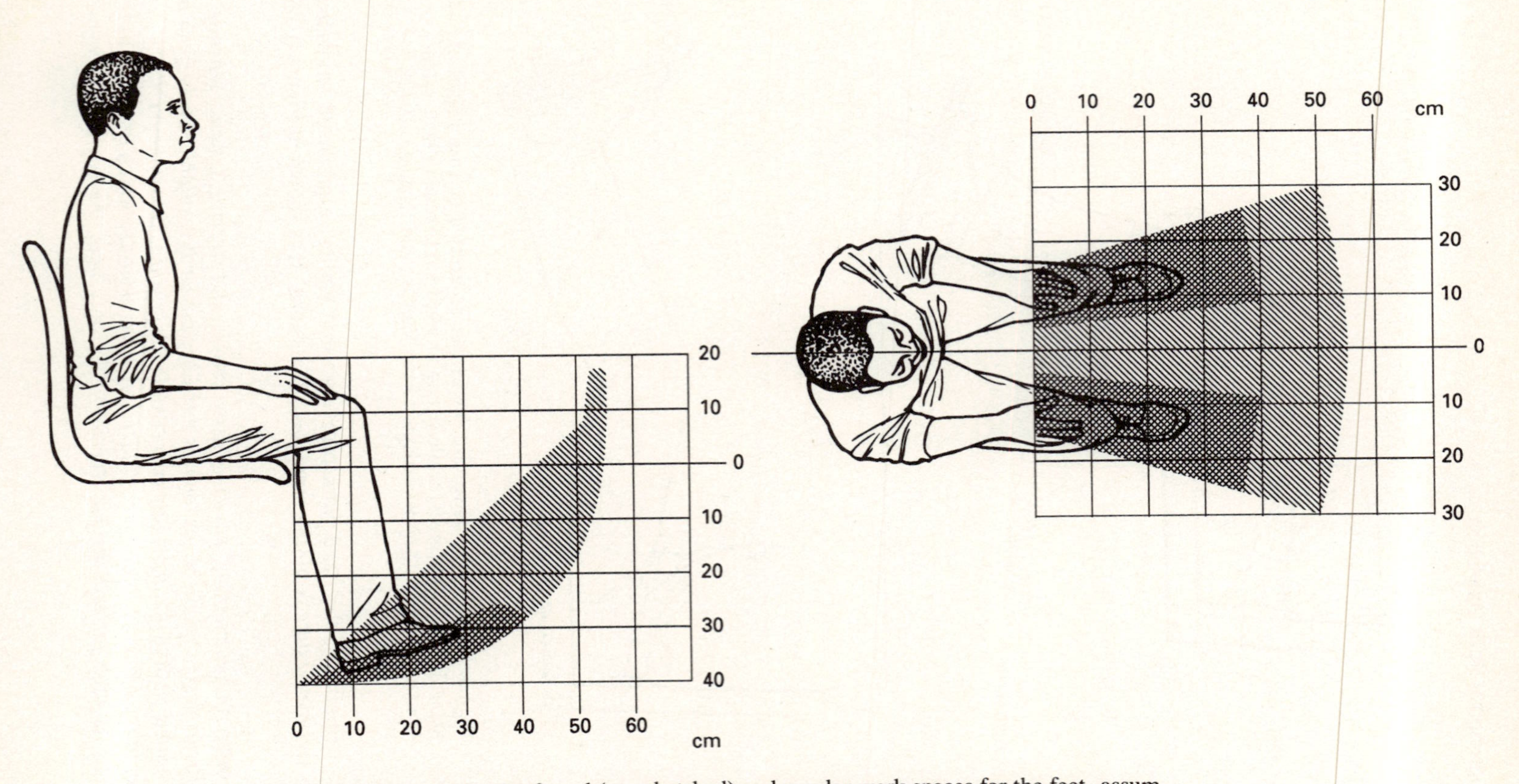

Figure 8-24. Preferred (crosshatched) and regular work spaces for the feet, assuming a seated operator.

that much more effort is required, but also this effort must be developed quickly and often in a body posture that is not favorable for such large exertion, such as with a strongly bent knee.

DESIGNING FOR HAND USE

The human hand is able to perform a large variety of activities, ranging from those that require fine control to others that demand large forces. (But the feet and legs are capable of more forceful exertions than the hand; see above.)

One may divide hand tasks in this manner:

- fine manipulation of objects, with little displacement and force. Examples are writing by hand, assembly of small parts, adjustment of controls.
- fast movements to an object, requiring moderate accuracy to reach the target but fairly small force exertion there. An example is the movement to a switch and its operation.
- frequent movements between targets, usually with some accuracy but little force; such as in an assembly task, where parts must be taken from bins and assembled.
- forceful activities with little or moderate displacement (such as with many assembly or repair activities, for example when turning a hand tool against resistance).
- forceful activities with large displacements (e.g., when hammering).

Accordingly, there are three major types of requirements: for accuracy, for strength exertion, and for displacement. For each of these, certain characteristics of hand movements can be described, if one starts from a "reference position" of the upper extremity: the upper arm hangs down; the elbow is at right angle, hence the forearm horizontal, and extended forward; and the wrist is straight. In this case, the hand and forearm are in a horizontal plane at approximately umbilicus height.

For *accurate* and *fast movements*, Fitts' law provides guidance (see Chapter 3): the smaller the distance traveled and the larger the target, the more accurate is a fast movement. Thus, the fingers are able to perform the fastest and most accurate motions. This is followed by movements of the forearm only, meaning (since the upper arm is fixed) that either (a) the forearm does a horizontal rotating sweep about the elbow (in fact, about the shoulder joint in which the upper arm twists), or (b) the forearm flexes/extends in the elbow joint. The least accurate and the most time-consuming movements are those in which the upper arm is pivoted out of its vertical reference location. If the hand must be displaced only by short distances from its reference position, pure forearm movements are preferable to those in which the upper arm is also

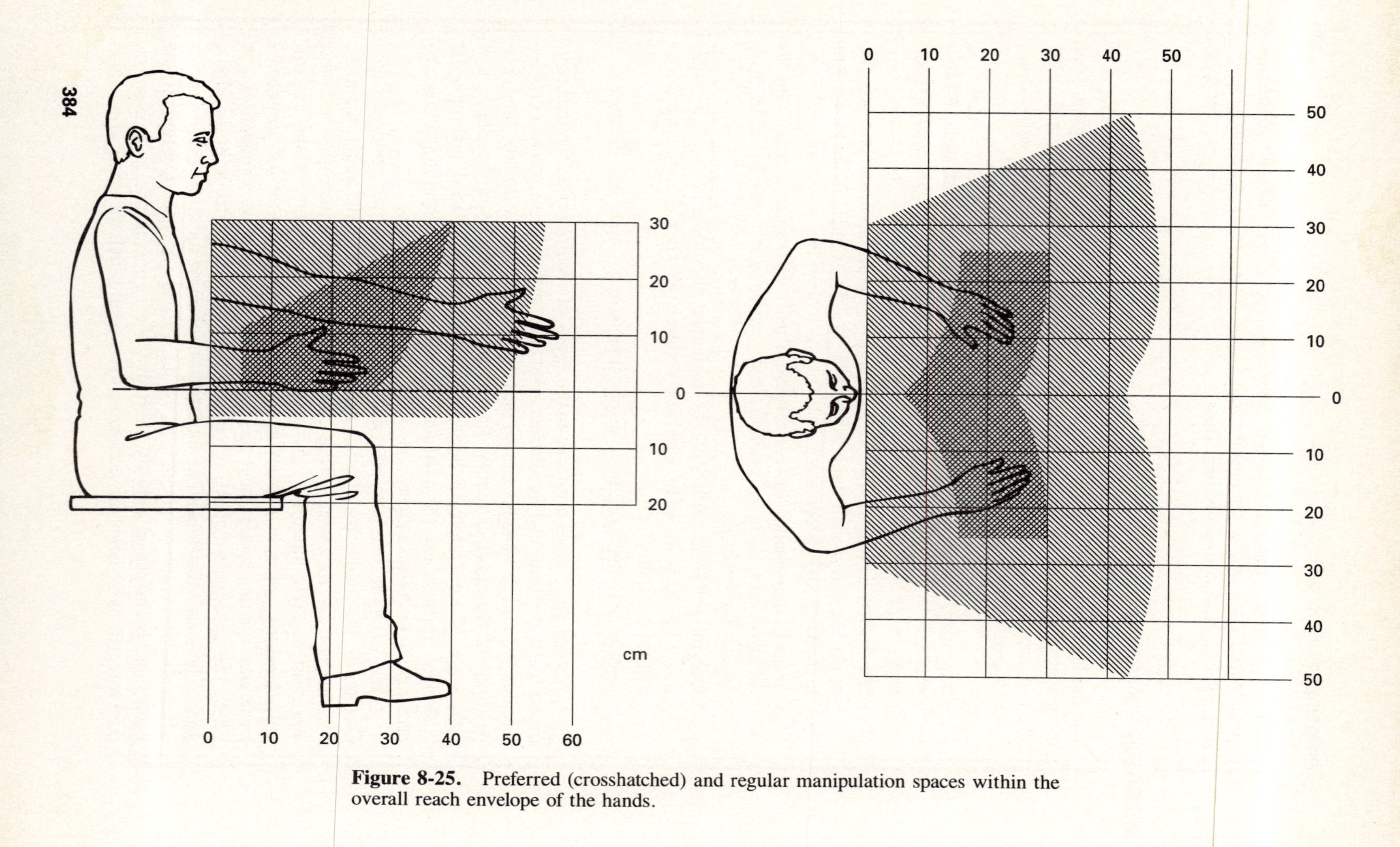

Figure 8-25. Preferred (crosshatched) and regular manipulation spaces within the overall reach envelope of the hands.

moved out of its hanging position. This establishes the "preferred manipulation space" mentioned in the first chapter (Das and Grady, 1983; Farley, 1955; Herzberg, Daniels, and Churchill, 1954; Konz and Goel, 1969; Squires, 1956). Its location is sketched in Figure 8-25.

Exertion of force with the hands is a more complex matter. Of the digits, the thumb is the strongest and the little finger the weakest. The gripping and grasping strengths of the whole hand are larger, but depend on the coupling used between the hand and the handle—see Figure 8-26. The forearm can develop fairly large twisting torques. Large force and torque vectors are exertable with the elbow bent at about a right angle, but the strongest pulling/pushing forces toward/away from the shoulder can be exerted with the extended arm, provided that the trunk can be braced against a solid structure. Torque about the elbow depends on the elbow angle, as depicted in Figure 8-27. Thus, forces exerted with the arm and shoulder muscles are largely determined by body posture and body support, as shown in Figure 8-28. Likewise, finger forces depend on the finger-joint angles, as listed in Tables 8-3 and 8-4.

Table 8-5 provides detailed information about manual strength capabilities measured in male students and machinists. Female students developed between 50 and 60 percent the digit strengths of their male peers, but 80 to 90 percent in "pinches" (Williams, 1988). Yet, the data presented here or found elsewhere must be applied with much caution, because they are likely to have been determined on different subject groups, with different techniques, and under different physical and psychological conditions than in the specific application case. Furthermore, the users may be fatigued or may be particularly trained or motivated, with possibly major effects on strength—see the discussion of "strength measurement" in Chapter 1.

Designing Hand Tools

Hand tools need to fit the contours of the hand; they need to be held securely with suitable wrist and arm postures; they must utilize strength and energy capabilities without overloading the body. Hence, the design of hand tools is a complex ergonomic task.

Use of hand tools is as old as mankind. It developed from simple beginnings—using a stone, bone, or piece of wood that fitted the hand and served the purpose—to the purposeful design of modern implements (such as the screwdriver, cutting pliers, or power saws) and of controls in airplanes and power stations (see Chapter 11 on controls in this book). Thus, a vast literature is available on tool design, summarized by Greenberg and Chaffin (1977), Fraser (1980), in a special issue of *Human Factors* (Vol. 28, No. 3, 1986), then by Freivalds (1987), Konz (1990), Chaffin and Andersson (1991), and Mital (1991).

The tool must fit the dimensions of the hand and utilize the strength and motion capabilities of the hand/arm shoulder system. Some dimensions of the

Coupling #1. Digit Touch:
One digit touches an object.

Coupling #2. Palm Touch:
Some part of the palm (or hand) touches the object.

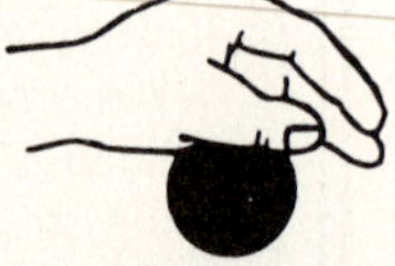

Coupling #3. Finger Palmar Grip (Hook Grip):
One finger or several fingers hook(s) onto a ridge, or handle. This type of finger action is used where thumb counterforce is not needed.

Coupling #4. Thumb–Fingertip Grip (Tip Pinch):
The thumb tip opposes one fingertip.

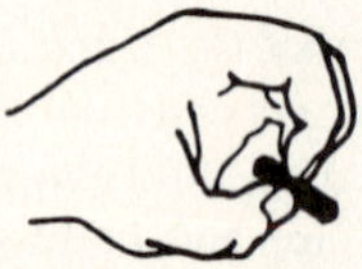

Coupling #5. Thumb–Finger Palmar Grip (Pad Pinch or Plier Grip):
Thumb pad opposes the palmar pad of one finger (or the pads of several fingers) near the tips. This grip evolves easily from coupling #4.

Coupling #6. Thumb–Forefinger Side Grip (Lateral Grip or Side Pinch):
Thumb opposes the (radial) side of the forefinger.

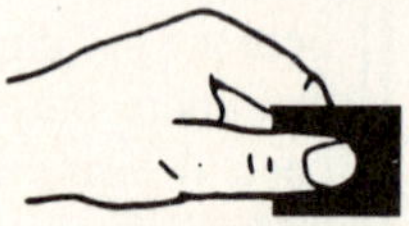

Coupling #7. Thumb–Two–Finger Grip (Writing Grip):
Thumb and two fingers (often forefinger and middle finger) oppose each other at or near the tips.

Coupling #8. Thumb–Fingertips Enclosure (Disk Grip):
Thumb pad and the pads of three or four fingers oppose each other near the tips (object grasped does not touch the palm). This grip evolves easily from coupling #7.

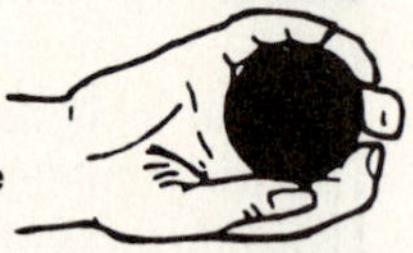

Coupling #9. Finger–Palm Enclosure (Collet Enclosure):
Most, or all, of the inner surface of the hand is in contact with the object while enclosing it. This enclosure evolves easily from coupling #8.

Coupling #10. Power Grasp:
The total inner hand surfaces is grasping the(often cylindrical) handle which runs parallel to the knuckles and generally protrudes on one or both sides from the hand. This grasp evolves easily from coupling #9.

Figure 8-26. Couplings between hand and handle (adapted with permission from "Coupling the Hand with the Handle: An Improved Notation of Touch, Grip, and Grasp" by Karl H. E. Kroemer, *Human Factors*, Vol. 28, No. 3, 1986).

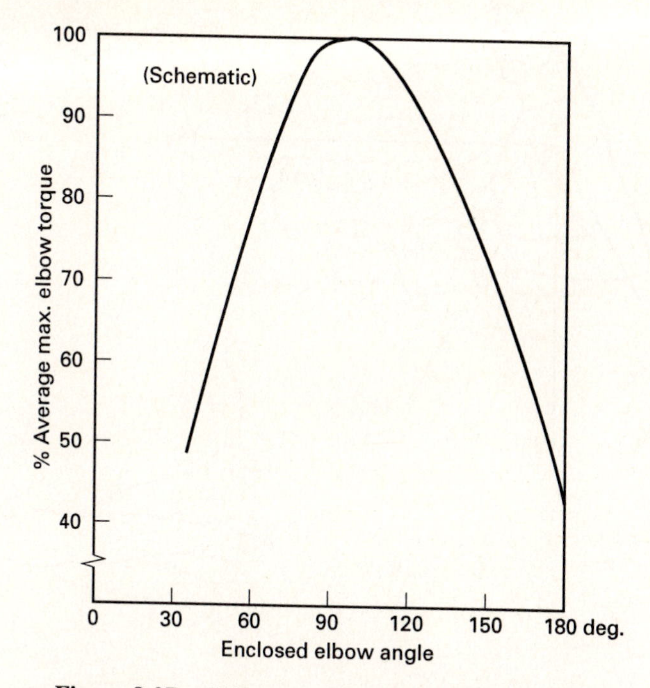

Figure 8-27. Effects of elbow angle on elbow torque.

human hand were given in the first chapter of this book. Further information can be found in the publications by Garrett, 1971; Gordon, Churchill, Clauser, et al. 1990; NASA/Webb, 1978; Wagner, 1988; and particularly by Greiner (1991).

Some hand tools require a fairly small force but precise handling, such as surgical instruments, screwdrivers used by optometrists, or writing instruments. Commonly, the manner of holding these tools has been called "precision grip." Other instruments must be held strongly between large surfaces of the fingers, thumb, and palm. Such holding of the hand tool allows the exertion of large forces and torques, hence has commonly been called "power grasp." Yet, there are many transitions from merely touching an object with a finger (such as pushing a button) to pulling on a hooklike handle, from holding small objects between the fingertips to transmitting large energy from the hand to the handle. One attempt to systematically arrange the various couplings of hand with handle is shown in Figure 8-26.

For the touch-type couplings (numbers one through six in Figure 8-26) relatively little attention must be paid to fitting the surface of the handle to the touching surface of the hand. Yet, one may want to put a slight cavity into the surface of a pushbutton so that the fingertip does not slide off; to hollow out the handle of a scalpel slightly so that the fingertips can hold on securely; to roughen the surface of a dentist's tool and not make it round in cross section,

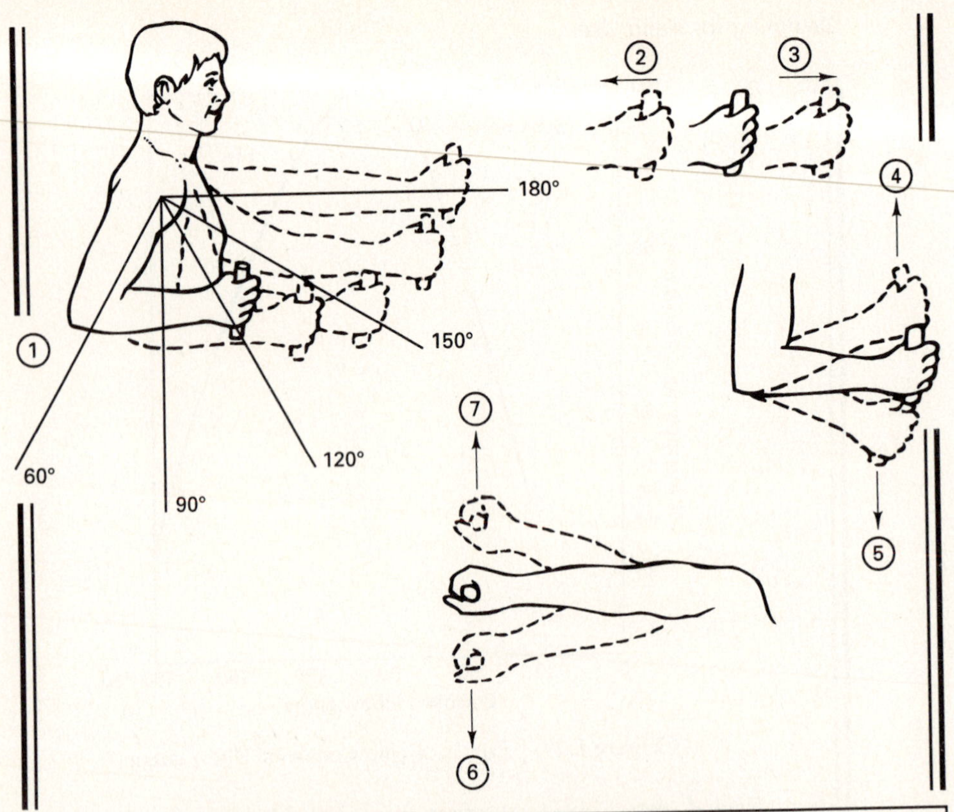

Fifth-percentile arm strength (N) exerted by sitting men													
(1)	(2)		(3)		(4)		(5)		(6)		(7)	
Degree of elbow flexation (deg)	Pull		Push		Up		Down		In		Out	
	Left	Right	L	R	L	R	L	R	L	R	L	R
180	222	231	187	222	40	62	58	76	58	89	36	62
150	187	249	133	187	67	80	80	89	67	89	36	67
120	151	187	116	160	76	107	93	116	89	98	45	67
90	142	165	98	160	76	89	93	116	71	80	45	71
60	116	107	96	151	67	89	80	89	76	89	53	71

Figure 8-28. Fifth-percentile arm strengths exerted by sitting men (adapted from MIL-HDBK 759).

which would allow it to turn in the hand. Thus, design details that facilitate holding onto the tool, moving it accurately, and generating force or torque, play important roles even for small hand tools.

These considerations of "secure tool handling" become even more important for the more powerful enclosure couplings (numbers eight, nine, and ten in Figure 8-26). These are typically used when large energies must be transmitted

TABLE 8-3. AVERAGE FORCES (AND STANDARD DEVIATIONS) IN NEWTON EXERTED BY NINE SUBJECTS IN FORE, AFT, AND DOWN DIRECTIONS WITH THE FINGERTIPS, DEPENDING ON THE ANGLE OF THE PROXIMAL INTERPHALANGEAL PIP JOINT

	PIP joint at 30 degrees			PIP joint at 60 degrees		
Direction:	Fore	Aft	Down	Fore	Aft	Down
DIGIT						
2 Index	5.4 (2.0)	5.5 (2.2)	27.4 (13.0)	5.2 (2.4)	6.8 (2.8)	24.4 (13.6)
2 n	4.8 (2.2)	6.1 (2.2)	21.7 (11.7)	5.6 (2.9)	5.3 (2.1)	25.1 (13.7)
3 Middle	4.8 (2.5)	5.4 (2.4)	24.0 (12.6)	4.2 (1.9)	6.5 (2.2)	21.3 (10.9)
4 Ring	4.3 (2.4)	5.2 (2.0)	19.1 (10.4)	3.7 (1.7)	5.2 (1.9)	19.5 (10.9)
5 Little	4.8 (1.9)	4.1 (1.6)	15.1 (8.0)	3.5 (1.6)	3.5 (2.2)	15.5 (8.5)

n: Nonpreferred hand

SOURCE: Kroemer, unpublished data.

TABLE 8-4. AVERAGE DIGIT POKE FORCES (MEANS AND STANDARD DEVIATIONS) IN NEWTON EXERTED BY 30 SUBJECTS IN DIRECTION OF THE STRAIGHT DIGITS (SEE ALSO TABLE 8-5.)

Digit	10 Mechanics	10 Male students	10 Female students
1, Thumb	83.8 (25.19) A	46.7 (29.19) C	32.4 (15.36) D
2, Index Finger	60.4 (25.81) B	45.0 (29.99) C	25.4 (9.55) DE
3, Middle Finger	55.9 (31.85) B	41.3 (21.55) C	21.5 (6.46) E

Entries with different letters are significantly different from each other ($p \leq 0.05$).

SOURCE: Williams, 1988.

between the hand and the tool. The design purpose is to hold the handle securely (without fatiguing muscles unnecessarily, and avoiding pressure points) while exerting linear force or rotating torque at the "working end" of the tool.

It is important to distinguish between the energy transmitted to the work object, and the energy transmitted between the hand and the handle. In many cases, the energy transmitted to the external object is not the same, in type, amount, or time, as generated between hand and handle. Consider, for example, the impulse energy transmitted by the head of a mallet compared to the way energy is transmitted between hand and handle, or the torque applied to a screw with the tip of a screwdriver compared to the combination of thrust and torque generated by the hand. Thus, the ergonomist must consider both the interface between tool and object, and the interface between tool and hand.

TABLE 8-5. FORCES OF DIGITS, GRIP AND GRASP FORCES EXERTED BY 21 MALE STUDENTS* AND BY 12 MALE MACHINISTS. MEANS (AND STANDARD DEVIATIONS) IN *N*

Couplings (see Figure 8-26)	Digit 1 (thumb)	Digit 2 (index)	Digit 3 (middle)	Digit 4 (ring)	Digit 5 (little)	
Push with digit tip in direction of the extended digit ("Poke")	91 (39)* 138 (41)	52 (16)* 84 (35)	51 (20)* 86 (28)	35 (12)* 66 (22)	30 (10)* 52 (14)	See also Table 8-4
Digit Touch (Coupling #1) perpendicular to extended digit.	84 (33)* 131 (42)	43 (14)* 70 (17)	36 (13)* 76 (20)	30 (13)* 57 (17)	25 (10)* 55 (16)	—
Same, but all fingers press on one bar	—	digits 2, 3, 4, 5 combined: 162 (33)				
Tip force (like in typing; angle between distal and proximal phalanges about 135 degrees)	—	30 (12)* 65 (12)	29 (11)* 69 (22)	23 (9)* 50 (11)	19 (7)* 46 (14)	—
Palm Touch (Coupling #2) perpendicular to palm (arm, hand, digits extended and horizontal)	—	—	—	—	—	233 (65)
Hook Force exerted with digit tip pad (Coupling #3, "Scratch")	61 (21) 118 (24)	49 (17) 89 (29)	48 (19) 104 (26)	38 (13) 77 (21)	34 (10) 66 (17)	all digits combined: 108 (39)* 252 (63)
Thumb-Fingertip Grip (Coupling #4, "Tip Pinch")	—	1 on 2 50 (14)* 59 (15)	1 on 3 53 (14)* 63 (16)	1 on 4 38 (7)* 44 (12)	1 on 5 28 (7)* 30 (6)	—
Thumb-Finger Palmar Grip (Coupling #5, "Pad Pinch")	1 on 2 and 3 85 (16)* 95 (19)	1 on 2 63 (12)* 34 (7)	1 on 3 61 (16)* 70 (15)	1 on 4 41 (12)* 54 (15)	1 on 5 31 (9)* 34 (7)	—
Thumb-Forefinger Side Grip (Coupling #6, "Side Pinch")	—	1 on 2 98 (13)* 112 (16)	—	—	—	—
Power Grasp (Coupling #10, "Grip Strength")	—	—	—	—	—	318 (61)* 366 (53)

SOURCE: Higginbotham and Kroemer, unpublished data.

Manually driven tools may be classified as follows, using in part Fraser's 1990 listing:

- Percussive (e.g., ax, hammer)—human task: swing and hold handle
- Scraping (saw, file, chisel, plane)—human task: push/pull and hold handle
- Rotating, boring (borer, drill, screwdriver, wrench, awl)—human task: push/pull, turn and hold handle
- Squeezing (pliers, tongs)—human task: press and hold handle
- Cutting (scissors, shears)—human task: pull and hold handle
- Cutting (knife)—human task: pull/push and hold handle

Note that in each case the operator must "hold" the tool.

Power-driven tools may use an electric power source (saw, drill, screwdriver, sander, grinder); compressed air (saw, drill, wrench); or smoothed internal combustion (chainsaw); or explosive power (bolter, cutter, riveter).

While using manual tools, the operator generates all the energy and therefore is always in full control of the energy exerted (with the exception of percussive tools, such as hammers or axes), while power-driven tools are usually held or moved while the energy is largely generated and applied to the outside by the auxiliary power. Yet, if that tool suddenly experiences resistance, the reaction force may directly affect the operator, often in terms of a jerk or impact which can exceed the person's abilities and lead to injuries. This occurs particularly often with chainsaws, power screwdrivers, and power wrenches (Mital, 1991). A major problem with many powered hand tools are impacts and vibrations transmitted to the operator, such as by jack hammers, riveters, powered wrenches, and sanders. Impacts and vibrations transmitted to the human, particularly if associated with improper postures, often lead to "overuse disorders" (discussed below).

Proper posture of the hand/arm system while using hand tools is very important. As a "handy rule," the wrist should not be bent but kept straight to avoid overexertion of such tissues as tendons, tendon sheaths, and compression of nerves and blood vessels (see below). Normally, the "grasp centerline" or "thrust line" of a straight handle is at about 70 degrees to the forearm axis—see Figure 8-29. For example, use of common straight-nose pliers often requires a strong bend in the wrist, and neither the direction of thrust nor the axis of rotation correspond with those of the hand and arm. "Bending the tool, not the wrist" improves that situation, as shown in Figure 8-30.

Another technique to avoid unsuitable postures and unnecessary muscle exertions in keeping the tool in the hand is to "form-fit" the handle to the human hand. Instead of using a straight, uniform surface (see Figure 8-31), it may be formed to fit the touching body parts, for example, the thenar pad and

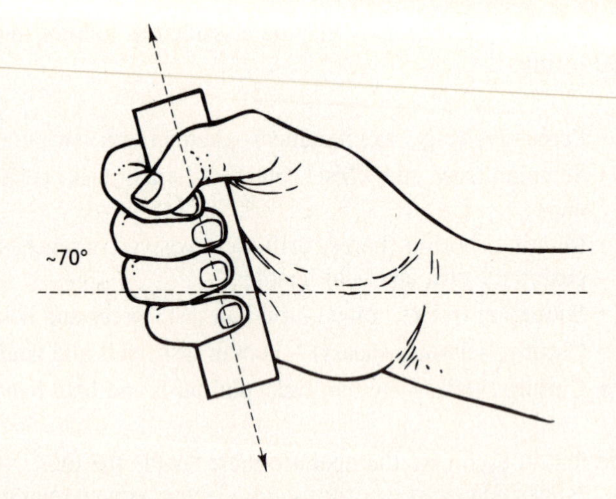

Figure 8-29. The natural angle between forearm and grasp center is about 70 degrees.

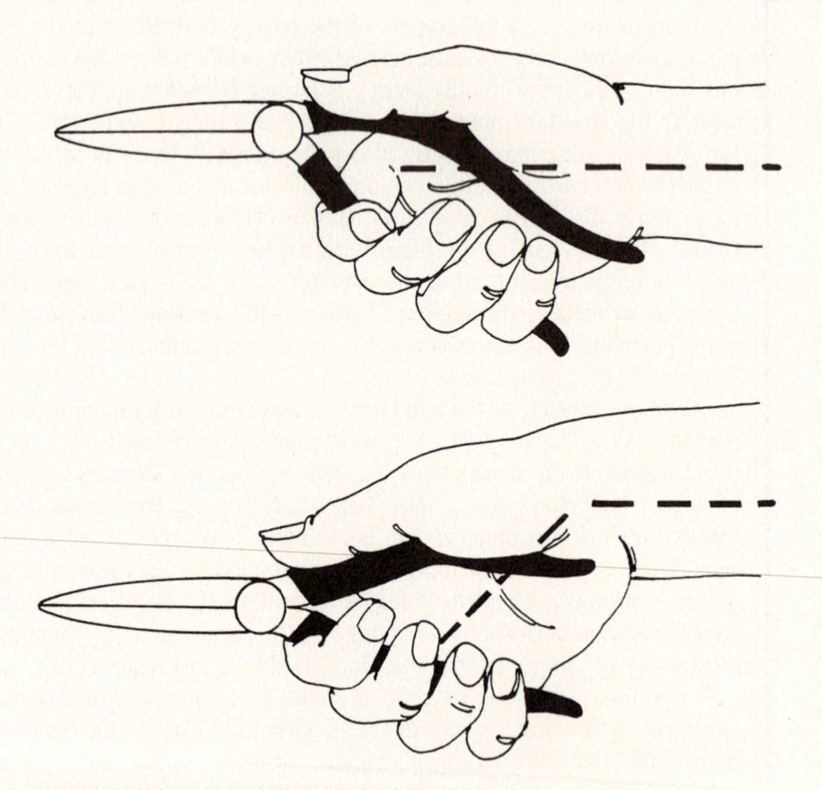

Figure 8-30. Common use of straight-nose pliers is often accompanied by a strong bend in the wrist (modified from NIOSH, 1973).

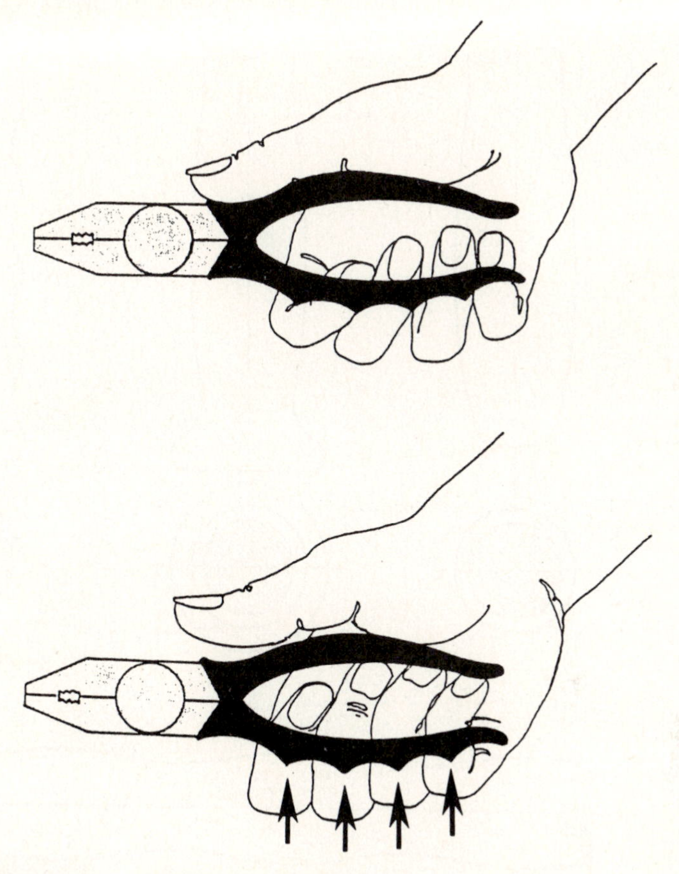

Figure 8-31. Form-fitting a handle can be helpful or painful (modified from NIOSH, 1973).

the rest of the palm, as shown in Figures 8-30, 8-32, and 8-33. Bulges and restrictions along the handle generate some form-fit, which, in addition to friction, prevents the handle from sliding out. Strong notchings or serrations, however, or other extreme form-fits can make the handle very uncomfortable, if it is held differently than anticipated by the tool designer.

Nine out of ten people, men or women, are right-handed (Greiner, 1991), and some tools are designed to fit only the right hand. Many tools can be used either with the left or right hand, but about one of ten persons prefers to use the left hand, and has better skills and more strength available there. Thus, it is advisable to provide, if needed, hand tools specifically designed for use with the left hand.

Form and surface treatment of the handle can be very important if dirt, dust, oil, or sweat changes the coefficient of friction between the handle and the hand. In such a case, special shapes and surface treatments must be consid-

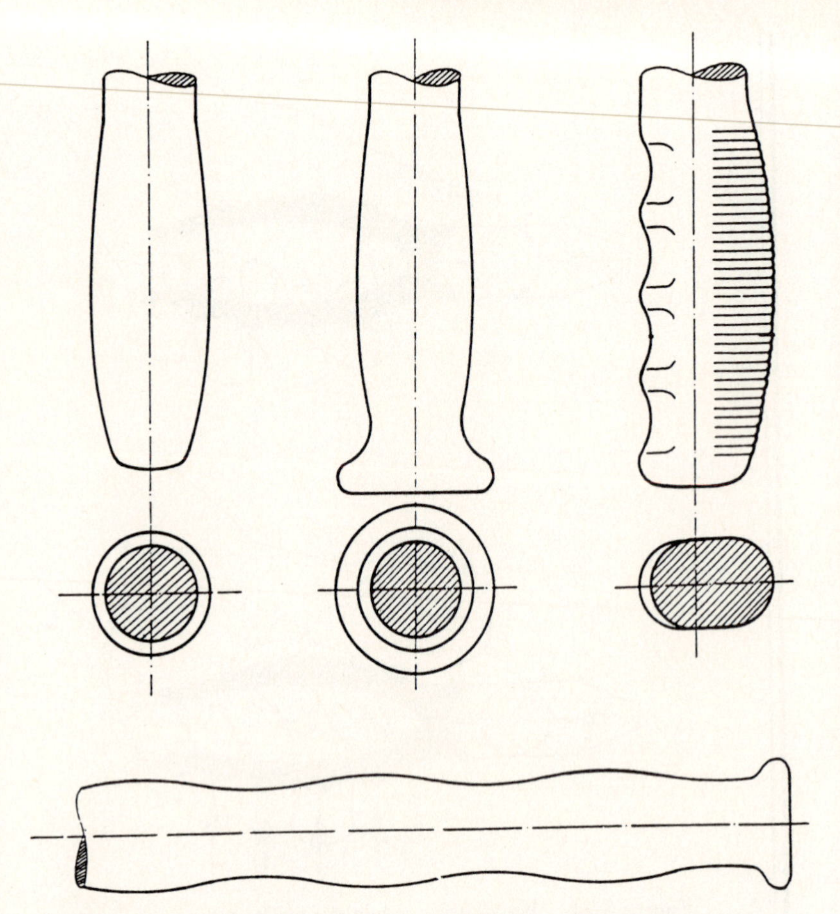

Figure 8-32. Suitable bulges and constrictions along the handle allow many hand positions without severe tissue compression. A flange at the end presents the hand from sliding off the handle.

ered, which are suitable either to keep these media away or to alleviate their effects.

Often, gloves or mittens are worn, either because of temperatures, hot or cold, that must be kept away from the skin, or because mechanical injuries must be avoided, or to alleviate the transmission of vibrations and shocks from the tool to the hand. Usually, wearing suitable gloves increases the friction between hand and handle (Buchholz, Frederick, and Armstrong, 1988). The "thrust force" T that can be developed on a handle is a function of the coefficient of friction between the handle and the hand (or the glove) and the "grasp force" G, which is perpendicular to T:

$$T = \mu G$$

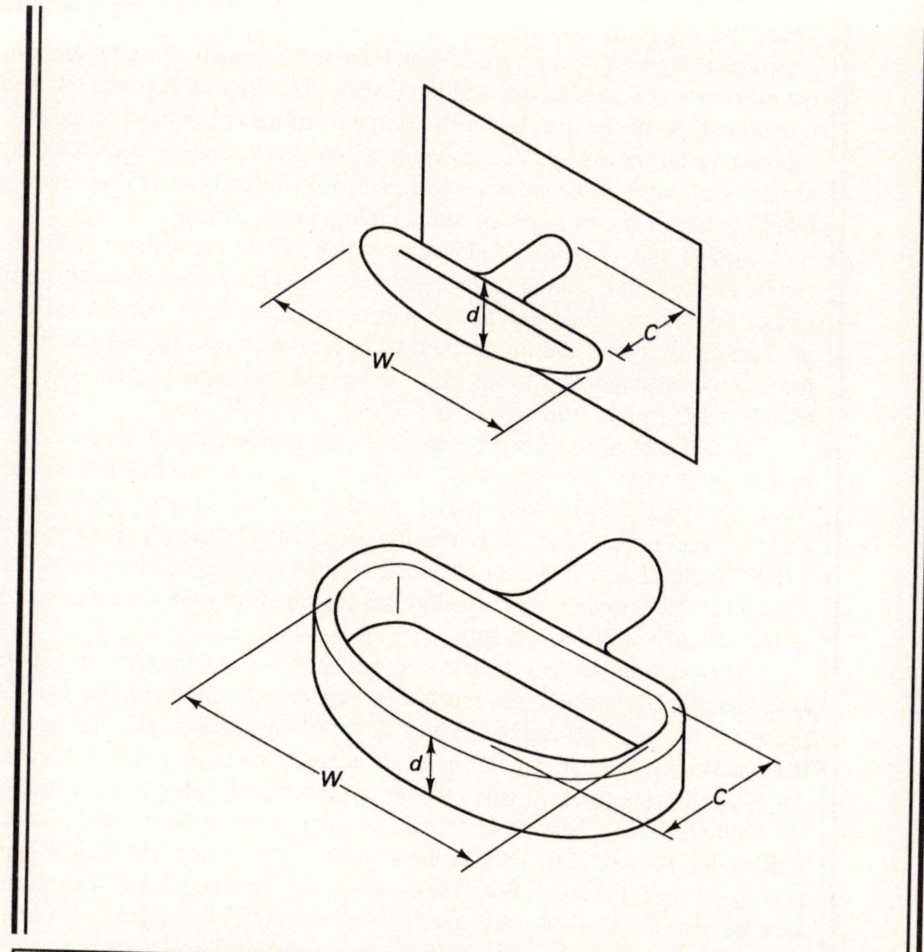

	W Width (cm)	d Thickness (cm)	C Clearance (cm)	D Displacement (cm)
Minimum	10	1,6	3,8	2,5
Maximum	13*	3,8	4,4**	Preferred 5

* Gloved operation of D handle
** Gloved operation of T handle

Figure 8-33. The T-handle is suitable for push and pull, the D-handle mostly for pulling (MIL-HDBK 759). Either can be used with the left or right hand, hence the thrust line is perpendicular to the handle.

Thus, the larger the grasping force G and the friction μ, the larger T. The grasp force depends on a person's "hand strength" (which is itself a function of the ratio between handle size and hand size). The friction depends on the texture and shape of the handle, on the skin conditions of the hand (e.g., soft vs. callous), or on the surface of a glove or mitten worn, and on intermediate materials, such as sweat or grease, which increase slipperiness, or dust, sand, and other abrasive materials, which make gliding more difficult.

Design and materials of gloves or mittens affect the amount of force that can be exerted. For example, gloves of space suits have been shown to strongly reduce not only mobility and tactile sensitivity of the hand, but also endurance in exerting strength, and the amount of force available to the outside, because the glove itself requires a strong effort to be bent and moved by the hand inside (Riley and Cochran, 1988; Roesch, 1986).

In terms of hand strength, obviously the biomechanical details discussed in Chapter 1 apply—in particular, whether intrinsic or extrinsic muscles of the hand are used, how well these muscles are developed, and the skill ("experience") of the user. It is also important to consider the duration of each grasp. As discussed in the muscle section, endurance is enhanced and fatigue reduced if the muscular effort is short and if it requires only a small percentage of the actually available strength.

The relation between handle size and hand size is important in two ways. If the handle is too small, not much force can be exerted, and large local tissue pressures might be generated (such as when writing with a very thin pencil). If the handle is too large for the hand, hand muscles must work at disadvantageous lever arms (such as when trying to squeeze the calking gun common in the United States). Numerous studies on grip strength have been conducted: Reith (1982) wrote a review. In most cases, more-or-less cylindrical handles have been used for tests. With them, diameters between 3 and 4 centimeters have been found to allow the largest grasp or compression force, with up to 6-centimeter diameters being suitable for persons with large hands. Yet, assessment of the grasp force is not as simple as one might believe, since the contribution of each digit, or of sections of the palm, should be measured separately (because each contributes its own "force times coefficient of friction"). Only a few studies shed light on the contributions of hand sections (e.g., Fransson and Winkel, 1991; Radwin and Oh, 1991), while mostly an overall, somehow averaged grip strength is measured.

The friction between hand and handle is largely determined by the surface texture of the handle and by its cross-sectional shape. As the terms indicate, "smooth" surfaces provide little friction, while it is difficult to slide on "rough" surfaces. Grooves, ridges, and serrations hinder sliding perpendicularly to them, the more so with more protruding and sharper edges. Of course, care must be taken not to exert too much pressure to damage hand tissues or even cut into the skin, a problem that can be alleviated by wearing gloves (which by themselves might increase the coefficient of friction). Thus, a proper balance must be found in the shaping of a handle between "easy sliding" (low coefficient of friction) and "mechanical interlocking" (coefficient of friction at unity value). Hence, depending on the conditions, various handle sizes and forms have been recommended, e.g., by Fraser (1980), Mital (1991), and Strasser (1991), and compiled in Vol. 28, No. 3, of *Human Factors,* 1986.

EXAMPLE

Design rules for hand tools

- Bend the tools, not the wrist.
- Push or pull in the direction of the forearm, with the handle directly in front of it.
- Provide good coupling between hand and handle by shape and friction.
- Avoid pressure spots and "pinch points."
- Round edges, pad surfaces.
- Avoid tools that transmit vibrations to your hand.
- Do not operate tools "frequently and forcefully" by hand: use a "*machine*" *instead.*

THE USE OF TABLES OF EXERTED TORQUES AND FORCES

There are many sources for forces and torques that operators can apply—see, e.g., the relevant tables and figures in this chapter. While these data indicate "orders of magnitude," the exact numbers should be viewed with great caution, because they were measured on various subject groups under widely varying circumstances—see, for example, Tables 8-3, 8-4, and 8-5. In many cases, it is advisable to take strength measurements on a sample of the intended user population to verify that a new design is operable.

Note that thumb and finger forces, for example, depend decidedly on "strength, skill, and training" of the digits as well as the posture of the hand and wrist (Hallbeck, 1989; Imrhan, 1991; Williams, 1988). Hand forces (and torques) also depend on wrist positions and on arm and shoulder posture. Exertions with arm, leg, and "body" (shoulder, backside) depend much on the posture of the body and on the support provided to the body (i.e., on the "reaction force" in the sense of Newton's third law) in terms of friction or bracing against solid structures. Figure 8-34 and Table 8-6 illustrate this: both were

Posture	Force-plate[1] height	Distance[2]	Force, N	
			Mean	SD
 1	50	80	664	177
	50	100	772	216
	50	120	780	165
	70	80	716	162
	70	100	731	233
	70	120	820	138
	90	80	625	147
	90	100	678	195
	90	120	863	141
	Percent of shoulder height		Both hands	
 2	60	70	761	172
	60	80	854	177
	60	90	792	141
	70	60	580	110
	70	70	698	124
	70	80	729	140
	80	60	521	130
	80	70	620	129
	80	80	636	133
	Percent of shoulder height			
 3	70	70	623	147
	70	80	688	154
	70	90	586	132
	80	70	545	127
	80	80	543	123
	80	90	533	81
	90	70	433	95
	90	80	448	93
	90	90	485	80
	Percent of shoulder height		Both hands	

Figure 8-34. Maximal static horizontal push forces of males (adapted from NASA, 1989).

	Force-plate[1] height	Distance[2]	Force, N	
			Mean	SD
			Both hands	
Force plate	100 percent of shoulder height	50	581	143
		60	667	160
		70	981	271
		80	1285	398
		90	980	302
		100	646	254
			Preferred hand	
		50	262	67
		60	298	71
		70	360	98
		80	520	142
		90	494	169
		100	427	173
		Percent of thumb-tip reach*		
	100 percent of shoulder height	50	367	136
		60	346	125
		70	519	164
		80	707	190
		90	325	132
		Percent of span**		

[1]Height of the center of the force plate – 20 cm high by 25 cm long – upon which force is applied.
[2]Horizontal distance between the vertical surface of the force plate and the opposing vertical surface (wall or footrest, respectively) against which the subjects brace themselves.

*Thumb-tip reach – distance from backrest to tip of subject's thumb as arm and hand are extended forward.
**Span – the maximal distance between a person's fingertips when arms and hands are extended to each side.

Figure 8-34. *(cont.)*

TABLE 8-6. HORIZONTAL PUSH AND PULL FORCES EXERTABLE INTERMITTENTLY OR FOR SHORT PERIODS OF TIME BY MALE SOLDIERS

Horizontal force*; at least	Applied with**	Condition (μ: coefficient of friction at floor)
100 N push or pull	both hands or one shoulder or the back	with low traction, $0.2 < \mu < 0.3$
200 N push or pull	both hands or one shoulder or the back	with medium traction, $\mu \sim 0.6$
250 N push	one hand	if braced against a vertical wall 51–150 cm from and parallel to the push panel
300 N push or pull	both hands or one shoulder or the back	with high traction, $\mu > 0.9$
500 N push or pull	both hands or one shoulder or the back	if braced against a vertical wall 51–180 cm from and parallel to the panel or if anchoring the feet on a perfectly nonslip ground (like a footrest)
750 N push	the back	if braced against a vertical wall 600–110 cm from and parallel to the push panel or if anchoring the feet on a perfectly nonslip ground (like a footrest)

*May be nearly doubled for two and less than tripled for three operators pushing simultaneously. For the fourth and each additional operator, not more than 75 percent of their push capability should be added.

**See Figure 8-34 for examples.

SOURCE: Adapted from MIL-STD 1472.

derived from the same set of empirical data but extrapolated to show the effects of

- friction at the feet
- body posture
- location of the point where force is exerted
- use of body parts

on horizontal push (and pull) forces applied by male soldiers.

DESIGNING FOR VISION

For many work tasks, the eyes must focus on the work object, or the tool, or must at least provide general guidance for the manipulation. This is often a problem in repair work, or in some assembly tasks, when either only a small opening is available for manipulation and vision, or where other objects may interfere with vision.

A particularly difficult ergonomic problem is often found at microscope workplaces. Many microscopes are so designed that the eye must be kept close to the ocular. This enforces maintenance of the same posture, often over extended periods. In addition, the microscope may be so designed or placed that the operator must bend head and neck to obtain proper eye location in relation to the eyepiece. Significant forward bends of the head and cervical column, exceeding 25 degrees from the vertical, are particularly stressful, because neck extensor muscles must be tensed to prevent the head from pitching forward even more. This "unbalanced" condition also affects the posture of the trunk; thus both neck and trunk muscles must be kept in tension over long periods. Consequently, complaints of microscope operators about pains and aches in the neck and back areas are frequent. Furthermore, some microscopes have hand-operated controls located high in front of the shoulders. This requires that the hands be lifted to that position and kept there, also requiring tension in muscles controlling arm and hand posture. Proper selection of microscopes that allow a variation of the eye position with respect to the eyepiece, the making of arrangements so that the operator need not bend forward, and the proper location of hand controls can alleviate many of these problems (Helander, Grossmith, and Prabhu, 1991; Krueger, Conrady, and Zuelch, 1989).

In general, visual targets that need close viewing should be placed in front of the operator (in or near the medial plane), 40 to 80 cm ("reading distance") away from the eyes. The angle of the line of sight (from pupil to target) is preferably between 20 to 60 degrees below the Ear-Eye Line plane (i.e., 0 to 40 degrees below the horizon if the head is held "straight up")—see the discussions in Chapters 4 and 9.

If the vision requirements are less stringent, especially if the person must look at the target only occasionally, the visual targets may be placed on a partial sphere surrounding the operator—see Figure 8-35.

AVOIDING OVERUSE DISORDERS IN SHOP AND OFFICE

In Australia during the early 1980s, an epidemic of so-called repetition strain injuries occurred in offices where keyboards were used. Figure 8-36 shows the frequency of injuries reported in Telecom Australia (Hocking, 1987). Similar

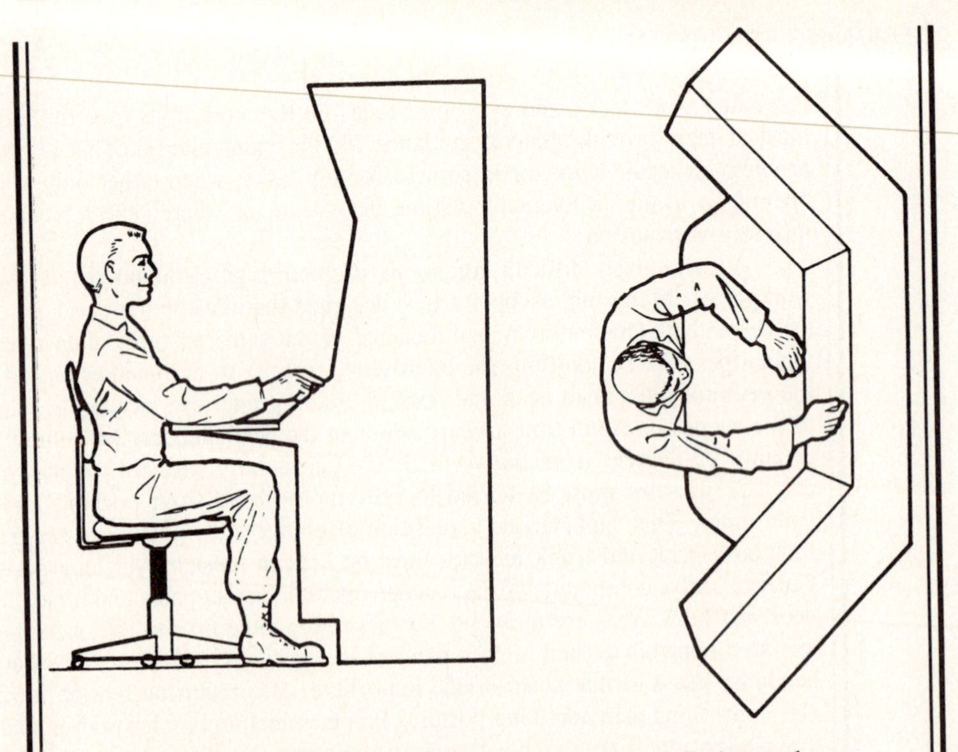

Figure 8-35. Console space suitable for placement of displays and controls (MIL-STD 1472).

events, though not so dramatic, have since occurred in Japan and the United States. Draganova reported in 1990 from Bulgaria that of 89 computer data-entry operators, 69 had musculo-skeletal complaints. Understanding of the injury mechanisms is necessary, so that they may be prevented by ergonomic measures at work.

An *overuse disorder* is the result of excessive use of a body element, often a joint, muscle, or tendon. In contrast to a single-event injury, called acute or traumatic, the overuse disorder stems from maintained efforts or from oft-repeated actions which are not injurious when occurring once or seldom, but whose time-related cumulative effects finally result in an injury. As discussed later, these effects are usually related to body posture or motion, energy or force exerted, and duration or repetitiveness. Different terms have been used to describe these phenomena, such as occupational overuse disorder (or injury, or syndrome); regional musculo-skeletal disorder; work-related disorder; repetitive stress or strain or motion injury; osteoarthrosis; rheumatic disease; or cumulative trauma disorder (Kroemer, 1992b; Putz-Anderson, 1988). In this text, the term "overuse disorder," OD, will be used.

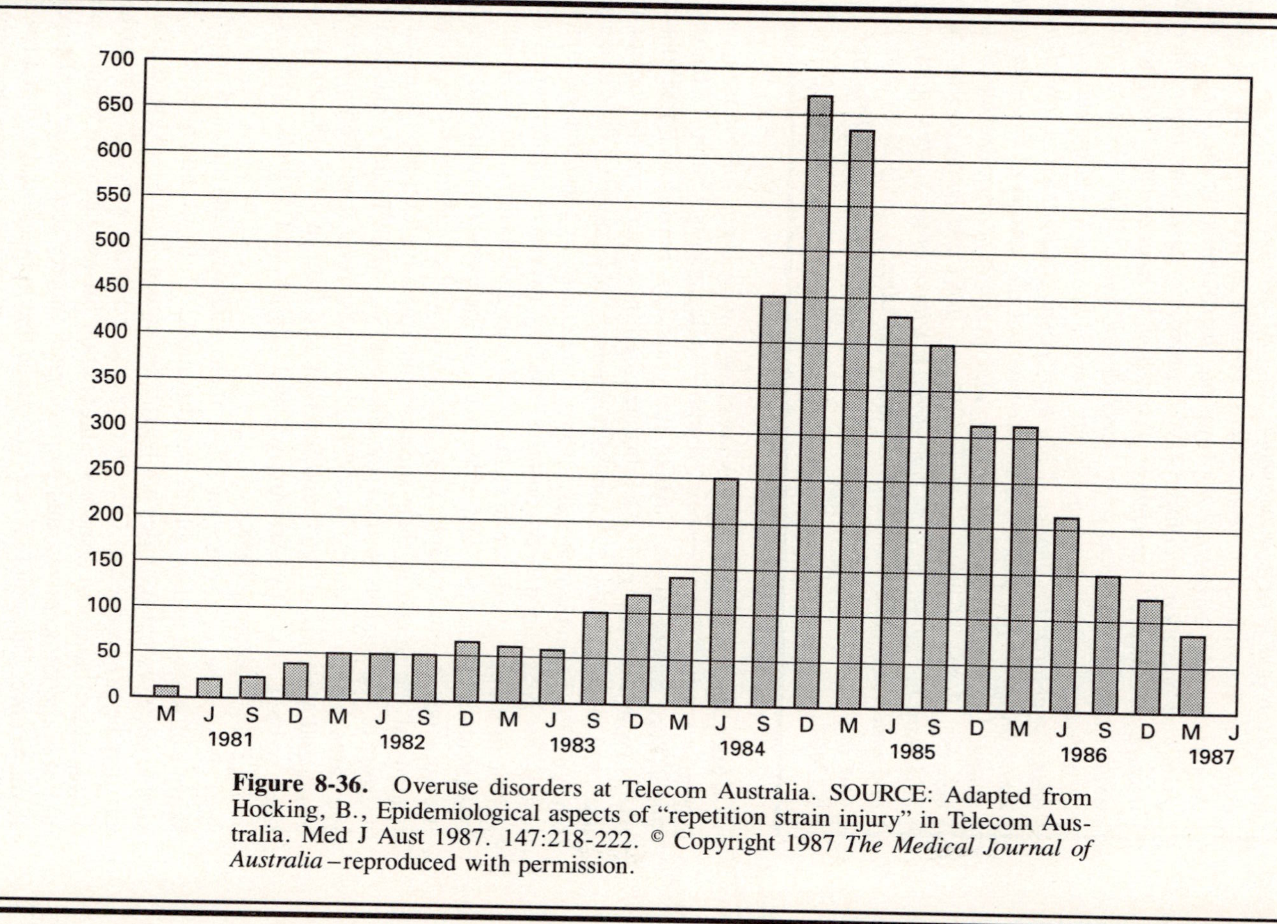

Figure 8-36. Overuse disorders at Telecom Australia. SOURCE: Adapted from Hocking, B., Epidemiological aspects of "repetition strain injury" in Telecom Australia. Med J Aust 1987. 147:218-222. © Copyright 1987 *The Medical Journal of Australia* – reproduced with permission.

"Steno-typists and other persons working with keyboards are often afflicted with disorders of tendons, sheaths, and synovial tissues of tendons, of the tendon and muscle attachments. . . . At this moment it is unknown what causes or aggravates these disorders. Possible sources may be, for example, the force needed to operate the keys, the displacement of the keys, or the frequency of operation. Hettinger (1957) attributes particular importance to the frequency. Practical experiences give reason to assume that "electrical" typewriters are advantageous over "mechanical" ones. With electrical machines, the typing frequency is certainly not lower, but the key displacement and the operational force are smaller.—The body posture is indicted in several publications. Inappropriate posture and extensive muscle tension of the arms are mentioned. . . , and the working position and direction of arms and hands are indicated. . . . The disadvantageous posture of arms and hands appears to be an important but generally little considered attribute of the work with typewriters and other keyboard machines, as indicated already in 1926 by Klockenberg and in 1931 by the Allgemeiner Freier Angestelltenbund" (Kroemer, 1964, p. 240).

Such disorders have been known for a long time, for example, as "washer women's sprain," "gamekeepers' thumb," "telegraphists' cramp," "trigger finger," "tennis" or "golfers' elbow." About 300 years ago, Bernadino Ramazzini described effects (which we now call ODs) that appeared in workers who did violent and irregular motions and assumed unnatural postures; he also reported ODs to occur among office clerks, believing that these events were caused by repetitive movements of the hands, by constrained body postures, and by excessive mental stress (Tichauer, 1973). In the twentieth century, many OD cases were reported from various kinds of agricultural, industrial, and office work. In the medical profession, ODs have long been diagnosed. For example, Gray in 1893 described inflammations of the extensor tendons of the thumb in their sheaths after excessive exercise; Hammer discussed diseases of the tendons in their sheaths in 1934. After Tanzer's classical report on the carpal tunnel syndrome in 1959, Phalen discussed it in 1966 and 1972, as did Posch and Marcotte in 1967, and Birkbeck and Beer in 1975. (Recent reviews have been written, e.g., by Armstrong, 1991; Ayoub and Wittels, 1989; Chatterjee, 1987; Putz-Anderson, 1988.)

While the occurrence of ODs, their diagnoses and medical treatment were fairly well known and established in the middle of the twentieth century, their relation to occupational activities has been hotly argued. There is evidence that individuals may be predisposed to OD—for instance, persons suffering from arthritis, diabetes, endocrinological disorders, and vitamin B6 deficiency. In women, pregnancy, use of oral contraceptives, and gynecological surgery seem to be statistically related to occurrence. One point of view maintains that in body "usage within reason," ODs should not occur in healthy persons and that, therefore, a connection between work and pathology is inadvertently introduced by a physician (Burnette and Ayoub, 1989;

Hadler, 1989, 1990). The fear of contracting an OD may lead to the condition that normally acceptable muscular discomfort thresholds are suddenly lowered, as in the so-called RSI epidemic of the 1980s in Australia, mentioned above. Some people who claim an OD are suspected to suffer from normal fatigue, to have compensation neurosis, to be victims of mass hysteria, or to malinger (Bammer and Martin, 1988; Caple and Betts, 1991; Hocking, 1987; Pickering, 1987). Yet, the prevalent position taken in the current literature is that occupational activities, especially repetitive ones, are causative, precipitating, or aggravating.

"Mountain Peaking Through Fog"

The appearance of health complaints related to cumulative trauma may be compared to a mountain. Its wide base is an accumulation of common everyday cases of tiredness, fatigue, uneasiness, and discomfort during or after a long day's work. The next higher layer consists of instances of occasional movement or postural problems beyond just weariness, often accompanied by small aches and pains that, however, disappear after a good night's rest. The narrower levels above are composed of cases of soreness, pain, and related persistent symptoms; they are present throughout most of the day and do not go away completely during the night or over a weekend. Above this layer, smaller again, is a layer of symptoms that make it difficult to continue related activities and that may lead to seeking advice from friends and co-workers as to how to alleviate the problems. The very pronounced symptoms and health complaints in the next higher level prompt discussions with nurses or physicians, who may recommend managerial and engineering changes at work. On top of this are disorders, injuries, and diseases that need medical attention and often cause short-term disability. At the peak are injuries and diseases that require acute medical treatment such as surgery. The very tip consists of a fortunately small number of disabilities which medical interventions cannot alleviate.

The top of this "mountain of problems" with its broad base and small point is visible above the "fog of psycho-social perception" that usually shrouds its base and lower sections. What becomes visible depends on the prevalent sensitivity to discomfort and pain, the existing willingness to disclose problems to supervisors and health-care givers, and the actual awareness level of society in recognizing existing problems. If the fog reaches high, only the peak of the mountain with the severe cases is visible. The lower the fog, the more problems become evident. In clear conditions, even the basic and most widespread layers of the "mountain of cumulative trauma" are in sight.

Figure 8-37 shows this analogy. The less serious concerns for health or work appear at the low levels. With increasing height, the risk to the person's health becomes more pronounced, and task performance is affected. Changes in engineering or managerial aspects of work, including work-rest arrangements, often can still alleviate the conditions. If persistent symptoms or acute aches and pains appear, medical advice is usually sought. At the high levels, pathological conditions exist that require medical intervention. If these are not completely successful, disability results.

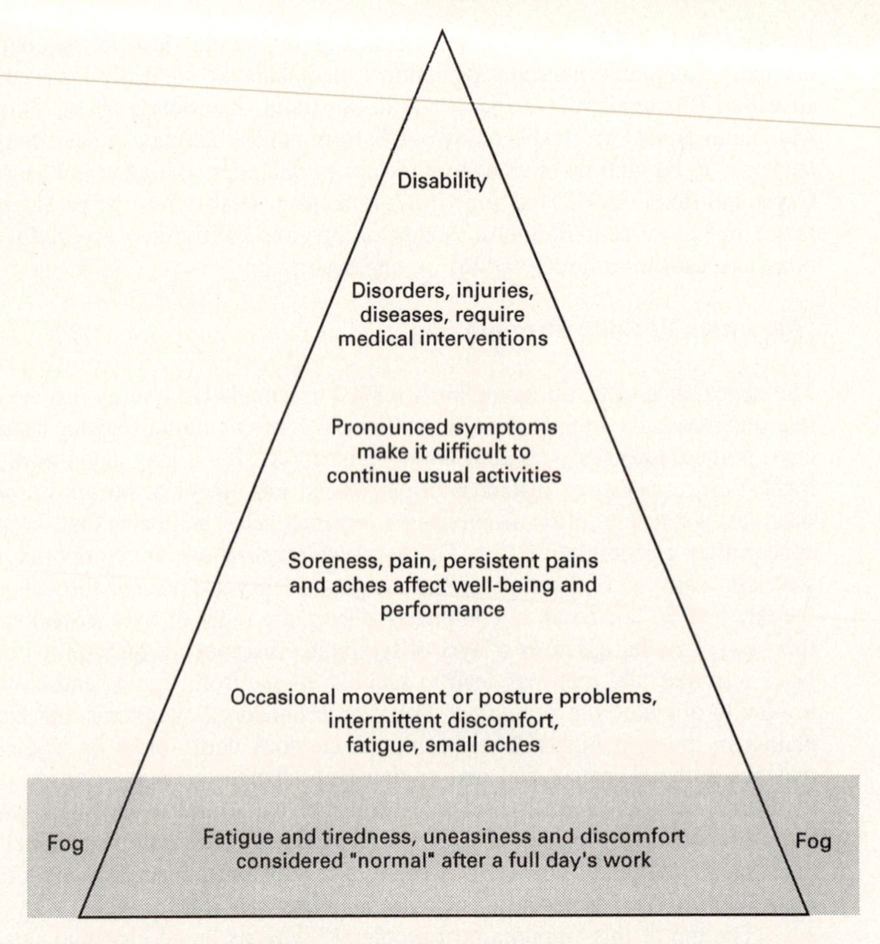

Figure 8-37. Analogy of the "mountain of cumulative trauma partly obscured by fog," the level of which indicates perception and awareness of symptoms.

Biomechanical Strains of the Body

EXAMPLE

Overuse disorders occur often in connective soft tissues, particularly to tendons and their sheaths. They may irritate or damage nerves and impede the blood flow through arteries and veins. They are frequent in the hand-wrist-forearm area (for example, in the "carpal tunnel") and in shoulder and neck. Repetitive loadings may even damage bone, such as the vertebrae of the spinal column.

Biomechanically, one can model the human body as consisting of a bony skeleton whose segment "links" connect in "joints" and are powered by muscles that bridge the joints—see Chapter 1. The muscle actions are "controlled" by the nervous and hormonal systems and "supplied" through a network of blood vessels, which also serves to remove metabolic byproducts such as CO_2, lactic acid, heat, and water (Kroemer, Kroemer, Kroemer-Elbert, 1990).

Bones provide the stable internal framework for the body. Bones can be shattered or broken through sudden impacts, and they can be damaged through continual stresses, such as in vibration. Bones are connected to each other and with other elements of the body through connective tissues: *cartilage*, *ligaments* and *tendons*. Tendons are often encapsulated by a fibrous tissue, the sheath, which allows gliding motion of the tendon, facilitated by synovia, a viscuis fluid that reduces friction with the inner lining of the sheath.

The network of *blood vessels* provides oxygen and nutrients through arteries to the working muscles and organs, and removes metabolic waste products from them—see Chapter 2.

Information from various sensors is sent through *nerves* of the feedback (afferent) pathways of the peripheral nervous system to the spinal cord and brain. Here, decisions about appropriate reactions and actions are made. Accordingly, signals are generated and sent through nerves of the feedforward (efferent) pathways to the muscles—see Chapter 3.

Body Components at Risk from ODs

While bones (except vertebrae) usually do not suffer from ODs, joints, muscles, and tendons and their related structures, as well as nerves and blood vessels, are at risk.

"Strains," inflammations, and "sprains." A "strain" is an injury to a muscle or tendon. Muscles can be stretched, which is associated with aching and swelling. A more serious injury is present when a group of fibers is torn apart. If blood or nerve supply is interrupted for an extended time, the muscle atrophies. *Tendons* contain collagen fibers, which neither stretch nor contract; if overly strained, they can be torn. Scar tissue may form, which creates chronic tension and is easily reinjured. Also, tendon surfaces can become rough, impeding their motion along other tissues. Gliding movement of a tendon in its *sheath*, caused by muscle contraction and relaxation, may be quite large—for example, 5 cm in the hand when a finger is moved from fully extended to completely flexed. Synovial fluid in the tendon sheath, acting as a lubricant to allow easy gliding, may be diminished, which causes friction between tendon and its sheath. First signs are feelings of tenderness, warmth, and pain, which may indicate inflammation.

Inflammation of a tendon or its sheath is a protective response of the body, its purpose being to limit bacterial invasion. The feeling of warmth and swelling stems from the influx of blood. The resulting compression of tissue produces pain. Movement of the tendon within its swollen surroundings is limited. Repeated forced movement may cause the inflammation of additional fiber tissue, which, in turn, can establish a permanent (chronic) condition.

A *bursa* is a fluid-filled sac lined with synovial membrane, a slippery cushion which prevents rubbing between muscle or tendon and bone. An often-used tendon, particularly if it has become roughened, may irritate its adjacent bursa, setting up an inflammatory reaction (similar to the inflammation in tendon sheaths) which inhibits the free movement of the tendon and hence reduces joint mobility.

When a joint is displaced beyond its regular range, fibers of a *ligament* may be stretched, torn apart, or pulled from the bone. This is called a "sprain," often resulting from a single trauma but also possibly caused by repetitive actions. Injured ligaments may take weeks or even months to heal, because their blood supply is poor. A ligament sprain can bring about a lasting joint instability and increase the risk of further injury.

Nerve compression. *Nerves* can also be affected by repeated or sustained pressure. Such pressure may stem from bones, ligaments, tendons, tendon sheaths, and muscles within the body, or from hard surfaces and sharp edges of work places, tools, and equipment. Pressure within the body can occur if the position of a body segment reduces the passage opening through which a nerve runs. Another source of compression, or an added one, may be irritation and swelling of other structures within this opening, often of tendons and tendon sheaths. The carpal tunnel syndrome (see later) is a typical case of nerve compression.

Impairment of a *motor nerve* reduces the ability to transmit signals to enervated muscle motor units. Thus, motor-nerve impairment impedes the controlled activity of muscles, and hence reduces the ability to generate force or torque for application to tools, equipment, and work objects. *Sensory-nerve* impairment reduces the information that can be brought back from sensors to the central nervous system. Sensory feedback is very important for many activities, because it contains information about force and pressure applied, position assumed, and motion experienced. Sensory-nerve impairment usually brings about sensations of numbness, tingling, or even pain. The ability to distinguish hot from cold may be reduced. Impairment of an *autonomic nerve* reduces the ability to control such functions as sweat production in the skin. A common sign of autonomic-nerve impairment is dryness and shininess of skin areas controlled by that nerve.

Blood-vessel compression. Compressing an artery reduces blood flow to the supplied area, which means reduced supply of oxygen and nutrients to such tissues as muscles, tendons, and ligaments; it also means diminished removal of metabolic byproducts, such as lactic acid. Vascular compression hence produces ischemia, which particularly limits the possible duration of muscular actions and impairs recovery of a "fatigued" muscle after activity. Such neurovascular compression is often found in the neck, shoulder, and upper arm regions.

Vibrations of body members, particularly of hand and fingers, may result in vasospasms that reduce the diameter of arteries. Of course, this impedes blood flow to the body areas supplied by the vessels, visible as "blanching" of the area, known particularly as the white-finger (or Raynaud's) phenomenon. Exposure to cold may aggravate the problem, because it can trigger vasospasms, particularly in the fingers. Associated symptoms include intermittent or continued numbness and tingling, with the skin turning pale and cold, and eventually loss of sensation and control. In the fingers, this condition is often caused by vibration transmitted from tools like pneumatic hammers, chainsaws, power grinders, power polishers. Frequent operation of keys on keyboards might be a source of vibration strain to the hand-wrist area.

Carpal Tunnel Syndrome

☞☞☞ *In 1959, Tanzer described several cases of carpal tunnel syndrome. Two of his patients had recently started to milk cows on a dairy farm, three worked in a shop in which objects were handled on a conveyor belt, two had done gardening with considerable hand weeding, one had been using a spray gun with a finger trigger. Two patients had been working in large kitchens with much stirring and "ladled soup" twice daily for about 600 students.* ☜☜☜

Among the best known ODs is the carpal tunnel syndrome (CTS), first described 125 years ago. In 1966 and 1972, Phalen published his "classical" reviews: he described the typical gradual onset of numbness in the thumb and the first two and a half fingers of the hand (supplied by the median nerve), with the little finger and the ulnar side of the ring finger unaffected. In 1975, Birkbeck and Beer described the results of a survey they made of the work and hobby activities of 658 patients who suffered from CTS. Four out of five patients were employed in work requiring light, highly repetitive movements of the wrists and fingers. In 1976, Posch and Marcotte analyzed 1,201 cases of CTS.

In 1964, Kroemer described the occurrence of ODs in typists and their possible relation to key force and displacement, key operation frequency, and posture of arms and hands. In 1980, Huenting, Grandjean, and Maeda found frequent impairments in hands and arms of operators of accounting machines; they believed that the disorders were related to working posture and key operation. In 1981, Cannon, Bernacki, and Walter described a case-control study on the factors associated with the onset of CTS. They linked the occurrence of this syndrome to diverse etiological factors, which included injury and illness, use of drugs, and hormonal changes in women. Yet: "Ergonomic theory relates the occurrence of the condition to repetitive movements of the wrists, performance of tasks with the wrists in ulnar deviation and chronic exposure to low-frequency vibrations" (p. 255). In 1983, the American Industrial Hygiene Association acknowledged the prevalence and importance of CTS by publishing Armstrong's *An Ergonomics Guide to Carpal Tunnel Syndrome,* which is called there an "occupational illness."

Thus, in the 1970s and early 1980s, CTS was well recognized as an often-occurring, disabling condition of the hand that can be caused, precipitated or aggravated by certain work activities in the office and on the shop floor. Of course, leisure activities may be involved as well. Critical activities include a flexed or hyperextended wrist, especially in combination with forceful exertions, in highly repetitive activities, and vibrations. Of course, case studies and epidemiological surveys have inherent attributes (such as the lack of a control group and existence of confounding variables) that make it difficult or impossible to "prove without a doubt" a connection between activities and disorders, but apparent correlations between the two strongly suggest a cause-effect relationship.

In 1963, Robbins described the anatomical conditions which explain the tunnel syndrome. On the palmar side of the wrist, near the base of the thumb, the carpal bones form a concave floor. It is bridged by three ligaments (the radial carpal, intercarpal, and carpometacarpal), which in turn are covered by the transverse carpal ligament, which is firmly fused to the carpal bones. Thus, a canal is formed by the carpal bones as the floor and sides, and ligaments cover it like a roof—see Figure 8-38. This structure is called the "carpal tunnel." Its cross section is roughly oval. Through the carpal tunnel, flexor tendons of the digits pass, also the median nerve and the radial artery. This crowded space is reduced if the wrist is flexed or extended, or pivoted ulnarly or radially. Also, swelling of the tendons and/or of their sheaths reduces the space.

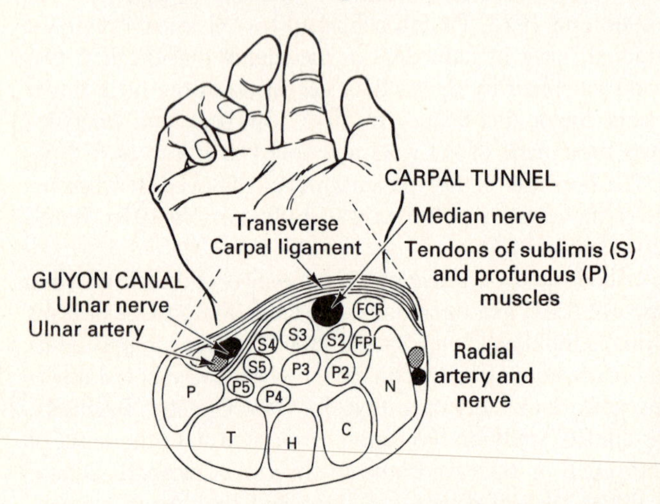

Figure 8-38. Schematic view of the carpal tunnel with the tendons of the superficial (S) and profound (P) finger flexor muscles, flexors of the thumb (FCR, FPL), nerves and arteries, carpal bones (P, T, H, C, N) and ligaments (adapted from Kroemer, 1989).

The median nerve, which passes through the carpal tunnel, innervates the thumb, much of the palm, and the index and middle fingers as well as the radial side of the ring finger. Outside pressure on the tunnel ligament, as well as tendon or sheath inflammation, reduces the tunnel space, as does wrist deviation from straight. This leads to pressure on the median nerve and/or blood vessels, resulting in the carpal tunnel syndrome. Compression of the passage containing the ulnar nerve and the ulnar artery can lead to the "guyon canal syndrome"; a similar event can occur to the radial nerve and artery.

APPLICATION

Occupational Activities and Related Disorders

Table 8-7 lists those conditions that are most often associated with overuse disorders (Kroemer, 1992b). Of course, this list is neither complete nor exclusive. New occupational activities occur, and several activities may be part of the same job.

OD-Prone Activities and Postures

The major activity-related factors in repetitive strain injuries are rapid often repeated movements, less frequent but more forceful movements, static muscle loading, and vibrations. Their negative effects are, or may be, aggravated by inappropriate postures.

Silverstein (1985) proposed that *high repetitiveness* be defined as a cycle time of less than 30 seconds, or as more than 50 percent of the cycle time spent performing the same fundamental motion. Silverstein also suggested that *high force* exerted with the hand, e.g., more than 45 N, may be a causative factor by itself. Posture may be highly important. For example, a "dropped" or an "elevated wrist"—see Figure 8-39—reduces the available cross section of the carpal tunnel, and hence generates a condition that may cause Carpal Tunnel Syndrome, particularly in persons with small wrists (Morgan, 1991).

Among the postural effects, maintained isometric contraction of muscles needed to keep the body or its part in position, or to hold a hand tool, or to carry an object, is often associated with a Carpal Tunnel Syndrome condition. Also, inward or outward twisting of the forearm with a bent wrist, a strong deviation of the wrist from the neutral position, and the pinch grip can be stressful. If muscles must remain contracted at more than about 15 to 20 percent of their maximal capability, circulation is impaired. This can result in tissue ischemia and delayed dissipation of metabolites, which constitute conditions of general physiological strain.

"Seven sins." There are seven conditions that specifically need to be avoided:

1. Job activities with many repetitions.
2. Work that requires prolonged or repetitive exertion of more than about one-third of the operator's static muscular strength available for that activity.

TABLE 8-7(a). COMMON ODs*

Disorder name	Description	Typical job activities
Carpal Tunnel Syndrome (writer's cramp, neuritis, median neuritis) (N)	The result of compression of the median nerve in the carpal tunnel of the wrist. This tunnel is an opening under the carpal ligament on the palmar side of the carpal bones. Through this tunnel pass the median nerve, the finger flexor tendons, and blood vessels. Swelling of the tendon sheaths reduces the size of the opening of the tunnel and pinches the medan nerve, and possibly blood vessels. The tunnel opening is also reduced if the wrist is flexed or extended, or ulnarly or radially pivoted.	Buffing, grinding, polishing, sanding, assembly work, typing, keying, cashiering, playing musical instruments, surgery, packing, housekeeping, cooking, butchering, hand washing, scrubbing, hammering.
Cubital Tunnel Syndrome (N)	Compression of the ulnar nerve below the notch of the elbow. Tingling, numbness, or pain radiating into ring or little fingers.	Resting forearm near elbow on a hard surface and/or sharp edge, also when reaching over obstruction.
DeQuervain's Syndrome (or Disease) (T)	A special case of tendosynovitis (see there) which occurs in the abductor and extensor tendons of the thumb, where they share a common sheath. This condition often results from combined forceful gripping and hand twisting, as in wringing cloths.	Buffing, grinding, polishing, sanding, pushing, pressing, sawing, cutting, surgery, butchering, use of pliers, 'turning' control such as on motorcycle, inserting screws in holes, forceful hand wringing.

Epicondylitis ("tennis elbow") (T)	Tendons attaching to the epicondyle (the lateral protrusion at the distal end of the humerus bone) become irritated. This condition is often the result of impacting of jerky throwing motions, repeated supination and pronation of the forearm, and forceful wrist extension movements. The condition is well known among tennis players, pitchers, bowlers, and people hammering. A similar irritation of the tendon attachments on the inside of the elbow is called medical epicondylitis, also known as "golfer's elbow."	Turning screws, small parts assembly, hammering, meat cutting, playing musical instruments, playing tennis, pitching, bowling.
Ganglion (T)	A tendon sheath swelling that is filled with synovial fluid, or a cystic tumor at the tendon sheath or a joint membrane. The affected area swells up and causes a bump under the skin, often on the dorsal or radial side of the wrist. (Since it was in the past occasionally smashed by striking with a bible or heavy book, it was also called a "bible bump.")	Buffing, grinding, polishing, sanding, pushing, pressing, sawing, cutting, surgery, butchering, use of pliers, 'turning' control such as on motorcycle, inserting screws in holes, forceful hand wringing.
Neck Tension Syndrome (M)	An irritation of the levator scapulae and trapezius group of muscles of the neck, commonly occurring after repeated or sustained overhead work.	Belt conveyor assembly, typing, keying, small parts assembly, packing, load carrying in hand or on shoulder.

*Type of disorder: N nerve, T tendon, M muscle, V vessel.

TABLE 8-7(b). COMMON ODs* (continued)

Disorder name	Description	Typical job activities
Pronator (Teres) Syndrome (N)	Result of compression of the median nerve in the distal third of the forearm, often where it passes through the two heads of the pronator teres muscle in the forearm; common with strenuous flexion of elbow and wrist.	Soldering, buffing, grinding, polishing, sanding.
Shoulder Tendonitis (rotator cuff syndrome or tendonitis, supraspinatus tendinitis, subacromial bursitis, subdeltoid bursitis, partial tear of the rotator cuff) (T)	This is a shoulder disorder, located at the rotator cuff. The cuff consists of four tendons that fuse over the shoulder joint where they pronate and supinate the arm and help to abduct it. The rotator cuff tendons must pass through a small bony passage between the humerus and the acromion, with a bursa as cushion. Irritation and swelling of the tendon or of the bursa are often caused by continuous muscle and tendon effort to keep the arm elevated.	Punch press operations, overhead assembly, overhead welding, overhead painting, overhead auto repair, belt conveyor assembly work, packing, storing, construction work, postal "letter carrying," reaching, lifting, carrying load on shoulder.
Tendonitis (tendinitis) (T)	An inflammation of a tendon. Often associated with repeated tension, motion, bending, being in contact with a hard surface, vibration. The tendon becomes thickened, bumpy, and irregular in its surface. Tendon fibers may be frayed or	Punch press operation, assembly work, wiring, packaging, core making, use of pliers.

	torn apart. In tendons without sheaths, such as within elbow and shoulder, the injured area may calcify.	
Tendosynovitis (tenosynovitis, tendovaginitis) (T)	This disorder occurs to tendons which are inside synovial sheaths. The sheath swells. Consequently movement of the tendon with the sheath is impeded and painful. The tendon surfaces can become irritated, rough, and bumpy. If the inflamed sheath presses progressively onto the tendon, the condition is called stenosing tendosynovitis. "DeQuervain's Syndrome" (see there) is a special case occurring in the thumb, while the "Trigger Finger" (see there) condition occurs in flexors of the fingers.	Buffing, grinding, polishing, sanding, punch press operation, sawing, cutting, surgery, butchering, use of pliers, 'turning' control such as on motorcycle, inserting screws in holes, forceful hand wringing.
Thoracic Outlet Syndrome (neurovascular compression syndrome, cervicobrachial disorder, brachial plexus neuritis, costocalvicular syndrome, hyperabduction syndrome) (V, N)	A disorder resulting from compression of nerves and blood vessels between clavicle and first and second ribs, at the brachial plexus. If this neurovascular bundle is compressed by the pectoralis minor muscle, blood flow to and from the arm is reduced. This ischemic condition makes the arm numb and limits muscular activities.	Buffing, grinding, polishing, sanding, overhead assembly, overhead welding, overhead painting, overhead auto repair, typing, keying, cashiering, wiring, playing musical instruments, surgery, truck driving, stacking, material handling, postal "letter carrying," carrying heavy loads with extended arms.

TABLE 8-7(c). COMMON ODs (concluded)

Disorder name	Description	Typical job activities
Trigger Finger (or Thumb) (T)	A special case of tendosynovitis (see there) where the tendon becomes nearly locked, so that its forced movement is not smooth but in a snapping or jerking manner. This is a special case of stenosing tendosynovitis crepitans, a condition usually found with digit flexors at the A1 ligament.	Operating trigger finger, using hand tools that have sharp edges pressing into the tissue or whose handles are too far apart for the user's hand so that the end segments of the fingers are flexed while the middle segments are straight.
Ulnar Artery Aneurysm (V, N)	Weakening of a section of the wall of the ulnar artery as it passes through the Guyon tunnel in the wrist; often from pounding or pushing with heel of the hand. The resulting "bubble" presses on the ulnar nerve in the Guyon tunnel.	Assembly work.
Ulnar Nerve Entrapment (Guyon Tunnel Syndrome) (N)	Results from the entrapment of the ulnar nerve as it passes through the Guyon tunnel in the wrist. It can occur from prolonged flexion and extension of the wrist and repeated pressure on the hypothenar eminence of the palm.	Playing musical instruments, carpentering, brick laying, use of pliers, soldering, hammering.
White Finger ("dead finger," Raynaud's Syndrome, vibration syndrome) (V)	Stems from insufficient blood supply bringing about noticeable blanching. Finger turns cold, numb, tingles, and sensation and control of finger movement may be lost. The condition is due to closure of the digit's arteries caused by vasospasms triggered by vibrations. A common cause is continued forceful gripping of vibrating tools, particularly in a cold environment.	Chain sawing, jack hammering, use of vibrating tool, sanding, paint scraping, using vibrating tool too small for the hand, often in a cold environment.

Figure 8-39. Dropped and elevated wrists. Note also the pressure at the edge of the table. Courtesy of Herman Miller Inc.

3. Putting body segments in extreme positions—see Figures 8-39 and 8-40.
4. Work that makes a person maintain the same body posture for long periods of time—see Figure 8-39.
5. Pressure from tools or work equipment on tissues (skin, muscles, tendons), nerves, or blood vessels—see Figures 8-40 and 8-41.
6. Work in which a tool vibrates the body, or part of the body.
7. Exposure of working body segments to cold, including air flow from pneumatic tools.

EXAMPLE

Forerunners of overuse disorders

- rapid and often-repeated actions
- exertion of finger, hand, or arm forces
- pounding and jerking
- contorted body joints
- polished-by-use sections of workplace, clothing; custom-made padding
- blurred outlines of the body owing to vibration
- the feeling of cold and the hissing sound of fast-flowing air

Figure 8-40. Contorted working posture at a punch press.

Figure 8-41. Edge of desk presses into forearm. Courtesy of Herman Miller Inc.

Stages of Disorders and Their Treatments

The clinical features of ODs are various, variable, and often confusing. The onset of these symptoms can be gradual or sudden. Three stages have been defined by Chatterjee (1987):

> Stage 1 is characterized by aches and "tiredness" during the working hours, which usually settle overnight and over days off. There is no reduction in work performance. This condition may persist for weeks or months, and is reversible.
>
> Stage 2 has symptoms of tenderness, swelling, numbness, weakness, and pain that start early in the workshift and do not settle overnight. Sleep may be disturbed, and the capacity to perform the repetitive work is often reduced. This condition usually persists over months.

Stage 3 is characterized by symptoms that persist at rest and during the night. Pain occurs even with nonrepetitive movements, and sleep is disturbed. The patient is unable to perform even light duties and experiences difficulties in daily tasks. This condition may last for months or years.

The early stage is often reversible through work modification and rest breaks. Exercise as a prophylactic measure is of questionable value (Silverstein, Armstrong, Longmate, and Woody, 1988; Williams, Smith and Herrick, 1989). In the later stages, the most important factor is abstinence from performing the causing motions, combined with rest. This may mean major changes in working habits and lifestyle. Further treatments are physiotherapy, drug administration, and other medical treatments including surgery. Medically, it is important to identify an OD case early, at a stage that allows effective treatment. Ergonomically, it is even more important to prevent injury by recognizing potentially injurious activities and conditions and alleviating them through work (re-)organization and work (re-)design.

APPLICATION

Ergonomic Means to Counter ODs

Of course, it is best to avoid conditions that may lead to an OD. Work object, equipment, and tools used should be suitably designed; instruction on and training in proper postures and work habits are important; managerial interventions such as work diversification (the opposite of job simplification and specialization), relief workers, and rest pauses are often appropriate. Of course, it is most important to "not repeat" possibly injurious motions and force exertions and to avoid unsuitable postures.

As a general rule, tools and tasks should be so designed that they can be handled without causing wrist deviations. The wrist should not be severely flexed, extended, or pivoted but in general remain aligned with the forearm. The forearm should not be twisted. The elbow angle should be varied. The upper arms usually should hang down along the sides of the body. The head should be held fairly erect. The trunk should be mostly upright when standing; when sitting for long periods, a tall, well-shaped backrest is desirable. Trunk rotation should not be required at work. There should be enough room for the legs and feet to allow standing or sitting comfortably. It is important that the postures of the body segments, and of the whole body, can and will be varied often during the working time.

Jobs should be analyzed for their posture, movement, and force requirements, using, e.g., the well-established industrial engineering procedure of "motion and time study" (Konz, 1990). Each element of the work should be screened for factors that can contribute to OD (Hahn, Chin, Ma, and Rebello, 1991; Jegerlehner, 1991). After the job analysis has been completed, workstations and equipment can be ergonomically engineered and work procedures organized to avoid stress on the operator's body. Table 8-8 (adapted from Kroemer, 1989) provides an overview of generic ergonomic measures to fit the job to the person.

Avoidance of OD, whether by redesign of an existing workstation or by appropriate planning of a new workstation, follows one simple generic rule: let the operator perform "natural" activities, i.e., those for which the human body is suited. Avoid highly repetitive activities and those in which straining forces or torques must be exerted or in which awkward posture must be maintained over prolonged time.

The general process of work, the particular hand tools to be used, or the parts on which work needs to be performed, should be designed—or, if found wanting, altered as needed—to fit human capabilities. The opposite way, i.e., selecting persons who seem to be especially able to perform work that most people cannot do, or letting several people work at the same workstation alternately so that nobody has to work long periods of time on the same job, involves basically inappropriate measures which should be applied only if no other solution can be found.

EXAMPLE

The overall principle is to fit the job to the person, not to attempt to fit persons to the job.

In addition to keeping the number of repetitions and the amounts of energy (force) small, the following posture-related measures should be considered:

1. Provide a chair with a headrest, so that one can relax neck and shoulder muscles at least temporarily;
2. Provide an armrest, possibly cushioned, so that the weight of the arms must not be carried by muscles crossing the shoulders and elbow joints;
3. Provide flat, possibly cushioned surfaces on which forearms may rest while the fingers work;
4. Provide a wrist rest for people operating traditional keyboards, so that the wrist cannot "drop" below the key level;
5. Round, curve, and pad all "edges" that otherwise might be point-pressure sources;
6. Select jigs and fixtures to hold workpieces in place, so that the operator does not have to hold the workpiece;

TABLE 8-8. ERGONOMIC MEASURES TO AVOID COMMON ODs

Disorder	Avoid in general	Avoid in particular	**Do**	**Design**
Carpal Tunnel Syndrome	Rapid, often repeated finger movements, wrist deviation	Dorsal and palmar flexion, pinch grip, vibrations between 10 and 60 Hz	use large muscles but infrequently and for short durations — let wrists be in the line with the forearm — let shoulder and upper arm be relaxed	the work object properly — the job task properly — hand tools properly ("bend tool, not the wrist") —
Cubital Tunnel Syndrome	Resting forearm on sharp edge or hard surface			
DeQuervain's Syndrome	Combined forceful gripping and hard twisting			
Epicondylitis	"Bad tennis backhand"	Dorsiflexion, pronation		
Pronator Syndrome	Forearm pronation	Rapid and forceful pronation, strong elbow and wrist flexion		
Shoulder Tendonitis, Rotator Cuff Syndrome	Arm elevation	Arm abduction, elbow elevation		
Tendonitis	Often repeated movements, particularly with force exertion; hard surface in contact with skin; vibrations	Frequent motions of digits, wrists, forearm, shoulder		

Tendosynovitis, DeQuervain's Syndrome, Ganglion	Finger flexion, wrist deviation	Ulnar deviation, dorsal and palmar flexion, radial deviation with firm grip	— let forearms be horizontal or more declined	round corners, pad — placed work object properly
Thoracic Outlet Syndrome	Arm elevation, carrying	Shoulder flexion, arm hyperextension		
Trigger Finger or Thumb	Digit flexion	Flexion of distal phalanx alone		
Ulnar Artery Aneurism	Pounding and pushing with the heel of the hand			
Ulnar Nerve Entrapment	Wrist flexion and extension	Wrist flexion and extension, pressure on hypothenar eminence		
White Finger, Vibration Syndrome	Vibrations, tight grip, cold	Vibrations between 40 and 125 Hz		
Neck Tension Syndrome	Static head posture	Prolonged static head/neck posture	alternate head/neck postures	

SOURCE: Adapted from Kroemer, 1989.

7. Select and place jigs and fixtures so that the operator can easily access the workpiece without "contorting" hand, arm, neck, or back;
8. Select bins and containers, and place them so that the operator can reach into them with least possible flexion/extension, pivoting, or twisting of hand, arm, and trunk;
9. Select tools whose handles distribute pressure evenly over large surfaces of the operator's digits and palm;
10. Select hand tools and working procedures that do not require "pinching";
11. Select the lightest possible hand tools;
12. Select hand tools that are properly angled so that the wrist must not be bent;
13. Select hand tools which do not require the operator to apply a twisting torque;
14. Select hand tools whose handles are so shaped that the operator does not have to apply much grasping force to keep it in place or to press it against a workpiece;
15. Avoid tools that have sharp edges, fluted surfaces, or other prominences that press into tissues of the operator's hand;
16. Suspend or otherwise hold tools in place so that the operator does not have to do so for extended periods of time;
17. Select tools that do not transmit vibrations to the operator's hand;
18. If the hand tool must vibrate, have energy absorbing/dampening material between the handle and the hand (yet, the resulting handle diameter should not become too large);
19. Make sure that the operator's hand does not become undercooled, which may be a problem particularly with pneumatic equipment;
20. Select gloves, if appropriate, to be of proper size, texture, thickness.

☞☞☞ *Three decades ago, Peres wrote:*

"It has been fairly well established, by experimental research overseas and our own experience in local industry, that the continuous use of the same body movement and sets of muscles responsible for that movement during the normal working shift (not withstanding the presence of rest breaks), can lead to the onset initially of fatigue, and ultimately of immediate or cumulative muscular strain in the local body area" (Peres, 1961, p. 1).

"It is sometimes difficult to see why experienced people, after working satisfactorily for, say, 15 years at a given job, suddenly develop pains and strains. In some cases these are due to degenerative arthritic changes and/or traumatic injury of the bones of the wrist or other joints involved. In other cases, the cause seems to be

compression of a nerve in the particular vicinity, as for example, compression of the median nerve in carpal tunnel syndrome. However, it may well be that many more are due to cumulative muscle strain arising from wrong methods of working . . ." (Peres, 1961, p. 11).

Research Needs

Our current knowledge about the relationships between activities (on the job or during leisure) and ODs is mostly limited to exertions of fairly large hand forces and their association with exertion frequencies and body postures. Yet, even for those gross muscular activities, some uncertainty about their causal relationships to ODs exist. Even if a causal relationship between job activity and disorder is presumed, the exact job factors are not well defined, nor are their critical threshold values generally known. For example, "forcefulness" of a job exertion may be measured statically (isometrically) or in dynamic terms. Silverstein's (1985) definitions are apparently mostly applicable to static exertions, while many industrial activities are in fact dynamic in nature. Some ODs are explicitly related to motion, i.e., dynamics—for example, as indicated by the term "repetitive motion injury." Exertion of energy in either static or in dynamic conditions establishes very different body strains (Kroemer, Marras, McGlothlin, McIntyre, and Nordin, 1990). Unfortunately, our current knowledge base seems to be largely dependent on the assumption of isometric muscle efforts, i.e., on a static condition.

Another major problem, both in concept and in application, is the "repetitiveness" of activities thought to be related with ODs. Repetitiveness expressed as "frequency of activity" presumes that the activity occurs at regular intervals, so that the occurrence may be described correctly by an "average number of exertions per time unit." Yet, this is not the case for activities that bunch together in some time periods, but occur seldom during the rest of the working time. It is unknown how an uneven distribution of activities over working time may be related to the occurrence of ODs.

Shop activities, usually with much muscle-strength exertion, have been mostly linked with ODs. In the office, operation of keys (such as on typewriters, word processing units, adding machines, telephone key sets) requires, per key activation, rather small energies to be exerted by the fingers, but the number of such activations per hour is often high, in some cases up to 20 thousand. This brings up the unresearched problem of interactions between energy, displacement, and repetitiveness of actions with respect to ODs. It is very likely that those interrelations are rather complex, and that they include factors beyond (static) force, frequency and body posture (Moore, Wells, and Ranney, 1991). Marras and Schoenmarklin (1991) have shown that certain wrist velocities and accelerations appear closely related to carpal tunnel syndrome. Armstrong, Werner, Warring, and Foulke reported in 1991 an experimental procedure to measure the pressure within the carpal tunnel.

NEED

What exactly are the interactions among activity components (such as force, displacement, acceleration, direction, repetition, posture) and their actual effects on body tissues? For example: Why does rapid and often repeated depressing of the fingertip (to move a key) cause a strain situation in the carpal tunnel? Is it the rubbing of tendons against their sheaths in terms of the magnitude of frictional force, the pressure exchanged between the surfaces, the heat energy generated, the tension within the tendon? How do these relate to the posture of the wrist? How is pressure generated within the carpal tunnel?

Clearly, much research must be conducted to identify the components of activities that may lead to ODs, and to understand why and how these physical events, singly or combined, overload body structures and tissues. When these relationships are understood, it should be possible to establish thresholds or doses for factors such as posture, force, displacement, their rates of occurrence and duration, etc., which separate suitable from harmful conditions. Until then, generic recommendations, as in this text, must do.

SUMMARY

The human body is "the measure" for workstations, equipment, and hand tools—see Table 8-9. One may assume different postures at work: sitting is generally preferred, but standing (and walking) allows one to cover more space and exert larger hand forces, while one can apply forces with the feet better while seated.

TABLE 8-9. FOLK NORMS OF MEASUREMENT

Inch	Breadth of thumb; length of distal phalanx of little finger
Phalanx	Length of distal phalanx of thumb; length of middle phalanx of middle finger (2 inches)
Hand	Width of palm across knuckles, length of index finger (2 phalanges, 4 inches)
Span (of hand)	Distance between tips of spread thumb and index finger (2 hands, 4 phalanges, 8 inches)
Foot	Length of foot (3 hands, 6 phalanges, 12 inches)
Ell	Length from elbow to tip of extended middle finger (3 spans, 6 hands, 24 inches)
Step	Distance covered by one step (4 spans, 16 phalanges)
Fathom	Distance between tips of fingers of the hands with arms extended laterally ("span akimbo") (3 steps)

SOURCE: Reprinted with permission from "Folk Norms and Biomechanics" by Rudolph J. Drillis, *Human Factors* Vol. 5, No. 5, 1963.

The hands are the body's primary "working ends," which operate with finest control directly in front of the body, at belly to chest height. Hand tools are often needed when finger manipulation alone is insufficient, but they should be so designed that they are helpful for the intended purpose, not stressful or even potentially damaging for the body. Foot operation of controls is more forceful, but frequent foot operation should be required only if the operator sits.

Overuse disorders are related to "unnatural" body postures, especially occurring in wrists, arms, shoulders, and neck; to often repeated activities, particularly if these require extensive body force and energy; and to vibrations, impacts, and cold. Although in many cases exact injury mechanisms are not yet well understood, rules and recommendations for design of proper equipment and its use are at hand.

CHALLENGES

Occasionally, one hears the argument that sitting may be more conducive to falling asleep on-the-job than standing.

Which are the structures that keep the spinal column in balance?

Consider the problems associated with drawing a person's attention to perceived working conditions in the course of an interview.

Consider the effects of trunk muscle use in sitting without a backrest.

How could the attributes of "sitting comfort" change with sitting time?

Can one sit too comfortably?

What changes would be needed in conventional lathe design to allow the operator to sit?

What makes work in tight, confined spaces so difficult?

Why is it difficult to exert large forces with the foot when one is standing?

Consider alternative design solutions for the foot controls in automobiles.

Consider alternative design solutions for the hand-control functions of flight direction and engine speed in an airplane.

What are the disadvantages of contouring hand tools to fit the hand closely?

Consider tests to assess the "usability" of gloves.

Which might be means to measure pressure distribution of surfaces that form-fit the human body, such as a hand tool, or a seat surface?

Often force is applied to an object, not in a continuous way, but in steps, such as in the breakaway force to set an object into motion, and then the force to keep it in motion. How could such force exertions be measured?

Consider the statement that "overexertion injuries should not occur in normal use of the body."

Is "social awareness" for problems at work detrimental or advantageous to productivity?

Which biomechanical models could be applied to explain cumulative trauma injuries to the hand-arm system as a result of manipulations?

Describe specialized job analyses to check for potential overuse disorders.

Under what conditions might it be admissible to select persons to do difficult jobs instead of fitting the job better to human capabilities?

Discuss the possible interactions of forcefulness, repetitiveness, and posture in the causation of cumulative trauma.

9

The Office (Computer) Workstation

OVERVIEW

The modern office, often equipped with computers, has little resemblance to the rooms in which male clerks made entries in ledgers or pinned letters a century ago. Yet, the idea of "sitting upright is healthy" had not changed until it became obvious that people in modern offices sit "any way they like"—apparently without bad health consequences.

Different working postures, use of computers in well-lit offices, and changed attitudes give reasons to rethink the recommendations for office furniture design and to facilitate appropriate motor, vision, and behavior interfacing.

INTRODUCTION

At the beginning of this century (when the clerks were male) it was common to stand while working in the office. Since then the concept has changed: it is now accepted that one should sit in the office. Yet, musculo-skeletal pain and discomfort (together with eye strain) constitute at least half, in some surveys up to 80 percent, of all subjective complaints and objective symptoms of computer operators, in both Europe and North America (Helander, 1988)—but it is no longer known how the standing clerks felt a century ago.

EXAMPLE

Grieco (1986) believes that HOMO ERECTUS is biomechanically designed for moving around, but neither for standing still nor for sitting still, and that the human spinal column, suitable for a body in four-legged motion, has not had enough time to adapt to the upright position used in bipedal locomotion. He also fears that the current rapid transformation to HOMO SEDENS will lead to serious maladaptation of the spine.

THEORIES OF "HEALTHY" SITTING

In the nineteenth century, several physiologists/orthopaedists (particularly Staffel, 1884) published theories about "proper" sitting postures and from them deduced recommendations for seat and furniture design. (For reviews, see for example Akerblom, 1948; Bradford and Prete, 1978; Grandjean, 1969; Grieco, 1986; Keegan, 1952; Kember, 1985; Kroemer and Robinette, 1968; Kroemer, 1988a, 1991; Mandal, 1982; Schoberth, 1962; and Zacharkow, 1988.) Their common concept was that "sitting with an upright trunk means sitting healthily." This idea had remained virtually unchallenged until recently, when various (occasionally even contradictory) theories about healthy postures while sitting at work have been published. Some of these have led to rather radical proposals for sitting and "seats" at work, such as "semisitting" on high forward-tilted supports with or without knee pads; saddle-type supports; and use of a "belly rest" instead of a backrest—see Figure 9-1.

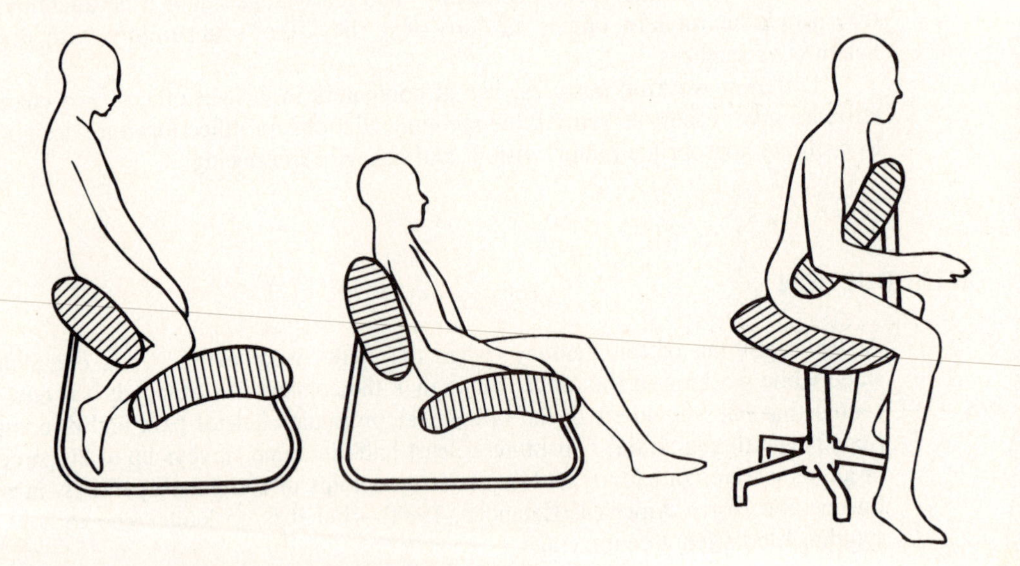

Figure 9-1. Modes of sitting.

Sitting "Upright"

The nineteenth-century concept was that the spinal column of the sitting person should be "erect" or "upright" (a contradiction in terms because it is actually curved, in the side view—see Chapter 1) similar to that of a "healthy normal upright standing" person. Special emphasis was placed on maintaining a "normal" lumbar lordosis. This posture was believed to put the least strain on the spinal column and its supportive structures, including the musculature. Also, this posture was considered "socially proper" (Grimsrud, 1990). Recommendations for the design and use of seats were based on this desired posture. These ideas about suitable posture and furniture were repeated in the literature, apparently without much questioning or contradiction, until the middle of the twentieth century. In retrospect, this is quite surprising, given the lack of experimental support for the appropriateness of the upright posture.

Standing or sitting "straight" means that, in the lateral view, the spinal column in fact forms an S-curve with forward bends (lordoses) in the neck and low back regions and a rearward bulge (kyphosis) in the chest region. While there appears to be no reason to doubt that this is—in the current evolutionary condition of the civilized human—a "normal" and hence desirable curvature of the spine, which keeps the trunk and its organs in acceptable order, it needs to be discussed how such a posture should be achieved, supported, or even enforced.

When one sits down on a hard flat surface, not using a backrest, the ischial tuberosities (inferior protuberances of the pelvic bones) act as fulcra around which the pelvic girdle rotates under the weight of the upper body. Since the bones of the pelvic girdle are linked by connective tissue to the thighs and the lower trunk, rotation of the pelvis affects the posture of the lower spinal column, particularly in the lumbar region (Keegan, 1952). If the pelvis rotation is rearward, the normal lordosis of the lumbar spine is flattened—see Figure 9-2. This was deemed highly undesir-

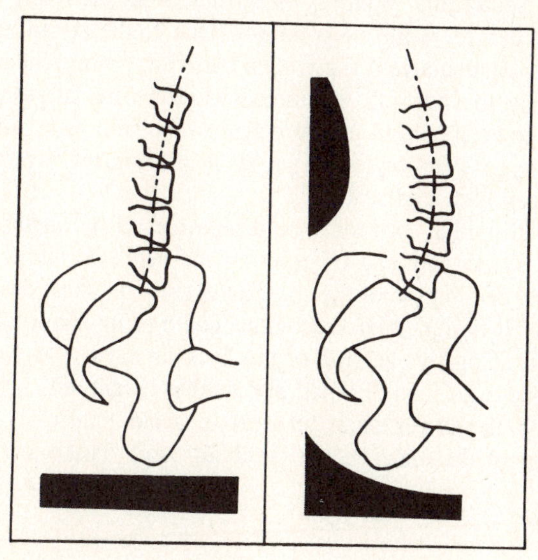

Figure 9-2. Positions of the pelvis and lumbar spine on a flat and on a declined seat surface (from Kroemer and Robinette, 1968).

able by orthopedists and physiologists. Hence, avoidance of backward rotation of the pelvis is a main purpose of many theories of seat design (Kroemer and Robinette, 1968).

Given the tissue connections between pelvis and thigh, particularly the effects of muscles spanning the hip joint or even both knee and hip joints (e.g. hamstrings, quadriceps, rectus femoris, sartorius, tensor fasciae latae, psoas major), the actual hip and knee angles also affect the location of the pelvis and hence the curvature of the lumbar spine. At a large hip angle, a forward rotation of the pelvis on the ischial tuberosities is likely, accompanied by lumbar lordosis. "Opening the hip angle" is an aim of some theories of seat design (Keegan, 1952). (These actions on the lumbar spine take place if associated muscles are relaxed; muscle activities or changes in trunk tilt can counter the effects.)

APPLICATION

In 1884, Staffel proposed a forward-declining seat surface to open up the hip angle and thus to bring about lordosis in the lumbar area. Eight decades later, this idea led to a seat pan design with an elevated rear edge which was popular in Europe in the 1960s. More recently, Mandal (1975, 1982) and Congleton et al. (1985) again promoted that the whole seat surface slope fore-downward. To prevent the buttocks from sliding down the forward-declined seat, the seat surface may be shaped to fit the human underside (Congleton), or one may counteract the downward-forward thrust either by bearing down on the feet (Mandal) or by propping knees or upper shins on special pads. One may call this posture "semisitting" or "semikneeling."

Anecdotal evidence and several studies (Bridger, Von Eisenhart-Rothe and Henneberg, 1989; Drury and Francher, 1985; Lander, Korbon, DeGood, and Rowlingson, 1987; Seidel and Windel, 1991) have indicated that some users like this support structure, while most do not. One of the advantages is the mobility of the trunk, particularly if there is no backrest; among the disadvantages are the tendency to slide off, counteracted either by fatiguing leg thrusts or by unpleasant or even painful pressure against the shin pads, and the difficulties often encountered when trying to get up in moving the legs out of the confined space between pads and seat.

Although the desired lumbar lordosis should be achieved by opening the hip angle to more than 90 degrees and by rotating the pelvis forward, most proponents of the forward declination of the seat pan deem a backrest desirable or necessary. In fact, a well-designed backrest alone could bring about lordosis of the lumbar spine by pushing this section of the back forward. Old wooden school benches simply had a horizontal wood slat lumbar height which forced the seated pupil to bend the lower back forward to avoid painful contact: hence, to the satisfaction of the teacher, the child sat "up". There are more subtle and agreeable ways to promote lumbar concavity: a fixed "Akerblom pad" (Akerblom, 1948) or inflatable lumbar cushions incorporated in the seat back of some car and airplane seats are examples.

Of course, one can shape the total backrest: Apparently independently from each other, experimental subjects of Ridder (1959) in the United States and of Grandjean and his co-workers (1963) in Switzerland found rather similar backrest shapes to be acceptable. In essence, these shapes follow the curvature of the rear side of the human body: at the bottom concave to accept the buttocks, above slightly convex to fill in the lumbar lordosis, then raising nearly straight but declined backward to support the thoracic area, but at top again convex to follow the neck lordosis. This shape (with more or less pronounced curvatures depending on the designer's assumptions about body size and body posture) has been used successfully for seats in automobiles, aircraft, passenger trains, and cars and for easy chairs. In the traditional office these "first-class" shapes were thoughtfully provided for managers, while other employees had to use simpler designs, such as the miserable small board attached to so-called secretarial chairs.

Experimental Studies

In addition to empirical studies, such as done by Ridder and Grandjean, many analytic experiments have been performed to measure physiologic responses of the human body to certain postures. Lundervold (1951) was apparently the first to record and interpret electromyograms (EMGs) of trunk muscles associated with seated positions. Based on these early studies, many EMG experiments with sitting persons have been performed (summarized by Chaffin and Andersson, 1984, 1991; Grieco, 1986; Winkel and Bendix, 1986a; Soderberg, Blanco, Consentino and Kurdelmeier, 1986—see also Chapter 1). EMGs show various activities of the five major trunk muscle pairs (erector spinae, latissimus dorsi, internal and external oblique, rectus abdominus) exerting pull forces along the length at the trunk and stabilizing the spinal column. Variations in postures, external loads, and backrests or other seat features influence these muscular activities (Nag, Chintharia, Saiyed, and Nag, 1986).

However, tension of muscles in the trunk does not seem to be very important for maintained "regular" seated postures, because the observed EMG activities indicate rather low demands on muscular capabilities (Andersson, Schultz and Ortengren, 1986), typically well below one-tenth of the maximal contraction capabilities. (Yet, muscular strains in the low back area increase substantially in unusual conditions and postures, such as leaning over the desk at which one is seated to lift an object such as a computer monitor with extended arms—see below.) The interpretation of these weak EMG signals is controversial, since one cannot necessarily assume that little muscle use (i.e., flat EMG signals) should be preferable over more extensive use. "Dynamic sitting" is desired by some physiologists and biomechanicists to obtain suitable muscle tone and training as well as for electrolyte and fluid balance (Grieco, 1986; Kilbom, 1986); for intervertebral disk metabolism (Hansson, 1986); to improve the circulation of blood (Winkel and Bendix, 1986b), preventing blood pooling (Thompson, Yates and Franzen, 1986). These considerations would encourage occasional or repeated bursts of muscular activities while sitting. Maintenance of the same posture over long periods via continued muscle tension (even at the low levels just mentioned) becomes uncomfortable and should be avoided by introducing "rest" periods or physical activities and exercises.

Muscular tensions observed in the shoulder/neck region appear, in contrast to those observed in the lower trunk, to be more critical. Tension and pain in the neck area are among the most-mentioned health complaints of computer operators (Sauter, Schleifer, and Knutson, 1991), apparently much more related to vertical tilt ("nodding") in the mid-sagittal plane than to sideways (horizontal) motions (Collins, Brown, Bowman and Carkeet, 1990). The relative intensity of muscle tension often is considerably higher than the 10 or less percent reported for lower trunk muscles, and often must be maintained over long periods of time when the head needs to be kept in a fixed relation to the visual object (Ekholm, Schuldt, and Harms-Ringdhal, 1986; Harms-Ringdhal and Ekholm, 1986; Hansson and Attebrant, 1986; Zwahlen, Hartmann and Kothari, 1986). Intensity, frequency, and length of time of such muscle contractions can generate intense discomfort, pain, and related musculo-skeletal health complaints, which may persist over long periods.

Other analytical studies addressed the pressure in the intervertebral disks dependent on trunk posture. Disk compression and trunk muscle activities are related: the stability of the stacked vertebrae is achieved by contraction of muscles (primarily latissimi dorsi, erectores spinae, obliques, recti abdominus) which generate vertical force vectors in the trunk. Together, these pull "down on the spine" and keep the vertebrae aligned on top of each other. (Each vertebra rests upon its lower one, cushioned by the spinal disk between their main bodies, and also supported lateral-posteriorly in the two facet joints of the articulation processes—see Chapters 1 and 2.) Since the downward pull of the muscles generates disk and facet joint strain (in response to upper body weight and to external forces), one should expect close relationships between trunk muscle activities and disk pressures. In the 1960s, experiments were performed in Scandinavia where pressure transducers were placed into spinal disks. These experiments showed that the amount of intradisk force in the lumbar region was dependent upon trunk posture and body support, as illustrated in Figures 9-3 and 9-4. (For a compilation and review see Chaffin and Andersson, 1984.)

When standing at ease, the forces in the lumbar spine were in the neighborhood of 330 newtons, as Figure 9-3 shows. This force increased by about 100 N when sitting on a stool without a backrest. It made little difference if one sat erect with the arms hanging, or relaxed with the lower arms on the thighs. (Sitting relaxed but letting the arms hang down increased the internal force to nearly 500 N.)

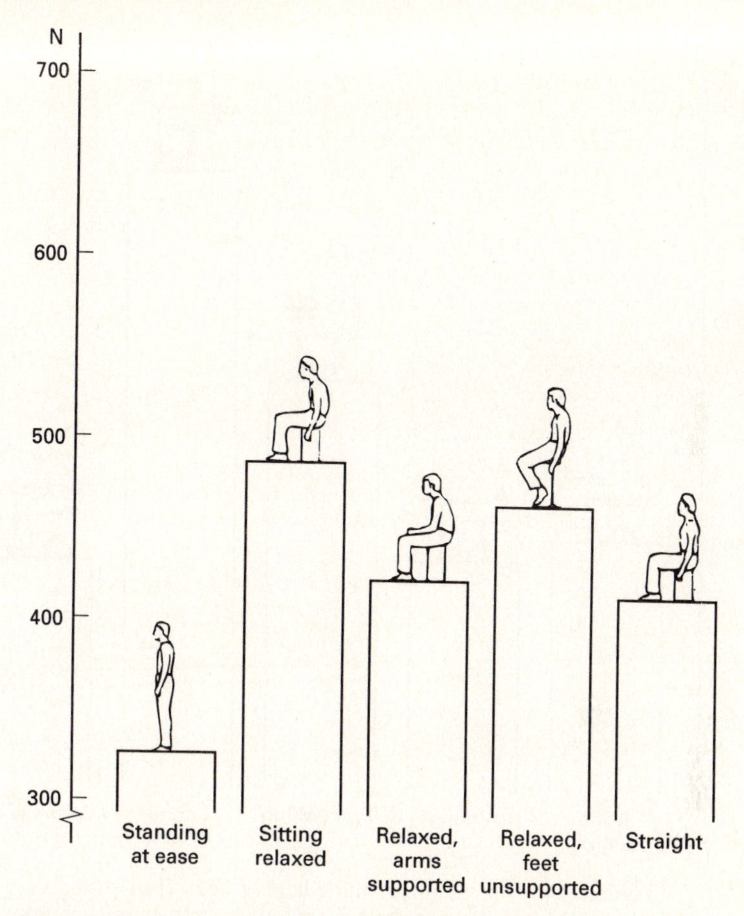

Figure 9-3. Forces in the third lumbar disk when standing, or when sitting on a stool without backrest (with permission from Chaffin and Andersson, 1984).

Thus, there was an increase in spinal compression force in the lumbar region when sitting down from standing; but the differences among sitting postures were not very pronounced. About the same force values were measured in persons sitting on an office chair with a small lumbar support. Sitting with the arms hanging, writing with the arms supported on the table, and activating a pedal led to forces around 500 N. The spinal forces were increased by typing, when the forearms and hands were lifted to keyboard height. (A further increase was seen when a weight was lifted in the hands with forward-extended arms.) None of these postures made use of the backrest. However, if one leaned back decidedly over the small backrest, and let the arms hang down, the internal compression forces reduced to approximately 400 N.

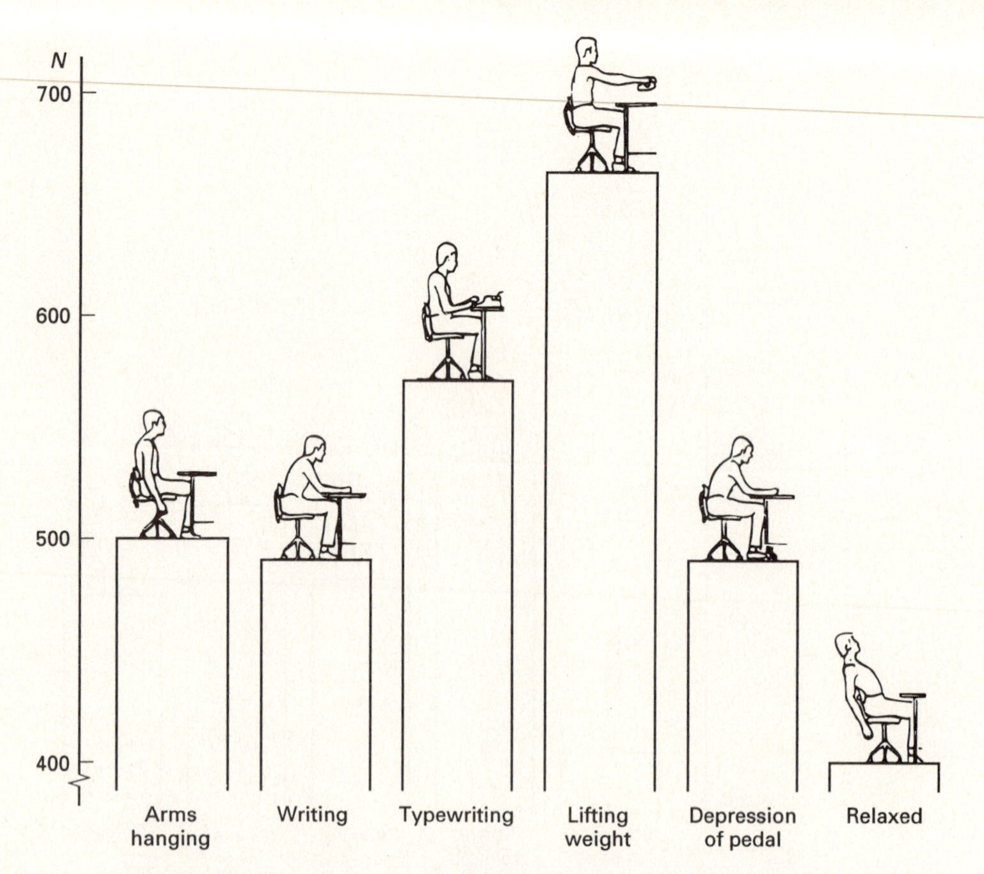

Figure 9-4. Forces in the third lumbar disk when sitting on an office chair with a small lumbar backrest at a desk (with permission from Chaffin and Andersson, 1984).

Figure 9-5 shows the effect of backrest use even more dramatically. When the backrest is upright, it cannot support the body, and disk forces between about 350 and 660 N may occur. Declining the straight backrest behind vertical brings about decreases in internal force, because part of the upper body weight is now transmitted to the backrest and hence does not rest on the spinal column. An even more pronounced effect can be brought about by making the backrest protrude toward the lumbar lordosis. A protrusion of 5 centimeters nearly halves the internal disk forces from the values associated with the flat backrest; protrusions of 4 to 1 cm in the lumbar region bring about proportionally smaller effects.

In a series of more recent studies (summarized by Andersson, Schultz and Ortengren, 1986), disk pressures were measured and calculated for various desk tasks and for sitting and standing postures. It was concluded that "In a well-designed chair the disk pressure is lower than when standing" (p. 1113).

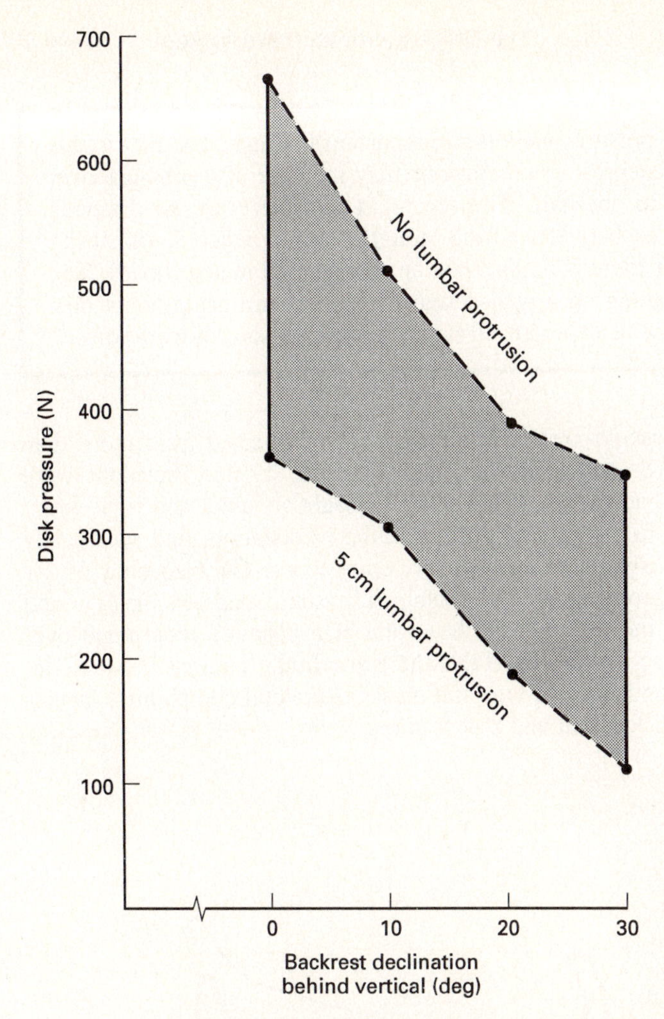

Figure 9-5. Disk forces (L3/4) depending on backrest angle and lumbar pad size. SOURCE: Adapted from *Occupational Biomechanics* by Chaffin, D.B. and Andersson, G.J.B. Copyright © 1984 by John Wiley & Sons, Inc. Reprinted by permission of John Wiley & Sons, Inc.

If the backrest consists only of a small lumbar board, a dramatic beneficial effect requires that one nearly "drapes oneself on it" by leaning backward over it, as depicted in Figure 9-4. Even a large backrest that could support the total trunk is nearly useless when put upright, but highly beneficial when declined backward behind vertical (Corlett and Eklund, 1984). Its positive effects are enhanced if it is shaped to bring about the S-curve of the spinal column (Branton, 1984), particularly lumbar lordosis.

Relaxed leaning against a declined backrest is the least stressful sitting posture. This is a condition that is often freely chosen by persons working in the office if there is a suitable backrest available: ". . . an impression which many observers have already perceived when visiting offices or workshops with VDT workstations: Most of the operators do not maintain an upright trunk posture. . . . In fact, the great majority of the operators lean backwards even if the chairs are not suitable for such a posture" (Grandjean, Huenting, and Nishiyama, 1984, 100–101).

EXAMPLE

Experiments on sitting postures yield three important findings. The first is that sitting without use of backrest or of armrests may increase disk pressure over standing by one-third to one-half. The second is that there are no dramatic disk-pressure differences between sitting straight, sitting relaxed, or sitting with supported arms, if there is no backrest or only a small lumbar board. The third finding is that leaning on a well-designed backrest can bring about disk pressures that are as low as experienced by a standing person, or even lower.

Statements about musculo-skeletal discomfort were collected from more than 900 computer operators (Sauter, Schleifer, and Knutson, 1991). Prevalent were problems in the neck and back areas, followed by the right shoulder and right wrist. This is apparently related to the general posture at the workstation, and specifically to the preferred use of the right hand for keyboard entry work. The frequency of low back, neck, and shoulder problems may be explained by static loads on shoulder and neck muscles induced by the demands on head and arm postures, continued over long periods. The study also demonstrated that there are direct relations between design features of the work and workstations and musculo-skeletal complaints, including the effects of keyboard location and seat features.

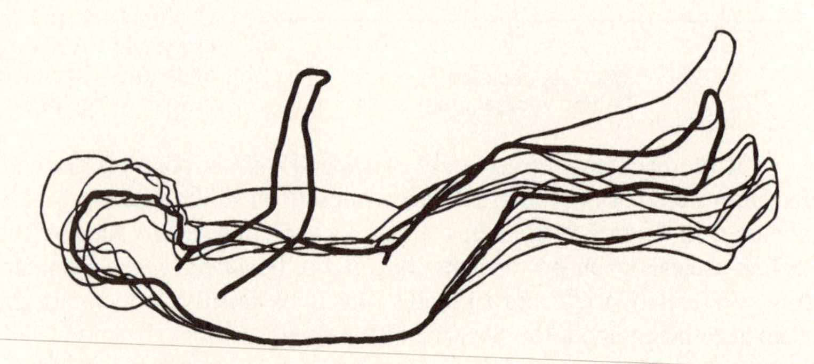

Figure 9-6. Relaxed underwater body postures (with permission from Lehmann, 1962), superimposed with the relaxed posture in weightlessness (from NASA/Webb, 1978).

"FREE POSTURING"

Allowing persons freely to select their posture has led in two instances to surprisingly similar results. In 1962, Lehmann showed the contours of five persons "relaxing" under water. Sixteen years later, relaxed body postures were observed in astronauts and reported by NASA (1978)—see Figure 9-6. The similarity between the postures under water and in space is remarkable. One might assume that, in both cases, the sum of all tissue torques around body joints has been nulled. Incidentally or not, the shape of "easy chairs" is quite similar to the contours shown in both figures. Some "executive" computer workstations, sketched in Figure 9-7, use similar support shapes, but it is doubtful that such postures and furniture will be used widely in regular offices.

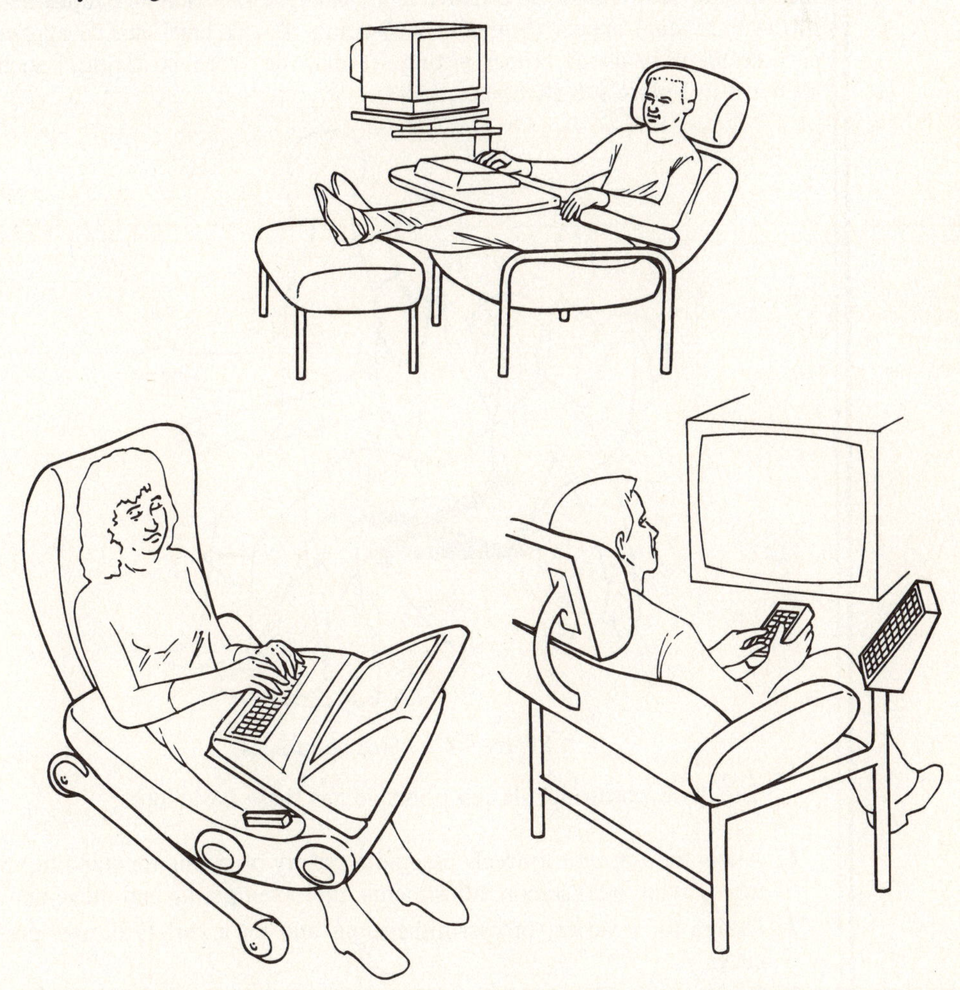

Figure 9-7. Body postures in "executive" computer workstations.

APPLICATION

Currently, no "one simple" theory about the proper, healthy, comfortable, efficient, etc., sitting posture at work prevails (Bendix, 1991; Kroemer, 1991; Lueder, 1983; Verbeek, 1991). With the idea abolished that everybody should sit upright, and that furniture should be designed to this end, the general tenet is that many postures may be comfortable (healthy, suitable, efficient, etc.), depending on one's body, preferences, and work activities (see Chapter 8). Consequently, it is now generally presumed that furniture should allow many posture variations and permit easy adjustments in its main features, such as seat height and angle, backrest position, or knee pads and footrests; and that the computer workstation should also allow easy variations in the location of the input devices and of the display. Thus, change, variation and adjustment to fit the individual appear central to well-being. If any label can be applied to current theories about proper sitting, it may be "free posturing," such as sketched in Figure 9-8 (Kroemer, 1988).

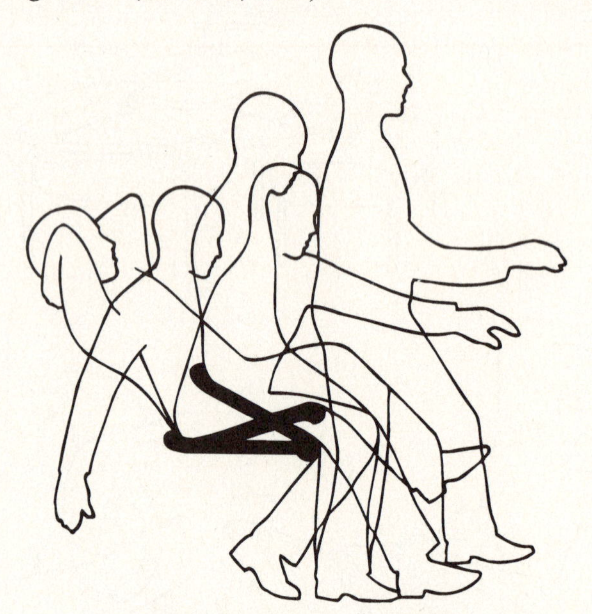

Figure 9-8. "Free posturing."

The "free-posturing" design principle has these basic ideas:

1. Allow the operator to freely assume a variety of sitting (or standing) postures, make workstation adjustments, and even get up and move about.
2. Design for a variety of user dimensions and for a variety of user preferences.
3. New technologies develop quickly and should be usable at the workstation. (For example, radically new keyboards and input devices, including

voice recognition, may be available soon; display technologies and display placement are undergoing rapid changes.)

EXAMPLE

"Egyptian tomb reliefs illustrate clearly that, even at that time, dignified man had to sit with back straight, thighs horizontal, and lower legs vertical. How was it, in spite of all the evidence against it being either comfortable or natural, that this stilted posture came to be accepted as standard, whether for sitting on a throne, for dining, for contemplation in the privacy of the boudoir, or for working in an office?" (In the editorial introduction of the March 1986 issue of *Ergonomics*.)

ERGONOMIC DESIGN OF THE OFFICE WORKSTATION

Successful ergonomic design of the workstation in the office depends on proper consideration of several interrelated aspects, sketched in Figure 9-9. The work task, the work posture, and the work activities all interact and are influenced by the workstation conditions, including furniture and other equipment, and by the environment. All of these must "fit" the person. Work postures determine to a large extent the person's well-being (as do attitude and social relations—Carayon-Saintfort, Smith and Lim, 1991; Smith, Carayon-Saintfort and Yang, 1991), and of course the work activities determine the work output. Naturally,

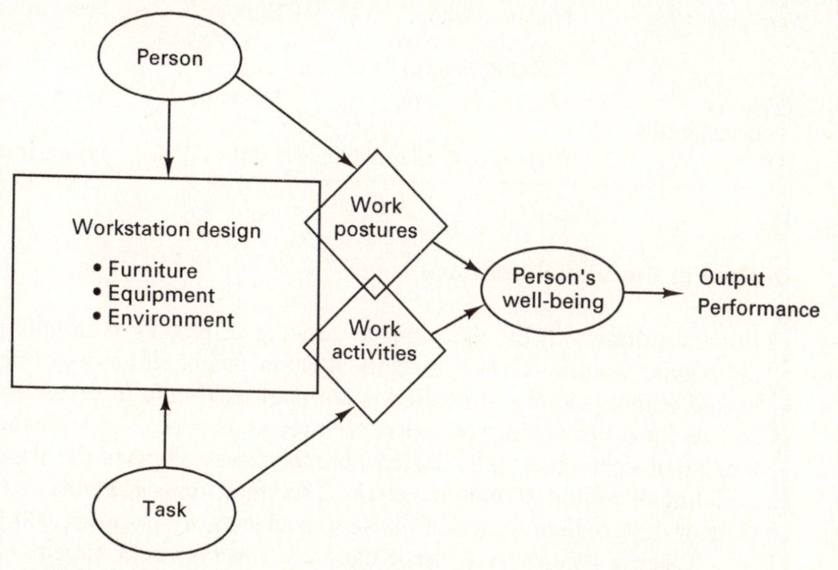

Figure 9-9. Interactions among person, task, workstation design, and performance.

the person's well-being also influences performance, and many people "feel good" if they are satisfied with their performance. Thus, there are many interactions in the relations among person, task, conditions, and results of the effort.

There are several "links" between a person and the job. The first is the "visual interface": one must look at the written material, the keyboard, and the computer screen, or the printed output. The second is "manipulation": the hands operate keys, a mouse or other input devices; manipulate pen, paper, telephone. A third link may exist between the feet and controls operated with them, such as when starting and stopping a dictation machine. The intensities of the visual and motor requirements depend on the specific job—see Table 9-1. Finally, the body and the seat (if one is used) are linked at the undersides of the thighs and buttocks, and at the back when a backrest is used. These four interface areas all have major impacts on the proper design of the office workstation, and its components.

TABLE 9-1. LINKING THE HUMAN WITH THE COMPUTER IN DIFFERENT WORK TASKS

Task	Visual requirements	Motor input requirements	Continuity of requirements
Data entry	High (source and screen)	High (keyboard)	Few interruptions
Data acquisition	High (screen)	Medium (keyboard)	Varies
Word processing	High to medium (source, screen and keyboard)	High (keyboard)	Few interruptions
Interactive communication	Medium (screen)	Medium	Varies
CAD	High (screen and source)	Low	Frequent interruptions

Designing the Visual Interface

In conventional offices, the paper for writing or reading is usually placed on the regular working surface, roughly at elbow height. If an object needs to be looked at more closely, it is lifted to a proper relation to the eyes. An inclined surface for easier reading often has been recommended, making its angle with the line of sight closer to 90 degrees, but the disadvantage of the sloped surface is sliding or rolling of work materials. Therefore, one usually has a horizontal table or desk surface on which the various objects are placed at will.

When a *typewriter* is used, there are several visual targets. One is the "printing" area, the platen on a conventional typewriter, the display area on electronic typewriters. This is usually only a few centimeters above the keyboard, and located fairly well with respect to human vision capabilities (distance) and preferences (line-of-sight angle), although its display surface (subtended angle) is very small. A problem has been, in many cases, the placement of a *source document* from which text is copied. To place it horizontally on a horizontal surface is quite often uncomfortable and makes exact reading difficult. Therefore, various types of document holders have been used that put the source document more nearly vertical and closer to the eyes. Still, often this document holder is placed far to one side, requiring a twisted body posture and lateral head and eye movements—see Figure 9-10.

Figure 9-10. "The computer user syndrome" (with permission from Grant, 1990).

Proper placement of all visual targets together is often a problem. First, with the large increase in the number of keys from the typewriter to most *computer keyboards*, many operators have reverted to scanning the keys, while previously typists were able to do their job without looking at the keys. Thus, the computer keyboard is a rather large, nearly horizontal visual target area, usually placed on the work surface. Second, there is the *display* area of the computer monitor (the "screen"), commonly placed in front of the operator at about right angles to the line of sight. Third, there is often some sort of *source document* from which information must be gleaned. Its placement causes problems like those encountered by typists, but in some cases that source document is fairly large, such as a drawing used in computer-aided design.

The placement problem is mostly one of available space within the center of the person's *field of view*. Research has shown that people prefer to look downward to close visual targets at angles between 20 and 60 degrees below the Ear-Eye plane, as discussed in Chapter 4; this is between 0 and 40 degrees below the horizon in a conventional sitting posture (Hill and Kroemer, 1986). They do so by inclining the head forward and by rotating the eyeballs downward, instead of looking straight ahead. As discussed in Chapter 4, this appears to be a "natural way" of focusing at a near target with least effort; the more upward one has to tilt one's eyes, the more difficult it becomes to focus. Thus, putting the monitor up high behind the keyboard (often by placing it on the CPU box of the computer or on a so-called monitor stand) is rather uncomfortable for the viewer. Instead of building a "monitor tower," the screen should be located immediately behind the keyboard, with its lower edge as near as possible to the rear section of the keyboard. A printer may be placed directly on top of the monitor, and a source document be placed next to it—see Figure 9-11. More refined solutions may incorporate all three visual areas in one setup: one such proposed solution is shown in Figure 9-12.

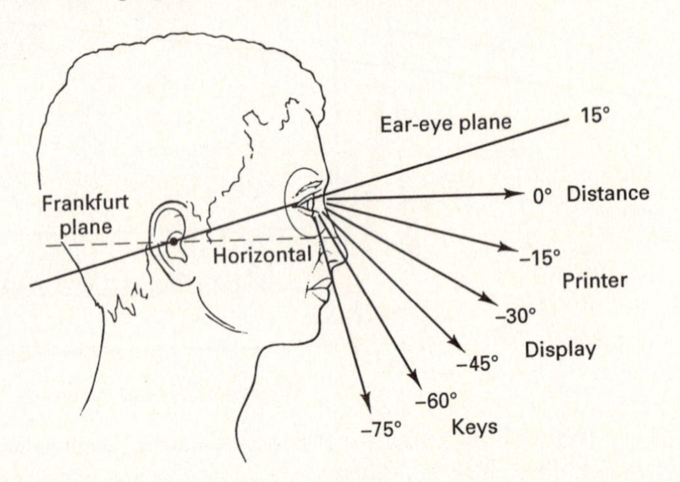

Figure 9-11. Suggested angles of the line of sight to observe keyboard, screen, and printer with current computer workplace technology. (Note that this is a compromise solution for imperfect components.)

The selection of a good display is discussed in Chapter 11 on "controls and displays" in this book.

Proper office illumination. For the engineer, the most important design factors are: illumination, luminance, and luminous contrast; and how they are distributed—see Chapter 4.

Illumination is the amount of lighting falling on a surface. The light may come from the sun or from luminaires (lamps).

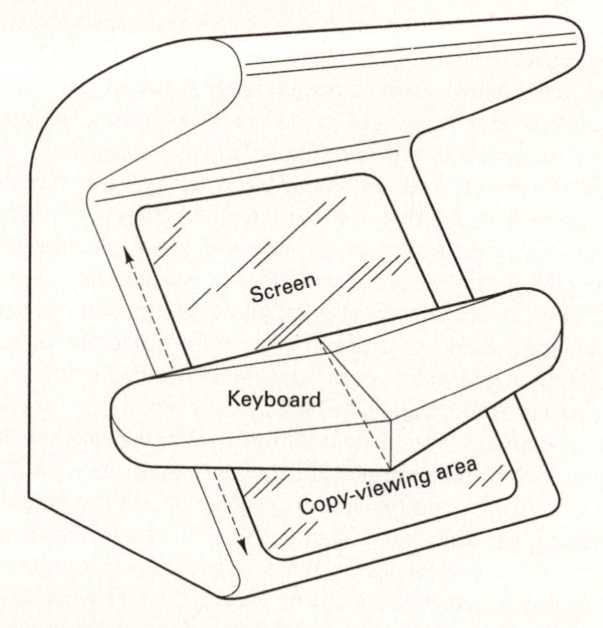

Figure 9-12. Proposed combination of screen and source display and keyboard (with permission from Grant, 1990).

Luminance is the amount of light reflected or emitted from a surface. Light may be reflected from ceiling, wall, or tabletop, or it may be emitted from a VDT screen surface.

Luminous contrast ratio describes the difference between the luminance values of two adjacent areas, assuming a defined boundary between them.

Of the three, *illumination* is the best known and the least useful phenomenon. The sun, the sky, or a lamp send visible energy (light) into the surroundings. Yet, what is emitted is of little direct consequence, because what "meets the eyes" is luminance (unless one stares directly at the light source): examples are looking at a piece of paper, seeing the surroundings about us, or viewing a computer screen.

Note that for most artificial light sources, particularly for candescent ones, usually the energy consumption (e.g., in watt) is given. However, what counts (with respect to illumination) is the energy emitted from the light source. For example, candescent light bulbs transform only about half of the energy consumed into light energy, while the remainder is converted into heat. (This explains why light bulbs can heat up a room considerably.) With age, most luminaires become even less efficient, lightwise.

The quantity that counts for human vision is the *luminance* experienced on the retina. This results from light sent (emitted or reflected) from the visual target and from other visual surroundings in our field of view. In rooms, most light energy reaching the eyes is reflected from surfaces, mostly walls and ceilings (with "glaring" reflections from "shiny" objects). To avoid specular ("directed") glare, surfaces can be made matte by giving them a rough surface that reflects incoming light in various directions. (The exceptions are mirrors and other polished surfaces that reflect incoming light at the same angle at which it was received. Mirrors have a background coating that absorbs little light but reflects nearly all of the incoming energy.) To describe the (reflected) luminous level, one uses the "reflectance" value (generally in percent) that indicates the portion of incident light that is reflected.

Contrast is the property that allows the human eye to distinguish between two adjacent areas, or objects. The larger the luminous contrast between two (reflecting or emitting) areas, and the better defined their common boundary, the easier it is to distinguish them.

The *distribution* of light within a room depends on the amount of light in general, on the location of light sources (luminaires) and the direction of light flow from them, and on the reflectances of ceiling, walls, and other surfaces. In general, the reflectances and colors of the room surfaces should be chosen so that there is a continuous decrease in reflectance from the ceiling to the floor. The ceiling reflectance should be about 80 to 90 percent, for instance accomplished by white paint. The walls should have reflectances of 40 to 60 percent, which corresponds to bright beige, yellow or green. The floors should have a reflectance of approximately 20 to 40 percent, such as by medium blue-green or brown-beige colors. Furniture and equipment surfaces should reflect at 25 to 45 percent.

One speaks of "indirect (or reflected) glare" at the computer workstation if a light source is reflected from the screen surface into the eyes, in the same way as a window or a lamp or a white shirt is reflected in a mirror. The reflected bright spot or surface on the screen hinders the attempt of the eyes to see the displayed "characters," e.g., letters, numbers, or lines. The contrast between the character and its background is reduced by the reflection, which makes it difficult to discern characters. This can lead to "eye fatigue" (a rather undefined but descriptive and popular term—National Research Council, 1983). A similar irritating condition is called "direct glare": it is caused by an intensive light source (the sun, light coming from a window, or a bright lamp) that shines directly into the eyes of the computer operator, thus generating high illumination of the retina, where rods and cones are trying to discern contrasts between characters and their background on the screen.

There are two other glare conditions, less well defined but common: one is "washout" or "veiling" of the contrast on the screen caused by high ambient illumination (this is a form of indirect glare), the other is "stray glare" caused by reflection from shiny surfaces in the field of vision (this is a form of direct glare).

Windows, so much preferred by many people, often generate difficult light-engineering conditions. On a bright day, they are sources of high intensity and strongly directed light, both able to generate direct or indirect glare.

To control the light coming from a window, several means can be employed:

- Coloring the window panes dark, to reduce the amount of illumination coming through.
- Use of louvers ("mini blinds"). Either horizontal or vertical louvers can be used, with vertical blinds best when the sun is low, and horizontal blinds when the sun is high. In proper adjustment, blinds can screen out the direct sun rays but still allow people to look out through the window. Also, blinds can be removed if not needed, for example when there is no sunshine. Light-colored louvers are also advantageous at night, when the darkness outside would turn the windows into dark surfaces reflecting on the inside like mirrors.
- Curtains can be used to hinder direct sunlight from entering the room. Light-colored curtains have little absorption of sun energy, and hence do not heat up and then send warmth into the room, but they are highly luminous and thus may generate a source of high light intensity. Therefore, dark-colored curtains are, with respect to light control, preferred.

Figure 9-13 shows three kinds of visibility problems that can be generated by improper illumination systems. A light source A positioned behind the operator can generate "specular" glare on the computer screen. A luminaire B positioned above the operator may "wash out" the contrast on the screen by creating a veiling luminance, and may cause reflections from the table and keyboard. A light source C positioned in front of the operator may cause "direct" glare by shining into the eyes. Figure 9-13 also shows the seated operator from behind, with three luminaires positioned either above the operator or to the side. Luminaire B, above the person, may cause reflections from a shiny workstation surface, while luminaires D and E, placed to the sides, do not cause any problems for this operator.

Another solution is to lower the position of luminaires and to direct some, most, even all of the emitted illumination upward. In this case, the workstation may be partly illuminated by direct light from the luminaire, and/or partly, mostly, or entirely by indirect light reflected from the ceiling. This has two advantages:

- The luminaire moved closer to the workstation generates defined areas of direct illumination.
- The illumination reflected from the ceiling does not create specular reflections.

One may illuminate only certain objects in the office with high intensity, such as the source document. This allows easier reading while keeping the overall illumination low. To do so, several arrangements can be used: the task

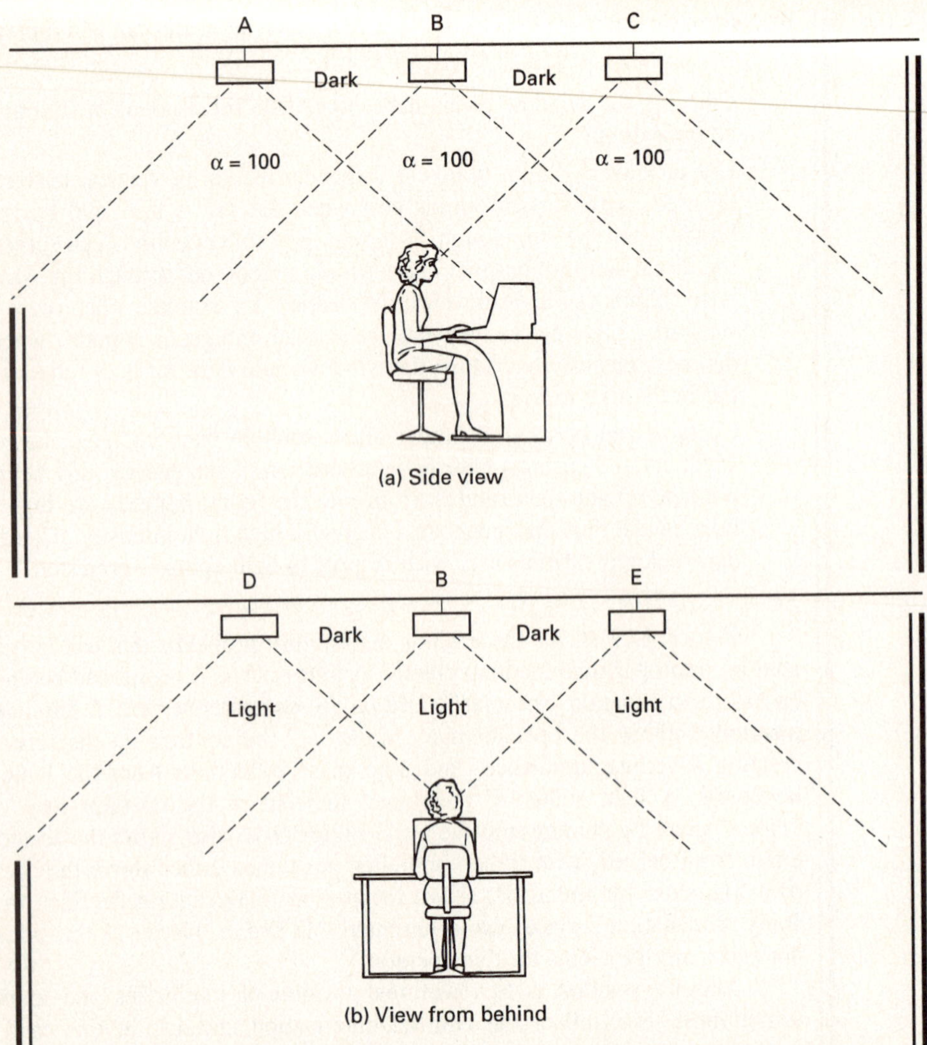

Figure 9-13. Luminaire A may generate specular reflections, B may wash out screen contrast, and C may cause direct glare. Luminaire B, positioned above the operator, may cause reflections from the workstation surface. Locations D and E are more suitable for the operator (adapted with permission from Helander, 1982).

luminaire may be mounted in the ceiling and shine a spotlight at the target; or, more commonly, a lamp is placed near the source document and directed at it. Care should be taken that the light source, or the lighted surface, does not generate direct or indirect glare for operator and neighbors.

Screen Filters and Treatments. The purpose of filters placed on the screen is to improve character-screen contrast and to reduce reflections. Ambient illumination passes through the filter before it reaches the screen surface, and then passes through the filter again on its way out. In each pass, light energy is lost. However, light from display characters passes through the filter only once, and therefore character luminance is less affected than screen-reflected ambient luminance, and hence the contrast increases. Yet, lowering the character luminance decreases visibility to some extent.

If *colored filters* are used, they should match the color of the screen characters. For example, a green filter should be used with green characters. This enhances luminous contrast but does not reduce the luminance of the characters.

Micro-louver or *micro-mesh filters* allow only light parallel to the openings to pass through, thus preventing much of the ambient illumination from getting to the screen. This improves contrast, owing to the reduction of veiling reflection. One disadvantage of micro-mesh filters is that they collect dirt, which reduces the amount of light that can pass through. Micro-louver filters are often embedded in plastic, which solves the dirt problem but may create specular reflections. Both filters have the disadvantage that the viewing angle of the operator toward the screen is restricted by the filter properties. *Antireflection coating* is used on certain filters, similar to the coating of camera lenses. Commonly, a quarter-wavelength coating is employed, which reduces specular reflections from the surface of the filter. The main disadvantages of coatings are that they can be degraded by fingerprints, lose effectiveness when aging, and are fairly expensive.

Specular reflections can also be reduced by making the screen surface matte, for example by etching. This eliminates the shiny mirrorlike surface and makes the image of a reflected object more diffused. Unfortunately, it also reduces the contrast between the characters and the background and makes the character edges less defined.

A *hood* is often used, which protrudes from the top and from the sides of the screen toward the operator. It can be effective in shielding from ambient illumination and hence reduces reflections on the screen. However, it is often difficult to position without casting a shadow on the screen from the edge of the hood. Also, a hood may make the operator feel as if "looking into a tunnel."

Some computer operators try to wear sunglasses at work. This does not increase contrast but instead reduces the luminous level of both the characters and the background alike. Visibility is, therefore, reduced.

Sometimes it is easy to position a *shield* or *screen* between luminaire and video display—see Figure 9-14. This is akin to using a small private office, illuminationwise, where there are no disturbing reflections from light sources other than those used at one's own will.

The condition of dark characters on a light background is often called "reversed video" in the United States, because most early displays used light characters on a dark background. The significant advantage of the light back-

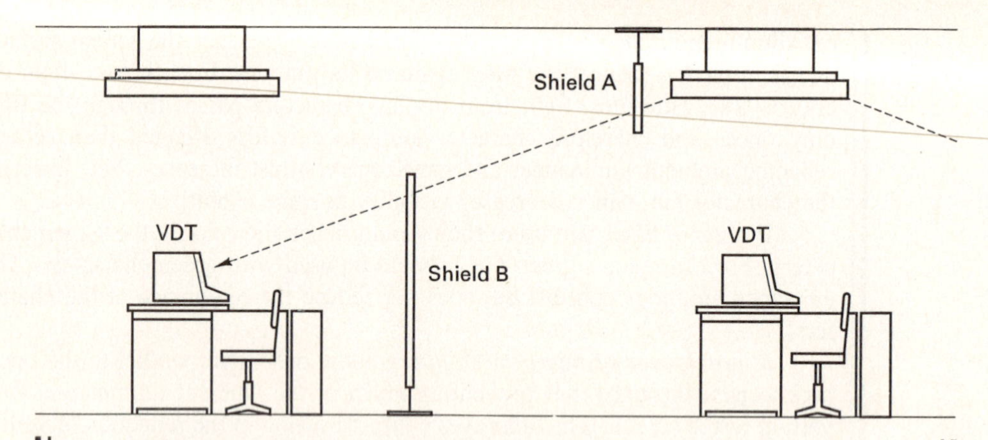

Figure 9-14. Shields used to keep light from directly shining onto a CDT (adapted with permission from Helander, 1982).

ground is that it acts less like a mirror with its disturbing reflections. The disadvantage is increased sensitivity to flicker and the need to make characters up to 40 percent wider than in light-on-dark displays.

In summary: it is best to use proper light sources, and to install them thoughtfully. This eliminates the sources of direct and indirect glare. The second-best solution is to intercept the flow of light from a source to the eyes of the operator or to the screen. Yet, such "interdiction of transmission" is usually more costly, and less effective. The third and least desirable solution is to apply means at the display surface to reduce screen reflections and to improve contrast, because most such treatments result in reduced visibility owing to loss of light energy. Table 9-2 summarizes the possible measures and their advantages and disadvantages.

Lenses to correct vision defects. Computer operators may experience vision difficulties and "eye strain": after a vision test they often find that corrective lenses are needed to compensate for eye deficiencies. But to use corrective lenses incorrectly can generate new problems. This is particularly the case with so-called reading glasses. These are ground for a viewing distance of about 40 centimeters and for a downward tilt of the line of sight of approximately 25 degrees (i.e., close to what Hill and Kroemer recommended in 1986), but many visual targets in the computer area are placed further away—particularly the screen, which is commonly behind the keyboard. If such visual target is beyond the focusing distance, one is tempted to "squint the eyes" while trying to focus, or to move the head forward to bring it closer to the correct focusing distance. The first attempt may lead to "eye fatigue," the second to improper neck posture and muscle tension. This is even more pronounced if one wears bifocals or trifocals, where the lowest section is meant for reading. In this case, one is likely to tilt the head backward in order to get the display on the screen on the "line of sight," which is predetermined by the glasses to be

TABLE 9-2. MEASURES FOR REDUCING SCREEN REFLECTIONS

Intervention	Advantage	Disadvantage
At the source:		
1. Covering windows		
(a) Dark film applied	Reduces veiling and specular reflections	Difficult to look out
(b) Louvers or mini-blinds	Exclude direct sunlight, reduce veiling and specular reflections	Must be readjusted in order to keep out light rays or to look out
(c) Curtains	Reduce veiling and specular reflections	Difficult to look out
(d) Cover windows permanently	Eliminates outside light	Not liked by most employees
2. Selecting and placing luminaries		
(a) Control of location and and direction of illumination	Reduces veiling reflections, may eliminate specular reflections	None
(b) Indirect lighting	Reduces specular reflections, allows larger number of workstations per square unit	None
(c) Task illumination	Increases visibility of source document	None (if properly arranged)
Between source and workstation:		
3. Shield or screen between luminairs or windows and workstation	Reduces reflections	Might create the impression of an "isolated" workplace
At the workstation:		
4. Readjust or relocate entire workstation	Eliminates all reflections	May need more space, disturb office design
5. Tilt screen	Eliminates specular reflection	May force operator to assume awkward posture
6. Use neutral density (gray) filter	Reduces veiling reflection, increases character contrast	Reduces character luminance
7. Use color filter (same color as characters)	Reduces veiling reflection, increases character contrast	Less character luminance
8. Use micro-mesh, microlouver filter	Reduces veiling reflection, increases character contrast	Operator must look directly onto the screen; embedded filter may get dirty
9. Use polaroid filter	Reduces veiling reflection, increases character contrast,	Decreases character luminance
10. Use quarter-wavelength reflection coating	Eliminates specular reflection	Expensive, difficult to maintain
11. Put matte finish on screen surface	Decreases specular reflection	Increases character edge spread (fuzziness), increases veiling reflection
12. Use screen hood	Reduces veiling and specular reflections	Difficult to avoid shadow on the screen; television feeling
At the operator:		
13. Wear tinted glasses	None	Reduces visibility

SOURCE: Adapted from Helander, 1982.

downward, with respect to the head. Ensuing severe tilting of the head often gives rise to muscular tension and severe headaches (Collins, Brown, Bowman, and Carkeet, 1990). When looking at the keyboard or a source document on the table, the head and neck are severely bent—see Figure 9-15. The solution is to wear "full-size" corrective lenses ground for the correct viewing distance, even if this blurs the impression of objects further away. Musicians, for example, habitually have their glasses ground to a distance of about one meter so that they can focus on their sheet music. Figure 9-16 shows that they still have a dilemma: they can see the notes clearly, but not the conductor.

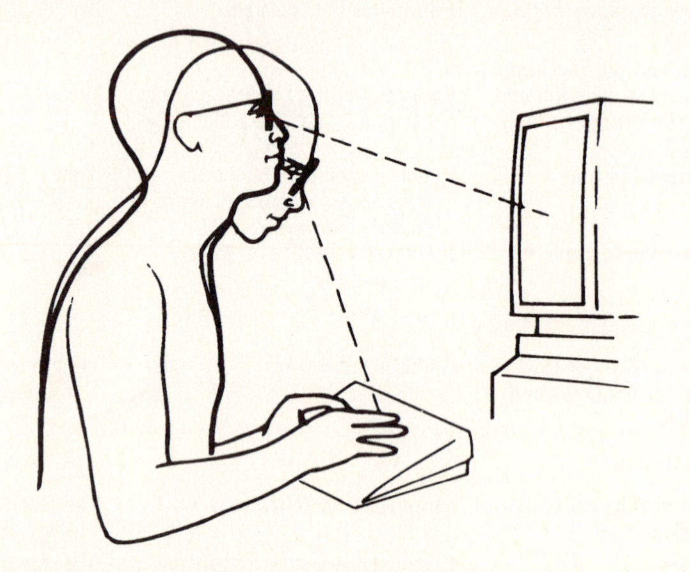

Figure 9-15. Excessive head and neck postures when looking through "reading glasses" at monitor or keyboard.

Designing the Motoric Interface

Unfortunately, the traditional typewriter keyboard is still used, largely unchanged, as the major input device for computers. The conventional QWERTY keyboard has several unergonomic features, such as that letters which frequently follow each other in common text (such as q and u in English) are spaced apart on the keyboard. This was originally done so that mechanical type bars might not entangle if struck in rapid sequence. Another characteristic is that the "columns" of keys run diagonally from left to right, which was also necessary on early typewriters due to the mechanical constraints of the type bars. Yet, the keys are arranged in straight sideways "rows," which the fingertips are not. The keyboard must be operated with pronated hands ("thumbs down"), owing to the horizontal arrangement of the rows of keys.

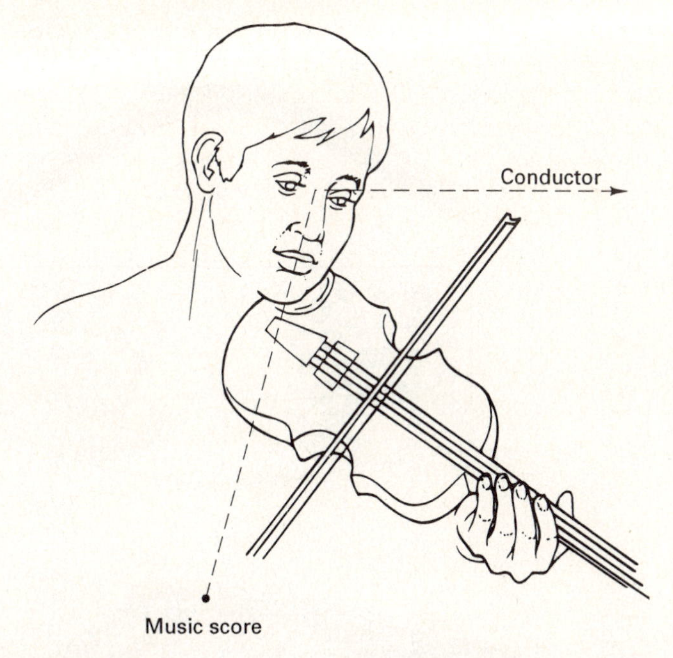

Figure 9-16. "The musician's malady" (with permission from Grant, 1990).

Furthermore, there is a large number of keys, of which one must be correctly selected so that the desired character be produced. This requires, cognitively, that a difficult multichoice decision be made, followed by motorically complex use of muscles to move the finger to the proper key.

"Mechanical" keyboards had strong key resistances and required large key displacements. Hence, it was suspected that weak fingers, particularly the little ones, were overworked. Thus, many recommendations for improvements of the traditional keyboard have been proposed in the past: for overviews, see for example, Klockenberg (1926); Kroemer (1972, 1990); Alden, Daniels, and Kanarick (1972); and Noyes (1983a). Suggested improvements include relocation of the letters on the keyboard and new geometries of the keyboard, such as curved arrangements of the keys. It was also suggested to divide the keyboard into one half for the left hand and one for the right, so arranged that the center sections are higher than the outsides, thus avoiding the pronation of the hand required on the flat keyboard; and to use two keys simultaneously (chording) to generate one character (Noyes, 1983b; Gopher and Raij, 1988; Keller, Becker, and Strasser, 1991). A new idea is to use chording in combination with keys that have three status conditions, i.e., ternary instead of the usually binary keys. The "ternary chord keyboard" requires only very few keys, such as one for each finger, to generate possibly thousands of different characters

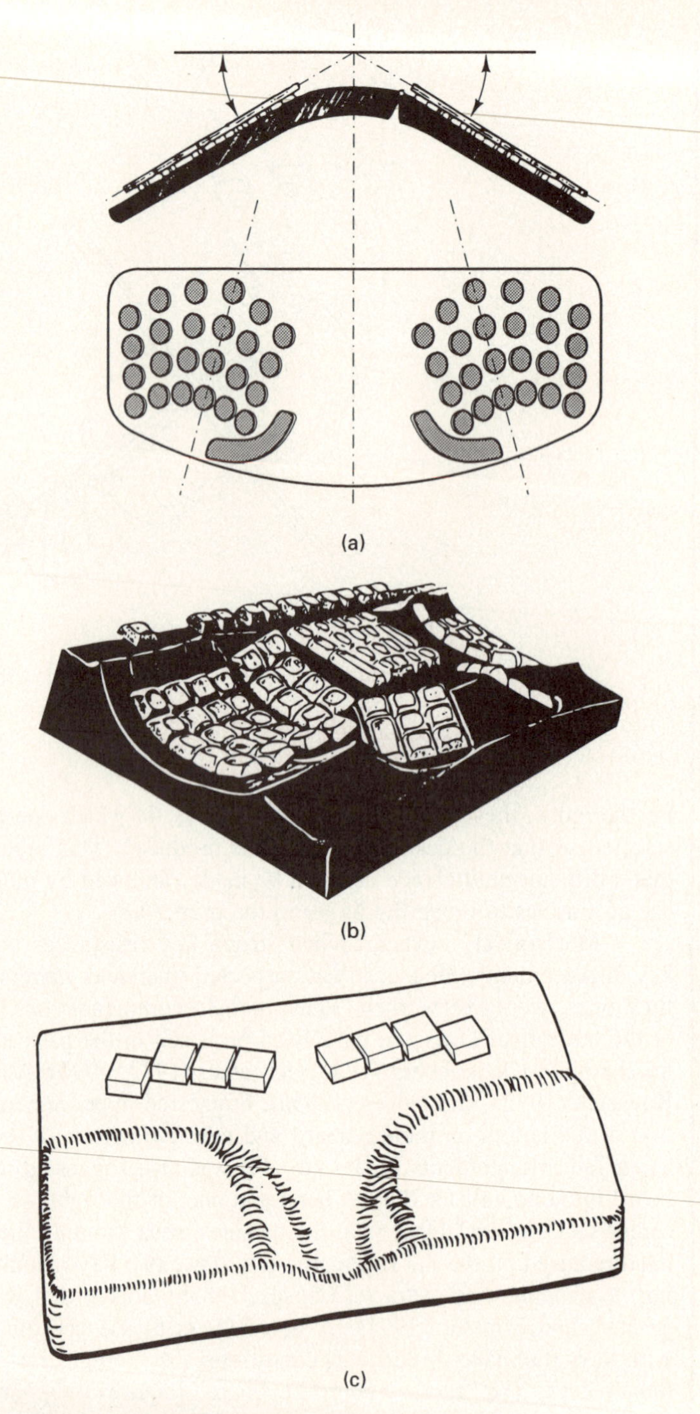

Figure 9-17. Examples of keyboards: (a) K-keyboard®, (b) Maltron®, and (c) the Ternary Chord Keyboard®.

(Kroemer, 1991). Examples of several keyboard developments are shown in Figure 9-17; further information on keyboards and other entry devices is presented in Chapter 11 on "controls."

Small keyboards can be placed nearly anywhere, at the user's convenience, such as on one's lap (which was quite difficult with the traditional large keyboard but was done nevertheless by some operators), or they may be incorporated in an armrest, in a glove, or even in the shell of a space suit. New developments may radically change the nature and appearance of keyboards, and hence allow new body and hand postures.

Foot-operated controls should be able to be placed at will within the reach space of the foot, or they should provide so much contact surface with the foot that it can be placed freely on it, thus allowing changes of posture.

The human is the most important component of the system, since she or he drives the output. Hence, the human must be accommodated first: the design of the workplace components should fit all operators, and allow many ideosyncratic variations in working posture. The myth of "one healthy upright posture, good for everybody, anytime" must be abolished.

Designing the Sit-Down Workstation

One of the first steps in designing office furniture for human use is to establish the main clearance and external dimensions; they derive from the anthropometric data of the office workers. There are, of course, many ways to do so; among them are "three alternative strategies" to determine major equipment dimensions (Kroemer, 1985). The first strategy assumes that all components are adjustable in height. The adjustability range for the seat height is established first; then heights for equipment supports (primarily the keyboard, display, and other working surfaces) are calculated. The second strategy assumes that the height of the major work surface (table height in traditional offices) must be fixed, but that seat height and display height are adjustable. The third strategy presumes a fixed seat height, but support and display height are adjustable.

The results of calculations for "conventional" furniture (using data on U.S. civilians) are compiled in Table 9-3. It contains the height adjustments necessary to fit about 90 percent of the U.S. civilian population by excluding only females who are smaller than the 5th percentile in the relevant dimension, and males who are larger than the 95th percentile.

In the first strategy, seat height is determined from popliteal height, adding 2 centimeters for heels. This results in a range of seat heights above floor adjustable from 37 cm to 50 cm, thereby accommodating the 5th-percentile female through the 95th-percentile male users. Then thigh thickness is added to calculate the necessary clearance height underneath the support structure; also adding 2 cm for the thickness of the support structure results in support surface heights of 53 cm to 70 centimeters.

TABLE 9-3. ADJUSTMENT RANGES FOR VDT WORKSTATION HEIGHTS IN CM ABOVE THE FLOOR

Height of . . . above floor	First strategy	Second strategy	Third strategy
	All adjustable	Support for keyboard fixed, all other adjustable	Seat fixed, all other adjustable
Seat Pan	37 to 50	50 to 55	Fixed at 50
Support Surface for keyboard or other data entry device	53 to 70	Fixed at 70	65 to 70
Center of Display	93 to 122	106 to 127	106 to 122
Footrest	Not needed	0 to 18	0 to 13

The next step is to determine eye height above the seat pan, considering actual trunk and head posture. From this the center height of the display is determined, using values for the preferred viewing distance and the preferred angle of sight (Kroemer, 1985c; Hill and Kroemer, 1986.) Accordingly, the height of the center of the computer display should be between 93 and 122 centimeters above the floor. A footrest is not needed in this design approach.

In the second design strategy, the support for the keyboard (the "table") is held fixed at 70 centimeters above the floor, so that even the tallest (95th-percentile) users fit underneath. If the seat heights are adjusted according to thigh thickness, many persons will need footrests. Also, the display must be arranged slightly higher than in the previous strategy, where most people used lower seats and keyboard heights.

The third strategy starts with the assumption that seat height is fixed at 50 centimeters above the floor to accommodate even persons with long legs; persons with shorter legs need to use footrests. Following the same logic as before results in support surfaces for the keyboard and for the display at intermediate height ranges.

Each of these strategies brings about design solutions with specific advantages and disadvantages. The first strategy, requiring complete adjustability, will easily accommodate all persons and does not require footrests. However, the adjustment ranges needed for seat pan height, support surface height, and display height are the largest of all strategies.

The second approach allows support surface height ("table height") to be kept constant for all workstations, but requires the tallest seat height, yet with relatively little adjustment needed. Footrests are often necessary, and to considerable heights. The display needs to be adjusted up and down considerably, although slightly less than in the first design strategy.

The third design approach allows using chairs of constant height but requires widespread use of footrests, which, however, need not be as high as in the second design strategy. Of course, the support surfaces for keyboard and display must be independently adjustable, but the adjustment ranges needed for each are the narrowest of all approaches.

Any of these approaches will provide "fit" to nearly all computer users but does not assume or require certain postures, such as upright trunk or horizontal forearms; instead, each solution allows much freedom in sitting how one likes, from bending forward to leaning back, having the legs at will within the leg room provided.

APPLICATION

DESIGN PRINCIPLES

Unconventional workplace layout and individual use of equipment and furniture require flexibility in work organization and management attitudes; indeed, providing freedom for individual variations from the conventional norm acknowledges that persons in the office differ in their physiques and work preferences. The ergonomic design of workstations (with or without computers), their adjustability, and their proper use affect, via many and often subtle interactions, the person's well-being and the related work performance.

The output of the human-computer system is driven by the human. The main interactions are through eyes, ears, hands, and feet. Thus, the body posture is largely determined by visual targets (screen, source document, keyboard) and by input devices (keyboard, mouse, etc.), and, of course, by the seat. Ill-designed and ill-arranged computer workstations lead to health complaints and attitude problems, while "ergonomic" conditions further well-being, both physically and psychologically.

The phantom of the "average person sitting upright with right angles at elbows, hips, and knees" must be abolished and replaced by a design model that incorporates the actual range of body sizes and of working postures, and their large variations reflecting individual sitting and standing preferences.

Long-term work with computers, particularly if it consists of extensive periods of data entry, requires ergonomic measures to assure that the job is healthy, satisfying, and productive. Several ways exist to achieve these goals. The first concerns the work environment and the work equipment (Grandjean, 1986; Kroemer, 1988). The second considers job content and work organization. The third utilizes technological progress to change the work altogether.

Environment

The *work environment* in the computer office should be designed along the same guidelines that apply to any office, but special attention is needed regarding the lighting, the furniture, and the computer equipment used—see Chapters 4, 5, and 8 for specific information.

The *illumination* level at the computer workstation is lower than required in ordinary offices. The reasons are that reflections on the screen (glare) must be avoided, and that the screen is itself a source of light which must be frequently viewed by the operator. Therefore, the general room illumination should be between 300 and 700 lx, with the lower levels appropriate when the hard copy (paper source document) used is of high quality (good contrast). If difficulties in reading the source document exist at low-level illumination, a task lamp may be used that shines exclusively upon the copy. The distribution of the illumination should be fairly constant throughout (with the exception of the spotlighted area) and be either diffuse or so directed that reflections on the screen are avoided. Wall and ceiling colors affect absorption/reflection of light but otherwise are a matter of personal preference (Kwallek and Lewis, 1990).

Noise is usually not a major concern, since most computers operate at fairly low sound levels. However, a cooling fan or transformer may need attention. Background noise, and noise interference from other workstations or equipment, need to be considered. Altogether, the general recommendation for offices is to keep the sound levels as low as possible, such as 60 dBA.

Temperature can be a problem in computer workstations, because some computer equipment generates and emits heat. Altogether, the same requirements as in other offices apply, meaning that the effective temperature should be in the low 20°C region, the relative humidity around 50 percent, and the air movement small, such as 0.5 m/s. Small deviations from these values should have no effects on output and well-being of the office workers, but complaints about the office being too hot or too cold are frequent in the United States.

Static electricity may also be a problem, not only because it can be irritating and may cause skin rashes in some operators, but also because it attracts dust to the screen.

The *office layout* should follow sound design principles, providing enough room and "privacy" to the individual. Personal preferences and job attributes may suggest either a separate room or cubicle, or an open layout ("office landscape") to facilitate communication with coworkers.

Design for Change

It is important to provide the opportunity and means to change body posture frequently during the work period. Maintaining a given posture, even if it appears comfortable in the beginning, becomes stressful as time goes on. Changes in posture are necessary, best combined with a brief period of physical activity. Hence, the computer workstation should be so designed, and be so easily adjustable, that the operator can assume a variety of postures. This, obviously, requires a chair which is comfortable in many positions.

To permit changes in hand/arm and eye locations as well, the input device (e.g., keyboard) should be movable within the workspace. Also, one should be able to move the display screen to various heights, which requires an easily adjustable, spring-supported or motor-driven suspension system of the support surface.

All components of the VDT workstation must "fit each other," and each must suit the operator. This requires easy adjustability. Figure 9-18 sketches

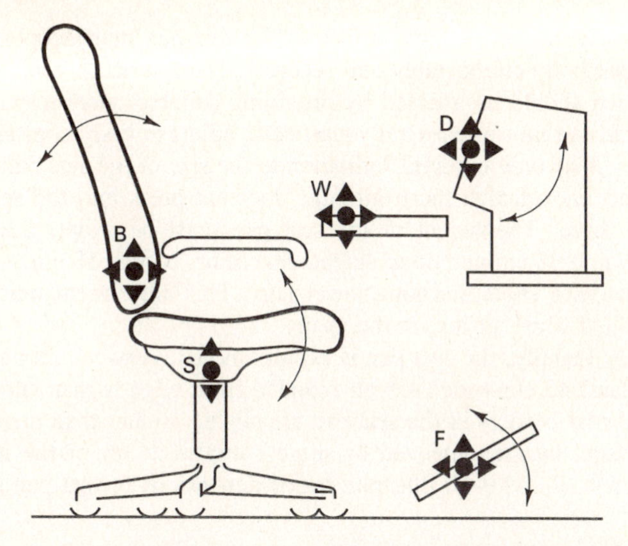

Figure 9-18. Adjustments of the components of a computer workstation.

various adjustment features which allow matching seat height with the height of support of the input devices or table, possibly while using a footrest, and positioning the monitor on its support.

Furniture

The *furniture* at the computer workstation consists primarily of the seat, the support for the data entry device, and the support for the display and a working surface. All of these should be independently adjustable in height, and the screen should be at proper viewing distance, usually at 40 to 50 cm from their eyes, located so that one looks slightly down at it. Recommended height adjustment ranges and other workstation characteristics are shown in Table 9-4 (Kroemer, 1988; Tougas and Nordin, 1987). Unusual designs (such as "stools" for semisitting) may be acceptable or comfortable for some individuals but should not be generally prescribed (Blair, 1991; Seidel and Windel, 1991).

TABLE 9-4. HEIGHT ADJUSTMENT RANGES ABOVE THE FLOOR IN CM FOR COMPUTER WORKSTATIONS

Seat pan	37 to 50
Keyboard support	53 to 70
Screen center	93 to 122
Work surface	53 to 70

Seat pan. The surface of the seat pan must support the weight of the upper body comfortably and securely. Hard surfaces generate pressure points, which should be avoided by providing suitable upholstery, cushions, or other surfaces that can elastically/plastically adjust to body contours.

The only inherent limitation to the size of the seat pan is that it should be short enough that the front edge does not press into the sensitive tissues near the knee. The height of the seat pan must be widely adjustable, preferably down to 38 cm and up to at least 50 cm or, better, 58 cm to accommodate persons with short and long lower legs. This adjustment must be easily accomplished while sitting on the chair.

Usually, the seat pan is essentially flat, between 38 and 42 cm deep, and at least 45 cm wide. A well-rounded front edge is mandatory. Often, the side and rear borders of the seat pan are slightly higher than the central parts of the surface, usually achieved by more compressibility of the inner sections. Figures 9-19 and 9-20 illustrate major dimensions of seat pan and backrest.

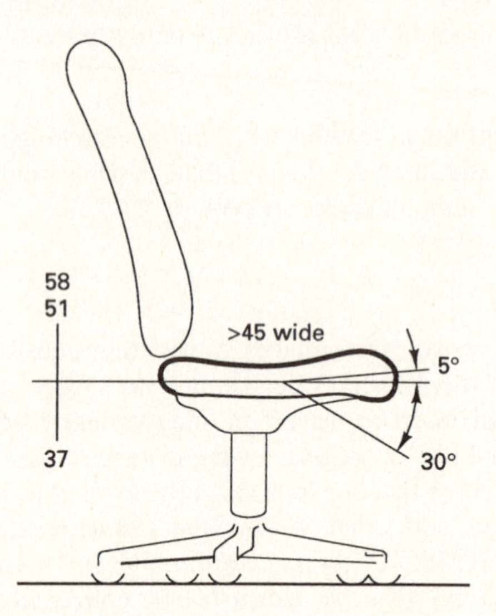

Figure 9-19. Essential dimensions of the seat.

In the side view, the seat pan is often essentially horizontal, but tilting it slightly backward or forward is usually perceived as comfortable and desirable. Seat pans that are higher in their rear portion and lower at the front facilitate "opening the hip angle." "Semisitting" on a distinctly forward-declined seat surface is comfortable for some persons. To counteract the forward/downward thrust along this declined surface, shin pads may be incorporated in the structure. Some inclined seat surfaces are shaped to fit the human underside to counteract slippage.

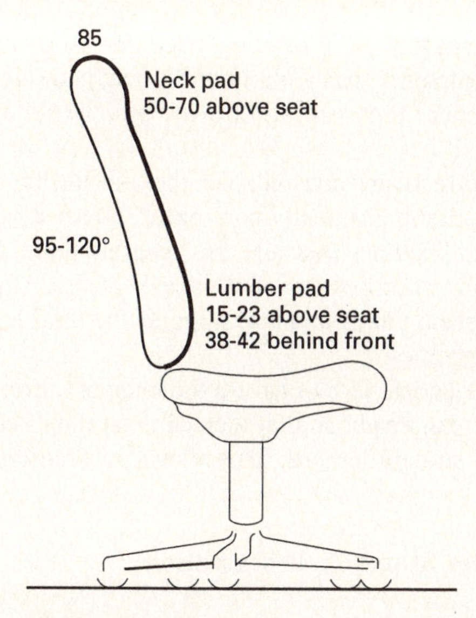

Figure 9-20. Essential dimensions of the backrest.

Backrests. Backrests serve two purposes: to carry some of the weight of upper trunk, arms, and head; and to allow muscle relaxation. Both purposes can be fulfilled only when the trunk reclines on the backrest.

The backrest should be as large as can be accommodated at the workplace. This means up to 85 cm high, and at least 30 cm wide. It should provide support from the head-neck on down to the lumbar region. For this purpose, in side view it is usually shaped to follow the back contours, specifically in the lumbar and the neck regions. An adjustable pad for the lumbar lordosis (e.g., an inflatable cushion) is appreciated by many users. The lumbar pad should be adjustable from 15 to 23 cm, the cervical pad 50 to 70 cm above the seat surface.

The angulation of the backrest must be easily adjustable while seated. It should range from slightly behind upright (95 deg from horizontal) to 30 deg behind vertical (120 deg), with further declination for rest and relaxation desirable. Whether or not the seat-back angle should be mechanically linked to the seat-pan angle is apparently a matter of personal preference. (Note that seat stability must be guaranteed, even if the backrest is strongly declined.)

Armrests. Armrests allow supporting the weight of hands, arms, and even portions of the upper trunk and head. Thus, armrests are useful, though often used only for short periods. They must be well located and of suitable load-bearing surface. Adjustability in height, width, and possibly direction within rather small limitations may be desirable. However, armrests can also hinder moving the arm, pulling the seat up to a workstation, or getting in and out of the seat. In these cases, having short armrests, or none, is appropriate.

Footrests. A prevailing need for footrests usually indicates that the height adjustments, particularly of the seat pan, are not sufficient for the seated person. Hence, presence of footrests usually indicates deficient workplace design.

If footrests are unavoidable, they should be high enough that the sitting person has the thighs nearly horizontal. Footrests should not consist of a single bar or a small surface (because this severely limits the ability of the sitting person to change the posture of the legs): instead, the footrest should provide a support surface that is about as large as the total leg room available in the normal work position.

Some people like to have a foot support, even if it is just a simple rail, at nearly seat-pan height so that they can rest their heels on it while extending the legs nearly straightforward. This allows an occasional "straightening of the elevated legs."

Designing the Stand-Up Workstation

Another way to change working posture is to allow the computer operator, at his or her own choosing, to work for some period of time while standing up. Such stand-up workstations can often use a spare computer in the office, to which work activities can be switched from the sit-down workstation for a while; or one may stand while reading, or writing, or telephoning. Stand-up workstations should be adjustable to have the input device at approximately elbow height when standing, i.e., between 90 and 120 centimeters. As in the sit-down workstation, the display should be located close to the other visual targets, such as source document and keyboard. If the surface is used for reading or writing, it may be slightly declined. A footrest at about two-thirds knee height (approximately 30 centimeters) is often welcomed so that the person can prop one foot up on it temporarily. This brings about changes in pelvis rotation and in spine curvature.

Data Entry Devices

The entry device is often a *keyboard*. The usual binary "tapping" keyboards are essentially flat in their key tops, with the rows of keys increasing in height by about 15 degrees, the further away from the operator they are—see ANSI Standard 100 (Human Factors Society, 1988). However, other keyboards have been recommended, particularly those that are formed to comply with the natural rotation of the wrist (actually: of the forearm), and the movement of the fingertips. Special keyboards use only a few specialized keys, in some cases arranged in unconventional manners and operated in chords—see the keyboard section in Chapter 10.

Other entry devices may be stationary, such as joysticks and trackballs. Among the movable input devices are the mouse, which is rolled on a horizontal surface; and the light pen, which is pointed by hand to targets on the screen. Each of these may be appropriate for given tasks. For more information, check the section on controls in Chapter 10.

Display Screen

The *screen* of the display unit, particularly its optical quality, is of major ergonomic importance. In general, the screen should provide a stable image (which neither flickers nor jitters), showing characters of good contrast against the background. Dark characters on light background are often preferred. The displayed characters and symbols should be of clean fonts with clear edge definition, of appropriate size, and suitably spaced. (For more information, see Table 9-5 and ANSI Standard 100.) Luminance level ("brightness") and contrast should be adjustable by the user. Use of more than two colors should be considered cautiously, since some colors pose visual problems, and because a large variety of colors displayed on the same screen may be more irritating than useful (Collins, Brown, Bowman and Carkeet, 1990; Umbers and Colliler, 1990). For more information, check the section on displays in Chapter 10.

TABLE 9-5. RECOMMENDATIONS FOR CHARACTERS ON DISPLAYS

Character height	min:	16 minutes of arc
	preferred:	20 to 22 minutes
	max:	45 minutes
Character height/width	min:	1/0.5
	preferred:	1/0.7 to 1/0.9
Character stroke width	min:	1/12 of height
Spacing between characters	min:	10% of height
Spacing between words	min:	one width
Spacing between lines	min:	double the stroke width, or 15% of height (whichever is greater)

SOURCE: Excerpted from ANSI 100, Human Factors Society, 1988.

Job Content and Work Organization

Many persons like to have *autonomy* in performing their work, to take responsibility for its quality and quantity, and to control their timing. Most prefer to perform larger tasks from beginning to end instead of simply doing specialized tidbits. The ability to receive direct feedback about work performance, preferably by reviewing one's own work daily, supplemented by constructive and positive comments from the supervisor, contributes essentially to the feeling of achievement and satisfaction. Within the limitations set by the requirement that certain work needs to be done, the operator should be free to distribute the workload, both in amount and in pace, according to one's own preferences and needs. Communication with colleagues and social relations are essential, although at individually varying intensities. Isolating people or submitting them to cold, formal relationships is usually detrimental to well-being and performance.

The organization of working time, particularly the provision of changes in work and of rest pauses, is important for many computer tasks. Most people are bored by repetitive, monotonous, and continuous tasks; instead, varying tasks of different lengths should be provided, so that the computer operator has occasion and cause to shift from one task to another, to move away from the computer for periods of time, to get or take away materials, do something else, or simply take a break. The "recovery value" of many short rest pauses is larger then that of a few long breaks.

Changes Through Technological Developments

Some of the current computer work tasks probably should not be imposed on humans, such as the prolonged simple input of numbers or of word texts. Not only is this boring, but repetitive finger movements may be a source of cumulative trauma disorders, such as tendonitis in hands or arms or carpal tunnel syndrome—see Chapter 8. Repetitive entry should be automated, e.g. by machine recognition of characters or voice. Feedback from the computer about its memory content can also be given to the operator through acoustical or sensory means other than just by display on the screen.

Clever software offers possibilities to facilitate the task of the computer operator, such as automated programs for spelling, stringing of characters, algorithms that check and indicate outliers in data or unusual events or repetitive occurrences. Certainly, a wide variety of opportunities exist to change and, one hopes, to improve and facilitate the work of the computer operator.

SUMMARY

The former simplistic ideal of sitting upright as "the" healthy body posture has been superseded by the recognition that many different postures may be suitable, and that postural change is important. This makes the design of office furniture much more challenging, particularly since other techniques than the old-fashioned QWERTY keyboard are suitable for interaction with the computer. The visual link between the eyes and the display monitor must be carefully designed: it is easier to look "down" at the display than to look at it straight ahead, or look up to it. Glare, either direct or reflected, should not strike the eyes.

While the design of the computer office is very important, so is its proper adjustment and use. It is often necessary to be reminded that one must change body posture, that one should occasionally stretch and reach and move about (see Figure 9-21), avoid assuming contorted positions (see Figure 9-22), and avoid excessive keyboarding. Proper work content, organization, and habits are personal challenges.

Figure 9-21. Occasional (but not habitual) reaching and stretching are desirable (courtesy of Herman Miller).

Figure 9-22. Contorted postures can lead to overuse disorders (courtesy of Herman Miller).

CHALLENGES

What is so bad about standing, or walking around, while doing office work?

What are some reasons for recommending an "upright" trunk while sitting?

Is there a physiological reason to provide only a low backrest to "typists"?

What are some appealing factors, and unappealing conditions, associated with office work?

Which are the advantages of having uniform heights for (a) seats, (b) work surfaces?

Which control functions, and data input functions, could conceivably be shifted from the fingers to other body parts?

What speaks against the computer "talking back" at the operator, instead of simply displaying information on the screen?

Why is changing one's posture desirable, as opposed to maintaining a comfortable posture for a long period?

Under what circumstances may the following links between the operator and the computer be the main determiners of posture: (a) vision, (b) keyboarding, (c) other finger-control interactions?

Which are the main arguments pro or con individual offices versus multiperson offices?

Should one strive to have windows in the office, in spite of the difficulties of controlling illumination?

Should eye examinations, and provision of suitable corrective lenses, be of more concern to the individual or to the employer?

10

"Handling" Loads

OVERVIEW

We all "handle" loads daily. We lift, hold, carry, push, pull, lower—while moving, packing, and storing objects. The material may be soft or solid, bulky or small, smooth or with corners and edges; it may come as bags, boxes, containers; with or without handles. We may handle objects occasionally or repeatedly; during leisure activities, but often as part of our occupational work. On the job, ergonomic design of material, containers, and workstations can help to avoid overexertions and injuries, as should instructions and training on how to "lift properly." For some jobs, selection of persons who are physically capable of strenuous material handling may be considered.

INTRODUCTION: STRAINS ASSOCIATED WITH LOAD HANDLING

Manipulation of even lightweight and small objects can strain us because we have to stretch, move, bend, or straighten out body parts, using fingers, arms, trunk and legs. Heavy loads pose additional strain on the body owing to their weight or bulk, or lack of handles.

Material handling is among the most frequent and the most severe causes of injury all over the world (Buis, 1990; Davis, 1985; Evans, 1990; ISO, 1988; Gilad and Kirschenbaum, 1986; National Academy of Science, 1985; NIOSH, 1985; Kroemer, McGlothlin, and Bobick, 1989). The direct and indirect cost are enormous, and the human suffering associated with, for example, low back injuries, immeasurable.

Exerting force and energy in lifting an object with the hand(s) strains hands, arms, shoulders, the trunk, and, if one stands, also the legs. The same parts of the musculo-skeletal system are under stress in lowering, in pushing and pulling, but directions and magnitudes of the external and internal force and torque vectors are different. The primary area of physiological and biomechanical concern has been the low back, particularly the disks of the lumbar spine. Thus, the operative words in the literature have been "low back pain related to lifting." Yet, when one considers the musculo-skeletal structures within the trunk (see Chapter 2), a variety of elements may be strained either singly or combined: the spinal vertebrae, primarily their disk or facet joints; connective tissue such as ligaments and cartilage, and muscles with their tendons. These all may experience "insults," sprains, or trauma. Compression strain of discs and vertebrae has been primarily studied, but tension is the primary loading of elastic elements such as muscles and connective tissues. Tension strains can be in form of linear elongation, or of bending movements, or of twisting torque. All structures may be subject to shear.

The loading of the body may come from activities done on external objects, such as lifting, lowering, pushing, pulling, carrying, or holding; thus, the strains may be static or dynamic, of fast or slow onset and of short or long duration. They may be single or several events; if the same or similar strains reappear, that repetition may be at regular or irregular timing, and the strains may be of similar or dissimilar magnitudes. Body structures are also strained by just moving one's own body (according to Newton's second law: force = mass × acceleration) or simply by maintaining a posture by muscle tension without external load. Any longitudinal contraction of the trunk muscles compresses the spinal column, which is (with some help from intrabdominal pressure—see Chapter 2) the only load-bearing structure of the trunk.

Musculo-skeletal strain may be experienced in sports, most obvious in weight lifting. Other sources of stresses are leisure and occupational activities. If done on the job, the activities are often labeled "manual material(s) handling" (MMH). In the course of such material handling, one may either exert energy intentionally toward an outside object, or the body may be subjected to unexpected energies, such as in catching a falling object or by accidents, such as slipping and falling.

☞☞☞ *Since the Latin word manus means hand, the term "manual material handling" is a perfect tautology.* ☜☜☜

Aside from sports, the literature has dwelt primarily on the MMH event "lifting," i.e. moving an object "by hand" from a lower position to a higher one; and on pushing and pulling, carrying, and holding. In terms of the internal structures, emphasis has been for several decades on the lumbar spine, particularly the lumbar discs, and on intraabdominal pressure. Occasionally in the past, but much more frequently in recent years, connective tissues have been considered, particularly the major longitudinal muscles and the spinal ligaments.

The following review of the literature summarizes the present knowledge in these areas and, based thereon, shows implications for "ergonomic interventions" to prevent overexertions of material handlers.

ASSESSING BODY CAPABILITIES RELATED TO MATERIAL HANDLING

Handling material requires exerting energy or force to lift, push, pull, carry, hold; that is move objects or keep them from moving. The energy needed to do these tasks must be generated within the body and exerted in terms of force to the outside object. Thus, in the past, research has been directed at the energy generation within the body (see the section on metabolic processes in Chapter 2) or was concerned with the external forces applied to the object handled. The metabolic research dealt with handling activities lasting for several hours; the force-directed research considered mostly one-time or just a few efforts.

If the human must lift (lower, push, pull, or carry) material over many hours in repetitive activities involving the whole body (or large sections thereof), then the ability to do so is likely to be limited by *metabolic and circulatory capabilities*. Given the energetic inefficiency of the body, moving the body in this way taxes body abilities usually to such an extent that fairly little external load may be moved, because so much energy is spent on moving body parts. This fact became obvious in the development of the 1981 "Lifting Guide" (NIOSH, 1981): MMH activities that are performed several times a minute, over hours, strain mostly metabolic and circulatory functions.

On the other hand, if very high force must be exerted just once, such as in lifting a heavy object, then indeed the ability to *generate large force* once is the limiting factor. This experience was apparently the reason why, in the past, guidelines were used that tried to determine the acceptable lifting task by establishing an upper weight limit (ILO, 1988). Of course, to set a weight limit for objects to be handled is not reasonable and prudent, because one might exert a large force even to a fairly small mass, if much acceleration is applied (Newton's second law). Generating a one-time upward force as needed to lift a heavy object does strain many musculoskeletal components of the body.

Most material handling work lies between the two extremes of enduring activities and one-time efforts. The assessment of human abilities to do heavy material movement has been primarily through biomechanical and psychophysical methods (Ayoub, 1991, 1992; Ayoub and Mital, 1989; Chaffin and Andersson, 1984; NIOSH, 1981; Kroemer, McGlothlin, and Bobick, 1989). Psychophysics bridge the range between assessments of metabolic capabilities and of one-time muscle strength.

The *psychophysical approach* relies on the assumption that the human can sense and integrate the perception of strain on all body functions and capabilities, be they metabolic, circulatory, muscular, or connective tissue-related, or on the bony structures. Judgment of the perceived strain contains an overall assessment of acceptability, suitability, and willingness to perform. And, indeed, while we have great difficulty in selecting the functions of the body upon which to base our judgments, we judge throughout our life. Psychophysical methods have become an important part of research regarding human material handling capabilities (Genaidy, Asfour, Mital, and Waly, 1990; Pope, Frymoyer, and Andersson, 1984; Karwowski, 1991; Snook, 1978; Snook and Cierello, 1991).

Biomechanical methods to evaluate strains and capabilities of the body have become widely used in the last few decades. Initial interest concerned mainly the spinal column, particularly with respect to the responses of the vertebrae and the vertebral disks to compression (Burns, Kaleps, and Kazarian, 1984; Kazarian, 1981; Kazarian and Graves, 1977; Schultz and Andersson, 1981.) Calculation of compression strain is complicated by the fact that the human spine is not a straight column but anterior-posteriorly curved even when "erect," and that this curvature changes with different trunk postures (Adams and Hutton, 1986; Aspden, 1988; Hutton and Adams, 1982). Diurnal variations (Adams, Dolan, and Hutton, 1987), as well as daily activities (Eklund and Corlett, 1984; Helander and Quance, 1990) change the mechanical parameters of the spinal unit. The strains (measured, modeled, and calculated) in the spinal column were initially assumed to be static or nearly so (Chaffin, 1981, 1987; Chaffin and Andersson, 1984; Jaeger and Luttmann, 1986; NIOSH, 1981). Dynamics were introduced into the models only slowly because of the complexity of the involved strains and of their calculation (Ayoub and Mital, 1989; Chaffin and Andersson, 1991; Jaeger and Luttmann, 1989; Marras, 1988). Early models were descriptive, mathematically expressing observed behavior, and did not include much analytical understanding of the sources and existing quantities of body strain; but analysis and modeling of the strains in the trunk and spinal column have made great advances (Ayoub and Mital, 1989; Bryant, Stevenson, French, Greenhorn, Andrew, and Deakin, 1990; Lee, Chaffin, Waikar, and Chung, 1989) with particular interest directed at the activation of the various muscles within the trunk in movement, such as lifting (Ayoub, 1992; Asfour, Genaidy, and Mital, 1988; Bush-Joseph, Schipplein, Andersson, and Andriacchi, 1988; Gagnon and Smyth, 1992; Marras, Ferguson, and Simon, 1990; Marras and Rangarajulu, 1987; Reilly and Marras, 1989; Sommerich and Marras, 1992).

The muscles that develop force vectors running between the inferior and the superior parts of the trunk (in essence, they try to "pull the shoulders onto the hips") are the right and left latissimus dorsi, right and left erector spinae, right and left internal and external obliques, and the right and left rectus abdominus (Schultz and Andersson, 1981). Their locations are sketched in Figure 10-1.

Marras and Reilly (1988) inserted wire electrodes in each of these muscle groups and measured their activities via EMG during controlled conditions of trunk motion. In this manner, they were able to identify which muscles were involved, and the sequence and intensity in which these muscles were called to action. The researchers found that muscle recruitment is quite different, both in quality and quantity, depending on whether the muscles hold the trunk in a static configuration or if various kinds of trunk motions are performed.

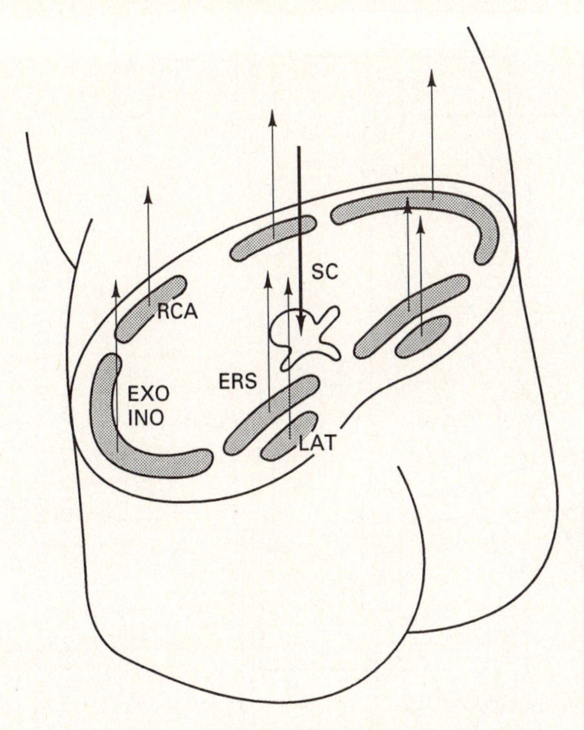

Figure 10-1. Sketch indicating the pull forces generated by the trunk muscles (RCA, rectus abdominus; EXO, INO, external and internal obliques; ERS, erector spinae; LAT, latissimus dorsi) resulting in spinal compression (SC). (Modified from Schulz and Andersson, 1981.)

Depending on the lifting task to be performed, particularly its height, distance from the body, speed, and sideways displacement (asymmetry), different sequences and magnitudes of muscle activation were observed. Figure 10-2 shows the events that occurred in an isovelocity trunk motion from a forward bent posture to upright, a 67-degree motion at half the subjectively possible speed. In this case, the latissimus dorsi, erector spinae, and external oblique muscles were activated first, together with a build-up of intraabdominal pressure; that pressure was the last event to subside after the motion.

In these studies it was also found that the role of the intraabdominal pressure (IAP) depends very much on the trunk loading and motion conditions. It had been assumed that the pressure column between the inferior and superior diaphrams of the trunk cavity would contribute toward carrying the load generated either by external forces while lifting, or by the reactive longitudinal muscle tensions within the trunk. It had been thought that, at least in static conditions, the IAP could transmit some of the load, thus relieving the spinal column from having to transmit all compression strain. Estimates had been that the IAP would carry up to 15 percent; in fact, European recommendations for suitable lifting conditions had largely relied on observing the development of that pressure (Freivalds, 1989). Yet, newer research findings showed that the contribution of the intraabdominal pressure column to load bearing depends very much on the given conditions of static or dynamic trunk activities (Mairiaux and Malchaire, 1988; Marras and Reilly, 1988; McGill and Norman, 1987).

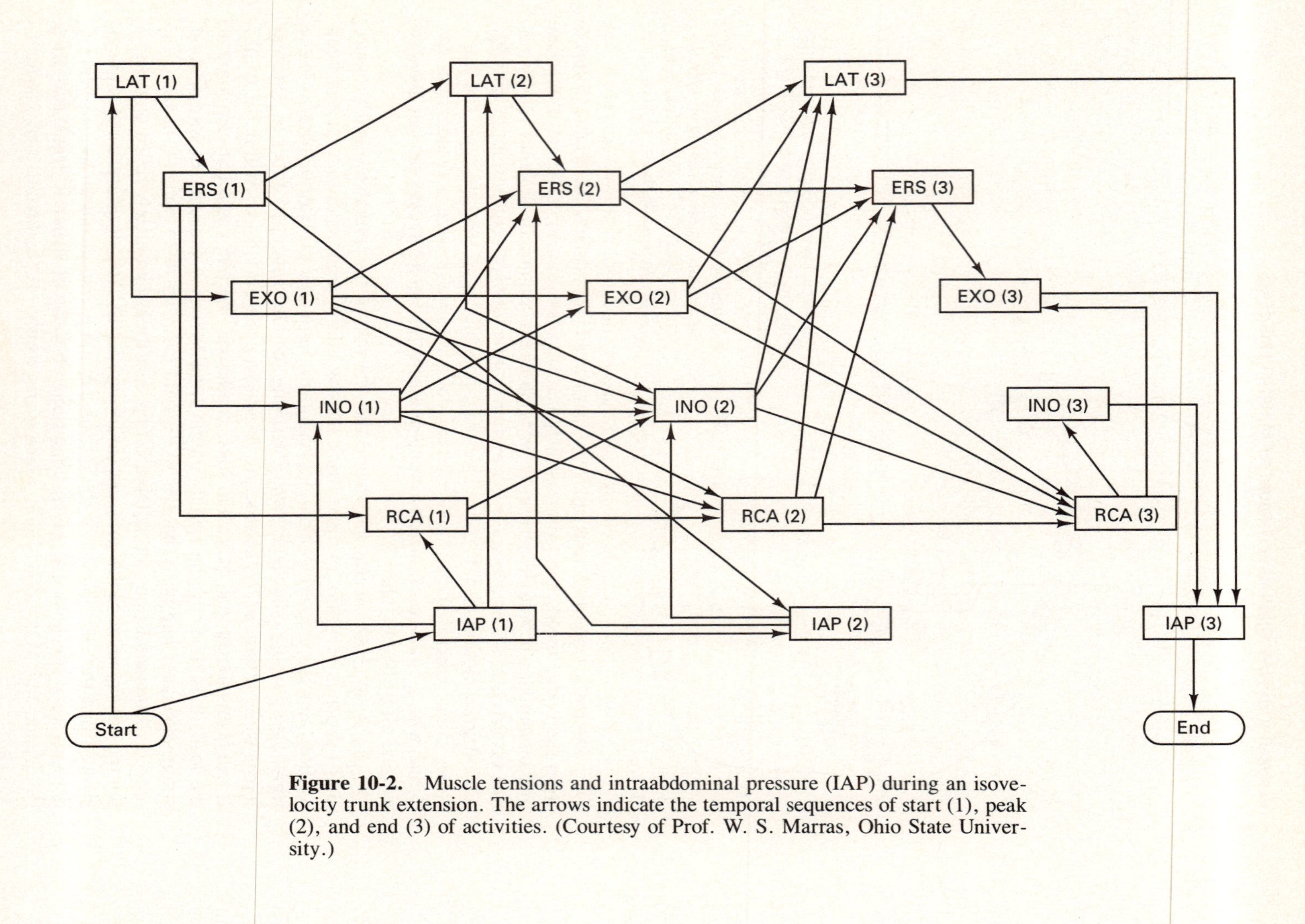

Figure 10-2. Muscle tensions and intraabdominal pressure (IAP) during an isovelocity trunk extension. The arrows indicate the temporal sequences of start (1), peak (2), and end (3) of activities. (Courtesy of Prof. W. S. Marras, Ohio State University.)

Initially, much research on lifting capabilities was done assuming that the person would perform this activity directly in front of the trunk. This has been called a "symmetrical" lift, because all activities occurring in sagittal planes, such as of the left and right hand, are mirror images reflected in the medial (mid-sagittal) plane. Of course, many actual activities are not symmetrical but involve activities to one or the other side of the body, requiring a twisting motion about spinal joints. It was apparent that the ability to perform such nonsymmetric activities is usually lower than for lift work directly in front of the body (NIOSH, 1981); however, the observations were usually empirical, until recording of electromyographic signals of the trunk muscles provided more detailed knowledge about the involvement of the various muscles within the trunk (Marras and Mirka, 1989, 1990; Ferguson and Marras, 1991; Sommerich and Marras, 1992). Analytic models of the contribution of muscles and the related loading of the spinal column are being developed using information of that kind.

Another aspect of biomechanical loading of the body relates to the "preparedness" of the person to perform a material handling activity. Usually, the person knows in advance that an activity is to be performed and has sufficient time to prepare for that activity. Yet, this is not always true; for example, a piece of material may suddenly fall from a shelf toward the operator, who had expected to pull it further before it began to move. Such sudden unexpected loadings of the body may lead to an overexertion injury (Marras and Reilly, 1988; Stobbe and Plummer, 1988).

Most of the research has been performed for "lifting" activities, where the force vector in the hands of the operator is vertical. Yet, many material handling activities involve horizontal pushing and pulling of a load. Following the pioneering research of Snook (1978), several researchers have described the abilities to perform such work (Imrhan and Ayoub, 1988; Lee, Chaffin, Waikar, and Chung, 1989; Snook and Ciriello, 1991).

APPLICATION

Back Injury and Pain

Injury occurs if the limits of maximal strain of the tissues (bone, cartilage, ligaments, muscles) are exceeded. This may happen in a single strenuous effort, an accidental trauma. However, often repeated loadings add up to a cumulative overloading. In this case, the person repeatedly insulting the back does not have to disrupt the normal pattern of work until finally onset of pain signals the accumulated injury.

A major difficulty in recognizing and analyzing the cause of a back injury is that it may happen without generating any pain. This is so because neither the facets of the apophyseal joints nor the intervertebral disks seem to have pain-sensitive nerves. Thus, the three load-bearing elements (two facet joints and one disk) of each spinal unit can indeed be injured without pain sensation.

Clinical evidence shows that old but stable fractures are commonly found in people who have no recorded history of injury.

If the disks or facet joints are repeatedly injured, degenerative changes may set in. However, radiography has shown evidence of disk or joint degeneration to be as frequent in persons who have low back pain as in those who do not suffer from it. Mechanical derangements of the intervertebral joint, such as a decrease in disk height, or a change in the positions of the components of the facet joints, may produce clinical symptoms long after the acute phase of the injury is over. Thus, there are many cases in which the onset of pain was delayed for a day or more after a known injury occurred (Andersson, 1991).

☞☞☞ *In many cases, low back pain cannot be traced to one specific incident of overexertion.* ☜☜☜

The actual reasons for back pain are quite often not clear, partly because of the lack of nerves at the facet and disk. Back pain may be caused by damage to the bony sections of the vertebra, but more likely by degeneration or deformation of the disk, or by microfractures and bone scarring of the cartilage endplates. Bulging or rupture of the disk can cause nerve root compression, or distortion of the ligaments around the joints. While it was believed that a large portion of all chronic low back pain may be diskogenic, muscular or ligament tissue problems near the spine may be just as frequent and important (Andersson, 1991; Snook, 1988, 1991). A large variety of individual factors have been associated with the likelihood of an incidence of a back injury and with its severity, particularly in its lumbar section. Among these factors are: degenerative disk disease, congenital anamolies, spondylolisthesis, differences in leg length, and previous injuries. Whether and to what extent these conditions contribute, and how they can be diagnosed and treated, has been a topic of often heated discussions for a long time, particularly between physicians and chiropractors.

Moderate activity increases the nutrition of the spinal disks, mostly by passive diffusion aided by the pumping action of compressing and releasing the disk tissue, and by bending and unbending. Rest and inactivity inhibit healing of strained disk tissue. After rising from bed rest in the morning, bending strains the fluid-filled and flexible disk tissue about three times more than later during the day (A. Nachemson, paper presented on March 5, 1989, Occupational Orthopaedics, American Academy of Orthopaedic Surgeons, New York, NY).

About every second person returns to work within one week after an incidence of low back pain (LBP). With increasing duration of the absence, the successful return becomes less likely; yet altogether, about 9 out of 10 LBP sufferers eventually try to return to work, some to medically recommended "restricted work," i.e. work that puts reduced strain on the back. The likelihood of successful return in patients above 45 years old is only about half that of persons younger than 24 years. Youthful workers have more frequent but less severe back disorders than older workers. The largest incidence rates for females are between ages 24 and 34, for males between 20 and 24 years. Altogether, males and females are equally often affected by LBP; however, given that fewer female workers than male workers are employed in the United States, more compensation cases are reported for males than for females (Andersson, 1991). Only a few patients with unrelenting LBP eventually need surgery. While acute LBP is often related to some kind of identifiable body strain, chronic pain (lasting three months or longer) becomes disassociated from its physical basis and instead associated with emotional distress (Snook, 1988, 1991).

Snook (1988) listed the following deterents to returning to work:

Associated with the worker:
- malingering
- psychological disability
- illness behavior

Associated with management:
- lack of follow-up or encouragement
- requirement to be 100 percent rehabilitated
- no work modification

Associated with the union:
- rigid work rules
- referrals to "friendly" physicians
- referrals to lawyers

Associated with the practitioner:
- inappropriate treatment
- ineffective treatment
- prolonged disability

Associated with the lawyer:
- lump-sum settlements instead of rehabilitation

In summary: the state of the science in determining the ability of the human body, particularly of the trunk, specifically of the low back region, to perform strenuous work in manual material handling is very much in flux. Physiological findings have been well integrated in the knowledge base, and psychophysical assessments are greatly contributing toward the overall assessment of one's willingness to perform certain activities without perceived overexertion. The newer field of biomechanics, although limited in its range, is quickly developing a large number of procedures to measure mechanical strains within the body, and to incorporate the various findings into models of strains and capabilities (see Chapter 8).

PERSONNEL TRAINING

There are three major ergonomic approaches for safer and more efficient manual material handling: personnel training, personnel selection, and job design. The first two fit the person to the job, the third fits the job to the person. For each, knowledge of the operator's capabilities and limitations is necessary. Thus, the research and knowledge base overlaps in all procedures.

Since "lifting" of loads is associated with the largest portion of all back injuries, training in "safe" lifting procedures has been advocated and conducted for decades. Studies have indicated that approximately one-half of compensable low back injuries are associated with lifting. Industrial experience has also identified lifting as a major cause of back injury. Hence, "training how to lift safely" has been targeted to persons performing manual material handling activities in industry, and to certain industries and jobs.

Training is expected to reduce injuries expressed in severity and frequency; to develop specific material handling skills; to further awareness and self-responsibility; to improve specific physical fitness characteristics; etc. Various instructional styles and media have been used. Training, either generic or customized, has been done in a single session or in various combinations of repetitions; at various times during employment; at the job site, in classrooms, or in outside workshops. Participants in these training efforts have been selected at random, or chosen according to risk, previous injuries, age, etc.; they were volunteers or they included all employees. Theoretically speaking: a large variety of experimental variables (independent, controlled, dependent) were used, with various experimental designs and experimental treatments.

With such a large variety of training approaches, a review (Kroemer, 1992) of previous procedures and their results is both complex and revealing.

What Are Proper Lifting Techniques?

Instructions on "how to lift" are meant to affect lifting behavior and to reduce the likelihood of an overexertion strain or injury.

In the laboratory, it has been shown that proper training regimes can increase the ability for lifting. Sharp and Legg (1988) reported that after four weeks of training, initially inexperienced lifters increased their work output significantly while maintaining their energy expenditures. The improvement was attributed to better skill through improved neuromuscular coordination and to possible increases of muscular endurance. Genaidy, Gupta, and Alshedi (1990) used six weeks of training and also found that muscular endurance, muscular strength, and cardiovascular endurance were improved. Yet, very few tightly controlled studies with large numbers of industrial material handlers have been performed, and the validity of laboratory findings for industrial environments has not been established. Intensive training can lead to muscle soreness which is apparently related to damage to muscle tissue—although not all tissue damage results in pain (Byrnes and Clarkson, 1986).

EXAMPLE

Many lifting instructions include the admonition "keep the back straight." However, what this means is that one should maintain the "natural curvature" of the "standing" spine, particular its lumbar lordosis. This imprecise use of words has led to much confusion.

From the 1940s on, the advocated method was the straight-back/bent-knees lift, where workers lowered themselves to the load by bending the knees and then lifted by using the leg muscles. Yet, biomechanical and physiological research has shown that the leg muscles used in this lift do not always have the needed strength; also, awkward and stressful postures may be assumed if one tries to enforce this technique under unsuitable circumstances, e.g. when the object is bulky. Hence, the straight-back/bent-knees action evolved into the "kinetic" lift, in which the back is kept "mostly" straight while the knees are unbent; but feet, chin, arm, hand, and torso positions are prescribed (International Labour Office, 1974; National Safety Council, 1971). Another variant was the "free-style lift" (new techniques and names appear and disappear frequently), which Garg and Saxena (1985) found to be better for male (but not for female) workers than the straight-back/bent-knees technique. In some contrast to a 1974 study by Park and Chaffin (1974), Garg and Herrin (1979), using calculated compressive force in the lumbosacral spine as critical measure, concluded that the stoop lift (with a bend at the waist and straight knees) may be superior to the squat (bent knees) position in some situations. However, Andersson and Chaffin in 1986 again advocated the squat posture with flat back.

Jones had found in 1972 that no single lifting method is best for all situations. Therefore, training of proper lifting technique is an area of confusion: what method should be taught? The fact that there is no one agreed-upon nor consistently appropriate method is probably one reason for the inconclusive findings regarding success or failure of teaching people "how to lift safely."

Unsuccessful training. Reviews by Brown (1972; 1975) and Yu, Roht, Wise, Kilian, and Weir (1984) detected no significant reductions in back injury due to education over four decades. Brown (1975) extracted from a questionnaire with 509 respondents that attendance at back-injury prevention lectures did not lead to a reduced incidence. Snook, Campanelli, and Ford (1980) compared the number of back injuries in companies that conducted training programs in "safe lifting" with the number of injuries in companies that did not have such programs: the numbers did not differ significantly. Although neither the quality nor the content of the programs was investigated, the general conclusion was that training (as done then) was not an effective preventive program for low back injuries (Dehlin, Hedenrud, and Horal, 1976; Stubbs, Buckle, Hudson and Rivers, 1983; Wood, 1987). Several studies on nurses did not find effects on the incidence of low back injury after receiving repeated instruction on lifting procedures; the principles taught were seldom used

(Harber, Billet, Shimozaki and Vojtecki, 1988; St.-Vincent, Tellier, and Lortie, 1989).

EXAMPLE

Scholey and Hair (1989) reported that 212 physical therapists, involved in back care education, had the same incidence, prevalence, and recurrence of back pain as a carefully matched control group.

Successful training. Hayne (1981) suggested that the three essential components of a training program are "knowledge, instruction, and practice," but neither provided sufficient information on the contents of such a program nor indicated how to make reliable evaluations of its effectiveness. Davies (1978) discussed three "carefully designed and properly carried out" (p. 176) training programs that were followed by decreases in back-injury incidence, unfortunately, he did not describe the specific criteria for "careful" and "proper"; therefore, the positive qualities of these programs cannot be determined. Miller (1977) reported success in decreasing the frequency of back injuries by using a five-minute slide/cassette program, a film, and posters to emphasize the theme "when you lift, bend your knees." A cohort was not observed. Hall (1973) advocated "clean-lifting" (a form of straight back/bent knees) based on his own personal experience with a bad back, but only presented anecdotal evaluative information.

Most training targets the individual worker or groups of workers. Other use of training is in educating supervisors, health and safety professionals, and management personnel in awareness of MMH problems, in ergonomic job design principles, and in how to respond to low back pain and injury once it has occurred. Such supervisor and manager training is probably of high importance for an effective injury prevention program.

Statistically, back pain has been related strongly to job satisfaction (Hultman, 1987) and attitude (Biering-Sorensen and Thomsen, 1986; Gentry, Show, and Thomas, 1977). This finding generates important questions regarding psychosociological aspects on and off the job which, so far, have found few answers.

"Back schools." The back-school concept is often traced to Fahrni (1975), who suggested education as a "conservative" treatment (as opposed to the "radical" treatment by surgery) for low back pain patients. He began using this treatment as early as 1958; other schools have since developed and in the 1980s became popular for rehabilitation of back-injured patients by health-care providers, physicians, nurses, and physical therapists. This training approach emphasizes knowledge, awareness, and attitude change by instructing the individual in anatomy, biomechanics, and injuries of the spine. Such programs may also provide information on stress management, drug use and abuse, vocational guidance, and other topics (Pope, 1987; Snook and White, 1984). The important goal is to encourage the individual to take responsibility for his or her own health by means of proper nutrition, physical fitness, and awareness of the effects of posture and movement on the back.

Controlled research concerning back schools has been limited. Back-school patients expressed increased understanding of their own back problem and a feeling of increased control over pain (Morris and Randolph, 1984). A study on the effectiveness of back schools was conducted by Bergquist-Ullman and Larsson (1977): low-back-pain patients from a plant in Sweden were randomly assigned to one of three treatments, i.e., back school, physical therapy, and placebo (short-wave diathermy). The results indicated that back school and physical therapy treatments were equally effective in reducing the number of days needed for pain relief, achieved by both treatments in fewer days than with the placebo. Although there were no differences between the back-school and therapy groups, the authors said that the back schools are more economical because patients are treated in groups rather than one-to-one. Fisk, DiMonte, and Courington (1983, p. 21) suggested that it may be the "influence of authority figures and the promotion of self-sufficiency" that account for the finding of no difference, rather than the different training contents.

Typical industrial case studies involving the use of a training program to change attitudes were reported by Fitzler and Berger (1982, 1983). In addition to instituting new safety rules and regulations and some ergonomic job redesign, an in-house back injury treatment program informed employees on spinal anatomy and physiology, posture, and appropriate exercises. Reported results include 75 percent reduction in compensation costs and 50 percent reduction in lost work days per injury. Training as part of a back-injury treatment program was deemed a great success but, unfortunately, a control group was not used. Results from a study of 3424 employees of the Boeing Company indicated that there were no significant differences, neither in the occurrence of back pain nor in the number of lost work days, between (healthy) employees who attended back school and a control group that did not (Nordin, Frankel, and Spengler, 1981; cited in Snook and White, 1984).

Several published case studies, however, report success. Melton (1983) described a study where 1,500 employees in eight industries received one hour of instruction on back-injury prevention. The training, which was "customized" for each industry, was conducted in small groups at the job sites. The training was followed by a six-month campaign of posters, booklets, and cards given out with paychecks. A year after the training, safety directors and some employees were randomly sampled and asked to complete a questionnaire. The safety directors reported a 40 percent reduction in lost work days in the year following the training. Three companies also reported decreases in medical insurance expenses. However, there was an increase in the number of reported injuries, but Melton attributed this to a change in management attitude and a willingness on the part of employees to report back injuries early.

Another training program was conducted at a company with 24,000 employees (Tomer, Olson and Lepore, 1984). Although it is not clear whether all employees participated, many attended training classes. Their content included anatomy, physiology, and kinesiology of the back, posture, physical fitness exercise program, good work practices, and a discussion of off-job practices that might cause back injuries. Refresher training that included new material on information on nutrition, physical fitness, and stress management was conducted seven months after the initial training. Five months later, a comparison of statistical loss data before and after training indicated a two-thirds reduction in the yearly average total costs for back injuries, and a return of training-dollar investment of better than 6 to 1. Two other studies discussed by Snook and White (1984) reported similar reductions in the number of back injuries and associated costs that were attributed to back-school education. Morris (1985) also presented a case with similar results: in the two years preceding a preventive educational program, a company paid over $190,000 in back-injury costs, but only $7,000 during the twenty months after training-program implementation. However, there was a production cutback during some of that period that might account for some of the decrease (Anderson, 1989). In none of these cases was a control group present. Also, there is no information about long-term effects.

The key factors for motivating compliance in efforts to prevent back injuries are

- recognizing the problem and the employee's personal role in it,
- gaining knowledge about the spine,
- understanding the relevance of the problem by seeing the working environment on slides or videotapes,
- being able to alter the behavior by learning "proper" techniques,
- practicing these new techniques, and
- following up the program at the workplace, as well as training new employees and using the program in safety meetings (Morris, 1984).

The concept of the back school (whether preventive or postinjury) with its emphasis on attitude change and awareness to encourage compliance with proper procedures at work is an interesting approach which goes beyond the traditional "safe lifting" teaching. However, it is very difficult to make any kind of reliable statement on the effectiveness of back schools from the literature currently available. One can hardly draw generalizable conclusions from case studies like those presented above. Clearly, in many cases (especially when published) the management is happy with the results and convinced of the program's effectiveness, even if its success may in reality be a "Hawthorne effect." (This is the positive result when persons are "treated" with actions that themselves may be ineffective, but the "subjects" react positively to the show of concern and interest—Roethlisberger and Dickson, 1939.)

☞☞☞ ***The Hawthorne effect—"bread and butter" for consultants***

Consultants called by management to improve (sufficiently bad) conditions of manual material activities will be successful if the improvement strategies are well-intended and well-directed by involving the material handlers. If, after the campaign, the level of performance sags again, as is to be expected, management is likely to call upon the same consultants again because "they were successful previously." ☜☜☜

Unfortunately, almost all studies and intervention results reported in the literature are "scientifically deficient" in that they do not provide evidence of a study design that allows the assessment of reliability and validity of the intended outcome of the experimental treatment. Granted, it is difficult (because of work interference, time needed and expenses) to conduct a field experiment in which one varies only the independent variable and excludes confounding variables and uncontrollable interferences. Still, including a control group in the experimental design is often feasible, and it would allow the evaluation of the claimed effect of the experimental training treatment (Cook and Campbell, 1979; ILO, 1974; Smith, 1976).

Only a few reports on the effectiveness of training mention control groups; these report on "back-school therapy" successfully applied to patients with back pain (Bergquist-Ullman and Larsson, 1977; Carlton, 1987; Lankhorst, Van de Stadt, Vogelaar, Van der Korst and Prevo, 1983; Moffett, Chase, Portek and Ennis, 1986). Generally, however, there is a dearth of publications concerning the proof of effectiveness of training programs through the use of control groups, or by other generally accepted procedures.

EXAMPLE

In 1987, a letter was sent to all 24 American "back schools" (listed on page 62 of the *Journal of Occupational Health and Safety,* February 1987 edition) asking them to relay the results of controlled training approaches which would substantiate their claims of success. Twelve back schools responded in writing, all claiming success, but none offered proof via a cohort/control group study (K. H. E. Kroemer).

"Fitness" and "flexibility" training. Another approach to training workers in prevention of low back injury is that of physical fitness training. Manual material handling is a physical job, and it is reasonable to assume that many aspects of physical fitness, such as musculo-skeletal strength, aerobic capacity, flexibility, etc. may be associated with the ability to perform MMH tasks without injury. Exercise has been used in the treatment of back injury for many years, although its exact role and effectiveness are not completely understood (Jackson and Brown, 1983; Kraus, Melleby and Gaston, 1977; Morris and Randolph, 1984).

☛☛☛ *Despite the well-known practice of "Japanese workers doing their calisthenics" the use of fitness training as a preventive program is rare in North America.* ☚☚☚

Musculo-skeletal strength is one of the aspects of physical fitness that is generally believed to be related to back injury. Experience indicates that the occurrence of musculo-skeletal injuries in weaker workers is up to three times greater than for stronger workers (Chaffin, Herrin and Keyserling, 1978), but there is also the unexpected finding that "strong" persons may be more often injured than their weaker colleagues (Battie, Bigos, Fisher, Hansson, Jones, and Wortley, 1989). Although the concept of strength training within an industrial environment as a means of injury prevention is occasionally mentioned, no major research literature on this topic seems to exist.

"Flexibility," particularly of the trunk, appears to be needed for bending and lifting activities (Jackson and Brown, 1983) which are part of MMH tasks. Locke (1983) suggested that a series of 25 stretching exercises be used as a means of warm-up and as a starting point for fitness improvement. Yet, Nordin (1991) stated that flexibility measures have been found to be poor predictors of back problems.

Chenoweth (1983a, 1983b) reported that an industrial fitness program was highly successful with volunteering participants, as compared to their control-group worker collegues. Unfortunately, the control group seems not to have been selected from persons who had volunteered to participate in the program. An empirical study investigated the relationship between fitness and low back injury of 1,652 Los Angeles County firefighters to determine the relationships between five strength and fitness measures with the occurrence of back injury over a three-year period (Cady, Bischoff, O'Connel, Thomas and Allan, 1979a, 1979b). Individuals were rated to be on one of three levels of fitness (high, middle, or low) based on measurements of flexibility, isometric lifting strength, recovery heart rate, blood pressure, and endurance. The results show that the fittest firefighters had the lowest percentage of back injuries, i.e., 0.8 percent; the middle group had 3.2 percent injuries, and the least fit group 7.1 percent injuries; but the fittest had the most severe injuries. This study has often been used to suggest that physical fitness may help to prevent back injuries, but Nordin (1991) cited two more recent longitudinal studies that did not show an association between fitness level (measured via maximal oxygen uptake) and reported back pain.

Regaining and improving "fitness," including flexibility, while recovering from a back injury or other disability has always been of concern to patients and to their nurses and physical therapists. This back-school concept has been incorporated in "work hardening" (see below), where specific body abilities deemed necessary to perform the job are improved through purposeful designed exercises. Fitness training for prevention in back injury is viewed with great interest; however, as in the other two training approaches discussed, evidence is insufficient at this time to allow making sound judgments on the effectiveness of this approach in general, or on specific programs.

Training: What and How

Content. A basic question that has not yet been answered is: "What to teach?" Certainly, the content of a training course is dependent on the aims of the training. Previous training efforts have generally fallen into three areas, as discussed above:

- training of specific lifting techniques, i.e. skill improvement,
- teaching biomechanics, awareness of and self-responsibility for back injuries, thereby changing attitudes, and
- training the body via physical fitness so that it is less susceptible to injury.

Although the aim of injury prevention is the same in each case, the methods of how to achieve that aim are quite different. The traditional approach of training in specific lifting techniques alone does not appear effective; probably because there is no one technique that is appropriate for all lifts. The method of preventing injury by increasing knowledge of the body and promoting attitude changes so that workers feel responsible for their bodies may be quite applicable in MMH. However, exactly what should be taught, and how, is still open. What method is most effective? How much knowledge is needed?

Another key to awareness and attitude change may be the attention paid to MMH problems by management, supervisors, or training instructors: making employees aware of "authority concern" is an underlying theme in the training received from back schools. (Perhaps this is the only element necessary if a Hawthorne effect is deemed sufficient success.)

Research regarding the role of physical training as a preventive method should provide useful information. For example, would "mobility" or "strength" training of workers effectively reduce back injury? Are "Asian-style" or "warm-up" exercises before the work shift beneficial?

Beyond these approach-specific research questions are the corollary problems of implementing two or more approaches in training. For example, how exactly does physical fitness interact with attitude change? In many cases, the programs presented in the literature contain elements of knowledge and attitude change, physical exercises, and specific lifting techniques. It is currently not clear if this combined approach is the best, or even effective at all.

Criteria. Another major and basic question is how to judge the effectiveness of any given training. The most commonly used methods rely on "objective" data derived from company records on cost and productivity, and from medical records to compare quantitative measures "before and after" training. The use of company loss data is most common, but sometimes the exact meaning of these numbers is not clear, particularly if other actions take place during the period of data recording that may have had effects on the loss statistics. Apparently, there is not enough standardization among companies in the definition of terms, or in the actual derivation of the statistics, to allow reliable industrywide comparisons. "Subjective" data result from asking trainees or managers questions such as, "Was training worthwhile?" The value of such judgments is uncertain.

Retention/refreshment. It would not make sense to do long-term evaluations of training if the students never learned the material in the first place. How can we measure whether the workers have actually absorbed what was presented in training sessions? Some test or demonstration as a criterion measure of learning directly following the training could be used as an evaluative tool, but what measure to use and how well one must do on the measure to be considered trained (or having a changed attitude) has not been discussed in the literature.

Another unanswered question concerns the retention of information by the trainees after training has been completed. Was there sufficient original learning? It may have been good immediately following instruction, but why then is there usually an increase in injuries (or whatever measure used) as time passes? Is this a reflection of good training that is forgotten over time, or was the training not "good"? To increase retention of information is a traditional concern. Can interference be reduced? Is transfer easily made from the learning environment to the work environment? Refresher courses (of some kind) probably should be offered, yet the time intervals between training need to be established.

Instructional style and media. Once it has been determined what should be taught, attention must be paid to the formation of the course itself. It seems from the literature that many courses are taught in a lecture format, with practice of techniques at some time during the session. Films and videotapes, audiotapes, posters, and cards in paycheck envelopes have also been used. Programmed instruction, computer-aided instruction, or interactive video are at hand. The (relative) effectiveness of these methods has not been determined (Druckman and Bjork, 1991). Is self-pacing an appropriate technique, and if so, is it more effective than other alternatives?

Where should the sessions be held? Would a classroom with desks and chairs be more appropriate than a lecture hall? The worksite appears most appropriate, but it may not be suitable for instructional purposes. Is it best to train employees working together as a group, or shall the group be split up? Of course, in any situation, there are practical constraints that might limit the ability to implement training ideals, but currently information is not available on which to base sound judgment.

Training "customized" for certain industries or jobs. Low back injuries are a major health problem among industrial workers. They are the most frequent and most expensive musculo-skeletal disorder in the United States, accounting for at least 25 percent of all compensable overexertion injuries. Every second worker is said to experience, at some time in life, back pain severe enough to require medical treatment (Snook, 1991; Yu, Roht, Wise, Kilian, and Weir, 1984).

Back pain has been related to weak trunk musculature, muscular fatigue, degenerative diseases, and improper posture and inappropriate lifting techniques; however, the majority of industrial back injuries has not been associated with objective pathological findings (Taylor, 1987; White and Gordon, 1982), and about every second back pain episode cannot be linked to a specific incident (Pope, 1987; Rowe, 1983).

A variety of actions and events at work have been associated with low back injuries. In the mining industry, overexertion, slips/trips/falls, and jolts in vehicles are the most frequently mentioned events (Bobick and Gutman, 1989). Back injuries have been directly associated with lifting (37 to 49 percent of the cases), pulling (9 to 16 percent), pushing (6 to 9 percent), carrying (5 to 8 percent), lowering (4 to 7 percent), bending (12 to 14 percent), twisting (9 to 18 percent)—the percentages vary considerably among industries and occupations. Industries with the largest incidence ratios for compensation claims for back injuries have been construction and mining. Occupations with high ratios were among garbage collectors, truck drivers, nurses, and, of course, material handlers.

☞☞☞ *The goal of training for prevention of back injuries is changing attitudes, not just the teaching of specific lifting techniques, to achieve self-responsibility "by modifying self-perception and attitudes so that patients view themselves as the primary agents . . . " of their own health (Fisk, DiMonte and Courington, 1983, p. 21).* ☜☜☜

One problem in MMH training is that broad generalizations concerning MMH tasks across all work environments may not be possible. Material handling task characteristics and requirements differ much among industries (e.g., tire-making, mining, nursing) as well as within one industry, depending on the specific job task, handling aids and equipment available, successful implementation of worker selection, and ergonomic job design. Therefore, it is a question how training recommendations are applicable across industrial settings. Even the group characteristics of material handlers in different industries might be important in designing a training program; for example, hospital workers might have higher educational skills than heavy-industry workers. Women employees may be predominant in a given industry or occupation, which might influence MMH training, because, on the whole, women are about two-thirds as strong as men. It is not known how personal or task-specific characteristics should influence training.

In the United States, "only" about two of every one hundred employees report a back injury per year. This poses another problem regarding the effectiveness and cost of back-care instructions. Of the actually reported injuries, about every tenth is serious, yet these few serious injuries cause the by far largest portion of the total cost. Hence, if one wanted to prevent specifically these serious injuries, two of every one thousand employees would be the target sample, while all one thousand must be in the educational program. Even if one wanted to address all persons who may suffer from any kind of back problem, about twenty out of one thousand, this is still a rather expensive approach, which may not appear cost-effective to the administrator.

Ethical considerations. There are ethical considerations in the area of MMH training and research. Workers do get injured and experience pain; so who shall receive training, everyone or just high-risk individuals? When conducting research, a no-treatment control group is often used. Yet, what are the responsibilities of the researcher in choosing whether to train or not train people in ways that might help prevent back injury? In the case of using physical-fitness training for injury prevention, there is the potential for injury from the training itself. Therefore, should training be mandatory or voluntary? If voluntary, will the training be given during working hours or on the worker's own time? Some individuals might feel that "mandatory" fitness training and nutrition guidelines, for example, decree a change in lifestyle that they may not want. How far can employers go?

☞☞☞ *The "willingness to exercise" has recently been investigated by questioning 444 employees of a Canadian power company who had agreed to take a fitness test. The prevalent determiners, on or off the job, are apparently*

- *the already existing habit of exercising: those who exercise are likely to continue; those who don't probably won't.*
- *the perceived obstacles to exercising: if it is difficult to arrange, or to succeed, one is likely not to try, or to drop out.*
- *the attitude: if exercise is considered positive and worthwhile, it is likely to be pursued.*

Unfortunately, this disappointing result appears to be true for all population subgroups tested so far (Godin and Gionet, 1991). ☜☜☜

❑ APPLICATION

Summary Conclusions of the Review of Training

This review covered about 100 published sources; yet, the available literature seems very spotty: it neither addresses all topics of interest, nor does it indicate many clear results. The first deficiency relates to the multifaceted magnitude of the problem, the second also to deficiencies in the "experimental design and control" of training approaches. Jensen (1985) reviewed specifically the literature on theories and processes of training. He developed a "block-diagram analysis" which illuminates the systematic sequence of needs assessment, training-program selection, pre- and post-training assessment, and program improvement. This systematic approach shown in Figure 10-3 helps in the evaluation of reliability, significance, and validity of past and future training for safe manual material handling.

Pizatella, McGlothlin and Putz-Anderson (1989, p. 107) wrote: "Contrary to the popular conception that lower back pain always originates from an identifiable trauma or injury, many of these cases have nonspecific origins, develop gradually over time, and result in a diffuse set of subjective complaints. The frequent absence of any discernible organic or physical pathology and apparent similarities in the personality profiles and psychosocial histories of many back pain sufferers have led researchers and rehabilitation practitioners alike to consider the role of psychological components in the back pain syndrome. . . . This does not imply that such cases involve deliberate faking or malingering but recognizes that factors such as personality structure, coping style, family support, and job satisfaction can act as contributory agents in the recognition, interpretation and reporting of back pain in the etiologic states and as predictors of recovery in the rehabilitation stage . . . ."

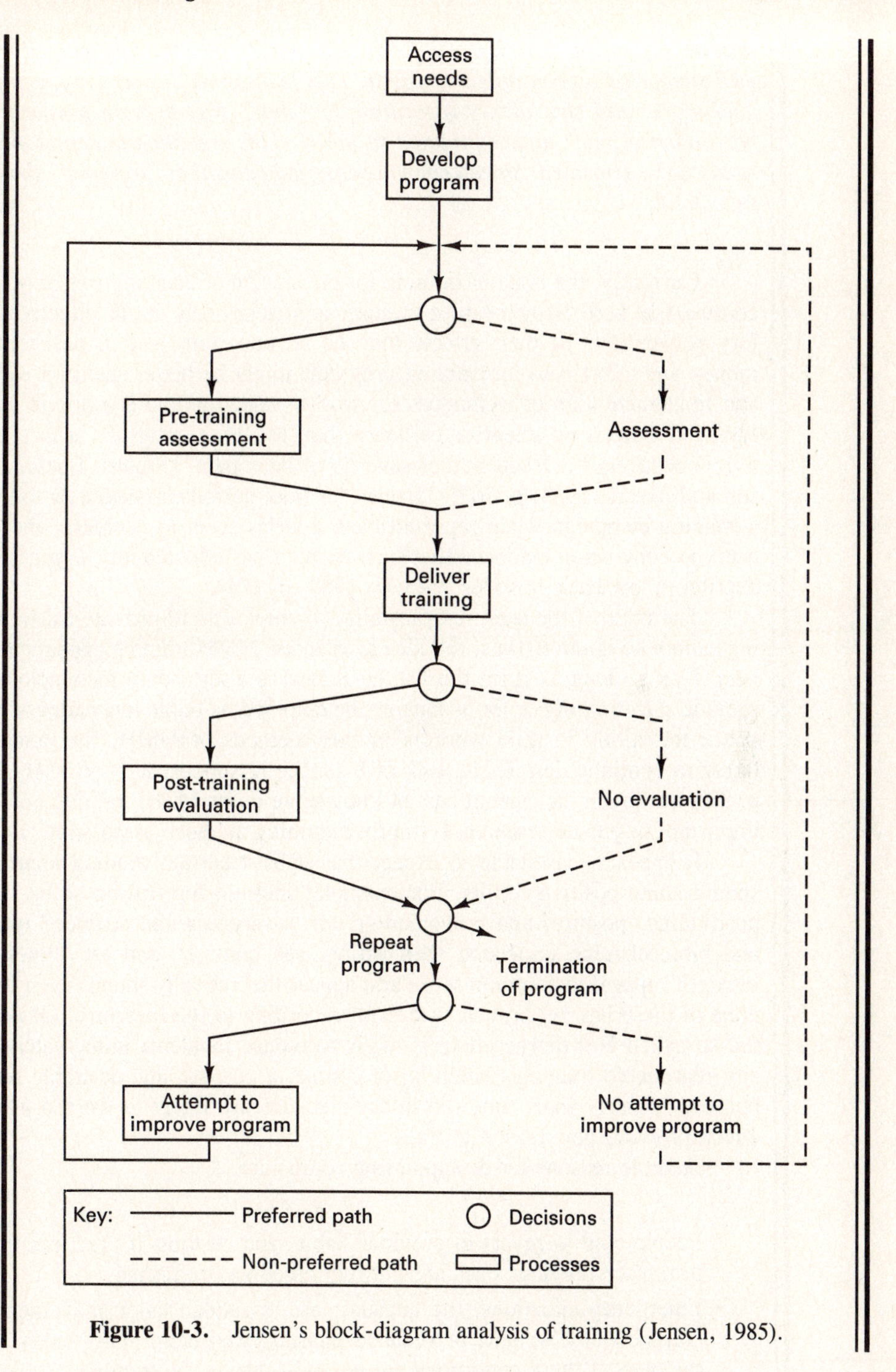

Figure 10-3. Jensen's block-diagram analysis of training (Jensen, 1985).

☞☞☞ *Given the scarcity of information, hardly any training guidelines are well supported by controlled research. This leaves much room for speculation, guesswork, and charlatanry regarding the "best" way to train people for the prevention of back injuries related to MMH. This condition is deplorable and needs to be remedied, since common sense indicates that "training" should be successful.* ☜☜☜

Currently, the issue of training for prevention of back injuries in MMH is confused, at best. Some or most training approaches may not be effective in injury prevention, or their effects may be so uncertain and inconsistent that money and effort paid for training programs might be better spent on research and implementation of techniques for worker selection and ergonomic job design. "[T]here is no scientific evidence that [MMH training] is, in fact, effective in reducing the frequency or severity of back pain" (Stubbs, Buckle, Hudson and Rivers, 1983, p. 767). "Except for brief periods immediately following a training campaign when reported back injuries seem to decrease, there has been no convincing evidence that such training has effected any significant reduction in low back episodes" (Rowe, 1983, p. 101).

The issue of the legal responsibility of employers to provide MMH training cannot be ignored; thus, the idea to abandon MMH training appears unrealistic. "Yet so long as it [in the United States] is a legal duty for employers to provide such training or for as long as the employer is liable to a claim of negligence for failing to train workers in safe methods of MMH, the practice is likely to continue despite the lack of evidence to support it." (NIOSH, 1981, p. 99). So, given the current lack of knowledge in the MMH training area, it is important to pursue research to improve training as much as possible.

It appears reasonable to expect that at least certain training approaches should show positive results. For example, training for "lifting skills" (body positioning, posture, and movements), for "awareness and attitude" (physics and biomechanics associated with lifting, self-control), and for "fitness and strength" appeal to common sense and appear theoretically sound, even though none of these has yet proven successful according to the literature. Of course, the so-called Hawthorne effect is likely to reduce incidents immediately after any reasonable training (which is, of course, a positive and desirable result), but after a fairly short time the injury statistics are likely to worsen again—which provides new reason to train.

Possible reasons for disappointing results are:

- People tend to revert to previous habits and customs if practices trained to replace previous ones are not reinforced and refreshed.
- Emergency situations, the unusual case, a sudden quick movement, increased body weight, or reduced physical well-being may overly strain the body if these conditions did not exist during the training.
- If the job requirements are stressful, "doctoring the symptoms" such as

behavior modification will not eliminate the inherent risk. Designing a safe job is fundamentally better than training people to behave safely.

Nevertheless, several general recommendations should help to reduce the risk of overexertion injuries when lifting. These are:

Leglift versus backlift. The "leg lift" has been heavily promoted in training, as opposed to the "back lift." It is indeed, normally, better to straighten the bent legs while lifting rather than unbending the back. But leg lifts can be done only with certain types of loads, either those of small size which fits between the legs, or those having "two handles" between which one can stand (two small suitcases instead of one big case). Large and bulky loads can usually not be lifted by unbending the knees without body contortion; if one attempts to do so, one may in fact stress the spine disks more than when also slightly unbending the back. Hence, proper task and material design is necessary to permit leg lifts.

Rules for "safe" lifting. There are no comprehensive and sure-shot rules for "safe lifting," which is a complex combination of moving body segments, changing joint angles, tightening muscles, and loading the spinal column.

Here are some guidelines for proper lifting:

+1. Design manual lifting (and lowering) out of the task and workplace. If it needs to be done by a person, perform it between knuckle and shoulder height.

+2. Be in good physical shape. If not used to lifting and vigorous exercise, do not attempt to do difficult lifting or lowering tasks.

+3. Think before acting. Place material conveniently. Make sure sufficient space is cleared. Have handling aids available.

+4. Get a good grip on the load. Test the weight before trying to move it. If it is too bulky or heavy, get a mechanical lifting aid, or somebody else to help, or both.

+5. Get the load close to the body. Place the feet close to the load. Stand in a stable position, have the feet point in the direction of movement.

+6. Involve primarily straightening of the legs in lifting.

There are some things to avoid:

−1. Do NOT twist the back, or bend sideways.

−2. Do NOT lift or lower, push or pull, awkwardly.

−3. Do NOT hesitate to get help, either mechanical or by another person.

−4. Do NOT lift or lower with arms extended.

−5. Do NOT continue heaving when the load is too heavy.

These rules and recommendations should not only be part of a worker training program but also tell the engineer and manager how to design the job. For example, the loads should be

- of proper size to be lifted by "straightening the knees."
- placed at proper height to be handled "in front of the trunk."
- of proper form and shape so that one can get "a good grip."

Obviously, the topic of training MMH techniques to prevent injury, particularly to the back, has not been studied in a cohesive or systematic manner. The overall impression is that many health and safety professionals, industrial engineers and managers have accepted the general idea of training workers in proper lifting and handling techniques and of improving "awareness and attitude," but have not yet clearly determined what content and which media to use nor made evaluations of long-term effectiveness and worth of the various training approaches. Much "applied research" needs to be done until training theory and practice are integrated and successful (Cannon-Bowers, Tannenbaum, Salas, and Converse, 1991).

PERSONNEL SELECTION BY PHYSICAL TESTING

While training is one approach to "fit the person to the job," the other is the selection of suitable persons, i.e., the screening of individuals with the purpose of placing those on strenuous jobs who can do them safely. This screening may be done either before employment, before placement on a new job, or on occasion of routine examinations during employment.

A basic premise of personnel selection by physical characteristics is that the risk of overexertion injury from manual material handling decreases as the handler's capability to perform such activity increases. This means that a test should be designed to allow judgment about the match between a person's capabilities for manual material handling, and the actual MMH demands of the job. Hence, this matching process requires that one knows, quantitatively, both the job requirements and the related capabilities (to be tested) of the individual. (Of course, if the job requirements are excessive, they should be lowered before any matching is attempted.)

Capability Limitations

The human body must maintain the balance between external demands of the work and of related internal capacities. The body is an "energy factory," converting chemical energy derived from nutrients into externally useful physical energy. Final stages of this metabolic process take place at skeletal muscle, which needs oxygen transported from the lungs by the blood. The blood flow also removes byproducts generated in the energy conversion, such as carbon dioxide, water, and heat, which are dissipated in the lungs, where oxygen is absorbed into the blood. Heat and water are also dispelled through the skin (sweat). The blood circulation is powered by the heart. (See the first chapters of this book for more details.)

Thus, the pulmonary system, the circulatory system, and the metabolic system establish *central limitations* of a person's ability to perform strenuous work.

A person's capability may be limited also by muscular strength, by the ability for movement in body joints, and in manual material handling often by the stress responses of the spinal column. As discussed earlier, these are *local limitations* for the force or work that a person can exert.

Hence, either central or local limitations may determine the individual performance capability. While handling material, the force or torque exerted with the hands must be transmitted through the body, that is via wrists, elbows, shoulders, trunk, hips, knees, ankles, and feet to the floor. In this chain of force vectors, the weakest link determines the capability of the whole body to do the job. If muscles are weak, or if they have to pull under mechanical disadvantages, the available handling force is reduced. Often, the lumbar section of the spinal column is the weakest link in this kinematic chain: muscular or ligamental strain or painful displacements of the vertebrae and/or of the intervertebral disks may limit a person's ability for material handling.

Assessment Methods

Various methods have been developed to assess an individual's capabilities for performing specified handling tasks (Himmelstein and Andersson, 1988). Such assessments rest on a fundament of (by nature general) epidemiological or of more specific etiological information.

The *medical examination* primarily identifies persons with physical impairments or diseases. While x-ray examinations are now sparingly used, the "physical" screens out the medically unfit on the basis of their physiological and orthopaedic traits. Unless specific job requirements are known to the testing physician or nurse, they must evaluate the person's capabilities against "generic" job demands. Hence, the examination is often not a specific match of capabilities (actually, of capability limitations) and job demands.

The *physiological examination* usually identifies individual limitations in central capabilities, e.g., of pulmonary, circulatory or metabolic functions. This provides essential criteria if these functions are indeed highly taxed by the material handling job, such as in frequent movements with heavy loads and much body involvement—however, this is now seldom the case in technically developed industries.

The *biomechanical examination* addresses mechanical functions of the body, primarily of the musculo-skeletal type, e.g., load-bearing capacities of the spine, or muscle strength exertable in certain postures or motions. Biomechanical methods rely on explicit models of such body functions, which is both their strength and weakness: The results are only as good (reliable and valid) as the underlying models.

The *psychophysical examination* addresses all (local or central) functions strained in the test: hence, it may include all or many of the systems checked via medical, physiological, or biomechanical methods. It filters the strain experienced through the sensation of the subject who "rates the perceived exertion" (Borg, 1982). In tests of maximal voluntary exertions, the subject decides how much strain is "acceptable" under the given conditions, e.g., how large an exertion will in fact be performed voluntarily.

The physician is expected to perform essentially all of these investigations during the medical examination (Dukes-Dobos, 1989), with the additional task of relating them to general and specific (but often unspecified) job demands to determine whether a proper match exists.

Screening Techniques

Several screening techniques exist which (should) serve to select persons who are able to perform defined material handling activities with no or acceptably small risks of overexertion injuries; i.e., whose capabilities match the job demands with a safety margin. In the past, managers, foremen, or physicians had to rely on many intuitive, experience-based guesses; only recently a number of preemployment or placement tests have been developed based on models and methods just discussed. Primarily, these tests differ in the techniques used to generate the external stresses that strain the (local and central) function capabilities to be measured. They must be judged against the criteria validity, reliability, and feasibility.

The biomedical ("strength") tests currently in use rely on either static or dynamic stressing of the subject:

Static techniques require the subject to exert isometric muscle strength against an external measuring instrument. Since muscle length does not change, there is no displacement of body segments involved; hence, no time derivatives of displacement exist. This establishes a mechanically and physiologically simple case which allows straightforward measurement of isometric muscle strength capability. A standardized procedure (Caldwell, Chaffin, Dukes-Dobos, Kroemer, Laubach, Snook, and Wasserman, 1974) has been widely accepted for such static strength measurement, which in turn has become part of well-established screening techniques claimed to be effective (Ayoub, Selan, and Jiang, 1984; Anderson and Catterall, 1987; Laughery, Jackson, and Fontenelle, 1988; Chaffin, 1981; Kumar, Chaffin, and Redfern, 1988; NIOSH, 1981).

Dynamic techniques appear more relevant to actual material handling activities (Ayoub and Mital, 1989; Stevenson, Bryant, Greenhorn, and Thomson, 1989), but they are also more complex because of the large number of possible displacements and of their time derivatives velocity, acceleration, and jerk—see the first chapter in this book. Hence, most current dynamic testing techniques employ either one of two ways to generate dynamic test stresses.

In the *isovelocity* (or isokinematic, often falsely called "isokinetic") technique, the subject moves the limb (or trunk) at constant angular velocity about a specific joint (usually knee, hip, shoulder, or elbow) while exerting maximal voluntary torque. Several kinds of test equipment are on the market. They are so designed that the exerted torque can be monitored continuously throughout the angular displacement (Marras, King, and Joynt, 1984). Since this equipment is rather costly, simplified versions are available where a handle is being moved at constant speed while angular joint speeds are not specified. (Note that in the initial and final portions of the test, limb speed is not constant—see the discussion in the first chapter of this text.)

The *isoinertial technique* requires the subject to move a constant mass (weight) between two defined points. The maximal load that can be lifted (or lowered, carried, or held) by the subject is the measure of that person's capability, while the forces or torques actually exerted depend (according to Newton's second law) on mass and acceleration. This rather realistic technique is part of many currently employed tests.

APPLICATION

Discussion of screening techniques. The main advantage of the *static* (isometric) techniques is the conceptual simplicity: putting a person into a few "frozen" positions and measuring the force which she or he can generate in one given direction over a period of just a few seconds is indeed easily understood. Furthermore, since no displacement takes place (muscle lengths and joint locations do not change) the physical conditions are very simple, since all time derivatives of displacement are zero. This allows the use of rather simple instruments and permits relatively easy control of the experimental condition. It is for these and other reasons that static strength measurement was initially suggested by Chaffin, Herrin, and Keyserling (1978), established by NIOSH in 1981, and since strongly promoted by Chaffin (1991) as a first analytical assessment of individual capabilities. Static strength testing was a significant step forward from the earlier simplistic assumption of single "safe" loads which children, women, and men supposedly could lift without danger of overexertion, but it has been shown to have rather low predictive power for dynamic tasks, with R^2 often well below 0.5 (Aghazadeh and Ayoub, 1985; Ayoub, Mital, Asfour, and Bethea, 1980; Kamon, Kiser, and Pytel, 1982; Kroemer, 1982; 1983a; McDaniel, Skandis, and Madole, 1983; Nordin, 1991). This unsatisfactory relationship between static testing and dynamic performance capability has triggered research to develop more suitable dynamic measuring techniques (Ayoub, Jiang, Smith, Selan, and McDaniel, 1987).

Apparently, the striking advantage of any suitable *dynamic* measuring technique is the better similarity to actual material handling, which is usually done in motion, not motionless. The testing difficulty lies in the fact that body members can be moved in a variety of dynamic conditions, depending on the actual path of motion, the time consumed, accelerations and decelerations applied to different masses involved, etc. Overcoming this difficulty by reducing the first derivative of displacement, velocity, to a constant is characteristic of *isovelocity* (falsely called "isokinetic") measurement techniques. Their main attraction lies in the hope that after measuring isovelocity strengths around all body joints (wrist, elbow, shoulder, spine, hip, knee, ankle) involved in the task, one may use this information about body component capabilities to synthesize the total capability of the body. Unfortunately, the pleasure of the analytical scientist is also the nightmare of the engineer and of the accountant (when seeing the price tag for the needed equipment). Simplifying the equipment so that, e.g., only a handle is moved at constant speed gives away most of the significant advantages and attractions of the analytical technique.

This leaves for consideration the apparently oldest and most practical test of lifting: to actually lift a load. To have somebody "try it" is certainly a realistic test of lifting capability: it assesses all involved capabilities or limitations,

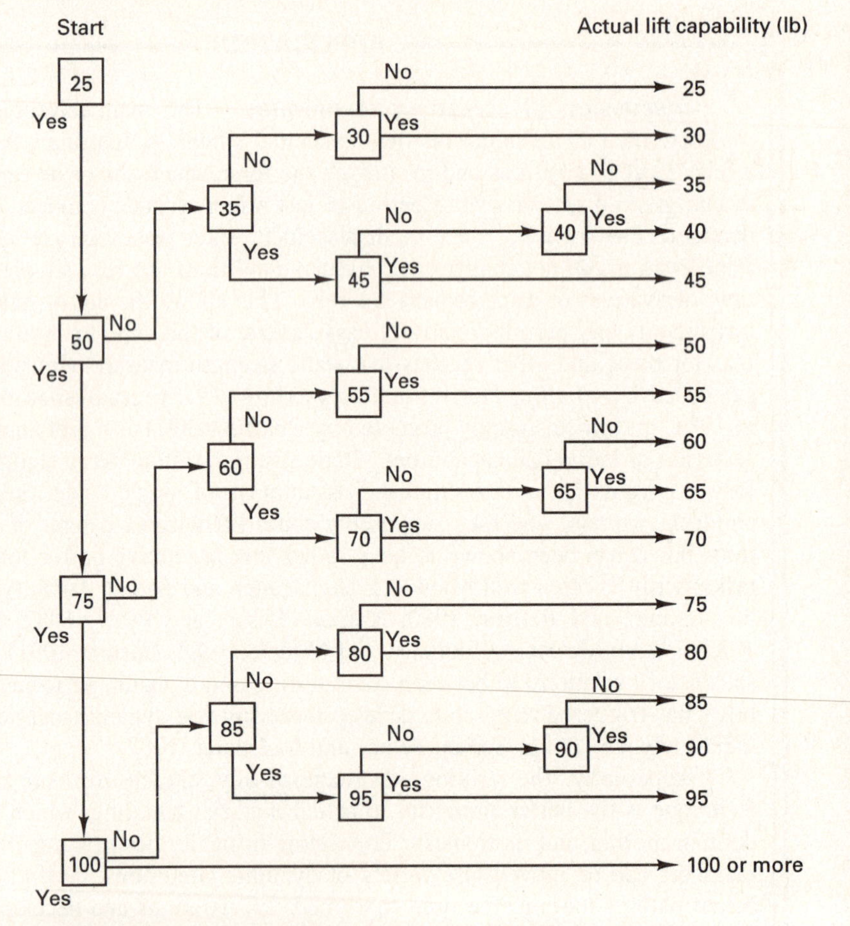

Figure 10-4. Sequence to determine actual lifting capability used in "LIFTEST" (Kroemer, 1982).

though it does not provide analytical information about the contributions of individual muscles, muscle groups, lever arms, etc. The incremental *isoinertial* test is easily understood, easily executed, easily controlled; safe, reliable and valid; and has been used to test millions of U.S. military recruits (McDaniel, Skandis, and Madole, 1983; Teves, Wright, and Vogel, 1985). The isoinertial lift test equipment consists essentially of two upright channels that guide a carriage to which weights are attached at the rear. The subject attempts to lift the carriage by its handles, not knowing what weight is attached. The experimenter increases the weight until the subjects reaches the maximum which he or she can lift. A strategy of increasing/decreasing the loads to determine a person's lifting capacity used in LIFTEST (Kroemer, 1982) is shown in Figure 10-4. The scheme of adding 25 lb, subtracting 15, adding 10, and subtracting 5 lb takes less than 10 minutes (including rest periods) to determine the actual lifting ability to the nearest 5 lb. The equipment is inexpensive and robust. Such an isoinertial test is an efficient, reliable, and realistic approach to assess an individual's capability to lift.

As in all areas of developing scientific and applied knowledge, the "state-of-the-art" changes, often rapidly. With respect to testing for lift strength, static testing was current just a few years ago, but is becoming outdated by dynamic tests which, although feasible, have not been generally validated yet.

It was pointed out in the section on muscle strength (in Chapter 1 of this book) that all maximal exertions depend on the subject's motivation or will to indeed give a "maximal" exertion, and that this depends on the internal feedback of sensory information about the strains felt in the body while exerting. Therefore, the outcomes of *all* testing procedures are strongly yet imperceptibly affected by the testee's evaluation of the experienced body strain and the associated motivation to exert more or less effort. As such, all tests are "psychophysical" regardless of whether they are static or dynamic; in fact, personal assessment is the overriding factor in strength testing. Using a mathematical model to predict "biomechanical" lifting ability, Waikar, Aghazadeh, and Parks (1991) found that the subjectively acceptable weight of an object to be lifted was below that predicted via the model. The reason may be that the model used was unrealistic (see the discussion of "models" in Chapter 7) or that, indeed, the subjective assessment of one's own body strain is the best (or safest) method to determine what is suitable or not. This thought gives some credence to the "psychophysical technique" of assessment, such as used prominently by Snook and co-workers (1978, 1991). They gave their subjects monetary incentives to work as hard as they could but instructed them to do so without becoming unusually tired, weakened, overheated, or out of breath.

Validity. A major problem in all current techniques of lift-strength testing is that of validity, meaning the predictive power of the tests for "real-life" ("true working") conditions. Given that static (isometric) strength testing can no longer be considered a sufficient measure of an individual's ability to perform lifting tasks, isoinertial strength testing appears to be a better personnel selection procedure. While its face validity is high, rigorous assessments of the validity of the isoinertial testing has not been performed yet (Fernandez, Ayoub, and Smith, 1991). To do so is straightforward and inexpensive. A typical scenario includes one large cohort group of material handlers whose capabilities are assessed with an isoinertial test. While no job assignments are made, the workers' work performance including incidence statistics is observed over a sufficient period of time, say one year. Validity is demonstrated if those persons who tested better also perform better, i.e., have fewer overexertion-related incidents.

Ethical and Legal Considerations

A major concern in all tests is to ensure that the tested individuals are not unfairly treated as a result of the testing. In essence, there is the problem of balancing the needs of the system (as perceived by the employer) and of individuals (on whom this system finally depends). The test criteria to be met must be representative of the actual and necessary requirements on the job. The testing procedures must be valid and reliable. A further problem is that both criteria and procedures are based on group data and do not consider individual variations (Webb and Tack, 1988). Although the specific legal requirements are different from country to country, in general tests should not discriminate against any given group. Yet, for example, as a group women are "weaker" than men (Astrand and Rodahl, 1986; NASA/Webb, 1978); younger employees have been found to have more frequent but often lighter injuries than older employees (Bigos, Spengler, Martin, Zeh, Fisher, and Nachemson, 1986). Also, rather low correlations exist between the occurrence of low back pain and body weight, static strength, or postural deficiencies (scoliosis, lordosis, kyphosis, uneven leg length) unless extreme (Andersson, 1991). Therefore, testing an *individual* to match her or his special capabilities with specified job requirements appears to be a desirable solution.

APPLICATION

II ERGONOMIC DESIGN

How the job is designed determines the stress imposed by the work on the human material handler. The size of the handled object, its weight, whether it has handles or not; the layout of the task, the kinds of body motions to be performed, the muscular forces and torques to be exerted, the frequency of these efforts, the organization of work and rest periods, and other engineering and managerial aspects determine whether a job is well designed or not, if it is safe, efficient, and agreeable to the operator.

In the following, facility layout, the environment, and equipment will be discussed.

Facility Layout

The layout of the overall work facility contributes toward safe and efficient material transfer. Organizing the flow of material in general, and designing it carefully in detail, determines the involvement of people and how they must handle the material.

In the real world, one encounters either one of two situations: a facility exists, it must be used, and the building and its interior layout cannot be changed significantly. By small changes, one must make the best use of what there is. In the second case, one can plan and design a new facility and its details according to the process at hand, to suit the persons and the production.

The opportunity to "do it right at the planning stage" is most desirable, allowing the closest approximation to the optimal solution. Striving for the best (possible) solution is also the purpose for modification of a given facility. Therefore, the ideal case will be used here to guide, even when only modifications may be possible.

It is the purpose of facility layout, or facility improvement, to select the most economical, safest, and efficient design of building, department, and workstation (Konz, 1990). Of course, specific details depend on the overall process.

A facility with well-planned material flow has short and few transportation lines. Reduction of material movement through proper facility layout can lower the cost of material transport considerably, which usually accounts to 30 to 75 percent of total operating cost, and is even higher in some instances. Unfortunately, material movement adds no value to the product, but is dangerous. Hence, reduction of material handling is both a major safety and a cost factor.

Process vs. product layout. There are two major design strategies: "process layout" and "product layout." In the first case, all machines or processes of the same type are grouped together, such as all heat treating in one room, all production machines in another section, and all assembly work in a different division. Figures 10-5 and 10-6 are examples of "process layout."

There is a major advantage to this design strategy. Quite different products or parts may flow through the same workstations, keeping machines busy. But much floor space is needed, and there are no fixed flow paths. Process layout requires relatively much material handling. It is worthwhile to study Figures 10-5 and 10-6 carefully to determine what improvements should be made in each case to improve the conditions depicted.

In contrast, in "product layout" all machines, processes, and activities needed for the work on the same product are grouped together. This results in short throughput lines, and relatively little floor space is needed. However, the layout suits only the specific product, and breakdown of any single machine or of special transport equipment may stop everything. Altogether, product layout is advantageous for material handling because routes of material flow can be predetermined and planned well in advance.

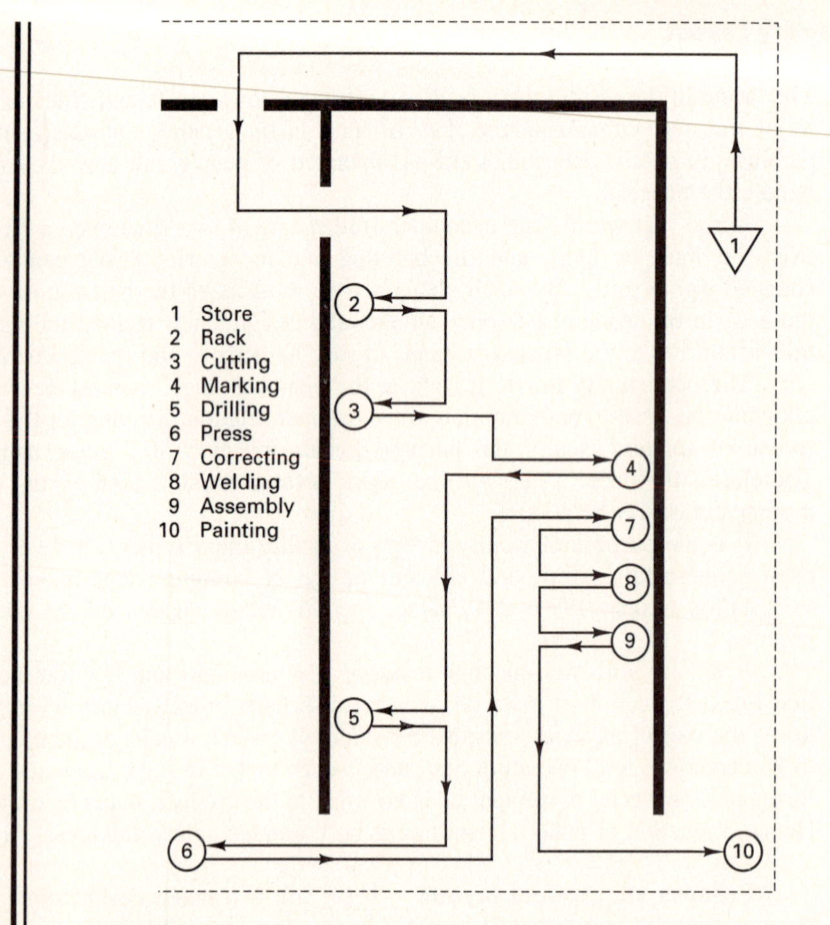

Figure 10-5. Flow diagram of bus seat production. (Adapted from ILO, 1974.)

Flow charts and diagrams. The Figures 10-5 and 10-6 show how easy it is to describe events and activities with simple sketches, symbols, and words.

The *flow diagram* is a picture or a sketch of the activities and events. It indicates their sequence, and where they take place.

The *flow chart* is a listing or table of the same activities and events, which can easily be put together and stored on a computer. It indicates their durations and provides detailed information on related facts or conditions. The events depicted in Figure 10-6 are charted in Figure 10-7.

Flow diagrams or charts may indicate activities performed on a given material: this is called a "material (centered) diagram," or chart. If it shows the activities of one person, it is called an "operator (centered) diagram" or chart.

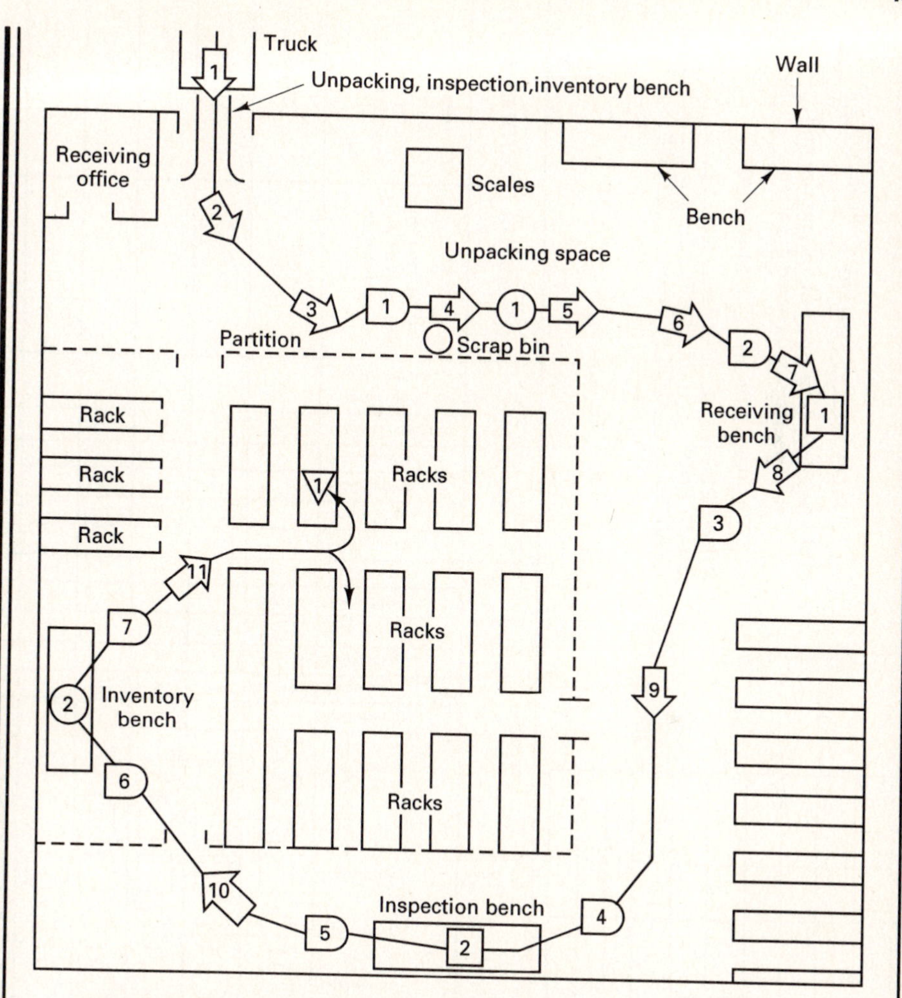

Figure 10-6. Flow diagram of receiving, inspection, inventorizing, and storage: original set-up. (Adapted from ILO, 1974.)

Making a flow chart and setting up a flow diagram are simple, straightforward tasks that can be done by even a lay person. Chart and diagram help to understand what is done, when it is done, how long it takes, what is involved, and which hazards are present. Many industrial engineering texts explain in detail how to make flow charts and flow diagrams, and how to interpret the results for ergonomic improvements of the working conditions.

An application of flow diagramming for a receiving facility was shown in Figure 10-6. Even a cursory look shows many workstations, many long transports, many delays—altogether, a waste of time, space, and personnel, combined with many possible sources of injuries caused by manual material movement. Figure 10-7 details all the actions that are necessary to perform the

SUMMARY	PRESENT		PROPOSED		DIFFERENCE	
	No.	Time	No.	Time	No.	Time
○ Operations	2	8				
⇨ Transportations	11	26				
□ Inspections	2	35				
D Delays	7	85				
▽ Storages	1	2				
Distance travelled	61 m		m		m	
Number of hazards: High	0					
Number of hazards: Medium	8					
Number of hazards: Low	8					

FLOW PROCESS CHART

No. 1
Page 1 of 2

JOB Receive, check, inspect, inventorize and storage of parts received in cartons.

☐ OPERATOR OR ☒ MATERIAL

CHART BEGINS 9:15 a.m.
CHART ENDS 11:31 a.m.
CHARTED BY KHEK Date 10/16/89

METHOD PRESENT OR PROPOSED	ACTIVITY					FACTS						HAZARDS						COMMENTS	CONTROL ACTION
Describe in Detail Each Shop	Operation	Transport	Inspection	Delay	Storage	Distance m	Time min	Weight kg	Size m.m.m	Freq/shift	# of people	Falling mat	Sharp edges	Pinch points	Hazard mat	Manual handling	Overall rating High/med/low	Identify Important Aspects, Particularly Hazard Potentials	Eliminate Combine Simplify Personal Protection etc.
1 Carton lifted from truck; placed on inclined roller conveyor		1				2	2	75	0.7 0.7 0.5	4 0 0	2	√				√	L	Back injury possible	Change
2 Slid on conveyor		2				6	1	▪	▪	▪	2	√					L		Combine
3 Stacked on floor		3				6	4	▪	▪	▪	2			√			L		Eliminate
4 Await unpacking				1		–	32										/		Eliminate
5 Unstacked		4				1	2	▪	▪	▪	2	√		√			M		Eliminate
6 Lid removed, shipment papers removed	1					–	3	▪	▪	▪	2		√	√			M		Change
7 Place on hand truck		5				1	1	▪	▪	▪	2	√		√			M	Back injury possible	Eliminate

Figure 10-7. Flow chart of the set-up shown in Figure 10-6.

No.	Details	Symbol														
8	Trucked to receiving bench	⇨ 6	9	4	▪	▪	▪	1	√					L		Combine
9	Await removal from truck	D 2	–	10										/		Eliminate
10	Placed on bench	⇨ 7	1	1	▪	▪	▪	2			√			L	Back injury possible	Eliminate
11	Parts taken from carton, checked, replaced	□ 1	–	15	▪	▪	▪	1		√	√			M		Simplify
12	Carton placed on hand truck	⇨ 8	1	1	▪	▪	▪	2	√		√			M	Back injury possible	Eliminate
13	Await transport	D 3	–	5										/		Eliminate
14	Trucked to inspection bench	⇨ 9	17	6	▪	▪	▪		√					L		Eliminate
15	Await inspection	D 4	–	10										/		Eliminate
16	Parts removed, inspected for function, replaced	□ 2	–	20	▪	▪	▪	3		√	√			M		Simplify
17	Await transport	D 5	–	5										/		Eliminate
18	Trucked to inventory bench	⇨ 10	9	5	▪	▪	▪	1	√					L		Eliminate
19	Await inventorizing	D 6	–	15										/		Eliminate
20	Inventized, removed, numbered, replaced	○ 2	–	5	▪	▪	▪	1	√	√	√			M		Change
21	Await transport	D 7	–	8										/		Eliminate
22	Transport to storage rack	⇨ 11	7	4	▪	▪	▪	2	√					L		Change
23	Carton stored	▽ 1	1	2	▪	▪	▪	2	√		√			M		Personnel protection equipment
24																
25																
26																

Figure 10-7. *(cont.)*

work under the original conditions. It shows that in order to do just two operations and two inspections, eleven transports are needed and seven delays are encountered. The material travels 61 meters, mostly by human hand, and it takes nearly $2\frac{1}{2}$ hours to store the material. Sixteen hazards exist.

Figure 10-8 shows a simple solution. Unpacking, inspection, and inventory taking now are all combined in one location, close to the entry. The partition is largely removed and the transport lines have become, through use of a conveyor, short and simple. Listing the activities of the ergonomically designed work in a new flow chart indicates that the time needed to place the material in storage, and the distance traveled by the material, both have been cut in half. The human labor has become much less intensive and dangerous.

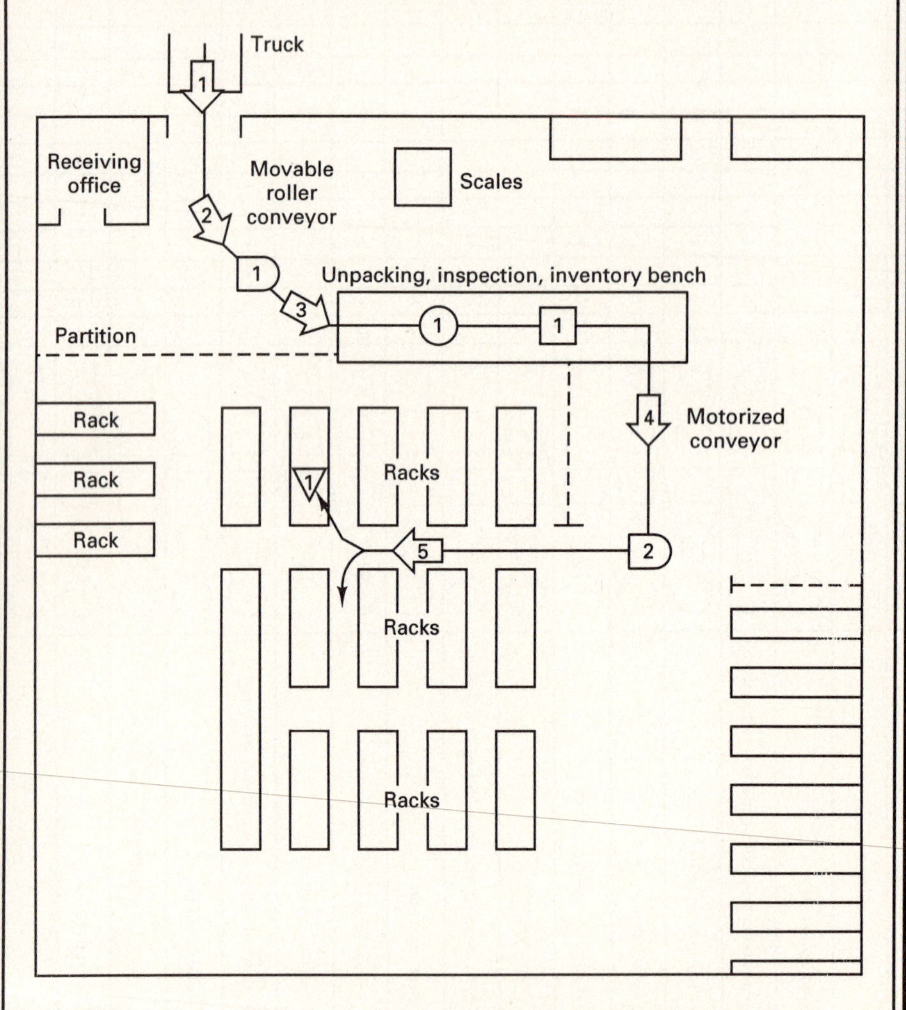

Figure 10-8. Flow diagram of improved set-up. (Adapted from ILO, 1974.)

The Work Environment

The work environment can be made to contribute to safe manual material activities if it is well designed and maintained.

The *visual environment* should be well lit, clean, and uncluttered; allowing good depth perception and discrimination of visual details, of differences in contrast, and of colors.

The *thermal environment* should be within zones comfortable for the physical work, usually in the range of about 18° to 22°C. Thermal stress resulting from conditions that are too hot or too cold may contribute to material handling safety problems.

The *acoustical environment* should be agreeable, with sound levels preferably below 75 dBA. Warning sounds and signals indicating unusual conditions should be clearly perceptible by the operator. High noise conditions can contribute to an overall straining of the operator, and hence affect safety of material handling.

For more information in environmental conditions, see Chapters 4 and 5.

Good housekeeping helps to avoid injuries. Safe gripping of the shoes on the floor, or good support from the chair when seated, are necessary conditions for safe material handling. Poor coupling with the floor can result in slipping, tripping, or misstepping. Floor surfaces should be kept clean to provide a good coefficient of friction with the shoes (Chaffin, Woldstad, and Trujillo, 1992). Clutter, loose objects on the floor, dirt, spills, etc. can reduce friction and lead to slip and fall accidents—see Figure 10-9.

Equipment

Equipment may provide assistance at the workplace to the material handler, or equipment may do the actual transportation—see Figures 10-10 through 10-12.

Equipment for assistance at the workplace includes:

- Lift tables, hoists, cranes
- Ball transfer tables, turntables
- Loading/unloading devices.

Transportation equipment includes:

- Nonpowered dollies, walkies, and trucks
- Powered walkies, rider trucks, tractors
- Conveyors of many kinds, trolleys
- Overhead and mobile cranes.

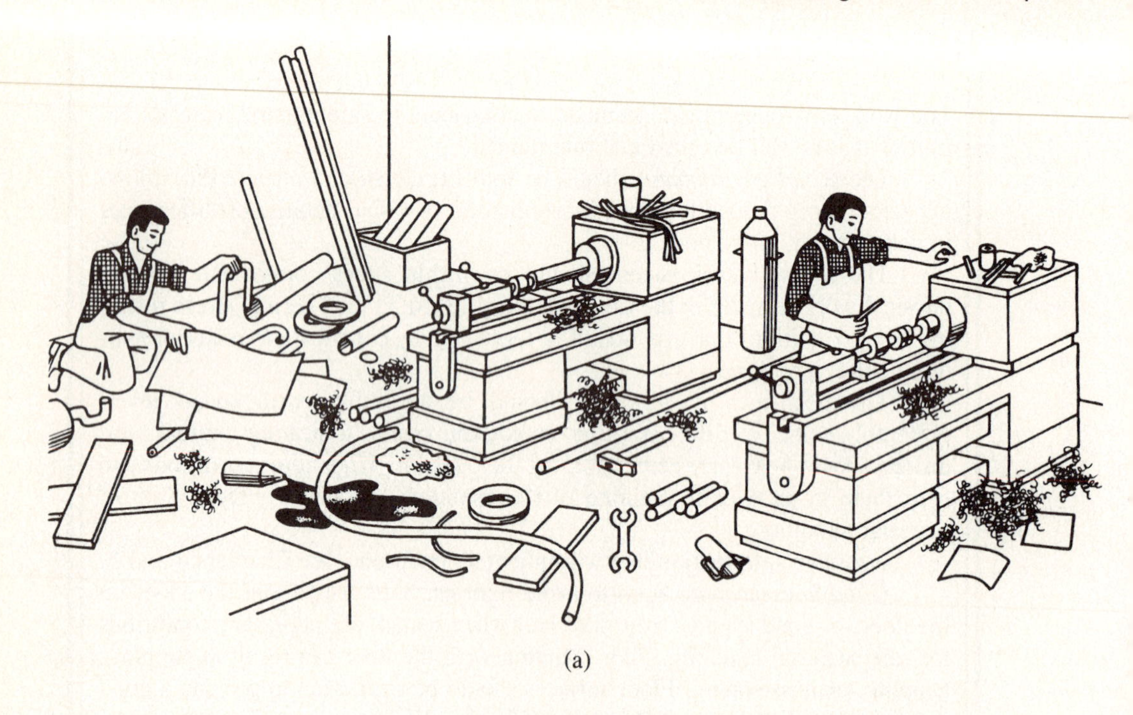

(a)

(b)

Figure 10-9. (a) Bad and (b) good housekeeping. (Courtesy of International Labor Office, 1988.)

Figure 10-10. Simple carts and dollies to roll materials instead of carrying them.

Obviously, several of these can be used both at the workplace and for in-process movement, such as hoists, conveyors, and trucks.

Finally, there is a group of material movement equipment primarily used at receiving and in warehousing. This includes, e.g.:

- Stackers
- Reach trucks
- Lift trucks
- Cranes
- Automated storage and retrieval systems.

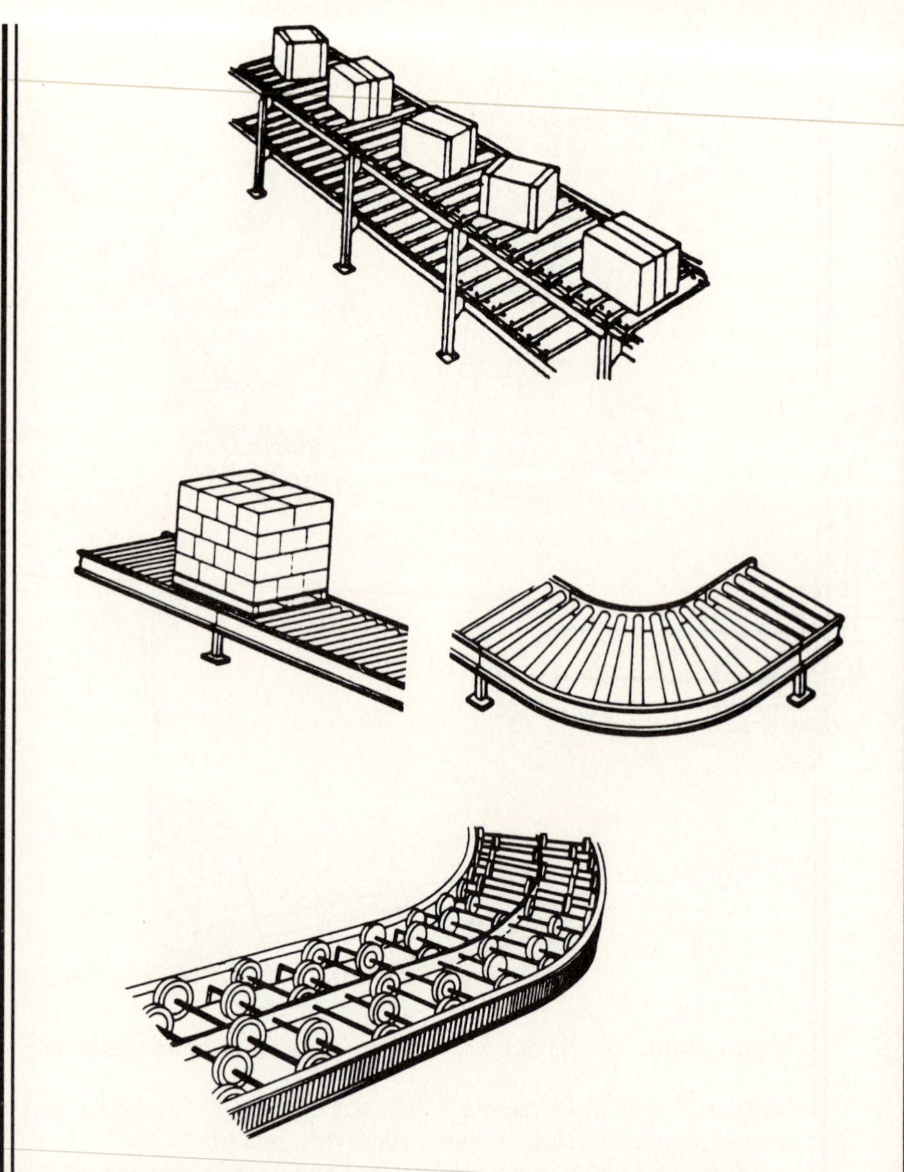

Figure 10-11. Conveyors on which one can push objects easily.

Will the equipment be used? All equipment mentioned above can take over the requirements of holding, carrying, pushing, pulling, lowering, and lifting of materials which would otherwise be performed with the hands of a person. However, whether this will indeed be done by machines depends, besides economical considerations, on the layout and organization of the work itself. For example: will an operator use a hoist to lift material to be fed into a

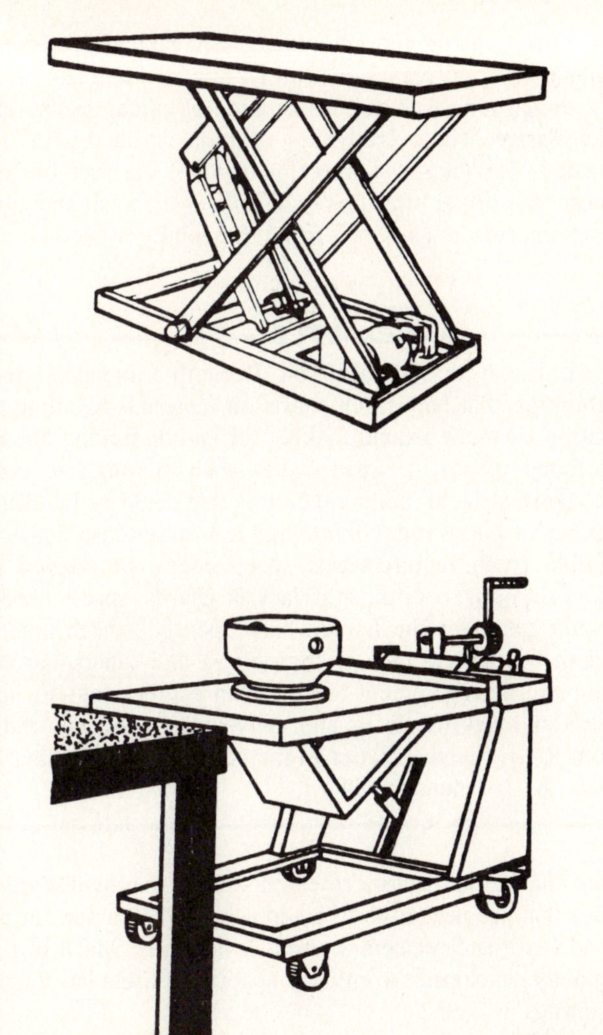

Figure 10-12. Lift tables.

machine if this is time-consuming, or awkward? Will a lift table be installed next to an assembly workstation if this means removal or relocation of another workstation in order to make sufficient room (Bobick and Gutman, 1989)?

Obviously, facility layout as well as workplace design must be suitable for the use of equipment. Furthermore, the operator must be convinced (trained) that it is worthwhile to go through the effort of using a hoist instead of heaving the material by hand.

Ergonomic design of equipment. Equipment must not only be selected to be able to perform the material handling job, but also to "fit" the human operator. Ease of use must be considered together with the safety of personnel working with or along the equipment. Unfortunately, some material movement equipment such as cranes and hoists, powered and hand trucks, and conveyors show an alarming lack of consideration of human factors and safety principles in their design. The overall worst example is found with "forklifts," which, when lifting, carrying and lowering a load, obstruct the operator's view. Furthermore, the driver often has only little space to sit and is subjected to vibrations and impacts transmitted from the rolling wheels—see Figure 10-13.

EXAMPLE

"I was shocked, dismayed, and perturbed. Recently I attended a regional industrial exhibition that had an emphasis on materials-handling equipment. I intentionally went around looking for bad or lacking human engineering. I found plenty . . . inadequate labels; wrong-size controls; lack of shape, position, color coding; controls that could be inadvertently actuated; absence of guard rails; unintelligible instructions; slippery surfaces; impossible reach requirements; sharp edges; unguarded pinch-points; extreme strength requirements; lack of guards; spaces needed for maintenance too small for the human hand; poorly located emergency switches; and so on, and so on! . . . Spacecraft and supersonic aircraft and missile monitoring equipment need human engineering; so, too, do hydraulic hoists and forklift trucks and conveyor systems and ladders." (Excerpts from R. B. Sleight's letter to the editor of the *Human Factors Society Bulletin*, p. 7, January 1984.)

Ergonomics and human-factors research work has provided information on design features that are needed to fit equipment to the human, in particular in order to provide safe and efficient working conditions. Much of this information was originally developed for military applications, but has found its way into industrial settings as well.

Overall dimensions and space requirements can be derived from human body and reach dimensions, as demonstrated in Chapter 8. Purposeful application of such information ensures, for example, that a driver compartment of a lift truck accommodates all driver sizes, or that operators must not strain themselves in trying to reach an object hanging from an overhead conveyor, or to grasp material in the far bottom corner of a transport bin. Of course, one should never design for the "average person," because this ghost does not really exist. Instead, one must design for body size ranges, such as from the 5th to the 95th percentile.

Handles on containers or tools should be so designed and oriented that hand and forearm of the operator are aligned. Do not force the operator to work with a bent wrist: "Bend the handle, not the wrist." Carpal tunnel syndrome and other cumulative trauma problems are less likely to occur if a person can work with a straight wrist—see Chapter 8.

Figure 10-13. Typical forklift truck. While it has a cage to protect the driver, the lift mechanism makes it very difficult to see what is ahead, even if there is no load.

The handle should be of such shape and material that squeezing forces are distributed over the largest possible palm area. The handle diameter should be in the range of $2\frac{1}{2}$ to 5 centimeters. Its surface should be slip resistant (Freivalds, 1987a). Coupling with the hands is best by protruding handles or by gripping-notch types, while hand-hold cutouts or drawer-pull types are acceptable (Deeb, Drury, and Pizatella, 1987). If they are on a boxlike object (of about 40 × 40 × 40 cm, weighing between 9 and 13 kg), one should arrange the location of handles according to the scheme shown in Figure 10-14: the right side of the box is divided into nine areas, with area #5 in the center. (The same numbering system is on the left side of the box, with #1 again on top and close to the body.) The best or at least suitable handle locations are listed in Table 10-1. The worst case is not to have any handles (Drury, Deeb, Hartman, Woolley, Drury, and Gallagher, 1989).

Containers and trays must be designed to have the proper weight, size, balance, and coupling with the human hand. The container should be as light as possible to add little to the load of the material. Its size and handle arrange-

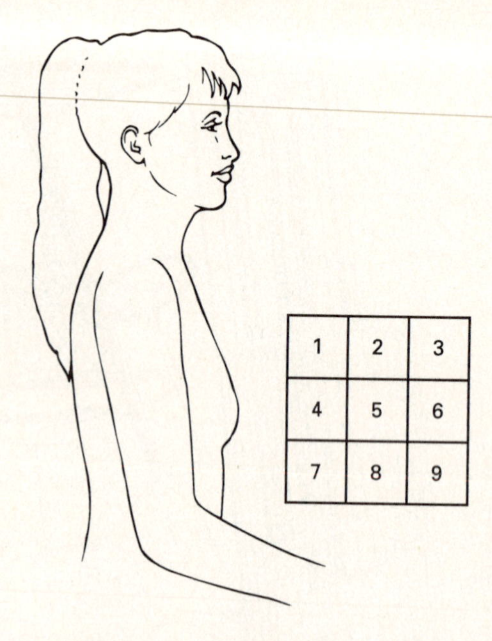

Figure 10-14. Scheme to describe handle locations in a box-type container. Each side is divided into nine regions, with #1 on top and closest to the body.

TABLE 10-1. PREFERRED HANDLE LOCATIONS (SEE FIGURE 10-14) AT 40 × 40 × 40 CM BOXES

	For lifting and lowering in front of the body	For work with sideways twisting of the body
Best	2/2	6/8, 8/8 for heavy loads
Acceptable	3/8, 6/8	3/8, 2/2

Handles or cutouts may be angled (against horizontal): 2/2 at 83 degrees, 3/8 at 76 degrees, 6/8 at 60 degrees, 8/8 at 50 degrees.

SOURCE: From Drury, Deeb, Hartman, Woolley, Drury and Gallagher, 1989.

ment should be so that the center of the loaded tray is close to the body. It should be well balanced, with its weight centered between, but below the hand-holds. Big, heavy, pliable "bags" are usually more difficult to handle than boxes or trays.

Written instructions and labels are not necessary if equipment and operation are designed to be "perfectly obvious." However, when instructions are needed, labeling should be done according to these rules (see also the appropriate section in Chapter 11 on controls and displays):

- Write the instructions in the simplest most direct manner possible.
- Give only the needed information.
- Describe clearly the required action. Never mix different instruction categories such as operation, maintenance, warning.

- Use familiar words.
- Be brief, but not ambiguous.
- Locate labels in a consistent manner.
- Words should read horizontally, not vertically.
- Label color should contrast with equipment background.

Ergonomic Design of Workstation and Work Task

The most effective and efficient way to reduce MMH injuries is to design equipment ergonomically, i.e. so that job demands are matched to human capabilities.

"Designing the job to fit the human" can take several approaches. The most radical solution is to "design out" manual material movement by assigning it to machines: *no people involved, no people at risk*. If people must be involved, load weight and size shall be kept small, best accompanied by ergonomic design of the work task, i.e. by selecting the proper type of material handling movements (e.g., horizontal push instead of vertical lift) and their frequency of occurrence. The location of the object with respect to the body is very important: best between hip and shoulder height, directly in front of the body so as to avoid twisting or bending the trunk—see Figures 10-15 and 10-16. The object itself is important, of course, regarding its bulk, its pliability (e.g., firm box versus pliable bag), and its ability to be grasped securely (shape, handles). Naturally, the workplace itself must be well designed and maintained: proper working height; material provided in containers from which it can be removed easily (Figure 10-17); nonslip floor, clean, quiet, orderly, suitable climate . . . all ergonomic conditions contribute to avoiding stress, physical overexertions and accidents.

Figure 10-15. Store material at proper height. (Modified from International Labor Office, 1988.)

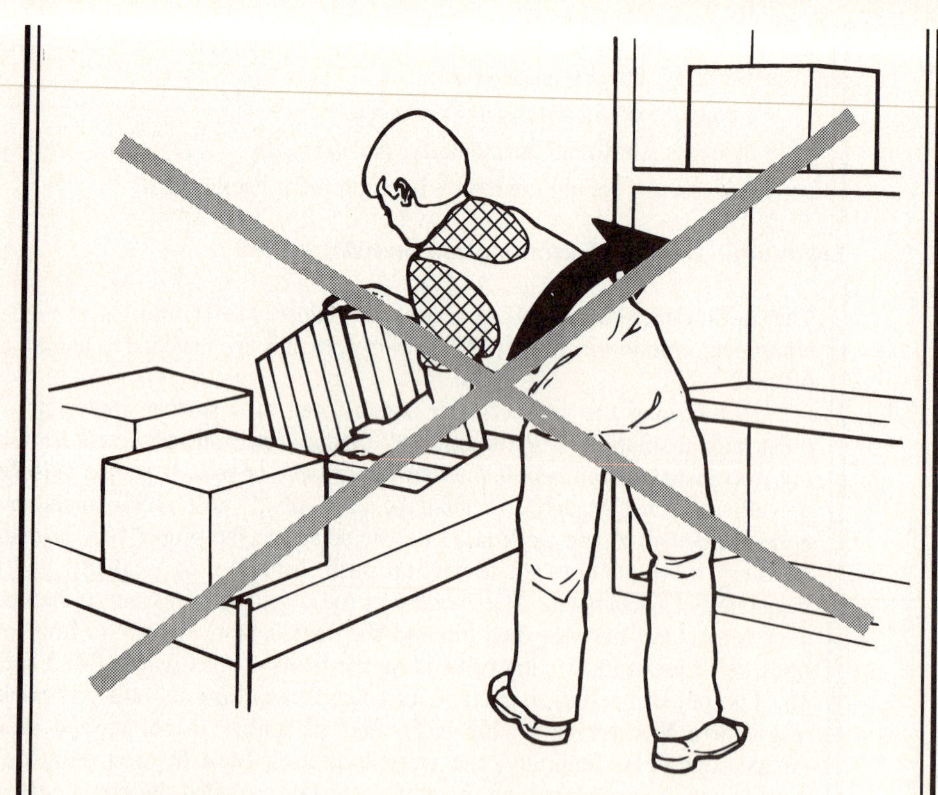

Figure 10-16. Avoid twisting the trunk. (Modified from International Labor Office, 1988.)

Permissible Loads for Manual Material Handling

Only a few decades ago it was believed that certain set weights could be lifted safely by men or women, or children. This simplistic idea is no longer legitimate, nor acceptable. Using epidemiological, medical, physiological, biomechanical, and administrative approaches, new knowledge on human capabilities for manual material handling has been gained, depending on such variables as frequency, location, direction, and other details of the MMH activities. Before describing two major sets of recommendations, it should be stressed they are based on assumptions and approaches that need refinement and further evaluation (Ayoub, 1991; Ayoub and Mital, 1989; Ayoub, Selan, and Jiang, 1986; Chaffin and Andersson, 1991; Ciriello, Snook, Blick, and Wilkinson, 1990; Dolittle, 1989; Gagnon and Smith, 1992; Noone and Mazumdar, 1992; Marras, 1988; Pope, Frymoyer, and Andersson, 1984; Potoin, Norman, Eckenrath, McGill, and Bennett, 1992; Sanders and Grieve, 1992; Sommerich and Marras, 1992; Snook, 1987).

Figure 10-17. Deliver and store containers at proper working height.

☞☞☞ *In many countries, specific weights have been set which one adult male worker is allowed to manually lift and carry. Examples of such maximal weights are 100 kg in Greece (meat or slaughter animals carried), 90 kg in Pakistan, and 50 kg in Czechoslovakia and the Philippines. Some countries have qualified their weight limits by provisos. In Brazil, 60 kg may be carried over distances up to 60 meters, while 40 kg may be lifted by one worker. "Workers may be allocated to lifting, carrying and transport of loads only in accordance with their constitution and physical strength" (Austria, 1983). "No worker in an undertaking shall be required to lift or carry a load which is so heavy that it could injure him (her)" (New Zealand, 1981). In Egypt, women are prohibited to do the work of loading and unloading of goods at docks, wharves, ports, and warehouses. In Hungary, boys under 14 years of age may not be employed in loading and transport of goods. In Cameroon, females under 18 years of age may not carry loads by means of hand barrows and two-wheeled carts (International Labour Office, 1988).* ☜☜☜

Limits for Lifting and Lowering

In 1981, a panel of experts prepared for the U.S. National Institute of Occupational Safety and Health a "Work Practices Guide for Manual Lifting" (NIOSH, 1981). For the first time, this document contained distinct recommendations for acceptable masses to be lifted that differed from the previous assumptions that one could establish *one* given weight each for men, women, or children which would be "safe" to lift. In this 1981 guide, two different threshold curves were established: the lower, called "Action Limit" (*AL*) was thought to be safe for 99 percent of working men and 75 percent of women in the United States. The *AL* values were dependent on the starting height of the load, the length of its upward path, its distance in front of the body, and the frequency of lifting. If the existing weight was above the *AL* value, engineering or managerial controls had to be applied to bring the load value down to the acceptable limit. However, under no circumstances was continuation of the lifting task allowed if the existing load was three times larger than the action limit values. This threshold was called the "Maximum Permissible Load" (*MPL*).

A decade later, NIOSH revised the technique for assessing overexertion hazards of manual lifting (Putz-Anderson and Walters, 1991). This new document no longer contains two separate weight limits, but has only one "Recommended Weight Limit" (*RWL*). It represents the maximal weight of a load that may be lifted or lowered by about 90 percent of American industrial workers, male or female, physically fit and accustomed to physical labor.

The 1991 equation used to calculate the *RWL* resembles the 1981 formula for *AL*, but includes new multipliers to reflect asymmetry and the quality of hand-load coupling. Yet, the 1991 equation allows as maximum a "Load Constant" (*LC*) (permissible only under the most favorable circumstances) with a value of 23 kg (51 lb) which may not be exceeded under any foreseeable circumstances. This is quite a reduction from the maximal 40 kg in the 1981 NIOSH guidelines.

The following assumptions and limitations apply:

- The equation does NOT include safety factors for such conditions as unexpectedly heavy loads, slips, or falls, or for temperatures outside the range of 19°C (66°F) to 26°C (79°F) and for humidity outside 35 to 65 percent.
- The equation does NOT apply to one-handed tasks while seated or kneeling, or to tasks in a constrained workspace.
- The equation assumes that other manual handling activities and body motions requiring high energy expenditure such as in pushing, pulling, carrying, walking, climbing, or static efforts as in holding, are less than 20 percent of the total work activity for the work shift (Waters, 1991).
- The equation assumes that the worker/floor surface coupling provides a coefficient of static friction of at least 0.4 between the shoe sole and the standing surface.
- The equation may be applied under the following circumstances:

Lifting or lowering tasks, i.e., the acts of manually grasping and moving an object of definable size without mechanical aids to a different height level.

The time duration of such an act is normally between two and four seconds. The load is grasped with both hands.

The motion is smooth and coninuous.

The posture is unrestricted (see above).

The foot traction is adequate (see above).

The temperature and humidity are moderate (see above).

The horizontal distance between the two hands is no more than 65 cm (25 in.).

To help apply the recommended weight limit, a "Lifting Index" (*LI*) is calculated: $LI = L/RWL$, with L the actual load. If *LI* is at or below 1 (one), no action must be taken. If *LI* exceeds 1, the job must be ergonomically redesigned.

For these conditions, NIOSH (Waters, Putz-Anderson and Garg, 1991) provides an equation for calculating the Recommended Weight Limit (*RWL*):

EXAMPLE

$$RWL = LC*HM*VM*DM*AM*FM*CM$$

LC is the "Load Constant" of 23 kg (51 lb).

Each multiplier can assume values between zero and one:

HM represents the "Horizontal Multiplier," where *H* is the horizontal location (distance) of the hands from the midpoint between the ankles at the start and at the end points of the lift.

VM is the "Vertical Multiplier," where *V* is the vertical location (height) of the hands above the floor at the start and end points of the lift.

DM is the "Distance Multiplier," where *D* is the vertical travel distance from the start to the end points of the lift.

AM is the "Asymmetry Multiplier," where *A* is the angle of asymmetry, i.e., the angular displacement of the load from the medial (mid-sagittal plane) which forces the operator to twist the body. It is measured at the start and end points of the lift, projected onto the floor.

FM is the "Frequency Multiplier," where *F* is the frequency rate of lifting, expressed in lifts per minute. It depends on the duration of the lifting task.

CM is the "Coupling Multiplier," where *C* indicates the quality of coupling between hand and load.

The following values are entered in the equation for *RWL*:

	METRIC	U.S. CUSTOMARY
LC = Load Constant =	23 kg	51 lb
HM = Horizontal Multiplier =	$25/H$	$10/H$
VM = Vertical Multiplier =	$1 - (0.003\|V - 75\|)$	$1 - (0.0075\|V - 30\|)$
DM = Distance Multiplier =	$0.82 + (4.5/D)$	$0.82 + (1.8/D)$
AM = Asymmetry Multiplier=	$1 - (0.0032A)$	$1 - (0.0032A)$
FM = Frequency Multiplier (see listing below)		
CM = Coupling Multiplier (see listing below).		

These variables can have the following values:

H is between 25 cm (10 in.) and 63 cm (25 in.) Although objects can be carried or held closer than 25 cm in front of the ankles, most objects that are closer cannot be lifted or lowered without encountering interference from the abdomen. Objects farther away than 63 cm (25 in.) cannot be reached and cannot be lifted or lowered without loss of body balance, particularly when the lift is asymmetrical and the operator is small.

V is between 25 cm (10 in.) and (175 − V) cm [(70 − V) in.]. For a lifting task, D is equal to $V_{end} - V_{start}$; for a lowering task, D equals $V_{start} - V_{end}$.

A is between 0° and 135°.

F is between one lift or lower every five minutes (over a working time of eight hours) to 15 lifts or lowers every minute (over a time of one hour, or less), depending on the vertical location V of the object. Table 10-2 lists the Frequency Multipliers, *FM*.

C is between 1.00 ("good") and 0.90 ("poor"). The effectiveness of the coupling may vary as the object is being lifted or lowered: a "good" coupling can quickly become "poor." Three categories are defined in detail in the NIOSH publication and result in the following listing of values for the Coupling Multiplier, *CM*:

COUPLINGS	$V < 75$ cm (30 in.)	$V \leq 75$ cm (30 in.)
Good	1.00	1.00
Fair	0.95	1.00
Poor	0.90	0.90

TABLE 10-2. FREQUENCY MULTIPLIERS FM

Frequency, lifts/min	Work duration (continuous) ≤ 8 hr		≤ 2 hr		≤ 1 hr	
	$V < 75$*	$V \geq 75$	$V < 75$	$V \geq 75$	$V < 75$	$V \geq 75$
0.2	0.85	0.85	0.95	0.95	1.00	1.00
0.5	0.81	0.81	0.92	0.92	0.97	0.97
1	0.75	0.75	0.88	0.88	0.94	0.94
2	0.65	0.65	0.84	0.84	0.91	0.91
3	0.55	0.55	0.79	0.79	0.88	0.88
4	0.45	0.45	0.72	0.72	0.84	0.84
5	0.35	0.35	0.60	0.60	0.80	0.80
6	0.27	0.27	0.50	0.50	0.75	0.75
7	0.22	0.22	0.42	0.42	0.70	0.70
8	0.18	0.18	0.35	0.35	0.60	0.60
9	0	0.15	0.30	0.30	0.52	0.52
10	0	0.13	0.26	0.26	0.45	0.45
11	0	0	0	0.23	0.41	0.41
12	0	0	0	0.21	0.37	0.37
13	0	0	0	0	0	0.34
14	0	0	0	0	0	0.31
15	0	0	0	0	0	0.28
>15	0	0	0	0	0	0

* Values of V are in cm.

SOURCE: From Putz-Anderson and Waters, 1991.

Limits for lifting, lowering, pushing, pulling, and carrying. In 1978, Snook published extensive tables of loads and forces found acceptable by male and female workers for continuous manual material handling jobs. These data were first updated in 1983 by Ciriello and Snook, finally revised in 1991 by Snook and Ciriello. The following prerequisites apply:

- Two-handed symmetrical material handling in the medial (mid-sagittal) plane, i.e., directly in front of the body; yet, a light body twist may occur during lifting or lowering.
- Moderate width of the load, such as 75 cm or less.
- Good couplings, i.e., of hands with handles, shoes with floor.
- Unrestricted working postures.
- Favorable physical environment, such as about 21°C at a relative humidity of 45 percent.
- Only minimal other physical work activities.
- Material handlers who are physically fit and accustomed to labor.

There are several differences in the format of Snook and Ciriello's recommendations compared to the NIOSH guidelines. The NIOSH values are unisex, while the Snook and Ciriello data are separated for female and males.

The Snook and Ciriello (1991) data are also grouped with respect to the percentage of the worker population to whom the values are acceptable. In the excerpts reprinted here, only those values are listed that are said to be acceptable to 50 percent, or 75 percent, or 90 percent of the worker population. (See the original tables for more information.)

Lifts are subdivided into three different height areas: floor to knuckle height, between knuckle and shoulder heights, and shoulder to overhead reach heights.

Tables 10-3, 10-4, 10-5, 10-6, and 10-7 show, in abbreviated form, Snook's and Ciriello's 1991 recommendations for suitable loads and forces in lifting, lowering, pushing, pulling, and carrying. Their original tables should be consulted for more information.

The data do not indicate individual capacity limits; rather, they represent the opinions of more than 100 experienced material handlers as to what they would do willingly and without overexertion.

It is of interest to note that, similar to NIOSH recommendations, the data in Snook and Ciriello's (1991) study also indicate that lack of handles reduces the loads that people are willing to lift and lower by an average of about 15 percent. If the objects become so wide or so deep as to be difficult to grasp, the lifting and lowering values are again considerably reduced. If several of the material handling activities occur together, the most strenuous task establishes the handling limit: look in the tables for the lowest percentage acceptance of a given set of parameters to find the most limiting task condition.

TABLE 10-3. MAXIMAL ACCEPTABLE LIFT WEIGHTS (KG)

Width (a)	Distance (b)	Percent (c)	Floor level to knuckle height One lift every								Knuckle height to shoulder height One lift every								Shoulder height to overhead reach One lift every							
			5	9	14	1	2	5	30	8	5	9	14	1	2	5	30	8	5	9	14	1	2	5	30	8
			sec			min				hr	sec			min				hr	sec			min				hr
	Males																									
		90	9	10	12	16	18	20	20	24	9	12	14	17	17	18	20	22	8	11	13	16	16	17	18	20
34	51	75	12	58	18	23	26	28	29	34	12	16	18	22	23	23	26	29	11	14	17	21	21	22	24	26
		50	17	20	24	31	35	38	39	46	15	20	23	28	29	30	33	36	14	18	21	26	27	28	31	34
	Females																									
		90	7	9	9	11	12	12	13	18	8	8	9	10	11	11	12	14	7	7	8	9	10	10	11	12
34	51	75	9	11	12	14	15	15	16	22	9	10	11	12	13	13	14	17	8	8	9	11	11	11	12	14
		50	11	13	14	16	18	18	20	27	10	11	13	14	15	15	17	19	9	10	11	12	13	13	14	17

(a) Handles in front of the operator (cm)

(b) Vertical distance of lifting (cm)

(c) Acceptable to 50, 75, or 90 percent of industrial workers

SOURCE: From Snook and Ciriello, 1991.

Conversion

1 kg = 2.2 lb

1 cm = 0.4 in.

TABLE 10-4. MAXIMAL ACCEPTABLE LOWER WEIGHTS (KG)

			Knuckle height to floor level								Shoulder height to knuckle height								Overhead reach to shoulder height							
			One lower every								One lower every								One lower every							
Width (a)	Distance (b)	Percent (c)	5 sec	9 sec	14 sec	1 min	2 min	5 min	30 min	8 hr	5 sec	9 sec	14 sec	1 min	2 min	5 min	30 min	8 hr	5 sec	9 sec	14 sec	1 min	2 min	5 min	30 min	8 hr
	Males																									
		90	10	13	14	17	20	22	22	29	11	13	15	17	20	20	20	24	9	10	12	14	16	16	16	20
34	51	75	14	18	20	25	28	30	32	40	15	18	21	23	27	27	27	33	12	14	17	19	22	22	22	27
		50	19	24	26	33	37	40	42	53	20	23	27	30	35	35	35	43	16	19	22	24	28	28	28	35
	Females																									
		90	7	9	9	11	12	13	14	18	8	9	9	10	11	12	12	15	7	8	8	8	10	11	11	13
34	51	75	9	11	11	13	15	16	17	22	9	11	11	12	14	15	15	19	8	9	10	10	12	13	13	16
		50	10	13	14	16	18	19	20	27	11	13	13	14	16	18	18	22	10	11	11	12	14	15	15	19

(a) Handles in front of the operator (cm)
(b) Vertical distance of lowering (cm)
(c) Acceptable to 50, 75, or 90 percent of industrial workers

Conversion
1 kg = 2.2 lb
1 cm = 0.4 in.

SOURCE: From Snook and Ciriello, 1991.

TABLE 10-5. MAXIMAL ACCEPTABLE PUSH FORCES (N)

	Height (a)	Percent (b)	One 2.1 meter push every 6 sec	12 sec	1 min	2 min	5 min	30 min	8 hr		Height (a)	Percent (b)	One 30.5 meter push every 1 min	2 min	5 min	30 min	8 hr
										INITIAL FORCES							
		90	206	235	255	255	275	275	334			90	167	186	216	216	265
Males	95	75	275	304	334	334	353	353	432		95	75	206	235	275	275	343
		50	334	373	422	422	442	442	530			50	265	294	343	343	432
		90	137	147	167	177	196	206	216			90	118	137	147	157	177
Females	89	75	167	177	206	216	235	245	265		89	75	147	157	177	186	206
		50	196	216	245	255	285	294	314			50	177	196	206	226	255
										SUSTAINED FORCES							
		90	98	128	159	167	186	186	226			90	79	98	118	128	157
Males	95	75	137	177	216	216	245	255	304		95	75	108	128	157	177	206
		50	177	226	275	285	324	335	392			50	147	167	196	226	265
		90	59	69	88	88	98	108	128			90	49	59	59	69	88
Females	89	75	79	108	128	128	147	157	186		89	75	79	88	88	98	128
		50	108	147	177	177	196	206	255			50	98	118	118	128	167

(a) Vertical distance from floor to hands (cm)

(b) Acceptable to 50, 75, or 90 percent of industrial workers

Conversion

$1\ kg_f = 2.2\ ib_f = 9.81\ N$

1 cm = 0.4 in.

SOURCE: From Snook and Ciriello, 1991.

TABLE 10-6. MAXIMAL ACCEPTABLE PULL FORCES (**N**)

	Height (a)	Percent (b)	One 2.1 meter pull every 6 sec	12 sec	1 min	2 min	5 min	30 min	8 hr
					INITIAL PULL				
		90	186	216	245	245	265	265	314
Males	95	75	226	265	304	304	314	324	383
		50	275	314	353	353	383	383	461
		90	137	157	177	186	206	216	226
Females	89	75	157	186	206	216	245	255	265
		50	186	226	245	255	285	294	314
					SUSTAINED PULL				
		90	98	128	157	167	186	196	235
Males	95	75	128	167	206	216	245	255	294
		50	157	206	255	265	304	314	363
		90	59	88	98	98	108	118	137
Females	89	75	79	118	128	128	147	157	196
		50	98	147	157	167	186	196	245

(a) Vertical distance from floor to hands (cm)

(b) Acceptable to 50, 75, or 90 percent of industrial workers

Conversion

1 kg_f = 2.2 lb_f = 9.81 N

1 cm = 0.4 in.

SOURCE: From Snook and Ciriello, 1991.

TABLE 10-7. MAXIMAL ACCEPTABLE CARRY WEIGHTS (KG)

	Height (a)	Percent (b)	One 2.1 meter carry every 6 sec	12 sec	1 min	2 min	5 min	30 min	8 hr
		90	13	17	21	21	23	26	31
Males	79	75	18	23	28	29	32	36	42
		50	23	30	37	37	41	46	54
		90	13	14	16	16	16	16	22
Females	72	75	15	17	18	18	19	19	25
		50	17	19	21	21	22	22	29

(a) Vertical distance from floor to hands (cm)

(b) Acceptable to 50, 75, or 90 percent of industrial workers

Conversion

1 kg = 2.2 lb

1 cm = 0.4 in.

SOURCE: From Snook and Ciriello, 1991.

If actual loads or forces exceed table values, engineering or administrative controls should be applied. Snook believes that industrial back injuries could be reduced by about one-third if the loads could be eliminated that lie above the values acceptable to 75 percent of the material handlers.

Comparing Recommendations

The recommendations by NIOSH (1981), Putz-Anderson (1991) and Snook and Ciriello (1991) provide information regarding what can be presumed to be suitable or unacceptable conditions for manual material activities in the United States.

For new systems, the data are initial planning guides for material handling conditions which either can be performed by persons or should be assigned to machines. Following these guides, the designer is able to avoid potentially injurious requirements and can establish both safe and economical conditions.

For the evaluation of existing material handling systems, one can compare actually existing job requirements with the table data in order to seek out those task demands that are likely to exceed human capabilities. Then the working conditions should be changed by engineering intervention (i.e. automation, mechanization, or lowering of demands), and/or managerial intervention (i.e. worker selection and training, and rest pauses or job rotation). Clearly, the engineering intervention is preferable, because it eliminates the source of the risk.

Of course, it is interesting to compare the recommendations. In general, one finds the 1981 NIOSH Action Limits for lifting to be quite similar to the data recommended by Snook in 1978 for 90th-percentile male workers. (This is not surprising, because Snook's lift data were part of the basic information used to develop the NIOSH guidelines.) In some cases, however, particularly at extreme working conditions, substantial discrepancies between the two sets of recommendations exist. In case of conflicting recommendations, it would be prudent to use the lower value.

Assessments of "suitable" efforts in manual material handling have been developed using four different sets of criteria:

1. Physiological (PH), mostly metabolic and circulatory strains.
2. Psychophysical (PP), mostly subjective assessments of what one is willing to do.
3. Intraabdominal pressure (IAP).
4. Biomechanical (BM), mostly assessing compression values in the lumbar area of the back, usually based on model calculations.

From these criteria sets, suitable loadings have been deduced, variously labeled as "maximum acceptable load (*MAL*)," "force limit," "weight limit," or "action limit (*AL*)." Their comparisons are difficult because of differences in the respective underlying concepts, assumptions, procedures, and units. Regardless of these concerns, the user of that information needs to compare the resulting recommendations:

- Garg and Ayoub (1980) found that recommendations resulting from physiological (PH), psychophysical (PP), and biomechanical (BM) approaches were not in general agreement. Specifically, lower recommended lift weights resulted from PP than BM (as was found in 1991 by Waikar, Aghazadeh, and Parks); and lower lift loads at greater lift frequencies resulted from PH than PP.

- NIOSH (1981) did not explicitly compare the outcomes of the different methods, but combined three (excluding IAP) into recommendations of its own. Freivalds (1989) compared the U.S. NIOSH recommendations with those in the U.K. derived from IAP for lifting and found differences in their recommendations. The greatest disparity was at small loads that are lifted far away from the body: altogether, the NIOSH recommendations were lower than the IAP values. Lower limits were also recommended by NIOSH for frequent lifting tasks.
- Nicholson (1989) compared lifting recommendations derived by PP, BM, and IAP criteria. The IAP recommendations were somewhat higher at working heights above shoulder, but often much lower for work below shoulder height, than the PP data. The PP recommendations compared reasonable well with BM recommendations, but the IAP recommendations were only about half the BM values.
- Nicholson also compared pushing and pulling recommendations: For pushing, PP values were in good agreement with IAP recommendations, but for pulling the IAP data were considerably higher than PP recommendations.

APPLICATION

The Effects of Posture

In the preceding considerations it was assumed that the material handler was free to assume any body postures suitable for the job. Yet, there are conditions in which only limited room is available, such as in underground mines (Gallagher, Marras, and Bobick, 1988; Gallagher and Unger, 1990) or in aircraft cargo holds (Stalhammar, Leskinen, Kuorinka, Gautreau, and Troup, 1986). Handling material in stooped, kneeling, bent, sitting, supine, and prone positions has been measured in the laboratory (Chomcherngpat, Mandhani, Lum, and Martin, 1989; Nag, 1991; Sims and Graveling, 1988; Smith, Ayoub, Selan, Kim, and Chen, 1989). These studies have shown that restrictions imposed on the body posture reduce the ability to lift, often severely. Of course, the magnitude of reduction depends very much on the given conditions, not only with respect to the posture, but also in regard to task parameters, such as the distance and the direction in which the object must be moved.

Helping Each Other

If a load appears to be too heavy or large to be handled alone, one might try to call another person to help—see Figure 10-18. This is quite often the case in hospitals, where nurses must lift and move patients (Vojtecky, Harber, Sayre, Billet, and Shimozaki, 1987). How much can be gained by helping each other? A second person might help to stabilize and balance a load which otherwise is difficult to handle (Johnson and Lewis, 1989), but the strengths of two persons

Figure 10-18. It may be easier to have somebody help handling an object than try to do it alone. (Modified from International Labor Office, 1988.)

do not simply add up, partly because the timing of efforts is not exactly the same, particularly if the load is difficult to grasp and handle. Even under conditions of good coordination and suitable placement of hands and feet, at best 90 percent but usually only about 80 percent of the sum of lift strengths of two or three persons can be exerted in isometric exertion. If the effort is dynamic instead of static, two persons can together generate only about two-thirds of their combined single strengths, even less if three persons cooperate (Karwowski 1988; Karwowski and Pongpatanasuegsa, 1988; Oriet and Dutta, 1989).

Use of "Pressure Belts"

The intraabdominal pressure is believed to help support the curvature of the spinal column and to relieve it from some of the compression force generated by the outside load, and by the body weight, in lifting or lowering of loads (see Chapter 1 and Figure 10-2). To maintain this internal pressure might be helped by an external wrapping around the abdominal region, making the walls of the pressure column stiffer. Weight lifters use wide and contoured belts. Based on their habits it has been advocated that people who do heavy manual material handling should also wear such belts. Several studies have been performed, with varying conclusions (Andersson, 1991; Harman, Rosenstein, Frykman, and Nigro, 1989; Krag, Pope, and Gilbertson, 1984; Lander, Simonton, and Giacobbe, 1990; McGill, Norman, and Sharratt, 1990; Walsh and Schwartz, 1990). Altogether, the use of such belts for industrial material handling does not seem to be an effective way of preventing overexertion injuries: even competitive weight lifters do suffer back injuries, belts might restrict mobility, and they may provide a false feeling of security.

Activities Other Than Industrial Lifting and Carrying

Past research has predominantly dealt with lifting of loads, often in fact considering only the ability to generate an upward-directed static force. Information for "lifting" does not always relate well to the opposite activity, lowering. Aside from static (isometric) horizontal forces, either as push or pull (see Chapter 8), little information about actual material handling is available beyond that reported by Snook (1978) and Snook and Ciriello (1991), discussed above. Their information on industrial carrying assumes that the load is carried by the hands, but there are other techniques of carrying loads that are hardly ever mentioned in the literature, yet commonly used. One is to carry loads on or in the arms, often by clasping the load and pressing it against the trunk. This everyday technique is not easily described in biomechanical terms, and the associated muscular effort and energy expenditure depend very much on the actual clasping manner used, and on the load carried. The other common but seldom described technique is to support a load on the hips, such as a mother does when carrying her small child.

Loads can be carried in many different ways. The technique that is most appropriate depends on many variables: the amount (weight) of load, its shape and size, rigidity/pliability, provision of handholds or points of attachment, the bulkiness/compactness of the load. What is the best technique also depends on the distance of carry, on whether the path is straight or curvy, flat or inclined, with or without obstacles; whether one can walk freely or must duck (because of limited space) or hide, such as a soldier who wants not to be detected. Many of these aspects have not been formally investigated, but some have been the object of studies (Ayoub and Smith, 1988; Datta and Ramanathlan, 1971; Haisman, 1988; Kirk and Schneider, 1991; Mello, Damokosh, Reynolds, Witt, and Vogel, 1988). The existing results and experiences may be summarized as follows:

In general, the load should be carried near to the mid axis of the body, near waist height. The further away from there, either toward the feet or toward one side, the more demanding the carrying becomes. Carrying a medium load, 25 to 30 kg, distributed on chest and back is the least energy-consuming method of carry (but the load on the front and sides must be kept small in size and very close to the body as not to hinder movements). Carrying the load on the back, or well distributed across both shoulders and the neck, also costs fairly little energy. Carrying the load in one hand is quite fatiguing and stressful, particularly for the muscles of the hand, shoulder, and back. Although carrying a load in one hand is biomechanically and physiologically disadvantageous, it is often done because of the convenience of quickly grasping and securely holding and releasing an object, and carrying it over short distances.

However, if one is used to doing so, carrying loads of proper sizes and weights on the head is also suitable. While the load is quite a distance away from the center of body mass, it is right on top of it. Thus, carrying a load on the head requires much balancing skill and a healthy spine, but does not demand much energy beyond that needed to move the body.

The suitable technique of load carrying also depends on the terrain to be covered. Walking on a smooth, solid surface is least energy demanding—See Chapter 2, Table 2–2. The effort increases as one walks on a dirt road, through light brush, on hard-packed snow, through heavy brush, or on a swampy bog. Walking in loose sand is about double as energy demanding as walking on smooth blacktop. In soft snow of approximately 20 cm depth, the energy demand is about three times that of walking on blacktop, and about four times when the snow is 35 cm deep (Duggan and Haisman, 1992; Pandolf, Givoni, and Goldman, 1977; Pandolf, Haisman and Goldman, 1976).

Table 10-8 provides a survey of the different ways of carrying a 30-kg load. Of the conditions investigated, carrying the load equally distributed in both hands was most energy demanding, while carrying it evenly distributed on chest and back was least demanding. Although few measured data are available, one should expect that proper positioning of the load on the trunk, including waist and hip, might be quite suitable. For very small loads, the location probably has little effect on energy demands, while convenience and availability of space may be the determining factors: one might place items into pockets along the thighs, around the waist, on the chest, or on the sleeves. With respect to muscular fatigue, "anecdotal evidence" indicates that carrying heavy loads in the hands (such as logs of wood, or suitcases) is very fatiguing for the muscles keeping the fingers closed and for the muscles crossing the shoulders. Also, stabilizing a load with the elevated hands and arm, such as when carrying on the head, on the back, or on the shoulders, can force termination of the carrying due to local muscle fatigue, not due to energy exhaustion in general. Fatigue may also occur when the load is placed asymmetrically, such as carried in one hand, on one shoulder, or on the back or chest alone. In these cases, the one-sided loading must be counteracted by muscles, which consequently are likely to fatigue if pulled into action for long periods of time, or at high percentages of their total strength.

Another major problem of load carrying is that of local discomfort, often brought about by ischemia, i.e., lack of blood supply due to compression of tissues. This is the case, for example, if a small hand grip cuts into the tissues of the hand, or if a narrow strap across the shoulder compresses the tissue.

TABLE 10-8. TECHNIQUES FOR CARRYING LOADS.

	Estimated energy expenditure for carrying 30 kg on straight flat path (kcal/min)	Estimated muscular fatigue	Local pressure and ischemia	Stability of loaded body	Special aspects	
In one hand	?	Very high	Very high	Very poor	Load easily manipulated and released	Suitable for quick pick-up and release; for short-term carriage of even heavy loads.
In both hands, equal weights	Very high, about 7	High	High	Poor		
Clasped between arms and trunk	?	?	?	?	Compromise between hand and trunk use	
On head. supported with one hand	Fairly low, about 5	(High if hand guidance needed)	?	Very poor	May free hand(s); strongly limits body mobility; determines posture; pad is needed	If accustomed to this technique, suitable for heavy and bulky loads.
On neck	Medium, about 5.5, therpa-type strap around forehead	?	?	Poor	May free hand(s); affects posture	
On one shoulder	?	High	Very high	Very poor	May free hand; strongly affects posture	Suitable for short-term transport of heavy and bulky loads.
Across both shoulders by yoke, held with one hand	High, about 6.2.	?	High	Poor	May free hand(s); affects posture	Suitable for bulky and heavy loads; pads and means of attachment must be carefully provided.

On back	Medium 5.3 backpack; 5.9 for bag held in place with hands	Low	?	Poor	Usually frees hands; forces forward trunk bend; skin cooling problem	Suitable for large loads and long-time carriage. Packaging must be done carefully, attachment means shall not generate areas of high pressure on body.
On chest	?	Low	?	Poor	Frees hands; easy hand access; reduces trunk mobility; skin cooling problem	Very advantageous for small loads that must be accessible.
Distributed on chest and back	4.8, lowest	Lowest	?	Good	Frees hands; may reduce trunk mobility; skin cooling problem	Very advantageous for loads that can be divided/distributed; suitable for long-time carriage.
At waist, on buttocks	?	Low	?	Very good	Frees hands; may reduce trunk mobility	Around waist for smaller items, distributed in pockets or by special attachments; superior surface of buttocks often used to partially support backpacks.
On hip	?	Low	?	Very good	Frees hands; may affect mobility	Often used to prop large loads temporarily.
On legs	?	High	?	Good	Easily reached with hands; may affect walking	Requires pockets in garments and/or special attachments.
On foot	Highest	Highest	?	Poor	Usually not useful	

In general, load carrying is an inefficient method of "doing external work": if the load is very small, the energy required to move the body is very high in relation to the energy imparted on the object. This phenomenon is well explained by the low "energy efficiency" of the body, meaning that large amounts of energies must be input in the body to generate fairly small external work—see Chapter 2. On the other extreme, to carry heavy loads (with the possible exception of skillfully balancing a load on the head) is a highly fatiguing effort for the muscles employed, accompanied by their low energy efficiency. Thus, the human body is, in general, not a good "vehicle" for transporting loads over large distances and long periods of time. Yet, the human's ability to perform finely controlled and carefully executed actions allows the manipulation of precious and fragile loads.

General principles for the design of load carrying are:

- use large muscle groups to avoid fatigue.
- keep the load close to the body, for the sake of stability and minimal effort.
- keep the load close to the center of mass of the human body to maintain a low center of gravity.
- avoid pressure points and other highly concentrated loads on body tissues.
- ensure adequate freedom of movement for the body in general, and the arms and trunk specifically.

Moving patients. Working as a nurse is an occupation with high risks for lower back injuries. Nurses and their aides must often lift and move "precious loads": patients. These have asymmetrical shapes, varying sizes and weights, and no handholds. They are often difficult to move because of limpness, they may be reluctant to be moved or uncooperative because of pain and discomfort. They must be moved in, from, or to beds, which, because of their size and height, may force nurses to assume awkward positions. Often, one nurse alone cannot move the patient, because this would generate a high risk of overexertion, particularly of back injury. Therefore, it is often advisable to call helpers; yet, as discussed above, the ability of teams of persons to lift is not simply the sum of each team member's individual lift capacity, but considerably less.

These problems have given rise to repeated attempts to redesign hospital beds, especially to provide means for raising, lowering or tilting. External lifting aids of the hoist type have not become generally accepted because of their awkwardness, the need to move them from bed to bed, and their unpleasant appearance and high price.

Gurneys, on the other hand, are more amenable for ergonomic design because they are not used for patient bedding, but rather as a means for short-term transportation. Thus, gurneys are often adjustable in height and narrow, traits that facilitate patient handling. Among the ideas to help move the patient laterally on a gurney is to do so by smooth gliding instead of lifting, which can be facilitated by having a fairly "slippery" gurney surface (possibly by using rollers, as on conveyors) and by using some kind of a "stiff bed sheet with handles."

Ergonomic Rules for Industrial Manual Handling Tasks, Particularly Lifting of Loads

Although general rules may not cover all specific cases, and there are always the proverbial exceptions, the following guidelines are useful:

Ground Rule: Eliminate the manual task if at all possible. (*No exposure, no injury.*)

Rule #1: Reduce size, weight and forces.

Rule #2: Provide good handholds.

Rule #3: Keep the object close to the body, always in front; don't twist.

Rule #4: For lifting, keep the trunk up and the knees bent.

Rule #5: Minimize the distance through which the object must be moved.

Rule #6: Move horizontally, not vertically: Convert lifting and lowering to pushing, pulling, or carrying. (*Provide material at the proper working height.*)

Rule #7: All movement should be smooth and planned.

Rule #8: Don't lift (lower) anything that must be lowered (lifted) later.

If material handling tasks cannot be avoided altogether, take measures to facilitate the task, make it less hazardous and strenuous—see Figures 10-19 and 10-20.

The following guidelines first address the (most hazardous) tasks of lifting and lowering, then carrying and holding, and finally pushing and pulling.

Eliminate the need to lift or lower manually by

- supplying the material at the working height, perhaps by raising (or lowering) the operator or the work area.

If elimination of lifting or lowering is not feasible, use for example:

- lift table, lift platform, elevated pallet, lift truck.
- crane, hoist, elevating conveyor.
- work dispenser, gravity dump, gravity chute, etc.

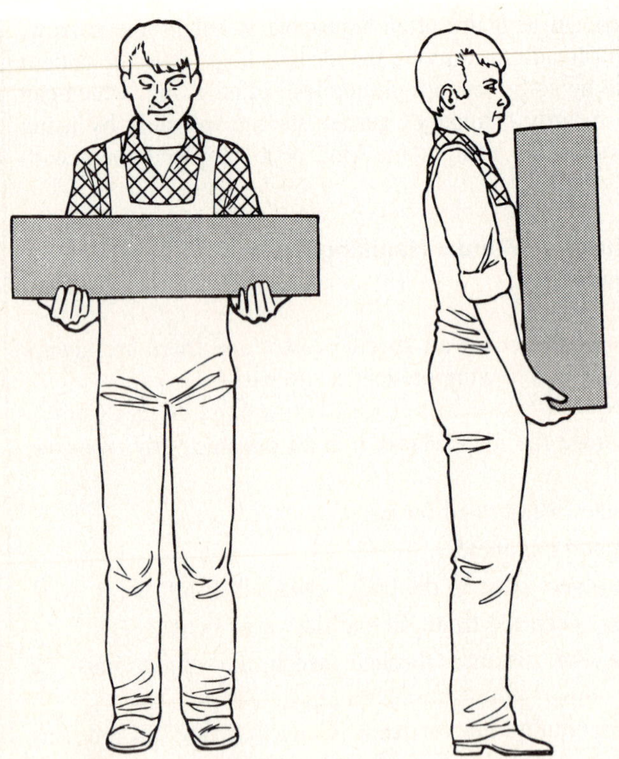

Figure 10-19. Keep the load close to the body.

Eliminate the need to carry through conversion to pushing or pulling by the use of

- conveyor, cart, dolly, truck; tables or slides between workstations, and rearrangement of workplace.
- increase the "unit weight" so that it must be handled mechanically (e.g., palletized loads).

If elimination of carrying is not feasible:

- reduce the weight by:

  using lighter material for the object.
  reducing the capacity of the container.
  reducing the weight of the container itself.
  reducing the size of the object.

- assign two workers to the job.

Figure 10-20. If lifting from the floor, keep the load between the legs, never in front of the knees.

- reduce the carrying distance by:

  changing the layout of the workplace.
  getting the operation closer to the previous or following operation.
  using conveyors.

Eliminate the need for manual holding by using:

- automatic feed and unload systems, jigs, fixtures, support stands and tables.

If elimination of the manual holding is not feasible:

- reduce the weight of the object to be held,
- reduce the time the object must be held,
- hold the object in front of the trunk and as close to the trunk as possible.

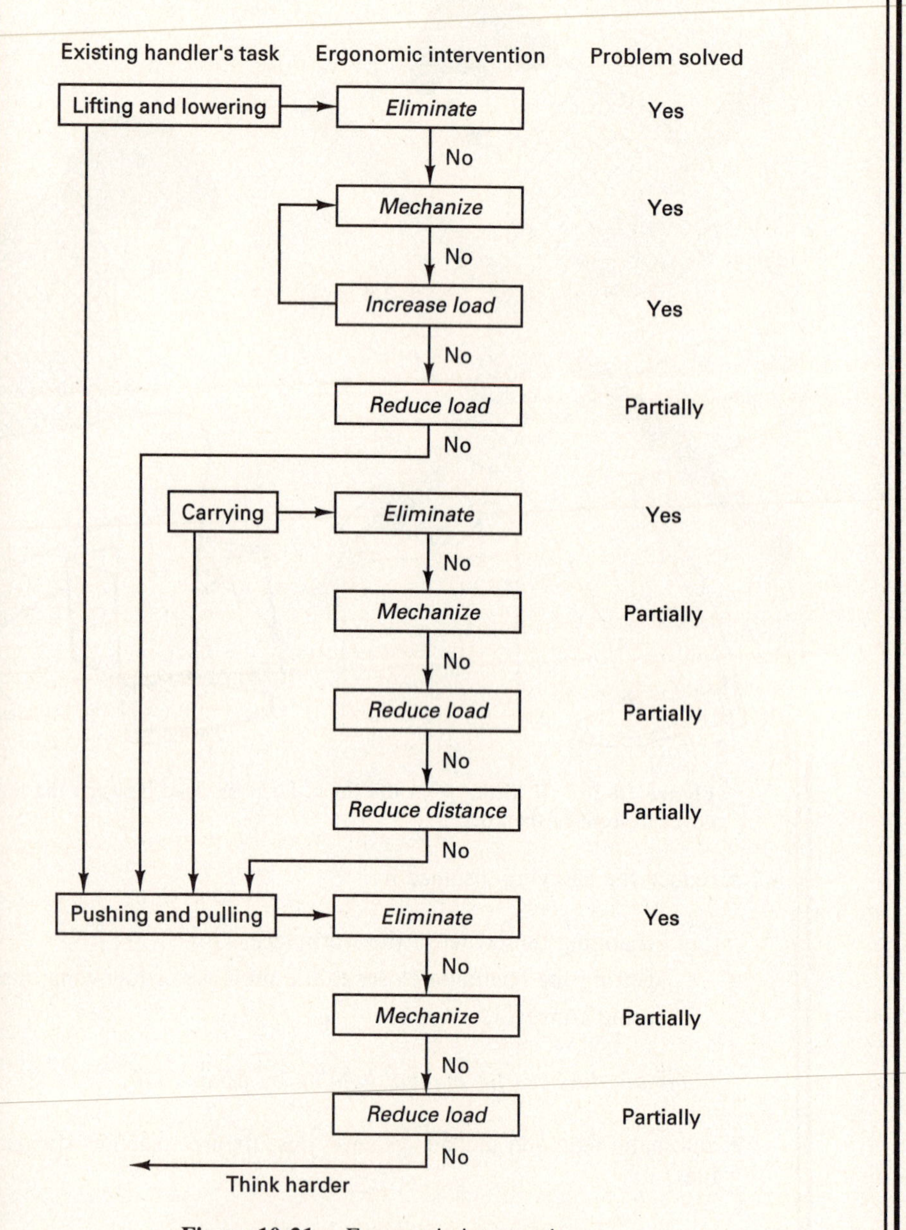

Figure 10-21. Ergonomic interventions.

Eliminate the need for pushing or pulling by use of

- conveyor, lift truck, powered truck, slide, chute, etc.

If elimination of pushing or pulling is not feasible, reduce the force required by

- reduction of weight or size of the load.
- using ramps, conveyor, dolly or truck, wheels and casters, air bearings, good maintenance of equipment and floor surface.

Figure 10-21 summarizes these ergonomic rules and their results.

SUMMARY

Avoidance of unnecessary strain, overexertions, and injuries in manual material movement is necessary for ethical and economic reasons. Basically, two major approaches exist: one is to fit the person to the job, through training or selection; the other is to fit the job to the person through ergonomic design.

Training for "safe" manual material handling has been attempted in many different ways. Training relies on the assumption that there are "safe" procedures that can be identified, taught, and followed. Unfortunately, no single training technique submitted to scientific scrutiny has yet been proven enduringly successful. Nevertheless, many claims of (often short-lived) improvements are being made. Study design and evaluation techniques are at hand to assess which training techniques are successful.

Personnel selection relies on the assumption that a "stronger" worker would be less susceptible to overexertion than a weaker colleague. While static strength measurements have been fairly successful during the last decade to assess individual capabilities, actual job demands of material handling generally are better reflected in dynamic test exertions. Such dynamic strength tests have been developed recently and applied to large numbers of military personnel. These techniques are now available for application in industry.

Ergonomic design of load, task, equipment, and workstation, including the work environment and the total facility, appears to be the most successful single approach. Human engineering solutions can avoid or remove causes of possible overexertion injuries, thus generating fundamentally safer, more efficient, and more agreeable working conditions: ergonomic design does not just "doctor the symptoms," as worker selection and training do. Even if (as Snook cautioned) two-thirds of all incidences may be unavoidable, reduction of suffering and expenses by one-third would be a major success of ergonomic design.

Developing successful personnel selection and training procedures and combining them with ergonomic design of task, workplace, and equipment should help to get the problem of overexertions and injuries of material handlers under control. In fact, the combination of all three approaches should provide the highest probability of success, and the best efficiency; this comprehensive approach has been reported to be successful in health care facilities (Aird, Nyran, and Roberts, 1988), where various and difficult manipulations have to be done on precious loads.

Both workers and managers are important players—one cannot do without the other. While material handlers are the direct recipients of proper ergonomic measures, the manager must fully understand and support them. Physical work, such as manual material movement, is accompanied by physical exertion of the body, energy expenditure, force generation, and accompanying fatigue and aches. Low back pain symptoms, for example, are likely to appear in nearly everybody's life, whether one works in the shop, in an office, or at home. As Snook (1988) convincingly explained, when workers experience low back pain during work, management should not allow adversary situations to develop which are likely to result in prolonged disability. Instead, understanding and acceptance of low back problems, early interventions, good follow-up and communication, and return-to-work-early programs may prevent, alleviate, or shorten disability.

CHALLENGES

What determines the time-dependent forces and impulses that must be applied to a load? How do these forces and impulses strain the body?

Why is the low back so frequently overexerted? What are the relations between force and torque in the spinal column?

What relationships exist among compressive forces, shear forces, bending, and twisting torques?

How are these transmitted by joints, bones, and disks, and by the ligaments and muscles that attach to vertebrae?

What is the role of skill development (experience) in load handling? Which traits can be specifically assessed using the biomechanical approach?

Under what conditions would physiological functions (such as metabolism and muscular efforts) establish limitations for a load handler's capacity for handling loads?

Why does contracting longitudinal trunk muscles load the spinal column?

How does "asymmetrical lifting" influence the use of different muscle groups and the loading of the spinal column?

What is the role of intraabdominal pressure in lifting tasks, and in pushing and pulling as well as carrying?

What effects might the use of "lifting belts" have on one's attitude and ability in material handling?

Would "staying in practice" keep an aging worker safe from a lifting injury? What can be done to have injured workers return to their jobs?

What needs to be done to make "training for safe material handling" more effective?

Which lifting techniques are appropriate for all tasks and conditions?

How can the success of training be assessed?

Is "work hardening" a promising approach?

How much intrusion on one's lifestyle should one accept to become "fit for material handling on the job"?

What are the responsibilities of the employer, and the employee, to promote noninjurious material handling on the job?

Should training for proper material handling be done during working hours?

Should an employer provide exercise facilities for the employees?

Is there anything wrong with use of the Hawthorne effect to reduce overexertion injuries?

Why is there not *one* safe load?

Which are appropriate means to select persons who are capable of handling heavy loads, or handling loads frequently?

Which medical examination techniques can be employed to select persons healthy enough to become material handlers?

How well can static strength testing predict a person's ability to perform load handling?

Which are the major ways for ergonomic design of the work process to reduce manual labor?

Should one introduce new symbols (to be used in flow charts) to identify details of load handling?

How can one redesign the fork lift truck to provide better vision and riding conditions for the operator?

Which are the major procedural differences underlying the recommendations by Snook and by NIOSH?

Which loads are particularly suitable to be carried close to the chest, and/or on the back?

Which ergonomic means can be devised to facilitate the moving of patients in hospitals?

11

Selection, Design, and Arrangement of Controls and Displays

OVERVIEW

Many traditional "knobs and dials" were researched in the 1940s and 50s, thus well-proven design recommendations are available. Unfortunately, they rely more on design and use experiences than on known psychological rules and principles.

Advances in technology as well as voiced user complaints and preferences have brought up new issues, related particularly to computer data entry and information display. Also, labeling and warnings have become of ergonomic importance.

INTRODUCTION

"Controls," called "activators" in ISO standards, transmit inputs to a piece of equipment. They are usually operated by hand or foot. The results of the control inputs are shown to the operator either in terms of "displays" or "indicators" or by the ensuing actions of the machine.

The 1940s and 50s are often called the "knobs and dials era" in human factors engineering because much research was performed on controls and displays. Thus, this topic is rather well researched, and summaries of the earlier findings were compiled, e.g., by Van Cott and Kinkade (1972), Woodson (1981), and McCormick and Sanders (1982). Military and industry standards (for example, MIL STD 1472 and HDBK 759; SAE J 1138, 1139, and 1048; HFS/ANSI 100) have established detailed design guidelines. These are matter-of-fact practical rules, concerning already existent kinds of controls and displays used in well-established designs and western stereotypes. Yet, overriding general "laws" based on human motion or energy principles, or on perception and sensory processess, are not usually known or stated. Thus, the current rules for selection and design are likely to change with new kinds of controls and displays, and new circumstances and applications.

NEED

Most recommendations for selection and arrangements of controls and displays are purely empirical and apply to existing devices and western stereotypes. Hardly any "general laws" are known that reflect human perceptual, decision-making, or motoric principles which would guide the ergonomics of control and display engineering.

APPLICATION

CONTROLS

Sanders and McCormick (1987) distinguish among the following control actions:

Activate or shut down equipment, such as with an ON-OFF key lock;

Make a "discrete setting," such as making a separate or distinct adjustment like selecting a TV channel;

Make a "quantitative setting," such as selecting a temperature on a thermostat (this is a special case of a "discrete" setting);

Apply "continuous control," such as steering an automobile;

"Enter data," as on a computer keyboard.

Control Selection

Controls shall be selected for their "functional usefulness." This includes:

- The control type shall be compatible with stereotypical or common expectations (e.g., use a pushbutton or a toggle switch to turn on a light, not a rotary knob).
- Control size and motion characteristics shall be compatible with stereotypical experience and past practice (e.g., have a fairly large steering wheel for two-handed operation in an automobile, not a small rotary control).
- Direction of operation shall be compatible with stereotypical or common expectations (e.g., ON control is pushed or pulled, not turned to the left).

- Operations requiring fine control and small force shall be done with the hands, while gross adjustments and large forces are usually exerted with the feet.
- The control shall be "safe" in that it will not be operated inadvertently or operated in false or excessive ways.

☞☞☞ *There are few "natural rules" for selection and design of controls. One is that hand-operated controls are expected to be used for fine control movements; in contrast, foot-operated controls are usually reserved for large force inputs and gross control. Yet, consider the pedal arrangements in modern automobiles, where vital and finely controlled operations, nowadays requiring fairly little force, are performed with the feet on the gas and brake pedals. Furthermore, the movement of the foot from the accelerator to the brake is very complex. Probably no topic has received more treatment in course projects and master theses than this against-all-rules but commonly used arrangement.* ☜☜☜

Compatibility of control-machine movement. Controls shall be selected so that the direction of the control movement is compatible with the response movement of the controlled machine, be it a vehicle, equipment, component, or accessory. Table 11-1 lists such compatible movements.

The term "compatibility" usually refers to the context (Sanders and McCormick, 1987) or situation where an association appears manifest or intrinsic, for example locating a control next to its related display. Yet, other relationships depend on what one has learned, what is commonly used in one's civilization. In the "western world," red is conceived to mean danger or stop, and green to mean safe and go. Such a relationship is called a population stereotype, probably learned during early childhood (Ballantine, 1983), which may differ from one user group to another. For example, in Europe, a switch is often toggled downward to turn on a light, while in the United States the switch is pushed up (Kroemer, 1982). Leaving the western world, one may encounter quite different stereotypical expectations and uses. For example, in China some durable conventions exist that are different from those in the United States (Courtney, 1988; Chapanis, 1975).

Control actuation force or torque. The force or torque applied by the operator for the actuation of the control shall be kept as low as feasible, particularly if the control must be operated often. If there are jerks and vibrations, it is usually better to stabilize the operator than to increase the control resistance in order to prevent uncontrolled or inadvertent activation.

[Note that in the following recommendations, often values for (tangential) "force" at the point of application instead of "torque" are given, even for some rotational operations. This is usually the most practical information.]

TABLE 11-1. CONTROL MOVEMENTS AND EXPECTED EFFECTS

	Direction of control movement											
Effect	Up	Right	Forward	Clockwise	Press*, Squeeze	Down	Left	Rearward	Back	Counter-clockwise	Pull**	Push**
On	1	1	1	1	2	—	—	—	—	—	1	—
Off	—	—	—	—	—	1	2	2	—	1	—	2
Right	—	1	—	2	—	—	—	—	—	—	—	—
Left	—	—	—	—	—	—	1	—	2	—	—	—
Raise	1	—	—	—	—	—	—	2	—	—	—	—
Lower	—	—	2	—	—	1	—	—	—	—	—	—
Retract	2	—	—	—	—	—	—	1	—	—	2	—
Extend	—	—	1	—	—	2	—	—	—	—	—	2
Increase	2	2	1	2	—	—	—	—	—	—	—	—
Decrease	—	—	—	—	—	2	2	1	—	2	—	—
Open Valve	—	—	—	—	—	—	—	—	—	1	—	—
Close Valve	—	—	—	1	—	—	—	—	—	—	—	—

Note: 1 = most preferred; 2 = less preferred.

* With trigger-type control.

** With push-pull switch.

SOURCE: Modified from Kroemer, 1988d.

TABLE 11-2. CONTROL-EFFECT RELATIONS OF COMMON HAND CONTROLS

Effect	Keylock	Toggle switch	Push-button	Bar knob	Round knob	Thumbwheel Discrete	Thumbwheel Continuous	Crank	Rocker switch	Lever	Joystick or ball	Legend switch	Slide*
Select ON/OFF	1	1	1	3	—	—	—	—	1	—	—	1	1
Select ON/ STANDBY/OFF	—	2	1	1	—	—	—	—	—	1	—	1	1
Select OFF/MODE 1/ MODE 2	—	3	2	1	—	—	—	—	—	1	—	1	1
Select one function of several related functions	—	2	1	—	—	—	—	—	2	—	—	—	3
Select one of three or more descrete alternatives	—	—	—	1	—	—	—	—	—	—	—	—	1
Select operating condition	—	1	1	2	—	—	—	—	1	1	—	1	2
Engage or disengage	—	—	—	—	—	—	—	—	—	1	—	—	—
Select one of mutually exclusive functions	—	—	1	—	—	—	—	—	—	—	—	1	—
Set value on scale	—	—	—	—	1	—	2	3	—	3	3	—	1
Select value in discrete steps	—	—	1	1	—	1	—	—	—	—	—	—	1

Note: 1 = most preferred; 3 = least preferred.

* Estimated, no experiments known.

SOURCE: Modified from Kroemer, 1988d.

Control-effect relationships. The relationships between the control action and the resulting effect shall be made apparent through common sense, habitual use, similarity, proximity and grouping, coding, labeling, and other suitable techniques, discussed later. Certain control types are preferred for specific applications. Table 11-2 helps (together with Tables 11-1 and 11-3) in the selection of hand controls.

Continuous versus detent controls. Continuous controls shall be selected if the control operation is anywhere within the adjustment range of the control; it does not need to be "set" in any given position. Yet, if the control operation is in discrete steps, these shall be marked and secured by "detents" or "stops" in which the control comes to rest.

Standard practices. Unless other solutions can be demonstrated to be better, the following rules apply to common equipment:

1. One-dimensional steering is by a steering wheel.
2. Two-dimensional steering is by a joystick or by combining levers, wheel, and pedals.
3. Primary vehicle braking is by pedal.
4. Primary vehicle acceleration is by a pedal or lever.
5. Transmission gear selection is by lever or by legend switch.
6. Valves are operated by round knobs or T-handles.
7. Selection of one (of two or more) operating modes can be by toggle switch, pushbutton, bar knob, rocker switch, lever, or legend switch.

Table 11-3 provides an overview of controls suitable for various operational requirements.

Arrangement and Grouping of Controls

Several "operational rules" govern the arrangement and grouping of controls.

"Locate for the Ease of Operation." Controls shall be oriented with respect to the operator. If the operator has different positions (such as in driving and operating a backhoe), the controls and control panels shall "move with the operator" so that in each position their arrangement and operation is the same for the operator.

"Primary Controls First." The most important and most frequently used controls shall have the best positions with respect to ease of operation and reaching.

"Group Related Controls Together." Controls that have sequential relations, that are related to a particular function, or that are operated together,

TABLE 11-3. SELECTION OF CONTROLS

Small operating force	
2 discrete positions	Keylock, hand-operated Toggle switch, hand-operated Pushbutton, hand-operated Rocker switch, hand-operated Legend switch, hand-operated Bar knob, hand-operated Slide, hand-operated Push-pull switch, hand-operated
3 discrete positions	Toggle switch, hand-operated Bar knob, hand-operated Legend switch, hand-operated Slide, hand-operated
4 to 24 discrete positions, or continuous operation	Bar knob, hand-operated Round knob, hand-operated Joystick, hand-operated Continuous thumbwheel, hand-operated Crank, hand-operated Lever, hand-operated Slide, hand-operated Track ball, hand-operated Mouse, hand-operated Light pen, hand-operated
Continuous slewing, fine adjustments	Crank, hand-operated Round knob, hand-operated Track ball, hand-operated
Large operating force	
2 discrete positions	Push button, foot-operated Push button, hand-operated Detent lever, hand-operated
3 to 24 discrete positions	Detent lever, hand-operated Bar knob, hand-operated
Continuous operation	Hand wheel, hand-operated Lever, hand-operated Joystick, hand-operated Crank, hand-operated Pedal, foot-operated

SOURCE: Modified from Kroemer, 1988.

shall be arranged in "functional groups" (together with their associated displays). Within each functional group, controls and displays shall be arranged according to operational importance and sequence.

"Arrange for Sequential Operation." If operation of controls follows a given pattern, controls shall be arranged to facilitate that sequence. The common arrangements are left-to-right (preferred) or top-to-bottom, as in print in the western world.

"Be Consistent." The arrangement of functionally identical or similar controls shall be the same from panel to panel.

"Dead-Man Control." If the operator becomes incapacitated and either lets go of a control, or continues to hold onto it, a "dead-man control" design shall be utilized which either returns the system to a noncritical operation state or shuts it down.

"Guard Against Accidental Activation." Numerous ways to guard controls against inadvertent activation may be applied, such as putting mechanical shields at or around them, or requiring critical forces or torques (see later). Note that most will reduce the speed of operation.

"Pack Tightly But Do Not Crowd." Often it is necessary to place a large number of controls into a limited space. Table 11-4 indicates the minimal separation distances for various types of controls.

TABLE 11-4. MINIMAL SEPARATION DISTANCES (IN MM) FOR HAND CONTROLS*

	Keylock	Bar knob	Detent thumbwheel	Push-button	Legend switch	Toggle switch	Rocker switch	Knob	Slide switch
Keylock	25	19	13	13	25	19	19	19	19
Bar Knob	19	25	19	13	50	19	13	25	13
Detent Thumbwheel	13	19	13	13	38	13	13	19	13
Pushbutton	13	13	13	13	50	13	13	13	13
Legend Switch	25	50	38	50	50	38	38	50	38
Toggle Switch	19	19	13	13	38	19	19	19	19
Rocker Switch	19	13	13	13	38	19	13	13	13
Knob	19	25	19	13	50	19	13	25	13
Slide Switch	19	13	13	13	38	19	13	13	13

* The given values are measured edge-to-edge with single controls in their closest positions. For arrays of controls, in some cases larger distances are recommended—see Boff and Lincoln (1988).

Control Design

The following descriptions, together with Tables 11-5 through 11-19 and Figures 11-1 through 11-14, present design guidance for various detent and continuous controls. The first set ("keylocks" through "Alphanumeric keyboards") consists of "detent" controls.

Keylock. *Application*—Keylocks (also called key-operated switches) are used to prevent unauthorized machine operation. Keylocks usually set into ON and OFF positions; they are always distinct.

Design Recommendations—Design recommendations are given in Figure 11-1 and Table 11-5. Other recommendations are as follows:

1. Keys with teeth on both edges (preferred) should fit the lock with either side up. Keys with a single row of teeth should be inserted into the lock with the teeth pointing up.
2. Operators should normally not be able to remove the key from the lock unless turned OFF.
3. The ON and OFF positions should be labeled.

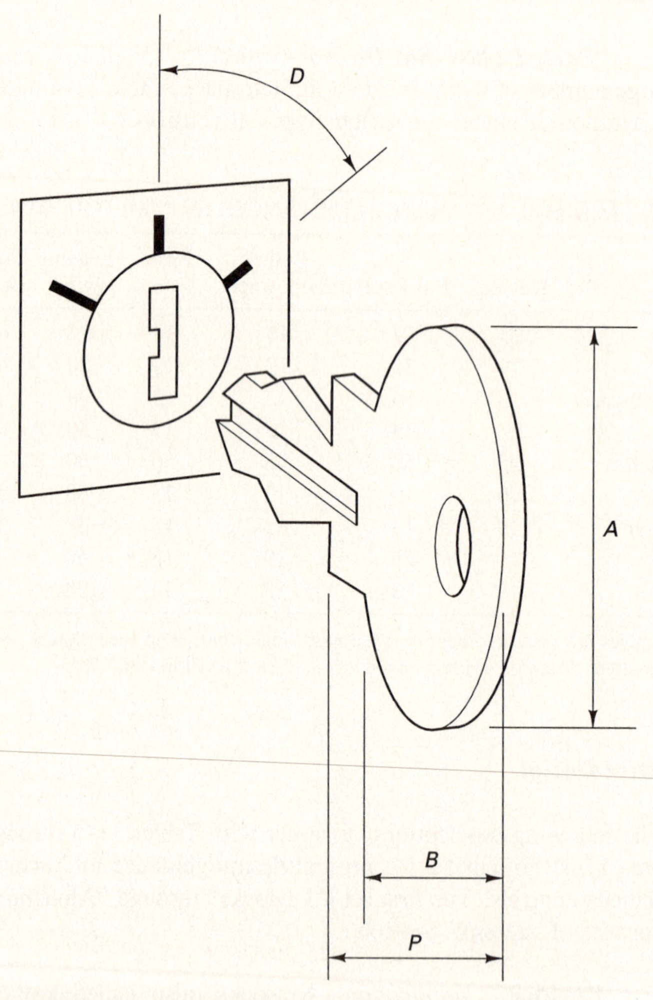

Figure 11-1. Keylock. (Modified from MIL-HDBK 759.)

TABLE 11-5. DIMENSIONS OF A KEYLOCK

	A Height (mm)	*B* Width (mm)	*P* Protrusion (mm)	*D* Displacement (degrees)	*S** Separation (mm)	*R*† Resistance (Nm)
Minimum	13	13	20	45	25	0.1
Preferred	—	—	—	—	—	—
Maximum	75	38	—	90	25	0.7

Note: Letters correspond to measurements illustrated in Figure 11-1.

* Between closest edges of two adjacent keys.

† Control should "snap" into detent position and not be able to stop between detents.

SOURCE: Modified from MIL-HDBK 759.

Bar knob. *Application*—Detent bar knobs (also called rotary selectors) should be used for discrete functions when two or more detented positions are required.

Shape—Knobs shall be bar-shaped with parallel sides, and the index end shall be tapered to a point.

Design Recommendations—Design recommendations are illustrated in Figure 11-2 and listed in Table 11-6.

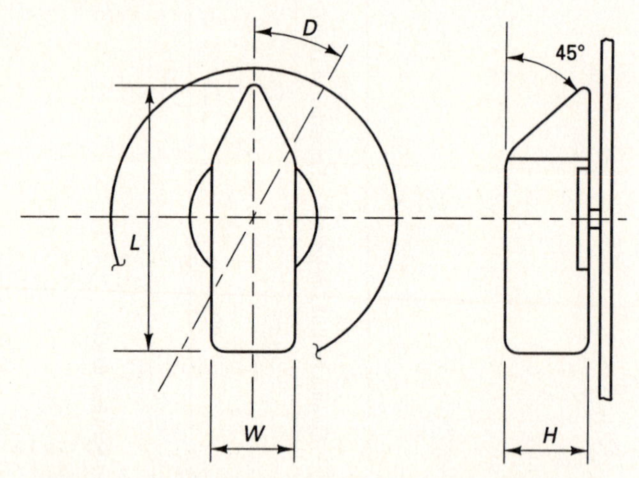

Figure 11-2. Bar knob. (Modified from MIL-HDBK 759.)

TABLE 11-6. DIMENSIONS OF A BAR KNOB

	L Length (mm)	*W* Width (mm)	*H* Height (mm)	*R*† Resistance (Nm)	*D* Displacement (degrees)	*S*, Separation: One hand, Random operation (mm)	*S*, Separation: Two hands, Simultaneous operation (mm)
Minimum	25 38*	13	16	0.1	15 30**	25 38*	75 100*
Preferred	—	—	—	—	—	—	125
Maximum	100	25	75	0.7	90	50 63*	150 175*

Note: Letters correspond to measurements illustrated in Figure 11-2.

* If operator wears gloves.

† High resistance with large bar knob only. Control should snap into detent position and not be able to stop between detents.

** For blind positioning.

SOURCE: Modified from MIL-HDBK 759.

Detent thumbwheel. *Application*—Detent (or: discrete) thumbwheels for discrete settings may be used if the function requires a compact input device for discrete steps.

Design Recommendations—Design recommendations are illustrated in Figure 11-3 and listed in Table 11-7.

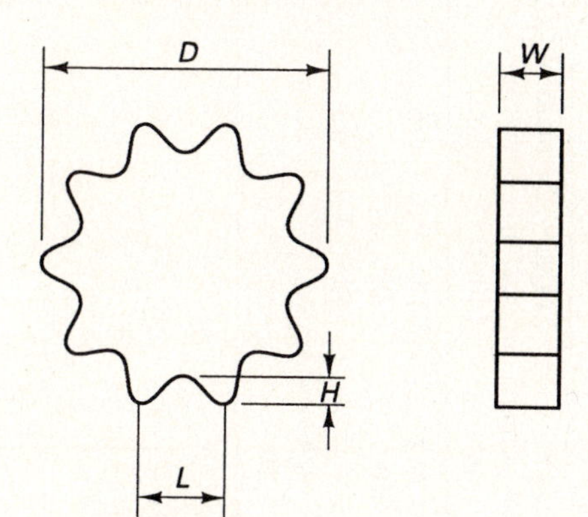

Figure 11-3. Discrete thumbwheel. (Modified from MIL-HDBK 759.)

TABLE 11-7. DIMENSIONS OF A DISCRETE THUMBWHEEL

	D Diameter (mm)	*W* Width (mm)	*L* Through distance (mm)	*H* Through depth (mm)	*S* Separation side-by-side (mm)	*R** Resistance (N)
Minimum	38	3	11	3	10	0.2
Preferred	—	—	—	—	—	—
Maximum	65	—	19	13	—	0.6

Note: Letters correspond to measurements illustrated in Figure 11-3.

* Control should snap into detent position and not be able to stop between detents.

SOURCE: Modified from MIL-HDBK 759.

Pushbutton. *Application*—Pushbuttons should be used for single switching between two conditions, for entry of a discrete control order, or for release of a locking system (e.g., of a parking brake). Pushbuttons can be used for momentary contact or for sustained contact.

Design Recommendations—Design recommendations are given in Figure 11-4 and listed in Table 11-8. Other recommendations are as follows:

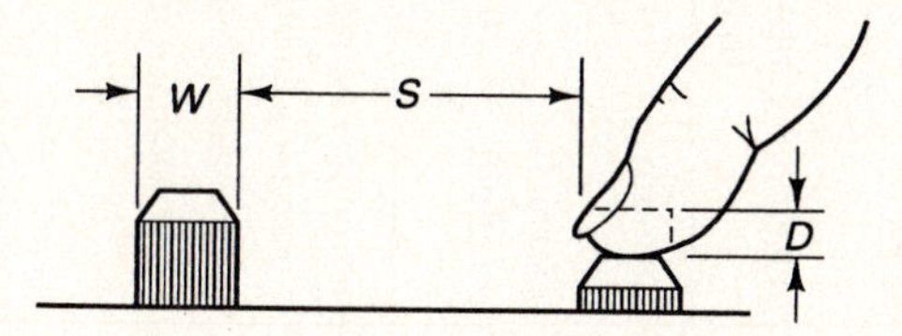

Figure 11-4. Pushbutton. (Modified from MIL-HDBK 759.)

TABLE 11-8. DIMENSIONS OF A PUSHBUTTON

	W, Width of square or diameter			*R*, Resistance					*S*, Separation			
									Single finger			
Operation	Fingertip (mm)	Thumb (mm)	Palm of hand (mm)	Little finger (N)	Other finger (N)	Thumb (N)	Palm of hand (N)	*D*** Displacement (mm)	Single operation (mm)	Sequential operation (mm)	Different fingers (mm)	Palm or thumb (mm)
Minimum	10 13*	19	25	0.25	0.25	1.1	1.7	3.2 16	13 25*	6	6	25
Preferred	—	—	—	—	—	—	—	—	50	13	13	150
Maximum	19	—	—	1.5	11.1	16.7	22.2	6.5 20*	—	—	—	—

Note: Letters correspond to measurements illustrated in Figure 11-4.

* If operator wears gloves.

** Depressed button shall stick out at least 2.5 mm.

SOURCE: Modified from MIL-HDBK 759.

Shape—The pushbutton surface should normally be concave (indented) to fit the finger. When this is impractical, the surface shall provide a high degree of frictional resistance to prevent slipping. Yet, the surface may be convex for operation with the palm of the hand.

Positive Indication—A positive indication of control activation shall be provided (e.g., snap feel, audible click, or integral light).

Push-pull switch. Push-pull controls have been used for discrete settings, commonly ON and OFF; intermediate positions have been occasionally employed, e.g., for the air-gasoline mixture of a combustion engine. They generally have a round flange under which to "hook" the fingers. Their diameter should be not less than 19 mm, protruding at least 25 cm from the mounting surface; the separation between adjacent controls should be at least 38 mm. There should be at least 13 mm displacement between the settings (MIL STD 1472D).

Legend switch. *Application*—Detent legend switches are particularly suited to display qualitative information on equipment status that requires the operator's attention and action.

Design Recommendations—Design recommendations are given in Figure 11-5 and listed in Table 11-9.

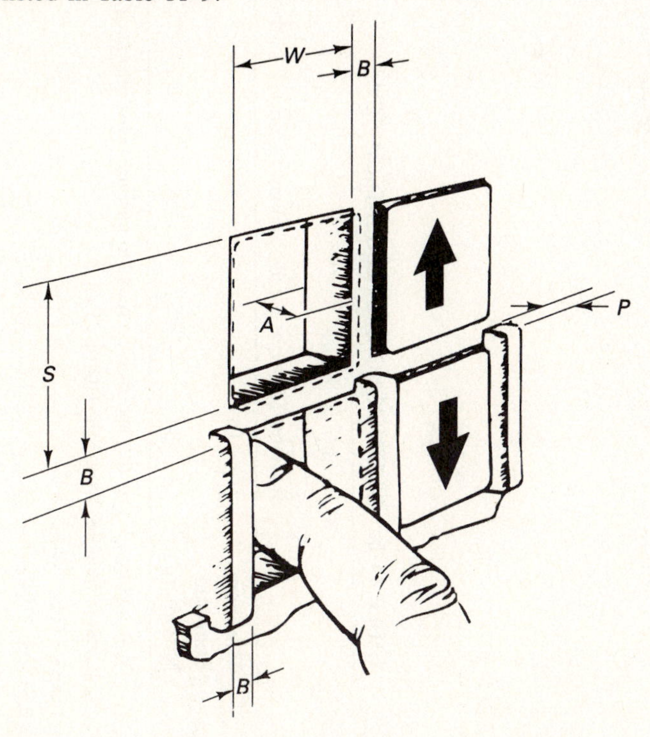

Figure 11-5. Legend switch. (Modified from MIL-HDBK 759.)

TABLE 11-9. DIMENSIONS OF A LEGEND SWITCH

	W, Width of square or diameter (mm)	D Displacement (mm)	B Barrier width (mm)	P Barrier protrusion (mm)	R Resistance (N)
Minimum	19	3	3	5	0.28
Preferred	—	—	—	—	—
Maximum	38	6	6	6	11

Note: Letters, W, A, B_w, and B_d correspond to measurements illustrated in Figure 11-5.

SOURCE: Modified from MIL-HDBK 759.

Legend switches should be located within a 30°-cone along the operator's line of sight.

Toggle switch. *Application*—Detent toggle switches may be used if two discrete positions are required. Toggle switches with three positions shall be used only where the use of a bar knob, legend switch, array of push buttons, etc., is not feasible.

Design Recommendations—Design recommendations are given in Figure 11-6 and listed in Table 11-10, and are as follows:

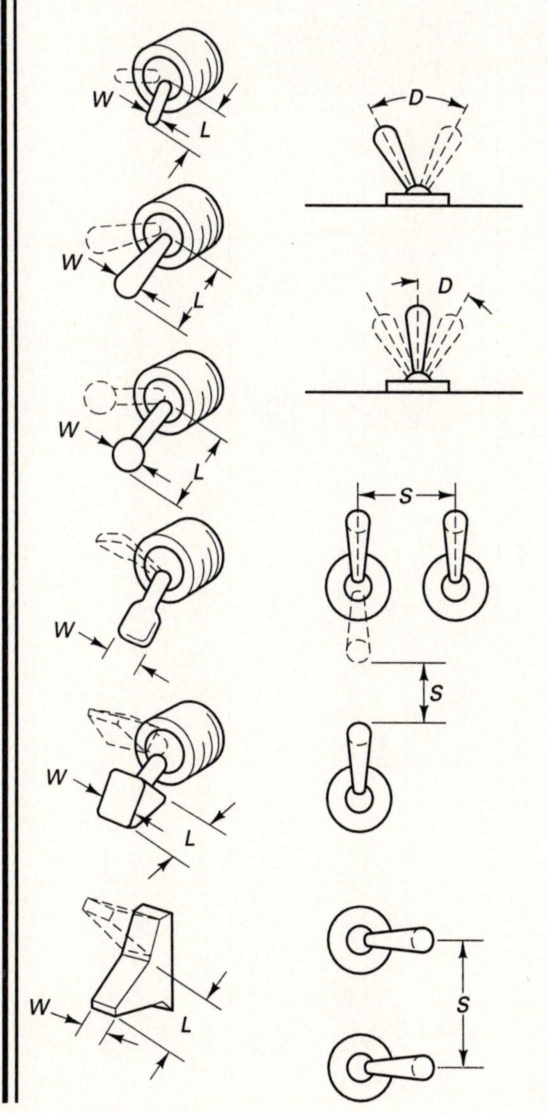

Figure 11-6. Toggle switch. (Modified from MIL-HDBK 759.)

TABLE 11-10. DIMENSIONS OF A TOGGLE SWITCH

				D, Displacement		S, Separation				
						Horizontal array, vertical operation				
						Single finger				
	L Arm length (mm)	W, Tip width or diameter (mm)	R† Resistance (N)	2 Positions (degrees)	3 Positions (degrees)	Random operation (mm)	Sequential operation (mm)	Several fingers simultaneous operation (mm)	Vertical array horizontal operation (mm)	Toward each other tip-to-tip other (mm)
Minimum	9.5 38*	3	2.8	25	18	19	13	19 32*	25 38*	25
Preferred	—	—	4.5	—	25	50	25	—	—	—
Maximum	50	25	11	120	60	—	—	—	—	—

Note: Letters correspond to measurements illustrated in Figure 11-6.

* If operator wears gloves.

† Control should snap into detent position and not be able to stop between detents.

SOURCE: Modified from MIL-HDBK 759.

Orientation—Toggle switches should be so oriented that the handle moves in a vertical plane, with OFF in the down position. Horizontal actuation shall be employed only if compatibility with the controlled function or equipment location is desired.

Rocker switch. *Application*—Detent rocker switches may be used if two discrete positions are required. Rocker switches protrude less from the panel than do toggle switches.

Design Recommendations—Design recommendations are given in Figure 11-7 and listed in Table 11-11.

Orientation—Rocker switches should be so oriented that the handle moves in a vertical plane, with OFF in the down position. Horizontal actuation shall be employed only if compatibility with the controlled function or equipment location is desired.

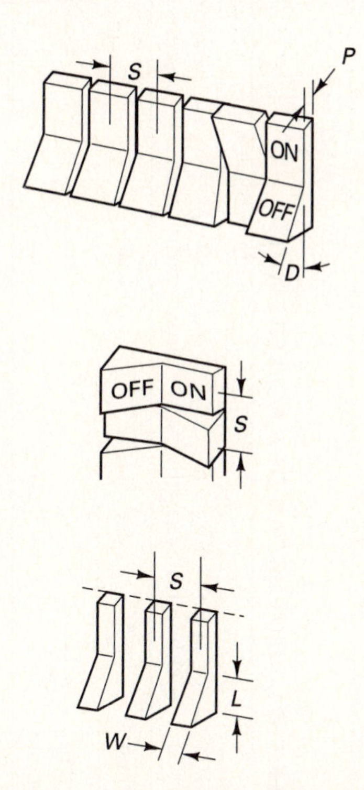

Figure 11-7. Rocker switch. Narrow switch (bottom) is especially desirable for tactile definition when gloves are worn. (Modified from MIL-HDBK 759.)

TABLE 11-11. DIMENSIONS OF A ROCKER SWITCH

	W Width (mm)	*L* Length (mm)	*D* Displacement (degrees)	*P* Protrusion depressed (mm)	*S* Separation center to center (mm)	*R* Resistance (N)
Minimum	6.5	13	30	2.5	19 32*	2.8
Preferred	—	—	—	—	—	—
Maximum	—	—	—	—	—	11.1

Note: Letters *D*, *H*, *L*, *S*, and *W* correspond to measurements illustrated in Figure 11-7.

* If operator wears gloves.

SOURCE: Modified from MIL-HDBK 759.

Sets of Numerical Keys

Two different kinds of numerical key sets are widely used.

One, usually associated with telephones, is arranged thus:

1 2 3
4 5 6
7 8 9
0

The other arrangement is called the calculator key set:

7 8 9
4 5 6
1 2 3
0

Keying performance may be more accurate and slightly faster with the telephone key set than with the calculator arrangement (Conrad and Hull, 1968; Lutz and Chapanis, 1955).

Alphanumeric Keyboards

On the original typewriter keyboard, developed in the 1860s, the keys were arranged in alphabetic sequence in two rows. The QWERTY layout, patented in 1879, was adopted after many modifications as an international standard in 1966. It has since been universally used on typewriters, then on computers and on many other input devices.

The letter-to-key allocation on the QWERTY has many, often obscure, reasons. Letters which frequently follow each other in English text (such as q and u) were spaced apart so that the mechanical type bars might not entangle if struck in rapid sequence. The "columns" of keys assigned to separate fingers run diagonally across the keyboard. This was also done originally due to mechanical constraints of the type bars. The keys are arranged in straight "rows," which the fingertips are not. Obviously, the original QWERTY keyboard was designed considering mostly mechanics, not "human factors." Many attempts were made to improve typing performance by changing the keyboard layout. They include relocating keys within the standard keyset layout, or changing the keyboard layout—for example, by breaking up the columns and rows of keys, by dividing the keyboard into separate sections, by adding sets of keys, etc.. (For reviews, see, for example, Alden, Daniels, and Kanarick, 1972; Gilad and Pollatsheck, 1986; Gopher and Raij, 1988; Hirsch, 1970; Kroemer, 1972, 1992; Lithrick, 1981; Martin, 1949; Michaels, 1971; Norman and Fisher, 1982; Noyes, 1983a; Seibel, 1972.) A terminology for describing major design features of keyboards is shown in Figure 11-8 with some descriptors taken from ANSI 100 (Human Factors Society, 1988).

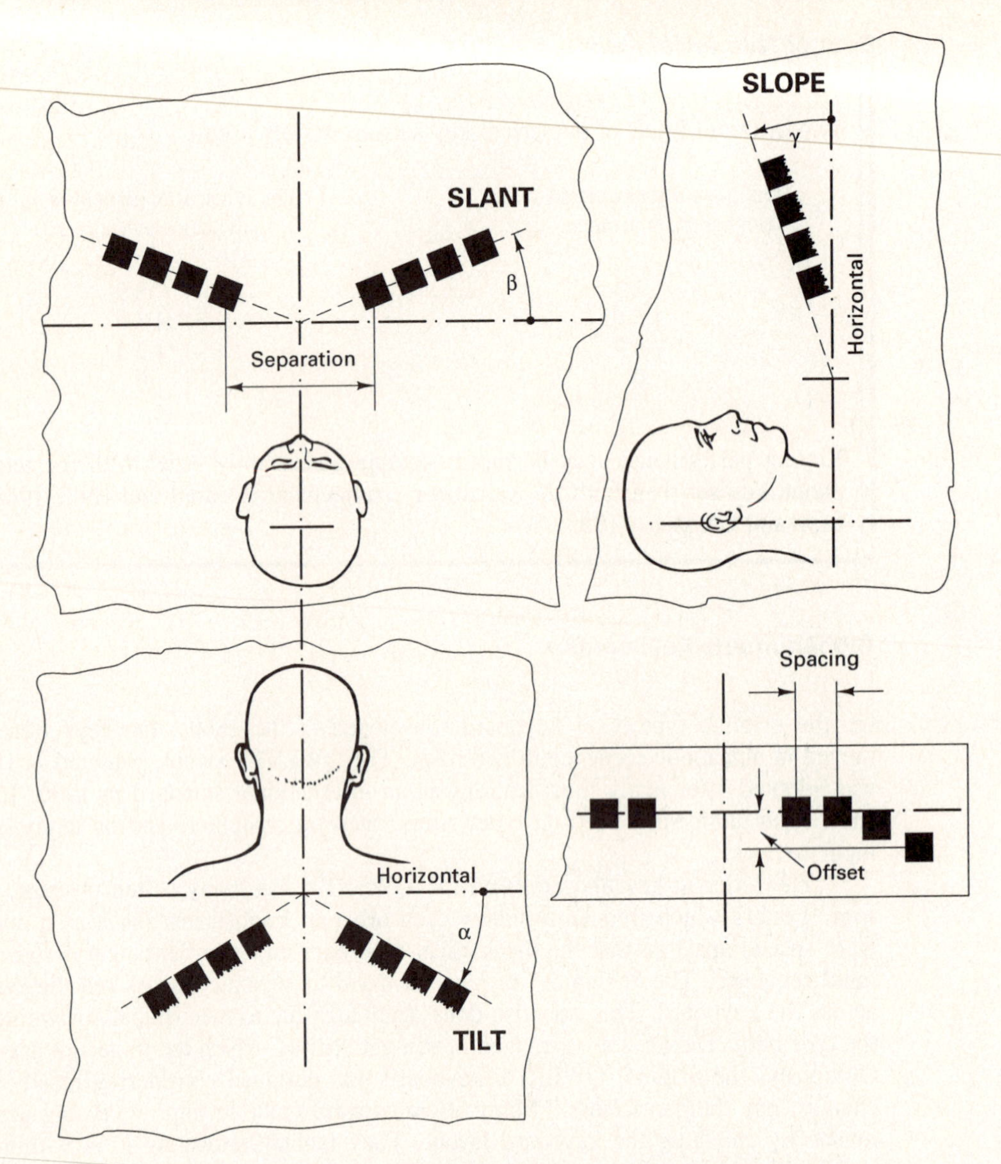

Figure 11-8. Terms to describe major design factors of keyboards.

APPLICATION

The dynamics of force-displacement characteristics of keys are important for the user, but they are difficult to measure, because the procedures commonly used rely on static force applications. This is not indicative of the actual dynamic operation, which varies from operator to operator (Sind, 1989). Furthermore, the technology of key switches is rapidly developing. Yet, there appears to be consensus that for "full-time keying," keys with some displacement (about 2 mm) and a "snap-back resistance" or an audible "click" signal (where key activation is felt by reduced key resistance, or heard) are preferable (Monty, Snyder, and Birdwell, 1983; Sind, 1989; Cushman and Rosenberg, 1991). Instead of having separate single keys, some keyboards consist of a membrane which, when pressed down in the correct location, generates the desired letter with virtually no displacement of the keyboard. A major advantage of the membrane is that dust or fluids cannot penetrate; however, many typists dislike it. In experiments, experienced users showed better performance with conventional keys than with the membrane keyboard, but the differences were not large and diminished with practice (Loeb, 1983; Sind, 1989). For "part-time" data input, keys with little or no displacement such as membrane keyboards may be acceptable.

Instead of simply improving conventional key and keypad designs, more radical proposals have recently appeared. These include the use of chording instead of single key activations for generating letters, words, or symbols. This has been used, for example for mail sorting or on-the-spot recording of verbal discussions (Gopher and Raij, 1988; Keller, Becker, and Strasser, 1991; Noyes, 1983b). While traditional keys are "tapped down" for activation, and have only two states, ON and OFF, other proposals use ternary keys that are "rocked" forth and back with the fingertips, where fingers are inserted into well-like switches which can be tapped, rocked, and moved sideways; or where finger movements are registered by means of instrumented gloves (Kroemer, 1992).

Traditionally, computer entries have been made by mechanical interaction between the operator's fingers and such devices as keyboard, mouse, trackball, or light pen. Yet, there are many other means to generate inputs. Voice recognition is one fairly well known method, but others can be employed that utilize, for example,

- hands and fingers for pointing, gestures, sign language, tapping, etc;
- arms for gestures, making signs, moving or pressing control devices;
- feet for motions and gestures, for moving and pressing of devices;
- the legs, also for gestures, moving and pressing of devices;
- the torso, including the shoulders, for positioning and pressing;
- the head, also for positioning and pressing;
- the mouth for lip movements, use of the tongue, or breathing such as through a blow/suck tube;
- the face for making facial expressions;
- the eyes for tracking;
- combinations and interactions of these different inputs could be used, such as in "ballet positions" (Jenkins, 1991).

Of course, the method selected must be clearly distinguishable from "environmental clutter," which may be defined as "loose energy" that interferes with sensor pickup. The ability of a sensor to detect the input signals depends on the type and intensity of the signal generated. For example, it may be quite difficult to distinguish between different facial expressions, while it is much easier to register displacement of, or pressure on, a sensor. Thus, the use of other than conventional input methods depends on the state of technology, which includes the tolerance of the system to either missed or misinterpreted input signals.

Pointing with the finger or hand, for example, is an attractive solution. Sensors are able to track position, the direction of pointing, and possibly movements. Sensing could be done with either a relative or an absolute reference. The sensing could be continuous or discrete in one, two, or three dimensions. Pointing, for example, is a common means to convey information; a large number of characters could be generated by it. Pointing is a dynamic activity; motion (or position) could be sensed either on the finger or hand directly, or by means of "landmarks," such as reflective surfaces, for example, attached to rings worn. Thus, pointing is a natural, easily learned and controlled activity, which could be observed fairly easily. However, there is the possibility of missed inputs, misinterpreted inputs, and of fatigue or overuse syndromes when many pointing actions are executed quickly and repeatedly (Jenkins, 1991).

While these ideas may be far-fetched, even current keyboards impose quite different motoric tasks on the hands of the user than did previous mechanical designs, which had much key displacement and large keying forces. Use of a smaller set of keys (such as with chording) would reduce the required hand travel and allow resting of the wrists on suitable pads. This might alleviate some of the problems associated with cumulative trauma disorders (discussed in Chapter 8) by virtue of reduced energy requirements and improved operator posture (Kroemer, 1992; Rose, 1991).

APPLICATION

The following set of control descriptions concerns continuous operations.

Knob. *Application*—Continuous knobs (also called round knobs or rotary controls) should be used when little force is required and when precise adjustments of a continuous variable are required. If positions must be distinguished, an index line on the knob should point to markers on the panel.

Design Recommendations—Design recommendations are given in Figure 11-9 and Table 11-12. Within the range specified in the table, knob size is relatively unimportant—provided the resistance is low and the knob can be easily grasped and manipulated. When panel space is extremely limited, knobs should be small and should have resistances as low as possible without permitting the setting to be changed by vibration or by inadvertent touching.

Knob Style—Unless otherwise specified, control knobs shall conform to the guidelines established in *Military Standards,* e.g., MIL-STD 1472 or MIL-HDBK 759.

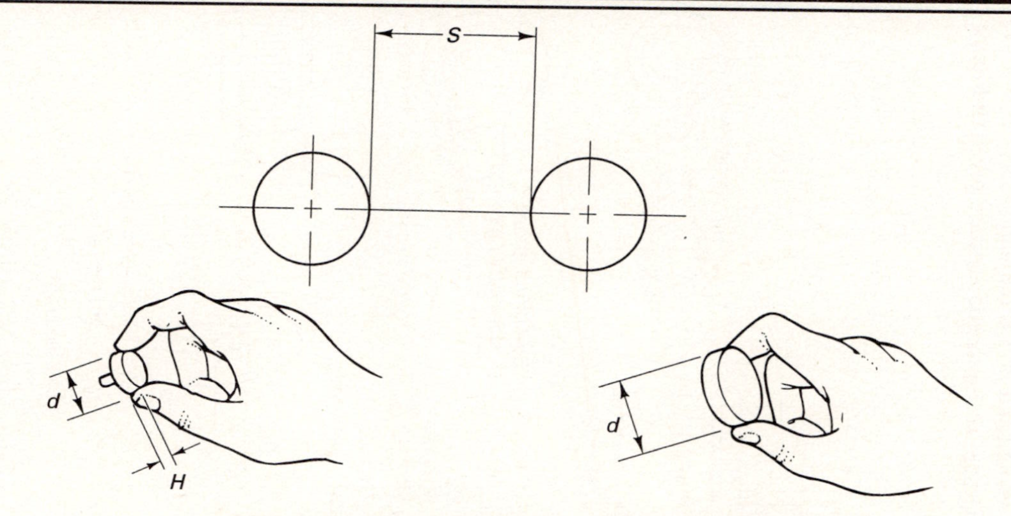

Figure 11-9. Knob. (Modified from MIL-HDBK 759.)

TABLE 11-12. DIMENSIONS OF A KNOB

	H	*D*, Diameter		*T*, Torque		*S*, Separation	
	Height (mm)	Fingertip grip (mm)	Thumb and finger grasp (mm)	Up to 25 mm in diameter (N m)	Over 25 mm in diameter (N m)	One hand (mm)	Two hands simultaneously (mm)
Minimum	13	10	25	—	—	25	50
Preferred	—	—	—	—	—	50	125
Maximum	25	100	75	0.03	0.04	—	—

Note: Letters correspond to measurements illustrated in Figure 11-9.

SOURCE: Modified from MIL-HDBK 759.

Crank. *Application*—Continuous cranks should be used primarily if the control must be rotated many times. For tasks involving large slewing movements or small, fine adjustments, a crank handle may be mounted on a knob or handwheel.

Grip Handle—The crank handle shall be designed so that it turns freely around its shaft, especially if the whole hand grasps the handle.

Design Recommendations—Design recommendations are given in Figure 11-10 and Table 11-13.

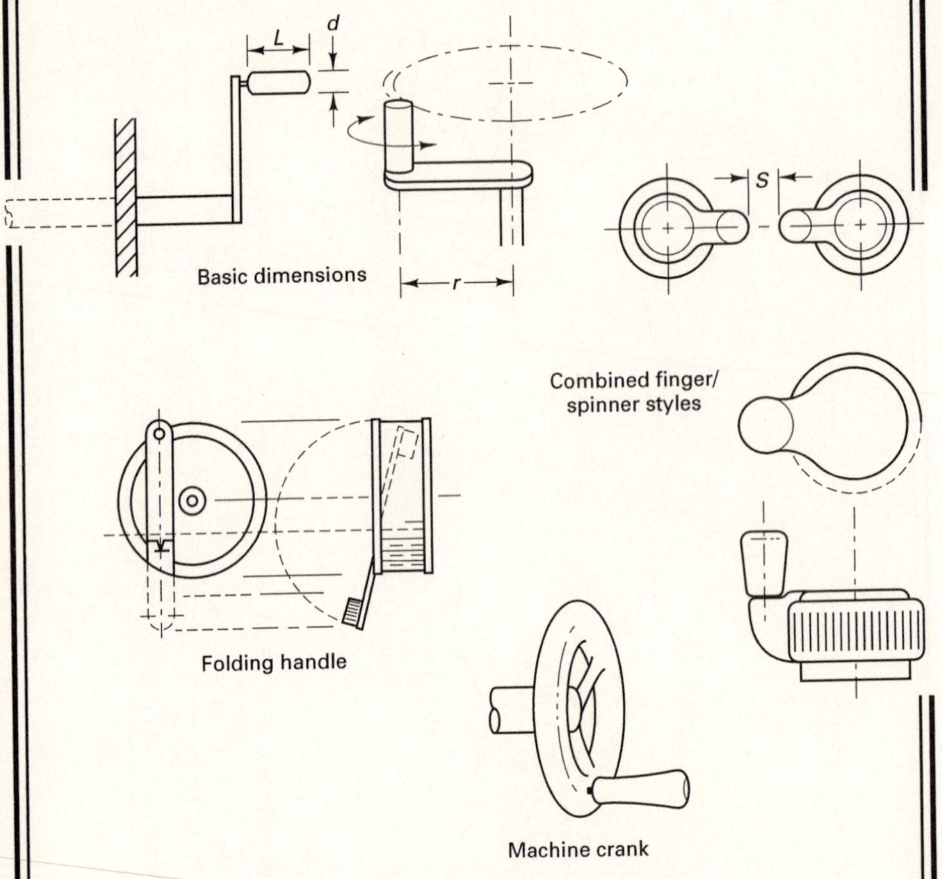

Figure 11-10. Crank. (Modified from MIL-HDBK 759.)

Handwheel. *Application*—Continuous handwheels that are designed for nominal two-handed operation should be used when the breakout or rotation forces are too large to be overcome with a one-hand control—provided that two hands are available for this task.

Knurling—Knurling or indentation shall be built into a handwheel to facilitate operator grasp.

TABLE 11-13. DIMENSIONS OF A CRANK

	Operated by finger and wrist movement (Resistance below 22 N)					Operated by arm movement (Resistance below 22 N)				
	L Length (mm)	d Diameter (mm)	r, Turning radius Below 100 RPM (mm)	r, Turning radius Above 100 RPM (mm)	S Separation (mm)	L Length (mm)	d Diameter (mm)	r, Turning radius Below 100 RPM (mm)	r, Turning radius Above 100 RPM (mm)	S Separation (mm)
Minimum	25	9.5	38	13	75	75	25	190	125	75
Preferred	38	13	75	58	—	95	25	—	—	—
Maximum	75	16	125	115	—	—	38	510	230	—

Note: Letters correspond to measurements illustrated in Figure 11-10.

SOURCE: Modified from MIL-HDBK 759.

Spinning Handle—When large displacements must be made rapidly, a spinner (crank) handle may be attached to the handwheel when this is not overruled by safety considerations.

Design Recommendations—Design recommendations are given in Figure 11-11 and listed in Table 11-14.

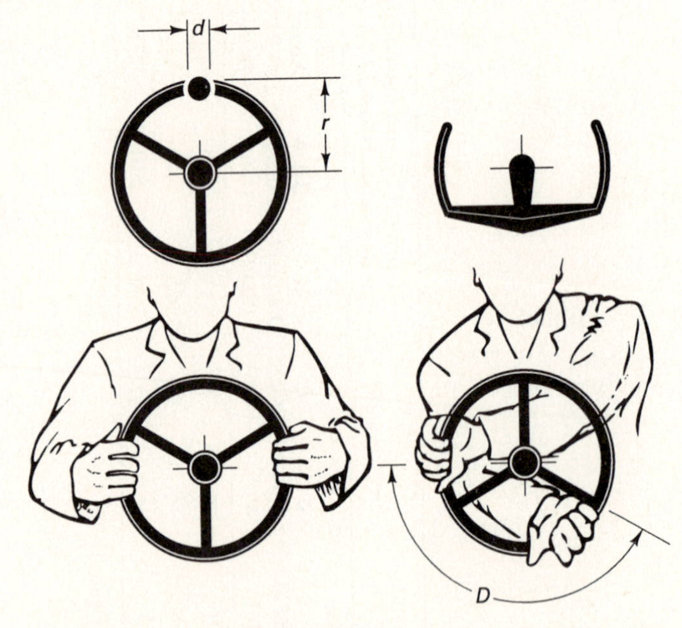

Figure 11-11. Handwheel. (Modified from MIL-HDBK 759.)

TABLE 11-14. DIMENSIONS OF A HANDWHEEL

	r, Wheel radius					
	With powersteering (mm)	Without powersteering (mm)	*d* Rim diameter (mm)	Tilt from vertical (degrees)	*R* Resistance (N)	*D* Displacement both hands on wheel (degrees)
Minimum	175	200	19	30 Light vehicle	20	—
Preferred	—	—	—	—	—	—
Maximum	200	255	32	45 Heavy vehicle	220	120

Note: Letters correspond to measurements illustrated in Figure 11-11.

SOURCE: Modified from MIL-HDBK 759.

Lever. *Application*—Continuous levers may be used when large force or displacement is required at the control and/or when multidimensional movements are required. There are two kinds of levers, often called "joysticks" or simply "sticks":

- A "force joystick," which does not move (is "isometric") but transmits control imputs according to the force applied to it.
- A "displacement joystick" (often falsely called "isotonic"), which transmits control inputs according to its spatial position, or movement direction, or speed.

The force-controlled lever is especially suitable when "return-to-center" of the system after each control input is necessary. The displacement lever is appropriate when control in two or three dimensions is needed, particularly when accuracy is more important than speed.

Design Recommendations—Design recommendations are given in Figure 11-12 and listed in Table 11-15. Other recommendations are as follows:

Limb Support—When levers are used to make fine or continuous adjustments, support shall be provided for the appropriate limb segment:

- for large hand movements: elbow support;
- for small hand movements: forearm support;
- for finger movements: wrist support.

Coding—When several levers are grouped in proximity to each other, the lever handles shall be coded (see below).

Labeling—When practicable, all levers shall be labeled (see below) with regard to function and direction of motion.

Elastic Resistance—For joystick controls, elastic resistance that increases with displacement may be used to improve "stick feel."

High-Force Levers—For occasional or emergency use, high-force levers may be used. They shall be designed to be either pulled up or pulled back toward the shoulder, with an elbow angle of 150° (±30°). The force required for operation shall not exceed 190 N. The handle diameter shall be from 25 to 38 mm, and its length shall be at least 100 mm. Displacement should not exceed 125 mm. Clearance behind the handle and along the sides of the path of the handle shall be at least 65 mm. The lever may have a thumb-button or a clip-type release.

Continuous thumbwheel. *Applications*—Thumbwheels for continuous adjustments may be used as an alternative to round knobs if a compact thumbwheel is beneficial.

Design Recommendations—Design recommendations are given in Figure 11-13 and Table 11-16.

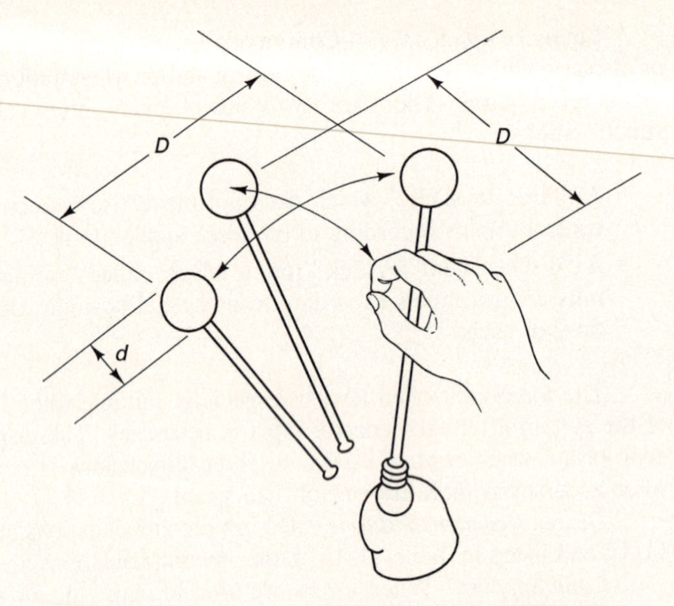

Figure 11-12. Lever. (Modified from MIL-HDBK 759.)

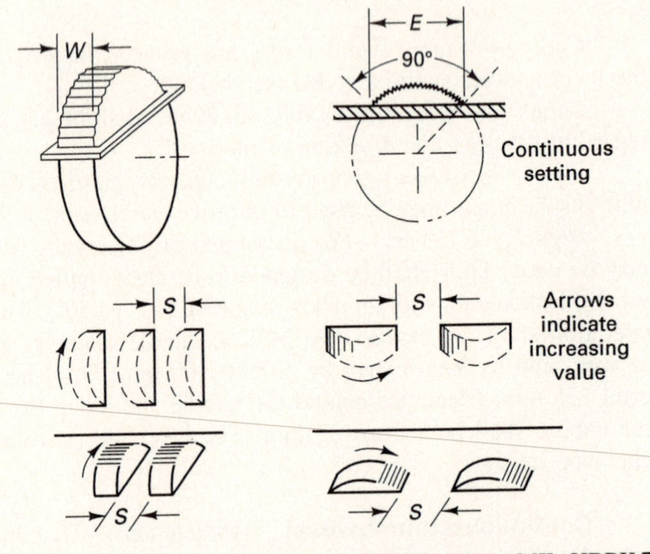

Figure 11-13. Continuous thumbwheel. (Modified from MIL-HDBK 759.)

TABLE 11-15. DIMENSIONS OF A LEVER

	d, Diameter		*R*, Resistance				*D*, Displacement		*S*, Separation	
			Fore-aft		Left-right					
	Finger grip (mm)	Hand grip (mm)	One hand (N)	Two hands (N)	One hand (N)	Two hands (N)	Fore-aft (mm)	Left-right (mm)	One hand (mm)	Two hands (mm)
Minimum	13	32	9	9	9	9	—	—	50*	75
Preferred	—	—	—	—	—	—	—	—	—	—
Maximum	75	75	135	220	90	135	360	970	100*	125

Note: Letters *A* and *D* correspond to measurements illustrated in Figure 11-12.

*About 25 mm if one hand usually operates two adjacent levers simultaneously.

SOURCE: Modified from MIL-HDBK 759.

TABLE 11-16. DESIGN RECOMMENDATIONS FOR A CONTINUOUS THUMBWHEEL

	E Rim exposure (mm)	*W* Width (mm)	*S*, Separation		*R* Resistance (N)
			Side-by-side (mm)	Head-to-foot (mm)	
Minimum	—	—	25 38*	50 75*	—
Preferred	25	3.2	—	—	—
Maximum	100	23	—	—	3.3†

Note: Letters correspond to measurements illustrated in Figure 11-13.

* If operator wears gloves.

† To minimize danger of inadvertent operation.

SOURCE: Modified from MIL-STD 759.

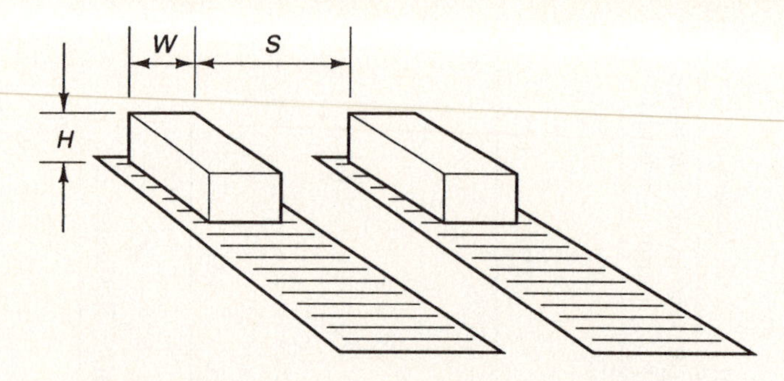

Figure 11-14. Slide. (Modified from MIL-HDBK 759.)

TABLE 11-17. DIMENSIONS OF A SLIDE OPERATED BY BARE FINGER

	W Width (mm)	*H* Height (mm)	*S* Separation (mm)	*R* Resistance (N)
Minimum	6	6	19* 13† 16**	3
Maximum	25	—	50* 25† 19**	11

Note: Letters correspond to measurements illustrated in Figure 11-14.

* Single-finger operation.

† Single-finger sequential operation.

**—simultaneous operation by different fingers.

SOURCE: Modified from MIL-STD 1472D.

Slide. Slide switches are used to make continuous settings, for example in music mix-and-control stations. Design recommendations are from Cushman and Rosenberg (1991) and MIL-STD 1472D. They are compiled in Figure 11-14 and Table 11-17.

Computer Input Devices

In addition to the traditional "key" (discussed above), a variety of other input devices has been used, such as the trackball, mouse, light pen, graphic tablet, touch screen, or stick (see above). These are used to move a cursor, manipulate the screen content, point (insert or retrieve information), digitize information, etc. Many tasks change with new technology and software. Therefore, any compilations current at the time (such as by Boff and Lincoln, 1988; Cushman and Rosenberg, 1991) become outdated quickly, and the designer must follow the current literature closely (see, for example, Chase and Casali, 1991; Han, Jorna, Miller, and Tan, 1990; Epps, 1987).

TABLE 11-18. DIMENSIONS OF A FOOT-OPERATED SWITCH FOR TWO DISCRETE POSITIONS

	W Width or diameter (mm)	*D* Displacement (mm)				*F* Resistance (N)	
		Operation by regular shoe	Operation by heavy boot	Operation by ankle flexion only	Operation by whole leg movement	Foot does not rest on control	Foot rests on control
Minimum	13	13	25	25	25	18	45
Maximum	—	65	65	65	100	90	90

SOURCE: Modified from MIL-STD 1472D.

TABLE 11-19. DIMENSIONS OF A PEDAL FOR CONTINUOUS ADJUSTMENTS (E.G., ACCELERATOR OR BRAKE)

			D, Displacement (mm)				*R*, Resistance (N)				Edge-to-edge separation, *S*	
	H Height or depth (mm)	*W* Width (mm)	Operation with regular shoe	Operation with heavy boot	Operation by ankle flexion only	Operation by whole leg movement	Foot does not rest on pedal	Foot rests on pedal	Operation by ankle flexion	Operation by whole leg movement	Random operation by one foot	Sequential operation by one foot
Minimum	25	75	13	25	25	25	18	45	—	45	100	50
Maximum	—	—	65	65	65	180	90	90	45	800	150	100

SOURCE: Modified from MIL-STD 1472D.

Foot-Operated Controls

Foot-operated switches may be used if only two discrete conditions—such as ON and OFF—need to be set. Recommended design parameters are compiled in Table 11-18.

In vehicles and cranes (seldom in other applications) pedals are used for continuous adjustments, such as in automobile speed control. Recommended design dimensions are listed in Table 11-19.

Remote Control Units

Remote control units, usually simply called "remotes," are small hand-held control panels which are manipulated at some distance from a computer, with which they communicate by cable or radiation.

Remote control units to be used with TV and radio equipment in the home got so difficult to understand and operate in the 1980s that they became almost proverbial for useless gadgetry. In 1991, a simple and inexpensive remote control unit "to replace all remotes" came on the market and was an immediate commercial success.

In the robotic industry, remote control units (often called "teach pendants") are used by a technician to specify the point in space to which the robot effector (its tool or gripper) must be moved to operate on a workpiece. This can be done in two ways: one is to take the three-dimensional coordinates of that point from a drawing or a computer-aided design program, and simply enter these into the remote unit. The other is to use the robot as its own measuring device by positioning the effector in the desired place, with all joint angles properly selected, and then simply to command the robot computer to record these positions internally.

Many human engineering issues are pertinent: one is the need of the human operator to see, at least initially, the exact location of the robot effector in relation to the workpiece. This may be difficult to do—for instance, because of lacking illumination—and it may be dangerous for the operator who must step into the operating area of the machine. A second ergonomic concern is proper design and operation of the controls. Joysticks are preferable in principle because they allow easy commands for movement in a plane; pushbuttons, the more common solutions, permit control only in linear or angular direction. Yet, pushbuttons are easier to protect from inadvertent operation and from damage when the unit is dropped (Parsons, 1988, 1991). The proper arrangements of sticks or buttons in arrays on the manipulation surface of the remote is similar to the design aspects discussed earlier in this chapter. Particular attention must be paid to the ability to immediately stop the robot in emergencies, and to move it away safely from a given position, even by an inexperienced operator. Reports about injuries to the robot operator or bystanders, or damage to property, by incorrect remote operation of the robot are frequent (Etherton and Collins, 1990).

APPLICATION

EXAMPLE

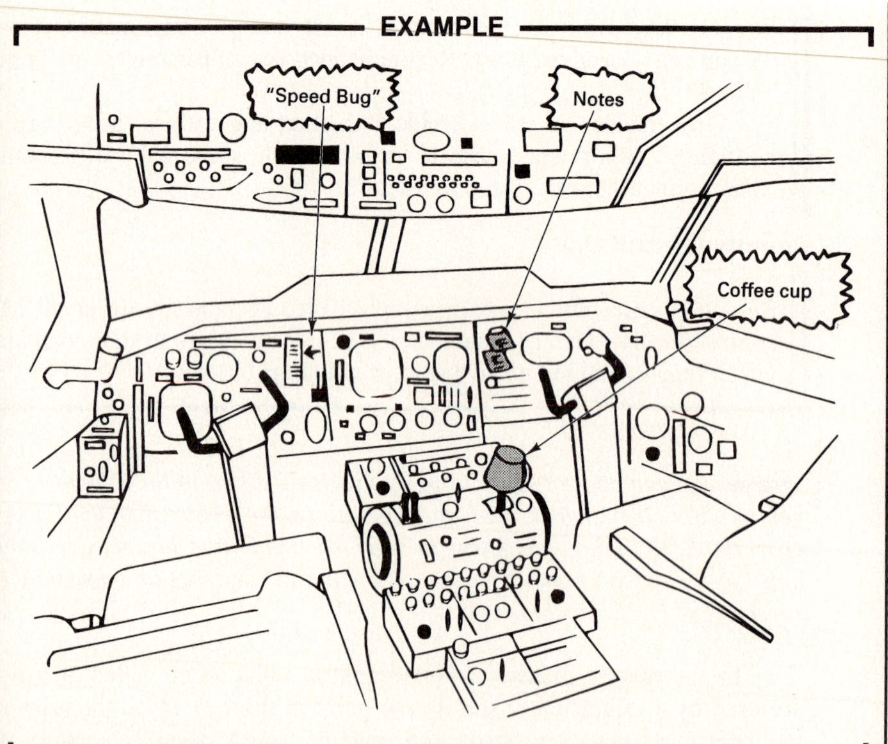

"Informal coding" used by aircraft crews (with permission from Norman, 1991).

Coding. There are numerous ways to help identify hand-operated controls, to indicate the effects of their operation, and to show their status. Throughout the system, the coding principles shall be uniform. The major coding means are:

- *Location*. Controls associated with similar functions shall be in the same relative location from panel to panel.
- *Shape*. Shaping controls to distinguish them can appeal to both the visual and tactation senses. Sharp edges shall be avoided. Various shapes and surface textures have been investigated for diverse uses. Figures 11-15 through 11-20 show examples.
- *Size*. Up to three different sizes of controls can be used for discrimination by size. Controls that have the same function on different items or equipments shall have the same size (and shape).

Figure 11-15. Examples of shape-coded aircraft controls. (Jenkins, 1953.)

- *Mode of operation*. One can distinguish controls by different manners of operation, such as push, turn, and slide. If the operator is not familiar with the control, a false manner of operation may be tried first, which is likely to increase operation time,
- *Labeling*. While proper labeling (see below) is a secure means to identify controls, this works only if the labels are in fact read and understood by the operator. The label must be placed so that it can be read easily, is well illuminated, and is not covered. Yet, labels take time to read. Trans-illuminated ("back-lighted") labels, possibly incorporated into the control, are often advantageous.
- *Color*. Most controls are either black (number 17038 in FED-STD 595) or gray (26231). For other colors, the following may be selected: red (11105, 21105, 31105, 14187); orange-yellow (13538, 23538, 33538); or white (17875, 27875, 37875). Use blue (1523, 25123) only if an additional color is absolutely necessary. Note that the use of color requires sufficient luminance of the surface.
- *Redundancy*. Often, coding methods can be combined, such as location, size, and shape; or color and labeling. This provides several advantages. One is that the combination of codes can generate a new set of codings. Second, it provides multiple ways to achieve the same kind of feedback,

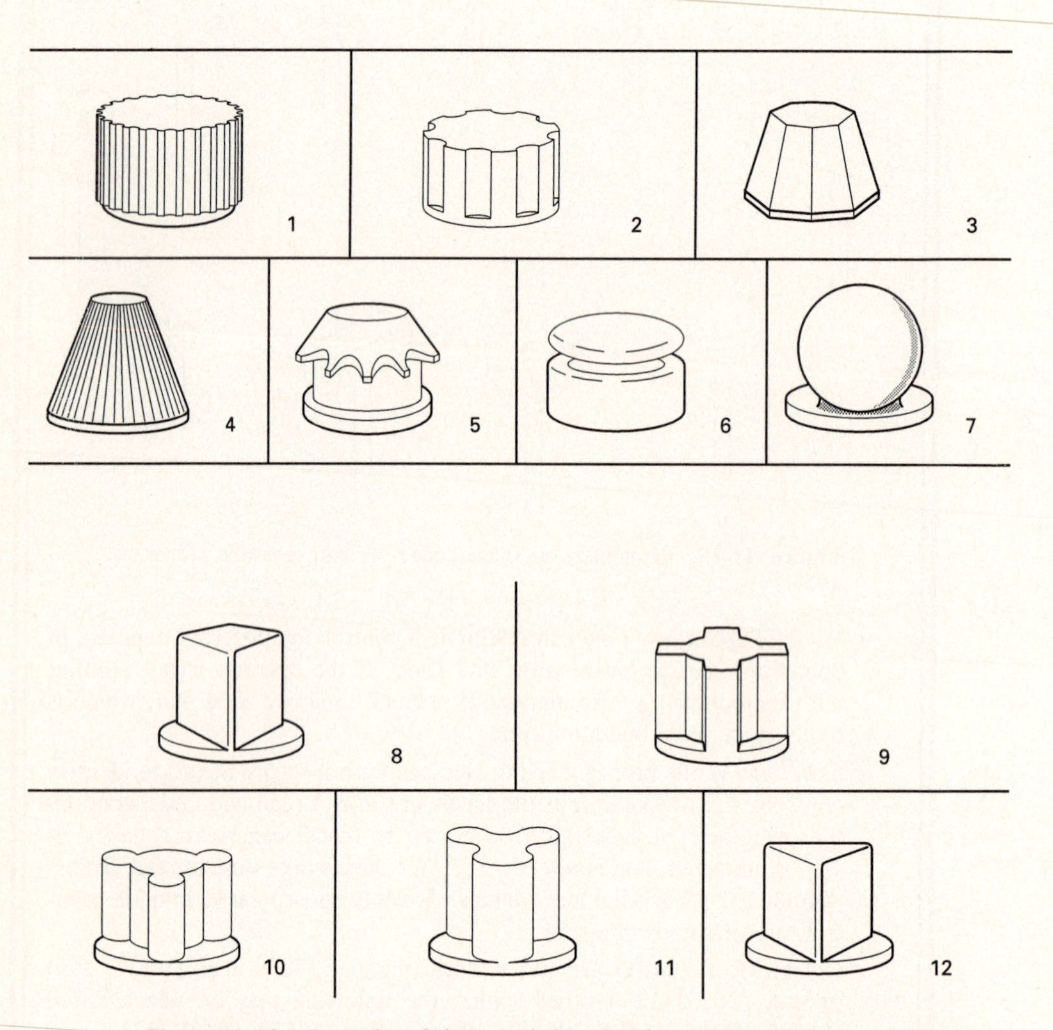

Figure 11-16. Examples of shape-coded knobs of approximately 2.5 cm diameter, 2 cm height. (Hunt and Craig, 1954.) Numbers 1 through 7 are suitable for full rotation (but: do not combine 1 with 2, 3 with 4, 6 with 7). Numbers 8 through 12 are suitable for partial rotation. Recommended combinations are 8 with 9, 10 with 11, 9 with 12 (but: do not combine 8 with 12, 9 with 10 or 11, 11 with 3 or 4).

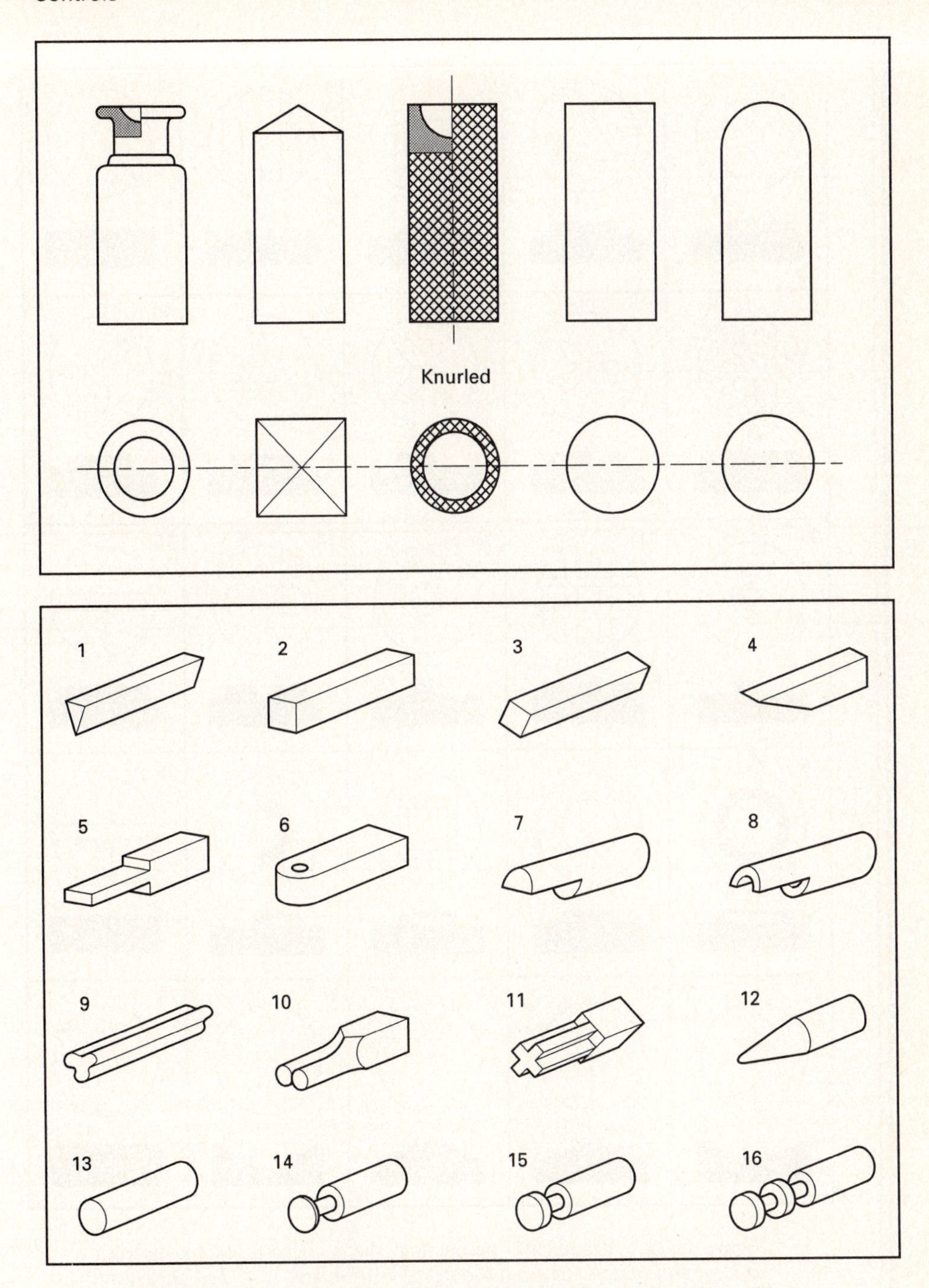

Figure 11-17. Examples of shape-coded toggle switches (top: Stockbridge, 1957; bottom: Green and Anderson, 1955) of approximately 1 cm diameter, 2.2 cm height.

Figure 11-18. Examples of pushbuttons shape-coded for tactile discrimination (with permission from Moore, 1974). Shapes 1, 4, 21, 22, 23, and 24 are best discriminable by bare-handed touch alone, but all shapes were on occasion confused.

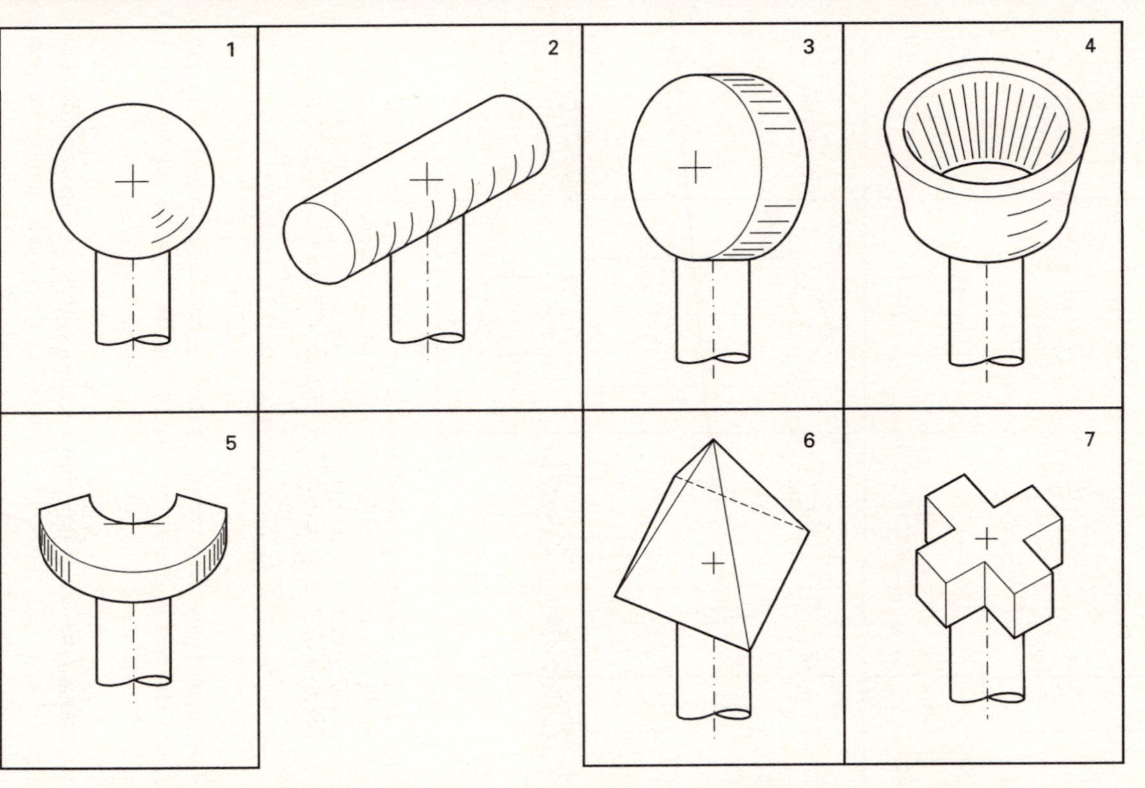

Figure 11-19. Shapes proposed (by the first author in 1980) for use on finger-operated controls of mining equipment. RECOMMENDED FOR CONCURRENT USE: *Two handles:* 1 & 2, 1 & 5, 1 & 6, 1 & 7, 2 & 3, 2 & 4, 2 & 5, 2 & 6, 2 & 7, 3 & 5, 4 & 6, 3 & 7, 4 & 5, 4 & 6, 4 & 7. *Three handles:* 1, 2, 6; 1, 2, 7; 2, 3, 6; 2, 3, 7; 3, 5, 6. *Four handles:* 1, 2, 3, 6; 1, 2, 3, 7; 2, 3, 4, 6; 2, 3, 4, 7. *Five handles:* 1, 2, 4, 5, 6. AVOID COMBINATIONS: 1 & 3, 3 & 4, 5 & 7.

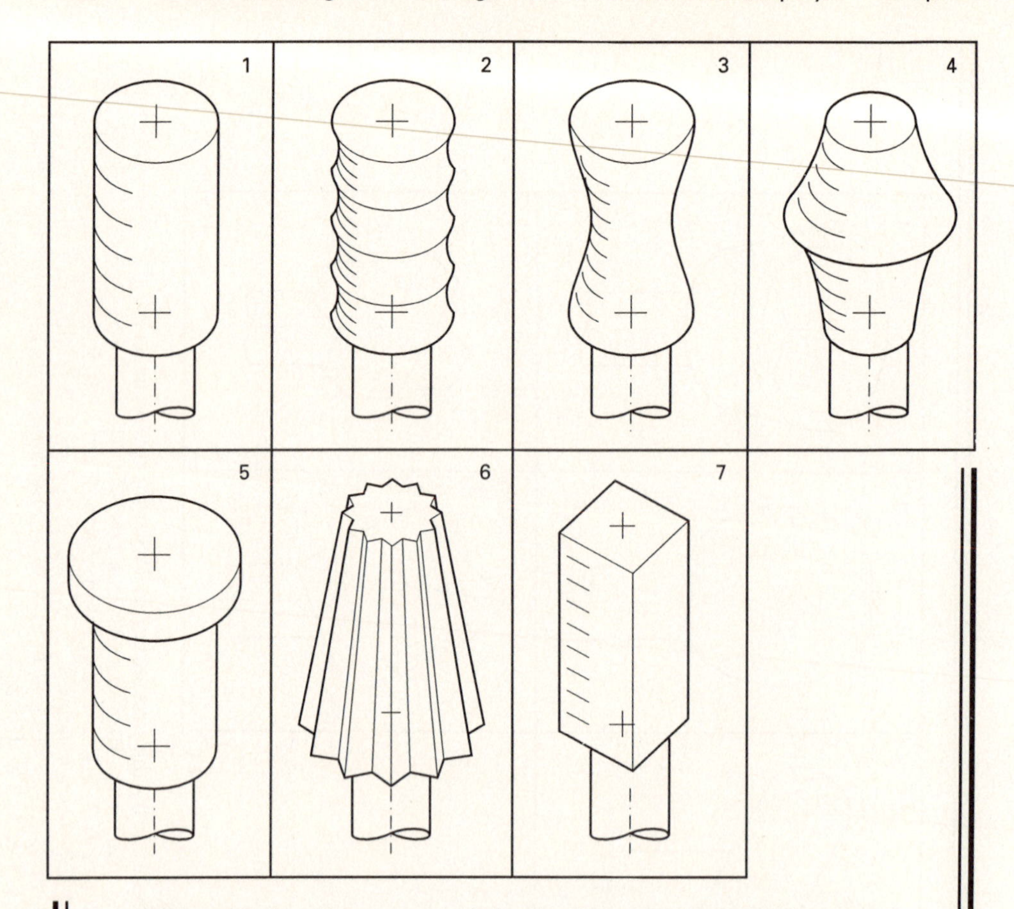

Figure 11-20. Shapes proposed (by the first author in 1980) for use on lever handles on mining equipment. RECOMMENDED FOR CONCURRENT USE: *Two handles:* 1 & 5, 1 & 6, 1 & 7, 2 & 6, 2 & 7, 3 & 4, 3 & 6, 3 & 7, 4 & 5, 4 & 6, 4 & 7, 5 & 6, 5 & 7. *Three handles:* 1, 2, 6; 1, 2, 7. *Four handles:* 1, 2, 4, 5; 1, 2, 4, 6; 1, 2, 4, 7. *Five handles:* 1, 2, 3, 4, 6; 1, 2, 3, 4, 7. AVOID COMBINATIONS: 2 & 5, 3 & 5, 6 & 7.

called redundancy: for example, if there is no chance to look at a control, one can still feel it; one knows that on a traffic light, the top signal is red, the bottom green.

The various types of codings have certain advantages and disadvantages, as listed in Table 11-20. Table 11-21 indicates the largest number of coding stimuli that can be used together.

Preventing accidental activation. Often it is necessary to prevent accidental activation of controls, particularly if this might cause injury to persons, damage to the system, or degradation of important system functions.

TABLE 11-20. ADVANTAGES AND DISADVANTAGES OF CODING TECHNIQUES

Advantages	Location	Shape	Size	Mode of Operation	Labeling	Color
Improves visual identification	X	X	X		X	X
Improves nonvisual identification (tactual and kinesthetic)	X	X	X	X		
Helps standardization	X	X	X	X	X	X
Aids identification under low levels of illumination and colored lighting	X	X	X	X	(When transilluminated)	(When transilluminated)
May aid in identifying control position (setting)		X		X	X	
Requires little (if any) training: is not subject to forgetting					X	
Disadvantages						
May require extra space	X	X	X	X	X	
Affects manipulation of the control (ease of use)	X	X	X	X		
Limited in number of available coding categories	X	X	X	X		X
May be less effective if operator wears gloves		X	X	X		
Controls must be viewed (i.e., must be within visual areas and with adequate illumination present)					X	X

SOURCE: Modified from Kroemer, 1988d.

TABLE 11-21. MAXIMAL NUMBER OF STIMULI FOR CODING

Visual stimuli	
Light intensity ("brightness")	2
Color of surfaces	9
Color of lights (lamps)	3
Flash rates of lights	2
Size	3
Shape	5
Auditory stimuli	
Frequency	4
Intensity ("loudness")	3
Duration	2

SOURCE: Adapted from information compiled by Cushman and Rosenberg, 1991.

There are various means to prevent accidental activation, some of which may be combined:

- Locate and orient the control so that the operator is unlikely to strike it or move it accidentally in the normal sequence of control operations.
- Recess, shield, or surround the control by physical barriers.
- Cover or guard the control.
- Provide interlocks between controls so that either the prior operation of a related control is required, or an extra movement is necessary to operate the control.
- Provide extra resistance (viscous or coulomb friction, spring-loading, or inertia) so that an unusual effort is required for actuation.
- Provide a "locking" means so that the control cannot pass through a critical position without delay.

Note that these means usually slow down the operation, which may be detrimental in an emergency case. Coding of foot-operated controls is difficult, and no comprehensive rules are known. Operation errors may have grave consequences—for example, in automobile driving (Rogers and Wierwille, 1988).

DISPLAYS

Displays provide the operator with necessary information about the status of the equipment. Displays are either visual (lights, scales, counters; CRT, flat panels) or auditory (bells, horns, recorded voice); or tactile (shaped knobs, Braille writing). Labels and instructions/warnings are special kinds of displays.

The "four cardinal rules" for displays are:

1. *Display only that information which is essential for adequate job performance.*
2. *Display information only as accurately as is required for the operator's decisions and control actions.*
3. *Present information in the most direct, simple, understandable, and usable form possible.*
4. *Present information in such a way that failure or malfunction of the display itself will be immediately obvious.*

Selecting either an auditory or visual display depends on conditions and purpose. The objective may be to provide

Status information—the current state of the system, the text input into a word processor, etc.

Historical—information about the past state of the system, such as the course run by a ship.

Predictive—such as the future position of a ship, given certain steering settings.

Instructional—telling the operator what to do, and how to do something.

Commanding—giving directions or orders for a required action.

An auditory display is appropriate if the environment must be kept dark; the operator moves around; the message is short, is simple, requires immediate attention, deals with events in time.

A visual display is appropriate if the environment is noisy; the operator stays in place; the message is long, is complex, will be referred to later, deals with spatial location.

Types of visual displays. There are three basic types of visual displays: the "check" display indicates whether or not a given condition exists (*example:* a green light to indicate normal functioning). The "qualitative" display indicates the status of a changing variable or the approximate value, or its trend of change (*example:* a pointer within a "normal" range). The "quantitative" display shows exact information that must be ascertained (*examples:* find your location on a map; read text or a drawing on a CRT) or indicates an exact numerical value that must be read (*example:* a clock). Overall guidelines are:

- Arrange displays so that the operator can locate and identify them easily without unnecessary searching.
- Group displays functionally or sequentially so that the operator can use them easily.

- Make sure that all displays are properly illuminated or luminant, coded, and labeled according to their function.

EXAMPLE

In some existing designs, it is advisable to reduce the information content. For example, a quantitative indicator may be reduced to a qualitative one by changing a numerical display of temperature to indicate simply "too cold," "acceptable," or "too hot."

TABLE 11-22. CODING OF INDICATOR LIGHTS

	Color			
Size/Type	Red	Yellow	Green	White
13 mm diameter or smaller/steady	Malfunction; action stopped; failure; stop action	Delay; check; recheck	Go ahead; in tolerance; acceptable; ready	Function or physical position; action in progress
25 mm diameter or larger/steady	Master summation (system or subsystem)	Extreme caution (impending danger)	Master summation (system or subsystem)	
25 mm diameter or larger/flashing	Emergency condition (impending personnel or equipment disaster)			

SOURCE: Modified from MIL-HDBK 759.

Light signals. Signals by light (color) are often used to indicate the status of a system (such as ON or OFF) or to alert the operator that the system, or a portion thereof, is inoperative and that special action must be taken. Common light (color) coding systems are (see Table 11-22):

- A *white* signal has no correct/wrong implications but may indicate that certain functions are ON.
- A *green* signal indicates that the monitored equipment is in satisfactory condition and that it is all right to proceed. For example, a green display

may provide such information as "go ahead," "in tolerance," "ready," "power on," etc.

- A *yellow* signal advises that a marginal condition exists and that alertness is needed, that caution be exercised, that checking is necessary, or that an unexpected delay exists.
- A *red* signal alerts the operator that the system or a portion thereof is inoperative and that a successful operation is not possible until appropriate correcting or overriding action has been taken. Examples for a red-light signal are to provide information about "malfunction," "failure," "error," and so on.
- A *flashing red* signal denotes an emergency condition that requires immediate action to avert impending personal injury, equipment damage, and the like.

Emergency signals. An emergency alert is best by auditory warning signal (see below) accompanied by a flashing light. (The operator may acknowledge the emergency by turning off one signal.)

Visual Warning Indicators. The warning light should be within 30 degrees of the operator's normal line of sight. The emergency indicator should be larger than general status indicators. The luminance contrast C with the immediate background (see Chapter 4) should be at least 3 to 1. Flash rate should be between three and five pulses per second, with "on" time about equal to "off" time. If the "flashing" device should fail, the light shall remain on steadily; warning indicators must never turn off merely because a flasher fails. A "word" warning (such as DANGER-STOP) should be used, if feasible.

Complex displays. More involved displays provide information that is either of a qualitative kind or that actually provides exact quantitative information. Some displays provide information about special settings or conditions, or indicate the difference between an expected and the actual condition. For these purposes, usually four different kinds of displays are used: moving pointer (fixed scale), moving scale (fixed pointer), counters, or "pictorial" displays. Table 11-23 lists these four kinds of displays and their relative advantages and disadvantages.

For a quantitative display it is usually preferable to use a moving pointer over a fixed scale. The scale may be straight (either horizontally or vertically), curved, or circular.

Scales should be simple and uncluttered; graduation and numbering of scales should be done such that correct readings can be taken quickly. Numerals should be located outside the scale markings so that they are not obscured by the pointer. The pointer, on the other side of the scale, should end with its tip directly at the markings. Figure 11-21 provides related information.

TABLE 11-23. CHARACTERISTICS OF DISPLAYS

Use	Moving Pointer	Moving Scale	Counters	Pictorial Displays
Quantitative information	*Good* Difficult to read while pointer is in motion	*Fair* Difficult to read while scale is in motion	*Good* Minimum time and error for exact numerical value, but difficult to read when moving	*Fair* Direction of motion/scale relations sometimes conflict, causing ambiguity in interpretation
Qualitative information	*Good* Location of pointer easy; numbers and scale need not be read; position changes easily detected	*Poor* Difficult to judge direction and magnitude of deviation without reading numbers and scale	*Poor* Numbers must be read; position changes not easily detected	*Good* Easily associated with real-world situation
Setting	*Good* Simple and direct relation of motion of pointer to motion of setting knob; position change aids monitoring	*Fair* Relation to motion of setting knob may be ambiguous; no pointer position change to aid monitoring; not readable during rapid setting	*Good* Most accurate monitoring of numerical setting; relation to motion of setting knob less direct than for moving pointer; not readable during rapid setting	*Good* Control-display relationship easy to observe

Tracking	*Good* Pointer position readily controlled and monitored; simplest relation to manual control motion	*Fair* No position changes to aid monitoring; relation to control motion somewhat ambiguous	*Poor* No gross position changes to aid monitoring	*Good* Same as above
Difference estimation	*Good* Easy to calculate positively or negatively by scanning scale	*Fair* Subject to reversal errors	*Poor* Requires mental calculation	*Good* Easy to calculate either quantitatively or qualitatively by visual inspection
General	Requires largest exposed and illuminated area on panel; scale length limited unless multiple pointers used	Saves panel space; only small section of scale need be exposed and illuminated; use of tape allows long scale	Most economical of space and illumination; scale length limited only by available number of digit positions	Picture/symbols need to be carefully designed and pretested

SOURCE: Modified from MIL-HDBK 759.

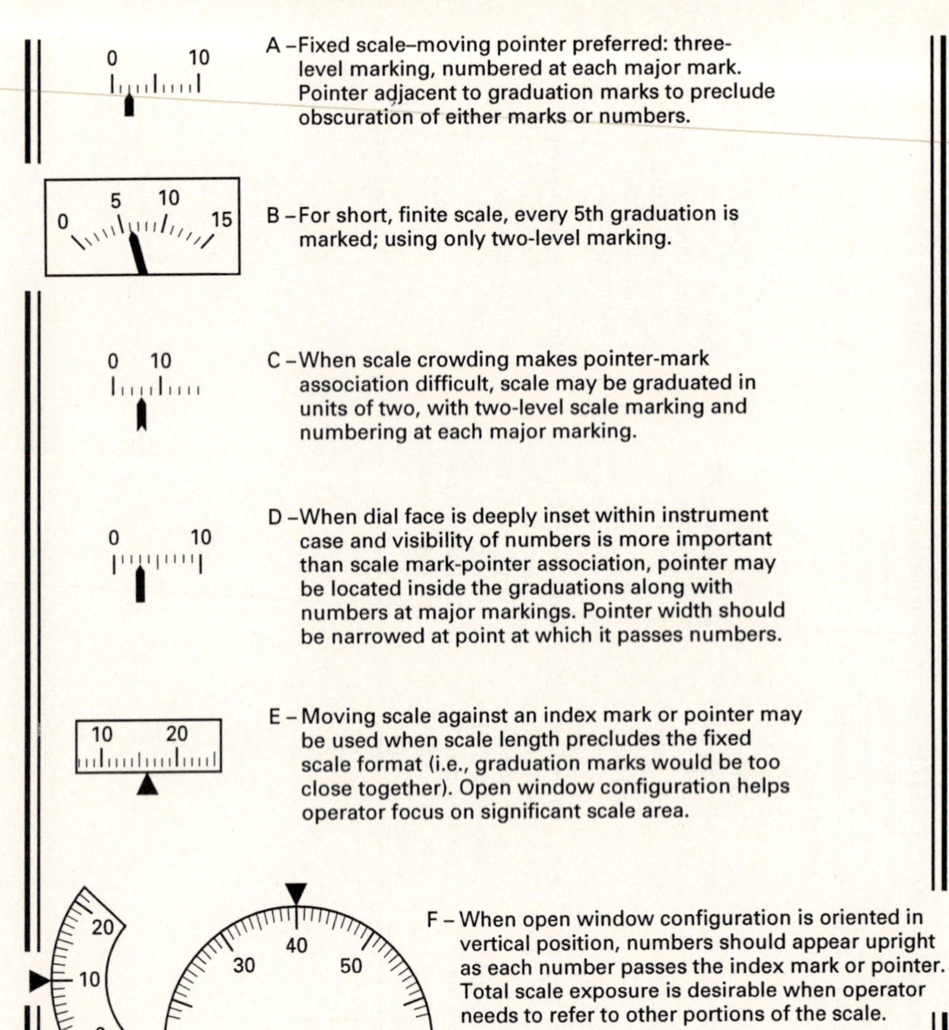

Figure 11-21. Scale graduation, pointer position, and scale numbering alternatives. (Modified from MIL-HDBK 759.)

The scale mark should show only such fine divisions as the operator must read. All major marks shall be numbered. Progressions of either 1, or 5, or 10 units between major marks are best. The largest admissible number of unlabeled minor graduations between major marks is nine, but only with a minor tick at 5. Numbers should increase as follows: left-to-right; bottom-to-top; clockwise. Recommended dimensions for the scale markers are presented in Figure 11-22. The dimensions shown there are suitable even for low illumination.

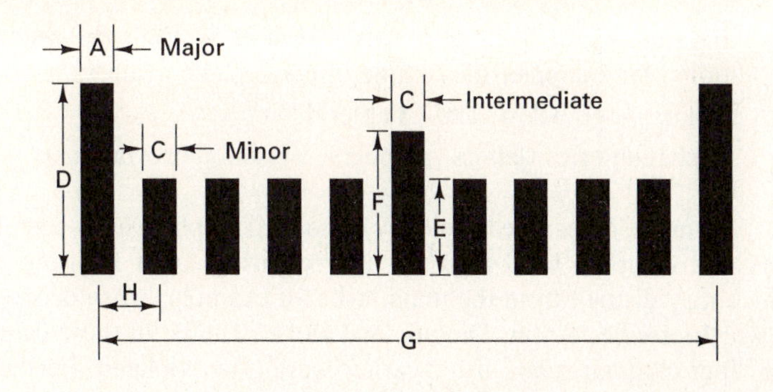

Dimension (in mm)	Viewing distance (in mm) 710	910	1525
A (Major index width)	0.89	1.14	1.90
B (Minor index width)	0.64	0.81	1.37
C (Intermediate index width)	0.76	0.99	1.63
D (Major index height)	5.59	7.19	12.00
E (Minor index height)	2.54	3.28	5.44
F (Intermediate index height)	4.06	5.23	8.71
G (Major index separation between midpoints)	17.80	22.90	38.00
H (Minor index separation between midpoints)	1.78	2.29	.381

Minimum scale dimensions suitable even for low illumination

Figure 11-22. Scale marks. (Reprinted from MIL-HDBK 759.)

Electronic displays. In the 1980s and 90s, "mechanical" displays (such as actual pointers moving over printed scales) were increasingly replaced by "electronic" displays with computer-generated images, or solid-state devices such as light-emitting diodes. Whether a mechanical or electronic display, the displayed information may be coded by the following means:

Shape—straight, circular, etc.
Shades—black and white or gray.
Lines and crosshatched patterns.

Figures, pictures, or pictorials—such as symbols or various levels of abstractions, for example, the outline of an airplane against the horizon.

Colors—see below, and Chapter 4.

Alphanumerics (letters, numbers, words, abbreviations).

In many cases, electronically generated displays have been fuzzy, overly complex and colorful, hard to read, and requiring exact focusing and close attention which may distract from the main task, for example, when driving a car (Wierwille, 1992; Wierwille, Antin, Dingus, and Hulse, 1988). In these cases, the first three of the "four cardinal rules" listed earlier were often violated. Furthermore, many electronically generated pointers, markings, and alphanumerics did not comply with established human engineering guidelines, especially when generated by line segments, scan lines, or dot matrices. Although some designs are "acceptable" according to current technology and user tolerance (Cushman and Rosenberg, 1991), requirements are changing and innovations in display techniques rapidly develop. Hence, printed statements (even if current and comprehensive when they appear, such as by Cushman and Rosenberg, 1991; Kinney and Huey 1990; Krebs, Wolf, and Sandrig, 1978; Silverstein and Merrifield, 1985; Woodson, 1981) are becoming obsolete quickly and none are given in this text. One must closely follow the latest information appearing in technical publications to stay up-to-date in this field.

EXAMPLE

"Consider the digital watch and digital speedometer. Although both are more accurate and easier to read than their analog counterparts, many people . . . prefer products with analog displays. One reason is that analog displays also provide information about deviations from reference values. For example, a watch with an analog display also shows the user how many minutes remain before a specific time. Similarly a speedometer with an analog display shows the driver of a car both the car's speed and how far it is above (or below) the posted speed limit" (Cushman and Rosenberg, 1991, pp. 92–93).

Electronic displays are widely used with computers, in cockpits of airplanes and automobiles, and in cameras. Critical physical aspects are size, luminance, contrast, viewing angle and distance, and color.

The overall quality of electronic displays has often been found wanting; viewers usually state that it is more difficult to read the same text from a CRT than from print on paper. However, the image quality of the compared text carriers must be comparable: a suitable metric is the *modulation transfer function,* MTF (Snyder, 1985). It describes the resolution of the display using a spatial sine-wave test signal. For the same MTF area (MTFA), i.e., for comparable image quality, one can expect the same reading times whether hard copy (photograph, printed material) or soft copy (electronic display) is used. Reading time and subjective impression of quality vary with image quality, largely independent of the display technology used (Jorna and Snyder, 1991). Yet, readers have many other criteria than just "reading time" to make use and preference statements (Dillon, 1992).

"Monochrome" displays have only one color, preferably near the middle of the color spectrum (see Chapter 4), i.e., white ("achromatic"), green, yellow, amber, orange rather than blue or red. While measured performance with each of the colors appears similar, personal preferences exist which make it advisable to provide a set with switchable colors (Cushman and Rosenberg, 1991).

If several colors appear on the ("chromatic") display, they should be easily discriminated. It is best to display simultaneously not more than three or four colors, including red, green, yellow or orange, cyan or purple, all strongly contrasting with the background; more may be used if necessary, if the user is experienced, and if the stimuli subtend at least 45′ of visual angle (Cushman and Rosenberg, 1991; Kinney and Huey, 1990).

There is a set of phenomena that have relevance to displaying visual information:

- Albney effect — Desaturating a colored light (by adding white light) may introduce a hue shift.
- Assimilation — A background color may appear to be blended with the color of an overlying structure (e.g., alphanumeric characters). The effect is the opposite of color contrast.
- Bezold-Brucke effect — Changing the luminance of a colored light is usually accompanied by a change in perceived hue.
- Chromostereopis — Highly saturated reds and blues may appear to be located in different depth planes (i.e., in front of or behind the display plane). This may induce visual fatigue and feeling of nausea or dizziness.
- Color adaptation — Prolonged viewing of a given color reduces an observer's subsequent sensitivity to that color. As a consequence, the hues of other stimuli may appear to be shifted.
- Color "blindness" — About 10 percent of all western males (1 percent of females) have hereditary color deficiencies, mostly such that they see colors (especially reds and greens) differently or less vividly.
- Color contrast — The hue of an object (e.g., a displayed symbol) is shifted toward the complementary of the surround. The effect is the opposite of assimilation.
- Desaturation — Desaturating highly saturated colors may reduce the effects of adaptation, assimulation, chromostereopsis, and color contrast but increase the Albney effect.

- Liebmann effect — Removing all luminance contrast from a color image may produce subjectively fuzzy edges.
- Receptor distribution — The distribution of cones (color-sensitive receptors) on the retina is uneven. Near the fovea there is most sensitivity to red and green (particularly under high illumination); at the periphery more sensitivity to blue and yellow (see Chapter 4).

APPLICATION

In spite of the many variables that (singly and interacting with each other) affect the use of complex color display, Cushman and Rosenberg (1991) compiled the following guidelines for use of color in displays:

- Limit the number of colors in one display to four if users are inexperienced or if use of the display is infrequent. No more than seven colors should ever be used.
- The particular colors chosen should be widely separated from one another in wavelength to maximize discriminability. Colors that differ from one another only with respect to the amount of one primary color (e.g., different oranges) should not be used.
- Suggested combinations:

  (1) green, yellow, orange, red, white
  (2) blue, cyan, green, yellow, white
  (3) cyan, green, yellow, orange, white

  Avoid:

  (1) reds with blues
  (2) reds with cyans
  (3) magentas with blues

In general, avoid displaying several highly saturated, spectrally extreme colors at the same time.

- Red and green should not be used for small symbols and small shapes in peripheral areas of large displays.
- Blue (preferably desaturated) is a good color for backgrounds and large shapes. However, blue should not be used for text, thin lines, or small shapes.
- Using opponent colors (red and green, yellow and blue) adjacent to one another or in an object/background relationship is sometimes beneficial and sometimes detrimental. No general guidelines may be given.
- The color of alphanumeric characters should contrast with that of the background.

- When using color, use shape or brightness as a redundant cue (e.g., all yellow symbols are triangles, all green symbols are circles, all red symbols are squares, etc.). Redundant coding makes the display much more acceptable for users who have color deficiencies.
- As the number of colors is increased, the sizes of color-coded objects should also be increased.

Location of displays. Displays should be oriented within the normal viewing area of the operator, with their surfaces perpendicular to the line of sight. The more critical the display, the more centered it should be within the operator's central cone of sight. (Avoid glare—see Chapters 4 and 9.)

A group of pointer instruments can be so arranged that all pointers are aligned under normal conditions. If one of the pointers deviates from that normal case, its displacement from the aligned configuration is particularly obvious. Figure 11-23 shows examples of such arrangements.

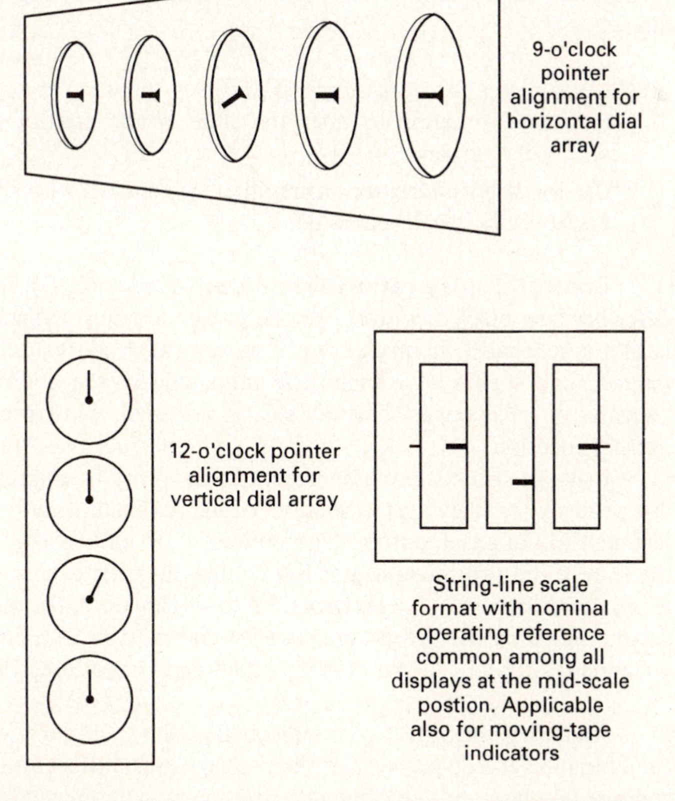

Figure 11-23. Aligned pointers for rapid check-reading. (Modified from MIL-HDBK 759.)

Control-display assignments. In many cases, instruments are set by controls, which should be located in suitable positions close to each other so that the control setting can be done without error, quickly, and conveniently. Proper operation is facilitated by selecting suitable control/display placements. Popular expectancies of relationships exist, but they are often not strong and may depend on the user's background and culture. The assignment is clearest when the control is directly below or to the right of the display. Figure 11-24 shows examples of suitable arrangements.

☞☞☞ *A frequent course assignment for students of ergonomics is to design a stove top in which the controls for four burners are located so that their relations are unambiguous. No solution has ever been found which combines satisfactorily good assignment, safety of operation, and easy cleaning.* ☜☜☜

Expected movement relationships are influenced by the types of controls and displays. When both are congruous, i.e., if both are linear or rotary, the stereotype is that they move in corresponding directions, e.g., both up or both clockwise. When the movements are incongruous, their preferred movement relationship can be taken from Figure 11-25. The following rules apply generally:

Gear-slide ("Warrick's rule"): a display (pointer) is expected to move in the same direction as does the slide of the control close to ("geared with") the display.

Clockwise-for-increase: turning the control clockwise should cause an increase in the displayed value.

Control/display ratio. The *control/display* (C/D) ratio (or: CD gain) describes how much a control must be moved to adjust a display: the C/D ratio is like a gear ratio. If much control movement produces only a small display motion, one speaks of a high C/D ratio, and of the control as having low "sensitivity." The opposite ratio, D/C, is also used: this expression resembles a transfer function.

Usually, two distinct movements are involved in making a setting: first a fast primary (or: slewing) motion to an approximate location, then a fine adjustment to the exact setting. The optimal C/D ratio is that which minimizes the sum of the two movements. For continuous rotary controls, the C/D ratio is usually 0.08 to 0.3, for joysticks 2.5 to 4. However, the most suitable ratios depend much on the given circumstance and must be determined for each application (Boff and Lincoln, 1988; Arnaut and Greenstein, 1990).

Auditory signals. As indicated earlier, auditory signals are better suited than visual displays when the message must attract attention. Therefore, auditory displays are predominantly used as warning devices, especially when the message is short or simple, often together with a flashing light.

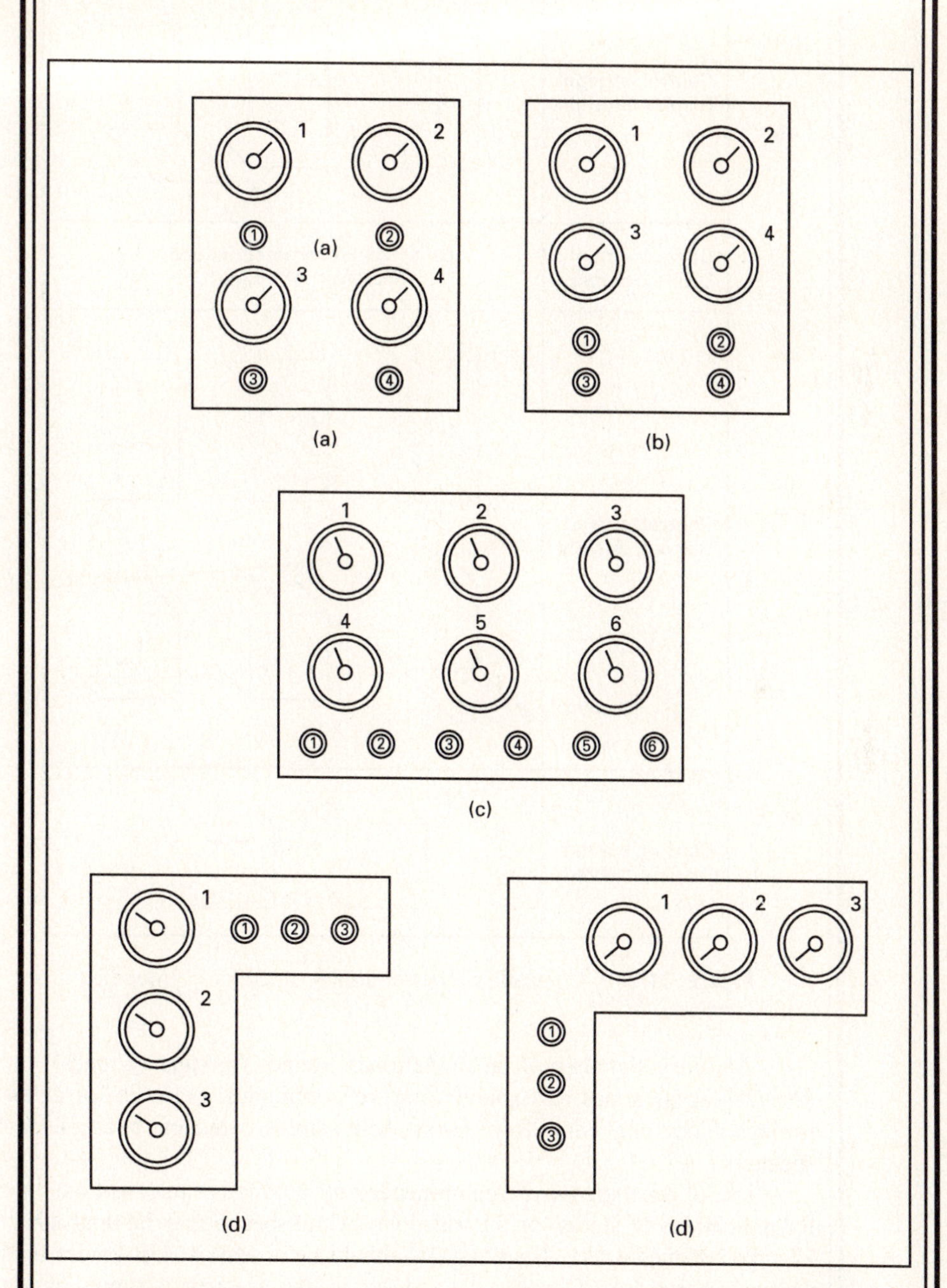

Figure 11-24. Control-display relationships. (Adapted from MIL-HDBK 759.)

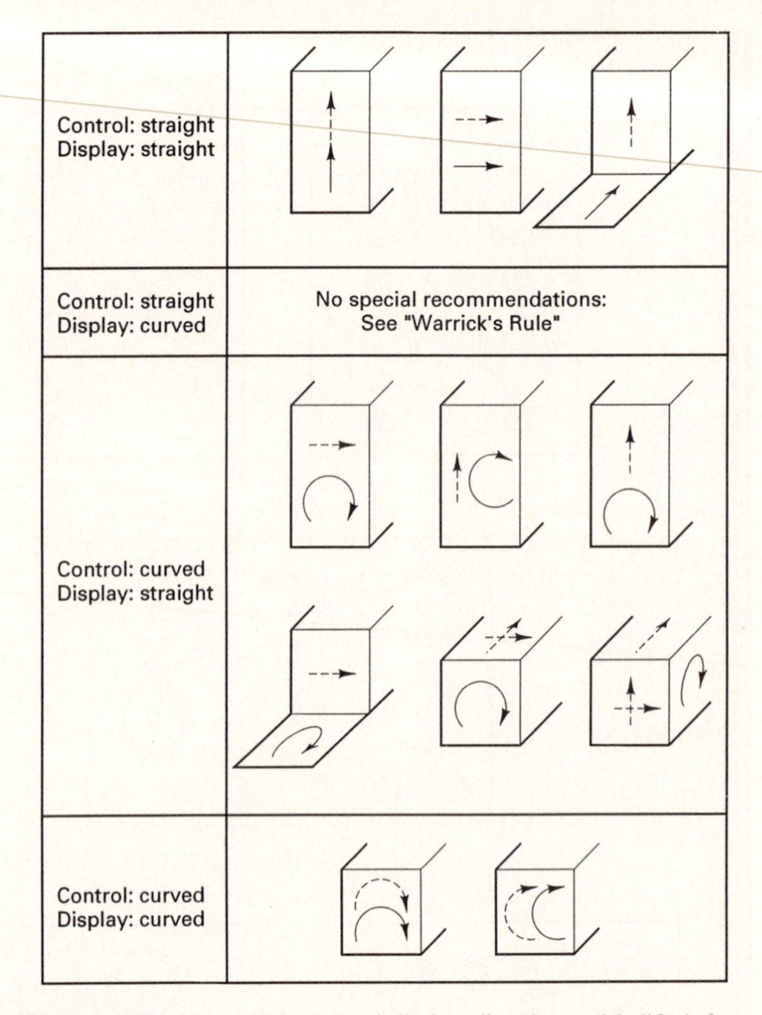

Figure 11-25. Compatible control-display directions. (Modified from Loveless, 1962.)

Auditory signals may be single tones, sounds (mixture of tones), or spoken messages. Tones and sounds may be continuous, periodic, or at uneven timings. They may come from horns, bells, sirens, whistles, buzzers, or loudspeakers.

Use of *tonal signals* is recommended for qualitative informations, such as for indications of status, or for warnings (while speech may be appropriate for all types of messages). Tonal signals should be at least 10 dB louder than the ambient noise (see Chapter 4) and be in the frequency range of 400 to 1,500 Hz; if the signal undulates or warbles, 500 to 1,000 Hz. Buzzers may have frequencies as low as 150 Hz and horns as high as 4,000 Hz. The tonal signal can be made more conspicuous by increasing its intensity, by interrupting it repeatedly, or by changing its frequency. For example: increase from 700 to 1,700 Hz in 0.85 s, be silent for 0.15 s (cycle time one second), then start over, etc. Where specific warning sounds are needed, people can identify the following as different:

1,600 ± 50 Hz tone, interrupted at a rate of 1 to 10 Hz;

900 ± 50 Hz steady tone, plus 1,600 ± 50 Hz tone interrupted at a rate of 0 to 1 Hz;

900 ± 50 Hz steady tone;

900 ± 50 Hz steady tone, plus 400 ± 50 Hz tone interrupted at a rate of 0 to 1 Hz;

400 ± 50 Hz tone, interrupted at a rate of 1 to 10 Hz.

Word messages may be prerecorded, digitized, or synthesized speech. The first two techniques are often used (as for telephone answering or "talking products") and are characterized by good intelligibility and natural sound. Synthesized speech uses compositions of phonemes; the result does not sound natural but is effective in converting written text to speech, and may sound startling, thus attracting attention to the message.

LABELS AND WARNINGS

Labels. Ideally, no label should be required on equipment or control to explain its use. Often, however, it is necessary to use labels so that one may locate, identify, read, or manipulate controls, displays, or other equipment items. Labeling must be done so that the information is provided accurately and rapidly. For this, the following guidelines apply:

- *Orientation*. A label and the information printed on it shall be oriented horizontally so that it can be read quickly and easily. (Note that this applies if the operator is used to reading horizontally, as in "western" countries.)
- *Location*. A label shall be placed on or very near the item that it identifies.
- *Standardization*. Placement of all labels shall be consistent throughout the equipment and system.
- *Equipment Functions*. A label shall primarily describe the function ("what does it do") of the labeled item.
- *Abbreviations*. Common abbreviations may be used. If a new abbreviation is necessary, its meaning shall be obvious to the reader. The same abbreviation shall be used for all tenses and for the singular and plural forms of a word. Capital letters shall be used, periods normally omitted.

- *Brevity*. The label inscription shall be as concise as possible without distorting the intended meaning or information. The texts shall be unambiguous, redundancy minimized.
- *Familiarity*. Words shall be chosen, if possible, that are familiar to the operator.
- *Visibility* and *Legibility*. The operator shall be able to be read easily and accurately at the anticipated actual reading distances, at the anticipated worst illumination level, and within the anticipated vibration and motion environment. Important are: contrast between the lettering and its background; the height, width, strokewidth, spacing, and style of letters; and the specular reflection of the background, cover, or other components.
- *Font* and *Size*. Typography determines the legibility of written information; it refers to style, font, arrangement, and appearance.

Font (typeface) should be simple, bold, and vertical, such as Futura, Helvetica, Namel, Tempo, and Vega. Note that most electronically generated fonts (such as by LED, LCD, matrix) are generally inferior to printed fonts; thus, special attention must be paid to make these as legible as possible.

Recommended *height of characters* depends on the viewing distance, e.g.,

Viewing distance 35 cm, suggested height 22 mm.
Viewing distance 70 cm, suggested height 50 mm.
Viewing distance 1 m, suggested height 70 mm.
Viewing distance 1.5 m, suggested height at least 1 cm.

The *ratio of strokewidth to character height* should be between 1:8 to 1:6 for black letters on white background, and 1:10 to 1:8 for white letters on black background.

The *ratio of character width to character height* should be about 3:5.

The *space between letters* should be at least one stroke width.

The *space between words* should be at least one character width.

For continuous text, mix upper- and lower-case letters. (For labels, use upper-case letters only.) For VDTs, see ANSI 100 (Human Factors Society, 1988, or newer edition).

Warnings. Ideally, all devices should be safe to use. In reality, often this cannot be achieved through design. In this case, one must warn users of danger associated with product use and provide instructions for safe use to prevent injury or damage.

It is preferable to have an "active" warning, usually consisting of a sensor that notices inappropriate use, and of an alerting device that warns the human of an impending danger. Yet, in most cases, "passive" warnings are used, usually consisting of a label attached to the product and of instructions for safe use in the user manual. Such passive warnings rely completely on the human to recognize an existing or potential dangerous situation, to remember the warning, and to behave prudently.

Factors that influence the effectiveness of product warning information have been compiled by Cushman and Rosenberg (1991). They are listed in Table 11-24.

TABLE 11-24. FACTORS THAT INFLUENCE THE EFFECTIVENESS OF PRODUCT WARNING INFORMATION

Situation	Warning Effectiveness	
	Low[1]	High[2]
User is familiar with product	X	
User has never used product before		X
High accident rate associated with product		X
Probability of an accident is low	X	
Consequences of an accident are likely to be severe		X
User is in a hurry	X	
User is poorly motivated	X	
User is fatigued or intoxicated	X	
User has previously been injured by product		X
User knows good safety practices		X
Warning label is adjacent to the hazard		X
Warning label is very legible and easy to understand		X
Active warnings alert user only when some action is necessary		X
Active warnings frequently give false alarms	X	
Product is covered with warning labels that seem inappropriate	X	
Warning contains only essential information		X
Source of warning information is credible		X

SOURCE: Reprinted with permission from Cushman and Rosenberg, 1991.

[1] Low probability of behavioral change.

[2] High probability of behavioral change.

Labels and signs for passive warnings must be carefully designed by following the most recent government laws and regulations, national and international standards, and the best applicable human engineering information. Warning labels and placards may contain text, graphics, and pictures—often graphics with redundant text. Graphics, particularly pictures and pictograms, can be used by persons with different cultural and language backgrounds, if these depictions are selected carefully. (For recent overviews, see, e.g., Cushman and Rosenberg, 1991; Lehto and Clark, 1991; Lehto and Miller, 1986; Miller and Lehto, 1986; Ryan, 1991.) One must remember, however, that the

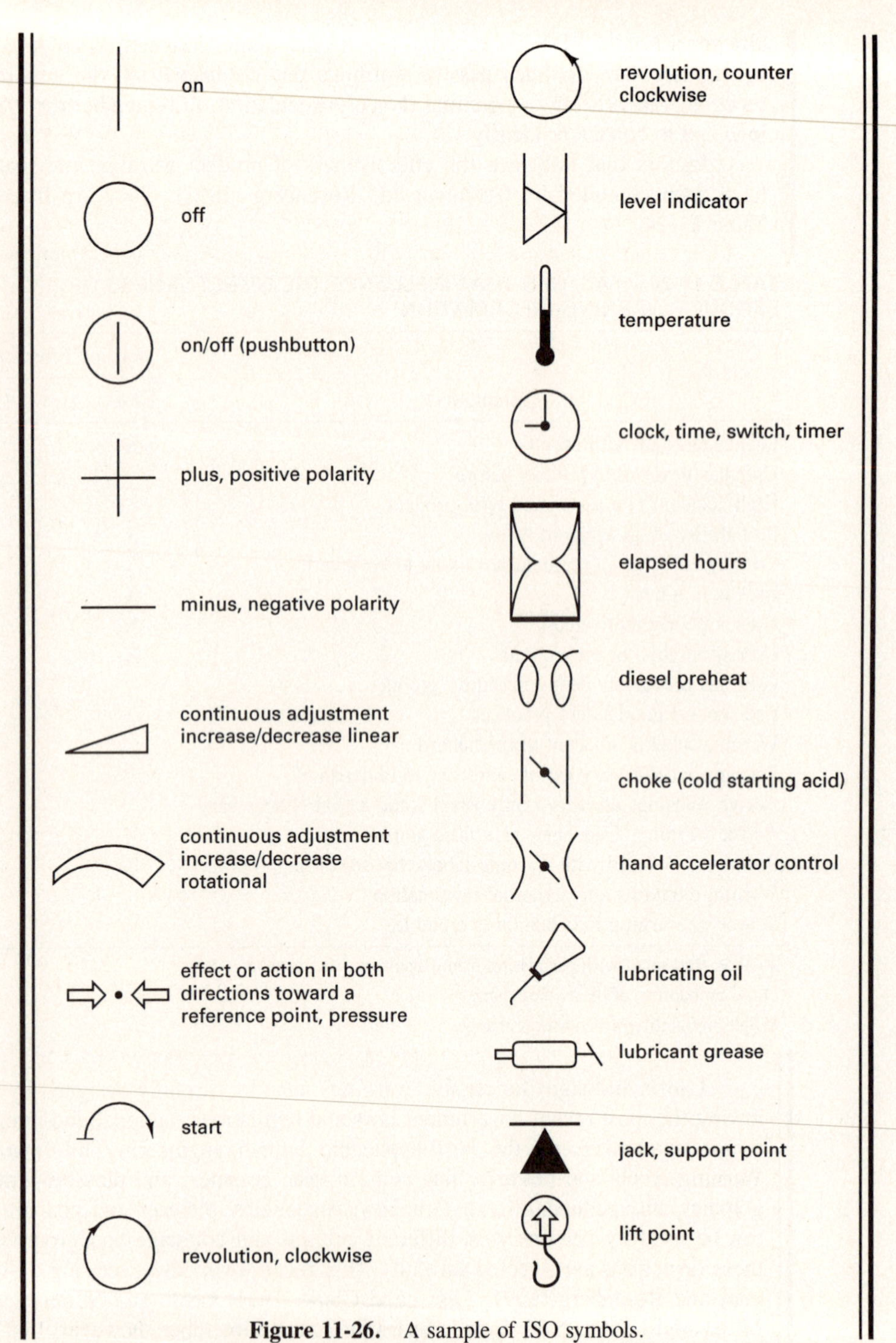

Figure 11-26. A sample of ISO symbols.

design of a "safe" product is much preferable to applying warnings to an inferior product. Furthermore, users of different ages and experiences, and users of different national and educational backgrounds, may have rather different perceptions of dangers and warnings (Chapanis, 1975; Tajima, Asao, Hill, and Leonard, 1991).

Symbols. Symbols ("icons") are simplified drawings of objects or abstract signs, meant to identify an object, warn of a hazard, or indicate an action. They are common in public spaces, automobiles, computer displays, and maintenance manuals. The Society of Automotive Engineers (SAE) and the International Standardization Organization (ISO), for example, have developed extensive sets of symbols and established guidelines for developing new ones. Some of the symbols for use in vehicles, construction machinery, cranes, and airport handling equipment are reproduced in Figure 11-26. Note that both abstract symbols (e.g., "ON") and simplified pictorials may require learning or the viewer's familiarity with the object (hourglass for "elapsed hours"). If one must develop new symbols, the cultural and educational background of the viewer must be carefully considered: many symbols have ancient roots and many invoke unexpected and unwanted reactions (Frutiger, 1989).

The following guidelines have been adapted from those of ISO Technical Committee 145 (dated October 29, 1987):

- Symbols should be graphically simple, clear, distinct, and logical to enhance recognition and reproduction.
- Symbols should incorporate basic symbol elements which can be used alone or combined as necessary into a logical symbolic language which, if not immediately obvious, is at least readily learned.
- Graphical clarity should prevail in disputes with logical consistency, because no symbol is recognizable, no matter how logical, if it cannot be distinguished from other symbols.
- A minimum of detail should be included; only details which enhance recognition should be allowed, even if other details are accurate renditions of the machine or equipment.

SUMMARY

Although psychological rules of "how the human best perceives displayed information and operates controls" are generally missing, there are many well-proven design recommendations for traditional controls and dial-type displays. These are presented, in much detail, in this chapter.

With the widespread use of the computer, new input devices are emerging, such as mouse, trackball, or touch screen. The classic typing keyboard has mushroomed: a redesign is urgently needed. Probably other key designs, possibly new input methods such as by voice, should replace the monstrous 100+ key arrangement that apparently overloads many users—see the section on overuse disorders in this book.

Display technology has also changed. Electronic displays, many of which caused "eye and head aches," can be made in a quality similar to hard copy. Use of color, if properly done, makes visual information transmission interesting and easy.

Warning signals, instructions on how to use equipment, and warnings to avoid misuse and danger are important, both ergonomically and juristically.

CHALLENGES

What are the implications for future control and display developments, if so few general rules exist that explain current usages?

Should one try to generate the same "population stereotypes" throughout the world?

Are there better means to control the direction of an automobile than by the conventional steering wheel?

Is it worthwhile to look for different keyboard and key arrangements if voice activation might become generally used in a decade, or two decades?

Why is it difficult to compare the "usability" of CRT displays with that of information displayed on paper?

12

Designing for Special Populations

OVERVIEW

Usually, one designs for a "regular adult" population in the age range of about 20 to 50 years. Of these people, anthropometry, biomechanics, physiology, and psychology, attitudes and behavior are fairly well known. This is the group of most interest to industry and to society as "movers" and contributors to the gross national product. Yet, other large population groups are of specific concern: pregnant women, children, the aging, and the disabled. These need special ergonomic attention, but much information is missing or incomplete.

BACKGROUND

As discussed in the first chapter, fairly complete anthropometric information is available mostly for military populations. Adult civilians are seldom measured as a large group, and their body dimensions must be inferred from those of soldiers. Fortunately, their body dimensions as well as their physical and psychological capabilities and traits are not very different; therefore, the ergonomic principles that help us design for civilian adults also apply to the military, and vice versa.

There are systematic differences among adult men and women—for instance, in body sizes and in physical capabilities. Yet, in general, one can design nearly any workstation, any piece of equipment or tool so that it is usable by either women or men. In some cases, adjustability is needed, or one may have to provide objects in different dimension ranges. These adjustments and ranges, however, often are not gender-specific but are simply needed to fit different people.

Healthy adults in their 20s, 30s, and 40s are commonly considered "the norm." About them most ergonomic information is available, and for them most ergonomic efforts are made. One group, however, among adults needs specific attention: that of pregnant females. Therefore, a subsequent section of the text will be devoted to them.

There is another group that draws everyone's attention: infants and children. The younger they are, the more different they are from our ergonomic norm, the adult. Specific ergonomic information is available for small children, but little is systematically known about teenagers.

Aging people are another large group who need special ergonomic attention. Body size and posture, physical capabilities, and psychological traits change, for some over a short time, for some over decades. To accommodate their special traits during their later working years, their retirement, and their waning period poses challenging yet ethically satisfying tasks to ergonomists.

Another group of people, most but not all of them adults, also need special ergonomic consideration: the impaired and disabled. They differ from their peers in size, posture, and abilities. In many cases, proper ergonomic design of their environment and their equipment can help them to overcome their handicaps and live a satisfactory life.

SPECIAL DESIGNS FOR WOMEN AND MEN?

Obviously, tight clothing must be cut differently for men and women to fit their bodies. But must one design special workstations and tools to fit either gender because, as groups, males may do certain jobs better than females, and vice versa?

The following statements, taken from the literature, are based on "group average and ranges"—findings which may not be true for individuals. Furthermore, there are generally large overlaps between capabilities and traits of males and females—for example, in body size or in muscle strength.

Size and Strength

In body dimensions, men are generally taller, with women having relatively shorter legs but wider hips. In muscle strength and work output, women are weaker than men, usually developing between 60 and 90 percent of the male values (Astrand and Rodahl, 1986; Hettinger, 1961; NASA/Webb, 1978). However, women's leg strength is only marginally lower, and muscle tension developed per cross-section unit is equal to men's. Mobility in body joints is generally but marginally greater in women—see Chapter 1.

Sensory Abilities

In audition (hearing), girls and adult females have lower absolute thresholds for pure tones than boys and men. Aging men have larger hearing losses than aging women. In vision, both static and dynamic, boys and men have better acuity than girls and women. With increasing age, acuity declines earlier in females than in males. Females have more vision deficiencies.

In gustation (taste), women detect sweet, sour, salty, and bitter stimuli at lower concentrations than men; however, some studies have not substantiated these findings. In olfaction (smell), women can detect some substances more easily than men. Both taste and smell capabilities and preferences change within the female menstrual cycle and during pregnancy.

In cutaneous senses, little difference has been shown between sexes in the threshold for temperature sensation. Females feel less warm (comfortable) than males in environments of 19° to 38°C initially, but adapt to the surrounding temperature more rapidly than males. Females begin to sweat at higher temperatures than males, and acclimatize to work in severely hot conditions somewhat more slowly. Men may find it a bit easier to adjust to very hot and very cold conditions (Burse, 1979), but the differences are minor and conditional, not indicating true gender differences (Shapiro, Pandolph, Avellini, Pimental, and Goldman, 1981).

Regarding touch, the sensitivity to vibration is about the same in males and females. However, females are more sensitive to pressure stimuli on their body, except on the nose. Pain sensations (a complex and difficult topic for physiological and behavioral reasons) seem to be about the same in males and females (Baker, 1987).

Thus, there appear to be some differences in sensory functioning (see Chapter 4) between the sexes, but such differences usually are small.

Cognition and Intellectual Skills

Only a few subsets of this topic have received sufficient research to allow conclusive statements. In mathematical reasoning men usually perform better than women, who excel in fine coordination. In the section of the Scholastic Aptitude Test (SAT) concerning numerical judgment, relational thinking, and reasoning, at age 13, boys have about 30 points (of about 400) higher scores than girls; the higher the absolute score, the larger the difference between boys and girls (Anderson, 1987).

Motor Skills

Girls and women are more skillful than boys and men on perceptual and psychomotor tests such as color perception, aiming and dotting, finger dexterity, inverted alphabet printing, and card sorting. Males perform better in speed-related tasks, on the rotary pursuit apparatus, and in other simple rhythmic eye-hand skills (Noble, 1978).

Coping with Environmental Stress

The stress concept encompasses a wide variety of situations, including job pressures, marital and family tensions, and physical aspects of the environment such as climate and noise—see Chapters 4 and 5. Among these, few gender-specific differences have been demonstrated. While females may be slightly more sensitive to sound and noise, their impact on performance, on health, and on social behavior seems to be similar for both sexes. Women appear to be better able to cope with little personal space than men; when forced to be in "crowded quarters," men tend to maintain greater distances from others, and react more negatively when people invade their personal space (Greene and Bell, 1987). Redgrove (1976) suggested that women should be able to cope better with low arousal conditions of work ("monotony, boredom") while men should cope better with higher arousal conditions ("pressure").

Cyclical Variations

In their behavior under biological circadian variations, men and women react essentially the same—see Chapter 5. Of course, men do not have a menstrual cycle, hence the question arises whether changes in physiological capabilities during the menstrual cycle of women are significantly different from a "constant" level, if this may be postulated for men. During the menstrual cycle, women show distinct changes in basal temperature, in daily temperature variations, and of course in internal hormone production. While this may affect some nervous system activity and sensory functioning, which possibly can affect scores on specific sensitive tasks, there is no evidence of significant variation in overall performance or in higher cognitive functioning. In comparison to men, women in general maintain a similar performance level throughout in spite of pre-menstrual tension and pain, the occurrence of which is predictable and can be counteracted by medication and individual effort (Asdso, 1987; Patkai, 1985). Yet, in a study on hand-steadiness (related to shooting with a hand gun), women outperformed men consistently except during the week preceding their menses, while women taking oral contraception medication maintained a steady performance level, which was, however, below that of men (Hudgens, Fatkin, Billingsley, and Mazarczak, 1988).

Task Performance in General

The stereotypical question whether females or males "perform better" is, put in such a general way, unanswerable, because there are wide individual differences and there is no "one" performance. Yet, interest in gender-related differences in performance has apparently been around since the earliest times. These (presumed) differences are topics with emotional, social, and political overtones, often biased by the point of view that inferiority or superiority of one or the other gender should be established, or dismissed. One of the first unbiased collection of facts about behavior of the sexes was the H. Ellis book *Man and Woman,* published by Black in London in six editions between 1894 and 1930.

"Task performance" depends on various attributes: the overall nature of the task, the specific conditions under which the task must be performed, the requirements for specific capabilities, the attitudes of the person performing the task, the subjective value of the task for the person performing it, and the social and task-related goals of the individual.

Regarding specific activities, only in those that require very large muscular strength and power do men (on the average) develop higher output than women. However, even here individual capabilities, selection of teams, training and skill, age, or motivation, often contradict statements based on "averages": clearly, in many cases strong women can outperform weak men.

Regarding responses to mental workloads, Hancock, Rodenburg, Mathews, and Vercruyssen (1988) found that women appeared to be more strained by a repetitive low-demand task than their male colleagues. In contrast, McCright (1988) found in simulated job demands that females experienced less strain in the low-demand condition but higher strain under high demand, as compared to their male cohorts. In a series of tests in which performance and subjective workloads were assessed, women tended to perform slightly better than men on the majority of tasks, including grammatical reasoning, linguistic processing, mathematical processing, and memory search (Schlegel, Schlegel, and Gilliland, 1988).

Few statistics indicate differences in actual on-the-job performance. For example, Vail (1988) checked accidents in general aviation that could be tracked to pilot error for the years 1982 to 1985. Adjusting the data for the proportions for male and female pilots indicated that female pilots had distinctly fewer accidents, and much less serious accidents, than male pilots. Female pilots had fewer accidents caused by errors in judgment and decision, or in motor skills. The accident numbers due to weather and communication problems were approximately the same for males and females.

In summary, only when body size and body strength either favor or are necessary for task execution, do clear differences appear in "average" task performance between the gender-related body sizes. No other genetically based sex differences produce essential differences in performance of males and females. While there are some other subtle variations, such as in sensory capabilities, or in reactions to climatic stress, the differences between genders are by far overshadowed by inter-individual differences. Even within areas where differences—though small—have been found, such as in reaction to crowded conditions, or in fine manual skills, the question is unanswered as to whether these reflect biologically or genetically based differences, or are merely or mostly the results of sex roles acquired through group traditions in a cultural contexts and under social pressure.

APPLICATION

No Special Designs for Women and Men

As this review of capabilities and performance has revealed, there are no gender-specific traits that require, or justify, design of workstations or work tools especially for men or women. Proper adjustment ranges will fit all people, whether they happen to be female or male. Nevertheless, it is likely that certain jobs may be preferred by one group or the other, though probably not exclusively so; for example, tasks that require brute force may be done mostly by strong men (but not all men are strong).

DESIGNING FOR PREGNANT WOMEN

For compiling anthropometric and ergonomic data on pregnant British women, Pheasant had to rely on body dimensions measured on pregnant Japanese (Pheasant, 1986, pp. 178–179).

While pregnancy is one of the most common life events, surprisingly little systematic and scientific information is available in the literature for ergonomic purposes.

One set of information concerns the changes in body dimensions during pregnancy. These become apparent after two or three months, and most increase throughout the course of pregnancy. The most obvious changes are in body weight, abdominal protrusion, and circumference. Figure 12-1 illustrates some of these changes.

Many brochures and books in anatomy, obstetrics, and gynecology contain "normative" tables about changes in body dimensions with pregnancy, but apparently few data have been measured on large population samples in recent years. For example, in his attempt to compile data for the United Kingdom, Pheasant (1986) was able to find only one anthropometric survey of pregnant women, but that was done in Japan. Thus, he had to estimate the anthropometric changes of British women based on Japanese data. Fluegel, Greil, and Sommer (1986) reported the changes from the fourth month of pregnancy measured on 198 German women: waist circumference increased by 27 percent, weight by 17 percent, chest circumference by 6 percent, and hip circumference by 4 percent. For American women, similar increases have been reported with average increases of 17 percent for body weight, 8 percent for chest circumference, 4 percent for hip circumference, and 2 percent for abdominal circumference above the values measured 16 weeks after onset of pregnancy on 105 white women with an average age of 26 years. Women with heavy prepregnant bodies showed smaller percentage increases in the abdominal region than lighter women (Rutter, Haager, Daigle, Smith, McFarland, and Kelsey, 1984).

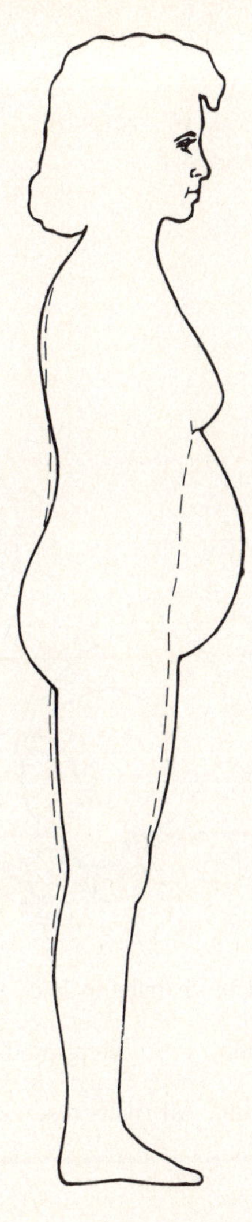

Figure 12-1. Changes in body dimensions and posture with pregnancy.

APPLICATION

In their attempt to design for crash protection in automobiles, Culver and Viano (1990) had to rely mostly on body dimensions derived from either older studies or biomechanical model estimates. Little change in shape and mass properties occurs during the first three months of pregnancy. However, from there on, biomechanically significant changes in abdominal depth and circumference take place, accompanied by a shift in the center of mass of the body. Culver and Viano present these changes in the form of ellipses—see Figure 12-2. These then allow estimates of the contact area between the woman's body and restraining devices, or automobile interior surfaces, in the case of a crash impact.

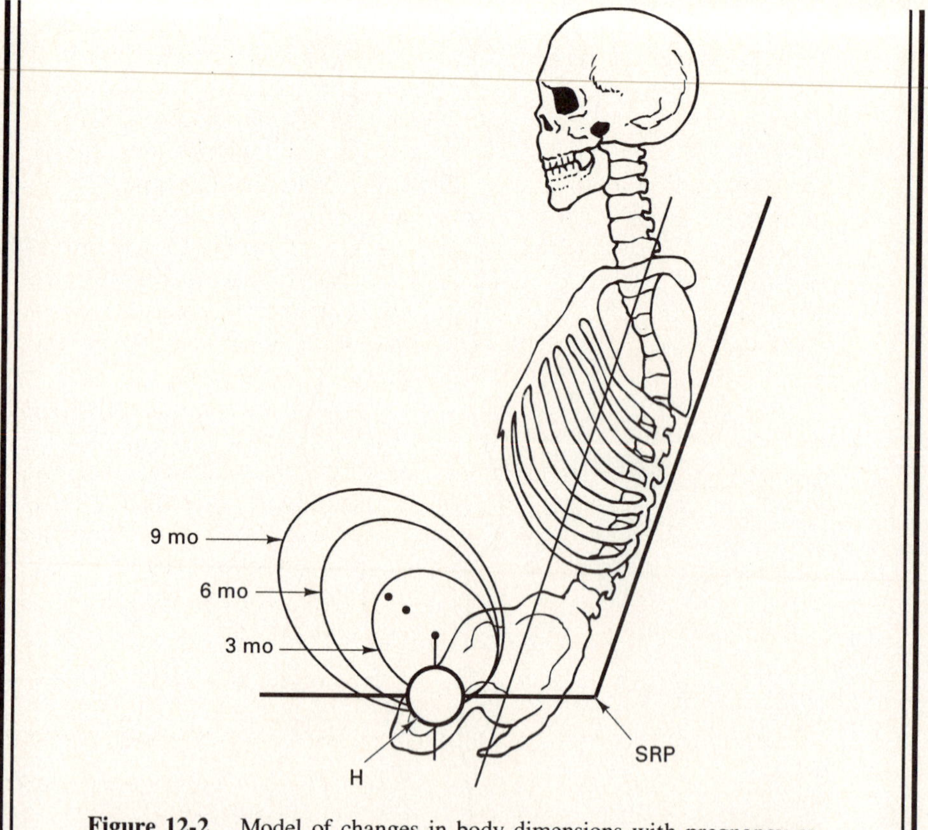

Figure 12-2. Model of changes in body dimensions with pregnancy related to hip joint (H) and seat reference point (SRP). SOURCE: Adapted from Culver and Viano, 1990. Reprinted with permission from *Human Factors*, Vol. 32, No. 6, 1990. Copyright 1990 by the Human Factors and Ergonomics Society, Inc. All rights reserved.

The increasing abdominal protrusion makes it increasingly difficult for women throughout their pregnancy to get as close to work objects as they can before and after pregnancy. The working area of the hands becomes smaller during pregnancy, and manipulating objects that are now further ahead of the spinal column generates an increased compression and bending strain on the spine and on ligaments and muscles in the back. This loading is, of course, also due to the increasing mass of the abdomen, and its increasing moment arm with respect to the spinal column. The variations in abdominal shape also change the body posture which, in the course of the pregnancy, assumes a backward pelvic rotation accompanied by forward movement of the trochanterion, and brings about a flattening of the lumbar lordosis.

These events explain, at least partly, the complaints of back strain and pain common with pregnancy (Boussena, Corlett, and Pheasant, 1982; Cherry, 1987; Fast, Shapiro, and Edmond, 1987).

Physical performance capabilities also change during pregnancy. Although there are great variations among individuals, with advancing pregnancy in general the ability decreases sharply to perform work that requires exertion of large energies, much mobility, or far reaches, particularly if repeated or over long periods (Cherry, 1987; Errkola, 1976). Nicholls and Grieve (1992) performed a survey of two hundred residents in London, UK, between 29 and 33 weeks pregnant. They were asked about their current performance of certain tasks compared to before becoming pregnant. The interviews used a five-point ordinal scale, and concerned 46 activities that were pre-selected for inclusion in the study. Of these 46 tasks, 32 were found to be significantly more difficult to perform during pregnancy than in prepregnancy. Among these, the following were considered the hardest: picking object up from floor; working at desk; walking upstairs; driving a car, getting in and out of car, using seat belts; ironing; reaching high shelves; using public toilets; and getting in and out of bed. The reasons perceived for the difficulties were related to back pain, reduced reach and clearance, feeling unstable, being fatigued, having reduced mobility, and having difficulties in seeing objects near the body. In general, (suitable) sitting was found less straining than standing.

APPLICATION

Many everyday tasks become more difficult with pregnancy. Yet, differences among individuals are striking, with some women finding only a few tasks more difficult and others finding nearly all activities harder to do.

These and other findings indicate clearly certain ergonomic measures that should be taken to accommodate pregnant women, either at the workplace, in transportation, or at home:

- Changing body dimensions, particularly increased abdominal protrusion, make it difficult to reach far objects. Thus, manipulation areas should be kept close to the body and possibly be somewhat elevated from their regular height.
- Work tasks should require as little force as possible, particularly in the vertical direction. Lifting of objects should be avoided.
- Suitable seats, easily adjustable by the woman, should be provided.
- Frequent rest periods should be allowed, to be freely selected by the woman.
- More space than usual should be allowed for moving around, and obstacles should be avoided, particularly low objects that might be difficult to see.

DESIGNING FOR CHILDREN

The time span between birth and early adulthood (at about 18 years) is characterized by very large changes in body dimensions, body strength, skill, and other physical and psychological variables. At birth we weigh about three and a half kilograms and are about 50 centimeters in length, of which the trunk represents about 70 percent. In the two decades that follow, body length increases three- to fourfold, our weight increases about twentyfold, and body proportions change drastically. The trunk accounts for just over 50 percent of stature when we are grown—see Figure 12-3. But these changes over time are individually quite different, and appear to be related not only to genetic factors but also to environmental variables. Thus, there is no "typical" boy or girl. In general, the rate of growth in boys is rapid during infancy (up to two years) and then declines until onset of puberty (at about 11 years), when growth becomes strong again, reaching its peak at about 14 years, and then again slowing. Final stature is attained in the early to middle 20s; yet, at 14 or 15 years, some boys have almost completed their growth while others are just beginning a strong growth phase. In girls, the puberty growth spurt begins earlier, at about 9 years of age, and is fastest at about 12 years; full adult stature is often complete at age 16. Hence, at about 11 to 13 years, many girls are taller than boys of the same age—see Table 12-1.

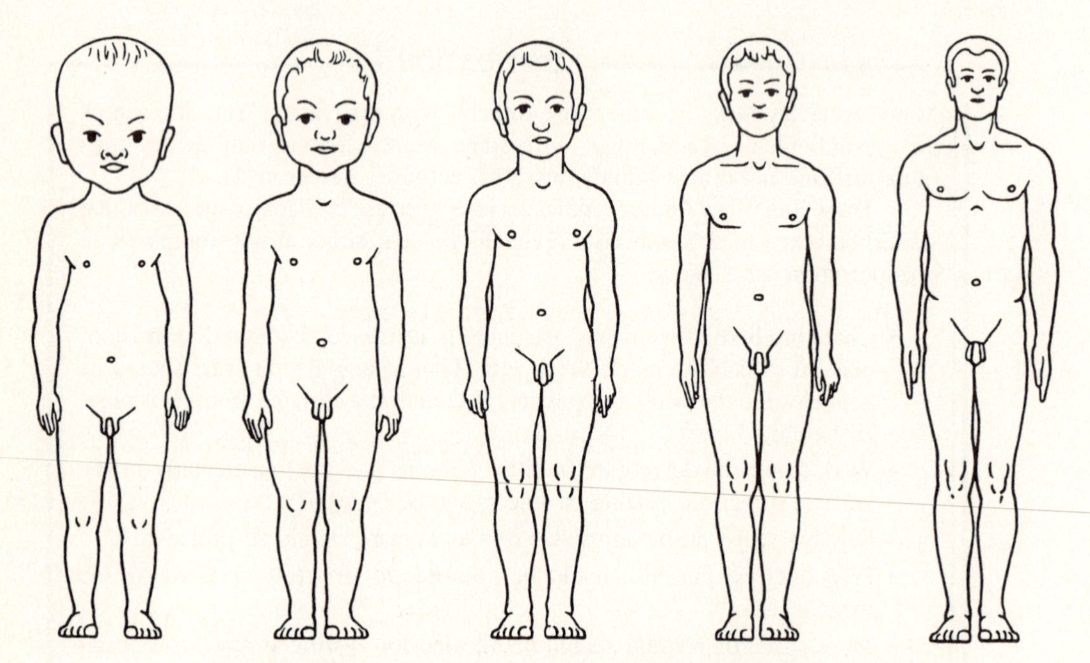

Figure 12-3. Changes in body proportions from birth to adulthood. (Modified from Fluegel, Greil, and Sommer, 1986.)

TABLE 12-1. CHILDREN'S DEVELOPMENTAL STAGES

Stage	Physical characteristics	Motor skills
	INFANCY AND TODDLERHOOD	
0 to 6 months	Oversized head, short stubby limbs	Reaches and grasps, sits with support
6 to 9 months	Increases in weight, hence plump appearance	Sits alone, stands with assistance
9 to 15 months	Flexible limbs	Crawls, walks, stands alone
after 18 months	Gradual appearance of neck, protruding abdomen	Wobbly, stiff, flat gait; climbs
2 to 3 years	Head has become smaller in proportion to body; less roundness; lean muscles; curved back; still protruding abdomen	Flexible at knees and ankles; can run, jump, kick, hop
	EARLY CHILDHOOD	
3 to 5 years	Rate of growth slows, body proportions change; loss of the babylike appearances, increase of muscle tissue; less back curvature	Masters walking; has smoother movements, better balance, turns corners; holds pencils and utensils
	MIDDLE CHILDHOOD	
6 to 12 years	Horizontal growth, gradual changes in physical appearance	Increases in running and jumping distance, accuracy, and endurance
	ADOLESCENCE	
12 to 18 years	Growth spurt peaks, hands and feet reach adult size, breasts develop in girls; body breadth increases; elongated trunks and legs for boys	Motor skills fully developed

Several "secular trends" have been observed in recent decades:

- increase in the growth rate of children; children seem to grow faster in earlier years.
- earlier onset of puberty, as indicated by menarche in girls and the adolescent growth spurt in both boys and girls.
- increase in the individually achieved, final adult stature (Fluegel, Greil, and Sommer, 1986; Lohman, Roche and Martorel, 1988; Pheasant, 1986).

Such information is often more anecdotal than based on large scientific surveys. For example, Konz (1991) reported that Japanese boys at the age of 13 years were in 1990 an average of 159 cm tall, an increase of 18 centimeters since 1950; similarly, the average stature of girls of 11 years of age was 146 cm in 1990, compared to 132 cm in 1950. Yet, a survey of more than 600 children in the Netherlands indicated significant differences among the body dimensions of children growing up in different provinces of Holland. Thus, information about children is not specific only to age groups, but also to areas of origin (Steenbekkers and Molenbroek, 1990). (In that respect, information on children is similarly variable, depending on the individual, as information on the aging, discussed below.)

Measurements on children are difficult to take, particularly if these are very young. They do not understand or follow instructions, and their attention span is very short. For example, the body length (stature) of babies and infants, up to the age of two years, is customarily measured with the child lying on its back; later, stature is taken when standing up. Given the many sources of variability, the data compiled in Tables 12-2 through 12-6 must be seen and used only as "examples in time and location" of anthropometric information on children. As these tables describe, rather large numbers of American and German children were measured, but those of British children estimated. The data are nearly a quarter of a century old: today's children may have different dimensions, and children from other areas and countries are likely to have different sizes.

TABLE 12-2. STATURE (IN CM) OF CHILDREN ESTIMATED FOR THE UNITED KINGDOM (UK; PHEASANT, 1986), MEASURED IN GERMANY (G; FLUEGEL, GREIL, AND SOMMER, 1986) AND IN THE UNITED STATES (USA; SNYDER, SPENCER, OWINGS, AND SCHNEIDER, 1975). STANDARD DEVIATIONS, IN PARENTHESES

Girls				Boys			
Age (years)	UK	G	USA	Age (years)	UK	G	USA
0		51.8	54.8 (3.6)	0		52.4	55.4 (4.0)
0.5		68.3	68.6 (2.3)	0.5		69.6	70.4 (2.4)
1		75.6	72.4 (2.9)	1		76.4	73.5 (3.2)
2	89	85.9	84.0 (3.4)	2	93	86.9	85.3 (3.4)
3	97	94.1	92.9 (4.4)	3	99	95.0	93.4 (3.9)
4	105	101.3	99.5 (4.3)	4	105	102.2	99.9 (3.8)
5	110	107.2	106.5 (4.7)	5	111	108.1	107.6 (5.0)
6	116	115.1	112.8 (5.0)	6	117	116.1	113.7 (4.8)
7	122	121.0	118.8 (5.0)	7	123	119.6	120.5 (4.7)
8	128	126.1	123.4 (5.3)	8	128	127.2	125.3 (5.8)
9	133	130.2	130.2 (5.9)	9	133	131.1	130.0 (5.8)
10	139	137.2	134.4 (6.1)	10	139	137.7	135.1 (6.3)
11	144	142.7	141.1 (6.8)	11	143	144.0	141.9 (5.3)
12	150	148.3	145.5 (6.5)	12	149	145.9	146.8 (7.1)
13	155	154.6	155.1 (6.2)	13	155	153.3	149.5 (7.8)
14	159	160.0		14	163	161.5	
15	161	162.2		15	169	166.5	
16	162	162.9		16	173	171.5	
17	162	163.5		17	175	173.6	
18	162	163.9		18	176	175.8	

APPLICATION

In the United States, a rather large number of children are injured every year by head, neck, and hand entrapment. Thus, the U.S. Consumer Product Safety Commission sponsored an anthropometric study of American children, performed and published by Schneider, Lehman, and Owings (1986). Their report not only provides related body dimensions, but also describes, in exemplary fashion, the sampling and measuring strategies that are suitable for gathering such information. Based on such data, various recommendations and regulations exist for the largest openings that still prevent children from moving through them, such as the distance between stakes of railings. Particularly critical dimensions are head breadth, chest depth, and hand clearance diameter. These have been compiled for American children in Table 12-3. Considering average and standard deviations values, the listing shows that up to the age of 12 years, girls have narrower heads, shallower chests, and smaller hand clearance diameters than boys. Thus, if openings are kept small enough to not let girls' bodies pass, boys should certainly not be able to squeeze through either.

For biomechanical design purposes, such as for restraint devices in automobiles, certain information is of importance, such as changes in body mass with age, as well as in the location of the center of mass of the body, standing

TABLE 12-3. AVERAGE VALUES(AND STANDARD DEVIATIONS) IN CM FOR HEAD BREADTH, CHEST DEPTH, AND HAND CLEARANCE DIAMETER FOR U.S. CHILDREN

Age (years)	Head breadth		Chest depth		Hand clearance diameter	
	Girls	Boys	Girls	Boys	Girls	Boys
0	10.3 (0.6)	10.4 (0.7)	9.0 (0.9)	9.3 (0.9)	3.21 (0.29)	3.33 (0.30)
0.5	11.4 (0.6)	11.7 (0.6)	9.9 (0.9)	9.9 (0.9)	3.55 (0.28)	3.72 (0.26)
1	12.3 (0.4)	12.6 (0.6)	10.4 (1.1)	11.0 (0.6)	3.86 (0.25)	4.14 (0.29)
2	13.0 (0.5)	13.3 (0.4)	11.3 (1.0)	11.6 (1.0)	4.10 (0.27)	4.24 (0.32)
3	13.3 (0.5)	13.5 (0.4)	11.8 (0.8)	12.0 (1.2)	4.30 (0.32)	4.51 (0.24)
4	13.5 (0.4)	13.8 (0.4)	12.2 (0.8)	12.5 (0.9)	4.50 (0.27)	4.57 (0.28)
5	13.6 (0.4)	14.0 (0.5)	12.7 (1.1)	13.0 (1.0)	4.66 (0.30)	4.82 (0.32)
6	13.7 (0.4)	14.0 (0.4)	13.2 (1.0)	13.3 (1.1)	4.79 (0.28)	4.99 (0.30)
7	13.9 (0.4)	14.2 (0.5)	13.5 (1.0)	14.1 (1.1)	5.01 (0.31)	5.16 (0.30)
8	14.0 (0.4)	14.2 (0.5)	13.7 (1.4)	14.3 (1.3)	5.08 (0.34)	5.28 (0.38)
9	14.1 (0.5)	14.3 (0.4)	14.4 (1.4)	14.8 (1.3)	5.22 (0.33)	5.42 (0.35)
10	14.1 (0.5)	14.4 (0.5)	14.7 (1.5)	15.2 (1.3)	5.42 (0.33)	5.56 (0.33)
11	14.2 (0.4)	14.6 (0.4)	15.7 (2.0)	16.2 (1.6)	5.60 (0.40)	5.85 (0.38)
12	14.5 (0.6)	14.5 (0.5)	16.2 (1.7)	16.8 (1.6)	5.82 (0.34)	6.03 (0.37)
13	14.6 (0.5)	14.5 (0.4)	17.9 (2.2)	17.2 (1.7)	6.16 (0.37)	6.06 (0.40)

SOURCE: Data excerpted from Snyder, Spencer, Owings, and Schneider, 1975.

TABLE 12-4. BODY MASS, AND LOCATION OF THE CENTER OF MASS FOR U.S. GIRLS AND BOYS; AVERAGE AND STANDARD DEVIATIONS

Age (years)	Body mass (kg)		Height (in percent of stature) of the center of mass of the body: Standing (above floor)		Seated (above seat)	
	Girls	Boys	Girls	Boys	Girls	Boys
0	4.6 (1.1)	4.8 (1.2)	59.4 (1.9)	58.5 (2.1)	50.2 (3.3)	48.0 (4.4)
0.5	6.7 (0.9)	7.4 (0.9)	58.1 (2.5)	59.1 (2.3)	47.1 (2.8)	46.6 (3.7)
1	8.9 (1.3)	9.5 (0.8)	58.1 (1.8)	58.5 (2.4)	44.6 (2.3)	45.6 (2.7)
2	11.2 (1.1)	12.2 (1.2)	57.5 (1.0)	57.5 (1.1)	41.3 (2.4)	39.3 (2.6)
3	12.8 (1.1)	14.2 (1.5)	59.3 (2.5)	58.9 (1.0)	39.1 (1.9)	37.6 (1.2)
4	15.4 (1.8)	15.8 (1.8)	58.8 (2.0)	59.7 (1.7)	37.9 (3.3)	37.2 (2.2)
5	17.7 (2.3)	18.3 (2.1)	59.3 (2.0)	58.9 (1.7)	35.3 (2.6)	36.6 (2.6)
6	19.3 (2.7)	20.8 (3.0)	59.3 (1.5)	59.1 (1.3)	34.0 (2.3)	35.0 (1.9)
7	21.8 (2.7)	23.2 (3.1)	58.6 (1.1)	58.7 (1.3)	33.3 (1.8)	33.1 (2.1)
8	24.2 (4.0)	25.3 (4.4)	58.0 (1.7)	58.6 (1.1)	32.2 (2.4)	32.3 (2.2)
9	27.7 (5.2)	27.7 (4.6)	58.0 (1.4)	57.9 (1.1)	30.7 (1.6)	32.1 (1.8)
10	30.6 (5.8)	30.4 (5.2)	57.5 (0.9)	58.0 (1.1)	30.2 (2.0)	31.1 (1.9)
11	34.4 (7.2)	35.4 (5.8)	57.4 (0.7)	57.7 (1.0)	29.5 (1.6)	30.0 (1.6)
12	38.1 (7.3)	38.8 (6.4)	57.4 (1.1)	57.8 (1.0)	29.4 (1.4)	30.1 (2.1)
13	48.0 (8.1)	40.7 (7.0)	57.4 (1.3)	58.0 (1.5)	29.2 (1.3)	29.7 (1.5)

SOURCE: Data excerpted from Snyder, Spencer, Owings, and Schneider, 1975.

or while sitting. Such information is compiled in Table 12-4. It shows that the location of the center of mass of the standing body, expressed in percent of body height above the floor, does not change much with increasing age, and is quite similar for girls and boys. For seated children, the relative height of the center of mass above the seat decreases with age if expressed in percent of stature, but again in fairly similar fashion for girls and boys. This information suggests that no distinction need be made between boys and girls for the design of body restraint systems.

Body dimensions of children are important for the design of furniture, particularly as used in schools. This poses problems, because children of different body sizes may be combined in the same rooms from kindergarten on up. Thus, tables and chairs of very different sizes should be made available to fit the different children. This is often difficult to do for a variety of organizational reasons. Provision of adjustable chairs and tables, for example, might appear a suitable solution, but especially young children might have great difficulties in adjusting that furniture to their size and liking.

While infants are still fairly uncoordinated and do not show great body strength (although their uncontrolled movements can inflict damage to others and themselves), body strength develops quickly during early and middle childhood.

Hand strength shows a strong positive correlation with age, while, at least in early years, there is little relation between strength, hand dominance, and gender. A variety of publications provide information on the strength capabilities of children (e.g., Ager, Olivett, and Johnson, 1984; Bowman and Katz, 1984; Burke, Tuttle, Thompson, Janney, and Weber, 1953; Imrhan, 1986; Lowrey, 1986; and Owings, Chaffin, Snyder, and Norcutte, 1975). Tables 12-5 and 12-6 are excerpts that indicate torque and force capabilities measured on American children between three and ten years of age. The tables are shown here for two reasons. First, they reflect how (average) strength increases with increasing age. Second, they show the very large interindi-

TABLE 12-5. AVERAGE TORQUES (AND STANDARD DEVIATIONS) IN NCM AROUND WRIST, ELBOW, AND KNEE EXERTED BY U.S. CHILDREN (GIRLS AND BOYS COMBINED)

Age (yrs)	Wrist		Elbow		Knee	
	Flexion	Extension	Flexion	Extension	Flexion	Extension
3	84 (47)	63 (22)	606 (156)	616 (111)	500 (197)	1673 (616)
4	122 (61)	61 (28)	731 (233)	724 (259)	468 (194)	1866 (710)
5	152 (79)	69 (30)	932 (319)	901 (285)	706 (351)	2301 (738)
6	224 (85)	90 (40)	1192 (299)	1034 (373)	956 (386)	2717 (961)
7	268 (105)	113 (47)	1687 (415)	1332 (441)	1175 (334)	3788 (1165)
8	352 (128)	122 (44)	2114 (506)	1612 (437)	1371 (564)	4762 (1391)
9	453 (188)	167 (74)	2248 (674)	1676 (527)	1986 (638)	5648 (1386)
10	434 (166)	164 (41)	2362 (603)	1596 (446)	2084 (842)	5553 (1826)
	$N = 211$	$N = 205$	$N = 495$	$N = 496$	$N = 267$	$N = 496$

SOURCE: Data excerpted from Owens, Chaffin, Snyder, and Norcutt, 1975.

TABLE 12-6. AVERAGE SIDE GRIP AND GRASP FORCES (AND STANDARD DEVIATIONS) IN *N* EXERTED BY U.S. CHILDREN (BOYS AND GIRLS COMBINED, $N = 227$)

Age (years)	Thumb-forefinger side grip* ("side pinch," see Fig. 8-26)	Power grasp ("grip strength")
3	18.6 (4.9)	45.1 (14.7)
4	26.5 (5.9)	57.9 (17.7)
5	31.4 (7.8)	71.9 (18.6)
6	38.3 (5.9)	89.3 (22.6)
7	41.2 (6.9)	105.0 (32.4)
8	47.1 (9.8)	124.6 (33.4)
9	52.0 (9.8)	145.2 (35.3)
10	51.0 (8.8)	163.8 (37.3)

*Pinch surfaces 20 mm apart.

SOURCE: Data Excerpted from Owens, Chaffin, Snyder, and Norcutt, 1975.

vidual differences: the coefficients of variations range from 30 to 60 percent, a wide variation indeed.

The tables combine the measurements taken on girls and boys, largely because at these age groups no systematic differences between the genders can be observed. The slight decrease from nine to ten years of age, visible in some of the data sets, is probably an artifact of relatively small subject numbers in the ten-year-old age group. Between 10 and 12 years of age, boys are usually, but not consistently, slightly stronger than girls. Hand preference, as determined both by handwriting and ball throwing, is not associated with strength up to the age of 12 years (Ager, Olivett, and Johnson, 1984; Bowman and Katz, 1984).

DESIGNING FOR THE AGING

☞☞☞ *In the United States, there is a curious use of terms: a "middle-aged" person becomes "older" at 45 years of age; "elderly" at 65 years; "old" as one reaches 75 years; and "very old" or "old old" if one lives beyond 85 years of age. Anthropometric and much demographic and capability-related information is usually collected in five-year intervals until the age of 65, and then just lumped together for the remaining years, with only occasionally a time marker set at 75 years of age.* ☜☜☜

Life expectancy of humans has changed dramatically. The average life span was less than 20 years in "prehistory" (until about 1000 B.C.), and increased into the low 20s in Ancient Greece (up to around the year 0). In the Middle Ages (about 1000 A.C.), life expectancies increased in Western Europe to the thirties and reached the low forties in the nineteenth century. In Colonial America, until about 1700, the average life span was 35 years. It increased to about 50 years by 1900, and to about 75 around 1990 in the United States. (Committee on an Aging Society, 1988; Kermis, 1984; Kelly and Kroemer, 1990). Life expectancies depend on genetic heritage, gender, climate, hygiene, nutrition, diseases, wars, and accidents.

The role of the aging person in society has been quite variable in different eras and in different regions. The aged person might be considered wise and experienced, a leader or advisor; or a useless and expensive appendix that is removed from societal life (as vividly portrayed by Margaret Mead, 1901–1978). Intermediate positions may be prevalent: consider, for example, the forced early cessation of flying duties by airline pilots; the common "going into retirement" at about age 65 for most occupations, or the late-life activities of some professors and politicians in America. What owes a person to society, or society to a person? How much care can be expected from relatives and friends, how much from society, and what can the aging person return as a favor? These general and many specific concerns have been addressed, in the United States, recently by the Committee on an Aging Society (1988) and by Czaja (1990), both under the aegis of the National Research Council.

Anthropometry

Measurement of body dimensions is done, in the aging population as in the younger adults, usually in a cross-sectional approach: one measures all available people and then lumps their measurements together within certain age brackets. This does not create a big problem in the "young adult" population, because dimensions do not change very much in the 20-to-40-year age span—see Chapter 1. However, this is a major problem in the description of the aging (as well as of children) for several reasons:

- Among the aging, some persons change dimensions rapidly within a few years—for example, in stature because of posture and shrinking thickness of spinal discs, or in weight because of changes in nutrition, metabolism, and health; and in musculature and strength because of changes in activity levels, habits, and health. Other people, in contrast, show little change over long periods.
- The age brackets used for surveys are rather wide, usually encompassing decades or even longer time spans, as opposed to the common five years in younger cohorts. Hence, people with very different dimensions are contained in each observation sample.

Thus, chronological age is not a good classifying criterion for the aging (or for children). They would be better described by a longitudinal procedure in which changes in body dimensions and capacities are observed within one individual over many years; yet, few such data are available (Annis, Case, Clauser, and Bradtmiller, 1991).

Anthropometric information reported in the literature on sufficiently large samples is, as a rule, the result of cross-sectional surveys. An excerpt from a recent compilation of available data is listed in Table 12-7. It exemplifies the current problems with anthropometric information: most of the samples are exceedingly small; surveys are done for a few age ranges only; there are no distinctions between ethnic origin, region, socioeconomic status, health or other attributes that are codeterminers of anthropometry. Given the limitations, in fact the paucity, of these data from the United States, one anticipates even poorer and less complete data from other regions of the earth. Thorough discussions of American demographic aspects, with respect to both current and future aging cohorts, have been provided by Annis et al. (1991), Serow and Sly (1988), and Soldo and Longino (1988).

The apparent height loss with age (of about 1 cm per decade) starting in the thirties may be a result of (a) flattening of the cartilaginous disks between the vertebrae; (b) a flattening or thinning of the bodies of the vertebrae; (c) a general thinning of all weight-carrying cartilages; (d) a change in the S-shape of the spinal column in the side view, particularly an increased kyphosis in the thoracic area (hump back); (e) in some cases, scoliosis, a lateral deviation from the straight line displayed by the

TABLE 12-7. ANTHROPOMETRIC DATA ON THE ELDERLY: MEANS (AND STANDARD DEVIATIONS)

Age range:	50–100[a]	60–69[b]	60–69[c]	65–69[d]	65–74[e]	65–90[f]	66–70[a]	70+[b]	70+[d]	70+[c]	72–91[e]	75–94[e]
Sample size:	822	43	72	24	72	184	169	12	20	28	130	40
Stature, against wall		172.8 (6.6)		171.9 (6.6)				171.5 (9.0)	170.4 (7.5)			
Stature, free standing	175.1 (8.9)		172.6 (6.4)	171.2 (6.6)		169.0			169.6 (7.6)	171.9 (8.4)	168.4 (5.3)	
Sitting height	79.9 (5.3)	90.8 (3.0)	90.8 (2.9)	90.0 (2.9)				89.5 (3.5)	89.0 (3.4)	89.8 (3.9)	88.3 (3.1)	
Knee height		53.9 (2.5)	53.6 (2.5)					53.5 (3.4)	53.2 (2.9)	53.7 (3.2)	53.8 (2.1)	
Popliteal height	42.1 (3.5)		42.1 (2.3)							42.1 (3.0)	44.0 (2.1)	
Thigh clearance height			19.7 (1.4)							14.8 (1.2)		
Hip breadth	37.4 (3.9)			36.0 (2.3)					35.8 (1.7)	37.8 (2.4)		
Bideltoid breadth			45.3 (2.4)	45.1 (2.1)					44.7 (1.6)	45.0 (1.7)	43.4 (2.3)	
Biacromial breadth			38.9 (1.7)							39.2 (1.8)	37.8 (1.6)	
Hand breadth	7.7 (0.6)		8.5 (0.4)	8.5 (0.4)					8.5 (0.4)	8.6 (0.4)	8.4 (0.4)	
Head breadth			15.5 (0.5)	15.5 (0.5)					15.5 (0.5)	15.5 (0.4)	15.4 (0.5)	
Foot breadth			9.8 (0.6)							9.9 (0.5)	10.0 (0.5)	
Head circumference			57.1 (1.4)	57.1 (1.3)					58.0 (1.4)	57.4 (1.6)	56.9 (1.8)	
Calf circumference			35.9 (2.5)	36.0 (2.9)					34.7 (2.1)	35.3 (2.2)	34.3 (2.7)	
Chest circ., resting			99.6 (7.1)	99.9 (6.3)					99.6 (5.5)	99.7 (5.9)	96.2 (7.6)	
Chest circ., maximum			101.8 (6.9)	101.7 (6.1)					101.5 (5.4)	101.7 (5.7)	98.7 (7.4)	
Chest circ., minimum			97.6 (7.2)	97.5 (6.5)					97.8 (5.6)	97.9 (6.0)	94.5 (7.6)	
Upper arm circumference			30.9 (2.7)	30.5 (2.6)					30.0 (2.4)	28.7 (2.8)		
Waist circumference			95.5 (9.3)	97.4 (8.9)					97.1 (8.0)	97.0 (7.6)		
Head length			19.6 (0.6)	19.6 (0.6)					19.5 (0.6)	19.7 (0.7)	19.7 (0.6)	
Hand length	17.5 (1.2)		18.9 (0.9)	18.9 (0.9)					18.8 (0.9)	19.0 (1.0)	18.8 (0.8)	

Buttock-knee length		58.6 (3.0)						58.4 (3.2)	59.1 (2.4)	
Buttock-popliteal length	46.3 (3.6)	48.2 (2.8)						48.1 (3.1)	47.2 (2.5)	
Elbow to middle finger length	44.2 (2.8)	46.8 (2.0)	46.8 (1.9)				46.6 (2.5)	46.9 (2.8)	46.4 (1.8)	
Shoulder to elbow length		37.3 (1.8)	37.4 (1.7)				37.0 (2.1)	37.4 (2.2)	36.9 (1.7)	
Forward reach		84.2 (3.7)						85.9 (5.4)	86.9 (3.8)	
Span		178.7 (7.5)	178.8 (7.5)				177.6 (9.0)	179.2 (9.9)	174.0 (7.0)	
Skinfold (triceps) (right)			1.1 (0.4)		1.2 (0.3)			0.9 (0.4)	1.1 (0.4)	
Skinfold (subscap.) (right)			1.7 (0.8)					1.5 (0.7)	1.6 (0.7)	
Foot height			26.3 (1.2)	26.4 (1.2)			26.5 (1.3)	26.8 (1.4)	26.0 (1.0)	
Weight (kg)	63.7	76.6 (1.1)	76.4 (1.0)	65.6 (11.6)	63.7		74.3 (0.9)	75.3 (9.0)	69.0 (10.5)	63.7 (11.7)
Grip strength (left) (N)		432 (88)				323 (58)		352 (88)	262 (80)	
Grip strength (right) (N)		461 (88)				370 (68)		412 (88)	283 (78)	

All measures in centimeters unless otherwise noted.

References:

[a] Molenbroek (1987; Netherlands; average of males and females).

[b] Borkan, Hults, and Glynn (1983; United States; males only).

[c] Damon et al. (1972; United States; males only).

[d] Friedlander et al. (1977; United States; males only).

[e] Dwyer et al. (1987; United States; average of males and females).

[f] Pearson, Bassey, and Bendall (1985; United States; average of males and females).

[g] Clement (1974; United States; males only).

SOURCE: Excerpted with permission from "Anthropometry of the Elderly: Status and Recommendations" by P. I. Kelly and K. H. E. Kroemer, *Human Factors*, Vol. 32, No. 5, 1990.

spinal column in the frontal view; and (f) possibly bowing of the legs and flattening of the feet (Barlow, Braid, and Jayson, 1990; Stoudt, 1981). As groups, American men usually have their largest body weights in their thirties, then lose weight with aging; American women are relatively light in their twenties, but then increase their weight with age, becoming heaviest, on average, at about sixty years (Annis et al., 1991).

☞☞☞ *Old age is not necessarily a condition of disability, but rather an increase in the probability of a number of small changes in performance parameters (Rabbitt, 1991, p. 776).* ☜☜☜

Changes in Biomechanics

In addition to the anthropometric changes that may occur with increasing age, there are numerous alterations in biomechanical features (discussed in Chapter 1). These include:

Bones. Particularly the long bones become larger in outer diameter, and larger in inner diameter (hollower), and larger pores appear. Total bone mass decreases. This is, together with a change in mineral content, the major component in age-related "osteoporosis": bones become stiffer and more brittle (Osthere and Gold, 1991). Women and persons who exercise little are more exposed to this development than men and active people. The changes in bone structure are associated with an increased likelihood of broken bones as a result of falls or other accidents in which sudden forces and impulses are exerted on the body. Injuries to the pelvic girdle, hip joint, or femur are particularly frequent in older women, followed by bone injuries to the shoulder and arms.

Joints. The lining of joints, the bony surfaces in joints, the supply of synovial fluids, the elasticity and resilience of joint capsules and ligaments are all reduced. This leads to reduced mobility in the joints, often associated with pain.

Muscles. While well-used muscles can retain their capabilities into advanced age, reduced use, often accompanied by decreased circulatory supply, generally leads to a loss of musculature and ensuing loss of strength capabilities.

Skills. While many or some skills can be retained, by practice and with good health, others deteriorate for the reasons discussed and possibly because of reductions in the performance of the central and peripheral nervous systems, and because of diminished circulatory and metabolic capabilities. The reductions are most likely in activities that require exertion of large energies or forces, often combined with endurance requirements and controlled through perceptual information, particularly of the visual and vestibular modes.

Manipulation capability can be considered a special subgroup of skills: it requires strength, mobility, and sensory control. In spite of the fact that manipulation skills are very important for many tasks (on the job, at home, or during leisure), unified and standardized measuring techniques are not at hand, although several attempts have been made—see Chapter 8 and the publications by Kroemer (1986a) and by Scott and Marcus (1991). Thus, little reliable information on hand capabilities exists for adults in general; and next to nothing is known for aging persons.

Changes in Respiration and Circulation

Respiratory capabilities reduce with increasing age mostly because the alveoli in the lungs are less able to perform the exchanges of gases, i.e., oxygen and carbon dioxide (see Chapter 2). Furthermore, the intercostal muscles and the chest diaphragm lose some of their ability to generate "breathing space" in the chest, hence vital capacity decreases. This is coupled with reduced blood flow, and possibly with emphysema, often resulting from smoking.

The elasticity of blood vessels seems to decrease. Resistance to blood passage in vessels may be increased due to deposits along their walls. Blood-cell production in the bone marrow is decreased. Thin aging people may have reduced volumes of body fluids.

The heart functions also change. The size of the heart may be reduced. Cardiac output is lower. Heart rate takes longer to return to resting level after the rate has been increased. Neural control of the heart may be impaired (Spence, 1989).

Changes in Nervous Functions

The ability to cope with the environment depends in large part on detecting, interpreting, and responding appropriately to sensory information. Both *sensation* (the reception of stimuli at sensors and the resulting neural inpulses in the afferent part of the neurons system) and *perception* (the interpretation of the stimuli—see Chapter 3) change with age. There appears to be a reduction of cells which, together with diminished arterial and venous flow in the blood vessels, changes the stimulation and conduction activities in the nervous system. This may lead to increased variability in reception and integration of, and hence response to, external and internal stimuli (Hayslip and Panek, 1989).

In the somesthetic system, the numbers of cells in the skin (dermis and epidermis) are decreasing, together with collagen and elastic fibers. Receptors such as Meissner's and Pacinian corpuscules decrease in number. The reduction of sensors and afferent fibers may be combined with reduced nerve conduction velocity.

Changes in Taste and Smell

The number of taste buds and the production of saliva in the mouth is often reduced. Fissuring of the tongue may occur. The sense of smell is often said to be diminished, although this finding is not uniform (Belsky, 1990; Hayslip and Panek, 1989; Kermis, 1984).

Changes in Visual Functions

Changes in visual functions are tied to many concurrent anatomical, physiological, and psychological processes that develop with age (Committee on Vision, 1987; Cowen, 1988; Sekuler, Kline, and Dismukes, 1982). Using the analogy of the camera, the human eye as a photographic device loses precision. Structures that bend, guide, and transform light (see Chapter 4) change, which reduces the amount of light reaching the retina, and defocuses the image projected to the retina. It becomes more difficult to focus on near objects, particularly if they are elevated or move fast (Heuer, Breuer, Roemer, Kroeger, and Knapp, 1991; Tyrrel and Leibowitz, 1990; Rabbitt, 1991).

One problem is the watering and tearing of eyes, due to accumulation of fluid on the outside. Water can affect the properties of the eye as a lens, and can simply be annoying. Baggy or droopy eyelids may reduce the amount of light reaching the cornea.

The cornea flattens, which limits the ability to focus. Fatty deposits reduce the transmission of light and scatter arriving light. The opening of the pupil gets smaller, which further reduces the amount of light entering the eyeball. This disorder (*senile miosis*) has the most serious effects in dim light. There is a possible benefit from the smaller diameter, however, similar to having a smaller aperture opening in a camera lens: the depth of field may be enhanced, meaning that objects both near and far are in better focus, although they appear dimmer.

A common problem in people over 40 years is hardening of the lens, which reduces its ability to become thicker and more rounded for focusing on near objects (*presbyopia*).

The young eye has a slightly yellow-tinted lens, which acts as an ultraviolet filter for the retina. The aging eye becomes more yellow because of the development of fluorescent chromophores of yellow color. A yellower lens is a stronger light filter, absorbing energy, hence raising the threshold for detection of light in general, and specifically absorbing some of the blue and violet wavelengths. This changes perception of colors: white objects appear yellow, blue is hard to detect, blue and green are difficult to distinguish. With increasing age, water-insoluble dry protein becomes more prevalent, and macromolecules appear. This decreases lens transparency and thus the amount of light transmitted to the retina. Dispersion of light rays at the macromolecules may act like a light veil through which one tries to see.

A large increase in insoluble proteins can "cloud" the lens of the eye (forming a *cataract*), which can occur at any age but is most common in the aging. Increased opacity distorts and decreases available light. The effects of a cataract on vision depend on its size, location, and density. A small cataract in the center of the lens is likely to affect vision far more than even a large cataract at the periphery. Cataracts can cause blurred or double vision, spots, difficulty in seeing at too little or too much illumination, change in the color of the pupil, and the sensation of having a film over the eye or looking through a waterfall (*cataract* is derived from the Latin word for waterfall).

The vitreous humor may yellow, increasing the problems already engendered by a yellowing lens. Liquid and gel portions may clump together, causing "floaters" or spots to appear in the field of vision. Also, pockets of liquids may form. Clumping and liquefaction together result in changing refraction, making the image formed on the retina less coherent. A sudden jolt or vibration may detach the posterior vitreous from the retina (probably as a result of macular edema), which brings about a severe vision impairment.

Light sensors, i.e., cones and rods at the retina, are reduced in number, particularly the cones at the fovea. This reduces the clarity of vision. Retinal pigment is reduced, a degenerative pigment (lipofuscin) begins to appear, and the retina becomes thinner, particularly at the periphery.

The natural fluids produced in the eye may not drain well but collect inside the eyeball, and the ensuing pressure (*glaucoma*) may eventually destroy fibers of the optic nerve. Thus, regular checkups for glaucoma are recommended after the age of 40 to detect its occurrence before damage occurs.

APPLICATION

Designing for the Aging Vision

Visual functions develop deficits with age in several basic areas: light sensitivity, near vision, depth perception, dynamic vision; and perception, visual search, and visual processing (Klosnik, Winslow, Kline, Rasinski, and Sekuler, 1988). Individuals develop visual impairments of different types and magnitudes, and at varying ages, but common experiences are difficulties in dimness, reading small print, distinguishing similar colors, or coping with glare. Most of these problems can be dealt with fairly simply, such as by providing proper corrective lenses, higher intensity of lighting, and increased color contrast, using large characters with high contrast against the background, repositioning a computer screen, or shielding bright lights in the field of view—discussed also in Chapter 4.

Red and yellow should be preferred to indicate color accents, because green and blue become difficult to distinguish with increasing age. In fact, all kinds of contrast and sensitivities are reduced, which reduces one's ability to perceive details of an object or scene, particularly at twilight. Recognizing details from a cluttered background becomes difficult: picking out individual faces in a crowd at dim light is nearly impossible, even reading a book with large print may be difficult if inadequate white space exists between the black letters.

For work at video workstations, separate optical lenses may have to be used for viewing the video display and for reading and writing tasks, particularly if the display is further away than "reading distance," i.e., beyond about 50 cm—see Chapter 9. Since low-contrast images are difficult to see, aging workers may have difficulty viewing the green-on-dark lettering of many displays. Bright reds and yellows are easier to distinguish and should be used instead of blues and greens.

Extra lighting (increased illuminance) can improve the visual ability of aging persons. But, since glare is a problem for many, spotlights and other bright light sources must be placed carefully. Aging people often find it more difficult to adapt to sudden changes in lighting, particularly to adapt from bright to dim conditions. Detection of targets on a cluttered background is facilitated if the light-on-dark contrast of the target is enhanced.

Ocular motility, the ability to move the eyeball through muscular actions, becomes impaired, both in performing quick movements and in turning the eye to an extreme angle. Pursuit movements are impaired in most aging persons, which limits both the ability to follow a target smoothly and to move the gaze to a target and to fixate it. Furthermore, the field of vision is reduced at the edges, particularly in the upward direction.

Many aging people drive automobiles, although they may avoid driving at night, on unfamiliar roadways, and in congested areas and times. To help them, one should avoid "light and color clutter," such as experienced in many commercial streets, where illumination and advertising lights compete with traffic lights. (In some European countries, no red or green lamps may be used near traffic lights.) Regarding the design of the "visual interior" of the automobile, the standard minimal illuminance of 300 cd m^{-2} on instruments should be exceeded to avoid "washout" in direct sunlight: for this, luminances of the order of 1,200 cd m^{-2} are required. The common minimal contrast ratio of five to one is sufficient for passive displays, but should be about twenty to one for moving displays. Regarding character size, 25 minutes of arc should be the minimum. Critical visual information should be displayed in the central area of view. Red versus blue or green are most easily distinguished, followed by yellow and white. These colors should be pure in saturation. Such design recommendations (see Chapter 11) not only help older drivers, but also are advantageous for younger eyes.

Changes in Hearing

The ability to hear decreases quite dramatically in the course of one's life, first at the high frequencies between 10 and 20 kHz, then down to about 8,000 Hz, an impairment typically recognized as "age-related hearing loss" (*presbycusis*). Yet, difficulties in hearing may extend into the lower frequencies, and they are often coupled with noise-induced hearing loss—see Chapter 4.

The changes start at the pinna, the outer ear, which becomes hard, inflexible, and may change in size and shape. Wax build-up is frequent in the ear canal. Often, the Eustachian tube becomes obstructed, leading to an accumulation of fluid in the middle ear. There may also be arthrosic changes in the joints of the bones (anvil, hammer, stirrup) of the middle ear, which, however, do not usually impair sound transmission to the oval window of the inner ear. There, atrophy and degeneration of hair cells in the basilar membrane of the cochlea occur. Deficiencies in the bioelectric and biomechanical properties of the inner ear fluid and mechanical degeneration of the cochlear partition occur, often together with a loss of auditory neurons. These degenerations cause either frequency-specific or more general deficiencies in hearing capabilities.

While it is estimated that 70 percent of all individuals over 50 years of age have some kind of hearing loss, the changes are individually quite different. Typically, in populations that do not suffer from industry or civilization-related noises, the hearing sensitivity in the higher frequencies is less reduced than in people from "developed" countries. Such changes related to the environment overlap with, and in some cases mask, age-dependent changes.

Loss of hearing ability in the higher frequencies of the speech range reduces the understanding of consonants that have such high-frequency components. This explains why older persons often are unable to discriminate between phonetically similar words, which may make it difficult to follow conversations in noisy environments. Severe hearing disorders may lead to speech disorders which, to some extent, may be psychologically founded: for example, if others have to speak loudly for you to hear, they may not want to interact with you because they may feel embarrassed. There may be hesitance to speak if it is uncertain what level of loudness is required and that one understands what is said.

APPLICATION

Ergonomic interventions are available. First, one can try to improve a person's ability to hear by providing "hearing aids" that amplify sounds for which hearing deficiencies exist. This, however, is difficult if the deficient areas are not exactly known (for example, because a person has not taken a hearing test recently), and the amplification of sounds might also enforce unwanted background noise. Improvement is very difficult if the hearing loss is due to destruction of structures of the middle and inner ear, and with current technology nearly impossible if the auditory nerves have been damaged. Further technical development may bring better help in the future. Ergonomic measures can improve the "clarity of the message," as discussed in Chapter 4: provision of sound signals that are easily distinguishable and of sufficient intensity, and avoidance of masking background sounds ("noise").

Another solution is to provide—at least some—information through other sensory channels, such as vision (e.g., present illustrations and written texts to accompany an auditory message) or to employ the taction sense, such as when using Braille.

Changes in Somesthetic Sensitivity

The somesthetic senses include those related to touch, pain, vibration, temperature, and motion—see Chapter 4. In spite of decades of research, there is deplorably little reliable quantitative information available about the changes, if any, in tactile sensitivity, including pain, with aging (Boff, Kaufman, and Thomas, 1986). Apparently, absolute thresholds increase, which may be associated with a loss of touch receptors, but that phenomenon and its explanation are still rather unresearched. There is a well-observed change in vibratory sensitivity, particularly in the lower extremities. This effect is used in the diagnosis of disorders of the nervous system. Among the possible explanations may be a reduction in blood supply to the spinal cord with ensuing damage to the nerve tracks, possibly a decline in the number of myelinated fibers in the spinal roots, or diminished blood flow to the peripheral structures of the body in general. Dietary deficiencies might also play a role.

Temperature sensitivity also seems to decline, but this may be partially offset by the often observed desire of elderly persons to be in warmer temperatures, outside or within a building. The observed differences in temperature behavior may be associated with a decline in the body's temperature regulation system.

Decrease in information from kinesthetic receptors, or decrease in the use of that information in the central nervous system, may contribute to the higher incidence of falls: with increasing age, one seems less able to perceive that one is being moved (such as in an automobile or airplane) or that one moves body parts. Again, surprisingly little is known in a systematic fashion. The same is true for pain sensitivity, which appears much reduced as one gets older. Possible explanations are reductions in the number of Meissner's corpuscles and other receptor organs in the skin, in the number of myelinated fibers in the peripheral nervous system, and in blood supply (Committee on an Aging Society, 1988; Hayslip and Panek, 1989).

Changes in Psychometric Performance

With aging, typically, reaction and response times to stimuli (see Chapter 4) increase. This may be partially explained by deficiencies in the sensory peripheral parts of the nervous system, delays in afferent provision of information to the central nervous system, and reduction of efficiency in the efferent part of the peripheral nervous system. Yet, successful performance involves perception, sensation, attention, short-term memory, decision making, intelligence, and personality as well as "motor behavior." Thus, while poor performance may be attributed to some of the more physiological factors, much depends on other processes "of the mind" which go beyond the scope of this text. Among the theories to explain changes in central functions are those of neural noise, expectancy or set theory, complexity and information overload theories, and rigidity hypotheses—see, e.g., publications by the Committee on an Aging Society (1988); Belsky (1990); Czaja (1990); Hayslip and Panek (1989); Kermis (1984); and literature in geriatric psychology, physiology, and sociology.

NEED

Aging of friends and relatives, and one's own aging, are of perpetual interest and concern. The discussion in this book of knowledge about changes with age, and related research, is by no means complete. Anecdotal observations and case studies abound in the literature, but systematic research and its compilation are scarce. The recent phenomenon of large numbers of older people, many still competent and politically and economically powerful, is prompting much research now.

Available research findings do not provide a complete picture. This is largely due to the fact that cross-sectional research is nearly meaningless: comparing the anthropometry or performance of persons of similar age provides only that information, if any. Chronological age is not a meaningful classifier. Better classification systems need to be established: the so-called biological age is one attempt in that direction, but its definition is rather difficult. Thus, a basic research task is to establish a suitable reference system, or reference systems, with proper scales and anchoring points.

It is obvious that many or almost all physical, perceptual, cognitive, and decision-making capabilities decline with age. Yet, some of those losses are slow and are not easily observed. Other capabilities and facilities decline fast, or they may deteriorate slowly at first and then quickly at some point in time, perhaps again stabilizing for a while. Some of these changes are independent of each other, but many are linked, directly or indirectly. Failing physical health may have effects on attitude and intentness, or failing eyesight might lead to a fall and serious injury with ensuing illness.

In addition, people have different coping strategies. One's failing ability to recall names, for example, may be overcome to some extent, and for some time, by developing mnemonic strategies. An aging person may reduce activity boundaries by maintaining only those with which one is comfortable and that one can competently handle. Maintaining physical or mental activities, perhaps purposefully so, can counteract reductions in facility significantly and for long periods of time.

Thus, in our current framework scaled by chronological age, one finds an immense variety of maintained, perhaps in some respect even increased, facilities, of decreasing capabilities, and of either slow or fast changes with time.

Much research needs to be done.

APPLICATION

Designing for the Older Worker

The U.S. Age Discrimination Acts have defined the "older" person as either age 40 or 45 and above. While one cannot set a particular year as the beginning of aging, there is no doubt that certain work tasks become more difficult as one approaches retirement. This includes strenuous physical exertions, such as moving heavy loads; tasks that require high mobility, particularly of the trunk and back; and work that requires high visual acuity and close focusing. On the other hand, there is at least "anecdotal evidence" that tasks which require patient and experience-generated skill may be performed better by at least some older workers than by young persons. Nearly all age-related difficulties can be overcome by proper workplace and tool design and selection, by proper arrangement of work procedures and tasks, by provision of working aids such as power-assisted tools or magnifying lenses, or by managerial measures such as assignment of proper work tasks and by provision of work breaks (Rice and Kemmerling, 1990; Kelly and Kroemer, 1990; Czaja, 1991).

Many of these ergonomic measures are logical results of the intents to counteract known deficiencies. Manipulation tasks are facilitated by providing proper work height, supports for arms and elbows, and provision of special hand tools which might be power-aided (see Chapter 10). Postural aids include the just-mentioned proper work height coupled with provision of a seat with appropriate dimensions and good adjustability. In essence the same recommendations for chair design exist whether it is used in the shop or office (see Chapter 9), but the shop seat must be especially sturdy and protected against soiling. Many sensory problems can be overcome by ergonomic interventions, as already discussed. For example: vision deficiencies can be counteracted by provision of corrective eye lenses, by intense and well-directed illumination, and by avoiding direct or indirect glare (see Chapter 4). If text must be read, either on paper or a dial, or on a computer screen, proper character size and contrast as well as carefully selected color schemes should be used (for example, avoid difficult visual discriminations of similar hues, particularly in the blue-green range). The auditory environment should be controlled to keep background noise at a minimum and to provide sufficient "penetration" of auditory signals that must be heard by selecting appropriate intensities and frequencies (avoid frequencies about 4,000 Hz)—see Chapter 4.

In essence, use of proper ergonomic measures, carefully selected and applied, is just "good human engineering," which would help *all workers of all age groups* but is of particular importance for the older worker.

Designing for the Aging Driver and Passenger

Depending on whose statistics one reads, aging drivers may have fewer or more traffic accidents, but the pattern of accidents with age indicates increasingly missed or misinterpreted perceptual cues, slow or false reactions to cues, and wrong motor actions such as accelerating instead of braking. There is a vast literature on the many traits of the aged driver, compiled by Barr and Eberhard (1991) and Eberhard and Barr (1992).

Recommendations for the design of the interior of automobiles have been discussed already. Regarding "public" transportation (planes, trains, subways, busses, trams, lifts, elevators, moving walkways, etc.), aging persons report the following problems:

- *Information* (cues and signs) indicating direction, location, and use, such as

  Where does the bus go that I see coming?
  Where does the tram stop?
  Where is the exit to Brown Street?
  Where are we?
  How do I buy a ticket?

  Many of these problems can be overcome by applying common "human engineering principles" such as using signs with good lettering (contrast, size, symbols), proper illuminance and avoidance of glare, auditory announcements that are timely and understandable, and redundant information such as showing the floor number in an elevator in large numerals and announcing it early over a public address system.
- *Ingress and egress, and body posture,* while using the transport system. Entrance/exit passages and steps are often difficult to negotiate and may require forceful and complex stepping, typically so in trams, trains, busses, or moving walkways. Uneven floors and damaged or misplaced floor coverings often pose problems in vehicles or hallways. Handholds used to be a problem in busses and trams, but most now provide a variety of hooks, columns, and hand grips.

- *Use of walking aids and wheelchairs.* Many aged people use canes or walkers, and some need wheelchairs. While a cane usually does not pose a problem in public transportation (in fact, it may alert other passengers to be considerate and helpful), walking aids and specifically wheelchairs are often not easily accommodated and may be a serious deterrent to use of public transportation.

☞☞☞ *"Human aging encompasses much more than physiologic change over the life course. Age-related changes are manifest across all aspects of life including physical, environmental, economic, and social aspects and mental well-being. Yet change along any dimension is not simply, or irrevocably, correlated with chronological age. The serious loss or compromise of capacity in one area, however, can accelerate the rate of decline in others. Such interactions are often complex. Poor health, for example, can require increased medical expenditures that divert income from other essential areas such as home upkeep or the purchase of food. Over time, such interactions can result in further erosion of functional capacity in the aging. Alternatively, a supportive social or physical environment may retard the rate of functional loss to some degree" (Soldo and Longino, 1988, p. 103).* ☜☜☜

DESIGNING THE HOME FOR THE AGING

A 50-year-old who purchases a home (or durable product) will be a rather different individual still using it ten, twenty, or thirty years later. A person's ability to live alone, to live in the home with some outside help, to be cared for at home, or to live in a care environment is a complex function of various abilities, or disabilities. Figure 12-4 presents an overview of common disorders among the aged, and how these problems can be alleviated, at least to some degree, by proper ergonomic measures.

Functional ability, or its opposite, dependency, is commonly assessed in terms of clusters of abilities: "instrumental activities of daily living" (IADL) or the less specific "activities of daily living" (ADL)—shown in Table 12-8. Numerous surveys of elderly persons have been done that use these two activity listings. The results have been used to classify persons into groups that require help and specifically designed environments (see, for example, Smith, 1990; Lawton, 1990; Clark, Czaja, and Weber, 1990; Committee on an Aging Society, 1988). While IADLs are rather practical and self-explanatory, they lack specificity and objectivity and are not easily scaled. One attempt to improve was to subdivide ADLs into more specific tasks such as lifting/lowering, pushing/pulling, bending/stooping, reaching (Clark, Czaja, and Weber, 1990). Further work in this direction could result in a list of basic demands and activities somewhat similar to the "motion elements" used in industrial engineering for method studies.

Problems/ Manifestations	Senescence	Arteriosclerosis	Hypertension	Parkinson's disease	Peripheral neuropathy	Drowsiness	Cataracts/glaucoma	Arthritis	Paget's disease	Osteoporosis	Low back pain	Bronchitis/emphysema	Pneumonia	Diabetes	Senile dementia
General debility	☆		☆	☆	☆			☆	☆	☆	☆	☆	☆	☆	
Mobility	☆	☆		☆	☆		☆	☆	☆		☆				
Posture	☆			☆			☆	☆	☆	☆	☆				
Pain		☆		☆	☆			☆	☆		☆				
Incoordination		☆		☆	☆		☆	☆							
Reduced sensory input	☆	☆			☆	☆	☆								
Loss of balance	☆	☆		☆			☆								
Reduced joint mobility								☆	☆	☆	☆				
Weakness in muscles	☆			☆	☆			☆							
Auditory disorders	☆					☆		☆						☆	☆
Locating body in space		☆					☆	☆							
Shortening of breath		☆		☆								☆	☆		
Deformity								☆	☆	☆					
Memory impairment		☆													☆
Visual problems	☆	☆					☆							☆	
Disorientation		☆		☆											☆
Loss of sensation	☆				☆	☆									
Cognition disturbance		☆		☆											
Incontinence		☆													☆
Speech disorders				☆		☆	☆								
Touch disabilities	☆				☆			☆						☆	

Figure 12-4. Problems arising from common age-related disorders. (Adapted from Kemmerling, 1991.)

TABLE 12-8. MEASURES OF DAILY LIVING

Instrumental activities of daily living (IADL)	Activities of daily living (ADL)
Managing money	Living
Shopping	Bed/chair transference
Light housework	Indoor and outdoor mobility
Laundry	Dressing
Meal preparation	Bathing
Making a phone call	Toileting
Taking medication	

To be in one's home has the major advantage of a "familiar" setting with all its physical and emotional implications. Unless by happenstance or foresight designed to be ergonomic, private homes usually need some adjustments to allow the aging inhabitant to perform all necessary activities, even with somewhat reduced sensory, motoric, and decision-making capabilities. In addition to passage areas, there are several rooms of particular concern. One is the kitchen, where one stores, prepares, and serves food. Often, it is also a phone-in message center. In the past, the kitchen was the woman's territory, but this is no longer true.

☞☞☞ *The first "scientific study" of kitchens was completed by Lillian Gilbreth in the 1920s. Her classical study relied mainly on work flow, and on the time and motion study methods that she pioneered together with her husband. Her re-designed kitchen reduced motions by nearly 50 percent.* ☜☜☜

APPLICATION

"Seven principles" derived from time and motion studies, augmented by ergonomic findings, apply to the kitchen:

1. One shall design for a small "work triangle," the corners of which are the refrigerator, sink, and stove or range.
2. The "work flow" for food preparation is to remove food from the refrigerator or cabinet, mix or otherwise prepare ingredients near the sink, cook on the range or in the oven, and serve. Kitchen components should facilitate that flow.
3. If there is "traffic flow" by others, it should not cut through the patterns of work triangle and work flow.
4. Items should be stored at the "point of first use," as determined by work "triangle" and "flow."
5. The "work space" for the hands should be at about elbow height, or slightly below. This facilitates manipulation and visual control. Counter and sink heights are derived from elbow height. Note that it might be advisable to consider walking aids, and that stools or chairs might be used.
6. The reaches to items stored in the kitchen should be at or slightly below eye height to allow for visual control and easy arm and shoulder motions.
7. The motion and working space should not be reduced or interrupted by doors of appliances and cabinets, hence they should open "outward" from the working person.

Another area of major ergonomic concern is the bathroom. It is one of the busiest and, unfortunately, most dangerous rooms in the house. Basic equipment includes a bathtub and/or shower, toilet, and lavatory. Furthermore, there are usually storage facilities for toiletries, towels, etc.

Bathtub and shower are the two common means for cleansing the whole body and are common accident sites. Their major danger stems from slipperiness between bare skin and floor or walls. The more dangerous of the two is the bathtub, because of its slanted surfaces and the high sides above which one has to step, a procedure difficult for most people and particularly so for the elderly, who may have balance and mobility deficiencies. Kira (1976) described the several techniques employed by most users getting in and out of a tub. They involve shifts in body weight: from the legs, to arms and legs used together, and finally to the buttocks while entering the tub; in leaving, these shifts occur in the reverse order. There is much potential for loss of balance and for slips and ensuing falls. While resting in the tub, the angle of the backrest and its slipperiness comprise the most critical design aspect. Proper hand rails and grab bars, within easy reach both for sitting and getting in and out, are of importance. The shower stall may also have a slippery floor to step on, but its lower enclosure rim makes it easier to move in and out.

Use of the control handles for hot and cold water is quite often difficult for aged persons, particularly when they are not at their familiar home and have to cope with different handle designs and movement directions and varying resistances. Both better design principles and standardization would be helpful—for example, in the mode and direction of control movement to regulate water temperature.

The wash basin may be difficult to use if it is too far away, as it may be if inserted in a cabinet, so that one cannot step close to it. The faucet often reduces the usable opening area of the wash basin. Proper height is important, as are the water controls.

The toilet is, obviously, of great importance in allowing one to expel body wastes and to keep the body openings and adjacent anatomical areas clean. Kira's classical study has provided much information about proper design, sizing, shaping, and location.

For western-style private homes, a variety of publications contains valuable ergonomic design recommendations—for example, by Singer and Graeff (1988). In other civilizations and other parts of the earth quite different customs and conditions exist, for which at present little ergonomic information appears available.

The Design of Nursing Homes

Severe functional disabilities are what we dread most when we think of "old age," because they are at the core of our fears about growing old. In the United States, currently only one of four people of age 65 has any problems negotiating life, but there is the likelihood that impairments are in progress. As more health deficiencies, mental problems, or functional disabilities occur, the elderly person first needs more help in his or her own home. Initially, that care may be privately secured, through friends and relatives or a hired person. For many, this is the beginning of a path that leads to a nursing home.

Many aged persons suffer from "therapeutic nihilism": they are prone to self-diagnosed treatable conditions, but have changed their attitude to accept aches, pains, and physical distress, normally considered a disease, as something "normal" at advanced age. They may not want to appear weak and discouraged and hence do not want to "bother" the caregiver, even if they are generally ill and in need of help (Belsky, 1990). This, of course, makes it particularly difficult to provide them with the help and care that they need and deserve.

Aged people are the most frequent consumers of physicians' services. Basically, it is a physician's job to diagnose an illness, make a medical intervention, and cause a cure. But this is mostly not the case with the elderly. The older one gets, the more likely one is to suffer from a chronic illness that cannot be cured, although occasionally it can be alleviated, or covered up, at least for a while. Fighting disability requires diagnostic strategies different from battling illness. They include techniques outside the physician's traditional realm of expertise (Belsky, 1990). Physicians who specialize in geriatric medicine collaborate with nurses, physical therapists, dietitians, psychologists, and ergonomists.

There are different kinds of institutions for the elderly who cannot stay at home. Some simply offer room, board, and personal care to the residents. Other are more like a hospital, offering intensive medical services to seriously ill people. Some cater to certain religious or ethnic groups; some freely accept people with Alzheimer's disease or those who are bedridden; others want only occupants who are not severely physically or mentally impaired. Despite this diversity, Belsky classified them by the intensity of care which they offer and by their mode of ownership. In the United States, nursing homes must be classified as offering either "intermediate" or "skilled" care in order to be reimbursed by Medicaid or Medicare.

☞☞☞ *The quality of nursing facilities varies widely from "home" to "snake pit" (Belsky, 1990, p. 107).* ☜☜☜

Residents who need ongoing assistance in functioning but not intensive care are in "intermediate" care facilities. Thus, the architectural and other ergonomic recommendations given earlier for the private home also apply to intermediate care facilities insofar as they facilitate the residents' efforts to look after themselves. However, in addition, design and organizational means must be considered that facilitate the caregivers' activities, such as easy access, cleaning, awareness of immediate help requirements, and emergency access. While it is important that aged people have as much personal freedom as possible, being in an institution limits their choices in the most basic aspects of life, such as where to live, when to get up or lie down, what to do, what meals to have. Home management should carefully provide various choices for the residents, keeping their interest in mind and not primarily organizational ease.

Much of that ability to control themselves, and their environment, does not exist for patients in what are usually called in the United States "skilled nursing care facilities." Their residents tend to be very old and in continual need of help and care. Owing to gender differences in longevity and types of illnesses and injuries, women are more likely than men to be disabled but not to have a life-threatening illness. Since nursing home care is, in the United States, largely financed by Medicaid, most residents are poor; furthermore, many are single, divorced, or widowed (Belsky, 1990). Regarding the architecture and interior design of nursing homes for people who need intensive care, some of the earlier detailed recommendations for ergonomic designs still apply; but now the aspects of providing 24-hour supervision and care, and possibly intensive medical treatment, prevail. Unfortunately, ergonomic information on architecture and interior design for nursing care facilities is piecemeal and incomplete, requiring much systematic research before encompassing design guidelines can be expected (Committee on an Aging Society, 1988; Meadows, 1988; Czaja, 1990).

☞☞☞ *With increased longevity, vulnerability among the aging has assumed growing importance.*

- *Changes among aging persons are not correlated with chronological age but exhibit variance.*
- *Changes among aging people are not manifest as a simple linear decline but show a variety of rates; change may even be arrested.*
- *Changes in function may produce different effects in the same person, and the variance among aged persons tends to increase with age.*
- *The rate of change can proceed along some dimensions relatively independently of change in others, but the serious loss or compromise of functional capacity in one area can accelerate the rate of decline in others.*
- *A supportive social or physical environment or positive change can retard the rate of functional loss to some degree (Committee on the Aging Society, 1988, p. 7).*

☜☜☜

ERGONOMIC DESIGN FOR DISABLED PERSONS

Many of the design considerations discussed in the foregoing section as applying to aging people are also relevant to disabled persons. This is particularly true for those design recommendations that help to alleviate, overcome, or sidestep impairments so as to attain the largest possible independence and everyday functioning capability.

EXAMPLE

At pedestrian street crossings, it is advantageous to combine traffic lights with acoustic signals that indicate to persons with impaired hearing and sight that they may or shall not cross the street. In addition to the familiar red and green lights, selected sounds help to indicate when it is safe to proceed. For example, a "cuckoo" sound may signal crossing in one direction and an electronic "chirp" indicate the orthogonal direction. This combination of light and sound has been found beneficial not only for aging and impaired persons, but also for other pedestrians (Szeto, Valerio, and Novak, 1981).

Description of "Disability" or "Handicap"

We are all only temporarily able-bodied: as children, we lack the strength and skill that we hope to acquire during adulthood; and while aging, we lose some of the facilities that we previously enjoyed. Many suffer from injuries or illnesses that deprive us of certain capabilities, often only for some period of time, but possibly forever. Estimates of the percentage of the working-age population with disabilities range from 8 to 17 percent in the United States, and the estimates for children and the elderly are also quite variable. Some of that diversity stems from different accepted definitions (Levine, Zitter, and Ingram, 1990; Gardner-Bonneau, 1990). The Americans with Disabilities Act (ADA), signed into law in the United States in 1990, defines disability as "a physical or mental impairment which substantially limits one or more of an individual's major activities of daily living such as walking, hearing, speaking, learning, and performing manual tasks." The Committee on National Statistics of the U.S. National Research Council, like the World Health Organization, defined disability as "any restriction or lack (resulting from an impairment) of ability to perform an activity in the manner, or in the range, considered normal." Since human activities are various, there are many different kinds of disabilities (Haber, 1990).

In this context, "impairment" is a chronic physiologic, psychological, or anatomical abnormality of bodily structure function caused by disease or injury. More specifically, "work disability" is a dysfunction in the vocation for which a person is trained, which is often selected early in adulthood and may be influenced by impairment existing during youth. "Handicap" is the social and economic disadvantage that results from impairment or disability. It may entail loss of income, social status, or social contacts (Levine, Zitta, and Ingram, 1990). Accordingly, there is a wide range for judgment regarding impairments, disabilities, and handicaps, both with respect to their extent and regarding specific age groups. For example, judged against "normal adults," children and many elderly persons could be described as disabled (Haber, 1990).

❑ **APPLICATION**

Ergonomic Means to Enable the Disabled

As just discussed, there is a large variety of disabilities. Faste (1977) attempted to classify these, as shown in Figure 12-5. The disabilities and impairments shown there may result from very dissimilar conditions. For example, "poor

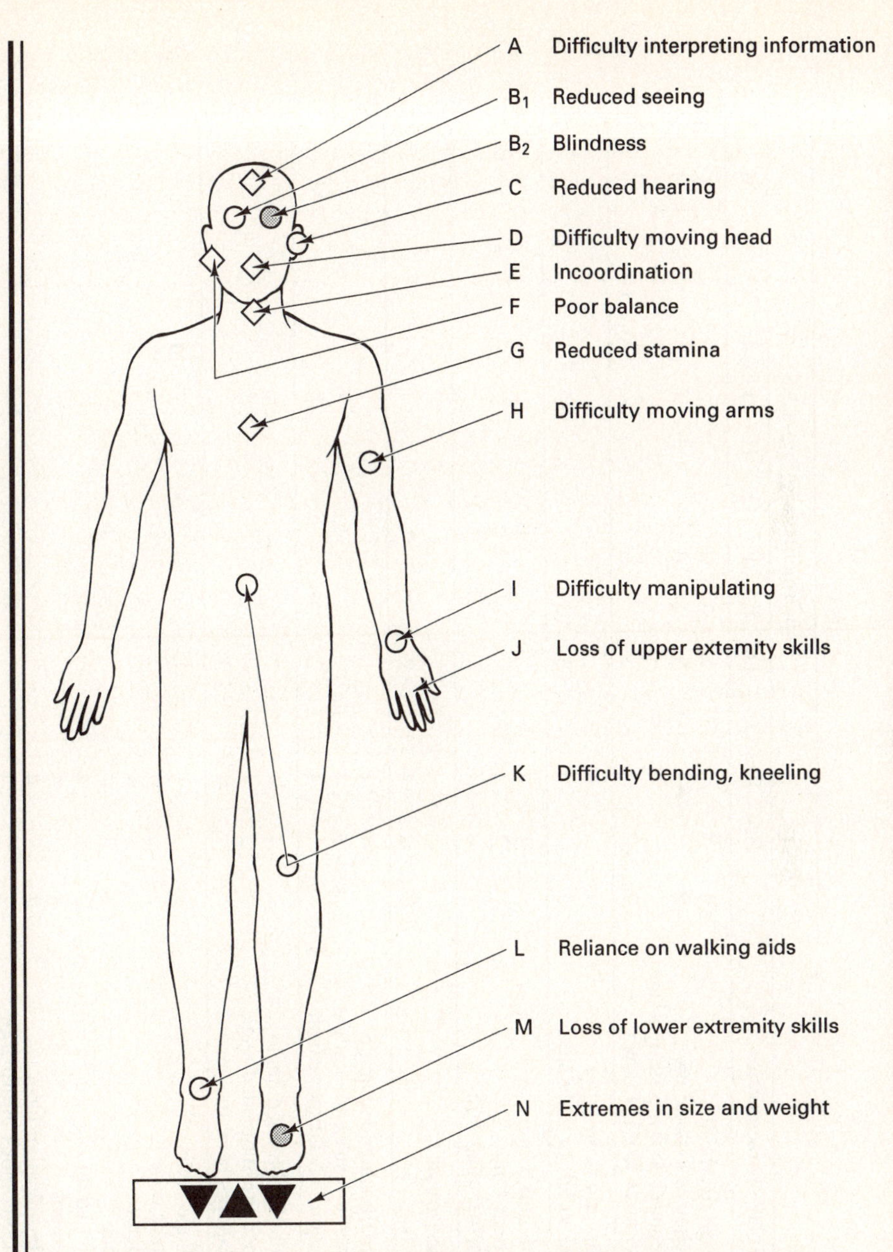

Figure 12-5. Impairment classification. (Adapted from Faste, 1977.)

balance" may result from conditions related to blood pressure, hemiplegia, paraplegia, amputation, multiple sclerosis, muscular dystrophy, cerebral palsy, Parkinson's disease, brain tumor, and other causes. The design matrix shown in Figure 12-6 relates display design issues to disabilities. Such charting identifies ergonomic challenges and indicates possible solutions.

For most impaired persons, a specific disability, or a combination of disabilities, can be compensated by ergonomic means. For example, one para-

		Impairments (see Fig. 12-5)														
Display characteristics		A	B_1	B_2	C	D	E	F	G	H	I	J	K	L	M	N
Vertical location	high overhead	3	4	4		4	2	4						3	3	1
	requires looking up	2	2	4		3	1	3						1		
	requires looking straight ahead			4												
	requires looking down	2	2	4		3							1			1
Horizontal location	directly in front			3												
	off to left or right side	3	2	4		3	3							2		
Viewing distance	about 0.5 m		1	4												
	about 1 m		2	4												
	farther than 1 m		4	4												
Orientation	horizontal			4												
	other	2	2	4		2										
Vertical size	small subtended angle		4	4												
	medium subtended angle		3	4												
	requires head movement			4		2										
Horizontal size	small subtended angle		4	4												
	medium subtended angle		2	4												
	requires head movement			4		3										
Content	shape code	3				1					1	4				
	color code	3		4												
	picture		1	4												
	map	3	2	4												
	pictogram	2	1	4												
	symbol	4		4												
	identification label	3	2	4												
	dichotomous information	2	2	4												

Figure 12-6. Effects of display characteristics on users with impairments. (Modified from Faste, 1977; Cushman and Rosenberg, 1991.)

	Display characteristics	Impairments (see Fig. 12-5)														
		A	B_1	B_2	C	D	E	F	G	H	I	J	K	L	M	N
	quantitative information	4	3	4												
	brief text	3	3	4												
	long text	4	4	4					1							
	audio cue supplement	2			2											
	audio cue only	3			3											
Exposure variables	used frequently	2														
	used occasionally	3	3	2												
	short viewing time	4	4	4												
	observer or display moving	4	4	4												
	dynamic display	3	4	4												
	interactive display	4	3	4	4						3	4				
Illumination	high contrast on display		1	4												
	low contrast on display	4	4	4												
	high contrast to surround			4												
	low contrast to surround	4	4	4												
	front lighted			4												
	translucent or back lighted			4												
	daylight		1	4												
	artificial light			4												
	glare present	4	4													
Other variables	legibility	4	4	4												
	readability	4	4	4												
	logic of location	4	2	4		1									1	
	logic of message content	4	3	4												
		A	B_1	B_2	C	D	E	F	G	H	I	J	K	L	M	N

Legend: 1—Potential problem; 2—Problem; 3—Severe problem; 4—Impossibility.

Figure 12-6. (*cont.*)

plegic may work mostly with a computer, and hence the operation and control of that equipment is of particular importance; serving customers at a sales or information counter may be the specific occupational concern. As Casali and Williges (1990) describe, computers can be particularly beneficial for a person with disabilities for most physical tasks. With the computer, one does not have to manipulate papers, files, or other printed information, and one need not be able to write, draw, or use a phone. There are now many software programs available that have been specifically developed for disabled individuals, such as to control the environmental conditions of a home or office (Griffith, 1991; Williges, Williges, and Elkerton, 1986). Yet, the computer workstation itself, particularly the current multi-key QWERTY keyboard, or a mouse, may be difficult or impossible for the impaired person to use. Thus, adaptive hardware is often needed, such as a stick (attached to head or hand, or held in the mouth) to press the keys.

A systematic process was described by Casali and Williges (1990) to determine the most appropriate aid for a disabled computer user. As shown in Figure 12-7, the approach relies on a database containing information about available hardware and software. The residual abilities of the client regarding computer use are assessed and compared with the task. After choosing a candidate solution, its usability must be tested, and changes and improvements introduced as needed.

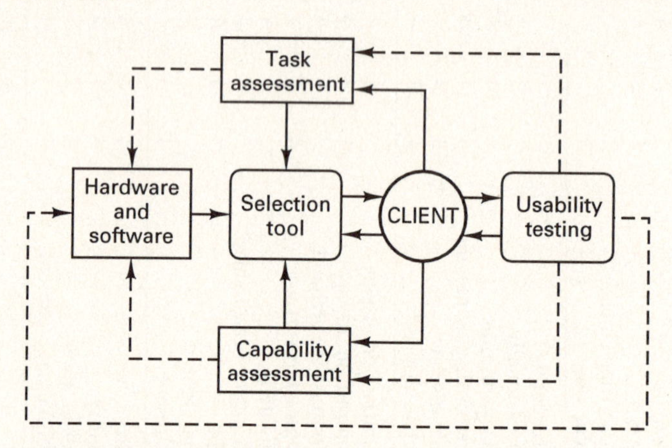

Figure 12-7. Systematic process to determine proper computer aids for an impaired client. (With permission from Casali and Williges, 1990.)

Testing devices and procedures have been developed to quantitatively measure the residual capabilities of impaired persons, such as the Basic Elements of Performance (Kondraske, 1988) or Available Motions Inventory (AMI) (Dryden and Kemmerling, 1990; Smith and Leslie, 1990). The AMI, for example, has a number of panels that contain switches, knobs, and other devices to be reached, turned, operated, and otherwise activated and manipulated. These panels are put into standardized positions within the work area, and each person's ability or disability to operate these as required is recorded, together with the time needed and the strength exerted, as appropriate. A large database is at hand that allows the comparison of the individual's performance with that of other persons, able or disabled. Thus, the individual capability or limitation for certain tasks of particular interest can be determined. This information facilitates a systematic approach to find work tasks and work conditions suitable for a given individual's capabilities, or to modify them to match the person's abilities.

Locomotion and transportation are serious problems for many disabled persons. Providing technical aids in terms of prostheses, crutches, and walkers has been an age-old concern. Wheelchairs have been described for decades; a comprehensive review covering the time prior to the mid-1980s was provided by Zacharkow (1988). He points out that, in addition to movement, the wheelchair should provide stable yet relaxed support, particularly avoiding pressure points and sores. Accordingly he presents specific recommendations regarding the seat, backrest, arm rest, leg support, etc. More recently, a vast number of reports on wheelchair developments and uses have been published, including client-propelled, attendant-pushed, and motor-driven chairs for indoors and outside use, even for racing (see, e.g., the journals of *Biomechanical Engineering*, *Prosthetics and Orthotics International, Rehabilitation & Progress Reports, Rehabilitation and Research Development*). Another fairly large research and development effort concerns the development of specialized engine-driven covered vehicles (e.g., Oldenkamp, 1990), or the adaptation of existing automobiles for use by disabled drivers or passengers (e.g., Koppa, 1990).

The use of ergonomic knowledge in rehabilitation engineering is widespread, ranging from wrist splints to artificial limbs, from walking aids to special automobiles. Technology for people with disabilities has advanced beyond gadgeteering and now depends increasingly on the development and implementation of sophisticated devices. These often include control actions stimulated by EMG signals, movements of unimpaired body parts (such as the jaw), voice control, or direction of gaze (Smith and Leslie, 1990). Yet, such assist devices need to be selected according to a variety of criteria, which include the following (Batavia and Hammer, 1990):

- *Affordability*—the extent to which the purchase, maintenance, and repair causes financial hardship to the consumer.
- *Dependability and durability*—the extent to which the device operates with repeatable and predictable levels of accuracy for extended periods of time.
- *Physical security*—the probability that the device will not cause physical harm to the user or other people.
- *Portability*—the extent to which the device can be readily transported to and operated in different locations.

TABLE 12-9. RANKING OF ASSIST DEVICES

	For the motion-impaired					For the blind			For the deaf			
	Wheel-chair	Typing system	Robotic arm	Environ. control	Phone system	Type reader	Recording system	Orientat. system	Alert system	Phone system	Speech recog.	Average ranking
Effectiveness	1	2	1	1	1	2	2	2	1	1	1	1.36
Affordability	4	5	4	6	6	1	1	1	2	2	2	3.09
Operability	2	1	2	2	2	5	7	10	5	3	3	3.82
Dependability	3	4	3	3	3	3	3	6	4	5	5	3.82
Portability	*	*	*	*	*	7	4	3	6	4	8	5.33
Durability	8	8	8	7	7	4	5	11	3	6	6	6.64
Compatibility	13	3	6	4	5	6	6	13	10	7	4	7.00
Flexibility	7	6	5	5	4	11	9	15	8	9	7	7.82
Ease of maintenance	6	9	11	9	11	8	8	8	7	10	9	8.73
Securability	*	*	*	*	*	13	11	7	11	12	14	11.33
Learnability	14	7	10	11	10	9	16	14	14	11	10	11.45
Personal acceptance	5	11	7	13	9	15	15	12	13	14	13	11.55
Physical comfort	10	13	13	14	8	12	13	9	17	8	11	11.64
Supplier repair	9	10	12	8	12	10	10	16	16	16	12	11.91
Physical security	11	15	9	10	15	17	17	4	12	13	16	12.64
Consumer repair	12	12	14	12	13	14	12	5	15	17	15	12.82
Ease of assembly	15	14	15	15	14	16	14	17	9	15	17	14.64

SOURCE: Adapted from Batavia and Hammer, 1990.

- *Learnability and usability*—the extent to which the consumer can easily learn to use a newly received device and can use it easily, safely, and dependably for the intended purpose.
- *Physical comfort and personal acceptability*—the degree to which the device provides comfort, or at least avoids pain or discomfort to the user, so that the person is attracted to use it in public or private.
- *Flexibility and compatibility*—the extent to which the device can be augmented by options and to which it will interface with other devices used currently or in the future.
- *Effectiveness*—the extent to which the device improves the user's capabilities, independence, and objective and subjective situation.

Table 12-9 provides a ranking of several assist devices according to these criteria by a panel of disabled consumers.

SUMMARY

In body sizes and in physical capabilities, there are systematic differences between adult men and women. Yet, in general, one can design nearly any workstation, any piece of equipment or tool, so that it can be used by either women or men. In some cases particular adjustability is needed, or one may have to provide objects in different dimension ranges; yet, these adjustments and ranges are generally not gender-specific but are simply needed to fit different people.

Commonly, one directs ergonomic efforts toward the effectiveness and well-being of the working-age population. However, there are large population groups that especially need ergonomic design. Among them are pregnant women, who during their pregnancy experience significant changes in body dimensions and in capabilities to perform demanding physical work. It is disappointing to note that fairly little systematic information on ergonomic design useful for pregnant women has been published in the scientific literature.

Infants, children, and adolescents are another very large group of "ergonomic customers." From birth to adulthood, body dimensions, physical characteristics, skill, intelligence, and attitude change enormously. Yet, descriptors of these developments are usually ordered by age (often stopping in the early teens), which is a rather meaningless classification because, at the same age, very large differences are likely to exist among individuals. This lack of information is counter to the strong economic interest in providing goods to the young population, be it in clothing, furniture, or toys. There also is a need for safeguarding and protecting children—for example, by properly designing cribs, railings, and toys, or by providing restraint devices to protect them in case of an automobile accident.

Like children, aging people are also usually grouped in chronological age brackets, which is a rather meaningless classification scheme because physical characteristics, capabilities, and attitudes vary widely among individuals. As with children, the development of capabilities (here usually a decline) can also rapidly or slowly alter within one individual over time, or may remain on about the same level for a while. There is great societal and individual interest in providing an ergonomically properly designed and maintained environment to the older population, be it at the workplace, in the home, or in a care facility.

Helping impaired people to overcome their handicaps through assistive devices is another major concern of ergonomics and rehabilitation engineering. Systematic procedures are at hand to determine the kind and extent of disabilities for specific activities, and often help can be provided either with mechanical devices or computers. Electronic devices require relatively little physical capabilities from the user, but can control and operate rather complex machinery and provide information storage as well as transfer to others and to the impaired person.

Thus, ergonomic knowledge and ergonomic procedures can be of help from childhood through adulthood into old age.

☞☞☞ *The ergonomic principles and techniques for architectural design and interior layout, for workplace, equipment and tools, are the same for "anybody"; they are just more critical and important for the less abled.* ☜☜☜

CHALLENGES

Most design is for "normal adults" in the age range of 20 to 40 years. Yet, is this a homogeneous group? Are there developments within a person during this age span?

Are there physiological reasons to presume that the musculature of women is per se weaker than that of men?

Should one expect that a summation of small differences in sensory functions between the sexes should make a difference for professional task performance?

Is it true that females show better finger dexterity than males?

Are there specific occupations that appear to be better suited for women, or men?

Why would pregnant women have difficulty seeing objects on the ground in front of them?

What kind of seats would particularly accommodate pregnant women?

Would an air bag promise to provide better forward protection in an automobile, compared to a seat-shoulder belt?

How would the proper design of restraint devices in automobiles be influenced by the age-related development of children?

Why is it so difficult to take anthropometric and biomechanical measurements on small children?

What kind of adjustment features might be suitable for furniture used in kindergarten and elementary school?

Which professional activities appear to be particularly affected by changes related to aging?

Are there any professions and activities for which aging people appear particularly suited?

Why are there so few longitudinal studies of people? What would it take to do such longitudinal studies?

Which physiological functions are likely to be maintained, into old age, by physical exercises?

What is the problem associated with sound amplification in most hearing devices?

Which would be suitable approaches to assess changes in somesthetic sensitivities with age?

Which tests may be included, possibly for all and not only for aging people, to qualify them for a "driver's license"?

To what extent would the architectural layout of a house for the aging differ from a layout suitable for younger people?

Consider classification schemes for the various "disabilities," with the intent to use these to develop aids and enablers.

Why are computers so particularly helpful for disabled persons?

Consider replacements for IADLs and ADLs.

Consider the various uses of wheelchairs, temporary or long-term, by persons with different impairments, and for different purposes. Can certain classification schemes be developed with regard to specific design solutions?

Postscript

HOW TO DO ERGONOMIC INTERVENTIONS AND TO MEASURE THEIR SUCCESS

Obviously, it is best to design things for human use, including complex human-technology systems, ergonomically correctly from the beginning. This is possible only for new things or systems. Most are already around, and we may find reasons to improve them: either because the "ergonomic expert" (often the user) knows better, or because problems in ease of use or performance become apparent. Figure E-1 presents a stepwise sequence of analyzing the problem(s), stating the improvement goals, establishing the improvement interventions (engineering or managerial, often both), and checking the success of the ergonomic improvement.

☞☞☞ *Red spots appeared on the skin of flight attendants of a U.S. airline during three months in 1980. One hundred and thirty-two cases were reported, with 91 flight attendants affected, almost all on flights between New York and Miami, and nine of ten on the same type of aircraft. In some cases, a burning feeling, nausea, and headaches were associated with the spots. Fear developed that the spots were caused by bleeding through the skin and might indicate a serious health hazard. An investigation ruled out contaminants, food poisoning, cosmetics used, faulty air conditioning—but detected that flakes of red paint came off the life vests used to demonstrate emergency procedures to the passengers (AIHA Journal, 1991, 42(4), 323–324).* ☜☜☜

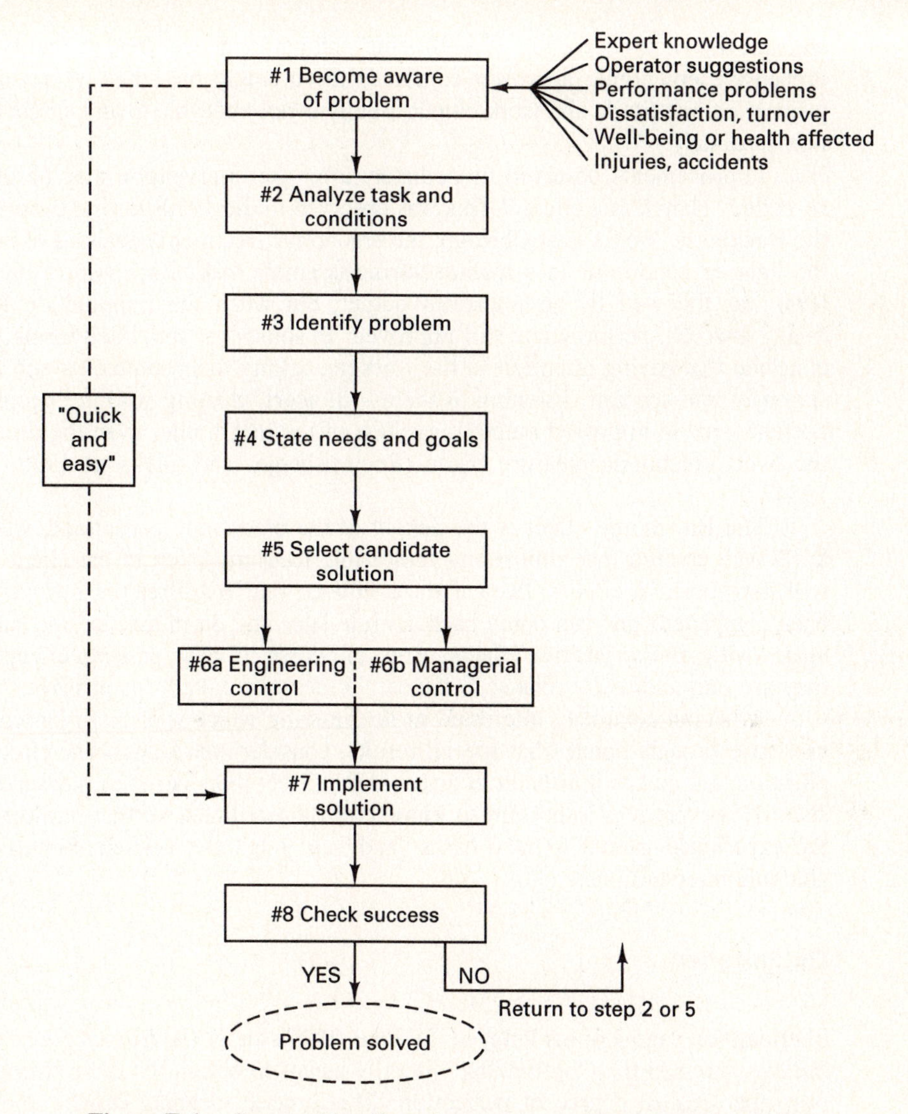

Figure E-1. Steps to identify and preempt or solve ergonomic problems.

A large variety of ergonomic procedures and techniques exist to improve working conditions. Taking computer workstations as an example—as discussed in Chapter 9—one can improve the layout of the room and workstation; improve the furniture (chair, support for the display, support for the input device, support for the source document); improve the lighting conditions and the physical climate; reduce unwanted sound; and consider the need for privacy. For "experimental cleanliness," one would like to introduce each intervention separately and then observe how it changes the person's feeling of well-being and performance. Yet, in reality, a comprehensive ergonomic approach combines these single measures. A resulting improvement in attitude and work output usually is not traceable to any single intervention separately.

Improvements observed immediately after the intervention may be due to the so-called "Hawthorne effect." This was observed in the 1920s during experiments in the Hawthorne Works near Chicago: the ergonomic treatment consisted of improving the lighting conditions in a manufacturing/assembly task. Each rise in the lighting level was followed by an improved output; but when the illumination level was finally lowered, performance still improved. In somewhat simplified terms, one may conclude that paying attention to the workers, taking their comments and activities seriously into account, listening to them—in short, treating working people as important—led to improved output regardless of the magnitude, even the direction, of the overt ergonomic measure taken (Roethlisberger and Dickson, 1939; Parsons, 1984).

The Hawthorne effect is the delight of the ergonomic consultant, who can be fairly well assured that almost any reasonable measure taken in the client's facility will have positive effects. Even if these effects wear out after the intervention has been completed, one can come back a while later and do in essence the same thing again, with similar positive effects. And, of course, to have positive effects, even if they are only indirectly related to the actual measures taken, "is positive."

A person genuinely interested in determining which specific ergonomic measures have brought about what special results, considers the Hawthorne effect a complication that makes it difficult to discover the direct links between measures and results. However, it is important to know these direct links, so that one can transfer the experience gained to new cases, and can judge the cost-effectiveness of ergonomic interventions.

Optimization

Sheridan (on page 4 of his letter in the June 1988 issue of the *Human Factors Society Bulletin)* stressed that "optimizing" literally means to obtain the most favorable solution, the greatest degree of perfection. This precise meaning can be applied to a given mathematical function, or to a (well-defined, small) set of objectives or events with respect to a well-defined criterion. But can we really claim to "optimize" hardware, software, human-machine interaction, or human behavior?

What we can do in ergonomics is to find a good solution, perhaps even an excellent one, within certain practical limitations. With respect to performance expectations, this often means a compromise, such as between system cost and quality. A typically difficult task is the design of a human-machine system for two distinct situations, such as exhaustingly high human performance in an emergency situation and low-stress activities during long-lasting routine operations. The proverbial airline pilot flies for hours in boring monotony which may be interrupted by seconds or minutes of highly exciting demands. Should we "optimize" the cockpit design for those seconds of panicky terror and make it as "good as possible" for routine operation?

Measuring Cost and Success

In industry, the effects of ergonomic work have been measured commonly in two categories: one is the "objective" measure of performance, work output, productivity. The other is the assessment of "subjective" attitudes regarding well-being, satisfaction, cooperative spirit, social interaction, etc., of the persons involved. For the assessment of attitudes, a large variety of tools and procedures is at hand, such as discussed by Muchinsky (1988).

The assessment of work output, performance, and productivity as a result of ergonomic intervention can be a fairly difficult task. In many work situations, the output depends not only on the person, or a team of persons, but also on circumstances over which one has little or no control, such as working at a speed that is determined by a machine, by the movement of an assembly line, or by the number of customers waiting. In such conditions, ergonomic improvements of the working situation are not likely to result in increased performance, measured in work units done per time. However, human-factors improvements may result in improved quality of work, even if the tempo is machine-paced, or in less strain on the person and in better attitude and well-being.

One common measure relies on productivity, which is defined as output, measured in quantity; divided by the input or effort that makes the output possible, also measured in quantity. Often, productivity is expressed in money values, meaning that both the input and the output are assessed in monetary units. The similar cost-benefit analysis may be a useful tool, but a comparison of costs and benefits is beset by serious methodological difficulties, which may require the analyst to make value-laden assumptions. Costs, like input, are easier to value quantitatively than benefits such as improved health, quality of life, and positive economic side effects that may defy accurate estimation. For example, the art of estimating the number of cancers or occupational diseases prevented, or of avoided injuries, is in its infancy. Even if we could accurately estimate the amount of disease or injury prevented, the task of expressing health benefits in terms of money is nearly impossible without making assumptions and setting boundary conditions that may be misleading or erroneous. For example, health benefits may accrue far into the future, and a human life or a lost limb do not have an established market value. One needs to consider such factors as how serious the hazard is, who the recipients of the costs and benefits are, and how informed and voluntary the risk assumption is (Ashford, 1982).

For these and other reasons it is nearly nonsensical to try to evaluate a change in social welfare in terms of dollars as a benefit-to-cost ratio, such as the number of fatalities prevented or injuries avoided, per dollar expenditure. While "cost-benefit," "economic efficiency," "cost-effectiveness" analyses, and other approaches indicate attempts at rational decision-making, these are much more complex than is often naively suggested.

RESEARCH AND PRACTICE

For the sake of argument let us assume there are indeed two extremes: (1) "pure" theory and related research; and (2) its opposite, "practice," the application of design solutions in the real world. (In fact, there is a large overlap on this *continuum* between the two opposites in which research concerns applications, and in which applications involve research.) The arguments have been going forth and back on the respective values of either research or application, and regarding whether and how research should be performed with an eye on application; or whether the designer and engineer should be expected to understand and evaluate theoretical data. *Bon mots* have been coined to describe the situation: for example, "A good theory assures applicability" or, "Behavioral research does not adequately support those who practice human factors." Thoughts, arguments, accusations, and defenses have been summarized and discussed by Boff (1988) and Chapanis (1991).

Use of New Information

The *usability* of research information depends, to a large extent, on the perception of its use value by the potential user. Research data can guide, confirm, deny, or inspire ideas and candidate design solutions ("informal hypotheses") conceived by the practitioner.

If the engineer or manager believes that research data are of value, then they will be sought, considered, and probably used.

Pressures of limited time and resources drive many designers to bias their decisions toward reduction of uncertainty and risk and toward the fastest solution. Hence, many "new" designs only adapt a previous design to a changed requirement, but are not original, i.e., do not demonstrate a new approach or technology. One takes a risk in deciding either to use new information to generate a new design, or not to search for new data (or not to use new data at hand) but to stay with a "proven" design. Calculation of that risk depends upon the perceived relevance of available scientific information; upon the designer's willingness to undergo the effort of obtaining and adapting it; and upon the consequences of an incorrect solution to the design problem. Uncertainty can be resolved by testing and evaluating a prototype—but only after it has been designed.

According to Boff, use of research information depends on its accessibility, interpretability, and applicability.

Accessibility refers to the ease, or difficulty, of searching for information and obtaining it for use. Where to look for the needed information (with the need often not yet clearly defined)? How to obtain it, such as from an obscure thesis or industry or military report? How to use it if declared confidential?

Interpretability refers to the ability of the nonspecialist to understand, evaluate, and decide to apply the information to the design project. Making the judgment about applying the data, or not, often requires considerable understanding of underlying theory and of complicated language, on the side of the applier. How to understand information if it is written in a foreign language, or presented in a difficult format?

Applicability involves risky judgment by the designer regarding the relevance, reliability, and generalizability of (or: ability to reduce and specify) theoretical information to the conditions of the application case. The problem is threefold, at least. First, research may address problems that have only some overlap with the practical question. Second, most experimental data are obtained under laboratory conditions which rigorously control experimental variables and exclude confounding influences in such a manner that the laboratory work becomes "unrealistic." Third, theoretical data may just be too complex, "too difficult" to use.

Theory Versus Practice?

With theory and application at the far ends of the conceptual continuum, it has often been argued that basic research is more easily generalizable than applied research. Supposedly the latter is more limited to specific practical conditions, while basic research provides general information which can apply to a wide range of applications. By improper or imprecise selection of key terms, one is often lured into believing that research results have a wide application. For example, research is often labeled by words referring to a general taxonomy, such as fatigue, noise, vigilance, stress, workload, etc. Yet such terms describe wide "conceptual variables" but do not identify specific independent or dependent variables researched.

Chapanis (1991) argues that even so-called basic studies are specific in their experimental design, in experimental procedures, and in selection and use of subjects so that the results of that research are indeed specific, and hence cannot be applied to a wide range of related topics. For example, studies concerning "noise" need to identify sound intensity, frequency, time duration, and combinations thereof. Likewise, research concerning "mental workload" considers such measurement variables as heart rate and eye blinks, tapping regularity and time estimation, and subjective ratings. All of these are, or may be, measures of certain aspects of workload, but they do not define workload, nor can the results of these studies be generalized to describe "the" workload.

Returning to the earlier assumed simple dichotomy of theory and research on one side, and practice and application on the other: obviously, research and theory usually do not provide "recipes for designers and managers." Of course, the intermediate approach of doing research to support design solutions (applied research) can provide such direct design guidance. But this guidance then applies, usually, only to the given problem. It is not generalizable, often not even remotely applicable, to other design challenges.

The Bridge

What is the solution? Obviously, what the practitioner must know today is often quite different from what the theoretician thinks should be researched tomorrow. Professional conferences, journals, textbooks, handbooks, computer models, or standards can help to bridge the gap. That bridge becomes wide and easily traveled if the researcher learns to express the findings in "usable language" and if the user learns to understand what the researcher says.

This book tries to provide such a bridge.

PART THREE: FURTHER INFORMATION

References

ACGIH (ed.). (1989). *Threshold Limit Values for 1980–90*. Cincinnati, OH: American Conference of Governmental Industrial Hygienists.

ACKERMAN, D. (1990). *A Natural History of the Senses*. New York: Random House.

ADAMS, M. A., and HUTTON, W. C. (1985). The Effect of Posture on the Lumbar Spine. *Journal of Bone and Joint Surgery, 67-B*(4), 625–629.

ADAMS, M. A., and HUTTON, W. C. (1986). Has the Lumbar Spine a Margin of Safety in Forward Bending? *Clinical Biomechanics, 1,* 3–6.

ADAMS, M. A., DOLAN, P., and HUTTON, W. C. (1987). Diurnal Variations in the Stresses on the Lumbar Spine. *Spine, 12*(2), 130–137.

AGER, C. L., OLIVETT, B. L., and JOHNSON, C. L. (1984). Grasp and Pinch Strength in Children 5 to 12 Years Old. *The American Journal of Occupational Therapy, 38,* 107–113.

AGHAZADEH, F., and AYOUB, M. M. (1985). A Comparison of Dynamic- and Static-Strength Models for Prediction of Lifting Capacity. *Ergonomics, 28,* 1409–1417.

AGHAZADEH, F., and JIANG, B. C. (1987). *Some Considerations in the Use of Isometric, Isoinertial and Isokinetic Strength Models for Predicting Lifting Capability*. Unpublished Report, Lubbock, TX: Texas Tech.

AIRD, J. W., NYRAN, P., and ROBERTS, G. (1988). Comprehensive Back Injury Prevention Program: An Ergonomics Approach for Controlling Back Injuries in Health Care Facilities. In F. Aghazadeh (ed.), *Trends in Ergonomics/Human Factors V* (pp. 705–712). Amsterdam: Elsevier.

AKERBLOM, B. (1948). *Standing and Sitting Posture*. Stockholm: Nordiska Bokhandeln.

ALDEN, D. G., DANIELS, R. W., and KANARICK, A. F. (1972). Keyboard Design and Operation: A Review of the Major Issues. *Human Factors, 14*, 275–293.

ALFREDSSON, L., AKERSTEDT, T., MATTSSON, M., and WILBORG, B. (1991). Self-Reported Health and Well-Being Amongst Night Security Guards: A Comparison With the Working Population. *Ergonomics, 34*(5), 525–530.

AL-HABOUBI, M. (1991). Anthropometric Study for the User Population in Saudi Arabia. In *Proceedings, 11th Congress of the International Ergonomics Association* (pp. 891–893). London: Taylor and Francis.

AMAR, J. and DUNOD, H. (1917). *Organization Physiologique du Travail*. Paris: Dumont.

AMOORE, J. (1970). *Molecular Basis of Odor*. Springfield, IL: Thomas.

ANDERSON, C. K. (1989). Strength and Endurance Testing for Pre-Employment Placement. In K. H. E. Kroemer, J. D. McGlothlin, and T. G. Bobick (eds.), *Manual Material Handling: Understanding and Preventing Back Trauma*. Akron, OH: American Industrial Hygiene Association.

ANDERSON, C. K., and CATTERALL, M. J. (1987). The Impact of Physical Ability Testing on Incidence Rate, Severity Rate, and Productivity. In S. S. Asfour (ed.), *Trends in Ergonomics/Human Factors IV*, (pp. 577–584). Amsterdam: Elsevier.

ANDERSON, C. K., and CHAFFIN, D. B. (1986). A Biomechanical Evaluation of Five Lifting Techniques. *Applied Ergonomics, 17*, 2–8.

ANDERSON, N. S. (1987). Cognition, Learning and Memory. Chapter 3 in M. A. Baker (ed.), *Sex Differences in Human Performance* (pp. 37–54). Chichester, UK: Wiley.

ANDERSSON, G. B. J. (1991). Low Back Pain. In *Proceedings, Occupational Ergonomics: Work Related Upper Limb and Back* Disorders (not paginated). San Diego, CA: American Industrial Hygiene Association, San Diego Section.

ANDERSSON, G. B. J., SCHULTZ, A. B., and ORTENGREN, R. (1986). Trunk Muscle Forces During Desk Work. *Ergonomics, 29*, 1113–1127.

ANDERSSON, K., KARLEHAGEN, S., and JONSSON, B. (1987). The Importance of Variations in Questionnaire Administration. *Applied Ergonomics, 18*, 229–232.

ANDRES, R. (1985). Impact of Age on Weight Goals. In *Proceedings, NIH Consensus Development Conference* (pp. 77–81). Beshesda, MD: National Institutes of Health.

ANDREWS, A. W. (1991). Hand-held Dynamometry for Measuring Muscle Strengths. *Journal of Human Muscle Performance, 1*, 35–50.

ANNIS, J. F., CASE, H. W., CLAUSER, C. E., and BRADTMILLER, B. (1991). Anthropometry of an Aging Work Force. *Experimental Aging Research, 17*, 157–176.

ANSON, J. G. (1982). Memory Drum Theory: Alternative Test and Explanations for the Complexity Effects on Simple Reaction Time. *Journal of Motor Behavior, 4*, 228–246.

Anthropology Research Project Inc., (ed.) (1985). *An Examination of Sample Specificity and Sub-Grouping in Regression Analyses*. Appendix S (Final Report, USAF Contract F33615-82-C-0510). Yellow Springs, OH: Author.

ANTONOVSKY, A. (1979). *Health, Stress, and Coping*. San Francisco, CA: Jossey-Bass.

APTEL, M. (1988). Comparison Between Required Clothing Insulation and That Actually Worn by Workers Exposed to Artificial Cold. *Ergonomics, 19*, 301–305.

ARMSTRONG, L. E. (1988). The Impact of Hyperthermia and Hypohydration on Circulation, Strength, Endurance and Health. *Journal of Applied Sport Science Research, 2*, 60–65.

ARMSTRONG, T. J. (1983). *An Ergonomics Guide to Carpal Tunnel Syndrome*. Akron, OH: American Industrial Hygiene Association.

ARMSTRONG, T. J. (1991). Work Related Cumulative Trauma Disorders. In *Proceedings, Occupational Ergonomics* (not paginated). San Diego, CA: American Industrial Hygiene Association, San Diego Section.

ARMSTRONG, T. J. and CHAFFIN, D. B. (1979). Some Biomechanical Aspects of the Carpal Tunnel. *J. Biomechanics, 12*, 567–570.

ARMSTRONG, T. J., WERNER, R. A., WARING, W. P., and FOULKE, J. A. (1991). Intra-carpal Canal Pressure in Selected Hand Tasks. In *Proceedings of the 11th Congress of the International Ergonomics Association* (pp. 156–158). London, UK: Taylor and Francis.

ARNAUT, L. Y., and GREENSTEIN, J. S. (1990). Is Display/Control Gain a Useful Metric for Optimizing an Interface? *Human Factors, 32*, 651–663.

ASCHOFF, H. (1981). *Handbook of Behavioral Neurobiology* (Vol. 4). New York, NY: Plenum.

ASFOUR, S. S., GENAIDY, A. M., and MITAL, A. (1988). Physiological Guidelines for the Design of Manual Lifting and Lowering Tasks: The State of the Art. *American Industrial Hygiene Association Journal, 49*, 150–160.

ASHFORD, N. A. (1982). Cost–Benefit Analysis: Can Balance Be Achieved? *Occupational Safety and Health, 51*, 10–12, 42, 44, 45.

ASHLEY-MONTAGU, M. F. (1960). *An Introduction to Physical Anthropology* (3d ed.). Springfield, IL: Thomas.

ASHRAE (ed.). (1981). *Thermal Environmental Conditions for Human Occupancy*. ANSI-ASHRAE Standard 55. Atlanta, GA: American Society of Heating, Refrigerating, and Air-Conditioning Engineers.

ASHRAE (ed.). (1985). *Physiological Principles for Comfort and Health*. Chapter 8 in 1985 Fundamentals Handbook. Atlanta, GA: American Society of Heating, Refrigerating, and Air-Conditioning Engineers.

ASIMOV, I. (1989). *Asimov's Chronology of Science and Discovery*. New York: Harper & Row.

ASPDEN, R. M. (1988). A New Mathematical Model of the Spine and its Relationship to Spinal Loading in the Workplace. *Applied Ergonomics, 19*, 319–323.

ASSO, D. (1987). Cyclical Variations. Chapter 4 in M. A. Baker (ed.), *Sex Differences in Human Performance* (pp. 55–80). Chichester, UK: Wiley.

ASTRAND, P. O., and RODAHL, K. (1977). *Textbook of Work Physiology* (2d ed.). New York: McGraw-Hill.

ASTRAND, P. O., and RODAHL, L. (1986). *Textbook of Work Physiology* (3d ed.). New York: McGraw-Hill.

AUNE, I. A., and JUERGENS, H. W. (1989). *Computer Models of the Human Body* (in German). (Ergonomics Studies, Report #27). Koblentz, Germany: Federal Office for Defense Technology and Acquisition.

AVOGARO, P. (1990). Alcohol: A Risk or Protective Factor in Aging. In F. Fabris, L. Pernigotti, and A. Ferrario (eds.), *Sedentary Life and Nutrition* (pp. 204–289). New York: Raven.

AYOUB, M. A. (1982). The Manual Lifting Problem: The Illusive Solution. *Journal of Occupational Accidents, 4*, 1–23.

AYOUB, M. A. and WITTELS, N. E. (1989). Cumulative Trauma Disorders. *International Review of Ergonomics, 2*, 217–272.

AYOUB, M. M. (1973). Workplace Design and Posture. *Human Factors, 15*, 265–268.

AYOUB, M. M. (1991). Determining Permissible Lifting Loads: An Approach. In *Proceedings of the 35th Annual Meeting* (pp. 825–829). Santa Monica, CA: Human Factors Society.

AYOUB, M. M., and KIM, H. K. (1989). A Look at the Approaches to Determine Lifting Capacity. In K. H. E. Kroemer, J. D. McGlothlin, and T. G. Bobick (eds.), *Manual Material Handling: Understanding and Preventing Back Trauma* (pp. 79–85). Akron, OH: American Industrial Hygiene Association.

AYOUB, M. M., and MILLER, M. (1991). Industrial Workplace Design. In A. Mital and W. Karwowski (eds.), *Workspace, Equipment and Tool Design* (pp. 67–92). Amsterdam, NL: Elsevier.

AYOUB M. M., and MITAL, A. (1989). *Manual Materials Handling*. London, UK: Taylor and Francis.

AYOUB, M. M., and SMITH, J. L. (1988). Manual Materials Handling in Unusual Postures: Carrying of Loads. In *Proceedings of the Human Factors Society 32nd Annual Meeting* (pp. 675–679). Santa Monica, CA: Human Factors Society.

AYOUB, M. M., JIANG, B. C., SMITH, J. L., SELAN, J. L., and MCDANIEL, J. W. (1987). Establishing a Physical Criterion for Assigning Personnel to US Air Force Jobs. *American Industrial Hygiene Association Journal, 45*, 464–470.

AYOUB, M. M., MITAL, A., ASFOUR, S. S., and BETHEA, N. J. (1980). Review, Evaluation, and Comparison of Models for Predicting Lifting Capacity. *Human Factors, 22*, 257–269.

AYOUB, M. M., SELAN, J. L., and JIANG, B. C. (1984). *A Mini-Guide for Manual Materials Handling*. Institute of Ergonomics Research. Lubbock, TX: Texas Tech University.

AYOUB, M. M., SELAN, J. L., and LILES, D. H. (1983). An Ergonomics Approach For the Design of Manual Materials Handling Tasks. *Human Factors, 25*, 507–515.

AYOUB, M. M., SELAN, J. L., and JIANG, B. C. (1986). Manual Materials Handling. Chapter 7.2 in G. Salvendy (ed.), *Handbook of Human Factors* (pp. 790–818). New York: Wiley.

Back Pain Monitor (1984). *Wausau Insurance Fights Client Back Injury Problems With Simple Program*. October Issue, 136–138.

BAKER, M. A. (ed.). (1987). *Sex Differences in Human Performance*. Chichester, UK: Wiley.

BALLANTINE, M. (1983). Well, How Do Children Learn Population Stereotypes? In *Proceedings of the Ergonomics Society's Conference* (p. 1). London, UK: Taylor and Francis.

BALOGUN, J. A. (1986). Optimal Rate of Work During Load Transportation on the Head and by Yoke. *Industrial Health, 24*, 75–86.

BAMMER, G., and MARTIN, B. (1988). The Arguments About RSI: An Examination. *Community Health Studies, 12*, 348–358.

BARLOW, A. M., and BRAID, S. J. (1990). Foot Problems in the Elderly. *Clinical Rehabilitation, 4*, 217–222.

BARNES, R. M. (1963). *Motion and Time Study* (5th ed.). New York: Wiley.

BARR, R. A., and EBERHARD, J. W. (eds.). (1991). Safety and Mobility of Elderly Drivers, Part 1. Special Issue of *Human Factors, 33*, 497–600.

BARTOL, A. M., HAZEN, V. L., KOWALSKI, J. F., MURPHY, B. P., and WHITE, R. P. (1990). *Advanced Dynamic Anthropometric Manikin (ADAM) Design Report* (Report AAMRL-TR-90-023). Wright-Patterson AFB, OH: Aerospace Medical Research Laboratory.

BASMAJIAN, J. V., and DELUCA, C. J. (1985). *Muscles Alive* (5th ed.). Baltimore, MD: Williams & Wilkins.

BATAVIA, A. I., and HAMMER, G. S. (1990). Toward the Development of Comsumer-based Criteria for the Evaluation of Assistive Devices. *Journal of Rehabilitation Research and Development, 27*, 425–436.

BATTIE, M. C., BIGOS, S. J., FISHER, L., HANSSON, T. H., JONES, M. E., and WORTLEY, M. D. (1989). Isometric Lifting Strength as a Predictor of Industrial Back Pain Reports. *Spine, 14*, 851–856.

BAUGHMAN, L. D. (1982). *Segmentation and Analysis of Stereophotometric Body Surface Data* (AFAMRL-TR-81-96). Wright-Patterson Air Force Base, OH: Air Force Aerospace Medical Research Laboratory.

BEECHER, R. M. (1986). Computer Graphics and Shape Diagnostics. In *Proceedings of the Human Factors Society Annual Meeting* (pp. 211–215). Santa Monica, CA: Human Factors Society.

BENDIX, T. (1991). Significance of Seated-Workplace Adjustments. In W. Karwowski and J. W. Yates (eds.), *Advances in Industrial Ergonomics and Safety III* (pp. 343–350). London, UK: Taylor and Francis.

BENDIX, T., WINKEL, J., and JESSEN, F. (1985). Comparison of Office Chairs with Fixed Forwards or Backwards Inclining, or Tiltable Seats. *European Journal of Applied Physiology and Occupational Physiology, 54*, 78–385.

BENEDICT, F. C., and CATHCARD, E. F. (1913). *Muscular Work, A Metabolic Study with Special Reference to the Efficency of the Human Body as a Machine*. Washington, D.C.: The Carnegie Institution of Washington.

BEN-SHAKHAR, G., and FUREDY, J. J. (1990). *Theories and Applications in the Detection of Deception*. New York: Springer.

BERANEK, L., BLAZIER, W., and FIGWER, J. (1971). Preferred Noise Criteria (PNC) Curves and Their Application to Rooms. *Journal Acoustical Society of America, 50*, 1223–1228.

BERG, V. J. (1987). *Review of the "Training" Literature on Manual Material Handling*. Unpublished Report, Industrial Ergonomics Laboratory, IEOR Department, Blacksburg, VA: Virginia Tech (VPI & SU).

BERGER, E. H., and CASALI, J. G. (1992, in press). Hearing Protection Devices. Chapter 8 in M. J. Crocker (ed.), *Handbook of Acoustics*. New York: Wiley.

BERGQUIST-ULLMAN, M., and LARSSON, U. (1977). Acute Low Back Pain in Industry. *Acta Orthopadica Scandinavica*, Supplement No. 170.

BERNS, T. A. R., and MILNER, N. P. (1980). TRAM: *A Technique for the Recording and Analysis of Moving Work Posture* (Report 80:23, pages 22–26). Stockholm, Sweden: Ergolab.

BESL, P., and JAIN, R. (1984). *Surface Characterization for Three-Dimensional Object Recognition in Depth Maps* (Center for Research on Integrated Manufacturing Report RSD-TR-20-84.) Ann Arbor, MI: University of Michigan.

BESL, P., and JAIN, R. (1985). Intrinsic and Extrinsic Surface Characteristics. In *IEEE Proceedings on Computer Vision and Pattern Recognition* (pp. 226–725). San Francisco, CA: Institute of Electrical and Electronics Engineers.

BIERING-SORENSEN, F., and THOMSEN, C. (1986). Medical, Social and Occupational History as Risk Indicators for Low-Back Trouble in a General Population. *Spine, 11*, 720–725.

BIGOS, S. J., SPENGLER, D. M., MARTIN, N. A., ZEH, J., FISHER, L., and NACHEMSON, A. (1986). Back Injuries in Industry: A Retrospective Study, III. Employee-related factors. *Spine, 11*, 252–256.

BIRCH, J. B. (1980). Some Convergence Properties of Iterated Re-weighted Least Squares in the Location Model. *Commun. Stat.*, B9, 359–369.

BIRKBECK, M. Q., and BEER, T. C. (1975). Occupation in Relation to the Carpal Tunnel Syndrome. *Rheumatology and Rehabilitation, 14*, 218–221.

BISHU, R. R., and DEEB, J. M. (1989). Optimum Handle Positions in a Lifting and Carrying Task. In A. Mital (ed.), *Advances in Industrial Ergonomics and Safety I* (pp. 707–714). London, UK: Taylor and Francis.

BLAIR, B. (1991). Evaluation of a Maldivian Chair. *Applied Ergonomics, 22*, 125–127.

BLANCHARD, B. S., and FABRYCKY, W. J. (1990). *Systems Engineering and Analysis*. Englewood Cliffs, NJ: Prentice Hall.

BLINKHORN, S. (1988). Lie Detection as a Psychometric Procedure. Chapter 3 in A. Gale (ed.), *The Polygraph Test. Lies, Truth, and Science* (pp. 29–39). London, UK: Sage.

BLUM, A. (1990). More Repetitive-Strain Injury Suits Predicted. *The National Law Journal* (July 23), 14.

BLUMENBACH, J. F. (1825). *A Manual of the Elements of Natural History*. London, UK: Simpkin and Marshall.

BOBICK, T. G., and GUTMAN, S. H. (1989). Reducing Musculo-Skeletal Injuries by Using Mechanical Handling Equipment. In K. H. E. Kroemer, J. D. McGlothlin, and T. G. Bobick (eds.), *Manual Material Handling: Understanding and Preventing Back Trauma* (pp. 87–96). Akron, OH: American Industrial Hygiene Association.

BOFF, K. R. (1988). The Value of Research Is in the Eye of the Beholder. *Human Factors Society Bulletin, 31*, 1–4.

BOFF, K. R., KAUFMAN, L., and THOMAS, J. P. (eds.). (1986). *Handbook of Perception and Human Performance*. New York: Wiley.

BOFF, K. R., and LINCOLN, J. E. (eds.) (1988). *Engineering Data Compendium: Human Perception and Performance*. Wright-Patterson AFB, OH: Armstrong Aerospace Medical Research Laboratory.

BOHANNON, R. W. (1991). Research Relevant to Human Muscle Performance. *Journal of Human Muscle Performance, 1*, 51–72.

BOHLE, P., and TILLEY, A. J. (1989). The Impact of Night Work on Psychological Well-Being. *Ergonomics, 34*, 1089–1099.

BORG, G. A. V. (1962). *Physical Performance and Perceived Exertion*. Lund, Sweden: Gleerups.

BORG, G. A. V. (1982). Psychophysical Bases of Perceived Exertion. *Medicine and Science in Sports and Exercise, 14*, 377–381.

BORRELLI, G. A. (1680). *De motu animalium*. Lugduni Batavorum.

BOTTOMS, D. J., and BUTTERWORTH, D. J. (1990). Foot Reach Under Guard Rails in Agricultural Machinery. *Applied Ergonomics, 21*, 179–186.

BOUDRIFA, H., and DAVIES, B. T. (1984). The Effect of Backrest Inclination, Lumbar Support and Thoracic Support on the Intra-Abdominal Pressure While Lifting. *Ergonomics, 27*, 379–387.

BOUSSENA, M., CORLETT, E. N., and PHEASANT, S. P. (1982). The Relation Between Discomfort and Postural Loading of the Joints. *Ergonomics, 25*, 315–322.

BOWMAN, O. J., and KATZ, B. (1984). Hand Strength and Prone Extension in Right-Dominant, 6 to 9 Year Olds. *The American Journal of Occupational Therapy, 38*, 367–376.

BOYD, P. R. (1982). *Human Factors in Lighting*. New York: MacMillan.

BRADFORD, P. and PRETE, B. (eds.). (1978). *Chair*. New York: Crowell.

BRAIN, W., WRIGHT, A., and WILKINSON, M. (1947). Spontaneous Compression of the Median Nerve in the Carpal Tunnel. *Lancet, 1*, 277–282.

BRANTON, P. (1984). Backshapes of Seated Persons—How Close Can the Interface Be Designed? *Applied Ergonomics, 15*, 105–107.

BRAUNE, W., and FISCHER, O. (1889). *The Center of Gravity of the Human Body as Related to the Equipment of the German Infantryman*. (In German.) (Abh. d. Math. Phys. Cl. d. k. Saechs. Gesell. d. Wissenschaften Leipzig, 1889, 15, pp. 561.) Translation in Human Mechanics (AMRL-TDR-63-123). Wright-Patterson AFB, OH: Aerospace Medical Research Laboratory, 1963.

BRENNAN, R. B. (1987). Trencher Operator Seating Positions. *Applied Ergonomics, 18*, 95–102.

BRESNAHAN, T. (1985). The Hazard Association Value of Safety Figures. *Professional Safety*, July issue, 27–31.

BRIDGER, R. S., EISENHART-ROTHE, C. V., and HENNEBERG, M. (1989). Effects of Seat Slope and Hip Flexion on Spinal Angles in Sitting. *Human Factors, 31*, 679–688.

BROUHA, L. (1967). *Physiology in Industry* (2d ed.). Riverside, NJ: Pergamon Press.

BROWN, J. R. (1972). *Manual Lifting and Related Fields: An Annotated Bibliography*. Toronto: Labour Safety Council of Ontario, Ontario Ministry of Labour.

BROWN, J. R. (1975). Factors Contributing to the Development of Low-Back Pain in Industrial Workers. *American Industrial Hygiene Association Journal, 36*, 26–31.

BRUNSWICK, E. (1956). *Perception and the Representative Design of Experiments* (2d ed.). Berkley, CA: University of California Press.

BRYANT, J. T., STEVENSON, J. M., FRENCH, S. L., GREENHORN, D. R., ANDREW, G. M., and DEAKIN, J. M. (1990). Four Factor Model to Describe an Isoinertial Lift. *Ergonomics, 33*, 173–186.

BUCHHOLZ, B., FREDERICK, L. J., and ARMSTRONG, T. J. (1988). An Investigation of Human Palmar Skin Friction and the Effects of Materials, Pinch Force Moisture. *Ergonomics, 31*, 317–325.

BUCHNER, H. J. (1983). *Finger Movement in the Sagittal Plane and the Mechanisms of Touch Control*. Unpublished MS thesis, The Ohio State University, Columbus, OH.

BUCKLE, P. W., and DAVID, G. C. (1989). Development of Anthropometric Selection Criteria for Airline Cabin Crew. In E. D. Megaw (ed.), *Contemporary Ergonomics* (pp. 320–325). London, UK: Taylor and Francis.

BUCKLE, P. W., DAVID, G. C., and KIMBER, A. C. (1990). Flight Deck Design and Pilot Selection: Anthropometric Considerations. *Aviation, Space, and Environmental Medicine, 61*, 1079–1084.

BUIS, N. (1990), Ergonomics, Legislation and Productivity in Manual Materials Handling. *Ergonomics, 33*, 353–359.

BURKE, W. E., TUTTLE, W. W., THOMPSON, C. W., JANNEY, C. D., and WEBER, R. J. (1953). The Relation of Grip Strength and Grip-strength Endurance to Age. *Applied Physiology, 5*, 628–630.

BURNETTE, J. I., and AYOUB, M. A. (1989). Cumulative Trauma Disorders. *Pain Management*, 196–209 and 256–264.

BURNS, M. L., KALEPS, I., and KAZARIAN, L. E. (1984). Analysis of Compressive Creep Behavior of the Vertebral Unit Subjected to a Uniform Axial Loading Using Exact Parametric Solution Equations of Kelvin-solid Models—Part I. Human Intervertebral Joints. *Journal Biomechanics, 17*, 113–130.

BURSE, R. L. (1979). Sex Differences in Human Thermoregulatory Responses to Heat and Cold Stress. *Human Factors, 21*, 687–699.

BURTON, K. (1991). Measuring Flexibility. *Applied Ergonomics, 22*, 303–307.

BUSH-JOSEPH, C., SCHIPPLEIN, O., ANDERSSON, G. B. J., and ANDRIACCHI, T. P. (1988). Influence of Dynamic Factors on the Lumbar Spine Moment in Lifting. *Ergonomics, 31*, 211–216.

BYRNES, W. C., and CLARKSON, P. M. (1986). Delayed Onset Muscle Soreness and Training. *Clinics in Sports Medicine, 5*, 605–614.

CADY, L. D., BISCHOFF, D. P., O'CONNELL, E., THOMAS, P. C., and ALLAN, J. (1979a). Strength and Fitness and Subsequent Back Injuries in Fire Fighters. *Journal of Occupational Medicine, 21*, 269–272.

CADY, L. D., BISCHOFF, D. P., O'CONNELL, E. R., THOMAS, P. C., and ALLAN, J. H. (1979b). Letters to Editor: Authors' Response. *Journal of Occupational Medicine, 21*, 720–725.

CAILLET, R. (1981). Low Back Pain (3d ed.). London: Davis.

CAIN, W. S., LEADERER, B. P., CANNON, L., TOSUN, T., and ISMAIL, H. (1987). Odorization of Inert Gas for Occupational Safety: Psychophysical Considerations. *American Industrial Hygiene Journal, 48*, 47–55.

CALDWELL, L. S., CHAFFIN, D. B., DUKES-DOBOS, F. N., KROEMER, K. H. E., LAUBACH, L. L., SNOOK, S. H., and WASSERMAN, D. E. (1974). A Proposed Standard Procedure for Static Muscle Strength Testing. *American Industrial Hygiene Association Journal, 35*, 201–206.

CAMERON, N. (1979). The Growth of London School Children 1904–1966: An Analysis of Secular Trend and Intra-county Variation. *Annals of Human Biology, 6*, 505–525.

CANNON, L. J., BERNACKI, E. J., and WALTER, S. D. (1981). Personnel and Occupational Factors Associated with Carpal Tunnel Syndrome. *Journal of Occupational Medicine, 23*, 225–258.

CANNON-BOWERS, J. A., and TANNENBAUM, S. I. (1991). Toward an Integration of Training Theory and Technique. *Human Factors, 33*, 281–292.

CAPLE, D. C., and BETTS, N. J. (1991). RSI—Its Rise and Fall in Telecom Australia 1981–1990. In *Proceedings of the 11th Congress of the International Ergonomics Association* (pp. 1037–1039). London: Taylor and Francis.

CARAYON-SAINTFORT, P., SMITH, M. J., and LIM, S. Y. (1991). Comparison of Objective and Subjective Ergonomic Evaluations of Office Environments and Workstations. In *Proceedings of the 11th Congress of the International Ergonomics Association* (pp. 768–770). London, UK: Taylor and Francis.

CARLTON, R. S. (1987). The Effects of Body Mechanics Instruction on Work Performance. *The American Journal of Occupational Therapy, 41*, 16–20.

CARVALHAIS, A. B., TEPAS, D. I., and MAHAN, R. P. (1988). Sleep Duration in Shift Workers. *Sleep Research, 17*, 109–124.

CASALI, J. G. (1989). Multiple Factors Affect Speech Communication in the Workplace. *Occupational Health and Safety*, July issue, 32–36, 37, 40–42.

CASALI, J. G., and FRANK, L. H. (1986). Perceptual Distortion and Its Consequences in Vehicular Simulation: Basic Theory and Incidence of Simulator Sickness. *Transportation Research Record*, 1059, 57–65.

CASALI, J. G., and PARK, M. Y. (1990). Attenuation Performance of Four Hearing Protectors Under Dynamic Movement and Different User Fitting Conditions. *Human Factors, 32*, 9–25.

CASALI, S. P., and WILLIGES, R. C. (1990). Databases of Accommodative Aids for Computer Users with Disabilities. *Human Factors, 32*, 407–422.

CASEY, S. M. (1989). Anthropometry of Farm Equipment Operators. *Human Factors Society Bulletin, 32*, 1–16.

CHAFFIN, D. B. (1981). Functional Assessment for Heavy Physical Labor. *Occupational Health and Safety*, 50, 24, 27, 32, 64.

CHAFFIN, D. B. (1987). Manual Materials Handling and the Biomechanics Bases for Prevention of Low-Back Pain in Industry—An Overview. *American Industrial Hygiene Association Journal, 48*, 989–996.

CHAFFIN, D. B. (1991). Occupational Ergonomics. In *Proceedings, Occupational Ergonomics: Work-Related Upper Limb and Back Disorders* (not paginated). San Diego, CA: American Industrial Hygiene Association, San Diego Section.

CHAFFIN, D. B., and ANDERSSON, G. B. J. (1984). *Occupational Biomechanics*. New York: Wiley.

CHAFFIN, D. B., and ANDERSSON, G. B. J. (1991). *Occupational Biomechanics* (2d ed.). New York: Wiley.

CHAFFIN, D. B., HERRIN, G. D., and KEYSERLING, W. M. (1978). Preemployment Strength Testing: An Updated Position. *Journal of Occupational Medicine, 20*, 403-408.

CHAFFIN, D. B., WOLDSTAD, J. C., and TRUJILLO, A. (1992). Floor/Shoe Slip Resistant Measurement. *American Industrial Hygiene Association J., 53*, 283–289.

CHANDLER, R. F., CLAUSER, C. E., MCCONVILLE, J. R., REYNOLDS, H. M., and YOUNG, J. W. (1975). *Investigation of Inertial Properties of the Human Body* (AMRL-TR-74-137). Wright-Patterson AFB, OH: Aerospace Medical Research Laboratory.

CHAPANIS, A. (ed.). (1975). *Ethnic Variables in Human Factors Engineering*. Baltimore: Johns Hopkins University Press.

CHAPANIS, A. (1991). To Communicate the Human Factors Message, *Human Factors Society Bulletin, 35*, 1–4.

CHARTERIS, J., SCOTT, P. A., and NOTTRODT, J. W. (1989). Metabolic and Kinematic Responses of African Women Headload Carriers Under Controlled Conditions of Load and Speed. *Ergonomics, 32*, 1539–1550.

CHASE, J. D., and CASALI, S. P. (1991). *A Comparison of Three Cursor Control Devices on a Cursor Control Benchmark Task* (Technical Report). Blacksburg, VA: Virginia Polytechnic Institute and State University, Human-Computer Interface Laboratory.

CHATTERJEE, D. S. (1987). Repetition Strain Injury—A Recent Review. *Journal of Society of Occupational Medicine, 37*, 100–105.

CHENOWETH, D. (1983a). Fitness Program Evaluation: Results with Muscle. *Occupational Health and Safety, 52*, 14–17 and 40–42.

CHENOWETH, D. (1983b). Health Promotion: Benefit vs. Cost. *Occupational Health and Safety, 52*, 37–41.

CHERRY, N. (1987). Physical Demands of Work and Health Complaints Among Women Working Late in Pregnancy. *Ergonomics, 30*, 689–701.

CHEVERUD, J., GORDON, C. C., WALKER, R. A., JACQUISH, C., KOHN, L., MOORE, A., and YAMASHITA, N. (1990). *1988 Anthropometric Survey of U.S. Army Personnel* (Technical Reports 90/031 through 036). Natick, MA: U.S. Army Natick Research, Development, and Engineering Center.

CHOMCHERNGPAT, C., MANDHANI, P., LUM, C., and MARTIN, C. (1989). Human Lifting Strength in Different Postures. In *Proceedings of the Human Factors Society 33rd Annual Meeting* (pp. 745–749). Santa Monica, CA: Human Factors Society.

CHRISTENSEN, C., CASALI, J. G., and KROEMER, K. H. E. (1984). A Methodology for Work Chair Assessment. In *Proceedings of the Human Factors Society 28th Annual Meeting* (pp. 832–835). Santa Monica, CA: Human Factors Society.

CHRISTENSEN, J. M. and TALBOT, J. M. (1986). Psychological Aspects of Space Flight. *Aviation, Space, and Environmental Medicine, 57*, 203–212.

CHRISTENSEN, J. M., TOPMILLER, D. A., and GILL, R. T. (1988). Human Factors Definitions Revisited. *Human Factors Society Bulletin, 31*, 7–8.

CHURCHILL, E. (1978). Statistical Considerations in Man-Machine Designs. Chapter IX in NASA-Webb (eds.), *Anthropometric Source Book*. Vol. 1, NASA Reference Publication 1024. (pp. IX-1–IX-62). Houston, TX: NASA.

CHURCHILL, E., CHURCHILL, T., and KIKTA, P. (1978). *Intercorrelations of Anthropometric Measurements: A Source Book for USA Data* (AMRL-TR-77-1). Wright-Patterson AFB, OH: Aerospace Medical Research Laboratory.

CIE (1951). *CIE Proceedings*, Volume 3. Paris: Bureau Central de la Commission Internationale de l'Eclairage.

CIRIELLO, V. M., and SNOOK, S. H. (1983). A Study of Size, Distance, Height, and Frequency Effects on Manual Handling Tasks. *Human Factors, 25*, 473–483.

CIRIELLO, V. M., SNOOK, S. H., BLICK, A. C., and WILKINSON, P. L. (1990). The Effects of Task Duration on Psychophysically-Determined Maximum Acceptable Weights and Forces. *Ergonomics, 33*, 187–200.

CLARK, F. J., and HORSCH, K. W. (1986). Kinesthesia. Chapter 13 in K. R. Boff, L. Kaufman, and J. P. Thomas, (eds.), *Handbook of Perception and Human Performance* (pp. 13.1–13.67). New York: Wiley.

CLARK, M. C., CZAJA, S. J., and WEBER, R. A. (1990). Older Adults and Daily Living Task Profiles. *Human Factors 32*, 537–549.

CLAYTON, G. D., and CLAYTON, F. E. (eds.) (1991). *Patty's Industrial Hygiene and Toxicology* (4th ed.). New York: Wiley.

COBLENTZ, A., IGNAZI, G., and MOLLARD, R. (1986). ERGODATA. *American Journal of Physiol. Anthropology, 69*, 188.

COBLENTZ, A., MOLLARD, R., and IGNAZI, G. (1991). Three-dimensional Face Shape Analysis of French Adults, and Its Application to the Design of Protective Equipment. *Ergonomics, 34*, 497–517.

COE, J. E., and FISHER, L. (1990). Occupational Carpal Tunnel Syndrome: Clinical Characteristics of an Affected Population and Utility of Screening Techniques. In *Abstracts, International Conference on Occupational Musculo-Skeletal Disorders and Prevention of Low Back Pain* (p. 23). Milan, Italy: University of Milan, Institute of Occupational Medicine.

COLLIGAN, M. J., and TEPAS, D. I. (1986). The Stress of Hours of Work. *Journal of the American Industrial Hygiene Association, 47*, 686–695.

COLLINS, M., BROWN, B., BOWMAN, K., and CARKEET, A. (1990). Workstation Variables and Visual Discomfort Associated with VDTs. *Applied Ergonomics, 21*, 157–161.

COLOMBINI, D., OCCHIPINTI, E., MOLTENI, G., GRIECO, A., PEDOTTI, A., BOCCARDI, S., FRIGO, C., and MENONI, O. (1985). Posture Analysis. *Ergonomics, 28*, 275–284.

COLQUHOUN, W. P. (1985). Hours of Work at Sea: Watch-keeping Schedules, Circadian Rhythms, and Efficiency. *Ergonomics, 28*, 637–653.

Committee on an Aging Society (ed.). (1988). *The Social and Built Environment in an Older Society*. Washington, DC: National Academy Press.

Committee on Government Operations (1991). *Confronting Repetitive Motion Illnesses in the Workplace*. Hearing Before the Employment and Housing Subcommittee (Chairman, Hon. Tom Lantos) of the Committee on Government Operations, House of Representatives, on March 28, 1992. Washington, DC: U.S. Government Printing Office.

Committee on Vision (ed.). (1987). *Work, Aging, and Vision*. Washington, DC: National Academy Press.

CONGLETON, J. J., AYOUB, M. M., and SMITH, J. L. (1985). The Design and Evaluation of the Neutral Posture Chair for Surgeons. *Human Factors, 27*, 589–600.

CONIGLIO, I., FUBINI, E., MASALI, M., MASIERO, C., PIERLORENZI, G., and SAGONE, G. (1991). Anthropometric Survey of Italian Population for Standardization in Ergonomics. In *Proceedings, 11th Congress of the International Ergonomics Association* (pp. 894–896). London, UK: Taylor and Francis.

CONRAD, R., and HULL, A. J. (1968). The Preferred Layout for Numerical Data Entry Key Sets. *Ergonomics, 11*, 165–173.

CONWAY, K., and UNGER, R. (1991). Ergonomic Guidelines for Designing and Maintaining Underground Coal Mining Equipment. In A. Mital and W. Karwowski (eds.), *Workspace, Equipment, and Tool Design* (pp. 279–302). Amsterdam: Elsevier.

COOK, T. D., and CAMPBELL, D. T. (1979). *Quasi-Experimentation: Design and Analysis Issues for Field Settings*. Chicago: Rand McNally.

COOK, T. M., and NEUMANN, D. A. (1987). The Effects of Load Placement on the EMG Activity of the Low Back Muscles During Load Carrying by Men and Women. *Ergonomics, 30*, 1413–1423.

CORLETT, E. N. (1990a). The Evaluation of Industrial Seating. Chapter 20 in J. R. Wilson and E. N. Corlett (eds.), *The Evaluation of Human Work* (pp. 500–515). London, UK: Taylor and Francis.

CORLETT, E. N. (1990b). Static Muscle Loading and the Evaluation of Posture. Chapter 22 in J. R. Wilson and E. N. Corlett (eds.), *The Evaluation of Human Work* (pp. 542–570). London, UK: Taylor and Francis.

CORLETT, E. N., and BISHOP, R. P. (1976). A Technique for Accessing Postural Discomfort. *Ergonomics, 19*, 175–182.

CORLETT, E. N., and EKLUND, J. A. E. (1984). How Does a Backrest Work? *Applied Ergonomics, 15*, 111–114.

CORLETT, E. N., and EKLUND, J. A. E. (1986). Change of Stature as an Indicator of Loads on the Spine. Chapter 20 in N. Corlett, J. Wilson, and I. Manenica (eds.), *The Ergonomics of Working Postures: Models, Methods and Cases* (pp. 232–242). London, UK: Taylor & Francis.

CORLETT, E. N., MADELEY, S. J., and MANENICA, I. (1979). Postural Targetting: A Technique for Recording Working Postures. *Ergonomics, 22*, 357–366.

COSTA, G., LIEVORE, F., CASALETTI, G., GAFFURI, E., and FOLKARD, S. (1989). Circadian Characteristics Influencing Inter-Individual Differences in Tolerance and Adjustment to Shiftwork. *Ergonomics, 32*, 373–385.

COURTNEY, A. J. (1988). Chinese Response Preferences for Display-Control Relationships. *Human Factors, 30*, 367–372.

COWEN, R. (1988). *Eyes on the Workplace*. Washington, DC: National Academy Press.

COX, T. (1990). The Recognition and Measurement of Stress: Conceptual and Methodological Issues. Chapter 25 in J. R. Wilson and E. N. Corlett (eds.), *The Evaluation of Human Work* (pp. 628–647). London, UK: Taylor and Francis.

CULVER, C. C., and VIANO, D. C. (1990). Anthropometry of Seated Women During Pregnancy: Defining a Fetal Region for Crash Protection Research. *Human Factors, 32*, 625–636.

CUSHMAN, W. H., and ROSENBERG, D. J. (1991). *Human Factors in Product Design*. Amsterdam: Elsevier.

CZAJA, S. J. (ed.). (1990). Aging. Special Issue, *Human Factors, 32*, 505–622.

CZAJA, S. J. (ed.). (1990). *Human Factors Research Needs for an Aging Population*. Washington, DC: National Academy Press.

CZAJA, S. J. (1991). Work Design for Older Adults. In A. Mital and W. Karwowski (eds.), *Workspace, Equipment, and Tool Design* (pp. 345–369). Amsterdam: Elsevier.

CZEISLER, C. A., DUMONT, M., and RICHARDS, G. S. (1990a). *Disorders of Circadian Function: Clinical Consequences and Treatment*. Paper presented at the NIH Consensus Development Conference on the Treatment of Sleep Disorders of Older People, Bethesda, MD.

CZEISLER, C. A., JOHNSON, M. P., and DUFFY, J. F. (1990b). Exposure to Bright Light and Darkness to Treat Physiologic Maladaptation to Night Work. *The New England Journal of Medicine, 322*, 1253–1259.

CZEISLER, C. A., KRONAUER, R. E., and ALLAN, J. S. (1989). Bright Light Induction of Strong (Type 0) Resetting of the Human Circadian Pacemaker. *Science, 244*, 1328–1332.

DAMON, A., STOUDT, H. W., and MCFARLAND, R. A. (1966). *The Human Body in Equipment Design*. Cambridge, MA: Harvard University Press.

DANIELS, L., and WORTHINGHAM, C. (1980). Muscle Testing (4th ed.). Philadelphia: Saunders.

DAS, B., and GRADY, R. M. (1983). The Normal Working Area in the Horizontal Plane: A Comparative Analysis Between Farley's and Squires' Concepts. *Ergonomics, 26*, 449–459.

DATTA, S. R., CHATTERJEE, B. B., and ROY, B. M. (1975). Maximum Permissible Weight to Be Carried on the Head by a Male Worker from Eastern India. *Journal of Applied Physiology, 38*, 132–135.

DATTA, S. R., and RAMANATHAN, N. L. (1971). Ergonomic Comparison of Seven Modes of Carrying Loads on the Horizontal Plane. *Ergonomics, 14*, 269–278.

DAVIES, B. T. (1978). Training in Manual Handling and Lifting. In. C. G. Drury (ed.), *Safety in Manual Materials Handling* (pp. 175–178). (DHEW (NIOSH) Publication No. 78-185). Cincinnati: NIOSH.

DAVIS, D. R., SHACKELTON, V. J., and PARASURAMAN, R. (1983). Monotony and Boredom. Chapter 1 in R. Hockey (ed.), *Stress and Fatigue in Human Performance* (pp. 1–33). Chichester, UK: Wiley.

DAVIS, P. R. (ed.). (1985). Industrial Back Pain in Europe. *Ergonomics, 28*, 1–405.

DAVY, D. T., KOTZAR, G. M., BROWN, R. H., HEIPLE, K. G., GOLDBERG, V. M., HEIPLE, JR., K. G., BEVILLA, J., and BURSTEIN, A. H. (1988). Telemetric Force Measurements Across the Hip After Total Arthroplasty. *Journal of Bone and Joint Surgery, 70(A)*, 45–50.

DAWS, J. (1981). Lifting and Moving Patients, a Revision of Training Programs. *Nursing Times*, 2067–2068.

DEBOUT, D. E., STORY, D., ROCA, J., HOGAN, M. C., and POOLE, D. C. (1989). Effects of Altitude Acclimatization on Pulmonary Gas Exchange During Exercise. *J. Applied Physiology, 67*, 2286–2295.

DEEB, J. M., DRURY, C. G., and PIZATELLA, P. (1987). Handle Placement on Containers in Manual Materials Handling. In *Proceedings, 9th International Conference on Production Research* (pp. 417–423). Amsterdam: Elsevier.

DEHLIN, O, BERG, S., HEDENRUD, B., ANDERSSON, G., and GRIMBY, G. (1978). Muscle Training, Psychological Perception of Work and Low-Back Symptoms in Nursing Aides. *Scandinavian Journal of Rehabilitation Medicine, 10*, 201–209.

DEHLIN, O., HEDENRUD, B., and HORAL, J. (1976). Back Symptoms in Nursing Aides in Geriatric Hospital. *Scandinavian Journal of Rehabilitation Medicine, 8*, 47–53.

Department of Defense (1980). *Anthropometry of U.S. Military Personnel* (metric). (DOD-HDBK-743) Washington, D.C.: U.S. Government Printing Office.

DERRICK, W. L. (1988). Dimensions of Operator Workload. *Human Factors, 30*, 95–110.

DIAMOND, J. M. (1988). Founding Fathers and Mothers. *Natural History, 97*(6), 10–15.

DICKINSON, C. E., CAMPION, K., FOSTER, A. F., NEWMAN, S. J., O'ROURKE, A. M. T., and THOMAS, P. G. (1992). Questionnaire Development: An Examination of the Nordic Musculoskeletal Questionnaire. *Applied Ergonomics, 23*, 197–201.

DILLON, A. (1992). Reading from Paper Verus Screens: A Critical Review of the Empirical Literature. *Ergonomics, 35*, 1297–1326.

DIONNE, E. D. (1984). Carpal Tunnel Syndrome, Part 1—The Problem. *National Safety News*, March issue, 42–45.

DOHERTY, E. T. (1991). Speech Analysis Techniques for Detecting Stress. In *Proceedings of the Human Factors Society 35th Annual Meeting* (pp. 689–693). Santa Monica, CA: Human Factors Society.

DOOLITTLE, T. L. (1989). Selection Standards for Manual Material Handling Tasks Using Dynamic Strength Tests. In A. Mital (ed.), *Advances in Industrial Ergonomics and Safety I* (pp. 515–521). London, UK: Taylor and Francis.

DOUWES, M., and DUL, J. (1991). Validity and Reliability of Estimating Body Angles by Direct and Indirect Observations. *Proceedings, 11th Congress of the International Ergonomics Association* (pp. 885–887). London, UK: Taylor and Francis.

DRAGANOVA, N. (1990). Musculo-skeletal Complaints in Some Professions. In *Abstracts, International Conference on Occupational Musculo-Skeletal Disorders and Prevention of Low Back Pain* (p. 21). Milan, Italy: University of Milan, Institute of Occupational Medicine.

DRERUP, B., and HIERHOLZER, E. (1985). Objective Determination of Anatomical Landmarks on the Body Surface: Measurement of the Vertebra Prominens from Surface Curvature. *J. Biomechanics, 18*, 467–474.

DRILLIS, R. J. (1963). Folk Norms and Biomechanics. *Human Factors, 5*, 427–441.

DRILLIS, R., and CONTINI, R. (1966). *Body Segment Parameters* (Report 1166-03). Office of Vocational Rehabilitation, Department of Health, Education and Welfare. New York: University School of Engineering and Science.

DRUCKMAN, D., and BJORK, R. A. (eds.). *In the Mind's Eye: Enhancing Human Performance*. Washington, DC: National Academy Press.

DRUCKMAN, D., and SWETS, J. A. (eds.). (1988). *Enhancing Human Performance: Issues, Theories, and Techniques*. Washington, DC: National Academy Press.

DRURY, C. G., and FRANCHER, M. (1985). Evaluation of a Forward Sloping Chair. *Applied Ergonomics, 16*, 41–47.

DRURY, C. G., DEEB, J. M., HARTMAN, B., WOOLLEY, S., DRURY, C. E., and GALLAGHER, S. (1989). Symmetric and Asymmetric Manual Materials Handling; Part 1: Physiology and Psychophysics. *Ergonomics, 32*, 467–489.

DRYDEN, R. D., and KEMMERLING, P. T. (1990). Engineering Assessment. In S. P. Sheer (ed.), *Vocational Assessment of Impaired Workers* (pp. 107–129). Aspen, CO: Aspen Press.

DUCHON, J., WAGNER, J., and KERAN, C. (1989). Forward Versus Backward Shift Rotation. In *Proceedings of the Human Factors Society 33rd Annual Meeting* (pp. 806–810). Santa Monica, CA.: Human Factors Society.

DUGGAN, A., and HAISMAN, M. F. (1992). Prediction of the Metabolic Cost of Walking With and Without Loads. *Ergonomics, 35*, 417–426.

DUKES-DOBOS, I. N. (1989). The Physician's Role as a Team Member in Preventing Low Back Injury. In Kroemer, K. H. E., McGlothlin, J. D., and Bobick, T. G. (eds.), *Manual Material Handling: Understanding and Preventing Low Back Trauma* (pp. 51–54). Akron, OH: American Industrial Hygiene Association.

DUMBLETON, J. H. (1988). The Clinical Significance of Wear in Total Hip and Knee Prostheses. *Journal of Biomaterials Applications, 3*, 3–32.

DUNHAM, R. B., PIERCE, J. L., and CASTANEDA, M. B. (1987). Alternative Work Schedules: Two Field Quasi Experiments. *Personnel Psychology, 40*, 215–242.

DUPUIS, H., and ZERLETT, G. (1986). Whole-Body Vibration and Disorders of the Spine. *International Archives of Occupational and Environmental Health, 59*, 323–336.

DURNIN, J. V. G. A., and PASSMORE, R. (1967). *Energy, Work, and Leisure*. London, UK: Heinemann.

DURNIN, J. V. G. A., and RAHAMAN, M. M. (1967). The Assessment of the Amount of Fat in the Human Body from Measurements of Skinfold Thickness. *British Journal of Nutrition, 21*, 681–689.

Eastman Kodak Company. (Vol. 1, 1983; Vol. 2, 1986). *Ergonomic Design for People at Work*. New York: Van Nostrand Reinhold.

EBERHARD, J. W., and BARR, R. A. (eds.). (1992). Safety and Mobility of Elderly Drivers, Part 2. *Special Issue of Human Factors, 34*, 1–65.

EDHOLM, O. G., and MURRELL, K. H. F. (1974). *The Ergonomics Research Society*. A History 1949 to 1970. (No location given): The Council of the Ergonomics Research Society.

EKHOLM, J., SCHULDT, K., and HARMS-RINGDAHL, K. (1986). Arm Suspension and Elbow Support in Sitting Work Postures: Effects on Neck and Shoulder Muscular Activity. In *Proceedings of the Conference "Work with Display Units"* (pp. 333–335). Stockholm: Swedish National Board of Occupational Safety and Health.

EKLUND, J. A. E. (1986). *Industrial Seating and Spinal Loading*. Doctoral Dissertation, University of Nottingham. Linkoeping, Sweden: University of Technology (ISBN 91-7870-14409).

EKLUND, J. A. E., and CORLETT, E. N. (1984). Shrinkage as a Measure of the Effect of Load on the Spine. *Spine, 9*, 189–194.

EKLUND, J. A. E., and CORLETT, E. N. (1986). Experimental and Biomechanical Analysis of Seating. Chapter 28 in N. Corlett, J. Wilson, and I. Manenica (eds.), *The Ergonomics of Working Postures: Models, Methods and Cases* (pp. 319–330). London, UK: Taylor and Francis.

EKMAN, G. (1964). Is the Power Law a Special Case of Fechner's Law? *Perception and Motor Skills, 19*, 730.

ELBERT, K. E. K. (1991). *Analysis of Polyethylene in Total Joint Replacement*. Unpublished doctoral dissertion, Cornell University, Ithaca, NY.

EL KARIM, M. A., SUKKAR, M. Y., COLLINS, K. J., and DORE, C, C. (1981). The Working Capacity of Rural, Urban and Service Personnel in the Sudan. *Ergonomics, 24*, 945–952.

ENANDER, A. (1987). Effects of Moderate Cold Performance on Psychomotor and Cognitive Tasks. *Ergonomics, 30*, 1431–1445.

ENOKA, R. M. (1988). *Neuromechanical Basis of Kinesiology*. Champaign, IL: Human Kinetics.

EPPS, B. W. (1987). A Comparison of Cursor Contol Devices on a Graphics Editing Task. In *Proceedings of the Human Factors Society 31st Annual Meeting* (pp. 442–446). Santa Monica, CA: Human Factors Society.

ERB, B. D., FLETCHER, G. F., and SHEFFIELD, T. L. (1979). Standards for Cardiovascular Exercise Treatment Programs. *Circulation, 59*(108A), 30–40.

ERRKOLA, R. (1976). The Physical Work Capacity of the Expectant Mother and Its Effect on Pregnancy, Labour and the Newborn. *International Journal of Obstetrics and Gynaecology, 14*, 153–159.

ETHERTON, J. R., and COLLINS, J. W. (1990). Working With Robots. *Professional Safety*, March Issue, 15–18.

EVANS, W. A. (1990). The Relationship Between Isometric Strength of Cantonese Males and the US NIOSH Guide for Manual Lifting. *Applied Ergonomics, 21*, 135–142.

FAHRINI, W. H. (1975). Conservative Treatment of Lumbar Disc Degeneration, Our Primary Responsibility. *Orthopedic Clinics of North America, 6*, 93–103.

FALLON, E. F., DILLON, A., SWEENEY, M., and HERRING, V. (1991). An Investigation of the Concept of Designer Style and its Implications for the Design of CAD Man-Machine Interfaces. In W. Karwowski and J. W. Yates (eds.), *Advances in Industrial Ergonomics and Safety III* (pp. 873–880). London: Taylor and Francis.

FARLEY, R. R. (1955). Some Principles of Methods and Motion Study as Used in Development Work. *General Motors Engineering Journal, 2*, 20–25.

FAST, A., SHAPIRO, M. D., and EDMOND, J. (1987). Low Back Pain in Pregnancy. *Spine, 12*, 368–371.

FASTE, R. A. (1977). New System Propels Design for the Handicapped. *Industrial Design*, 51–55.

FECHNER, G. T. (1860). *Sachen der Psychophysik*. Leipzig, Germany: Breitkopf und Hertel.

FERGUSON, D. (1981). Repetition Strain Injuries in Process Workers. *Medical Journal of Australia, 2*, 408.

FERGUSON, S. A., and MARRAS, W. S. (1991). Differences in Back Motion Characteristics as a Function of Task Direction. In *Proceedings of the Human Factors Society 35th Annual Meeting* (pp. 800–803). Santa Monica, CA: Human Factors Society.

FERNANDEZ, J. E., AYOUB, M. M., and SMITH, J. L. (1991). Psychophysical Lifting Capacity Over Extended Periods. *Ergonomics, 34*, 23–32.

FISHER, W., and TARBUTT, V. (1988). Some Issues in Collecting Data on Working Postures. In *Proceedings of the Human Factors Society 32nd Annual Meeting* (pp. 627–631). Santa Monica, CA: Human Factors Society.

FISK, J. R., DIMONTE, P., and COURINGTON, S. M. (1983). Back Schools: Past, Present and Future. *Clinical Orthopaedics, 179*, 18–23.

FITZLER, S. L., and BERGER, R. A. (1982). Attitudinal Change: The Chelsea Back Program. *Occupational Health and Safety, 52*, 24–26.

FITZLER, S. L., and BERGER, R. A. (1983). Chelsea Back Program: One Year Later. *Occupational Health and Safety, 52*, 52–54.

FLACH, J. M. (1989). An Ecological Alternative to Egg-Sucking. *Human Factors Society Bulletin*, 32, 4–6.

FLUEGEL, B., GREIL, H., and SOMMER, K. (1986). *Anthropologischer Atlas*. Berlin, Germany: Tribuene.

FOLKARD, S., and MONK, T. H. (eds.). (1985). *Hours of Work*. Chichester, UK: Wiley.

FOLKARD, S., MONK, T. H., and LOBBAN, M. C. (1979). Towards a Predictive Test of Adjustment to Shift Work. *Ergonomics, 22*, 79–91.

FOLKMAN, S., and LAZARUS, R. S. (1988). The Relationship Between Coping and Emotion: Implications for Theory and Research. *Social Science and Medicine, 26*, 309–317.

FOX, J. G. (1983). Industrial Music. In D. J. Oborne and M. M. Gruneberg (eds.), *The Physical Environment at Work* (pp. 221–226). New York: Wiley.

FRANSSON, C., and WINKEL, J. (1991). Hand Strength: The Influence of Grip Span and Grip Type. *Ergonomics, 34*, 881–892.

FRASER, T. M. (1980). *Ergonomic Principles in the Design of Hand Tools* (Occupational Safety and Health Series, No. 44). Geneva, Switzerland: International Labour Office.

FRASER, T. M. (1989). *The Worker at Work*. London, UK: Taylor and Francis.

FRAZER, L. (1991). Sex in Space. *Ad Astra, 3*, 42–45.

FREEMAN, R. L. (1990). Going Around and Around With CTDs: A Look at the Legal Issues. *Occupational Safety and Health Report* (13 June), 63–67.

FREIVALDS, A. (1987a). The Ergonomics of Tools. *International Reviews of Ergonomics, 1*, 43–75.

FREIVALDS, A. (1987b). Comparison of United States (NIOSH Lifting Guidelines) and European (ECSC Force Limits) Recommendations for Manual Work Limits. *American Industrial Hygiene Association Journal, 48*, 698–702.

FREIVALDS, A. (1989). Understanding and Preventing Back Trauma: Comparison of U.S. and European Approaches. In K. H. E. Kroemer, J. D. McGlothlin, and T. G. Bobick (eds.), *Manual Material Handling: Understanding and Preventing Back Trauma* (pp. 55–63). Akron, OH: American Industrial Hygiene Association.

FROEBERG, J. E. (1985). Sleep Deprivation and Prolonged Working Hours. Chapter 6 in S. Folkard and T. H. Monk (eds.), *Hours of Work* (pp. 67–76). Chichester, UK: Wiley.

FRONE, M. R., and MCFARLIN, D. B. (1989). Chronic Occupational Stressors, Self-Focused Attention, and Well-Being: Testing a Cybernetic Model of Stress. *Journal of Applied Psychology, 74*, 876–883.

FRUITIGER, A. (1989). *Signs and Symbols*. New York: Van Nostrand Reinhold.

FRYKMAN, P. N. (1988). Effects of Air Pollution on Human Exercise Performance. *Journal of Applied Sport Science Research, 2*, 66–71.

FULCO, C. S. (1988). Human Acclimatization and Physical Performance at High Altitude. *Journal of Applied Sport Science Research, 2*, 79–84.

GAGNON, M., and SMYTH, G. (1992). Biomechanical Exploration on Dynamic Modes of Lifting. *Ergonomics, 35*, 329–345.

GALANTE, J. O., LEMONS, J., SPECTOR, M., WILSON, P. D., and WRIGHT, T. M. (1991). The Biologic Effects of Implant Materials. *Journal of Orthopaedic Research, 9*, 760–775.

GALE, A. (ed.). (1988). *The Polygraph Test. Lies, Truth, and Science*. London, UK: Sage.

GALLAGHER, S., MARRAS, W. S., and BOBICK, T. G. (1988). Lifting in Stooped and Kneeling Postures: Effects on Lifting Capacity, Metabolic Costs, and Electromyography of Eight Trunk Muscles. *International Journal of Industrial Ergonomics, 3*, 65–76.

GALLAGHER, S., and UNGER, R. L. (1990). Lifting in Four Restricted Lifting Conditions. *Applied Ergonomics, 21*, 237–245.

GALLWEY, T. J., and FITZGIBBON, M. J. (1991). Some Anthropometric Measures on an Irish Population. *Applied Ergonomics, 22*, 9–12.

GAMBERALE, F., LJUNGBERG, A. S., ANNWALL, G., and KILBOM, A. (1987). An Experimental Evaluation of Psychophysical Criteria for Repetitive Lifting Work. *Applied Ergonomics, 18*, 311–321.

GARDNER-BONNEAU, D. J. (ed.). (1990). Assisting People with Functional Impairments. *Special Issue of Human Factors, 32*, 379–475.

GARG, A. (1983). Physiological Responses to One-handed Lift in the Horizontal Plane by Female Workers. *American Industrial Hygiene Association Journal, 44*, 190–200.

GARG, A. (1986). Maximum Acceptable Weights and Maximum Voluntary Isometric Strengths for Asymmetric Lifting. *Ergonomics, 29*, 879–892.

GARG, A., and AYOUB, M. M. (1980). What Criteria Exist for Determining How Much Load Can be Safely Lifted? *Human Factors, 22*, 475–486.

GARG, A., and BADGER, D. (1986). Maximum Acceptable Weights and Maximum Voluntary Isometric Strengths for Asymmetric Lifting. *Ergonomics, 29*, 879–892.

GARG, A., and HERRIN, G. D. (1979). Stoop or Squat: A Biomechanical and Metabolic Evaluation. *AIIE Transactions, 11*, 293–302.

GARG, A., and SAXENA, U. (1985). Physiological Stresses in Warehouse Operations with Special Reference to Lifting Technique and Gender: A Case Study. *American Industrial Hygiene Association Journal, 46*, 53–59.

GARG, A., SHARMA, D., CHAFFIN, D. B., and SCHMIDLER (1983). Biomechanical Stresses as Related to Motion Trajectory of Lifting. *Human Factors, 25*, 527–539.

GARRETT, J. W., and KENNEDY, K. W. (1971). *A Collation of Anthropometry* (AMRL-TR-68-1). Wright-Patterson Air Force Base, OH: Aerospace Medical Research Laboratories.

GARROW, J. S. (1980). Problems in Measuring Human Energy Balance. In J. M. Kinney (ed.), *Assessment of Energy Metabolism in Health and Disease* (p. 25). Columbus, OH: Ross Laboratories.

GENAIDY, A. M., and ASFOUR, S. S. (1987). Review and Evaluation of Physiological Cos Prediction Models for Manual Materials Handling. *Human Factors, 29*, 465–476.

GENAIDY, A. M., ASFOUR, S. S., MITAL, A., and WALY, S. M. (1990). Psychophysical Models for Manual Lifting Tasks. *Applied Ergonomics, 21*, 295–303.

GENAIDY, A. M., GUPTA, T., and ALSHEDI, A. (1990). Improving Human Capabilities for Combined Manual Handling Tasks Through a Short and Intensive Physical Training Program. *American Industrial Hygiene Association Journal, 51*, 610–614.

GENAIDY, A. M., and HOUSHYAR, A. (1989). Optimization Techniques in Occupational Biomechanics. In *Proceedings of the Human Factors Society 33rd Annual Meeting* (pp. 672–676). Santa Monica, CA: Human Factors Society.

GENTRY, W. D., SHOW, W. D., and THOMAS, M. (1977). Chronic Low Back Pain: A Psychological Profile. *Psychosomatics, 14*, 52–56.

GIL, H. J. C., and TUNES, D. B. (1977). Posture Recording: A Model for Sitting Posture. *Applied Ergonomics, 20*, 53–57.

GILAD, I., and KIRSCHENBAUM, A. (1986). About the Risks of Back Pain and Work Environment. *International Journal of Industrial Ergonomics, 1*, 65–74.

GILAD, I., and POLLATSCHEK, M. A. (1986). Layout Simulation for Keyboards. *Behaviour and Information Technology, 5*, 273–281.

GILLBERG, M. (1985). Effects of Naps on Performance. Chapter 7 in S. Folkard and T. H. Monk (eds.), *Hours of Work* (pp. 77–86). Chichester, UK: Wiley.

GODIN, G., and GIONET, N. (1991). Determinants of an Intention to Exercise of an Electric Power Commission's Employees. *Ergonomics, 34*, 1221–1230.

GOENEN, E., KALINKARA, V., and OEZGEN, O. (1991). Anthropometry of Turkish Women. *Applied Ergonomics, 22*, 409–411.

GOLDSTEIN, I. L. (1974). *Training: Program Development and Evaluation*. Monterey, CA: Brooks/Cole.

GOPHER, D., and RAIJ, D. (1988). Typing With a Two-Hand Chord Keyboard: Will the QWERTY Become Obsolete? *IEEE Transactions on Systems, Man, and Cybernetics, 18*, 601–609.

GORDON, C. C., CHURCHILL, T., CLAUSER, C. E., BRADTMILLER, B., MCCONVILLE, J. T., TEBBETTS, I., and WALKER, R. A. (1989). *1988 Anthropometric Survey of U.S. Army Personnel: Summary Statistics Interim Report* (Technical Report NATICK/TR-89-027). Natick, MA: United States Army Natick Research, Development and Engineering Center.

GOULD, S. J. (1981). *The Mismeasure of Man*. New York: Norton.

GOULD, S. J. (1988). A Novel Notion of Neanderthal. *Natural History, 97*, 16–21.

GRAEBER, R. C. (1988). Aircrew Fatigue and Circadian Rhythmicity. Chapter 10 in E. L. Wiener and D. C. Nagel (eds.), *Human Factors in Aviation* (pp. 305–344). San Diego, CA: Academic Press.

GRAF, M., GUGGENBUEHL, U., and KRUEGER, H. (1991). Movement Dynamics of Sitting Behaviour During Different Activities. In *Proceedings of the 11th Congress of the International Ergonomics Association* (pp. 15–17). London, UK: Taylor & Francis.

GRANDJEAN, E. (1963). *Physiological Design of Work* (in German). Thun, Switzerland: Ott.

GRANDJEAN, E. (ed.). (1969). *Sitting Posture*. London, UK: Taylor and Francis.

GRANDJEAN, E. (1973). *Ergonomics of the Home* (particularly pages 99–136). London, UK: Taylor and Francis.

GRANDJEAN, E. (1980). *Fitting the Task to the Man: An Ergonomic Approach* (1st ed.). London, UK: Taylor and Francis.

GRANDJEAN, E. (1986). *Ergonomics in Computerized Offices*. Philadelphia: Taylor and Francis.

GRANDJEAN, E. (1988). *Fitting the Task to the Man* (4th ed.). London, UK: Taylor and Francis.

GRANDJEAN, E., HUENTING, W., and NISHIYAMA, K. (1984). Preferred VDT Workstation Settings, Body Postures and Physical Impairments. *Applied Ergonomics, 15*, 99–104.

GRANT, A. (1990). Homo-Quintadus, Computers and Rooms (Repetitive Orcular Orthopedic Motion Stress). *Optometry and Vision Science, 67*(4), 297–305.

GREEN, B. F., and ANDERSON, L. K. (1955). The Tactual Identification of Shapes for Coding Switch Handles. *Journal of Applied Psychology, 39*, 219–226.

GREEN, R. A., BRIGGS, C. A., and WRIGLEY, T. V. (1991). Factors Related to Working Posture and its Assessment Among Keyboard Operators. *Applied Ergonomics, 22*, 29–35.

GREENBERG, L., and CHAFFIN, D. B. (1977). *Workers and Their Tools*. Midland, MI: Pendell.

GREENE, T. C., and BELL, P. A. (1987). Environment Stress. Chapter 5 in M. A. Baker (ed.), *Sex Differences in Human Performance* (pp. 81–106). Chichester, UK: Wiley.

GREENWOOD, K. M. (1991). Psychometric Properties of the Diurnal Type Scale of Torsvall and Akerstedt (1980). *Ergonomics, 34*, 435–443.

GREINER, T. M. (1991). *Hand Anthropometry of U.S. Army Personnel* (Technical Report TR-92/011). Natick, MA: U.S. Army Natick Research, Development and Engineering Center.

GREINER, T. M., and GORDON, C. C. (1990). *An Assessment of Long-Term Changes in Anthropometric Dimensions: Secular Trends of U.S. Army Males* (Natick/TR-91/006). Natick, MA: U.S. Army Natick Research, Development and Engineering Center.

GRIECO, A. (1986). Sitting Posture: An Old Problem and a New One. *Ergonomics, 29*, 345–362.

GRIFFIN, M. D., and FRENCH, J. R. (1991). *Space Vehicle Design*. Washington, DC: American Institute of Aeronautics and Astronautics, Inc.

GRIFFIN, M. J. (1990). *Handbook of Human Vibration*. San Diego, CA: Academic Press.

GRIFFITH, D. (1990). Computer Access for Persons Who are Blind or Visually Impaired: Human Factors Issues. *Human Factors, 32*, 467–475.

GRIMSRUD, T. M. (1990). Humans Are Not Created to Sit—And Why You Have to Refurnish Your Life. *Ergonomics, 33*, 291–295.

GUDJONSSON, G. H. (1988). How to Defeat the Polygraph Tests. Chapter 10 in A. Gale (ed.), *The Polygraph Test. Lies, Truth, and Science* (pp. 126–136). London: Sage.

GUIDI, M. A. (1989). *Equipment Needs to Support Human Performance in Long-Duration Spaceflight: A Review of the Literature* (Project Report IEOR 5614). Blacksburg, VA: Virginia Tech (VPI & SU).

GUIGNARD, J. C. (1985). Vibration. Chapter 15 in L. V. Cralley and L. J. Cralley (eds.), *Patty's Industrial Hygiene and Toxicology* (pp. 635–724). New York: Wiley.

GUYTON, A. C. (1979). *Physiology of the Human Body* (5th ed.). Philadelphia: Saunders.

HABER, L. D. (1990). Issues in the Definition of Disability and the Use of Disability Survey Data. In D. B. Levine, M. Zitter, and L. Ingram (eds.), *Disability Statistics: An Assessment* (pp. 35–71). Committee on National Statistics, National Research Council. Washington, DC: National Academy Press.

HACKETT, T. P., ROSENBAUM, J. F., and TESAR, G. E. (1988). Emotion, Psychiactric Disorders, and the Heart. In E. Braunwald (ed.), *Heart Disease—A Textbook of Cardiovascular Medicine* (pp. 1883–1900). Philadelphia: Saunders.

HADLER, N. M. (1978). Legal Ramifications of the Medical Definition of Back Disease. *Annals of Internal Medicine, 89*, 992–999.

HADLER, N. M. (1982). A Rheumatologist's View of the Back. *Journal of Occupational Medicine, 24*, 282–285.

HADLER, N. M. (1989). The Roles of Work and of Working in Disorders of the Upper Extremity. *Bailliere's Clinical Rheumatology, 3*, 121–141.

HADLER, N. M. (1990). Cumulative Trauma Disorders: An Iatragenic Concept. *Journal of Occupational Medicine, 32*, 38–41.

HAEGG, G. M., SUURKUELA, J., and LIEW, M. (1987). A Worksite Method for Shoulder Muscle Fatigue Measurements Using EMG, Test Contractions and Zero Crossing Technique. *Ergonomics, 30*, 1541–1551.

HAHN, H. A. (1986). *The Effects of Alcohol on Four Behavioral Processes: Perception, Mediation, Communication, and Motor Activity*. Unpublished doctoral dissertation, Blacksburg, VA: Virginia Tech (VPI & SU).

HAHN, K., CHIN, D., MA, P., and REBELLO, S. (1991). Cumulative Trauma Disorder Reductions in the Industrial Workplace—A Systems Approach. In W. Karwowski and J. W. Yates (eds.), *Advances in Industrial Ergonomics and Safety III* (pp. 147–154). Philadelphia: Taylor and Francis.

HAISMAN, M. F. (1988). Determinants of Load Carrying Ability. *Applied Ergonomics, 19*, 111–121.

HALL, H. W. (1973). "Clean" Versus "Dirty" Lifting, An Academic Subject for Youth. *American Society of Safety Engineers Journal, 18*, 20–25.

HALLBECK, M. S. (1990). *Biomechanical Analysis of Carpal Flexion and Extension*. Unpublished doctoral dissertation, Blacksburg, VA: Virginia Tech (VPI & SU).

HALLBERG, G. (1976). A System for the Description and Classification of Movement Behavior. *Ergonomics, 19*, 727–739.

HAMPSON, S. E. (1988). What are Truthfulness and Honesty? Chapter 5 in A. Gale (ed.), *The Polygraph Test. Lies, Truth, and Science* (pp. 53–64). London: Sage.

HAN, S. H., JORNA, G. C., MILLER, R. H., and TAN, K. C. (1990). A Comparison of Four Input Devices for the Macintosh Interface. In *Proceedings of the Human Factors Society 34th Annual Meeting* (pp. 267–271). Santa Monica, CA: Human Factors Society.

HANCOCK, P. A., and MESHKAT, N. (eds.). (1988). *Human Mental Workload*. Amsterdam, NL: Elsevier.

HANCOCK, P. A., RODENBURG, G. C., MATHEWS, W. D., and VERCRUYSSEN, M. (1988). Estimation of Duration and Mental Workload at Differing Times of Day by Males and Females. In *Proceedings of the Human Factors Society 32nd Annual Meeting* (pp. 857–861). Santa Monica, CA: Human Factors Society.

HANGARTNER, M. (1987). *Standardization in Olfactometry with Respect to Odor Pollution Control; Assessment of Odor Annoyance in the Community*. Presentations 87-75A.1 and 87-75B.3 at the 80th Annual Meeting of the APCA. New York: June 21–26.

HANSEN, C. P. (1989). A Causal Model of the Relationship Among Accidents, Biodata, Personality, and Cognitive Factors. *Journal of Applied Psychology, 74*, 81–90.

HANSSON, J. E., and ATTEBRANT, M. (1986). The Effect of Table Height and Table Top Angle on Head Position and Reading Distance. In *Proceedings of the Conference "Work With Display Units"* (pp. 419–422). Stockholm: Swedish National Board of Occupational Safety and Health.

HANSSON, T. (1986). Prolonged Sitting and the Back. In *Proceedings of the Conference "Work With Display Units"* (pp. 491–492). Stockholm: Swedish National Board of Occupational Safety and Health.

HARALICK, R. M., CHU, Y. H., WATSON, L. T., and SHAPIRO, L. G. (1984). Matching Wire Frame Objects from their Two-Dimensional Perspective Projects. *Pattern Recognition, 17*, 607–619.

HARALICK, R. M., and WATSON, L. T. (1981). A Facet Model for Image Data, Computer Graphics. *Image Processing, 15*, 113–129.

HARALICK, R. M., WATSON, L. T., and LAFFEY, T. J. (1983). The Topographic Primal Sketch. *International Journal of Robotics Research, 2*, 50–72.

HARBER, P., BILLET, E., SHIMOZAKI, S., and VOJTECKY, M. (1988). Occupational Back Pain of Nurses: Special Problems and Prevention. *Applied Ergonomics, 19*, 219–224.

HARLESS, E. (1860). The Static Moments of the Component Masses of the Human Body (in German). *Trans. Math. Phys. Royal Bavarian Acad. Sci. 8*, 69–96, 257–294.

HARMAN, E. A., ROSENSTEIN, R. M., FRYKMAN, P. H., and NIGRO, G. A. (1989). Effects of a Belt on Intra-abdominal Pressure During Weight Lifting. *Medicine and Science in Sports and Exercise, 21*, 186–190.

HARMS-RINGDHAL, K., and EKHOLM, J. (1986). Pain and Extreme Position of Lower Cervical Spine in Sitting Postures. In *Proceedings of the Conference "Work With Display Units"* (pp. 341–342). Stockholm: Swedish National Board of Occupational Safety and Health.

HARRISON, A. A., CLEARWATER, Y. A., and MCKAY, C. P. (1991). *From Antarctica to Outer Space: Life in Isolation and Confinement*. New York: Springer.

HASLAM, R. A., and PARSONS, K. C. (1988). Quantifying the Effects of Clothing on Models of Human Response to the Thermal Environment. *Ergonomics, 3*, 1787–1806.

HASLEGRAVE, C. M. (1979). An Anthropometric Survey of British Drivers. *Ergonomics, 22*, 145–154.

HASLEGRAVE, C. M. (1986). Characterizing the Anthropometric Extremes of the Population. *Ergonomics, 29*, 281–301.

HAY, J. G. (1973). *The Center of Gravity of the Human Body*. In *Kinesiology III* (p. 2044). Washington, DC: American Association for Health, Physical Education, and Recreation.

HAYNE, C. R. (1981). Lifting and Handling. *Health and Safety at Work, 3*, 18–21.

HAYSLIP, B., and PANEK, P. (1989). *Adult Development and Aging*. New York: Harper & Row.

HEILMAN, M. E., RIVERO, J. C., and BRETT, J. F. (1991). Skirting the Competence Issue: Effects of Sex-Based Preferential Selection on Task Choices of Women and Men. *Journal of Applied Psychology, 76*, 99–105.

HELANDER, M. G. (1981). *Human Factors/Ergonomics for Building and Construction*. New York, NY: Wiley.

HELANDER, M. G. (1982). *Ergonomic Design of Office Environments for Visual Display Terminals* (Report for NIOSH). Blacksburg, VA: Virginia Tech (VPI & SU).

HELANDER, M. G. (ed.). (1988). Handbook of Human-Computer Interaction. Amsterdam: North-Holland.

HELANDER, M. G., GROSSMITH, E. J., and PRABHU, P. (1991). Planning and Implementation of Microscope Work. *Applied Ergonomics, 22*, 36–42.

HELANDER, M. G., and QUANCE, L. A. (1990). Effect of Work-Rest Schedules on Spinal Shrinkage in the Sedentary Worker. *Applied Ergonomics, 21*, 279–284.

HERRIN, G. D., JARAIEDI, M., and ANDERSON, C. K. (1986). Prediction of Overexertion Injuries Using Biomechanical and Psychophysical Models. *American Industrial Hygiene Association Journal, 47*, 322–330.

HERRON, R. E. (1973). Biostereometric Measurement of Body Form. *Yearbook of Physical Anthropology 1972, American Association of Physical Anthropologists, 16*, 80–121.

HERTZBERG, H. T. E. (1968). The Conference on Standardization of Anthropometric Techniques and Terminology. *American Journal of Physical Anthropology, 28*, 1–16.

HERTZBERG, H. T. E. (1972). Engineering Anthropology. In H. P. Van Cott and R. G. Kinkade (eds.), *Human Engineering Guide to Equipment Design* (revised ed.). New York: McGraw-Hill.

HERTZBERG, H. T. E. (1979). *Engineering Anthropology: Past, Present and Potential*. In *The Uses of Anthropology* (pp. 184–204). Special Publication No. 11 of the American Anthropological Association.

HERTZBERG, H. T. E., DANIELS, G. S., and CHURCHILL, E. M. (1954). *Anthropometry of Flying Personnel—1950* (Report WADC-TR-52-321). Wright-Patterson AFB, OH: Wright Air Development Center.

HETTINGER, T. (1961). *Physiology of Strength*. Springfield, IL: Thomas.

HEUER, H., BRUEWER, M., ROEMER, T., KROEGER, H., and KNAPP, H. (1991). Preferred Vertical Gaze Direction and Observation Distance. *Ergonomics, 34*, 379–392.

HEUER, H., and OWENS, D. A. (1987). Variations of Dark Vergence as a Function of Vertical Gaze Deviation. *Investigative Ophthamology, 28*, 315.

HEUER, H., and OWENS, D. A. (1989). Vertical Gaze Direction and the Resting Posture of the Eyes. *Perception, 18*, 353–377.

HILL, S. G. (1985). *Review of the "Training" Literature on Manual Material Handling*. Unpublished Report, Industrial Ergonomics Laboratory, IEOR Department, Blacksburg, VA: Virginia Tech (VPI & SU).

HILL, S. G., and KROEMER, K. H. E. (1986). Preferred Declination and the Line of Sight. *Human Factors, 28*, 127–134.

HIMMELSKIN, J. S., and ANDERSSON, G. B. J. (1988). Low Back Pain: Risk Evaluation and Preplacement Screening. *Journal of Occupational Medicine, 3*, 255–269.

HIRSCH, R. S. (1970). Effect of Standard vs. Alphabetical Formats in Typing Performance. *Journal of Applied Psychology, 54*, 484–490.

HOCKING, B. (1987). Epidemiological Aspects of "Repetition Strain Injury" in Telecom Australia. *The Medical Journal of Australia, 147* (Sept.), 218–222.

HOFFMANN, E. R. (1991). Capture of Moving Targets: A Modification of Fitts' Law. *Ergonomics, 34*, 211–220.

HOFFMANN, E. R. (1991). A Comparison of Hand and Foot Movement Times. *Ergonomics, 34*, 397–406.

HOFFMAN, R. G., and POZOS, R. S. (1989). Experimental Hypothermia and Cold Perception. *Aviation, Space and Environmental Medicine, 60*, 964–969.

HOLEWIJN, M. (1990). Physiological Strain Due to Load Carrying. *European Journal of Applied Physiology and Occupational Physiology, 61*, 237–245.

HOLLAND, D. A. (1991). *Systems and Human Factors Concerns for Long-Duration Space Flight*. Unpublished MS thesis, Dept. of Systems Engineering, Blacksburg, VA: Virginia Tech (VPI & SU).

HOLMES, T. H., and RAHE, R. H. (1967). The Social Readjustment Rating Scale. *Journal of Psychosomatic Research, 11*, 213–218.

HOLZMANN, P. (1981). ARBAN—A New Method of Analysis of Ergonomic Effort. *Applied Ergonomics, 13*, 82–86.

HOOD, D. C., and FINKELSTEIN, M. A. (1986). Sensitivity to Light. Chapter 5 in K. R. Boff, L. Kaufman, and J. P. Thomas (eds.), *Handbook of Perception and Human Performance* (pp. 5.1–5.66). New York: Wiley.

HORNE, J. A. (1985). Sleep Loss: Underlying Mechanisms and Tiredness. Chapter 5 in S. Folkard and T. H. Monk (eds.), *Hours of Work* (pp. 53–65). Chichester, UK: Wiley.

HORNE, J. A. (1988). *Why We Sleep—The Functions of Sleep in Humans and Other Mammals*. Oxford, UK: Oxford University Press.

HORNE, J. A., and GIBBONS, H. (1991). Effects on Vigilance Performance and Sleepiness of Alcohol Given in the Early Afternoon ('Post Lunch') vs. Early Evening. *Ergonomics, 34,* 67–77.

HORNE, J., and OESTBERG, O. (1976). A Self-assessment Questionnaire to Determine Morningness-Eveningness in Human Circadian Rhythms. *International Journal of Chronobiology, 4,* 97–110.

HOUSE, L. H., and PANSKY, B. (1967). *A Functional Approach to Neuroanatomy (2nd ed.).* NY: McGraw-Hill.

HOWARD, I. P. (1986). The Vestibular System. In K. R. Boff, L. Kaufman, and J. P. Thomas (eds.), *Handbook of Human Perception and Human Performance* (pp. 11.1–11.30). New York: Wiley.

HUDGENS, G. A., FATKIN, L. T., BILLINGSLEY, P. A., and MAZURACZAK, J. (1988). Hand Steadiness: Effects of Sex, Menstrual Phase, Oral Contraceptives, Practice, and Handgun Weight. *Human Factors, 30*, 51–60.

HUENTING, H., GRANDJEAN, E., and MAEDA, K. (1980). Constrained Postures in Accounting Machine Operations. *Applied Ergonomics, 11,* 145–149.

HUISKES, R. (1985). Stress Analysis and Fixation Problems in Joint Replacement. In N. Berne, A. E. Engin, and K. M. Correia da Silva (eds.), *Biomechanics of Normal and Pathological Human Articulating Joints* (pp. 337–358). NATO ASI Series. New York: Plenum.

HULTMAN, G. (1987). The Healthy Back, Its Environment and Characteristics: A Pilot Study. *Ergonomics, 30,* 295–298.

Human Factors Society (ed.) (1988). *American National Standard for Human Factors Engineering of Visual Display Terminal Workstations* (HFS/ANSI 100). Santa Monica, CA: Human Factors Society.

HUMBOLDT, A. VON (1849). *Cosmos*. London: Bohn.

HUNT, DR. P., and CRAIG, D. R. (1954). *The Relative Discriminability of Thirty-one Differently Shaped Knobs* (WADC-TR-54-108). Wright-Patterson AFB, OH: Wright Air Development Center.

HUNT, S. R. (1987). Human Engineering for Space. Chapter 6.5 in G. Salvendy (ed.), *Handbook of Human Factors* (pp. 708–721). New York: Wiley.

HUNTER, K. I., and SHANE, R. H. (1979). Time of Death and Biorhythmic Cycles. *Perceptual and Motor Skills, 48,* 220.

HURRELL, J. J., and COLLIGAN, M. J. (1985). Alternative Work Schedules: Flextime and the Compressed Work Week. Chapter 8 in C. L. Cooper, and M. J. Smith (eds.), *Job Stress and Blue Collar Work* (pp. 131–148). New York: Wiley.

HUTTON, W. C. and ADAMS, M. A. (1982). Can the Lumbar Spine Be Crushed in Heavy Lifting? *Spine, 7,* 586–590.

ILO (ed.) (1974). *Introduction to Work Study*. Geneva, Switzerland: International Labour Office.

ILO (ed.) (1988). *Maximum Weights in Load Lifting and Carrying* (Occupational Safety and Health Series, No. 59). Geneva, Switzerland: International Labour Office.

IMRHAN, S. N. (1986). An Analysis of Finger Pinch Strength in Children. In *Proceedings of the Human Factors Society 30th Annual Meeting* (pp. 667–671). Santa Monica, CA: Human Factors Society.

IMRHAN, S. N. (1991). The Influence of Wrist Position on Different Types of Pinch Strength. *Applied Ergonomics, 22,* 379–384.

IMRHAN, S. N., and AYOUB, M. M. (1988). Predictive Models of Upper Extremity Rotary and Linear Pull Strength. *Human Factors, 30,* 83–94.

INGELS, N. B. (1979). *Molecular Basis of Force Development in Muscle*. Palo Alto, CA: Palo Alto Medical Research Foundation.

INGLEMARK, B. E., and LEWIN, T. (1968). Anthropometrical Studies on Swedish Women. *Acta Morphologica Neerlando-Scandinavica, III,* 145–166.

INTARANONT, K. (1991). Human Characteristics of Workers in Northeast Thailand. In *Proceedings, 11th Congress of the International Ergonomics Association* (pp. 888–890). London: Taylor and Francis.

INTARANONT, K., KHOKHAJAIKIAT, P., SOMNASANG, S., and ASVAKIAT, P. (1988). Anthropometry and Physical Work Capacity of Agricultural Workers in Northeast Thailand. In *Proceedings of the 10th Congress of the International Ergonomics Association* (pp. 215–217). (Cited by Intaranont, 1991.)

IRVINE, C. H., SNOOK, S. H., and SPARSHATT, J. H. (1990). Stairway Rises and Treads: Acceptable and Preferred Dimensions. *Applied Ergonomics, 21*(3), 215–225.

ISO (ed.) (1985). *Evaluation of Human Exposure to Whole-Body Vibration* (ISO Standard 2631). Geneva, Switzerland: International Organization for Standardization.

ISO (ed.) (1987). *Mechanical Vibration and Shock: Mechanical Transmissibility in the Human Body in the Z Direction* (ISO Standard 7962). Geneva, Switzerland: International Organization for Standardization.

ISO (ed.) (1989). *Hot Environments* (ISO Standard 7243). Geneva, Switzerland: International Organization for Standardization.

JACKSON, C. P., and BROWN, M.D. (1983). Is There a Role for Exercise in the Treatment of Patients with Low Back Pain? *Clinical Orthopaedics, 179,* 39–45.

JAEGER, M. (1987). Biomechanical Human Model for Analysis and Evaluation of the Strain in the Spinal Column While Manipulating Loads (in German). *Biotechnik Series 17,* No. 33. Duesseldorf, Germany: VDI Verlag.

JAEGER, M., and LUTTMANN, A. (1986). Biomechanical Model Calculations of Spinal Stress for Different Working Postures in Various Workload Situations. Chapter 15 in N. Corlett, J. Wilson, and I. Manenica (eds.), *The Ergonomics of Working Postures: Models, Methods and Cases* (pp. 144–423). London, UK: Taylor and Francis.

JAEGER, M., and LUTTMANN, A. (1989). Biomechanical Analysis and Assessment of Lumbar Stress During Load Lifting Using a Dynamic 19-Segment Human Model. *Ergonomics, 32,* 93–112.

JASCHINSKI-KRUZA, W. (1991). Eyestrain in VDU Users: Viewing Distance and the Resting Position of Ocular Muscles. *Human Factors, 33,* 69–83.

JEGERLEHNER, J. L. (1991). Ergonomic Analysis of Problem Jobs Using Computer Spreadsheets. In W. Karwowski and J. W. Yates (eds.), *Advances in Industrial Ergonomics and Safety III* (pp. 865–871). Philadelphia: Taylor and Francis.

JENIK, J., and NORTH, K. (1982). Somatography in Workspace Design. In R. Easterby, K. H. E. Kroemer, and D. B. Chaffin (eds.), *Anthropometry and Biomechanics* (pp. 215–224). New York: Plenum.

JENKINS, J. A. (1991). *Alternative Input Methods* (Report for ISE 5614). Blacksburg, VA: Virginia Tech (VPI & SU).

JENKINS, J. P. (ed.) (1991). *Human Performance for Long Duration Space Missions* (Final Report, NASA-SSTAC Ad Hoc Committee). Washington, DC: NASA.

JENKINS, W. L. (1953). *Design Factors in Knobs and Levers for Making Settings on Scales and Scopes* (WADC-TR-53-2). Wright-Patterson AFB, OH: Aero Medical Laboratory.

JENKINS, W. L., and JEGERLEHNER, J. L. (1991). Ergonomic Analysis of Problem Jobs Using Computer Spreadsheets. In W. Karwowski and J. W. Yates (eds.), *Advances in Industrial Ergonomics and Safety III* (pp. 865–871). Philadelphia: Taylor and Francis.

JENSEN, R. C. (1985). A Model of The Training Process Devised from Human Factors and Safety Literature. In R. E. Eberts and C. G. Eberts (eds.), *Trends in Ergonomics/ Human Factors II* (pp. 501–509). Amsterdam: Elsevier.

JIANG, B. C., and AYOUB, M. M. (1987). Modelling of Maximum Acceptable Load of Lifting by Physical Factors. *Ergonomics, 30,* 529–538.

JIANG, B. C., SMITH, J. L., and AYOUB, M. M. (1986). Psychophysical Modeling of Manual Materials-Handling Capacities Using Isoinertial Strength Variables. *Human Factors, 28,* 691–702.

JOHNSON, L. C., TEPAS, D. I., COLGUHOUN, W. P., and COLLIGAN, M. J. (eds.). (1981). *Biological Rhythms, Sleep and Shift Work*. New York: Spectrum.

JOHNSON, S. L., and LEWIS, D. M. (1989). A Psychophysical Study of Two-Person Manual Material Handling Tasks. In *Proceedings of the Human Factors Society 33rd Annual Meeting* (pp. 651–653). Santa Monica, CA: Human Factors Society.

JOHNSTON, W. W. (1982). Back Injuries: A Problem for Both Workers and Employers. *Ohio Monitor, 55,* 15.

JONES, D. F. (1972). Back Injury Research. *American Industrial Hygiene Association Journal, 33,* 596–602.

JONES, R. G. (1990). Worker Interdependence and Output: The Hawthorne Studies Reevaluated. *American Sociological Review, 55,* 176–190.

JORNA, G. C., MOHAGEG, M. F., and SNYDER, H. L. (1989). Performance, Perceived Safety and Comfort of the Alternating Tread Stair. *Applied Ergonomics, 20,* 26–32.

JORNA, G. C., and SYNDER, H. L. (1991). Image Quality Determines Differences in Reading Performance and Perceived Image Quality with CRT and Hard-Copy Displays. *Human Factors, 33*(4), 459–469.

JUDD, D. B. (1951). *Colorimetry and Artificial Daylight*. Report of the Technical Committee Number 7, International Commission on Illumination, 12th Session (pp. 1–60). Stockholm: International Commission on Illumination.

JUERGENS, H. W. (1984). Anthropometric Reference System. In H. Schmidtke (ed.), *Ergonomic Data for Equipment Design* (pp. 93–100). New York: Plenum.

JUERGENS, H. W., AUNE, I. A.,and PIEPER, U. (1990). *International Data on Anthropometry* (Occupational Safety and Health Series No. 65). Geneva: International Labour Office.

KAHN, J. F., and MONOD, H. (1989). Fatigue Induced by Static Work. *Ergonomics, 32,* 839–846.

KALEPS, I., CLAUSER, C. E., YOUNG, J. W., CHANDLER, R. F., ZEHNER, G. F., and MCCONVILLE, J. (1984). Investigation Into the Mass Distribution Properties of the Human Body and Its Segments. *Ergonomics, 27,* 1225–1237.

KAMARCK, T., and JENNINGS, J. R. (1991). Behavioral Factors in Sudden Cardiac Death. *Psychological Bulletin, 109,* 42–75.

KAMON, E., KISER, D., and PYTEL, J. (1982). Dynamic and Static Lifting Capacity and Muscular Strength of Steelmill Workers. *American Industrial Hygiene Association Journal, 43,* 853–857.

KANNER, A. D., COYNE, J. C., SCHAEFER, C., and LAZARUS, R. S. (1981). Comparison of Two Modes of Stress Measurement: Daily Hassels and Uplifts vs. Major Life Events. *Journal of Behavioral Medicine, 4,* 1–39.

KANTOWITZ, B. H., and SORKIN, R. D. (1983). *Human Factors: Understanding People-System Relationships*. New York: Wiley.

KARHU, O., KANSI, P., and KUORINKA, I. (1977). Correcting Working Postures in Industry: Practical Method for Analysis. *Applied Ergonomics, 18,* 199–201.

KARHU, O., KARKONEN, R., SORVALI, P., and VEPSALAINEN, P. (1981). Observing Working Postures in Industry: Examples of OWAS Application. *Applied Ergonomics, 12,* 13–17.

KARWOWSKI, W. (1988). Maximum Load Lifting Capacities of Males and Females in Teamwork. In *Proceedings of the Human Factors Society 32nd Annual Meeting* (pp. 680–682). Santa Monica, CA: Human Factors Society.

KARWOWSKI, W. (1989). Perception of Load Heaviness by Males. In K. H. E. Kroemer, J. D. McGlothlin, and T. G. Bobick (eds.), *Manual Material Handling: Understanding and Preventing Back Trauma* (pp. 9–14). Akron, OH: American Industrial Hygiene Association.

KARWOWSKI, W. (1991). Psychophysical Acceptability and Perception of Load Heaviness by Females. *Ergonomics 34,* 487–496.

KARWOWSKI, W., and KASDAN, M. L. (1988). The Partnership of Ergonomics and Medical Intervention in Rehabilitation of Workers with Cumulative Trauma Disorders of the Hand. In A. Mital and W. Karwowski (eds.), *Ergonomics in Rehabilitation* (pp. 35–53). Philadelphia: Taylor and Francis.

KARWOWSKI, W., and PONGPATANASUEGSA, N. (1988). Testing of Isometric and Isokinetic Lifting Strengths of Untrained Females in Teamwork. *Ergonomics, 31,* 291–301.

KAUFMAN, J. E., and HAYNES, H. (eds.). (1981). *IES Lighting Handbook, 1981*. Application Volume. New York: Illuminating Engineering Society of North America.

KAYIS, B., and OEZOK, A. F. (1991a). The Anthropometry of Turkish Army Men. *Applied Ergonomics, 22,* 49–54.

KAYIS, B., and OEZOK, A. F. (1991b). Anthropometric Survey Among Turkish Primary School Children. *Applied Ergonomics, 22,* 55–56.

KAZARIAN, L. (1981). Injuries to the Human Spinal Column: Biomechanics and Injury Classification. In D. I. Miller (ed.), *Injuries to the Spine* (pp. 297–352). Philadelphia: Franklin Institute Press.

KAZARIAN, L. and GRAVES, G. A. (1977). Compressive Strength Characteristics of the Human Vertebral Centrum. *Spine, 2,* 1–14.

KEEGAN, J. J. (1952). Alterations to the Lumbar Curve Related to Posture and Sitting. *Journal of Bone and Joint Surgery, 35,* 589–603.

KEELE, S. W. (1986). Motor Control. Chapter 30 in K. R., BOFF, L. KAUFMAN, and J. P. THOMAS (eds.) *Handbook of Human Perception and Human Performance* (pp. 30.1–30.60). New York: Wiley.

KELLER, E., BECKER, E., and STRASSER, H. (1991). An Objective Assessment of Learning Behavior With a Single-Hand Chord Keyboard for Text Inputs (in German). *Z. Arbeitswissenschaft, 45,* 1–10.

KELLY, P. L., and KROEMER, K. H. E. (1990). Anthropometry of the Elderly: Status and Recommendations. *Human Factors, 32,* 571–595.

KEMBER, P. (1985). *Bibliography: Anthropometry Related to Seating, 1975–1984*. Cranfield, UK: Ergonomics Laboratory, Cranfield Institute of Technology.

KEMMERLING, P. T. (1991). *Human Factors Engineering for the Disabled and Aging*. Course Information ISE 5654, Fall Semester, 1991. Blacksburg, VA: Virginia Tech (VPI & SU).

KEMMERLING, P. T., and DRYDEN, R. D. (1990). Designs, Paradigms, and the Changing Times. In *Proceedings of the International Ergonomics Association Conference on Human Factors in Design for Manufacturability and Process Planning* (pp. 181–184). Buffalo, NY: Helander, Dept. of IE, SUNYAB.

KENNEDY, R. S. (1991). *Long Term Effects of Rotating in an Artificial Gravity Environment*. Presentation at the Annual Aerospace Medical Association Meeting in Cincinnati, Ohio, May 5–7.

KERKHOF, G. (1985). Individual Differences and Circadian Rhythms. Chapter 3 in S. Folkard and T. H. Monk (eds.), *Hours of Work* (pp. 29–35). Chichester, UK: Wiley.

KEYSERLING, W. M. (1986a). Postural Analysis of the Trunk and Shoulder in Simulated Real Time. *Ergonomics, 29,* 569–583.

KEYSERLING, W. M. (1986b). A Computer-Aided System to Evaluate Postural Stress in the Workplace. *American Industrial Hygiene Association Journal, 47,* 641–649.

KHALIL, T. M., GENAIDY, A. M., ASFOUR, S. S., and VINCIGUERRA, T. (1985). Physiological Limits in Lifting. *American Industrial Hygiene Association Journal, 46,* 220–224.

KILBOM, A. (1986). Physiological Effects of Extreme Physical Inactivity. In *Proceedings of Work with Display Units* (pp. 486–489). Stockholm: Swedish National Board of Occupational Safety and Health.

KILBOM, A., PERSSON, J., and JONSSON, B. (1985). *Risk Factors for Work-Related Disorders of the Neck and Shoulder with Special Emphasis on Working Postures and Movement*. Zadar, Yugoslavia: International Symposium on the Ergonomics of Working Posture.

KING, A. I. (1984). A Review of Biomechanical Models. *Journal of Biomechanical Engineering, 106,* 104–124.

KING, A. I. and CHOU, C. C. (1976). Mathematical Modeling, Simulation and Experimental Testing of Biomechanical System Crash Response. *Journal Biomechanics, 9,* 301–317.

KINNEY, J. M. (ed.). (1980). *Assessment of Energy Metabolism in Health and Disease*. Columbus, OH: Ross Laboratories.

KINNEY, J. S., and HUEY, B. M. (eds.). (1990). *Application Principles for Multicolored Displays*. Washington, D.C.: National Academy Press.

KIRA, A. (1976). *The Bathroom*. New York: Viking.

KIRK, J., and SCHNEIDER, D. A. (1990). *Physiological and Perceptual Responses to Load Carrying in Female Subjects Using Internal and External Frame Backpacks.* (Technical Report TR-91/023). Natick, MA: United States Army Natick Research, Development, and Engineering Center.

KLEIN, B. P., JENSEN, R. C., and SANDERSON, L. M. (1984). Assessment of Workers' Compensation Claims for Back Strains/Sprains. *Journal of Occupational Medicine, 26,* 443–448.

KLOCKENBERG, E. A. (1926). *Rationalization of the Typewriter and of Its Use* (in German). Berlin: Springer.

KNAPIK, J. (1989). *Loads Carried by Soldiers: A Review of Historical, Physiological, Biomechanical, and Medical Aspects* (Technical Report T19-89). Natick, MA: United States Army Natick Research, Development, and Engineering Center.

KNAUTH, P., ROHMERT, W., and RUTENFRANZ, J. (1979). Systematic Selection of Shift Plans for Continuous Production with the Aid of Work-Physiological Criteria. *Applied Ergonomics 10,* 9–15.

KOBRICK, J. L., and FINE, B. J. (1983). Climate and Human Performance. In D. J. Osborne and M. M. Grunebert (eds.), *The Physical Environment at Work* (pp. 69–107). Chichester, UK: Wiley.

KOGI, K. (1985). Introduction to the Problems of Shiftwork. Chapter 14 in S. Folkard and T. H. Monk (eds.), *Hours of Work* (pp. 115–184). Chichester, UK: Wiley.

KOGI, K. (1991). Job Content and Working Time: The Scope for Joint Change. *Ergonomics, 34,* 757–773.

KONDRASKE, G. (1988). Rehabilitation Engineering: Towards a Systematic Process. *IEEE Engineering in Medicine and Biology Magazine, 10,* 11–15.

KONZ, S. (1983). *Work Design: Industrial Ergonomics* (2d ed.). Columbus, OH: Grid.

KONZ, S. (1985). *Facility Design.* New York: Wiley.

KONZ, S. (1990). *Work Design: Industrial Ergonomics.* Worthington, OH: Publishing Horizon.

KONZ, S. (ed.) (1991). Japanese Children. *Ergonomics, 34,* 971.

KONZ, S., and GOEL, S. C. (1969). The Shape of the Normal Work Area in the Horizontal Plane. *AIIE Transactions, 1,* 70–73.

KONZ, S. A., and MITAL, A. (1990). Carpal Tunnel Syndrome. *International Journal of Industrial Ergonomics, 5,* 175–180.

KOPPA, R. J. (1990). State of the Art in Automotive Adaptive Equipment. *Human Factors, 32,* 439–455.

KOSNIK, W., WINSLOW, L., KLINE, D., RASINSKI, K., and SEKULER, R. (1988). Visual Changes in Daily Life Throughout Adulthood. *Journal of Gerontology, 43,* 63–70.

KRAG, M. H., POPE, M. H., and GILBERTSON, L. G. (1984). Intra-abdominal Pressure: A Study of Its Role in Spine Biomechanics. In 1984 *Advances in Bioengineering* (pp. 125–126). New York: American Society of Mechanical Engineers.

KRAMER, A. F. (1991). Mental Workloads: A Review of Recent Papers. In D. L. Damos (ed.), *Multiple Task Performance* (pp. 279–328). London: Taylor and Francis.

KRAUS, H., MELLEBY, A., and GASTON, S. (1977). Back Pain Correction and Prevention. *New York State Journal of Medicine, 77,* 1335–1338.

KREBS, M. J., WOLF, J. D., and SANDRIG, J. H. (1978). *Color Display Design Guide* (Report ONR-CR-213-136-2F). Arlington, VA: Office of Naval Records.

KROEMER, K. H. E. (1964). On the Effect of the Spatial Position of Keyboards on Typing Performance (in German). *Int. Zeitschrift Angewandte Physiologie einschl. Arbeitsphysiol., 20,* 240–251.

KROEMER, K. H. E. (1967). *What One Should Know About Switches, Cranks, and Pedals* (in German). Frankfurt: Beuth.

KROEMER, K. H. E. (1970a). Human Strength: Terminology, Measurement and Interpretation of Data. *Human Factors, 12,* 279–313.

KROEMER, K. H. E. (1970b). Human Engineering the Keyboard. *Human Factors, 12,* 51–63.

KROEMER, K. H. E. (1971). Foot Operation of Controls. *Ergonomics, 14,* 333–361.

KROEMER, K. H. E. (1972). *Pedal Operation by the Seated Operator* (SAE Paper 72004). New York: Society of Automotive Engineers.

KROEMER, K. H. E. (1979). A New Model of Muscle Strength Regulation. In *Proceedings of the Annual Conference of the Human Factors Society, Boston, MA* (pp. 19–20). Santa Monica, CA: Human Factors Society.

KROEMER, K. H. E. (1981). Engineering Anthropometry: Designing the Work Place to Fit the Human. In *Proceedings of the Annual Conference of the American Institute of Industrial Engineers* (pp. 119–126). Norcross, GA: AIIE.

KROEMER, K. H. E. (1982). *Development of LIFTEST, A Dynamic Technique to Assess the Individual Capability to Lift Material* (Final Report, NIOSH Contract 210-79-0041). Blacksburg, VA: Ergonomics Laboratory, IEOR Department, Virginia Tech (VPI & SU).

KROEMER, K. H. E. (1983a). An Isoinertial Technique to Assess Individual Lifting Capability. *Human Factors, 25,* 493–506.

KROEMER, K. H. E. (1983b). Physiological Responses to Work: An Ergonomic Assessment. In *Proceedings of the Annual Professional Development Conference, American Society of Safety Engineers* (pp. 185–200). Park Ridge, IL: ASSE.

KROEMER, K. H. E. (1983c). Field Testing of Workers Involved in Material Handling. In *Proceedings of the Bureau of Mines Technology Transfer Symposia* (pp. 47–53). Washington, DC: U.S. Government Printing Office.

KROEMER, K. H. E. (1984a). Ergonomics of Manual Material Handling: Review of Models, Methods, and Techniques. In *Proceedings of the International Conference on Occupational Ergonomics,* Vol. 2, (pp. 56–60). Rexdale, Ontario: HFAC-IEA.

KROEMER, K. H. E. (1984b). *Ergonomic Seminars in Material Handling* (Final Report, NIOSH Contract 1349-01/02). Blacksburg, VA: Ergonomics Laboratory, IEOR Department, Virginia Tech (VPI & SU).

KROEMER, K. H. E. (1984c). *Ergonomics of VDT Workplaces* (American Industrial Hygiene Association Ergonomic Guide). Akron, OH: American Industrial Hygiene Association.

KROEMER, K. H. E. (1985a). Testing Individual Capability to Lift Material: Repeatability of a Dynamic Test Compared with Static Testing. *Journal of Safety Research, 16,* 1–7.

KROEMER, K. H. E. (1985b). *Personnel Selection for Material Handling.* Reprint 85-326, SME-AIME Fall Meeting. Littleton, CO: Society of Mining Engineers.

KROEMER, K. H. E. (1985c). Office Ergonomics: Work Station Dimensions. Chapter 18 in D. C. Alexander and B. M. Pulat (eds.), *Industrial Ergonomics* (pp. 187–201). Norcross, GA: Institute of Industrial Engineers.

KROEMER, K. H. E. (1985d). An Ergonomist's Perspective of Ski Facilities Design. In *Proceedings of the 5th Annual Vail Aerial Tramway and Safety Seminar* (pp. 39–43). Vail, CO: Vail Associates.

KROEMER, K. H. E. (1986a). *Review of the "Training" Literature on Manual Material Handling*. Unpublished Report, Industrial Ergonomics Laboratory, IEOR Department, Blacksburg, VA: Virginia Tech (VPI & SU).

KROEMER, K. H. E. (1986b). Human Muscle Strength: Definition, Generation and Measurement. In *Proceedings of the Human Factors Society Annual Meeting* (pp. 977–981). Santa Monica, CA: Human Factors Society.

KROEMER, K. H. E. (1986c). Coupling the Hand with the Handle. *Human Factors, 28,* 337–339.

KROEMER, K. H. E. (1987a). Matching Individuals to the Job Can Reduce Manual Labor Injuries. *Occupational Safety and Health News Digest, 3,* 4–7.

KROEMER, K. H. E. (1987b). Biomechanics of the Human Body. In G. Salvendy (ed.), *Handbook of Human Factors* (pp. 169–181). New York: Wiley.

KROEMER, K. H. E. (1987c). Ergonomics. Chapter 13 in B. A. Plog (ed.), *Fundamentals of Industrial Hygiene* (3d ed.) (pp. 283–334). Chicago, IL: National Safety Council.

KROEMER, K. H. E. (1988a). VDT Workstation Design. Chapter 23 in M. Helander (ed.), *Handbook of Human Computer Interaction* (pp. 521–539). Amsterdam: Elsevier.

KROEMER, K. H. E. (1988b). Ergonomic Seats for Computer Workstations. In F. Aghazadeh (ed.), *Trends in Ergonomics/Human Factors V* (pp. 313–320). Amsterdam: Elsevier.

KROEMER, K. H. E. (1988c). *Ergonomics Manual for Manual Material Handling* (5th ed). Radford, VA: Author.

KROEMER, K. H. E. (1988d). Ergonomics. In A. Plog (ed.), *Fundamentals of Industrial Hygiene* (pp. 183–334). Chicago: National Safety Council.

KROEMER, K. H. E. (1989a). Engineering Anthropometry. *Ergonomics, 32,* 767–784.

KROEMER, K. H. E. (1989b). Cumulative Trauma Disorders: Their Recognition and Ergonomic Measures to Avoid Them. *Applied Ergonomics, 20,* 274–280.

KROEMER, K. H. E. (1989c). Personnel Testing and Selection. In K. H. E. Kroemer, J. D. McGlothlin, and T. G. Bobick (eds.), *Manual Material Handling: Understanding and Preventing Back Trauma* (pp. 65–71). Akron, OH: American Industrial Hygiene Association.

KROEMER, K. H. E. (1989d). A Survey of Ergonomic Models of Anthropometry, Human Biomechanics, and Operator-Equipment Interfaces. In *Proceedings of the Human Factors Society 33rd Annual Meeting* (pp. 571–575). Santa Monica, CA: Human Factors Society.

KROEMER, K. H. E. (1989e). A Survey of Ergonomic Models of Anthropometry, Human Biomechanics, and Operator-Equipment Interfaces. In G. R. McMillan, D. Beevis, E. Salas, M. H. Strub, R. Sutton, and L. Van Breda (eds.), *Applications of Human Performance Models to System Design* (pp. 331–339). New York: Plenum.

KROEMER, K. H. E. (1990). Cumulative Trauma Disorders. *Applied Ergonomics, 20,* 274–280.

KROEMER, K. H. E. (1991a). Sitting at Work: Recording and Assessing Body Postures, Designing Furniture for Computer Workstations. In A. Mital and W. Karwowski (eds.), *Work Space, Equipment and Tool Design* (pp. 93–109). Amsterdam, NL: Elsevier.

KROEMER, K. H. E. (1991b). Experiments with the TCK—A Keyboard with Built-in Wrist Rest and Only Eight Keys. In W. Karwowski and J. W. Yates (eds.), *Advances in Industrial Ergonomics and Safety III* (pp. 537–542). London, UK: Taylor and Francis.

KROEMER, K. H. E. (1991c). Cumulative Trauma Disorders in Computer Operators. In *Proceedings of the 11th Congress of the International Ergonomics Association* (pp. 727–729). London: Taylor and Francis.

KROEMER, K. H. E. (1992a). Performance on a Prototype Keyboard with Ternary Chorded Keys. *Applied Ergonomics, 23,* 83–90.

KROEMER, K. H. E. (1992b). Avoiding Cumulative Trauma Disorders in Shop and Office. *J. Am. Industrial Hygiene Association* (in press).

KROEMER, K. H. E. (1993). Operation of Ternary Chorded Keys. Accepted for publication in *American Journal of Human-Computer Interaction.*

KROEMER, K. H. E., FATHALLAH, F. A., and LANGLEY, L. W. (1988). A New Keyboard with Chorded Ternary Keys. In *Proceedings of the Human Factors Society 32nd Annual Meeting* (pp. 1005–1008). Santa Monica, CA: Human Factors Society.

KROEMER, K. H. E., KROEMER, H. J., and KROEMER-ELBERT, K. E. (1986). *Engineering Physiology: Bases of Ergonomics* (1st ed.). Amsterdam: Elsevier.

KROEMER, K. H. E., KROEMER, H. J., and KROEMER-ELBERT, K. E. (1990). *Engineering Physiology: Bases of Ergonomics* (2d ed.). New York: Van Nostrand Reinhold.

KROEMER, K. H. E., and MARRAS, W. S. (1980). Toward an Objective Assessment of the Maximal Voluntary Contraction Component in Routine Muscle Strength Measurements. *European Journal of Applied Physiology, 45,* 1–9.

KROEMER, K. H. E., MARRAS, W. S., MCGLOTHLIN, J. D., MCINTYRE, D. R., and NORDIN, M. (1990). Assessing Human Dynamic Muscle Strength. *International Journal of Industrial Ergonomics, 6,* 199–210.

KROEMER, K. H. E., MCGLOTHLIN, J. D., and BOBICK, T. J. (eds.). (1989). *Manual Material Handling: Understanding and Preventing Back Trauma.* Akron, OH: American Industrial Hygiene Association.

KROEMER, K. H. E., and ROBINETTE, J. C. (1968). *Ergonomics in the Design of Office Furniture. A Review of European Literature* (AMRL-TR 68-90). Wright-Patterson AFB, OH. Also pubished with shortened list of references (1969) in International Journal of Industrial Medicine and Surgery, 38, 115–125.

KROEMER, K. H. E., SNOOK, S. H., MEADOWS, S. K., and DEUTSCH, S. (eds.). (1988). *Ergonomic Models of Anthropometry, Human Biomechanics, and Operator-Equipment Interfaces.* Washington, DC: National Academy Press.

KRUEGER, H., CONRADY, P., and ZUELCH, J. (1989). Work with Magnifying Glasses. *Ergonomics, 32,* 785–794.

KUBOVI, M. (1986). Perceptual Organization and Cognition, Overview. Section VI in K. R. Boff, L. Kaufman, and J. P. Thomas (eds.). *Handbook of Perception and Human Performance* (pp. vi.i–vi.9). New York: Wiley.

KUMAR, S. (1984). The Physiological Cost of Three Different Methods of Lifting in Sagittal and Lateral Planes. *Ergonomics, 27,* 425–433.

KUMAR, S., and DAVIS, P. R. (1983). Spinal Loading in Static and Dynamic Postures: EMG and Intra-Abdominal Pressure Study. *Ergonomics, 26,* 913–922.

KUMAR, S., CHAFFIN, D. B., and REDFERN, M. (1988). Isometric and Isokinetic Back and Arm Lifting Strengths: Device and Measurement. *Biomechanics, 21,* 35–44.

KUORINKA, I., JONSSON, B., KILBOM, A., VINTERBERG, H., BIERING-SORENSEN, F., ANDERSSON, G., and JORGENSEN, K. (1987). Standardized Nordic Questionnaires for the Analysis of Musculoskeletal Symptoms. *Applied Ergonomics, 18,* 233–237.

KVALSETH, T. O. (1991). Reaction Time and Stimulus Information. In *Proceedings of the 11th Congress of the International Ergonomics Association* (pp. 475–459). London: Taylor and Francis.

KWALLEK, N., and LEWIS, C. M. (1990). Effects of Environmental Colour on Males and Females: A Red or White or Green Office. *Applied Ergonomics, 21,* 275–278.

LADIN, Z., and WU, G. (1991). Combining Position and Acceleration Measurements for Joint Force Estimation. *Journal Biomechancis, 12,* 1173–1187.

LAMEY, J., AGHAZADEH, F., and NYE, J. (1991). A Study of the Static Anthropometric Measurements of Jamaican Agricultural Workers. In *Proceedings, 11th Congress of the International Ergonomics Association* (pp. 897–899). London, UK: Taylor and Francis.

LANDER, C., KORBON, G. A., DEGOOD, D. E., and ROWLINGSON, J. C. (1987). The Balans Chair and Its Semi-kneeling Position. An Ergonomic Comparison with the Conventional Sitting Position. *Spine, 12,* 269–272.

LANDER, J. E., SIMONTON, R. L., and GIACOBBE, J. F. K. (1990). The Effectiveness of Weight-Belts During the Squat Exercise. *Medicine and Science in Sports and Exercise, 22,* 117–124.

LANKHORST, G. J., VAN DE STADT, R. J., VOGELAAR, T. W., VAN DER KORST, J. K., and PREVO, A. J. H. (1983). The Effect of the Swedish Back School in Chronic Ideopathic Low Back Pain. *Scandanavian Journal of Rehabilitation and Medicine, 15,* 141–145.

LATECK, J. C., and FOSTER, L. W. (1985). Implementation of Compressed Work Schedules: Participation and Job Redesign as Critical Factors for Employee Acceptance. *Personnel Psychology, 38,* 75–92.

LAUGHERY, K. R., JACKSON, A. S., and FONTENELLE, G. A. (1988). Isometric Strength Tests: Predicting Performance in Physically Demanding Transport Tasks. In *Proceedings of the Human Factors Society 32nd Annual Meeting* (pp. 695–699). Santa Monica, CA: Human Factors Society.

LAVENDER, S. A., TISUANG, Y., HAFEZI, A., ANDERSSON, G. B. J., CHAFFIN, D. B., and HUGHES, R. E. (1992). Coactivation of the Trunk Muscles During Asymmetric Loading of the Torso. *Human Factors, 34,* 239–247.

LAVIE, P. (1985). Ultradian Cycles in Wakefulness. Chapter 9 in S. Folkard and T. H. Monk (eds.), *Hours of Work* (pp. 97–106). Chichester, UK: Wiley.

LAWTON, M. P. (1990). Aging and Performance of Home Tasks. *Human Factors, 32,* 527–536.

LAZARUS, R. S. and COHEN, J. B. (1977). Environmental Stress. In L. Altman and J. F. Wohlwill (eds.), *Human Behavior and the Environment: Current Theory and Research,* Vol. 2. New York: Plenum.

LAZARUS, R. S., and FOLKMAN, S. (1984). *Stress, Appraisal, and Coping.* New York: Springer.

LEE, C. H., HOSNI, Y. A., GUTHRIE, L. L., BARTH, T., and HILL, C. (1991). Design and Evaluation of a Work Seat for Overhead Tasks. In W. Karwowski and J. W. Yates (eds.), *Advances in Industrial Ergonomics and Safety III* (pp. 555–562). London, UK: Taylor and Francis.

LEE, K. S., CHAFFIN, D. B., WAIKAR, A. M., and CHUNG, M. K. (1989). Lower Back Muscle Forces in Pushing and Pulling. *Ergonomics, 32,* 1551–1563.

LEE, P., WEI, S., ZHAO, J., and BADLER, N. I. (1990). Strength Guided Motion. *Computer Graphics, 24,* 253–262.

LEE, Y. T. (1988). Toward Electronic Work Design. In *Proceedings of the Human Factors Society 32nd Annual Meeting* (pp. 622–626). Santa Monica, CA: Human Factors Society.

LEGG, S. J. (1985). Comparison of Different Methods of Load Carriage. *Ergonomics, 28,* 197–212.

LEGG, S. J., and PATEMAN, C. M. (1984). A Physiological Study of the Repetitive Lifting Capabilities of Healthy Young Males. *Ergonomics, 27,* 259–272.

LEHMANN, G. (1962). Praktische Arbeitsphysiologie (2d ed.). Stuttgart, Germany: Thieme.

LEHTO, M. R., and CLARK, D. R. (1991). Warning Signs and Labels in the Work Place. In A. Mital and W. Karwowski (eds.), *Workspace, Equipment and Tool Design* (pp. 303–344). Amsterdam: Elsevier.

LEHTO, M. R., and MILLER, J. M. (1986). *Warnings,* Volume I. Ann Arbor, MI: Fuller.

LESKINEN, T. P. J., STALHAMMAR, H. R., and KUORINKA, I. A. A. (1983). A Dynamic Analysis of Spinal Compression with Different Lifting Techniques. *Ergonomics, 26,* 595–604.

LEVINE, D. B., ZITTER, M., and INGRAM, L. (eds.). (1990). *Disability Statistics: An Assessment*. Committee on National Statistics, National Research Council. Washington, DC: National Academy Press.

LEVINE, M. D. (1971). Depression, Back Pain, and Disc Protrusion: Relationships and Proposed Psychophysiological Mechanisms. *Diseases of the Nervous System, 32,* 41–45.

LEWIN, T. (1969). Anthropometric Studies on Swedish Industrial Workers When Standing and Sitting. *Ergonomics, 12,* 883–902.

LI, C., HWANG, S., and WANG M. (1990). Static Anthropometry of Civilian Chinese in Taiwan Using Computer-Analyzed Photography. *Human Factors, 32,* 359–370.

LIND, A. R., and MCNICOL, G. W. (1968). Cardiovascular Responses to Holding and Carrying Weights by Hand and by Shoulder Harness. *Journal of Applied Physiology, 25,* 261–267.

LINDGREN, G. (1976). Height, Weight and Menarche in Swedish Schoolchildren in Relation to Socio-economic and Regional Factors. *Annals of Human Biology, 3,* 510–528.

LITTERICK, I. (1981). QWERTYUIOP–Dinosaur in a Computer Age. *New Scientist* (January), 66–68.

LOCKE, J. C. (1983). Stretching Away from Back Pain Injury. *Occupational Health and Safety, 52,* 8–13.

LOEB, K. M. C. (1983). Membrane Keyboards and Human Performance. *The Bell System Technical Journal, 62,* 1733–1749.

LOHMAN, T. G., ROCHE, A. F., and MARTOREL, R. (eds.) (1988). *Anthropometric Standardization Reference Manual*. Champaign, IL: Human Kinetics.

LOUHEVAARA, V., ILMARINEN, J., and OJA, P. (1985). Comparison of Three Field Methods for Measuring Oxygen Consumption. *Ergonomics, 28,* 463–470.

LOUPAJARVI, T. (1987). Workers' Education. *Ergonomics, 30,* 305–311.

LOVELESS, N. E. (1962). Direction-of-Motion Stereotypes: A Review. *Ergonomics, 5,* 357–383.

LOWREY, G. H. (1986). *Growth and Development of Children* (8th ed.). Chicago: Year Book Medical Publishers.

LUEDER, R. K. (1983). Seat Comfort. A Review of the Construct in the Office Environment. *Human Factors, 25,* 701–711.

LUEDER, R., and NORO, K. (eds.) (1992). *Hard Facts About Soft Mechanisms: Ergonomics of Seating*. London: Taylor and Francis.

LUNDERVOLD, A. (1951). Electromyographic Investigations of Position and Manner of Working in Typewriting. *Acta Physiologica Scandinavica, 24,* 84–104.

LUTZ, M. C., and CHAPANIS, A. (1955). Expected Location of Digits and Letters on Ten-Button Key Sets. *Applied Psychology, 39,* 314–317.

LYKKEN, D. T. (1988). The Case Against the Polygraph Test. Chapter 9 in A. Gale (ed.), *The Polygraph Test. Lies, Truth, and Science* (pp. 111–125). London, UK: Sage.

MAIRIAUX, P., and MALCHAIRE, J. (1988). Relation Between Intra-abdominal Pressure and Lumbar Stress: Effect of Trunk Posture. *Ergonomics, 31,* 1331–1342.

MALONE, R. L. (1991). *Posture Taxonomy*. Unpublished MS thesis, Department of Industrial and Systems Engineering, Blacksburg, VA: Virginia Tech (VPI & SU).

MANDAL, A. C. (1975). Work-Chair with Tilting Seat. *Lancet,* 642–643.

MANDAL, A. C. (1982). The Correct Height of School Furniture. *Human Factors, 24,* 257–269.

MARQUER, P., and CHALMA, M. C. (1961). The Development of Morphologic Traits as a Function of Age Among 2089 French Women Aged 20 to 91 Years (in French). *Bulletin et Memoires de la Societe d'Anthropologie de Paris, XI,* 1–78.

MARRAS, W. S. (1987). Trunk Motion During Lifting: Temporal Relations Among Loading Factors. *International Journal of Industrial Ergonomics, 1,* 551–562.

MARRAS, W. S. (1988). Predictions of Forces Acting Upon the Lumbar Spine Under Isometric and Isokinetic Conditions: A Model-Experiment Comparison. *International Journal of Industrial Ergonomics, 3,* 19–27.

MARRAS, W. S. (1989). Towards an Understanding of Internal Trunk Responses to Dynamic Trunk Activity. In K. H. E. Kroemer, J. D. McGlothlin, and T. G. Bobick (eds.), *Manual Material Handling: Understanding and Preventing Back Trauma* (pp. 23–33). Akron, OH: American Industrial Hygiene Association.

MARRAS, W. S., FERGUSON, S. A., and SIMON, S. R. (1990). Three Dimensional Dynamic Motor Performance of the Normal Trunk. *International Journal of Industrial Ergonomics, 6,* 211–224.

MARRAS, W. S., KING, A. I., and JOYNT, R. L. (1984). Measurement of Loads on the Lumbar Spine under Isometric and Isokinetic Conditions. *Spine,* 9, 176–188.

MARRAS, W. S., and MIRKA, G. A. (1989). Trunk Strength During Asymmetric Trunk Motion. *Human Factors, 31,* 667–677.

MARRAS, W. S., and MIRKA, G. A. (1990). Muscle Activities During Asymmetric Trunk Angular Accelerations. *Journal of Orthopaedic Research, 8,* 824–832.

MARRAS, W. S., and RANGARAJULU, S. L. (1987). Trunk Force Development During Static and Dynamic Lifts. *Human Factors, 29,* 19–29.

MARRAS, W. S., RANGARAJULU, S. L., and LAVENDER, S. A. (1987). Trunk Loading and Expectation. *Ergonomics, 30,* 551–562.

MARRAS, W. S., RANGARAJULU, S. L., and WONGSAM, P. E. (1987). Trunk Force Development During Static and Dynamic Lifts. *Human Factors, 29,* 19–29.

MARRAS, W. S., and REILLY, C. H. (1987). Internal Trunk-Loading Sequence Responses to Lifting Motions. In *Proceedings of the Human Factors Society, 31st Annual Meeting* (pp. 447–451). Santa Monica, CA: Human Factors Society.

MARRAS, W. S., and REILLY, C. H. (1988). Networks of Internal Trunk-Loading Activities Under Controlled Trunk-Motion Conditions. *Spine, 13,* 661–667.

MARRAS, W. S., and SCHOENMARKLIN, R. W. (1991). Wrist Motions and CTD Risk in Industrial and Service Environments. In *Proceedings of the 11th Congress of the International Ergonomics Association* (pp. 36–38). London, UK: Taylor and Francis.

MARRAS, W. S., WONGSAM, P. E., and RANGARAJULU, S. L. (1987). *Trunk Motion During Lifting: The Relative Cost* (Unpublished Report). Columbus: Ohio State University.aaae

MARTIN, B. J., ROLL, J. P., and GAUTHIER, G. M. (1986). Inhibitory Effects of Combined Agonist and Antagonist Muscle Vibration on H-Reflex in Men. *Aviation, Space, and Environmental Medicine, 57,* 681–687.

MARTIN, D. K., and DAIN, S. J. (1988). Postural Modifications of VDU Operators Wearing Bifocal Spectacles. *Applied Ergonomics, 19,* 293–300.

MARTIN, E. (1949). *The Typewriter and its Historical Development* (in German). Aachen, Germany: Basten.

MARTIN, P. E., and NELSON R. C. (1985). The Effect of Carried Loads on the Combative Performance of Men and Women. *Military Medicine, 150,* 357–362.

MARTIN, R. (1914). *Lehrbuch der Anthropologie* (1st ed.). Jena, Germany: Fischer.

MARUTA, T., SWANSON, D. W., and SWANSON, W. M. (1976). Pain as a Psychiatric Symptom: Comparison Between Low Back Pain and Depression. *Psychosomatics, 17,* 123–127.

MCCONVILLE, J. T., and CHURCHILL, E. (1976). *Statistical Concepts in Design* (AMRL-TR-76-79). Wright-Patterson Air Force Base, OH: Aerospace Medical Research Laboratory.

MCCONVILLE, J. T., CHURCHILL, T., KALEPS, I., CLAUSER, C. E., and CUZZI, J. (1980). *Anthropometric Relationships of Body and Body Segment Moments of Inertia* (AFAMRL–TR-80-119). Wright-Patterson AFB, OH: Aerospace Medical Research Laboratory.

MCCONVILLE, J. T., ROBINETTE, K. M., and CHURCHILL, T. (1981). *An Anthropometric Data Base for Commercial Design Applications* (Final Report, NSF-DAR-80 09 861). Yellow Springs, OH: Anthropology Research Project, Inc.

MCCORMICK, E. J., and SANDERS, M. S. (1982). *Human Factors in Engineering and Design*. New York: McGraw-Hill.

MCDANIEL, J. W. (1991). The Development of Computer Models for Ergonomic Accommodation. In A. Mital and W. Karwowski (eds.), *Workspace, Equipment, and Tool Design* (pp. 29–66). Amsterdam: Elsevier.

MCDANIEL, J. W., SKANDIS, R. J., and MADOLE, S. W. (1983). *Weight Lift Capabilities of Air Force Basic Trainees* (AFAMRL-TR-83-0001). Wright-Patterson Air Force Base, OH: Air Force Aerospace Medical Research Laboratory.

MCEVOY, G. M., and CASCIO, W. F. (1989). Cumulative Evidence of the Relationship Between Employee Age and Job Performance. *Journal Applied Psychology, 74,* 11–17.

MCGILL, S. M., and NORMAN, R. W. (1987). Reassessment of the Role of Intra-abdominal Pressure in Spinal Compression. *Ergonomics, 30,* 1565–1588.

MCGILL, S. M., NORMAN, R. W., and SHARRATT, M. T. (1990). The Effect of an Abdominal Belt on Trunk Muscle Activity and Intra-Abdominal Pressure During Squat Lifts. *Ergonomics, 33,* 147–160.

MCMILLIAN, G. R., BEEVIS, D., SALAS, E., STRUB, M. H., SUTTON, R., and VAN BREDA, L. (eds.). (1989). *Applications of Human Performance Models to System Design*. New York: Plenum.

McWright, A. (1988). Gender Differences in the Strain Responses to Job Demand. In *Proceedings of the Human Factors Society 32nd Annual Meeting* (pp. 853–856). Santa Monica, CA: Human Factors Society.

Meadows, S. (1988). The Wave of Innovation for an Aging Society: Enhancing Independent Living. In *Proceedings of the Human Factors Society 32 Annual Meeting* (p. 184). Santa Monica, CA: Human Factors Society.

Mebarki, B., and Davies, B. T. (1990). Anthropometry of Algerian Women. *Ergonomics, 33,* 1537–1547.

Medawar, P. B. (1944). Size, Shape and Age. Presented to D'Arcy Wentworth Thompson (pp. 155–187). Oxford, UK: Clarendon.

Meister, D. (1989). *Conceptual Aspects of Human Factors*. Baltimore: Johns Hopkins University Press.

Meister, D. (1990). Simulation and Modelling. Chapter 8 in Wilson and E. N. Corlett (eds.), *Evaluation of Human Work* (pp. 180–199). London, UK: Taylor and Francis.

Meister, K. J. (1990). A Few Implications of an Ecological Approach to Human Factors. *Human Factors Society Bulletin,* 33(11), 1–4.

Mellerowicz, H., and Smodlaka, V. N. (1981). *Ergometry*. Baltimore: Urban and Schwarzenberg.

Mello, R. P., Damokosh, A. I., Reynolds, K. L., Witt, C. E., and Vogel, J. A. (1988). *The Physiological Determinants of Load Bearing Performance at Different March Distances* (Technical Report No. T15–88). Natick, MA: U.S. Army Research Institute of Environment Medicine.

Melton, B. (1983). Back Injury Prevention Means Education. *Occupational Health and Safety, 52,* 20–23.

Mendenhall, G. S. (1977). *Carpal Tunnel Syndrome*. Pueblo, CO: Dynamic Communications.

Meredith, H. W. (1976). Findings from Asia, Australia, Europe and North America on Secular Change in Mean Height of Children, Youths and Young Adults. *American Journal of Physical Anthropology, 44,* 315–326.

Metropolitan Life Foundation (ed.) (1983). Comparison of 1959 and 1983 Metropolitan Height and Weight Tables. *Statistical Bulletin, 64,* 6–7.

Meyer, H. von (1863). *The Changing Locations of the Center of Gravity in the Human Body*. (In German, translation in) *Human Mechanics* (AMRL–TDR-63-123). Wright-Patterson AFB, OH: Aerospace Medical Research Laboratory, 1963.

Michaels, S. E. (1971). QWERTY vs. Alphabetic Keyboards as a Function of Typing Skills. *Human Factors, 13,* 419–426.

MIL-HDB 759, U.S. Army Missile Command (1981). *Human Factors Engineering Design for Army Material* (Metric). Philadelphia, PA: Naval Publications and Forms Center.

Millar, J. D. (1989). *Occupational Safety and Health Objectives for the Year 2000*. Centers for Disease Control, Atlanta, GA: Letter of 23 February 1989.

Miller, J. F., and Stamford, B. A. (1987). Intensity and Energy Cost of Weighted Walking vs. Running for Men and Women. *Journal Applied Physiology, 62,* 497–1501.

Miller, N. S., and Gold, M. S. (1991). *Alcohol*. New York: Plenum.

Miller, J. M., and Lehto, M. R. (1986). *Warnings,* Vol. II. Ann Arbor, MI: Fuller.

Miller, R. L. (1977). Bend Your Knees. *National Safety News, 115,* 57–58.

MILLER, R. J. (1990). Pitfalls in the Conception, Manipulation, and Measurement of Visual Accommodation. *Human Factors, 32,* 27–44.

MIL-STD 1472, U.S. Army Missile Command (1981). *Human Engineering Design Criteria for Military Systems, Equipment and Facilities*. Philadelphia: Naval Publications and Forms Center.

MINORS, D. S., and WATERHOUSE, J. M. (1981). *Circadian Rhythms and the Human*. Bristol, UK: Wright.

MINORS, D. S., and WATERHOUSE, J. M. (1987). The Role of Naps in Alleviating Sleepiness During an Irregular Sleep-wake Schedule. *Ergonomics, 30,* 1261–1273.

MITAL, A. (1987). Maximum Weights of Asymmetrical Loads Acceptable to Industrial Workers for Symmetrical Manual Lifting. *American Industrial Hygiene Association Journal, 48,* 539–544.

MITAL, A. (1991). Hand Tools: Injuries, Illnesses, Design, and Usage. In A. Mital and W. Karwowski (eds.), *Workspace, Equipment, and Tool Design* (pp. 219–256). Amsterdam: Elsevier.

MITAL, A., and KARWOWSKI, W. (eds.). (1991). *Workspace, Equipment and Tool Design*. Amsterdam: Elsevier.

MITAL, A., KARWOWSKI, W., MAZOUZ, A., and ORSARH, E. (1986). Prediction of Maximum Acceptable Weight of Lift in the Horizontal and Vertical Planes Using Simulated Job Dynamic Strengths. *American Industrial Hygiene Association Journal, 47,* 288–292.

MOFFETT, J. A. K., CHASE, S. M., PORTEK, I., and ENNIS, J. R. (1986). A Controlled Perspective Study to Evaluate the Effectiveness of a Back School in the Relief of Chronic Low Back Pain. *Spine, 11,* 120–121.

MONK, T. H. (1989). Shift Worker Safety: Issues and Solutions. In A. Mital (ed.), *Advances in Industrial Ergonomics and Safety I* (pp. 887–893). Philadelphia: Taylor and Francis.

MONK, T. H., and TEPAS, D. I. (1985). Shiftwork. Chapter 5 in Cooper and Smith, M. J. (eds.), *Job Stress and Blue Collar Work* (pp. 65–84). New York: Wiley.

MONK, T. H., and WAGNER, J. A. (1989). Social Factors Can Outweigh Biological Ones in Determining Night Shift Safety. *Human Factors, 31,* 721–724.

MONOD, H., and VALENTIN, M. (1979). The Predecessors of Ergonomy (in French). *Ergonomics, 22,* 673–680.

MONTY, R. W., and SNYDER, H. L. (1983). Keyboard Design: An Investigation of User Preference and Performance. In *Proceedings of the Human Factors Society 27th Annual Meeting* (pp. 201–205). Santa Monica, CA: Human Factors Society.

MONTEGRIFFO, V. M. E. (1968). Height and Weight of a United Kingdom Adult Population with a Review of Anthropometric Literature. *Annals of Human Genetics, 31,* 389–398.

MOOG, R. (1987). Optimization of Shift Work: Physiological Contributions. *Ergonomics, 30,* 1249–1259.

MOORE, A., WELLS, R., and RANNEY, D. (1991). Quantifying Exposure in Occupational Manual Tasks with Cumulative Trauma Potential. *Ergonomics, 34,* 1433.

MOORE, T. G. (1974). Tactile and Kinesthetic Aspects of Pushbuttons. *Applied Ergonomics, 5,* 66–71.

MORAY, N. (1988). Mental Workload Since 1979. *International Reviews of Ergonomics, 2,* 123–150.

MORETZ, S. (1987). How to Prevent Costly Back Injuries. *Occupational Hazards, 7,* 45–48.

MORGAN, S. (1991). Wrist Factors Contributing to CTS Can be Minimized, if Not Eliminated. *Occupational Health and Safety, 60,* 47–51.

MORRIS, A. (1984). Program Compliance Key to Preventing Low Back Injuries. *Occupational Health and Safety, 53,* 44–47.

MORRIS, A. (1985). Identifying Workers at Risk to Back Injury is Not Guess Work. *Occupational Health and Safety, 55,* 16–20.

MORRIS, A., and RANDOLPH, J. (1984). Back Rehabilitation Programs Speed Recovery of Injured Workers. *Occupational Health and Safety, 53,* 53–55, 64–68.

MORROW, D. Y., and JEROME, L. (1990). The Influence of Alcohol and Aging on Radio Communication During Flight. *Aviation, Space and Environmental Medicine, 61,* 12–20.

MOTOWIDLO, S. J., PACKARD, J. S., and MANNING, M. R. (1986). Occupational Stress: Its Causes and Consequences for Job Performance. *Journal Applied Psychology, 71,* 618–629.

MOUSTAFA, A. W., DAVIES, B. T., DARWICH, M. S., and IBRAHEEM, M. A. (1987). Anthropometric Study of Egyptian Women. *Ergonomics, 30,* 1089–1098.

MUCHINSKY, P. M. (1987). *Psychology Applied to Work* (2d ed.). Chicago: Dorsey.

MUNSEL, A. H. (1942). *Book of Color*. Baltimore: Munsell Color Book Corp.

MURRELL, K. F. H. (1969). *Ergonomics—Man and His Working Environment*. London: Chapman and Hall.

MYERS, R. H. (1986). *Classical and Modern Regression with Applications*. Boston: Duxbury.

MYERS, R. H., and WALPOLE, R. A. (1985). *Probability and Statistics for Engineers and Scientists* (3d ed.). New York: MacMillan.

NACHEMSON, A. (1987). Lumbar Intradiscal Pressure. In M. A. V. Jayson (ed.), *The Lumbar Spine and Back Pain* (3d ed.) (pp. 191–203). Edinburgh, UK: Churchill-Livingstone.

NACHEMSON, A. (1989). Individual Factors Contributing to Low Back Pain. Presented at the American Academy of Orthopaedic Surgeons, New York, May 1989.

NACHEMSON, A., and ANDERSSON, G. (1982). Classification of Low Back Pain. *Scandinavian Journal of Work and Environmental Health, 8,* 134–136.

NADEL, E. R. (1984). Energy Exchanges in Water. *Undersea Biomedical Research, 11,* 149–158.

NADEL, E. R., and HORVATH, S. M. (1975). Optimal Evaluation of Cold Tolerance in Man. Chapter 6A in S. M. Horvath, S. Kondo, H., Matsui, and H. Yoshimura (eds.), *Comparative Studies on Human Adaptability of Japanese, Caucasians, and Japanese American,* Vol. 1. Tokyo: Japanese Committee of International Biological Program.

NAG, P. K. (1991). Endurance Limits in Different Modes of Load Holding. *Applied Ergonomics, 22,* 185–188.

NAG, P. K., CHINTHARIA, S., SAIYED, S., and NAG, A. (1986). EMG Analysis of Sitting Work Postures in Women. *Applied Ergonomics, 17,* 195–197.

NAG, P. K., SEBASTIAN, N. C., and MAVLANKAR, M. G. (1980). Occupational Work Load of Indian Agricultural Workers. *Ergonomics, 23,* 91–102.

NASA (1989). *Man-Systems Integration Standards* (Revision A). (NASA-STD 3000). Houston, TX: L.B.J. Space Center, SP 34-89-230.

NASA/Webb (eds.) (1978). *Anthropometric Sourcebook* (3 volumes). (NASA Reference Publication 1024.) Houston, TX: LBJ Space Center.

National Academy of Sciences (1980). Recommended Standard Procedures for the Clinical Measurement and Specification of Visual Acuity (Report of Working Group 39, Committee on Vision). *Archives of Opthalmology, 41,* 103–148.

National Academy of Sciences (ed.) (1985). *Injury in America*. Washington, DC: National Academy Press.

National Institutes of Health (ed.) (1985). *Health Implications of Obesity*. Conference Statement, 5(9). Washington, DC: U.S. Government Printing Office.

National Institutes of Health (ed.) (1990). *Noise and Hearing Loss*. NIH Consensus Development Conference Statement, Vol. 8, No. 1. Bethesda, MD: National Library of Medicine, Office of Medical Applications of Research.

National Research Council (ed.) (1979). *Odors from Stationary and Mobile Sources*. Washington, DC: National Academic Press.

National Research Council (ed.) (1983). *Research Needs for Human Factors*. Washington, DC: National Academic Press.

National Research Council Committee on Vision (ed.) (1983). *Video Displays, Work and Vision*. Washington, DC: National Academy Press.

National Safety Council (1971). Human Kinetics in Lifting. *National Safety News* (June issue), 44–47.

NELSON, B. D., GARDNER, R. M., OSTLER, D. V., SCHULTZ, J. M., and LOGAN, J. S. (1990). Medical Impact Analysis for the Space Station. *Aviation, Space, and Environmental Medicine, 61*, 169–175.

NELSON, R. M. (1987). Prevention—A Government Perspective. *Ergonomics, 30*, 221–226.

NICE (1987). *Study of Body Dimensions of Workers* (in Thai). Bangkok: Department of Labour. (Cited by Intaranont, 1991.)

NICHOLLS, J. A., and GRIEVE, D. W. (1992). Performance of Physical Tasks in Pregnancy. *Ergonomics, 35*, 301–311.

NICHOLSON, A. S. (1989). A Comparative Study of Methods for Establishing Load Handling Capabilities. *Ergonomics, 32*, 1125–1144.

NICHOLSON, A. S. (1991). Anthropometry and Workspace Design. In A. Mital and W. Karwowski (eds.), *Workspace, Equipment and Tool Design* (pp. 3–28). Amsterdam: Elsevier.

NICOGOSSIAN, A. E., HUNTOON, C. L., and POOL, S. L. (1989). *Space Physiology and Medicine* (2nd ed.). Philadelphia: Lea and Febiger.

NIELSEN, R., GAVHED, D., and NILLSON, H. (1989). Thermal Function of a Clothing Ensemble During Work: Dependency on Inner Layer Fit. *Ergonomics, 32*, 1581–1594.

NIOSH (1981). *Work Practices Guide for Manual Lifting*. DHHS (NIOSH) Publication No. 81–122. Washington, DC: U.S. Government Printing Office.

NIOSH (1985). *Prevention of Musculo-Skeletal Disorders*. Draft Statement, National Symposium on the Prevention of Work-Related Disease and Injuries. Atlanta, GA: Centers for Disease Control.

NOBLE, C. E. (1978). Age, Race, and Sex in the Learning and Performance of Psycho-Motor Skills. In R. T. Osborne, C. E. Noble, and N. Weyl (eds.), *Human Virbration: The Biopsychology of Age, Race, and Sex* (pp. 287–378). New York: Academic Press.

NOONE, G., and MAZUMDAR (1992). Lifting Low-lying Loads in the Sagittal Plane. *Ergonomics, 35*(1), 65–92.

NORDIN, M. (1991). Worker Training and Conditioning. In *Proceedings, Occupational Ergonomics: Work Related Upper Limb and Back Disorders*. San Diego, CA: American Industrial Hygiene Association, San Diego Section.

NORMAN, D. A. (1991). Cognitive Science in the Cockpit. *CSERIAC Gateway, 2*, 1–6.

NORMAN, D. A., and FISHER, D. (1982). Why Alphabetic Keyboards Are Not Easy to Use: Keyboard Layout Doesn't Much Matter. *Human Factors, 24,* 509–519.

NORO, K. (1992). Construction of Parametric Model of Operator and Workstation. *Ergonomics, 35,* 661–676.

NOYES, J. (1983a). The Qwerty Keyboard: A Review. *International Journal of Man-Machine Studies, 18,* 265–281.

NOYES, J. (1983b). Chord Keyboards. Applied Ergonomics, 14(1), 55–59.

NYGREN, T. W. (1991). Psychometric Properties of Subjective Workload Measurement Techniques. *Human Factors, 33,* 17–33.

OBORNE, D. J. (1983). Vibration at Work. In D. J. Oborne and M. M. Gruneberg (eds.), *The Physical Environment at Work* (pp. 143–177). New York: Wiley.

O'BRIAN, R., and SHELTON, W. C. (1941). *Women's Measurements for Garment and Pattern Constructions* (U.S. Department of Agriculture, Publication No. 454). Superintendent of Documents, Washington, DC: U.S. Government Printing Office.

OCCHIPINTI, E., COLOMBINI, D., FRIGO, C., PEDOTTI, A., and GRIECO, A. (1985). Sitting Posture: Analysis of Lumbar Stresses with Upper Limbs Supported. *Ergonomics, 28,* 1333–1346.

OCCHIPINTI, E., COLOMBINI, D., and GRIECO, A. (1991). A Procedure for the Formulation of Synthetic Risk Indices in the Assessment of Fixed Working Postures. In *Proceedings, 11th Congress of the International Ergonomics Association* (pp. 3–5). London, UK: Taylor & Francis.

OCCHIPINTI, E., COLOMBINI, D., MOLTENI, G., MENONI, O., BOCCARDI, S., and GRIECO, A. (1988). Clinical and Functional Examination of the Spine in Working Communities: Occurrence of Alterations in the Male Control Group. *Clinical Biomechanics, 4* (no pagination).

O'DONNELL, R. D., and EGGLEMEYER, F. T. (1986). Workload Assessment Methodology. In K. R. Boff, L. Kaufman, and J. P. Thomas (eds.), *Handbook of Perception and Human Performance,* Vol. II (pp. 42.1–42.49). New York: Wiley.

OEZKAYA, N., and NORDIN, M. (1991). *Fundamentals of Biomechanics*. New York: Van Nostrand Reinhold.

OLDENKAMP, I. (1990). Comfort and Appearance. *Ergonomics, 33,* 413–420.

OLZAK, L. A., and THOMAS, J. P. (1986). Seeing Spatial Patterns. Chapter 7 in K. R. Boff, L. Kaufman, and J. P. Thomas (eds.), *Handbook of Perception and Human Performance* (pp. 7.1–7.56). New York: Wiley.

ONG, C. N., and KOGI, K. (1990). Shiftwork in Developing Countries: Current Issues and Trends. In A. J. Scott (ed.), *Shiftwork* (pp. 417–428). Philadelphia: Hanley and Belfus.

ONG, C. N., KOH, D., PHOON, W. O., and LOW, A. (1988). Anthropometrics and Display Station Preferences of VDU Operators. *Ergonomics, 31,* 337–347.

ORIET, L. P., and DUTTA, S. P. (1989). Investigations of Modelling Two-Worker Lifting Teams. In A. Mital (ed.), *Advances in Industrial Ergonomics and Safety I* (pp. 679–683). London, UK: Taylor and Francis.

OSTLERE, S. J., and GOLD, R. H. (1991). Osteoporosis and Bone Density Measurement Methods. *Clinical Orthopaedics, 271,* 149–163.

OWENS, D. A., and LEIBOWITZ, H. W. (1983). Perceptual and Motor Consequences of Tonic Vergence. In C. Schor and K. Ciuffreda (eds.), *Vergence Eye Movements: Basic and Clinical Aspects* (pp. 25–74). Boston, MA: Butterworths.

OWINGS, C. L., CHAFFIN, D. B., SNYDER, R. G., and NORCUTT, R. (1975). *Strength Characteristics of U.S. Children for Product Safety Design* (011903-F). Ann Arbor, MI: The University of Michigan.

PAAS, F. G. W., and ADAM, J. J. (1991). Human Information Processing During Physical Exercise. *Ergonomics, 33,* 1385–1397.

PANDOLPH, K. B., GIVONI, B., and GOLDMAN, R. F. (1977). Predicting Energy Expenditure with Loads While Standing or Walking Very Slowly. *Journal of Applied Physiology: Respiration, Environmental, and Exercise Physiology, 43,* 577–581.

PANDOLPH, K. B. (1983). Advances in the Study and Application of Perceived Exertion. *Exercise and Sport Sciences Review, 11,* 118–158.

PANDOLPH, K. B. (ed.). (1988). *Human Performance Physiology and Environmental Medicine at Terrestrial Extremes.* Indianapolis, IN: Benchmark.

PANJABI, M. M., GOEL, V., OXLAND, T., TAKATA, K., DURANCEAU, J., KRAG, M., and PRICE, M. (1992). Human Lumbar Vertebrae Quantitative Three-Dimensional Anatomy. *Spine, 17,* 299–306.

PAQUETTE, S. P. (1990). *Human Analogue Models for Computer-Aided Design and Engineering Applications* (Technical Report Natick/TR-90/954). Natick, MA: U.S. Army Natick Research, Development and Engineering Center.

PARK, K. S., and CHAFFIN, D. B. (1974). A Biomechanical Evaluation of Two Methods of Manual Load Lifts. *AIIE Transactions, 6,* 105–113.

PARKES, K. R. (1990). Coping, Negative Affectivity, and the Work Environment: Additive and Interactive Predictors of Mental Health. *Journal of Applied Psychology, 75,* 399–409.

PARSONS, H. M. (1974). What Happened at Hawthorne? *Science, 18,* 922–932.

PARSONS, H. M. (1988). Human Factors in Robot Design and Robotics. In D. J. Oborne (ed.), *International Reviews of Ergonomics,* Vol. 3 (pp. 151–176). London, UK: Taylor and Francis.

PARSONS, H. M. (1988). Robot Programming. In M. Helander (ed.), *Handbook of Human-Computer Interaction* (pp. 737–754). Amsterdam: Elsevier.

PARSONS, H. M. (1990). Assembly Ergonomics in the Hawthorne Studies. In *Proceedings of the International Ergonomics Association Conference on Human Factors in Design for Manufacturability and Process Planning* (pp. 299–305). Buffalo, NY: Helander, Dept. of IE, SUNYAB.

PARSONS, H. M. (1991). Remote Control Units for Industrial Robots. In M. Rahimi and W. Karwowski (eds.), *Human-Robot Interaction* (pp. 266–283). Basingtoke, UK: Taylor and Francis.

PARSONS, K. C. (1988). Protective Clothing: Heat Exchange and Physiological Objectives. *Ergonomics, 31,* 991–1007.

PATKAI, P. (1985). The Menstrual Cycle. Chapter 8 in S. Folkard and T. H. Monk (eds.), *Hours of Work* (pp. 87–96). Chichester, UK: Wiley.

PATTERSON, P., CONGLETON, J., KOPPA, R., and HUCHINGSON, R. D. (1987). The Effects of Load Knowledge on Stresses at the Lower Back During Lifting. *Ergonomics, 30,* 539–549.

PATTON, J. F. (1988). The Effects of Acute Cold Exposure on Exercise Performance. *Journal of Applied Sport Science Research, 2,* 72–78.

PEAY, J. M. (1983). *Back Injuries.* Bureau of Mines Technology Transfer Symposium, Reno, Nevada, Bureau of Mines Information Center, IC 8943: U.S. Department of Interior.

PENNINGTON, A. J., and CHURCH, H. N. (1985). *Food Values of Portions Commonly Used*. New York: Harper.

PERES, N. J. V. (1961). Process Work Without Strain. *Australian Factory, 1,* 1–12.

PERSINGER, M. A., COOKE, W. J., and JANES, J. T. (1978). No Evidence for Relationship Between Biorhythms and Industrial Accidents. *Perceptual and Motor Skills, 46,* 423–426.

PHALEN, G. S. (1966). The Carpal-Tunnel Syndrome: Seventeen Years' Experience in Diagnosis and Treatment of Six-Hundred Fifty-Four Hands. *The Journal of Bone and Joint Surgery, 48-A*(2), 211–228.

PHALEN, G. S. (1972). The Carpal-Tunnel Syndrome, Clinical Evaluation of 598 Hands. *Clinical Orthopaedics and Related Research, 83,* 29–40.

PHEASANT, S. (1986). *Bodyspace*. London, UK: Taylor and Francis.

PHILLIPS, C. A., and PETROFSKY, J. S. (eds.). (1983). *Mechanics of Skeletal and Cardiac Muscle*. Springfield, IL: Thomas.

PICKERING, T. (1987). "RSI": Putting the Epidemic to Rest. *The Medical Journal of Australia, 147,* 213–218.

PISONI, D. B., JOHNSON, K., and BERNACKI, R. H. (1991). Effects of Alcohol on Speech. In *Proceedings of the Human Factors Society 35th Annual Meeting* (pp. 694–698). Santa Monica, CA: Human Factors Society.

PIZATELLA, T. J., MCGLOTHLIN, J. D., and PUTZ-ANDERSON, V. (1989). An Overview of NIOSH Research Activities to Reduce Low Back Injuries. In K. H. E. Kroemer, J. D. McGlothlin, and T. G. Bobick (eds.), *Manual Material Handling: Understanding and Preventing Back Trauma* (pp. 103–108). Akron, OH: American Industrial Hygiene Association.

PLOG, B. A. (ed.). (1987). *Fundamentals of Industrial Hygiene* (3d ed.). Chicago: National Safety Council.

POKORNY, J., and SMITH, V. C. (1986). Colorimetry and Color Discrimination. Chapter 8 in K. R. Boff, L. Kaufman, and J. P. Thomas (eds.), *Handbook of Perception and Human Performance* (pp. 8.1–8.51). New York: Wiley.

POPE, M. H. (1987). The Biomechanical Basis for Early Care Programmes. *Ergonomics, 30,* 351–358.

POPE, M. H. (1987). Modification of Work Organization. *Ergonomics, 30,* 449–455.

POPE, M. H., FRYMOYER, J. W., and ANDERSSON, G. (eds.). (1984). *Occupational Low Back Pain*. Philadelphia: Praeger.

PORTER, J. M. CASE, K., and BONNEY, M. C. (1990). Computer Workspace Modelling. Chapter 19 in J. R. Wilson and E. N. Corlett (eds.), *Evaluation of Human Work* (pp. 472–499). London: Taylor and Francis.

POSCH, J. L., and MARCOTTE, D. R. (1976). Carpal Tunnel Syndrome, An Analysis of 1,201 Cases. *Orthopaedic Review, 5,* 25–35.

POTVIN, J. R., NORMAN, R. W., ECKENRATH, M. E., MCGILL, S. M., and BENNETT, G. W. (1992). Regression Models for the Prediction of Dynamic L4/L5 Compression Forces During Lifting. *Ergonomics, 35,* 187–201.

PRICE, A. D. F. (1990). Calculating Relaxation Allowances for Construction Operatives, Part 1: Metabolic Cost; Part 2: Local Muscle Fatigue. *Applied Ergonomics, 21,* 318–324.

PRICE, D. L. (1988). Effects of Alcohol and Drugs. In G. A. Peters and B. J. Peters (eds.), *Automotive Engineering and Litigation,* Vol. 2 (pp. 489–551). New York: Garland.

PRIEL, V. Z. (1974). A Numerical Definition of Posture. *Human Factors, 16,* 576–584.

PUNNETT, L., and KEYSERLING, W. M. (1987). Exposure to Ergonomic Stressors in the Garment Industry: Application and Critique of Job-Site Work Analysis Methods. *Ergonomics, 30,* 1099–1116.

PUTZ-ANDERSON, V. (1988). *Cumulative Trauma Disorders: A Manual for Musculoskeletal Diseases of the Upper Limbs*. London, UK: Taylor and Francis.

PUTZ-ANDERSON, V. and WATERS, T. (1991). Revisions in NIOSH Guide to Manual Lifting. Paper presented at the Conference, "A National Strategy for Occupational Musculoskeletal Injury Prevention," Ann Arbor, MI, April 1991.

PYTEL, J. L., and Kamon, E. (1981). Dynamic Strength Test as a Predictor for Maximal and Acceptable Lifting. *Ergonomics, 24,* 663–672.

QLOX, T. (1990). The Recognition and Measurement of Stress: Conceptual and Methodological Issues. Chapter 25 in J. R. Wilson and E. N. Corlett (eds.), *The Evaluation of Human Work* (pp. 628–647). London: Taylor and Francis.

QUINET, R. J., and HADLER, N. M. (1979). Diagnosis and Treatment of Backache. *Seminars in Arthritis and Rheumatism, 8,* 261–287.

RABBITT, P. (1991). Management of the Working Population. *Ergonomics, 34,* 775–790.

RADWIN, R. G., and OH, S. (1991). Handle and Trigger Size Effects on Power Tool Operation. In *Proceedings of the Human Factors Society 35th Annual Meeting* (pp. 843–847). Santa Monica, CA: Human Factors Society.

RAMSEY, J. D. (1983). Heat and Cold. Chapter 2 in R. Hockey (ed.), *Stress and Fatigue in Human Performance* (pp. 33–60). Chicester, UK: Wiley.

RAMSEY, J. D. (1987). Practical Evaluation of Hot Working Areas. *Professional Safety,* 42–48.

RAMSEY, J. D. (1990). Do WGBT Head Stress Limits Apply to Both Physiological and Psychological Responses? In *Proceedings, International Conference on Environmental Ergonomics IV* (pp. 132–133). Washington, DC: USAF Office of Scientific Research.

RASKIN, D. C. (1988). Does Science Support Polygraph Testing? Chapter 8 in A. Gale (ed.), *The Polygraph Test. Lies, Truth, and Science* (pp. 96–110). London, UK: Sage.

REBIFFE, R., ZAYANA, O., and TARRIERE, C. (1969). Determination of Optimal Zones in the Workplace for the Placement of Manual Controls (in French). *Ergonomics, 12,* 913–924.

RECHTSCHAFFEN, A., and KALES, A. (1968). *A Manual of Standardized Terminology, Techniques, and Scoring System of Sleep Stages in Human Subjects*. Los Angeles: UCLA Brain Information Services.

REDGROVE, J. A. (1976). Sex Differences in Information Processing: A Theory and its Consequences. *Journal of Occupational Psychology, 49,* 29–37.

REILLY, C. H., and MARRAS, W. S. (1989). Simulift: A Simulation Model of Human Trunk Motion. *Spine, 14,* 5–11.

REITH, M. S. (1982). *The Relationship of Isometric Grip Strength, Optimal Dynamometer Settings, and Certain Anthropometric Factors*. Unpublished MS thesis, Virginia Commonwealth University, Richmond, VA.

RICE, V. B., and KEMMERLING, P. T. (1990). The Impact of the Aging U.S. Population on the Workforce and on Workplace Design. In *Proceedings of the International Ergonomics Association Conference on Human Factors in Design for Manufacturability and Process Planning* (pp. 347–354). Buffalo: Helander, Dept. of IE, SUNYAB.

RIDDER, C. A. (1959). *Basic Design Measurements for Sitting* (Bulletin 616, Agricultural Experiment Station). Fayetteville: University of Arkansas.

RILEY, M. W., and COCHRAN, D. J. (1988). Ergonomic Aspects of Gloves: Design and Use. *International Reviews of Ergonomics, 2,* 233–250.

RIPPLE, P. H. (1952). Accommodative Amplitude and Direction of Gaze. *American Journal of Ophthalmology, 35,* 1630–1634.

ROBBINS, H. (1963). Anatomical Study of the Median Nerve in the Carpal Tunnel and Etiologies of the Carpal-Tunnel Syndrome. *Journal of Bone and Joint Surgery, 45-A,* 953–966.

ROBERTSON, C. D., and GRIFFIN, M. J. (1989). *Laboratory Studies of the Electromyographic Response to Whole-Body Vibration* (Technical Report#184). Southampton University, Southampton, UK: Institute of Sound and Vibration Research.

ROBINETTE, K. M., and MCCONVILLE, J. T. (1981). *An Alternative to Percentile Models.* (SAE Technical Paper 810217). Warrendale, PA: Society of Automotive Engineers.

RODGERS, S. R. (1986). *Personal Communication,* 24 March 1986.

ROEBUCK, J. A. (1993). Santa Monica: *Anthropometric Methods.* Human Factors Society.

ROEBUCK, J. A., KROEMER, K. H. E., and THOMSON, W. G. (1975). *Engineering Anthropometry Methods.* New York: Wiley.

ROEBUCK, J., SMITH, K., and RAGGIO, L. (1988). *Forecasting Crew Anthropometry for Shuttle and Space Station* (STS 88–0717). Downey, CA: Rockwell International.

ROESCH, J. R. (1986). *Hand Grip Performance with the Bare Hand in the Extravehicular Activity Glove.* Unpublished MS thesis, Virginia Tech (VPI & SU), Blacksburg, VA.

ROETHLISBERGER, F. J., and DICKSON, W. J. (1939). *Management and the Worker.* Cambridge, MA: Harvard University Press.

ROGERS, A. S., SPENCER, M. B., STONE, B. M., and NICHOLSON, A. N. (1989). The Influence of a 1-hr Nap on Performance Overnight. *Ergonomics, 32,* 1193–1205.

ROGERS, S. B., and WIERWILLE, W. W. (1988). The Occurrence of Accelerator and Brake Pedal Actuation Errors During Simulated Driving. *Human Factors, 30,* 71–81.

ROHMERT, W., and RUTENFRANZ, J. (eds.). (1983). *Practical Work Physiology* (in German) (3d ed.). Stuttgart, Germany: Thieme.

ROSA, R. R., and COLLIGAN M. J. (1988). Long Workdays Versus Rest Days: Assessing Fatigue and Alertness with a Portable Performance Battery. *Human Factors, 30,* 305–317.

ROSE, M. J. (1991). Keyboard Operating Posture and Actuation Force: Implications for Muscle Over-Use. *Applied Ergonomics, 22,* 198–203.

ROSS, L. E., and MUNDT, J. C. (1988). Multi-attribute Modeling Analyses of the Effects of a Low Blood Alcohol Level on Pilot Performance, *Human Factors, 30,* 293–304.

ROWE, M. L. (1969a). Low Back Pain in Industry: A Position Paper. *Journal of Occupational Medicine, 11,* 161–169.

ROWE, M. L. (1969b). Low Back Pain in Industry: An Updated Position. *Journal of Occupational Medicine, 13,* 476-478.

ROWE, M. L. (1983). *Backache at Work.* Fairport, NY: Perinton.

RUTTER, B. G., HAAGER, J. A., DAIGLE, G. C., SMITH, S., MCFARLAND, N., and KELSEY, N. (1984). Dimensional Changes Throughout Pregnancy: A Preliminary Report. *Carle Select Papers, 36,* 44–52.

RYAN, J. (1991). *Design of Warning Labels and Instructions.* New York: Van Nostrand Reinhold.

SACKS, O. (1990). *The Man Who Mistook his Wife for a Hat*. New York: Harper Perennial.

SAE (1973). *Subjective Rating Scale for Evaluation of Noise and Ride Comfort Characteristics Related to Motor Vehicle Tires* (SAE J1060, Recommended Practice 29.11). Detroit, MI: Society of Automotive Engineers.

SAE (1974). *Symbols for Motor Vehicles Controls* (SAE J1048). Cleveland, OH: Society of Automotive Engineers.

SAE (1977). *Driver Hand Controls Location for Passenger Cars, Multi-purpose Passenger Vehicles, and Trucks* (SAE J1138). Cleveland, OH: Society of Automotive Engineers.

SALAME, P. (1991). The Effects of Alcohol on Learning as a Function of Drinking Habits. *Ergonomics, 34,* 1231–1241.

SALTHOUSE, T. A. (1990). Influence of Experience on Age Differences in Cognitive Functioning. *Human Factors, 32,* 551–569.

SALVENDY, G. (ed.) (1987). *Handbook of Human Factors*. New York: Wiley.

SAMA'MUR, P. K. (1985). Anthropometric Data in Indonesian Working Populations in the Industrial Sector. In *Proceedings of the International Symposium on Ergonomics in Developing Countries* (pp. 92–100). Geneva, Switzerland: International Labour Office. (Cited by Intaranont, 1991.)

SANCHEZ, D., and GRIEVE, D. W. (1992). The Measurement and Prediction of Isometric Lifting Strength in Symmetrical and Asymmetrical Postures. *Ergonomics, 35,* 49–64.

SANDERS, M. S., and MCCORMICK, E. J. (1987). *Human Factors in Engineering and Design* (6th ed.). New York: McGraw-Hill.

SANDOVER, J. (1983). Dynamic Loading as a Possible Source of Low-Back Disorders. *Spine, 8,* 652–658.

SANDOVER, J. (1986). Vibration and People. *Clinical Biomechanics, I,* 150–159.

SANDOVER, J., and DUPUIS, H. (1987). A Reanalysis of Spinal Motion During Vibration. *Ergonomics, 30,* 975–985.

SARNO, J. E. (1977). Psychosomatic Backache. *The Journal of Family Practice, 5,* 353–357.

SAUTER, S. L., SCHLEIFER, L. M., and KNUTSON, S. J. (1991). Work Posture, Workstation Design, and Musculoskeletal Discomfort in a VDT Data Entry Task. *Human Factors, 32,* 151–167.

SAYER, J. R., SEBOK, A. L., and SNYDER, H. L. (1990). *Color-Difference Metrics: Task Performance Prediction for Multichromatic CRT Applications as Determined by Color Legibility* (pp. 265–268). Society for Information Display Digest.

SCHIPPLEIN, O. D., TRAFIMOW, J. H., and ANDERSSON, G. B. J. (1988). The Influence of Load on Lifting Technique. *Transactions of the Orthopedic Research Society,* 375.

SCHLEGEL, B., SCHLEGEL, R. E., and GILLILAND, K. (1988). Gender Differences in Criterion Task Set Performance and Subjective Rating. In *Proceedings of the Human Factors Society 32nd Annual Meeting* (pp. 848–852). Santa Monica, CA: Human Factors Society.

SCHMIDT, R. A. (1988). *Motor Control and Learning* (2d ed.). Champaign, IL: Human Kinetics.

SCHNECK, D. J. (1985). *Biomechanics of Striated Skeletal Muscle*. Santa Barbara, CA: Kinko.

SCHNECK, D. J. (1985). Deductive Physiologic Analysis in the Presence of "Will" as an Undefined Variable. *Mathematical Modelling, 2,* 191–199.

SCHNECK, D. J. (1990). *Engineering Principles of Physiologic Function*. New York: New York University Press.

Schneider, L. W., Lehman, R. J., Pflug, M. A., and Owings, C. L. (1986). *Size and Shape of the Head and Neck from Birth to Four Years* (Final Report CPSC-C-83-1250). Ann Arbor, MI: The University of Michigan, Transportation Research Institute.

Schoberth, H. (1962). *Sitting Posture, Sitting Damage, Furniture for Sitting* (in German). Berlin, Germany: Springer.

Scholey, M., and Hair, M. (1989). Back Pain in Physiotherapists Involved in Back Care Education. *Ergonomics, 32,* 179–190.

Schultz, A. B., and Andersson, G. B. J. (1981). Analysis of Loads on the Lumbar Spine. *Spine, 6,* 76–82.

Schultz-Johnson, K. (1990). Upper Extremity Factors in the Evaluation of Lifting. *Journal of Hand Therapy,* 72–85.

Scott, D., and Marcus, S. (1991). Hand Impairment Assessment: Some Suggestions. *Applied Ergonomics, 22,* 263–269.

Scripture, E. W. (1899). *The New Psychology*. New York: Charles Scribner's Sons.

Seibel, R. (1972). Data Entry Devices and Procedures. Chapter 7 in H. P. Van Cott and R. G. Kinkade (eds.), *Human Engineering Guide to Equipment Design* (pp. 312–344). Washington, DC: U.S. Government Printing Office.

Seidel, B., and Windel, A. (1991). Sitting on "Balans" Chairs—Subjective Assessment of the Comfort and the Effect on Frequencies of Complaints. In *Proceedings, 11th Congress of the International Ergonomics Association,* Vol. 3 (pp. 11–12). London, UK: Taylor & Francis.

Seidel, H. (1988). Myoelectric Reactions to Ultra-Low Frequency, Whole-Body Vibration. *European Journal of Applied Physiology, 57,* 558–562.

Seidel, H., and Heide, R. (1986). Long-Term Effects of Whole-Body Vibration: A Critical Survey of the Literature. *International Archives of Occupational and Environmental Health, 58,* 1–26.

Sekuler, R., Kline, D., and Dismukes, K. (eds.). *Aging and Human Visual Function.* New York: Liss.

Sells, S. B. (1966). A Model for the Social System for the Multiman Extended Duration Space Ship. *Aerospace Medicine, 37,* 1130–1135.

Selye, H. (1978). *The Stress of Life* (rev. ed.). New York: McGraw-Hill.

Serow, W. J., and Sly, D. F. (1988). The Demography of Current and Future Aging Cohorts. In Committee on an Aging Society (ed.), *America's Aging—The Social and Built Environment in an Older Society* (pp. 42–102). Washington, DC: National Academy Press.

Seven, S. A. (1989). Workload Measurement Reconsidered. *Human Factors Society Bulletin, 32,* 5–7.

Shackel, B. (1991). Ergonomics from Past to Future: An Overview. In Mikumashiro and E. D. Megaw (eds.), *Towards Human Work* (pp. 3–17). London, UK: Taylor and Francis.

Shackel, B., Chidsey, K. D., and Shipley, P. (1969). The Assessment of Chair Comfort. *Ergonomics, 12,* 169–306.

Shapiro, Y., Pandolf, K. B., Avellini, B. A., Pimental, N. A., and Goldman, R. F. (1981). Heat Balance and Transfer in Men and Women Exercising in Hot-Dry and Hot-Wet Conditions. *Ergonomics, 24,* 375–386.

Sharp, M. A., and Legg, S. J. (1988). Effects of Psychophysical Lifting Training on Maximal Repetitive Lifting Capacity. *American Industrial Hygiene Association Journal, 49,* 639–644.

SHERRICK, C. E., and CHOLEWIAK, R. W. (1986). Cutaneous Sensitivity. Chapter 12 in K. R. Boff, L. Kaufmann, and J. P. Thomas, (eds.), *Handbook of Perception and Human Performance* (pp. 12.1–12.58). New York: Wiley.

SIEKMANN, H. (1990). Recommended Maximum Temperatures for Touchable Surfaces. *Ergonomics, 21,* 69–73.

SILVERSTEIN, B. A. (1985). *The Prevalence of Upper Extremity Cumulative Trauma Disorders in Industry*. Unpublished doctoral dissertation, University of Michigan, Ann Arbor, MI.

SILVERSTEIN, B. A., ARMSTRONG, T. J., LONGMATE, A., and WOODY, D. (1988). Can In-plant Exercise Control Musculoskeletal Symptoms? *Journal of Occupational Medicine, 30,* 922–927.

SILVERSTEIN, B. A., FINE, L. J., and ARMSTRONG, T. J. (1987). Occupational Factors and Carpal Tunnel Syndrome. *American Journal of Industrial Medicine, 11,* 343–358.

SILVERSTEIN, L. D., and MERRIFIELD, R. M. (1985). *The Development and Evaluation of Color Display Systems for Airborne Applications* (Report DOT/FAA/PM-85-19). Washington, DC: Federal Aviation Administration and Naval Air Test Center.

SIMS, M. T., and GRAVELING, R. A. (1988). Manual Handling of Supplies in Free and Restricted Headroom. *Applied Ergonomics, 19,* 289–292.

SINCLAIR, D. C. (1973). Mapping of Spinal Nerve Roots, in A. Jarrett (ed.), *The Physiology and Pathophysiology of the Skin*, Vol. 2, p. 349. London, UK: Academic Press.

SIND, P. M. (1989). *The Effects of Structural and Overlay Design Paramenters of Membrane Switches on the Force Exerted by Users.* Unpublished doctoral dissertation, Blacksburg, VA: Virginia Tech (VPI & SU).

SING, L. (1988). Furniture Took Ages to Grow Legs. *China Daily,* October 31, 5.

SINGER, L. D., and GRAEFF, R. F. (1988). *A Bathroom for the Elderly* (Report on Grants 230-11-110H-150-8903081 and CAE-86-005-01). College of Architecture and Urban Studies. Blacksburg, VA: Virginia Tech (VPI & SU).

SMITH, D. B. D. (1990). Human Factors and Aging: An Overview of Research Needs and Application Opportunities. *Human Factors, 32,* 509–526.

SMITH, J. L., AYOUB, M. M., SELAN, J. L., KIM, H. K., and CHEN, H. C. (1989). Manual Materials Handling in Unusual Postures. In A. Mital (ed.), *Advances in Industrial Ergonomics and Safety I* (pp. 685–591). London, UK: Taylor and Francis.

SMITH, J. L., and JIANG, B. C. (1984). A Manual Materials Handling Study of Bag Lifting. *American Industrial Hygiene Association Journal, 45,* 505–508.

SMITH, L. A., and SMITH, J. L. (1984). Observations on In-House Ergonomics Training for First-line Supervisors. *Applied Ergonomics, 15,* 11–14.

SMITH, M. J., CARAYON-SAINTFORT, P., and YANG, C.-L. (1991). Stability of the Relationships Between Job Characteristics and Computer User Well-Being. In *Proceedings, 11th Congress of the International Ergonomics Association* (pp. 765–767). London, UK: Taylor and Francis.

SMITH, P. C. (1976). Behaviors, Results and Organizational Effectiveness: The Problem of Criteria. In M. D. Dunnette (ed.), *Handbook of Industrial and Organizational Psychology* (pp. 745–775). Chicago: Rand McNally.

SMITH, R. V., and LESLIE, J. H. (eds.) (1990). *Rehabilitation Engineering*. Boca Raton, FL: CRC Press.

SNOOK, S. H. (1978). The Design of Manual Handling Tasks. *Ergonomics, 21,* 963–985.

SNOOK, S. H. (1982). Low Back Pain in Industry. In A. A. White, III, and S. L. Gordon (eds.), *American Academy of Orthopaedic Surgeons Symposium on Idiopathic Low Back Pain* (pp. 23–38). St. Louis, MO: Mosby.

SNOOK, S. H. (1987). Approaches to Preplacement Testing and Selection of Workers. *Ergonomics, 30,* 241–247.

SNOOK, S. H. (1988a). *The Control of Low Back Disability: The Role of Management.* Presented at the American Industrial Hygiene Conference, San Francisco, CA.

SNOOK, S. H. (1988b). *Low Back Pain.* (Available from Dr. S. H. Snook, 54 Oakvale Rd., Framingham, MA 01701.)

SNOOK, S. H. (1991). Low Back Disorders in Industry. In *Proceedings of the Human Factors Society 35th Annual Meeting* (pp. 830–833). Santa Monica, CA: Human Factors Society.

SNOOK, S. H., CAMPANELLI, R. A., and FORD, R. J. (1980). *A Study of Back Injuries at Pratt-Whitney Aircraft.* Hopkinton, MA: Liberty Mutual Insurance Company Research Center.

SNOOK, S. H., CAMPANELLI, R. A., and HART, J. W. (1978). A Study of Three Preventive Approaches to Low Back Injury. *Journal of Occupational Medicine, 20,* 478–481.

SNOOK, S. H., and CIRIELLO, V. M. (1972). Low Back Pain in Industry. *American Society of Safety Engineers Journal, 17,* 17–23.

SNOOK, S. H., and CIRIELLO, V. M. (1991). The Design of Manual Handling Tasks: Revised Tables of Maximum Acceptable Weights and Forces. *Ergonomics, 34,* 1197–1213.

SNOOK, S. H., and WHITE, A. H. (1984). Education and Training. In M. H. Pope, J. W. Frymoyer, and G. Andersson (eds.), *Occupational Low Back Pain* (pp. 233–244). Philadelphia: Praeger.

SNYDER, H. L. (1985a). The Visual System: Capabilities and Limitations. Ch. 3 in L. E. Tannas (ed.), *Flat-Panel Displays and CRTs* (pp. 54–69). New York: Van Nostrand Reinhold.

SNYDER, H. L. (1985b). Image Quality: Measures and Visual Performance. In L. E. Tannas (ed.), *Flat Panel Displays and CRTs* (pp. 71–90). New York: Van Nostrand Reinhold.

SNYDER, R. G. (1975). Impact. Chapter 6 in J. F. Parker and V. R. West (eds.), *Bioastronautics Data Book* (pp. 221–295). NASA SP-3006. Washington, DC: U.S. Government Printing Office.

SNYDER, R. G., SCHNEIDER, L. W., OWINGS, C. L., REYNOLDS, H. M., GOLOMB, D. H., and SCHORK, M. A. (1977). *Anthropometry of Infants, Children, and Youths to Age 18 for Product Safety Design* (Final Report UM-HSRI-88-17). Ann Arbor, MI: The University of Michigan, Highway Safety Research Institute.

SNYDER, R. G., SPENCER, M. L., OWINGS, C. L., and SCHNEIDER, L. W. (1975). *Physical Characteristics of Children as Related to Death and Injury for Consumer Product Safety Design* (Final Report, UM-HSRI-BI-75-5). Ann Arbor, MI: The University of Michigan, Highway Safety Research Institute.

SODERBERG, G. L. (ed.) (1992). *Selected Topics in Surface Electromyography for Use in the Occupational Setting: Expert Perspectives* (DDHS-NIOSH Publication 91-100). Washington, DC: U.S. Department of Health and Human Service.

SODERBERG, G. L., BLANCO, M. K., COSENTINO, T., and KURDELMEIER, K. A. (1986). An EMG Analysis of Posterior Trunk Musculature During Flat and Anteriorly Inclined Sitting. *Human Factors, 28,* 483–491.

SOLDO, B. J., and LONGINO, C. F. (1988). *Social and Physical Environments for the Vulnerable Aged.* In Committee on an Aging Society (ed.), America's Aging—The Social and

Built Environment in an Older Society (pp. 103–133). Washington, DC: National Academy Press.

SOMMERICH, C. M., and MARRAS, W. S. (1992). Temporal Patterns of Trunk Muscle Activity Throughout a Dynamic, Asymmetric Lifting Motion. *Human Factors, 34,* 215–230.

SPAIN, W. H., EWING, W. M., and CLAY, E. (1985). Knowledge of Causes, Control Aids, Prevention of Heat Stress. *Occupational Health and Safety, 54,* 27–33.

SPELT, P. F. (1991). Introduction to Artificial Neural Networks for Human Factors. *Human Factors Society Bulletin, 34,* 1–4.

SPENCE, A.P. (1989). *Biology of Human Aging*. Englewood Cliffs, NJ: Prentice Hall.

SPITZER, H., and HETTINGER, T. (1958). *Tables of the Calorie Consumption with Physical Work* (in German). Darmstadt, Germany: REFA.

SQUIRES, P. (1956). *The Shape of the Normal Work Area* (Report 275). New London, CT: Medical Research Laboratory, Navy Department.

STAFF, K. R. (1983). *A Comparison of Range of Joint Mobility in College Females and Males*. Unpublished Master's Thesis, Texas A&M University, College Station, TX.

STAFFEL, F. (1884). On the Hygiene of Sitting (in German). *Zbl. Allgemeine Gesundheitspflege, 3,* 403–421.

STALHAMMAR, H. R., LESKINEN, T. P. J., KUORINKA, I. A. A., GAUTREAU, M. H. J., and TROUP, J. D. G. (1986). Postural Epidemiological and Biomechanical Analysis of Luggage Handling in an Aircraft Luggage Compartment. *Applied Ergonomics, 17,* 177–183.

STEENBEKKERS, L. P. A., and MOLENBROEK, J. F. M. (1990). Anthropometric Data of Children for Non-Specialist Users. *Ergonomics, 33,* 421–429.

STEGEMANN, J. (1981). *Exercise Physiology*. Chicago, IL: Yearbook Medical Publishers, Thieme.

STEGEMANN, J. (1984). *Physiology of Performance* (in German) (3d ed.). Stuttgart, Germany: Thieme.

STEKELENBURG, M. (1982). Noise at Work: Tolerable Limits and Medical Control. *American Industrial Hygiene Association Journal, 43,* 402–410.

STERNBACH, R. A., WOLF, S. R., MURPHY, R. W., and AKESON, W. H. (1973). Aspects of Chronic Low Back Pain. *Psychosomatics, 14,* 52–56.

STEVENS, S. S. (1957). On the Psychophysical Law. *Psychology Review, 64,* 151–181.

STEVENSON, J. M., ANDREW, G. M., BRYANT, J. T., GREENHORN, D. R., and THOMSON, J. M. (1989). Isoinertial Tests to Predict Lifting Performance. *Ergonomics, 32,* 157–166.

STEVENSON, J. M., BRYANT, J. T., FRENCH, S. L., GREENHORN, D. R., ANDREW, G. M., and THOMSON, J. M. (1990). Dynamic Analysis of Isoinertial Lifting Technique. *Ergonomics, 33,* 161–172.

STEVENSON, J., BRYANT, T., GREENHORN, D., SMITH, T., DEAKIN, J., and SURGENOR, B. (1990). The Effect of Lifting Protocol on Comparisons with Isoinertial Lifting Performance. *Ergonomics, 33,* 1455–1469.

STEVENSON, M. G., MAHER, K., MCPHEE, B. J., LONG, A. F., and LUSTED, M. (1991). Examination of Design Criteria Following an Ergonomic Evaluation of 210 Office Chairs. In *Proceedings, 11th Congress of the International Ergonomics Association* (pp. 900–902). London: Taylor and Francis.

STOBBE, T. J., and PLUMMER, R. W. (1988). Sudden-Movement/Unexpected Loading as a Factor in Back Injuries. In F. Aghazadeh (ed.), *Trends in Ergonomics/Human Factors V,* (pp. 713–720) Amsterdam: Elsevier.

Stockbridge, H. C. W. (1957). *Micro-shape-Coded Knobs for Post Office Keys* (Techn. Memo No. 67). London: Ministry of Supply.

Stolwijk, J. A. J. (1980). Partitional Calorimetry. In J. M. Kinney (ed.), *Assessment of Energy Metabolism in Health and Disease* (pp. 21–22). Columbus, OH: Ross Laboratories.

Stoudt, H. W. (1981). The Anthropometry of the Elderly. *Human Factors, 23,* 29–37.

Strasser, H. (1991). Different Grips of Screwdrivers Evaluated by Means of Measuring Maximum Torque, Subjective Rating and by Registering Electromyographic Data During Static and Dynamic Test Work. In W. Karwowski and J. W. Yates, (eds.), *Advances in Industrial Ergonomics and Safety III* (pp. 413–420). London, UK: Taylor and Francis.

Strasser, H., Keller, E., Mueller, K-W, and Ernst, J. (1989). Local Muscular Strain Dependent on the Direction of Horizontal Arm Movements. *Ergonomics, 32,* 899–910.

Stubbs, D. A., Buckle, P. W., Hudson, M. P., and Rivers, P. M. (1983). Back Pain in the Nursing Profession. II. The Effectiveness of Training. *Ergonomics, 26,* 767–779.

St.-Vincent, M., Tellier, C., and Lortie, M. (1989). Training in Handling. *Ergonomics, 32,* 191–210.

Swets, J. A., and Bjork, R. A. (1990). Enhancing Human Performance: An Evaluation of "New Age" Techniques Considered by the U.S. Army. *Psychological Science, 1,* 85–96.

Szeto, A. Y. F., Valerio, N. C., and Novak, R. E. (1991). Audible Pedestrian Traffic Signals. *Journal of Rehabilitation Research and Development, 28,* 57–78.

Szeto, A. Y. J., and Riso, R. R. (1990). Sensory Feedback Using Electrical Stimulation of the Tactile Sense. Chapter 3 in R. V. Smith and J. H. Leslie (eds.), *Rehabilitation Engineering* (pp. 29–78). Boca Raton, FL: CRC Press.

Tache, J., and Selye, H. (1986). On Stress and Coping Mechanisms. In C. D. Spielberger and I. G. Sarason (eds.), *Stress and Anxiety: A Sourcebook of Theory and Research,* Vol. 10 (pp. 3–24). Washington, DC: Hemisphere.

Tajima, N., Asao, L., Hill, G. W., and Leonard, S. D. (1991). Comparison of Meanings of Warnings in Japan and the USA. In W. Karwowski and J. W. Yates (eds.), *Advances in Industrial Ergonomics and Safety III* (pp. 739–741). London, UK: Taylor and Francis.

Tanner, J. M. (1962). *Growth at Adolescence.* Oxford, UK: Blackwell.

Tanner, J. M., and Whitehouse, R. H. (1976). Clinical Longitudinal Standards for Height, Weight, Height Velocity, Weight Velocity and Stages of Puberty. *Archives of Diseases in Childhood, 51,* 170–179.

Tanner, J. M., Whitehouse, R. H., and Takaishi, M. (1966). Standards from Birth to Maturity for Height, Weight, Height Velocity and Weight Velocity: British Children, 1965. *Archives of Diseases of Childhood, 41,* 454–471 and 613–635.

Tanzer, R. C. (1959). The Carpal-tunnel Syndrome—A Clinical and Anatomical Study. *American Journal of Industrial Medicine, 11,* 343–358.

Taylor, B. B. (1987). Low Back Injury Prevention Training Requires Traditional, New Methods. *Occupational Health and Safety,* 44–52.

Tebbetts, I., Churchill, T., and McConville, J. T. (1980). *Anthropometry of Women of the U.S. Army—1977* (TR-80-016). Natick, MA: Clothing, Equipment and Materials Engineering Laboratory, United States Army Natick Research and Development Command.

Tepas, D. I. (1985). Flexitime, Compressed Work Weeks and Other Alternative Work Schedules. Chapter 13 in S. Folkard and T. H. Monk (eds.), *Hours of Work* (pp. 147–164). Chichester, UK: Wiley.

TEPAS, D. I., and MAHAN, R. P. (1989). The Many Meanings of Sleep. *Work and Stress, 3,* 93–102.

TEPAS, D. I., and MONK, T. H. (1986). Work Schedules. Chapter 7.3 in G. Salvendy (ed.), *Handbook of Human Factors* (pp. 819–843). New York: Wiley.

TEVES, M. A., WRIGHT, J. E., and VOGEL, J. A. (1985). *Performance on Selected Candidate Screening Test Procedures Before and After Army Basic and Advanced Individual Training* (Technical Report No. T13/85). Natick, MA: United States Army Research Institute of Environmental Medicine,

THOMPSON, D. (1989). Reach Distance and Safety Standards. *Ergonomics, 32,* 1061–1076.

THOMPSON, F. J., YATES, B. J., and FRANZEN, O. G. (1986). Blood Pooling in Leg Skeletal Muscles Prevented by a "New" Venopressor Reflex Mechanism. In *Proceedings of the Conference "Work With Display Units"* (pp. 493–496). Stockholm: Swedish National Board of Occupational Safety and Health.

THURMAN, J. E., LOUZINE, A. E., and KOGI, K. (eds.). (1988). *Higher Productivity and a Better Place to Work*. Geneva, Switzerland: International Labour Office.

TICHAUER, E. R. (1973). *The Biomechanical Basis of Ergonomics*. New York: Wiley.

TICHAUER, E. T. (1978). *Ergonomic Aspects of Biomechanics*. Chapter 32 in The Industrial Environment—Its Evaluation and Control (ed: NIOSH). Washington, D.C.: U.S. Government Printing Office.

TOMER, G. M., OLSON, C. N., and LEPORE, B. (1984). Back Injury Prevention Training Makes Dollars and Sense. *National Safety News, 129,* 36–39.

TORSVALL, A., and AKERSTEDT, T. (1980). A Diurnal Type Scale: Construction, Consistency, and Validation in Shift Work. *Scandinavian Journal of Work, Environment and Health, 6,* 283–290.

TOUGAS, G., and NORDIN, M. C. (1987). Seat Features Recommendations for Workstations. *Applied Ergonomics, 18,* 207–210.

TRACY, M. F. (1990). Biomechanical Methods in Posture Analysis. Chapter 23 in J. R. Wilson and E. N. Corlett (eds.), *The Evaluation of Human Work* (pp. 571–604). London: Taylor and Francis.

TRACY, M. F., and CORLETT, E. N. (1991). Loads on the Body During Static Tasks: Software Incorporating the Posture Targeting Method. *Applied Ergonomics, 22,* 362–366.

TROTTER, M., and GLESER, G. (1951). The Effect of Aging Upon Stature. *American Journal of Physical Anthropology, 9,* 311–324.

TUREK, F. W. (1989). Effects of Stimulated Physical Activity on the Circadian Pacemaker of Vertebrates. In S. Daan and E. Gwinner (eds.), *Biological Clocks and Environmental Time* (pp. 135–147). New York: Guilford.

TYRRELL, R. A., and LEIBOVITZ, H. W. (1990). The Relation of Vergence Effort to Reports of Visual Fatigue Following Prolonged Near Work. *Human Factors, 32,* 341–357.

UMBERS, I. G., and COLLILER, G. D. (1990). Coding Techniques for Process Plant VDU Formats. *Applied Ergonomics, 21,* 187–198.

U.S. Army (1981a). MIL-HDBK 759. *Human Factors Engineering Design for Army Material* (Metric). Redstone Arsenal, AL: U.S. Army Missile Command.

U.S. Army (1981b). MIL-STD 1472. *Human Engineering Design Criteria for Military Systems, Equipment, and Facilities*. Redstone Arsenal, AL: U.S. Army Missile Command.

U.S. Congress, Office of Technology Assessment (ed.) (1991). *Biological Rhythms: Implications for the Worker*. Washington, DC: U.S. Government Printing Office.

VAIL, G. J. (1988). A Gender Profile: U.S. General Aviation Pilot-Error Accidents. In *Proceedings of the Human Factors 32nd Annual Meeting* (pp. 862–866). Santa Monica, CA: Human Factors Society.

VAN COTT, H. P., and KINKADE, R. G. (eds.) (1972). *Human Engineering Guide to Equipment Design* (rev. ed). Washington, DC: U.S. Government Printing Office.

VAN DER GRINTEN, M. P. (1991). Test-Retest Reliability of a Practical Method for Measuring Body Part Discomfort. In *Proceedings, 11th Congress of the International Ergonomics Association* (pp. 54–56). London, UK: Taylor and Francis.

VERBEEK, J. (1991). The Use of Adjustable Furniture: Evaluation of an Instruction Programme for Office Workers. *Applied Ergonomics, 22,* 179–184.

VERRIEST, J. P., TRASBOT, J., and REBIFFE, R. (1991). MAN3D: A Functional and Geometrical Model of the Human Operator for Computer Aided Ergonomic Design. In W. Karwowski and J. W. Yates (eds.), *Advances in Industrial Ergonomics and Safety III* (pp. 901–908). London: Taylor and Francis.

VERSACE, J. (1971). *A Review of the Severity Index* (SAE Paper 71088, pp. 771–796). Detroit: Society of Automotive Engineers.

VINCENTE, K. J. and HARWOOD, K. (1990). A Few Implications of an Ecological Approach to Human Factors. *Human Factors Society Bulletin, 33,* 1–4.

VOGEL, J. A., and SHARP, M. A. (1989). High Intensity Repetitive Lifting Capacity. In K. H. E. Kroemer, J. D. McGlothlin, and T. G. Bobick (eds.), *Manual Material Handling: Understanding and Preventing Back Trauma* (pp. 1–8). Akron, OH: American Industrial Hygiene Association.

VOJTECKY, M. A., HARBER, P. SAYRE, J. W., BILLET, E., and SHIMOZAKI, S. (1987). The Use of Assistance While Lifting. *Journal of Safety Research, 18,* 49–56.

VOLGER, A., ERNST, G., NACHREINER, F., and HAENECKE, K. (1988). Common Free Time of Family Members in Different Shift Systems. *Applied Ergonomics, 19,* 213–128.

WAGNER, C. (1974). Determination of Finger Flexibility. *European Journal of Applied Physiology, 32,* 259–278.

WAIKAR, A., LEE, K., AGHAZADEH, F., and PARKS, C. (1991). Evaluating Lifting Tasks Using Subjective and Biomechanical Estimates of Stress at the Lower Back. *Ergonomics, 34,* 33–47.

WALSH, N. E., and SCHWARTZ, R. K. (1990). The Influence of Prophylactic Orthoses on Abdominal Strength and Low Back Injury in the Workplace. *American Journal Physical Medicine and Rehabilitation, 69,* 245–250.

WARM, J. S., DEMBER, W. N., GLUCKMAN, J. P., and HANCOCK, P. A. (1991). Vigilance and Workload. In *Proceedings of the Human Factors Society 35th Annual Meeting* (pp. 980–981). Santa Monica, CA: Human Factors Society.

WASSERMAN, D. (1982). *Vibration White Finger Disease in U.S. Workers Using Pneumatic Chipping and Grinding Hand Tools* (Technical Report DHHS 82-118). Cincinnati, OH: National Institute for Occupational Safety and Health.

WASSERMAN, D. E. (1987). *Human Aspects of Occupational Vibrations*. Amsterdam: Elsevier.

WASSERMAN, D. E., PHILLIPS, C. A., and PETROFSKY, J. S. (1986). The Potential Therapeutic Effects of Segmental Vibration on Osteoporosis. In *Proceedings of the 12th International Congress on Acoustics*. Paper F2-1.

WATERS, T. R. (1991). Strategies for Assessing Multi-Task Manual Lifting Jobs. In *Proceedings of the Human Factors Society 35th Annual Meeting* (pp. 809–813). Santa Monica, CA: Human Factors Society.

WATSON, L. T., LAFFEY, T. J., and HARALICK, R. M. (1985). Topographic Classification of Digital Image Intensity Surfaces Using Generalized Splines and the Discrete Cosine Transformation. *Comput. Vision, Graphics, Image Processing, 29,* 143–167.

WEBB, P. (1985) *Human Calorimeters.* New York: Praeger.

WEBB, R. D. G., and TACK, D. W. (1988). Ergonomics, Human Rights and Placement Tests for Physically Demanding Work. In F. Aghazadeh (ed.), *Trends in Ergonomics/Human Factors V* (pp. 751–758). Amsterdam: Elsevier.

WEBER, E. H. (1834). *De pulse, resorptione, auditu et tactu.* Leipzig, Germany: Kochler.

WEBER, W., and WEBER, E. (1836). *Mechanics of the Human Walking Implements* (in German). Goettingen, Germany: no publisher given.

WEDDERBURN, A. A. I. (1987). Sleeping on the Job: The Use of Anecdotes for Recording Rare But Serious Events. *Ergonomics, 30,* 1229–1233.

WELLER, H., and WILEY, R. L. (1979). *Basic Human Physiology.* New York: Van Nostrand.

WELLS, L. H. (1963). Stature in Earlier Races of Mankind. Chapter 39 in D. Bothwell and E. Higgs (eds.), *Science in Archaeology.* London, UK: Thames and Hudson.

WESTHEIMER, G. (1986). The Eye as an Optical Instrument. Chapter 4 in K. R. Boff, L. Kaufman, and J. P. Thomas (eds.), *Handbook of Perception and Human Performance* (pp. 4.1–4.20). New York: Wiley.

WEVER, R. A. (1985). Men in Temporal Isolation: Basic Principles of the Circadian System. Chapter 2 in S. Folkard and T. H. Monk (eds.), *Hours of Work* (pp. 15–28). Chichester, UK: Wiley.

WHITE, A. A., and GORDON, S. L. (eds.). (1982). *American Academy of Orthopaedic Surgeons Symposium on Idiopathic Low Back Pain.* St. Louis, MO: Mosby.

WHITE, M. K., HODOUS, T. K., and VERCRUYSSEN, M. (1991). Effects of Thermal Environment and Chemical Protective Clothing on Work Tolerance, Physiological Responses, and Subjective Ratings. *Ergonomics, 34,* 445–457.

WICKENS, C. D. (1991). Terms in Transition: Analogous Systems. In *Proceedings of the Human Factors Society 35th Annual Meeting* (pp. 976–979). Santa Monica, CA: Human Factors Society.

WICKENS, C. D., and KRAMER, A. G. (1985). *Engineering Psychology.* Annual Review of Psychology. New York: Annual Reviews.

WICKENS, C. D., and YEH, Y. Y. (1983). The Disassociation of Subjective Ratings and Performance: A Multiple Resources Approach. In *Proceedings of the Human Factors 27th Annual Meeting* (pp. 244–248). Santa Monica, CA: Human Factors Society.

WIERWILLE, W. W. (1992). Visual and Manual Demands of In-Car Controls and Displays. In J. B. Peacock and W. Karwowski (eds.), *Automotive Ergonomics: Human Factors in the Design and Use of the Automobile* (in press). London, UK: Taylor and Francis.

WIERWILLE, W. W., ANTIN, J. F., DINGUS, T. A., and HULSE, M. C. (1988). Visual Attentional Demand of an In-Car Navigation Display System. In A. G. Gale, M. H. Freeman, C. M. Haslegrave, P. Smith, and S. P. Taylor (eds.), *Vision in Vehicles II* (pp. 307–316). Amsterdam: Elsevier.

WIERWILLE, W. W., and CASALI, J. G. (1983). A Validated Rating Scale for Global Mental Workload. In *Proceedings of the Human Factors Society 27th Annual Meeting* (pp. 129–133). Santa Monica, CA: Human Factors Society.

WIERWILLE, W. W., CASALI, J. G., CONNOR, S. A., and RAHIMI, M. (1985). Evaluation of the Sensitivity and Intrusion of Mental Workload Estimation Techniques. *Advances in Man-Machine Systems Research, 2,* 51–127.

WIERWILLE, W. W. RAHIMI, M., and CASALI, J. (1985). Evaluation of 16 Measures of Mental Workload Using a Simulated Flight Task Emphasizing Mediational Activity. *Human Factors, 27,* 489–502.

WIKER, S. F., CHAFFIN, D. B., and LANGOLF, G. D. (1989). Shoulder Posture and Localized Muscle Fatigue and Discomfort. *Ergonomics, 32,* 211–237.

WIKER, S. F., CHAFFIN, D. B., and LANGOLF, G. D. (1990). Shoulder Postural Fatigue and Discomfort. *International Journal of Industrial Ergonomics, 5,* 133–146.

WIKER, S. F., LANGOLF, G. D., and CHAFFIN, D. B. (1989). Arm Posture and Human Movement Capability. *Human Factors, 31,* 421–441.

WILLIAMS, A., REILLEY, T., CAMPBELL, I., and SUTHERST, J. (1988). Investigation of Changes in Responses to Exercise and in Mood During Pregnancy. *Ergonomics, 31*(11), 1539–1549.

WILLIAMS, T. L., SMITH, L. A., and HERRICK, R. T. (1989). Exercise as a Prophylactic Device Against Carpal Tunnel Syndrome. In *Proceedings of the Human Factors Society 33rd Annual Meeting* (pp. 723–727). Santa Monica, CA: Human Factors Society.

WILLIAMS, V. H. (1988). *Isometric Forces Transmitted by the Fingers: Data Collection Using a Standardized Protocol.* Unpublished MS thesis, Virginia Tech (VPI & SU), Blacksburg, VA.

WILLIGES, R. C. (1987). The Use of Models in Human-Computer Interface Design. *Ergonomics, 30,* 491–502.

WILLIGES, R. C., WILLIGES, B. H., and ELKERTON, J. (1987). Software Interface Design. In G. Salvendy (ed.), *Handbook of Human Factors* (pp. 1416–1449). New York: Wiley.

WILSON, J. R., and CORLETT, E. N. (1990). *Evaluation of Human Work.* London, UK: Taylor and Francis.

WINKEL, J., and BENDIX, T. (1986a). Macro- and Micro-Circulatory Changes During Prolonged Sedentary Work and the Need for Lower Limit Values for Leg Activity. In *Proceedings of the Conference "Work With Display Units"* (pp. 497–500). Stockholm: Swedish National Board of Occupation Safety and Health.

WINKEL, J., and BENDIX, T. (1986b). Muscular Performance During Seated Work Evaluated by Two Different EMG Methods. *European Journal of Applied Physiology, 55,* 167–173.

WINKEL, J., and JORGENSEN, K. (1986a). Evaluation of Foot Swelling and Lower-Limb Temperatures in Relation to Leg Activity During Long-Term Seated Office Work. *Ergonomics, 29,* 313–328.

WINKEL, J., and JORGENSEN, K. (1986b). Swelling of the Foot, Its Vascular Volume and Systemic Hemoconcentration During Long-Term Constrained Sitting. *European Journal of Applied Physiology and Occupational Physiology, 55,* 162–166.

WINTER, D. A. (1990). *Biomechanics and Motor Control of Human Behavior* (2d ed.). New York: Wiley.

WOLFF, J. (1892). *The Law of Bone Translation* (in German). Berlin, Germany: Hirschwald.

WOOD, D. P. (1987). Design and Evaluation of a Back Injury Prevention Program Within a Geriatric Hospital. *Spine, 12,* 77–82.

WOOD, E. H., CODE, C. F., and BALDES, E. J. (1990). Partial Supination Versus GZ Protection. *Aviation, Space, and Environmental Medicine, 61,* 850–858.

WOODSON, W. E. (1981). *Human Factors Design Handbook.* New York: McGraw-Hill.

WOODSON, W. E., and CONOVER, D. W. (1964). *Human Engineering Guide for Equipment Designers* (2d ed.). Berkeley: University of California Press.

WOODSON, W. E., TILLMAN, B., and TILLMAN, P. (1991). *Human Factors Design Handbook* (2d ed.). New York: McGraw-Hill.

WYSZECKI, G. (1986). Color Appearance. In K. R. Boff, L. Kaufman, and J. P. Thomas (eds.), *Handbook of Perception and Human Performance* (pp. 9.1–9.57). New York: Wiley.

YEH, Y. Y., and WICKENS, C. D. (1988). Dissociation of Performance and Subjective Measures of Workload. *Human Factors, 30,* 111–120.

YETTRAM, A. L., and JACKMAN, J. (1981). Equilibrium Analysis for the Forces in the Human Spinal Column and its Musculature. *Spine, 5,* 402–411.

YOULE, A. (ed.). (1990). *The Thermal Environment* (Technical Guide No. 8, British Occupational Hygiene Association). Leeds, UK: Science Reviews Ltd. and H and H Scientific Consultants Ltd.

YU, T., ROHT, L. H., WISE, R. A., KILIAN, D. J., and WEIR, F. W. (1984). Low-back Pain in Industry: An Old Problem Revisited. *Journal of Occupational Medicine, 26,* 517–524.

ZACHARKOW, D. (1988). *Posture: Sitting, Standing, Chair Design and Exercise.* Springfield, IL: Thomas.

ZEHNER, G. F. (1986). Three-Dimensional Summarization of Face Shape. In *Proceedings of the Human Factors Society 30th Annual Meeting* (pp. 206–210). Santa Monica, CA: Human Factors Society.

ZIOBRO, E. (1991). A Contactless Method for Measuring Postural Strain. In W. Karwowski and J. W. Yates (eds.), *Advances in Industrial Ergonomics and Safety III* (pp. 421–425). London, UK: Taylor and Francis.

ZWAHLEN, H. T., HARTMANN, A. L., and KOTHARI, N. (1986). How Much Do Rest Breaks Help to Alleviate VDT Operation Subjective Occular and Musculoskeletal Discomfort? In *Proceedings of the Conference "Work With Display Units"* (pp. 503–506). Stockholm: National Swedish Board of Occupational Safety and Health.

Glossary of Terms*

A

abduct—same as *pivot:* to move away from the body or one of its parts; opposed to *adduct*.

absolute threshold—the amount of stimulus energy necessary to just detect the stimulus. Also called *detection threshold,* or merely *threshold*. Often taken as the value at which some specified probability of detection exists, such as 0.50 or 0.75. See *differential threshold*.

absorbed power—the power dissipated in a mechanical system as a result of an applied force.

accelerance—the complex ratio of acceleration to force during simple harmonic motion. (Accelerance is the inverse of *apparent mass* or *effective mass*.)

acceleration—a vector quantity that specifies the rate of change of velocity.

accommodation—in vision, an adjustment of the curvature (thickness) of the lens of the eye (which changes the eye's focal length) to bring the image of an object into proper focus on the retina.

* Some of these terms have been compiled from various sources, particularly Griffin (1990).

achromatic—absence of chroma and color.

acoustics—the science of sound.

acromion—the most lateral point of the lateral edge of the scapula. Acromial height is usually equated with shoulder height.

action, activation (of muscle)—see *contraction*.

acuity—the visual ability to discriminate fine detail. (Visual acuity is often expressed in terms of the angle subtended, or physical size, of the smallest recognizable object.) See *contrast sensitivity; fovea; Landolt C; Snellen chart*.

adaptation—a change in sensitivity to the intensity or quality of stimulation over time. This may be an increase in sensitivity (such as in dark adaptation) or a decrease in sensitivity (such as with continued exposure to a constant stimulus).

adduct—same as *pivot*: to move toward the body; opposed to *abduct*.

admittance—the complex ratio of displacement to force in a vibrating mechanical system. The displacement and force may be taken at the same point or at different points in the same system during simple harmonic motion.

afferent—the conduction of nerve impulses from the sense organs to the central nervous system. See *efferent*.

amplitude—the maximum value of a quantity (often a sinusoidal quantity). (Also called *peak amplitude* or *single amplitude*.)

anastomosis—a connection between two blood vessels.

anatomy—the science of the structure of the body.

angular frequency—the product of the frequency of a sinusoidal quantity with 2π (in radians per second, rad s^{-1}).

ankylosis—stiffening or immobility of a joint as a result of disease with a fibrous or bony union across the joint.

annulus fibrosus—a ring of fiber which forms the circumference of an intervertebral disc.

antagonistic muscles—pairs of muscles which act in opposition to each other, such as extensor and flexor muscles.

anterior—at the front of the body; opposed to *posterior*.

anthropometric dummy—a physical model constructed to reproduce the dimensions and ranges of movement of the human body for a specified percentile of an identified population.

anthropometry—the measurement of human physical form (e.g., height, reach).

apparent mass—the complex ratio of force to acceleration during simple harmonic motion.

arithmetic mean—see: *mean*.

arousal—a general term indicating the extent of readiness of the body. See *autonomic nervous system*.

arteriole—a minute artery with a muscular wall.

artery—a blood vessel carrying blood away from the heart. See *vein*.

arthritis—a condition of joints characterized by inflammation. See *osteoarthritis*.

arthrosis—degeneration of a joint.

articular—relating to a joint.

atrophy—wasting of body tissue.

audio frequency—any normally audible frequency of a sound wave (e.g., 20–20,000 Hz).

audio-frequency sound—sound with a spectrum lying mainly at audio frequencies.

audiogravic illusion—the apparent tilt of the body related to an auditory stimulus when an observer is exposed either to linear translational acceleration or deceleration or to centripetal acceleration. See *oculogravic illusion*.

audiogyral illusion—apparent movement of a source of sound, which is stationary with respect to an observer, when the observer is rotated. See *oculogyral illusion*.

autonomic nervous system—a principal part of the nervous system which is mainly self-regulating. It includes the sympathetic nervous system (involved in arousal) and the parasympathetic nervous system (involved in digestion and maintenance of functions that protect the body). See *nervous system*.

average—see *mean*.

axilla—the armpit.

axis—one of three mutually perpendicular straight lines passing through the origin of a coordinate system.

axon—a nerve fiber which normally conducts nervous impulses away from a nerve cell and its dendrites to another nerve cell or to effector cells. Axons vary from about 0.25 to more than 100 μm in width and can be many centimeters long. Axons more than 0.5 μm thick are usually covered by myelin sheath. See *neuron; synapse*.

B

ballistic—relating to a motion which is largely predetermined at its onset.

bandpass filter—a filter which has a single transmission band extending from a lower cut-off frequency (not zero) to an upper cut-off frequency (not infinite). Used to remove unwanted low- and high-frequency oscillations.

bandwidth—the (nominal) bandwidth of a filter is the difference between the nominal upper and lower cut-off frequencies. The difference may either be expressed in hertz, or as a percentage of the passband center frequency, or as the interval between the upper and lower nominal cut-off frequencies in octaves.

bel—a unit of level when the base of the logarithm is 10. Used with quantities proportional to power. See *decibel*.

biceps brachii—the large muscle on the anterior surface of the upper arm, connecting the scapula with the radius.

biceps femoris—a large posterior muscle of the thigh.

biodynamics—the discipline to study (human) body motions in terms of a mechanical system.

biomechanics—the discipline to study the (human) body in terms of a mechanical system.

biopsy—removal of tissue from a living person for diagnostic examination.

bite bar—a bar, plate or mount which a person holds between the teeth of the upper and lower jaws so that there is no relative motion between the bar and the skull (head).

blood pressure—pressure exerted on the artery walls by the blood. Systolic blood pressure is the maximal pressure produced by contraction of the left ventricle; diastolic blood pressure is the minimal pressure when the heart muscle is relaxed. Typically the pressures are 120 mm Hg (systolic) and 80 mm Hg (diastolic), but values vary within and between individuals and depend on body posture, exercise, emotion, etc.

brachialis—the muscle connecting the shoulder and upper arm with the ulna, crossing both shoulder and elbow joints.

brain—that part of the central nervous system enclosed within the skull. See *nervous system; electroencephalogram*.

breakthrough—the error appearing in the output of a system controlled by a human operator as a result of vibration being transmitted via the limb of the operator. (Also called *vibration-correlated error* or *feedthrough*.)

brightness—individual perception of the intensity of a given visual stimulus.

bump—a mild form of mechanical shock.

bursitis—inflammation of a bursa. Bursae provide slippery membranes for tendons and ligaments to slide over bones. Bursitis may arise from infection or repeated pressure, friction or other trauma.

buttock protrusion—the maximal posterior protrusion of the right buttock.

C

cadaver—a dead body.

candela (cd)—the SI base unit of the luminous intensity of a light source.

canthus—juncture of the eyelids.

capillary—minute blood vessel. Capillaries connect the arterial system to the venous system.

cardiovascular—relating to the heart and the blood vessels or the circulation of blood.

carpal bones—the wrist bones: scaphoid, lunate, triquetral, pisiform, trapezium, trapezoid, capitate, and hamate bones.

carpal tunnel syndrome—pressure of swollen tissue on the median nerve where it passes through the carpal tunnel formed by the wrist bones.

carpus—the wristbones, collectively.

cartilage—the tough, smooth, white tissue (i.e., gristle) covering the moving surfaces of joints.

case-control study—in epidemiology, a study in which individual cases of disease are matched with individuals from a control group. See *cross-sectional study; cohort*.

Celsius—scale of temperature on which the melting point of ice is 0°C and the boiling point of water is 100°C.

center frequency—the center frequency of a passband filter is the geometric mean of the nominal cut-off frequencies (i.e., center frequency $= \sqrt{(f_1 f_2)}$, where f_1 and f_2 are the cut-off frequencies).

centigrade—former name for the Celsius scale of temperature.

central nervous system—part of the nervous system consisting of the brain and the spinal cord. See *cerebellum; cerebral cortex; cerebrum; nervous system*.

central tendency—a measure of the central tendency of a distribution of values as given by a typical value, such as the mean, geometric mean, median, or mode.

cerebellum—the posterior part of the brain consisting of two connected lateral hemispheres. The cerebellum is involved in muscle coordination and the maintenance of body equilibrium. See *central nervous system.*

cerebral cortex—the outermost layer of the cerebrum consisting of "grey matter." The cerebral cortex is involved in sensory functions, motor control and some "higher processes." See *central nervous system.*

cerebrum—the largest part of the brain consisting of two interconnected hemispheres with an inner core of "white matter" (myelinated fibers) and an outer covering of "grey matter" (unmyelinated fibers, the cerebral cortex). The cerebrum is involved in "higher mental activities," including the interpretation of sensory signals, reasoning, thinking, decision making, control of voluntary actions. See *nervous system; central nervous system; myelin.*

cervical—relating to the neck. See *vertebra.*

cervicale—the posterior protrusion of the spinal column at the base of the neck caused by the tip of the spinous process of the 7th cervical vertebra.

chroma—attribute of color perception that determines to what degree a chromatic color differs from an achromatic color of the same lightness.

chromatic or achromatic induction—a visual process that occurs when two color stimuli are viewed side by side, when each stimulus alters the color perception of the other. The effect of chromatic or achromatic induction is usually called *simultaneous contrast,* or *spatial contrast.*

chronic—of long duration with slow progress; not acute.

clavicle—the collarbone linking the scapular with the sternum.

clinical—the consideration of symptoms of a disease as opposed to the scientific observation of changes.

closed-loop system—a system whose output is fed back and used to manipulate the input quantity. Closed-loop systems are frequently called *feedback control systems.* See *open-loop system.*

closed system—a system isolated from all inputs from outside. See *open system.*

cochlea—coiled, snail-shaped, structure in the inner ear. The cochlea has a spiral canal making two and a half turns around a central core. The canal consists of three parallel fluid-filled canals separated from each other by Reissner's membrane and the basilar membrane. See *Corti organ.*

cognitive task—those aspects of a task involving mental processes.

cohort—a defined population group. The term is used in epidemiology to refer to a group followed prospectively in a cohort study.

collimation—to make rays of light parallel, i.e., as if they came from an object at infinite distance. This may be achieved with an optical device such as a lens or mirror.

color illuminant—color perceived as belonging to an area that emits light as primary source.

color object—color perceived as belonging to an object.

color, related—color seen in direct relation to other colors in the field of view.

color surface—color perceived as belonging to a surface from which the light appears to be reflected or radiated.

color, unrelated—color perceived to belong to an area in isolation from other colors.

compliance—the reciprocal of stiffness.

concentric (muscle effort)—shortening of a muscle against a resistance.

conditioned response—a response which anticipates an event. May refer to the response to a normally neutral stimulus which has been elicited by "conditioning."

condyle—articular prominence of a bone.

cone—receptor of light in the retina. Cones, which mediate color vision, are the only receptors located in the fovea. The density decreases toward the periphery of the retina. They function in daylight (photopic) viewing conditions. See *rod*.

confidence interval—for a normal distribution of measured data points, the range within which one value will lie with a given degree of probability.

confounded—the results of an experiment are confounded if the observed effect may be caused by more than one variable.

contraction—literally, "pulling together" the Z lines (that delineate the length of a sarcomere), caused by the sliding action of actin and myosin filaments. Contraction develops muscle tension only if external resistance against the shortening exists. Note that during an "isometric contraction" no change in sarcomere length occurs (a contradiction in terms!) and that in an "eccentric contraction" the sarcomere is actually lengthened. Hence, it is often better to use the term *action*, or *effort*, or *exertion* instead of contraction. See *muscle*.

contralateral—relating to the opposite side. See *ipsilateral*.

contrast—the difference in luminances I_{max} and I_{min} between two areas. Suitable expressions include $(I_{max} - I_{min})/(I_{max} + I_{min})$. Note that several nonequivalent equations are used in the literature to describe contrast.

contrast sensitivity—a measure of the ability of the visual system to detect variations in contrast. Contrast sensitivity is dependent on the angular size subtended by test objects. It is the dependent variable used when measuring visual performance as a function of spatial frequency. See *acuity*.

control group—a group of persons with characteristics similar to those of an experimental group that is not exposed to the conditions under investigation.

coordinate system—orthogonal system of axes to indicate the directions of motions or forces. By convention, biodynamic coordinate systems follow the "right hand rule": the positive directions of the x-, y-, and z-axes are designated by the directions of the first finger, the second finger, and the thumb, respectively, of the right hand.

Coriolis force—force that arises when a movement is made on a body which is undergoing rotational motion. The force arises from a cross-coupling of the motions; the resultant motion is called Coriolis acceleration.

Coriolis oculogyral illusion—the visual and postural illusion which occurs within a rotating environment when making head movements about an axis orthogonal to the axis of rotation. A small light at the center of rotation will appear to rise diagonally forward to the right if the head is tilted to the left while exposed to clockwise rotation.

cornea—transparent outer layer of the anterior portion of the eye.

coronal plane—same as frontal plane.

corpuscle—a term used to refer to an encapsulated nerve ending. See *Golgi-Mazzoni corpuscle; Golgi (tendon) organ; Krause's end bulb; Meissner's corpuscle; Merkel's disk; Pacinian corpuscle; Ruffini ending*.

correlation—in statistics, a relationship between two (or more) variables such that increases in one variable are accompanied by systematic increases or decreases in the other variable. See *correlation coefficient; multiple correlation; product-moment correlation; rank-order correlation*.

correlation coefficient—a number which expresses the degree and direction of a relationship between two (or more) variables. A correlation coefficient of -1.00 indicates perfect negative correlation; $+1.00$ indicates perfect positive correlation; 0 indicates no correlation. See *product-moment correlation*.

cortex—see *cerebral cortex*.

Corti organ—complex structure in the cochlea of the inner ear associated with hearing. Includes the basilar membrane and attached hair cells; sounds impinging on the tympanic membrane (i.e., ear drum) cause vibration of the ossicles (three small bones), which is transmitted to the basilar membrane causing movement and firing of the hair cells. See *cochlea; ear*.

Coulomb damping—the dissipation of energy that occurs when a particle in a vibrating system is resisted by a force the magnitude of which is a constant independent of displacement and velocity, and the direction of which is opposite to the direction of the velocity of the particle. Also called *dry friction damping*.

covariance—a measure of the extent to which changes in one variable are accompanied by changes in a second variable.

crest factor—the ratio of the peak value to the root-mean-square value of a quantity over a specified time interval.

criterion—a characteristic by which something may be judged (plural: *criteria*).

cross-axis coupling—the motion occurring in one axis due to excitation in an orthogonal axis.

cross-sectional study—in epidemiology, an investigation in which a group of persons is studied at one point in time. (Also called *prevalence study*.) See *cohort; case-control study*.

cutaneous sensory system—sensory systems with receptors in or near the skin. (Receptors include those responsible for the senses of touch, pressure, warmth, cold, pain, taste, and smell.) See *proprioception; corpuscle*.

cycle—the complete range of values through which a periodic function passes before repeating itself.

D

dactylion—the tip of the middle finger.

damper—a device used for reducing the magnitude of a shock or vibration by energy dissipation methods. Also called an *absorber*.

damping—the dissipation of energy with time or distance. See *Coulomb damping*.

danger—an unreasonable and unacceptable combination of hazard and risk. See *hazard, risk*.

dB—see *decibel*.

dB(A)—A-weighted sound pressure level.

decay time—for a shock pulse, the interval of time required for the value of the pulse to drop from some specified large fraction of the maximal value to some specified small fraction.

decibel—one-tenth of a bel. A level in decibels is 10 times the logarithm, to the base 10, of the ratio of powerlike quantities. The power level in decibels is $L_p = 10 \log_{10} (p^2/p_0^2) = 20 \log_{10} (p/p_0)$.

degeneration—degeneration of physiological tissues involves some form of deterioration in their condition or function. It may be caused by injury or disease processes. The function may be impaired or destroyed.

degrees of freedom—the number of independent variables in an estimate of some quantity.

dendrite—a branching treelike process involved in the reception of neural impulses from neurons and receptors. Dendrites are rarely more than 1.5 mm in length. See *axon*.

dependent variable—any variable the values of which are the result of changes in an independent variable.

dermis—layers of skin between the epidermis and subcutaneous tissue. The dermis contains blood and lymphatic vessels, nerves and nerve endings, glands and hair follicles.

deterministic function—a function whose value can be predicted from knowledge of its behavior at previous times.

diastole—the dilation of the heart during which the heart cavities fill with blood. See *systole*.

differential threshold—the difference in value of two stimuli which is just sufficient for their difference to be detected. Also called *difference threshold*. See *absolute threshold*.

digits—the fingers (not: the thumb) and toes.

disability—loss of function and learning ability.

disease—an interruption, cessation, or disorder of body function. A disease is identified by at least two of the following: an identifiable cause, a recognizable group of signs and symptoms, or consistent anatomical alterations. See *syndrome*.

displacement—a vector quantity that specifies the change of position of a body with respect to a reference frame.

displacement control—a control, such as a hand control, which can be operated by isotonic contraction of muscles. The control moves but offers the same resistance to the applied force, or torque, at all positions. Also called *isotonic control*.

distal—the end of a body segment farthest from the center of the body opposed to proximal.

dominant frequency—a frequency at which a maximum value occurs in a spectral-density curve.

dorsa—to the rear. See *posterior; palmar*.

dorsal—at the back, also at the top of hand or foot; opposed to *palmar, plantar,* and *ventral*.

double amplitude—see *peak-to-peak value*.

double-blind—an experimental procedure in which neither the subject nor the person administrating the experiment knows the crucial aspects of the experiment, e.g., whether a substance being administered is a placebo or a drug.

driving-point impedance—the complex ratio of the force to velocity taken at the same point in a mechanical system during simple harmonic motion.

dummy—test device or physical model simulating one or more of the anthropometric or dynamic characteristics of the human or animal body for experimental or test purposes.

duodenum—the first part of the small intestine, following the stomach.

dynamic—relating to the existence of motion due to forces; not static or at equilibrium. See *mechanics*.

dynamics—a subdivision of mechanics that deals with forces and bodies in motion.

dynamic stiffness—(i) the ratio of change of force to change of displacement under dynamic conditions; (ii) the complex ratio of force to displacement during simple harmonic motion.

E

ear—the organ of hearing. It consists of: (i) external ear (pinna and ear canal to tympanic membrane, i.e. 'ear drum'); (ii) middle ear (cavity beyond ear drum including the ossicles, i.e., the three bones: malleus, incus, and stapes); and (iii) inner ear (the cochlea, semicircular canals, utricle and saccule). See *vestibular system; Corti organ*.

ear-eye plane—a standard plane for orientation of the head, especially for defining the angle of the line of sight. The plane is established by a line passing through the right auditory meatus (ear hole) and the right external canthus (juncture of the eyelids), with both eyes on the same level. The EE plane is about 10 degrees more inclined than the Frankfurt plane. See *frontal plane, medial plane, transverse plane*.

eccentric (muscle effort)—lengthening of a resisting muscle by external force.

ECG—see *electrocardiogram*.

EEG—see *electroencephalogram*.

effective mass—see *apparent mass*.

effector—a muscle or gland, at the terminal end of an efferent nerve, which produces the intended response.

efferent—the conduction of nerve impulses from the central nervous system toward the peripheral nervous system (e.g., to the muscles). See *afferent*.

effort—see *contraction*.

EKG—see *electrocardiogram*.

electrocardiogram (ECG or EKG)—a graphical presentation of the time-varying electrical potential produced by heart muscle.

electroencephalogram (EEG)—the graphical presentation of the time-varying electrical potentials of the brain.

electromyogram (EMG)—the graphical presentation of the time-varying electrical potentials of muscle(s).

electrooculogram (EOG)—the graphical presentation of the time-varying electrical potentials of the eye muscle(s).

embolism—obstruction or occlusion of a vessel by some means.

EMG—see *electromyogram*.

empirical—based on observation and experiment rather than theory.

endocrine—relating to internal secretion of hormones by glands. The hormones are usually distributed through the body via the bloodstream.

endolymph—a clear fluid in the membraneous semicircular canals of the vestibular system. Flow of endolymph relative to the canals causes movement of the cupula and the firing of hair cells consistent with rotation of the head.

end organ—term used to refer to sensory receptors. See *corpuscle*.

EOG—see *electrooculogram*.

epicondyle—the bony eminence at the distal end of the humerus, radius, and femur.

epidemiology—the study of the prevalence and spread of disease in a community.

epidermis—the outer layer of the skin.

equivalent comfort contour—the magnitudes of vibration, expressed as a function of frequency, which produce broadly similar degrees of discomfort.

equivalent continuous A-weighted sound pressure level—value of the A-weighted sound pressure level of a continuous, steady sound that, within a specified time interval, T, has the same mean square sound pressure as a sound under consideration.

ergonomics—the discipline to study human characteristics for the appropriate design of the living and work environment. Also called human factors or human engineering, mostly in the United States.

etiology—a part of medical science concerned with the causes of disease. Also spelled *aetiology*.

exertion—see *contraction*.

experimental design—the plan of an experimental investigation. Experimental designs are aimed at maximizing the sensitivity and ease of interpretation of experimental measures and minimizing the influence of unwanted effects. A design is usually associated with some statistical measures used to test for significant results.

extend—to move adjacent body segments so that the angle between them is increased, as when the leg is straightened; opposed to *flex*.

extensor—a muscle whose contraction tends to straighten a limb or other body part. An extensor muscle is the antagonist of the corresponding *flexor* muscle.

external—away from the central long axis of the body; the outer portion of a body segment.

exteroception—the perception of information about the world outside the body. Involves the cutaneous sensory system and the senses of vision, hearing, taste and smell. See *proprioception; interoception*.

exteroceptor—any sensory receptor mediating exteroception.

extrinsic variable—a variable external to a subject the properties of which are not directly under the control of the subject.

ex vivo—outside the living body.

eye—see *cone; fovea; retina; rod*.

eye movements—see *saccade*.

F

facet—a small smooth area on a bone. Especially the superior and inferior articular facets of a vertebra: these connect adjoining vertebra forming small synovial joints restricting the relative movement between vertebrae.

false negative—a term applied to results at one extreme of a distribution which are beyond the level required for members of the group. The criterion falsely suggests that the members are not part of the group. See *false positive*.

false positive—a term applied to results at one extreme of a distribution which are beyond the level required for inclusion in some other group. The criterion falsely suggests that the members are part of the other group. See *false negative*.

farsightedness—an error of refraction, when accommodation is relaxed, in which the parallel rays of light from an object at infinity are brought to focus behind the retina. Also called *hyperopia* or *hypermetropia*.

fatigue—weariness resulting from bodily (or mental) exertion, reducing the performance. Condition can be removed and performance restored by rest.

Fechner's law—the psychological sensation P produced by a physical stimulus of magnitude I increases in proportion to the logarithm of the stimulus intensity: $P = k \log I$. See *Stevens' (power) law; Weber's law.*

feedback—input of some information to a system on the output of the system.

feedback control system—see *closed-loop system.*

feedthrough—see *breakthrough.*

femur—the thigh bone.

fiber—see *muscle.*

fibril—see *muscle.*

filament—see *muscle.*

filter—a device for separating oscillations on the basis of their frequency: it attenuates oscillations at some frequencies more than those at other frequencies. A filter may be mechanical, acoustical, electrical, analog or digital.

Fitts' law—the motion time to a target depends on the length of the motion path D and the size of the target w, i.e., on the difficulty of the activity: $MT = a + b \log_2 (2D/W)$.

flex—to move a joint in such a direction as to bring together the two parts which it connects, as when the elbow is being bent; opposed to extend.

flexor—a muscle whose contraction tends to flex. A flexor muscle is the antagonist of the corresponding extensor muscle.

force control(ler)—a control, such as a hand control, which can be operated by isometric contraction of muscles. The control does not move but responds to the applied force, or torque. Also called *isometric control.*

fovea—the fovea (centralis) of the eye is a small pit in the center of the retina which contains cones but no rods. When looking directly at a point, its image falls on the fovea. The fovea covers an angle of about 2°. Visual acuity is normally greatest for images on the fovea.

Frankfurt plane (falsely spelled Frankfort)—a standard plane for orientation of the head. The plane is established by a line passing through the right tragion (approximately, the ear hole) and the lowest point of the right orbit (eye socket), with both eyes on the same level. See *ear-eye plane; frontal plane; medial plane; transverse plane.*

free dynamic—in muscle strength context, an experimental condition in which neither displacement and its time derivatives, nor force, are manipulated as independent variables.

frequency—the reciprocal of the fundamental period. Frequency is expressed in hertz (Hz), a unit which corresponds to one cycle per second.

frequency weighting—a transfer function used to modify a signal according to a required dependence on vibration frequency.

frontal plane—any plane at right angle to the medial (mid-sagittal) plane dividing the body into anterior and posterior portions. Also called *coronal plane*. See *ear-eye plane; Frankfurt plane; medial plane; transverse plane*.

function disorder—any disorder for which there is no known organic pathology.

G

***g*, acceleration of gravity**—the acceleration produced by the force of gravity at the surface of the earth, 9.80665 m s^{-2}. G if used as unit.

gain—the amplification provided by a system.

galvanic skin response—a measure of the electrical characteristics of the skin, usually the electrical resistance. The resistance varies with emotional tension and other factors, an effect used in any polygraph, lie detector.

Gaussian distribution—see *normal distribution*.

geometric mean—the geometric mean of two quantities is the square root of the product of the two quantities.

glabella—the most anterior point of the forehead between the brow ridges in the mid-sagittal plane.

glenoid cavity—the depression in the scapula below the acromion into which fits the head of the humerus, forming the "shoulder joint."

gluteal furrow—the furrow at the juncture of the buttock and the thigh.

Golgi-Mazzoni corpuscle—an encapsulated nerve ending found in the dermis of the skin.

Golgi (tendon) organ—a proprioceptive sensory nerve ending mainly embedded within fibers of tendons at their junction with muscles. It is a "stretch" receptor activated by a change in tension between muscles and bone. See *corpuscle*.

H

habituation—reduction in human response to a stimulus as a result of cumulative exposure to the stimulus. Habituation is often assumed to involve activity of the central nervous system. See *adaptation*.

haemoglobin—the red respiratory protein of erythrocytes; consisting of globin, a protein, and an iron compound, haem.

hand-transmitted vibration (or shock)—mechanical vibration (or shock) applied or transmitted directly to the hand-arm system, usually through the hand or fingers. A common example is the vibration from the handles of power tools.

harmonic—a sinusoidal oscillation whose frequency is an integral multiple of the fundamental frequency. The second harmonic is twice the frequency of the fundamental, etc.

hazard—condition or circumstance that present an injury potential. See *danger, risk*.

head-up display—a fixed display presenting information in the normal line of sight of the observer, such as in a windscreen of an automobile or aircraft.

helmet-mounted display—a visual display mounted on a helmet and moving with the head.

hemoglobin—see *haemoglobin*.

hernia—protrusion of a part of the body through tissues normally containing it.

herniated disk—see *prolapsed disk*.

high-pass filter—a filter which has a single transmission band extending from some critical cut-off frequency of interest. Used to remove unwanted oscillations at low frequency.

H-point—the pivot point of the torso and thigh on two-dimensional and three-dimensional devices used to measure vehicle seating accommodation.

hue—attribute of color perception that uses color names and combinations thereof, such as bluish purple, yellowish green. The four unique hues are red, green, yellow, and blue, neither of which contains any of the others.

human analog (or surrogate)—in biodynamics, a body which has biodynamic properties representative of those of the human body. See *dummy*.

human factors (engineering)—see *ergonomics*.

humerus—the bone of the upper arm.

hyper-—over, above, exceeding, excessive.

hypertrophy—increase in bulk of some tissue in the body without an increase in the number of cells.

hypo-—under, below, diminished.

hypothesis—a supposition made as a starting point for reasoning or investigation, possibly without an assumption as to its truth. Hypotheses should be formulated such that they are amenable to testing by, for example, empirical research. See *theory*.

I, J

iliac crest—the superior rim of the pelvic bone.

illium—see *pelvis*.

impact—a single collision of one mass with a second mass.

impedance—the ratio of a harmonic excitation of a system to its response (in consistent units), both of which are complex quantities and both of whose arguments increase linearly with time at the same rate. See *mechanical impedance*.

impulse—the integral with respect to time of a force taken over a time during which the force is applied; often simply the product of force and time during which the force is applied.

independent variable—any variable whose values are independent of changes in the values of other variables. In an experiment, the variable which is manipulated so that its effect on one or more dependent variables can be observed.

inferior—below, lower, in relation to another structure.

infra-—below, low, under, inferior, after.

infrasonic frequency—any frequency lower than normally audible sound waves, e.g., below 20 Hz.

infrasound—sound with a spectrum lying mainly at infrasonic frequencies.

injury—damage to body tissue due to trauma.

inner ear—the innermost part of the ear containing the vestibular system and cochlea. Also called *labyrinth*.

innervation—(i) provision of an organ with nerves; (ii) nervous stimulation or activation of an organ.

inseam—a term used in tailoring to indicate the inside length of a sleeve or trouser leg. It is measured on the medial side of the arm or leg.

intelligence quotient (IQ)—a measure of intelligence defined as 100 times the mental age divided by the chronological age. (Average intelligence is therefore 100.)

inter-—between, among.

internal—near the central long axis of the body; the inner portion of a body segment.

interoception—the perception of information about the interior functioning of the body. Involves receptors in the viscera, glands, and blood vessels and the senses of hunger, thirst, nausea, etc. See *proprioception, exteroception*.

interoceptor—any sensory receptor mediating interoception.

intersubject variability—variability between subjects. See *intrasubject variability*

interval scale—a scale in which differences between intervals on the scale have quantitative significance; although the intervals between values are significant, the absolute values are not. Allowable arithmetical operations are addition and subtraction, but not multiplication or division, thus one cannot use proportions. The valid statistical operations include calculation of the mean value and of the standard deviation.

intervertebral disk—flexible pads between the vertebra. Disks have a soft jellylike core, the nucleus pulposus, enclosed by hard fibrous tissue which is attached to the bodies of the adjacent vertebrae. Disks comprise about 25 percent of the length of the vertebral column. See *prolapsed disk; vertebra.*

in toto—as a whole.

intra- —within, on the inside.

intrasubject variability—variability within a subject. See *intersubject variability*.

in vitro—in an artificial environment, as in a test tube.

in vivo—in the living body

involuntary action—an action not under voluntary control. Either a reflex action, a very well-learned action, or a normal action which is part of the genetic makeup of the body.

ipsilateral—relating to the same side. See *contralateral*.

ischaemia—see *ischemia*.

ischemia—a deficiency in blood supply to a part of the body; often as a result of the narrowing or complete blockage of an artery or arteriole.

ischial tuberosity—bony projection at the lower and posterior part of the hip bone. Located at the junction of the lower end of the body of the ischium and its ramus. When seated on a flat rigid surface, the contact pressure is usually greatest beneath the ischial tuberosities.

ischium—the dorsal and posterior of the three principal bones that compose either half of the pelvis.

iso—equal, the same.

isoacceleration—a condition in which the acceleration is kept constant.

isoforce—a condition in which the muscular force (tension) is constant. This term is equivalent to *isotonic*.

isoinertial—a condition in which muscle moves a constant mass.

isojerk—a condition in which the time derivative of acceleration, jerk, is kept constant.

isokinematic—a condition in which the velocity of muscle shortening (or lengthening) is constant. Depending on the given biomechanical conditions, this may not coincide with a constant angular speed of a body segment about its articulation.

isometric—a condition in which the length of the muscle remains constant.

isometric contraction—a muscular effort which causes tension but no movement.

isometric control—see *force control.*

isotonic—a condition in which muscle tension (force) is kept constant—see *isoforce*. (In the past, this term was occasionally falsely applied to any condition other than isometric.)

isotonic control—see *force control.*

jerk—a vector quantity that specifies the rate of change of acceleration.

joule (J)—the work done by a force of 1 N acting over a distance of 1 m. The joule is the SI unit of work. It is equivalent to 10^7 ergs and is the energy dissipated by 1 W in 1 s.

just noticeable difference (JND)—the difference between two stimuli which is just noticeable in some defined condition. See *Weber's law*.

K

kinematics—a subdivision of dynamics that deals with the motions of bodies, but not the causing forces.

kinesthetic—the feeling of motion, especially from the muscles, tendons, and joints. Also called *somatosensory*.

kinetics—a subdivision of dynamics that deals with forces applied to masses.

knuckle—the joint formed by the meeting of a finger bone (phalanx) with a palm bone (metacarpal).

Krause's end bulb—encapsulated corpuscle in parts of the skin, mouth, and other locations generally believed to be sensitive to cold.

kyphosis—backward (convex) curvature of the spine. See *lordosis, scoliosis*.

L

labyrinth—see *inner ear*. (Name is derived from the complex mazelike structure within the bone.)

Landolt C—the letter "C" presented at various orientations as a test of visual acuity. The basic letter is formed from a ring with a thickness subtending a 1′ arch with a gap subtending a 1′ arc. Larger and smaller rings are presented and acuity may be expressed either by the angle subtended by the gap when its location can be correctly identified, or as a ratio as for the Snellen chart. (Also called Landolt ring.) See *acuity*.

latency—the period of apparent inactivity between the time a stimulus is presented and the moment that a specified response occurs.

lateral—near or toward the side of the body; opposed to *medial*.

ligament—fibrous band between two bones at a joint. Ligaments are flexible, but inelastic.

lightness—visual perception of how much more or less light a stimulus emits in comparison to a "white" stimulus also contained in the field of view.

linear function—one variable is said to be a linear function of another variable if changes in the first variable are directly proportional to changes in the second variable.

linear system—a system in which the response is proportional to the magnitude of the excitation.

line of sight—the line connecting the point of fixation in the visual field with the center of the entrance pupil and the center of the fovea in the fixating eye.

longitudinal study—see *cohort*.

longitudinal wave—a wave in which the direction of displacement caused by the wave motion is in the direction of propagation.

lordosis—forward (concave) curvature of the lumbar spine. See *kyphosis; scoliosis*.

low-pass filter—a filter which has a single transmission band extending from zero frequency up to a finite frequency. (Used to remove unwanted high-frequency oscillations from a signal.)

lumbago—a term used to describe pain in the mid and lower back. May be associated with sciatica, or some combination of pulled muscles and sprained ligaments, or an unknown cause. See *vertebral column; sciatica; prolapsed disk*.

lumbar—part of the back and sides between ribs and the pelvis; loins. See *vertebra*.

lumen (lm)—the luminous flux emitted from a point source of uniform intensity of 1 candela into unit solid angle. The lumen is the SI unit by which luminous flux is evaluated in terms of its visual effect.

lux—the SI unit of illumination equal to 1 lm m^{-2}.

M

macula—the macula (acousticae) is the collective name for the two patches of sensory cells in the utricle and saccule of the vestibular system of the inner ear. See *vestibular system; otoliths.*

magnitude—measure of largeness, size, or importance.

malinger—to feign an illness, possibly for compensation, sympathy, or to avoid work.

malleolus—a rounded bony projection in the ankle region. The tibia has such a protrusion on the medial side, and the fibula one on the lateral side.

manikin—a two-dimensional or three-dimensional model having the visual appearance of the human body. See *dummy; human analog.*

masking—a phenomenon in which the perception of a normally detectable stimulus is impeded by a second stimulus. The second stimulus may be presented at a different point in the same sensory system to cause "lateral masking." Alternatively it may be presented at a different time; either before, for "forward masking," or after, for "backward masking," of the test stimulus.

matched groups—an experimental procedure in which groups of subjects are matched for variables which may affect results but which are not under study in the experiment.

maximal value—the value of a function when any small change in the independent variable causes a decrease in the value of the function.

mean—(i) the mean value of a number of discrete quantities is the algebraic sum of the quantities divided by the number of quantities; (ii) the mean value of a function $x(t)$ over an interval between t_1 and t_2 is given by

$$\overline{x} = \frac{1}{t_2 - t_1} \int_{t_1}^{t_2} x(t)\, d(t)$$

The mean is a measure of central tendency. Also called *arithmetic mean* or *average*. See *geometric mean; median; mode.*

mean-square value—the square of the root-mean-square value.

meatus—the juncture of the eyelids.

mechanical advantage—in biomechanical context, the lever arm (moment arm, leverage) at which a muscle works around a bony articulation.

mechanical impedance—the complex ratio of force to velocity where the force and velocity may be taken at the same or different points in the same system during simple harmonic motion.

mechanical system—an aggregate of matter comprising a defined configuration of mass, stiffness, and damping.

mechanics—the branch of physics that deals with forces applied to bodies and their ensuing motions.

mechanoreceptor—a receptor which responds to mechanical pressures such as touch receptor in the skin. See *cutaneous sensory system; corpuscle*.

medial—near, toward, or in the midline of the body; opposed to lateral.

medial plane—a vertical plane through the midline of the body which divides the body (in the anatomical position) into right and left halves. Also called mid-sagittal plane. See *ear-eye plane; Frankfurt plane; frontal plane; transverse plane*.

median—the value, in a series of observed values, which has exactly as many observed values above it as there are below. The median is the middle score in a distribution of scores ordered according to their magnitude. It is a primary measure of central tendency for skewed distributions. See *mean; mode*.

Meissner's corpuscle—specialized encapsulated nerve ending found in the papillae of the skin of the hand and foot, the front of the forearm, lips, and tip of the tongue. This corpuscle responds to pressure and vibration within a small area and is rapidly adapting. It is often a principal means of sensing vibration of the skin in the range 5–60 Hz, depending on conditions. Meissner's corpuscles are located in glabrous skin and are orientated perpendicular to the skin surface. Also called tactile corpuscles. See *Pacinian corpuscle*.

Merkel's disk—specialized free nerve ending found immediately below the epidermis and around the ends of some hair follicles. The corpuscle is believed to respond to pressure applied perpendicular to the skin at frequencies below about 5 Hz. See *Ruffini ending*.

metacarpal—one of the long bones of the hand between the carpus and the phalanges.

metacarpus—the five bones of the hand between the carpus and the phalanges.

mid-sagittal plane—same as medial plane. See *Frankfurt plane; frontal plane; transverse plane*.

minute (of arc) (′)—unit of angular measure equal to 1/60 of a degree.

mitochondrion—a small granular body floating inside a cell's cytoplasm. Contains a maze of tightly folded membranes within which oxygen and nutrients (brought in the blood by the circulatory system) are processed by enzymes to reform adenosine triphosphate, ATP, from ADP, adenosine diphosphate. (plural: mitochondria)

modality—in psychology and physiology, indicating a sensory system, e.g., visual modality.

mode—in statistics, the value of a series of values which is the most frequently observed. A measure of central tendency. See *mean; median*.

model—a representation of some aspect of an idea item or functioning.

modulation—the variation in the value of some parameter which characterizes a periodic oscillation.

monotonic—a relationship between two variables in which for every value of each variable there is only one corresponding value of the other variable. Graphical representation shows steadily rising or steadily falling curves.

morbidity—the prevalence of disease in a population

motion sickness—vomiting (emesis), nausea, or malaise provoked by actual or perceived motion of the body or its surroundings.

motivation—a desire, or an incentive, to achieve.

motor(ic)—in life sciences, a term used to refer to processes or anatomical areas associated with muscular action.

motor unit—all muscle filaments under the control of one efferent nerve axon.

multiple-correlation (*R*)—the relation between a dependent variable and two (or more) independent variables. See *correlation*.

multivariate—any procedure in which more than one variable is considered simultaneously.

muscle—a tissue bundle of fibers, able to contract or be lengthened. Specifically, striated muscle (skeletal muscle) that moves body segments about each other under voluntary control. There are also smooth and cardiac muscles.

muscle contraction—the result of contractions of motor units so that tension is developed between the origin and insertion of one muscle which is often accompanied by shortening of the muscle. See *contraction*.

muscle fibers—elements of muscle, containing fibrils.

muscle fibrils—elements of muscle fibrils, containing filaments.

muscle filaments—elements of muscle fibrils (polymerized protein molecules) capable of sliding along each other, thus shortening the muscle and, if doing so against resistance, generating tension. See *contraction*.

muscle spindle—an end organ in skeletal muscle which is sensitive to stretch of the muscle in which it is enclosed. Muscle spindles are spindle-shaped, taper at both ends, and lie parallel to regular muscle bundles.

myelin—fatty substance which encloses some nerve fibers. Myelinated fibers propagate nerve impulses faster than unmyelinated fibers.

myo—a prefix referring to muscle.

myopia—see nearsightedness.

N

narrow-band filter—a bandpass filter for which the passband width is relatively narrow, e.g., one-third octave or less.

natural frequency—a frequency of free vibration resulting from only elastic and inertial forces of a mechanical system. See *resonance*.

nearsightedness—inability to see distant objects distinctly owing to an error of refraction if accommodation is relaxed, in which parallel rays of light from an object at infinity are brought to focus in front of the retina. Also called *myopia*.

nerve—a whitish cord made up of myelinated and/or unmyelinated neural fibers and held together by a connective tissue sheath. Nerves transmit stimuli to the central nervous system from sensors (extero- and interoceptors) (i.e., afferent nerves) or in the reverse direction to effectors (i.e., efferent nerves). See *myelin; axon; dendrite; synapse; effector*.

nerve cell—see *neuron*.

nervous system—a system composed of neural tissue controlling the human body's structures and organs. May be subdivided in many ways, e.g., central nervous system and peripheral nervous system, or autonomic nervous system and somatic nervous system.

neuromuscular—relating to the relationship between nerve and muscle, especially the motor innervation of skeletal muscle.

neuron—the functional unit of the nervous system, consisting of cell body, dendrite, and axon. See *synapse*.

nociceptor—specialized nerve endings involved in the sensation of pain.

noise—(i) any disagreeable or undesired sound; (ii) sound, generally of a random nature, the spectrum of which does not exhibit clearly defined frequency components; (iii) electrical (or mechanical) oscillations of an undesired or random nature.

nominal scale—the simplest form of a scale: it is qualitative, consisting of a set of categories or labels. A nominal scale facilitates the identification of items, to determine whether they are equivalent, and to count them.

normal distribution—a normal distribution has a probability density function, $P(x)$, given by

$$P(x) = \frac{1}{\sigma\sqrt{(2\pi)}} e^{-x_p^2/2\sigma}$$

where σ is the standard deviation of the signal and x_p is the instantaneous magnitude. Also called *Gaussian distribution*.

nucleus pulposus—the soft fibrocartilage central portion of the intervertebral disc.

nystagmus—eye movements consisting of a slow drift followed by a rapid return.

O

objective—in life sciences, something which is real and measurable in physical units. The term is ill defined and has many uses. Often used in contrast to *subjective*.

occipital—the back of the head, skull, or brain.

occular—relating to the eye.

octave—the interval between two frequencies which have a frequency ratio of 2.

octave bandwidth filter—a bandpass filter for which the passband is one octave, i.e., the difference between the upper and lower cut-off frequencies is one octave.

oculogravic illusion—the apparent tilt of the body relative to the visual scene when an observer is exposed either to linear translational acceleration or deceleration or to centripetal acceleration. See *audiogravic illusion*.

oculogyral illusion—apparent movement of a point of light, which is stationary with respect to an observer, when the observer is rotated. See *audiogyral illusion*.

oculomotor—term used to refer to eye movements and their muscular control.

olecranon—the proximal end of the ulna.

omphalion—the centerpoint of the navel.

one-half octave—the interval between two frequencies which have a frequency ratio of $2^{1/2}$ (= 1/4142).

one-third octave—the interval between two frequencies which have a frequency ratio of $2^{1/3}$ (= 1.2599).

open-loop system—a system with no feedback.

open system—a system that is influenced by inputs from the outside. See *closed system*.

orbit—the eye socket.

ordinal scale—a scale in which items are placed in order according to a characteristic. This is the simplest quantitative scale: items can be ranked, and large values imply that there is more of the characteristic, but do not indicate how much more. Nonparametric statistics are based on ordinal scales. The only other valid statistical operations on ordinal data are the determination of medians, percentiles, interquartile range, and similar procedures.

orthogonal—(i) at right angles; (ii) sets of variables which are independent of each other.

os—bone.

oscillation—the variation, usually with time, of the magnitude of a quantity with respect to a specified reference when the magnitude is alternately greater and smaller than some mean value.

osteoarthritis, osteoarthrosis—degeneration of the joints with some loss of the low-friction cartilage linings and formation of rough deposits of bone. The term *osteoarthrosis* is preferred when there is no inflammation.

osteoporosis—the wasting, or atrophy, of bone. Bone becomes more porous, more brittle, changes its geometry. Often associated with aging, but also arises from immobilization of limbs and nutritional deficiencies.

otoliths—crystalline particles of calcium carbonate and a protein adhering to the gelatinous membrane of the maculae of the uticle and saccule.

P-Q

Pacinian corpuscle—specialized oval-shaped encapsulated nerve ending. This corpuscle responds to pressure over a diffuse area and is rapidly adapting. It is a principal means of sensing vibration (e.g., 45–400 Hz, depending on conditions). Pacinian corpuscles are located in the palmar skin of the hands, the plantar skin of the feet, on some tendons, ligaments, and other locations. They are the largest nerve endings found in the skin (up to about 2 mm in length) and consist of concentric layers looking like a minute onion. Also called *lamellated corpuscle*. See *Meissner's corpuscle*.

palmar—relating to the palm of the hand or sole of the foot. Also called *plantar* or *volar*. Opposed to *dorsal*.

palpation—examination by feeling with the hands.

parasympathetic nervous system—see *autonomic nervous system*.

patella—the kneecap.

pathology—the science of diseases.

peak-to-peak value—the algebraic difference between the extreme values of a quantity.

peak value—the maximal value of a quantity during a given interval. The peak value is usually taken as the maximum deviation of the quantity from the mean value.

pelvis—the basin-shaped ring of bone. The "pelvic girdle," consisting of ilium, pubic arch, and ischium, which compose either half of the pelvis.

perception—awareness of some event; process by which the mind refers its sensations to external objects as cause.

phalanges—bones of the fingers or toes. There are 14 on each hand; three for each finger and two for the thumb (singular: *phalanx*).

phase—the fractional part of a period (in radians or degrees) through which a sinusoidal motion has advanced from some reference time.

phon—unit of perceived loudness level. The loudness level of a sound is the sound pressure level of a 1,000-Hz pure tone judged by the listener to be equally loud.

physiology—the science of the normal functions and phenomena of living things.

pink (random) vibration—a random vibration which has an equal mean-square acceleration for any frequency band having a bandwidth proportioned to the center frequency of the band. The energy spectrum of pink random vibration as determined by octave, one-third octave, etc., filters has a constant value. See *white (random) vibration*.

pitch—(i) perceived sound depending on frequency; (ii) rotational motion about the *y* axis. See *coordinate system*.

pivot—rotate a body joint, same as *abduct* or *adduct*, respectively.

placebo—a substance with no (intended, known, medicinal) effect.

plane—see *ear-eye plane; Frankfurt plane; frontal plane; medial plane; transverse plane*.

plantar—relating to the sole of the foot. See *palmar*.

plasma—the fluid portion of blood or lymph.

plethysmograph—a device for the measurement and graphical representation of the changes in volume, often blood volume, of some part of the body.

popliteal—pertaining to the ligament behind the knee or to the part of the leg behind the knee.

posterior—at the back of the body; opposed to *anterior*.

predisposition—a condition of having special susceptibility to a condition or disease.

probability (*p*)—an expression of the likelihood of occurrence of an event; usually expressed as a ratio of the number of occurrences of all types of events considered.

product-moment correlation (*r*)—the linear correlation between two variables based on the calculation of the mean of the products of the deviations of each score from the mean value of each variable. Also called the *Pearson product-moment correlation.*

prolapsed disk—a protrusion of the intervertebral disk caused by the nucleus pulposus having been pushed out in a bulge. If the bulge is to the rear it may exert pressure on the spinal cord, if the bulge occurs sideways it may press on nerve fibers. Commonly called *slipped disk* or *herniated disk.*

prone—the position of the body when lying face downward. Opposite of *supine.*

proprioception—the perception of information about the position, orientation and movement of the body and its parts. See *interoception; exteroception.*

proprioceptor—any sensory receptor that mediates proprioception.

proximal—the end of a body segment nearest the center of the body; opposed to *distal.*

psychology—the science concerned with the behavior (including related mental and physiological processes) of humans and animals.

psychophysics—science concerned with the quantitative relations between the perception of stimuli and their physical characteristics.

pure tone—a sound whose characteristic (.e.g., pressure) varies sinusoidally with time.

R

radian (rad)—unit of angular measure equal to the angle subtended at the center of a circle by an arc the length of which is equal to the radius. (There are 2π radians in 360 degrees; 1 rad = 57.3 degrees.)

radius—the bone of the forearm on its thumb side.

random process—a set (ensemble) of time functions that can be characterized through statistical properties. Also called a *stochastic process.*

range—the interval between the highest and lowest scores in a distribution.

rank—in statistics, the position of a score relative to all other scores ordered according to their value. When ranked, scores form an *ordinal scale*.

rank-order correlation—a correlation between two variables based on the differences in the rank orders of the two variables. Also called *Spearman rank-order correlation*.

rate coding—the time sequence in which efferent signals arrive at a motor unit to cause contractions.

rating scale—a scale of words or numbers along some dimension which may be used to report subjective reactions. Many types of rating scale may be devised. Scales may be bipolar, giving the options "good" and "bad," or unipolar. The number of steps on the scale is often in the range of 5 to 9. See *scaling*.

ratio scale—a scale in which the ratios of values on the scale have quantitative significance. Ratio scales are the most powerful scales: they are used for most physical measurements. They have true zero values, and an item which has twice as much of the characteristic as another item is represented by a value twice as large. All valid mathematical operations can be performed on data from a ratio scale.

reaction time (*RT*)—the time between the presentation of a stimulus and the beginning of an observer's response. "Simple reaction time" is the time to make a single simple response to a single stimulus. "Choice reaction time" is the time involved when there are two or more stimuli and two or more corresponding possible responses to every stimulus.

reclining posture—a body posture at some angle between seated upright and being supine.

recruitment coding—the time sequence in which efferent signals arrive at different motor units to cause them to contract.

reflex—an involuntary reaction in response to a stimulus applied peripherally and transmitted to the nervous centers of the brain or spinal cord. A reflex involves a "receptor" sensitive to the stimulus, an "effector" responding to the stimulus, and a "reflex arc" between receptor and effector.

reliability—the reliability of a test or measurement method is given by the degree to which it produces similar values when applied repeatedly in the same test conditions. See *sensitivity; specificity; validity*.

repeatability—the extent to which the outcome of an experiment would recur if the experiment were repeated.

repetition—performing the same activity more than once.

repetitive strain injury—an injury which arises from repetitive strain when a single strain gives no observable signs of injury.

residual—in statistics, the part of the variance which cannot be attributed to the factors which have been considered.

resonance—exists in a system in forced oscillation when any change, however small, in the frequency of excitation causes a decrease in a response of the system. Exists at the natural frequency.

retina—the light-sensitive membrane covering the inner rear surface of the eye.

rhythmic—the same action repeated in equal intervals.

risk—probability of injury. See *danger, hazard*.

rms—see *root-mean-square value*.

rod—receptor of light in the retina. Rods function under conditions of low illumination (scotopic) vision. They do not mediate color vision, or give perception of fine detail. There are no rods in the fovea. See *cone*.

roll—rotational motion about the x axis. See *coordinate system*.

root-mean-square value (rms)—(i) the rms value of a set of numbers is the square root of the average of their squared values; (ii) the rms value of a function, $x(t)$, over an interval between t_1 and t_2 is the square root of the average of the squared values of the function over the interval.

$$\text{rms value} = \left[\frac{\int_{t_1}^{t_2} x(t)^2\, dt}{t_2 - t_1}\right]^{1/2}$$

Ruffini ending—specialized encapsulated nerve ending found in the dermis of hairy skin. This corpuscle responds to lateral stretching of the skin (and possibly warmth) and adapts slowly. See *Merkel's disk*.

S

saccade—a quick jump of the eyes from one fixation point to another.

saccule—the smaller of the two vestibular sacs in the membranous labyrinth of the inner ear. Like the utricle, the saccule contains a layer of receptors sensitive to translational forces arising from wither translational acceleration of the head or rotation of the head in an acceleration field. See *vestibular system; macula; utricle*.

sacrum—segment of the vertebral column forming part of the pelvis. Formed from the fusion of five sacral vertebrae. The sacrum articulates with the last lumbar vertebra, the coccyx, and the hip bone on either side.

sagittal—("in the line of an arrow shot from the bow") pertaining to the medial (mid-sagittal) plane of the body, or to a parallel plane.

saturation—attribute of color perception that determines the degree to which a chromatic color differs from an achromatic color regardless of lightness.

scalar—any quantity that is completely defined by its magnitude. See *vector*.

scale—a procedure for placing items, individuals, events or sensations in some series. Four types of scale are possible: *nominal scale, ordinal scale, interval scale, ratio scale*. See these terms.

scaling—in psychology the determination of (a point on) a scale along a psychological dimension (e.g., discomfort) which has a continuous mathematical relation to some physical dimension (e.g., vibration magnitude). See *rating scale; scale*.

scapula—the shoulder blade.

sciatica—pain in the area of the sciatic nerve.

scoliosis—a sideways curvature of the spine. See *kyphosis; lordosis*.

semantic scale—a set of words describing various degrees of a characteristic, formed in a nominal or interval scale. See *interval scale, nominal scale, rating scale, scale*.

semicircular canals—three small membranous ducts (tubes) which form loops of about two-thirds of a circle within each inner ear.

sensation—the reception of stimuli at the sensors and their translation into neural impulses; followed by *perception*. See *perception*.

sensitivity—(i) the sensitivity of a transducer is given by the ratio of a specified output quantity (e.g., electrical charge) to a specified input quantity (e.g., acceleration); (ii) in screening, the proportion of individuals with a positive test result for the effect that the test is intended to reveal. See *reliability; specificity; validity*.

sensor—a transducer (usually a cell, organ, or nerve ending) by which one becomes aware of events or conditions within or outside the body. See *corpuscle, exteroceptor, interoceptor, proprioceptor*.

sensorimotor—relating to the neural circuit from a receptor to the central nervous system and back to a muscle.

shock—mechanical shock exists when a force, position, velocity, or acceleration is suddenly changed so as to excite transient disturbances in a system.

sign—in medicine, any abnormality which is discovered by a physician during an examination of a patient. (A sign is an objective symptom of disease.) See *symptom*.

skeleton—in humans, all or part of the 206 separate bones in the body.

skin—the outer covering of the body consisting of epidermis, dermis, and subcutaneous tissues.

slipped disk—see *prolapsed disk*.

Snellen acuity—visual acuity measured using a standard chart containing rows of letters of graduated sizes, expressed as the distance in which a given row of letters is correctly read, compared to the distance at which the letters can be read by a person with clinically normal eyesight. For example, an acuity score of 20/50 indicates that the tested individual can read letters at the distance of 20 feet which a normally sighted person can read at 50 feet.

Snellen chart—chart used to obtain approximate measures of visual acuity. Letters of various sizes are read at a fixed distance D (often 20 ft or 6 m). Visual acuity is given by D/D', where D' is the distance at which the smallest letters should be read by a normal eye. Normal acuity may be expressed as 20/20 (in ft) or 6/6 (in m). 6/12 indicates that letters which should be identified at 12 m were only recognized at 6 m.

somatic—relating to the body.

somatosensory—see *kinesthetic*.

sound—(i) acoustic oscillation capable of exciting the sensation of hearing; (ii) the sensation of hearing excited by an acoustic oscillation.

spasm—an involuntary muscular contraction.

specificity—in screening, the proportion of individuals with a negative test result for what the test is intended to reveal (i.e., true negative results as a proportion of the total of true-negative and false-positive results). See *reliability, sensitivity; validity*.

spectrum—a description of a quantity as a function of frequency or wavelength.

sphyrion—the most distal extension of the tibia on the medial side under the malleolus.

spinal cord—the column of neural tissue running the length of the vertebral column down to the second lumbar vertebra.

spinal nerve—in humans, one of the 31 pairs of nerves coming from the spinal cord: eight cervical, twelve thoracic, five lumbar, five sacral, and one coccygeal.

spine—the stack of vertebrae. See *vertebral column*.

spinous (or spinal) process of a vertebra—the posterior prominence.

spondylosis—degeneration of vertebrae and their joints. Vertebral ankylosis, osteoarthritis of vertebrae. Narrowing of joint spaces, or bony outgrowths on vertebrae or along edges of degenerating disks. These outgrowths may press on nerves where they join the spinal cord. Some signs are apparent in most persons past middle age; they may cause stiffness and intermittent pain.

static—at rest or in equilibrium; not dynamic.

statics—a subdivision of mechanics that deals with bodies at rest.

statistical significance—the probability that an obtained result would have occurred if only chance factors were operating; hence the degree to which the result may be attributed to some systematic effect. A probability less likely than 5 cases out of 100 is often chosen as the significance level. This is expressed as a significance of 5 percent, or $p = 0.05$. A lower probability is then required (i.e., $p < 0.05$) if a result is not to be assumed to be a chance finding. The significance of a set of results is determined by statistical tests.

sternum—the breastbone.

Stevens' (power) law—the relationship between the magnitude P of a psychological sensation produced by a stimulus of magnitude, I, is given by $P = kI^n$. The value of n is a characteristic of the physical stimulus—for example, noise or vibration—while k depends on the units of measurement. See *Fechner's law; Weber's law*.

stiffness—the ratio of change of force (or torque) to the corresponding change in translational (or rotational) displacement of an elastic element.

stochastic process—see *random process*. (Derived from the Greek word for "guess.")

strain—(i) in engineering, a dimensionless value given by the ratio of the deformation caused by a stress to the size of the material being stressed; (ii) in life sciences, the experienced result of stress generated by physical or mental demands. See *stress; stressor*.

stress—(i) in engineering, the load applied to a material which causes a change in its dimensions, i.e., strain; (ii) in psychology, either the cause of strain or the resulting strain. See *stressor*.

stressor—a cause of strain, or stress. The term helps to remove the confusing use of the word "stress": the cause is a "stressor" and the effect may be called either the "stress" (psychological use) or the "strain" (engineering use).

stylion—the most distal point on the styloid process of the radius.

styloid process—a long, spinelike projection of a bone.

sub—a prefix designating below or under.

subjective—in life sciences, something which is dependent on an individual. See *objective; scale; scaling*.

superior—above, in relation to another structure; higher.

supine—the position of the body when lying face upward. (Opposite of *prone*.)

supra—prefix denoting above, superior, or in very large quantities.

sympathetic nervous system—see *autonomic nervous system*.

symptom—in medicine, an abnormality in function, appearance, or sensation.

synapse—the junction of one neuron with another, where nerve impulses are transmitted (chemically or electrically) from the axon of one neuron to the dendrites of another neuron.

syndrome—in medicine, a combination of signs and symptoms which collectively indicate a disease.

synergistic effect—the effect of a combination of two or more stressors which may be more than the arithmetic sum of the effects of the individual stressors.

systole—the contraction of the heart by which blood is circulated. See *diastole*.

T

tactile—relating to touch or the sense of touch. See *corpuscle; Meissner's corpuscle; Merkel's disk; Pacinian corpuscle; Ruffini ending*.

tarsus—the collection of bones in the ankle joint.

tendon—fibrous cord joining a muscle to a bone.

tendonitis—inflammation of a tendon.

tendon reflex—a reflex muscular contraction elicited by a sudden stretching of a tendon. Sensors in tendons sense any stretching and trigger a reflexive contraction of the muscle to oppose movement and maintain posture. The "knee jerk" caused by percussion of the tendon at the knee is a tendon reflex.

tenosynovitis—inflammation of a tendon and its sheath.

theory—a reasoned explanation of how something occurs, or may occur, but without absolute proof. A theory is more securely established than a hypothesis.

thoracic outlet syndrome—compression of fifth cervical and first thoracic nerves and the subclavian artery by muscles in the region of the first rib and clavicle.

threshold—see *absolute threshold; differential threshold*.

tibia—the main bone of the lower leg (shin bone).

tibiale—the uppermost point of the medial margin of the tibia.

time history—the magnitude of a quantity expressed as a function of time.

tinnitus—a subjective sensation of noises in the ear (ringing, whistling, booming, etc.) experienced in the absence of an external acoustic stimulus.

tissue—a collection of similar cells and their surrounding structures.

tolerance—the ability to endure a stimulus without harm.

tone—(i) in muscles, a state in which there is normal muscular tension with slight stretching of the muscles maintained by proprioceptive reflexes; (ii) in acoustics, a sound at one given frequency.

torsion—the act of twisting by the application of forces at right angles to the axis of rotation.

touch—the act of, or the sensation which arises from, contact between the body and an object.

tracking task—a task which involves continuous following of a target. The two principal types of tracking task are "pursuit tracking" in which movements of the target are directly indicated, and "compensatory tracking," in which the difference between actual and desired target locations is indicated.

tragion—the point located at the notch just above the tragus of the ear.

tragus—the conical eminence of the auricle (pinna, external ear) in front of the ear hole.

trait—a characteristic of a person.

transducer—a device designed to receive energy from one system and supply energy, of either the same or a different kind, to another in such a manner that the desired characteristics of the input energy appear at the output.

transfer function—a mathematical relation between the output (response) and the input (excitation) of a system.

transfer impedance—in a mechanical sense, the complex ratio of the force taken at one point in a mechanical system to the velocity taken at another point in the same system during simple harmonic motion.

transfer mobility—the complex ratio of the velocity taken at one point in a mechanical system to the force taken at another point in the same system during simple harmonic motion.

translation—the (linear, straight-line) movement of an object so that all its parts follow the same direction (i.e., movement without rotation.)

transmissibility—the nondimensional ratio of the response amplitude of a system in steady-state forced vibration to the excitation amplitude expressed as a function of the vibration frequency. The ratio may be one of forces, displacements, velocities, or accelerations.

transverse plane—a plane across the body at right angles to the frontal plane and the medial (mid-sagittal) plane. See *ear-eye plane, Frankfurt plane*.

trauma—an injury caused by harsh contact with an object. (From the Greek word for "wound.")

triceps—the muscle of the posterior upper arm crossing the elbow.

trochanterion—the tip of the bony lateral protrusion of the proximal end of the femur.

tuberosity—a (large) rounded prominence on a bone.

Type-I error—the erroneous rejection of a true hypothesis. An error which arises in statistical tests which is more likely when requiring a low level of significance, such as $p < 0.05$, before rejecting the null hypothesis.

Type-II error—the failure to reject a false hypothesis. An error which arises in statistical tests which is more likely when requiring a high level of significance, such as $p < 0.0001$, before rejecting the null hypothesis.

U-Z

ulna—the bone of the forearm on its little-finger side.

ultra-—prefix meaning beyond, extreme, or excessive.

ultrasonic frequency—any frequency higher than normally audible sound waves, e.g., above 20,000 Hz.

umbilicus—depression in abdominal wall where the umbilical cord was attached to the embryo.

utricle—the larger of the two vestibular sacs in the membranous labyrinth of the inner ear.

validity—in tests, a measure of how well the test assesses what it purports to assess. See *reliability; sensitivity; specificity*.

valsalva maneuver—forced expiratory effort with either closed glottis or closed nose and mouth. This may be used to inflate the eustachian tube; it may also increase tolerance to high magnitudes of maintained downward acceleration by preventing blood from returning to the heart from the head.

variable—see *dependent variable; independent variable*.

variance—the square of the standard deviation.

vascular—relating to, or containing, blood vessels.

vaso—combining form denoting blood vessel. Also *vas-* and *vasculo-*.

vasoconstriction—narrowing of a blood vessel.

vasodilation—enlargement of a blood vessel.

vasomotor—relating to the nerves which control the muscular walls of blood vessels.

vasospasm—contraction of the muscular walls of the blood vessels.

vector—a quantity that is completely determined by its magnitude and direction. See *scalar*.

vein—a blood vessel carrying blood toward the heart. See *artery*.

ventral—pertaining to the anterior side of the trunk.

venule—a minute vein.

vertebra—one of the bones of the spinal column. In humans there are 33 vertebrae: 7 cervical vertebrae; 12 thoracic vertebrae; 5 lumbar vertebrae; 5 sacral vertebrae (fused as one bone, the sacrum); and 4 coccygeal vertebrae (fused as one bone, the coccyx).

vertebral column—the series of 33 vertebrae that extend from the coccyx to the cranium; the backbone or spine.

vertex—the top of the head.

vertigo—an inappropriate sensation of movement of the body or the visual field caused by disturbance of the mechanisms responsible for equilibrium.

vessel—any duct, tube, or canal conveying body liquid, such as blood.

vestibular system—collective term for the three semicircular canals and the two vestibular sacs (utricle and saccule) within the labyrinth of the inner ear.

vestibule—any small cavity at the front of a canal; especially the middle part of the inner ear containing the utricle and saccule.

vibration—the variation with time of the magnitude of a quantity which is descriptive of the motion or position of a mechanical system, when the magnitude is alternately greater and smaller than some average value.

viscera—the internal organs of the body, including the digestive, respiratory, urogenital, and endocrine systems and the heart, spleen, etc.

viscosity—the resistance of a fluid to shear forces and, therefore, to flow.

volar—relating to the palm of the hand or sole of the foot. See *palmar, plantar*.

volition—conscious, voluntary action. See *reflex*.

Weber-Fechner law—the combination of Weber's law and Fechner's law.

Weber's law—the just-noticeable difference JND in the stimulus magnitude I is proportional to the magnitude of the stimulus. The relation may be expressed as $\Delta I/I$ = constant. See *Fechner's law; Stevens' law*.

white (random) vibration—a random vibration which has equal mean-square acceleration for any frequency band of constant width over the spectrum of interest. See *pink (random) vibration*.

***x* axis**—see *coordinate system*.

yaw—rotational motion about a *z* axis. See *coordinate system*.

***y* axis**—see *coordinate system*.

***z* axis**—see *coordinate system*.

Index

A

B

C

D

E

F

M

N

Q

R

S